Body Structures and Functions

Anatomy and Physiology

14th EDITION

ALIGNED WITH PRECISION EXAMS by youscience

Body Structures and Functions

Anatomy and Physiology

Ann Senisi Scott
Tara Pullman Bell

Australia • Brazil • Canada • Mexico • Singapore • United Kingdom • United States

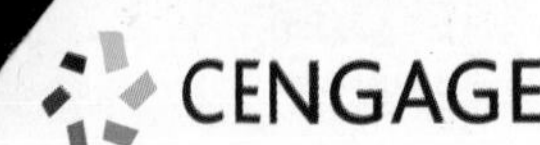

Acknowledgments
Grateful acknowledgment is given to the authors, artists, photographers, museums, publishers, and agents for permission to reprint copyrighted material. Every effort has been made to secure the appropriate permission. If any omissions have been made or if corrections are required, please contact the Publisher.

Cover: ©LEONELLO CALVETTI/SCIENCE PHOTO LIBRARY/Getty Images

National Geographic Learning | Cengage
200 Pier 4 Blvd., Suite 400
Boston, MA 02210

National Geographic Learning, a Cengage company, is a provider of quality core and supplemental educational materials for the PreK–12, adult education, and ELT markets. Cengage is a leading provider of customized learning solutions with employees residing in nearly 40 different countries and sales in more than 125 countries around the world. Find your local representative at **NGL.Cengage.com/RepFinder**

Visit National Geographic Learning online at **NGL.Cengage.com**

ISBN: 978-0-357-45754-2

Printed in the United States of America.

Print Number: 01
Print Year: 2021

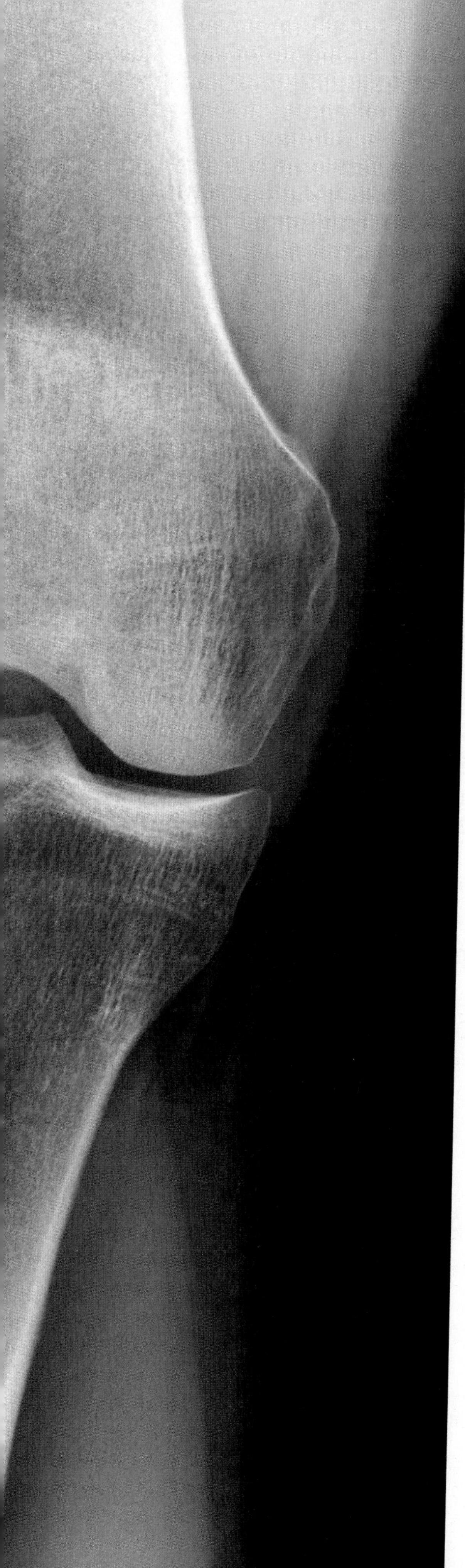

Contents

CHAPTER 5

CHAPTER 6

CHAPTER 7

CHAPTER 8

CHAPTER 9

CHAPTER 10

CHAPTER 11

CHAPTER 12

CHAPTER 13

CHAPTER 14

CHAPTER 15

CHAPTER 16

CHAPTER 17

CHAPTER 18

CHAPTER 19

CHAPTER 20

CHAPTER 21

CHAPTER 22

REPRODUCTIVE SYSTEM / 473

CHAPTER 23

GENETICS AND GENETICALLY LINKED DISEASES / 513

Spanish glossary available online

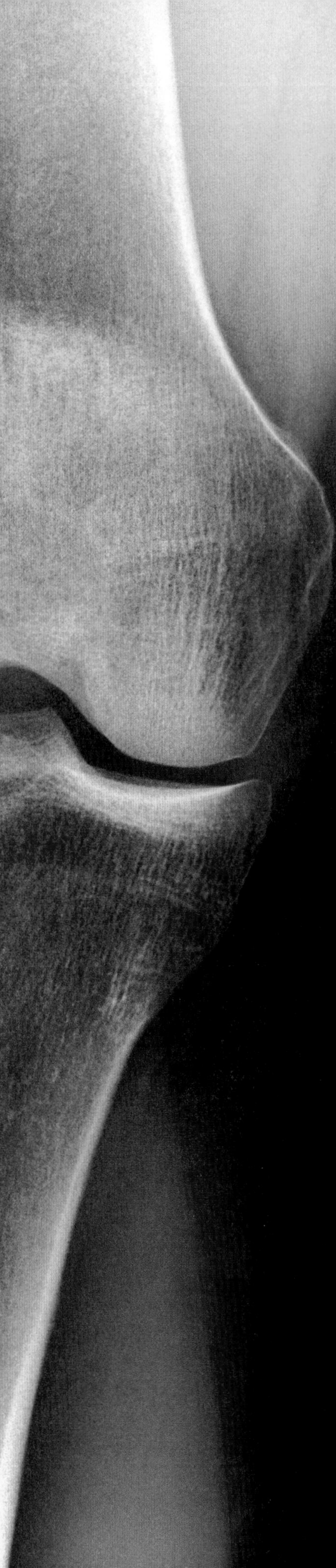

Preface

INTRODUCTION

The fourteenth edition of *Body Structures and Functions* has been revised to reflect the many changes that are occurring in today's health science and medical fields. The multiskilled health practitioner (MSHP) of today must know the structure and function of each body system as well as the common diseases. All disease and disorder content is integrated within each chapter as appropriate. *Body Structures and Functions*, together with *DHO Health Science* and *Medical Terminology*, provides a thorough introduction to the health sciences.

This book and the accompanying teaching materials are designed to facilitate learning. Review the introductory section "How to Study Using *Body Structures and Functions*."

KEY FEATURES

Key features in the Fourteenth Edition include the following:

- Phonetic pronunciations of key words are included in each chapter.
- The feature **One Body** outlines how each body system interacts with other body systems.
- The feature **Study Tools** directs learners to additional resources to enhance learning and assess mastery of the material.
- The new Chapter 1 discusses the foundations of science, proper lab practices, and how to communicate effectively as part of a scientific team.
- The **MindTap** platform provides easy access to the course materials online.
- New lab activities provide even more opportunities to investigate and apply chapter topics.

PHONETIC PRONUNCIATIONS OF KEY WORDS

Phonetic pronunciations of key words are included in each chapter in parentheses following the word. Pronounce the word by saying each syllable, placing more emphasis on the syllable in boldface capital letters. In the following example, the syllable *NAT* would receive more emphasis than the rest of the syllables would: anatomy (ah-**NAT**-oh-mee).

Most key word pronunciations contain only one syllable in boldface; however, some key words contain more than one. When a pronunciation contains more than one syllable in boldface, place *some* emphasis on the syllable in boldface *lowercase* letters and the *most* emphasis on the syllable in boldface *capital* letters. In the following example, the syllable *em* would receive some emphasis and the syllable *OL* would receive the most emphasis: embryology (**em**-bree-**OL**-oh-jee).

PRECISION EXAMS

This edition of *Body Structures and Functions* is aligned to Precision Exams' *Medical Anatomy and Physiology* exam. This exam is validated by industry, allowing students to earn a certification that connects skills taught in the classroom to a future career for a successful transition from high school to college and/or career. Working together, Precision Exams and National Geographic Learning, a part of Cengage, focus on preparing students for the workforce with exams and content that is kept up-to-date and relevant to today's jobs. To access a corresponding correlation guide, visit the accompanying Online Instructor Companion website for this title in NGLSync or at companion-sites.cengage.com. For more

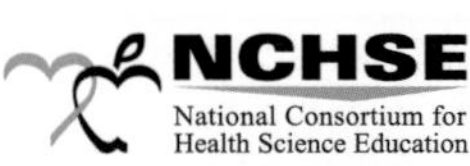

information on how to administer the *Medical Anatomy and Physiology* exam or to gain access to any of the 180+ Precision Exams, contact your local NGL/Cengage Sales Consultant. You can find your rep at ngl.cengage.com/repfinder.

MAJOR CHANGES TO THE FOURTEENTH EDITION

- Chapter 1: Scientific Foundations and Lab Practices—provides the history of science, details on the steps of the scientific method, proper lab practices and procedures, and how to communicate as a member of a scientific team.
- Chapter 2: Introduction to the Structural Units—includes new definitions for supine and prone as well as new information of the use of models and computers to reinforce learning objectives. There is updated information on biotechnology and nanotechnology in Medical Highlights.
- Chapter 3: Chemistry of Living Things—includes new information on homeostasis as it relates to acids and bases. New lab activities were also added to explore and investigate the structure of an atom and test for fats, proteins, and carbohydrates.
- Chapter 4: Cells—provides new information on congenital disorders relating to cell structure. This chapter also includes how the environment influences cell development, and new lab activities have been added to observe the effects of diffusion and active transport.
- Chapter 5: Tissues and Membranes—includes new information about the effects of congenital disorders and the environment on tissue and membrane development. There is updated information on tissue and organ transplant in Medical Highlights.
- Chapter 6: Integumentary System—includes new information on congenital deformities of the integumentary system as well as trauma and the environmental effects on the skin.
- Chapter 7: Skeletal System—includes a new lab activity to explore and describe joint movements. In addition, there is updated information in the Medical Highlights pertaining to damaged knees and hips, as well as new information on regenerative medicine. There is also new information on how trauma affects the skeletal system.
- Chapter 8: Muscular System—includes new information on torque, pressure, movement, and trauma. Force variance on the muscular system is also investigated in an all new lab activity.
- Chapter 9: Central Nervous System—includes new information on congenital nervous system disorders, motor neuron diseases, and how trauma affects the central nervous system. In addition, there is new information on the stages of Alzheimer's disease.
- Chapter 10: Peripheral Nervous System—includes new information on congenital disorders of the peripheral nervous system. In Medical Highlights, there is new information on monitored sedation.
- Chapter 11: Special Senses—provides new information on the congenital disorders of the eye, ear, nose, and tongue. There is also now treatment information provided for amblyopia and diplopia as well as how trauma can affect the special senses.
- Chapter 12: Endocrine System—provides new information about how trauma affects the endocrine system.
- Chapter 13: Blood—provides new information about how trauma affects blood.

- Chapter 14: Heart—provides new information about how trauma affects the heart.
- Chapter 16: The Lymphatic and Immune Systems—includes updated information on lymphedema and chemical responses. There is also new information on congenital disorders of the lymphatic system and the effect of trauma on the lymphatic system. In addition, a new lab activity explores how lymph fluid is drained from the body.
- Chapter 17: Infection Control and Standard Precautions—includes new information on parasitic infections, a new Medical Highlights explores COVID-19, and a new lab activity investigates the importance of contact tracing. There is updated information on MRSA, VRSA, VISA, and *C. diff.*
- Chapter 18: Respiratory System—includes new information on congenital disorders of the respiratory system and environmental causes of respiratory disease. There is also new information on trauma to the respiratory system and updated information in the Medical Highlights about sleep apnea technology and devices.
- Chapter 19: Digestive System—provides new information on congenital disorders of the digestive system and trauma's effect on the digestive system, as well as on the following conditions: canker sores, thrush, and salivary gland infection.
- Chapter 20: Nutrition—provides updates on binge eating and the treatment of eating disorders. There is new information on energy deficiencies and excesses, plant-based foods, and required nutrition labels on products.
- Chapter 21: Urinary System—includes new information on congenital anomalies of the kidneys and urinary system and how trauma can affect the urinary system.
- Chapter 22: Reproductive System—includes new information on congenital disorders of the female and male reproductive systems as well as the effect of trauma on the reproductive systems. There is also new information on ovarian cysts, polycystic ovarian syndrome (PCOS), Mycoplasma genitalium, and a new matching activity for assessment.
- Chapter 23: Genetics and Genetically Linked Diseases—provides updated information about the types of mutations as well as Down syndrome. In addition, there is new information on designer babies and stem cell treatments. The new lab activity explores genotype versus phenotype, common inherited traits, and investigates how common those traits are in the classroom setting.

MEDICAL HIGHLIGHTS

- Biotechnology and Nanotechnology (Chapter 2)
- Medical Imaging (Chapter 3)
- Stem Cells (Chapter 4)
- Tissue and Organ Transplant (Chapter 5)
- Hazards of the Sun (Chapter 6)
- RICE Treatment (Chapter 7)
- Procedures for Damaged Hips and Knees (Chapter 7)
- Orthopedic Diagnostic Technologies (Chapter 7)
- Massage Therapy and Health (Chapter 8)
- Specialized Brain Cells: Mirror Neurons (Chapter 9)
- Headaches (Chapter 9)
- Parkinson's Disease and Deep Brain Stimulation (Chapter 9)
- Types of Anesthesia (Chapter 10)
- Lasers (Chapter 11)
- Eye Surgery (Chapter 11)
- Hearing Aids (Chapter 11)
- Taste: Umami (Chapter 11)
- Hormone Imbalance: Mental Health (Chapter 12)
- Bone Marrow Transplant (Chapter 13)
- Diagnostic Tests for the Heart (Chapter 14)
- Pacemakers, Defibrillators, and Heart Pumps (Chapter 14)
- Mucosa-Associated Lymphoid Tissue (MALT) (Chapter 16)
- COVID-19 (Chapter 17)
- Changes Occurring in Infectious Diseases (Chapter 17)
- Sleep Apnea (Chapter 18)
- Pulmonary Function Tests (Chapter 18)
- Minimally Invasive Surgery: Laparoscopy (Chapter 19)

- Antioxidants (Chapter 20)
- Kidney Stone Removal (Chapter 21)
- Breast Self-Examination (Chapter 22)
- Testicular Self-Examination (Chapter 22)
- Treatment for Benign Prostatic Hypertrophy and Prostate Cancer (Chapter 22)
- Human Papillomavirus Vaccine (Chapter 22)

CAREER PROFILES

- Audiologist (Chapter 11)
- Cardiovascular Technologist and Technician/EKG Technician (Chapter 14)
- Certified Patient Care Technician (Chapter 16)
- Chiropractor (Chapter 8)
- Clinical Laboratory Technician and Clinical Laboratory Technologist (Chapter 13)
- Dental Hygienist, Dental Assistant, and Dental Laboratory Technician (Chapter 19)
- Dentist (Chapter 19)
- Dietitians and Nutritionist (Chapter 20)
- Electroneurodiagnostic Technician/EEG Technician (Chapter 9)
- Emergency Medical Technician and Paramedic (Chapter 14)
- Home Health Aide (Chapter 16)
- Licensed Practical Nurse (Chapter 15)
- Massage Therapist (Chapter 8)
- Medical Assistant (Chapter 22)
- Nursing Aide and Psychiatric Aide (Chapter 16)
- Optometrist and Dispensing Optician (Chapter 11)
- Orthotist and Prosthetist (Chapter 7)
- Physical Therapist and Physical Therapist Assistant (Chapter 7)
- Physician (Chapter 6)
- Radiologic Technologist (Chapter 3)
- Registered Nurse and Nurse Practitioner (Chapter 15)
- Respiratory Therapist (Chapter 18)
- Sports Medicine/Athletic Training (Chapter 8)

STUDENT WORKBOOK

The student workbook includes activities that focus on applied academics through practical application exercises, including multiple choice, fill in the blanks, matching, labeling, word puzzles, basic skill and theory to practice questions, and a Browse-the-Net feature.

THE ONLINE SOLUTION FOR CAREER AND TECHNICAL EDUCATION COURSES

CENGAGE | MINDTAP

MindTap for the fourteenth edition of *Body Structures and Functions* is the online learning solution for career and technical education courses that helps teachers engage and transform today's students into critical thinkers. Through paths of dynamic assignments and applications that you can personalize, real-time course analytics, and an interactive eBook, MindTap helps teachers organize and engage students. Whether you teach this course in the classroom or in hybrid/e-learning models, MindTap for *Body Structures and Functions* enhances the course experience with animations and videos; **Learning Lab Simulations**, which offer hands on decision making practice as students view real medical professionals working with patients; data analytics with engagement tracking; and student tools, such as flashcards, practice quizzes, auto-graded homework, and tests.

Teachers and students who have adopted MindTap can access their courses at nglsync.cengage.com.

ACCESS RESOURCES ONLINE, ANYTIME

Cognero®, Customizable Test Bank Generator is a flexible, online system that allows you to import, edit, and manipulate content from the text's test bank or elsewhere, including your own favorite test questions; create multiple test versions in an instant; and deliver tests from your learning management system, your classroom, or wherever you want.

Log on at nglsync.cengage.com, or companion-sites.cengage.com.

ABOUT THE AUTHORS

Ann Senisi Scott, RN, BS, MA, is the author of the thirteenth edition of *Body Structures and Functions*. Ann was previously the Coordinator of Health Occupations and Practical Nursing at Nassau Tech Board of Cooperative Education Services, Westbury, New York. As the Health Occupations Coordinator, she worked to establish a career ladder program from health care worker to practical nurse. Before becoming the administrator of these programs, she taught practical nursing for more than 12 years.

Tara Pullman Bell, M.Ed., is currently a CTE Campus Administrator for Cherry Creek School District in Greenwood Village, Colorado. She has a diverse background in multiple areas of the health sciences. Previously a member of the National Consortium for Health Science Education (NCHSE), Tara was also formerly the Health Science State Specialist for the Utah Board of Education, a health science teacher, HOSA Advisor, CTE coordinator, and the Program Director for Health Science and Public Safety for the Colorado Community College System.

ACKNOWLEDGMENTS

To complete this revision of the textbook, I have had the assistance of many people at Cengage Learning. I want to extend my thanks to Stefanie Boron, Gina Loverde Prokop, and Tara Pullman Bell for all their help in the preparation of this edition. You have provided so much encouragement and enthusiasm as you guided me through the fourteenth edition. Thanks to the reviewers who highlighted areas that needed additional information and corrections; your comments were invaluable. Thanks to Dr. Bryan Senisi, who assisted with some valuable insight into disease conditions.

In special memory of my husband, Wayne Scott, who was always my special mentor and reviewer through all the previous editions. In addition, special thanks to my family cheering section: Vincent, Margaret, Carolyn, Daniel, Michael, Kenneth, Leslie, Scotty, and their spouses.

To my grandchildren and future students: Have a love for learning, because it will bring much knowledge and rewards as you journey through life.

To the health care providers of tomorrow: Your knowledge will be an asset in the art of caring for the people entrusted to your care.

REVIEWERS

We are particularly grateful to the reviewers who continue to be a valuable resource in guiding this book as it evolves. Their insights, comments, suggestions, and attention to detail were very important in guiding the development of this textbook.

Dorothy J. Beaverson, RN, BSN
Health Science Instructor
Spencer County High School

Megan Duncan, MSN-Ed, RN
Nursing Education Specialist
NorthBay Healthcare

Melissa Legg, BSN, RN
Infection Control Coordinator
Medical Center of Aurora

Heather M. Matsuda, M. Ed., LAT
CTE Science
Clover Park School District

Ghadir Saidi, MS, CPhT
Program Coordinator/Instructor Pharmacy Technology
Southeastern College

Diana M. Sullivan, M. Ed.
Dental Program Director
Dakota County Technical College

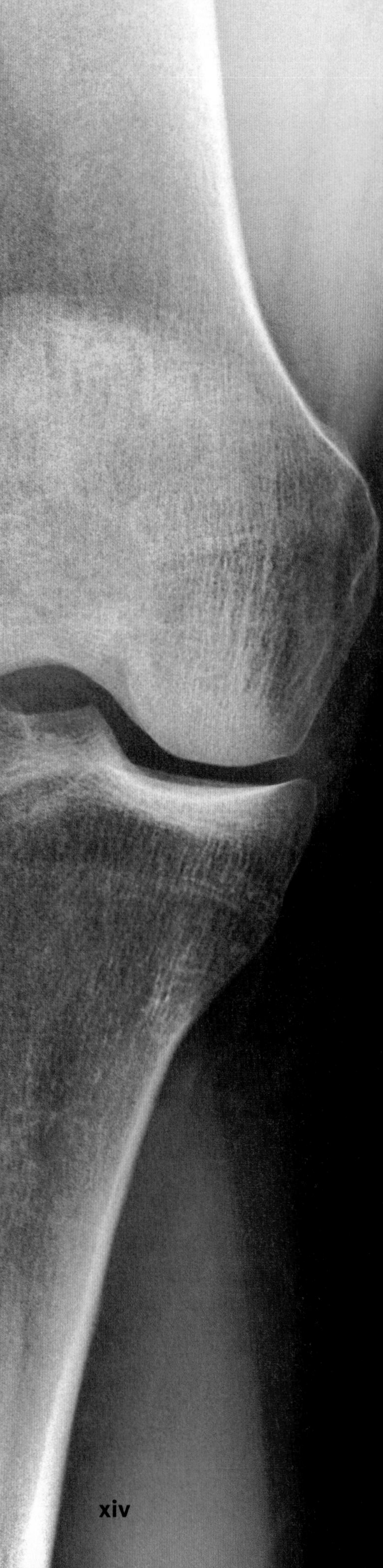

How to Study Using Body Structures and Functions

Preview the text before attempting to study the material covered in the individual chapters. By reviewing each section of this textbook, you will better understand its organization and purpose. Reading comprehension and long-term memory levels improve dramatically when you take the time to review the text and discover how it can help you learn. You can also review the chapters and use the online resources available through MindTap to learn and develop your understanding of the chapter topics.

To get the most from this course, take an active role in your learning by integrating your senses to increase your retention. You may want to

- *Visually* highlight important material.
- *Read* critically—turn headings, subheadings, and sentences into questions.
- *Recite* important material aloud to stimulate your auditory memory.
- *Draw* your own illustrations of anatomy or function processes and check them for accuracy.
- *Answer* (in writing or verbally) the review questions at the end of the chapter.

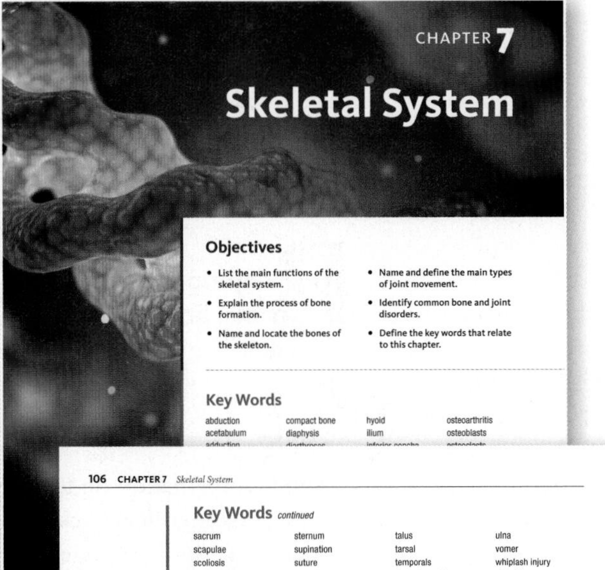

CHAPTER 7

Skeletal System

Objectives

- List the main functions of the skeletal system.
- Explain the process of bone formation.
- Name and locate the bones of the skeleton.
- Name and define the main types of joint movement.
- Identify common bone and joint disorders.
- Define the key words that relate to this chapter.

Key Words

abduction, acetabulum, compact bone, diaphysis, hyoid, ilium, osteoarthritis, osteoblasts

Each time you encounter a new chapter, preview it first to understand its overall structure. Review the **Objectives** presented at the beginning of each chapter to easily identify the key facts *before* you read the chapter. These objectives are also useful to review *after* you have completed a chapter. After reading a chapter, test yourself to see whether you can answer each objective. If you cannot, you will know exactly which areas to study again. The **Key Words** are listed at the beginning of each chapter, are highlighted in *red* within the chapter, and are also defined in the glossary.

Read the **main headings, subheadings,** and first sentence of each paragraph—these elements serve as the outline for the whole chapter. Be careful not to overlook the **illustrations, photographs,** and **tables,** which can help you comprehend difficult material.

Did You Know? boxes feature fun, interesting, trivialike facts to engage the learner.

Effects of Aging boxes are integrated within the chapters to highlight the changes that are associated with the body systems as we age.

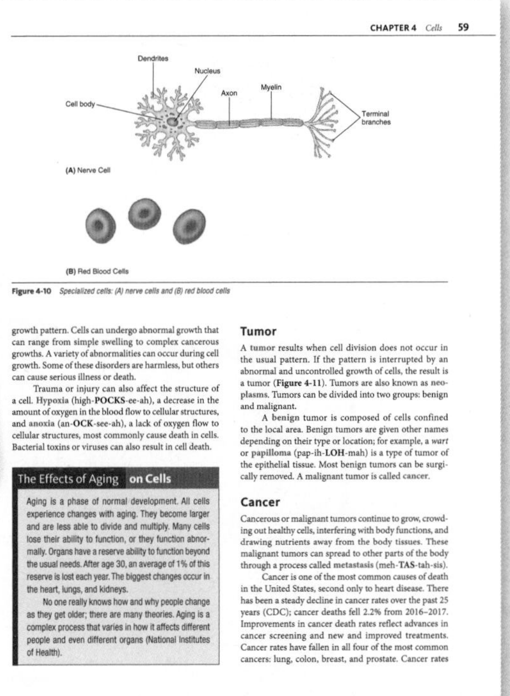

CHAPTER 4 Cells 59

Figure 4-10 Specialized cells: (A) nerve cells and (B) red blood cells

growth pattern. Cells can undergo abnormal growth that can range from simple swelling to complex cancerous growths. A variety of abnormalities can occur during cell growth. Some of these disorders are harmless, but others can cause serious illness or death.

Trauma or injury can also affect the structure of a cell. Hypoxia (high-POCKS-ee-ah), a decrease in the amount of oxygen in the blood flow to cellular structures, and anoxia (an-OCK-see-ah), a lack of oxygen flow to cellular structures, most commonly cause death in cells. Bacterial toxins or viruses can also result in cell death.

The Effects of Aging on Cells

Aging is a phase of normal development. All cells experience changes with aging. They become larger and are less able to divide and multiply. Many cells lose their ability to function, or they function abnormally. Organs have a reserve ability to function beyond the usual needs. After age 30, an average of 1% of this reserve is lost each year. The biggest changes occur in the heart, lungs, and kidneys.

No one really knows how and why people change as they get older; there are many theories. Aging is a complex process that varies in how it affects different people and even different organs (National Institutes of Health).

Tumor

A tumor results when cell division does not occur in the usual pattern. If the pattern is interrupted by an abnormal and uncontrolled growth of cells, the result is a tumor (Figure 4-11). Tumors are also known as neoplasms. Tumors can be divided into two groups: benign and malignant.

A benign tumor is composed of cells confined to the local area. Benign tumors are given other names depending on their type or location; for example, a *wart* or papilloma (pap-ih-LOH-mah) is a type of tumor of the epithelial tissue. Most benign tumors can be surgically removed. A malignant tumor is called cancer.

Cancer

Cancerous or malignant tumors continue to grow, crowding out healthy cells, interfering with body functions, and drawing nutrients away from the body tissues. These malignant tumors can spread to other parts of the body through a process called metastasis (meh-TAS-tah-sis).

Cancer is one of the most common causes of death in the United States, second only to heart disease. There has been a steady decline in cancer rates over the past 25 years (CDC); cancer deaths fell 2.2% from 2016–2017. Improvements in cancer death rates reflect advances in cancer screening and new and improved treatments. Cancer rates have fallen in all four of the most common cancers: lung, colon, breast, and prostate. Cancer rates

CASE STUDY

Your 80-year-old grandmother, Tess, fell off a step stool while putting dishes on a shelf and was unable to get up. She activated her medical lifeline, and the emergency medical team arrived at the scene. They noticed her right leg was abducted, and she was complaining of pain in her right leg and hip. Tess was taken to the emergency department, where an X-ray revealed that the neck of her right femur was fractured. Further X-rays revealed a reduced bone mass in her right hip, femur, and vertebrae. Surgery was done to repair the hip. Your grandmother is now recuperating and having physical therapy treatment daily.

1. What organ and body system were affected by the injury?
2. Name the type of tissue involved and the cells responsible for healing.
3. The physician says she will do an open reduction to repair Tess's hip. Explain the process of an open reduction.
4. What disease condition did the X-ray of the vertebrae reveal?
5. What is the significance of your grandmother's age and sex?
6. What other body systems may be affected by the fall?
7. What will the role of the physical therapist be in your grandmother's rehabilitation?
8. Name the test that might have revealed the disease condition.
9. What measures can be taken to prevent osteoporosis?
10. What limitations will your grandmoth... after her rehabilitation?

130 CHAPTER 7 *Skeletal System*

Medical Terminology

ab-	away from	kyph	humpback or hunchback
duc	move	-osis	process of
-tion	process	kyph/osis	process of being hunchbacked
ab/duc/tion	process of moving away from	lord	bending backward, swayback
ad-	to or toward	lord/osis	process of bending backward, inward curvature of the spine
ad/duc/tion	process of moving toward	meta	beyond
arthr	joint	meta/carp/al	pertaining to beyond the wrist, bones of the palm of the hand
-itis	inflammation	tars	ankle
arthr/itis	inflammation of a joint	meta/tars/al	pertaining to beyond the ankle bones of the sole of the foot
burs	small purselike sac	osteo/arthr/itis	inflammation of the joint
burs/itis	inflammation of a small sac	poro	pores in the bone
carp	wrist	-sis	abnormal condition
-al	pertaining to	osteo/poro/sis	abnormal condition of pores in the bone
carp/al	pertaining to the wrist	peri-	around
circum	around	peri/oste/um	presence of lining around the bone
circum/duc/tion	process of moving around	pronat	placing face down
end-	within	pronat/ion	process of being face down
oste	bone	rheumat	painful changes in the joints
-um	presence of	-oid	resembling
end/oste/um	presence of lining within the bone	rheumat/oid	resembling painful changes in the joints
extens	straightening	supinat	placing on the back
-ion	process of	supinat/ion	process of placing on the back
extens/ion	process of straightening		
	bend		
	process of bending		

Case Studies promote a real-world view of medical careers and encourage critical thinking.

Medical Terminology boxes introduce you to common medical prefixes and suffixes and how they work to form medical terms.

Career Profiles provide descriptions of many health professions in today's dynamic health and medical environment. These profiles describe the role of each professional, and may even provide you with insight into possible future career paths.

Medical Highlights provide information on technology, innovations, discoveries, and bioethical issues in research and medicine. These topics are based on current information obtained from research on various medical websites.

Review Questions will help you measure whether you have mastered the material you have covered. Questions in a variety of formats are presented to reinforce important information within each chapter. Also integrated here and in the workbook are applied academic activities for math, spelling, communication, and legal-ethical issues.

CHAPTER 7 *Skeletal System* 129

Career Profile

ORTHOTIST AND PROSTHETIST

Orthotists design, manufacture, and fit braces and other orthopedic devices such as artificial limbs, or prostheses.

Prosthetists measure, design, and fit or service prosthetic devices as prescribed by a physician.

Both of these fields require a master's degree in orthotics and prosthetics and a one-year residency program. After completion of the residency program, the candidate must pass a certification exam given by the American Board for Certification in Orthotics, Prosthetics, and Pedorthics. Employment is in hospitals, medical supply houses, and physicians' offices. The job outlook is expected to rise about 12% annually.

124 CHAPTER 7 *Skeletal System*

A hammertoe is a toe that is curled, or flexed, due to a bend in the middle joint of one or more toes. It may be caused by shoes that are too tight or heels that are too high. The longest of the four smaller toes may be forced against the front of the shoe, resulting in an unnatural bending of the toe with pain and pressure in the affected area. Proper footwear with a deep toe box and flexible material covering the toes may be helpful, as can using a special device ordered by a physician to wear in the shoe to help position the toe.

A whiplash injury is trauma to the cervical vertebrae, usually the result of an automobile accident. The force generated by a car's abrupt change in speed or direction whips the head, putting tremendous strain on the cervical spine and neck muscles. Treatment includes bracing, heat or cold, pain medication, and physical therapy. The goal of treatment is to alleviate pain and restore a normal range of motion; recovery time may be up to 6 weeks.

DISEASES OF THE BONES

Diseases of the Bones can be congenital, ranging from innocuous to severe, or they can evolve as a person begins to age.

Congenital Bone Disorders

The following are the common congenital disorders of the skeletal system:

- *Limb abnormalities:* This could be the absence of fingers or toes, or the partial or complete absence of an arm or leg.
- *Club foot:* The front part of the foot turns toward the inside of the heel. Treatment is the use of a cast or boot and physical therapy.
- *Congenital hip dislocation:* The hip and femur are underdeveloped, which leads to a dislocation of the hip. Treatment is the use of splints and sometimes surgery to allow the hip to develop properly.

Medical Highlights

RICE TREATMENT

RICE is the acronym for rest, ice, compression, and elevation, the recommended immediate treatment for bone, joint, and muscle injuries. Treatment that occurs in the first 24 to 72 hours after an injury can do a lot to relieve or even prevent aches and pains.

R = REST

Injuries heal faster if rested. Rest means staying off the injured body part. Using any part of the body increases the blood circulating to that area, which can cause more swelling of an injured part. In the case of an ankle sprain, there should be no weight bearing for at least the first 24 hours.

I = ICE

An ice pack should be applied to the injured area as soon as possible after the injury. Apply for 20 minutes every 4 hours during the first 48 hours. Skin treated with cold passes through four stages: cold, burning, aching, and numbness. Do not ice for more than 20 minutes at a time.

C = COMPRESSION

Compressing the injured area may squeeze some fluid and debris out of the injured area. Compression limits the ability of the skin and other tissues to expand and reduces internal bleeding. Apply an elastic bandage to the injured area, especially the foot, ankle, knee, thigh, hand, or elbow. Fill in the hollow areas with padding such as a washcloth or sock before applying the elastic bandage.

Caution: DO NOT apply an elastic bandage too tightly because this may restrict circulation. Leave fingers or toes exposed so possible skin color change due to restricted circulation can be observed. Compare the injured side with the uninjured side. Pale skin, numbness, pain, and tingling are signs of impaired circulation. Remove the elastic bandage immediately if any of these signs appear.

E = ELEVATION

Gravity slows the return of blood to the heart from the lower parts of the body. Once fluid gets to the hands or feet, the fluid has nowhere to go, and those parts of the body swell. Elevating the injured part, in combination with ice and compression, limits circulation to that area, which in turn helps limit internal bleeding and minimize swelling.

Source: http://athletics.mckenna.edu/sportsmedicine/rice.html; © Cengage Learning 2014

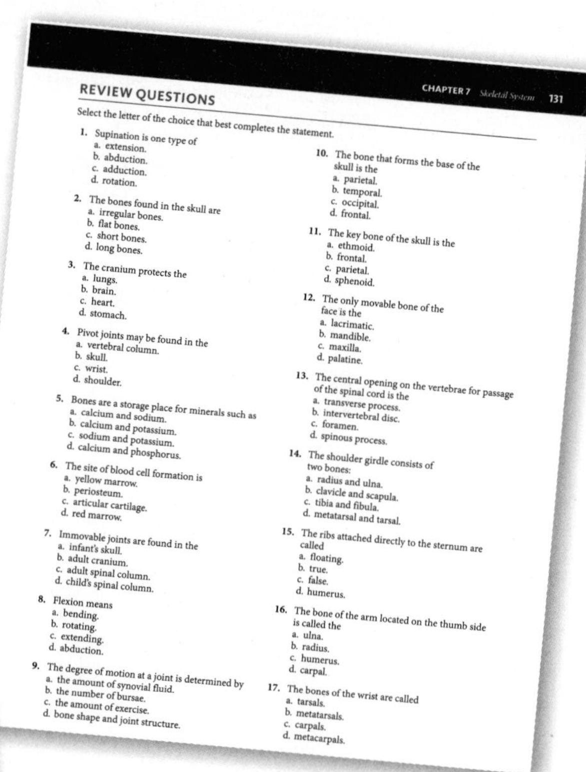

CHAPTER 7 *Skeletal System* 131

REVIEW QUESTIONS

Select the letter of the choice that best completes the statement.

1. Supination is one type of
   a. extension.
   b. abduction.
   c. adduction.
   d. rotation.
2. The bones found in the skull are
   a. irregular bones.
   b. flat bones.
   c. short bones.
   d. long bones.
3. The cranium protects the
   a. lungs.
   b. brain.
   c. heart.
   d. stomach.
4. Pivot joints may be found in the
   a. vertebral column.
   b. skull.
   c. wrist.
   d. shoulder.
5. Bones are a storage place for minerals such as
   a. calcium and sodium.
   b. calcium and potassium.
   c. sodium and potassium.
   d. calcium and phosphorus.
6. The site of blood cell formation is
   a. yellow marrow.
   b. periosteum.
   c. articular cartilage.
   d. red marrow.
7. Immovable joints are found in the
   a. infant's skull.
   b. adult cranium.
   c. adult spinal column.
   d. child's spinal column.
8. Flexion means
   a. bending.
   b. rotating.
   c. extending.
   d. abduction.
9. The degree of motion at a joint is determined by
   a. the amount of synovial fluid.
   b. the number of bursae.
   c. the amount of exercise.
   d. bone shape and joint structure.
10. The bone that forms the base of the skull is the
    a. parietal.
    b. temporal.
    c. occipital.
    d. frontal.
11. The key bone of the skull is the
    a. ethmoid.
    b. frontal.
    c. parietal.
    d. sphenoid.
12. The only movable bone of the face is the
    a. lacrimatic.
    b. mandible.
    c. maxilla.
    d. palatine.
13. The central opening on the vertebrae for passage of the spinal cord is the
    a. transverse process.
    b. intervertebral disc.
    c. foramen.
    d. spinous process.
14. The shoulder girdle consists of two bones:
    a. radius and ulna.
    b. clavicle and scapula.
    c. tibia and fibula.
    d. metatarsal and tarsal.
15. The ribs attached directly to the sternum are called
    a. floating.
    b. true.
    c. false.
    d. humerus.
16. The bone of the arm located on the thumb side is called the
    a. ulna.
    b. radius.
    c. humerus.
    d. carpal.
17. The bones of the wrist are called
    a. tarsals.
    b. metatarsals.
    c. carpals.
    d. metacarpals.

Lab Activities incorporate an element of interactivity to the content, further enhancing comprehension.

Phonetic pronunciations of key words in each chapter are in parentheses following the key word. Pronounce the word by saying each syllable, placing more emphasis on the syllable in boldface capital letters.

Media Links direct you to Online Resources that include PowerPoint presentations and 3D animations.

Study Tools alert you to additional resources to help you understand the material.

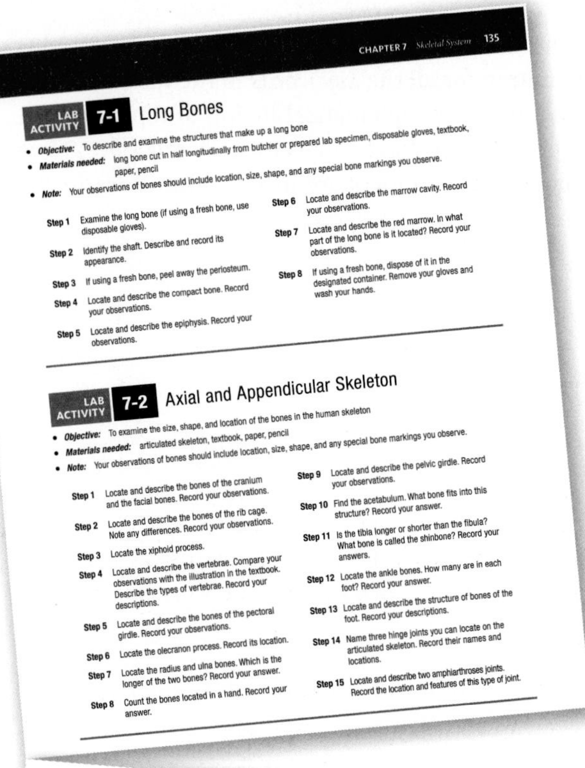

LAB ACTIVITY **7-1** Long Bones

- **Objective:** To describe and examine the structures that make up a long bone
- **Materials needed:** long bone cut in half longitudinally from butcher or prepared lab specimen, disposable gloves, textbook, paper, pencil
- **Note:** Your observations of bones should include location, size, shape, and any special bone markings you observe.

Step 1 Examine the long bone (if using a fresh bone, use disposable gloves).
Step 2 Identify the shaft. Describe and record its appearance.
Step 3 If using a fresh bone, peel away the periosteum.
Step 4 Locate and describe the compact bone. Record your observations.
Step 5 Locate and describe the epiphysis. Record your observations.
Step 6 Locate and describe the marrow cavity. Record your observations.
Step 7 Locate and describe the red marrow. In what part of the long bone is it located? Record your observations.
Step 8 If using a fresh bone, dispose of it in the designated container. Remove your gloves and wash your hands.

LAB ACTIVITY **7-2** Axial and Appendicular Skeleton

- **Objective:** To examine the size, shape, and location of the bones in the human skeleton
- **Materials needed:** articulated skeleton, textbook, paper, pencil
- **Note:** Your observations of bones should include location, size, shape, and any special bone markings you observe.

Step 1 Locate and describe the bones of the cranium and the facial bones. Record your observations.
Step 2 Locate and describe the bones of the rib cage. Note any differences. Record your observations.
Step 3 Locate the xiphoid process.
Step 4 Locate and describe the vertebrae. Compare your observations with the illustration in the textbook. Describe the types of vertebrae. Record your descriptions.
Step 5 Locate and describe the bones of the pectoral girdle. Record your observations.
Step 6 Locate the olecranon process. Record its location.
Step 7 Locate the radius and ulna bones. Which is the longer of the two bones? Record your answer.
Step 8 Count the bones located in a hand. Record your answer.
Step 9 Locate and describe the pelvic girdle. Record your observations.
Step 10 Find the acetabulum. What bone fits into this structure? Record your answer.
Step 11 Is the tibia longer or shorter than the fibula? What bone is called the shinbone? Record your answers.
Step 12 Locate the ankle bones. How many are in each foot? Record your answer.
Step 13 Locate and describe the structure of bones of the foot. Record your descriptions.
Step 14 Name three hinge joints you can locate on the articulated skeleton. Record their names and locations.
Step 15 Locate and describe two amphiarthroses joints. Record the location and features of this type of joint.

ANATOMY AND PHYSIOLOGY

Anatomy and physiology are branches of a much larger science called biology (bye-**OL**-oh-jee). Biology is the study of all forms of life. Biology studies microscopic one-celled organisms, multicelled organisms, plants, animals, and humans.

Anatomy (ah-**NAT**-oh-mee) studies the shape and structure of an organism's body and the relationship of one body part to another. The word *anatomy* comes from the Greek *ana*, meaning "apart," and *temuein*, "to cut"; thus, the acquisition of knowledge on human anatomy comes basically from dissection. However, one cannot fully appreciate and understand anatomy without the study of its sister science, physiology (fiz-ee-**OL**-oh-jee). Physiology studies the function of each body part and how the functions of the various body parts coordinate to form a complete living organism. Any abnormal change in a structure or function that produces symptoms is considered a disease (diz-**EASE**).

Branches of Anatomy

Anatomy is subdivided into many branches based on the investigative techniques used, the type of knowledge desired, or the parts of the body under study.

1. **Gross anatomy** is the study of large and easily observable structures on an organism. This is done through dissection and visible inspection with the naked eye. In gross anatomy, the different body parts and regions are studied with regard to their general shape, external features, and main divisions.
2. **Microscopic anatomy** refers to the use of microscopes to enable one to see the minute details of organ parts. Microscopic anatomy is subdivided into two branches. One branch is *cytology* (sigh-**TOL**-oh-jee), which is the study of the structure, function, and development of cells that comprise the different body parts. The other subdivision is *histology* (hiss-**TOL**-oh-jee), which studies the tissues and organs that make up the entire body of an organism.
3. **Developmental anatomy** studies the growth and development of an organism from fertilization to maturity, also called embryology (em-bree-**OL**-oh-jee).
4. **Comparative anatomy** is when the different body parts and organs of humans can be studied with regard to similarities to and differences from others in the animal kingdom. Humans are just one of the many animals found in the animal kingdom.
5. **Systematic anatomy** is the study of the structure of various organs or parts that comprise a particular organ system. Depending on the particular organ system under study, a specific term is applied; for example:
   a. **Dermatology** (der-mah-**TOL**-oh-jee)—study of the integumentary system (skin, hair, and nails)
   b. **Endocrinology** (en-doh-krin-**OL**-oh-jee)—study of the endocrine or hormonal system
   c. **Neurology** (new-**ROL**-oh-jee)—study of the nervous system

ANATOMICAL TERMINOLOGY

In the study of anatomy and physiology, special words are used to describe the specific location of a structure or organ, or the relative position or direction of one body part to another. The initial reference point used is the anatomical position. In the **anatomical position**, a human being is standing erect, with face forward, arms at the side, and palms forward (**Figure 2-1**).

Terms Referring to Location or Position and Direction

See **Figure 2-2** and **Figure 2-3**.

- **Anterior** or **ventral** means "front" or "in front of." For example, the knees are located on the anterior surface of the human body. A ventral hernia may protrude from the front or belly of the abdomen.
- **Posterior** or **dorsal** means "back" or "in back of." For example, human shoulder blades are found on the posterior surface of the body.
- **Supine** (su-**PINE**) refers to lying on one's back, with face and torso facing up.
- **Prone** (pron) refers to lying flat, with face and chest down.
- **Cephalic** (seh-**FAL**-ick) and **caudal** (**KAWD**-al) refer to direction: Cephalic means "skull" or "head end" of the body; caudal means "tail end." For example, a blow to the skull is a cephalic injury that may increase cranial pressure and cause headaches. Caudal anesthesia is injected in the lower spine.
- **Superior** means "upper" or "above another," and **inferior** refers to "lower" or "below another."

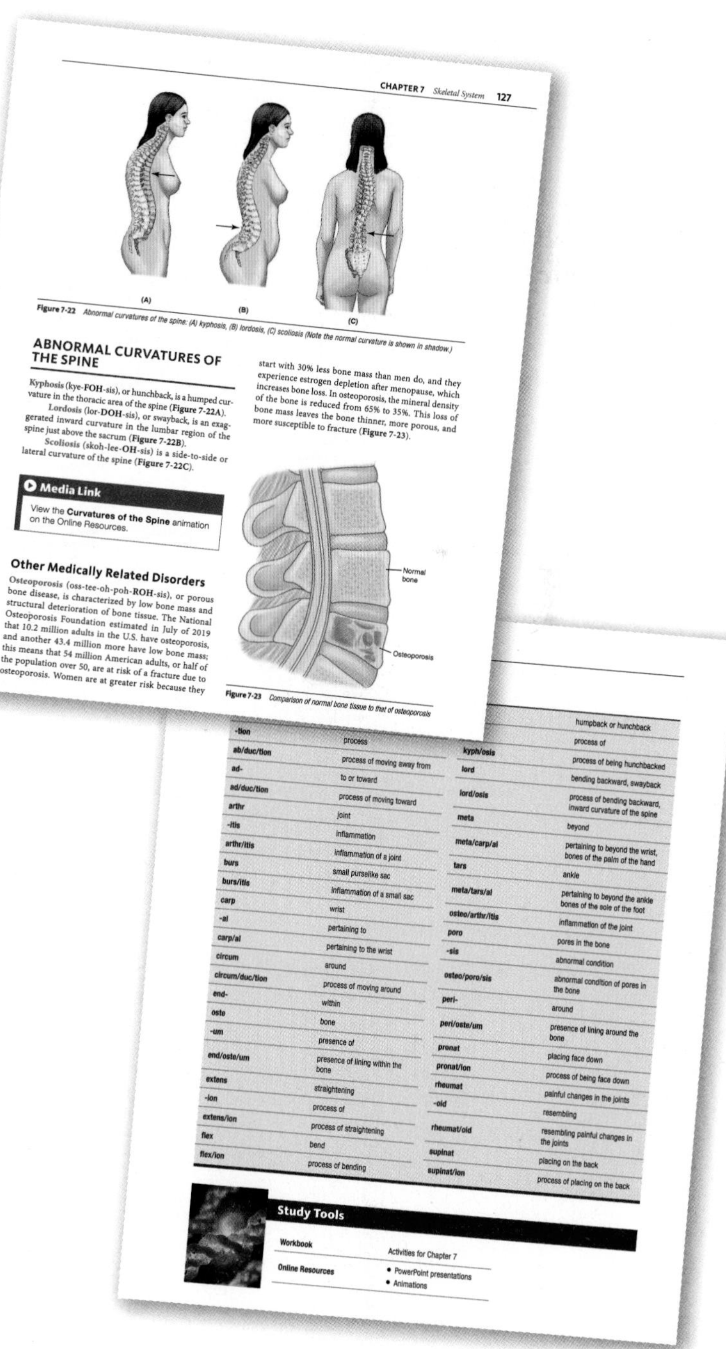

Figure 7-22 *Abnormal curvatures of the spine: (A) kyphosis, (B) lordosis, (C) scoliosis (Note the normal curvature is shown in shadow.)*

ABNORMAL CURVATURES OF THE SPINE

Kyphosis (kye-**FOH**-sis), or hunchback, is a humped curvature in the thoracic area of the spine (**Figure 7-22A**).

Lordosis (lor-**DOH**-sis), or swayback, is an exaggerated inward curvature in the lumbar region of the spine just above the sacrum (**Figure 7-22B**).

Scoliosis (skoh-lee-**OH**-sis) is a side-to-side or lateral curvature of the spine (**Figure 7-22C**).

Media Link

View the **Curvatures of the Spine** animation on the Online Resources.

Other Medically Related Disorders

Osteoporosis (oss-tee-oh-poh-**ROH**-sis), or porous bone disease, is characterized by low bone mass and structural deterioration of bone tissue. The National Osteoporosis Foundation estimated in July of 2019 that 10.2 million adults in the U.S. have osteoporosis, and another 43.4 million more have low bone mass; this means that 54 million American adults, or half of the population over 50, are at risk of a fracture due to osteoporosis. Women are at greater risk because they start with 30% less bone mass than men do, and they experience estrogen depletion after menopause, which increases bone loss. In osteoporosis, the mineral density of the bone is reduced from 65% to 35%. This loss of bone mass leaves the bone thinner, more porous, and more susceptible to fracture (**Figure 7-23**).

Figure 7-23 *Comparison of normal bone tissue to that of osteoporosis*

-tion	process		humpback or hunchback
ab/duc/tion	process of moving away from		process of
ad-	to or toward	**kyph/osis**	process of being hunchbacked
ad/duc/tion	process of moving toward	**lord**	bending backward, swayback
arthr	joint	**lord/osis**	process of bending backward, inward curvature of the spine
-itis	inflammation	**meta**	beyond
arthr/itis	inflammation of a joint	**meta/carp/al**	pertaining to beyond the wrist, bones of the palm of the hand
burs	small purselike sac	**tars**	ankle
burs/itis	inflammation of a small sac	**meta/tars/al**	pertaining to beyond the ankle bones of the sole of the foot
carp	wrist	**osteo/arthr/itis**	inflammation of the joint
-al	pertaining to	**poro**	pores in the bone
carp/al	pertaining to the wrist	**-sis**	abnormal condition
circum	around	**osteo/poro/sis**	abnormal condition of pores in the bone
circum/duc/tion	process of moving around	**peri-**	around
end-	within	**peri/oste/um**	presence of lining around the bone
oste	bone	**pronat**	placing face down
-um	presence of	**pronat/ion**	process of being face down
end/oste/um	presence of lining within the bone	**rheumat**	painful changes in the joints
extens	straightening	**-oid**	resembling
-ion	process of	**rheumat/oid**	resembling painful changes in the joints
extens/ion	process of straightening	**supinat**	placing on the back
flex	bend	**supinat/ion**	process of placing on the back
flex/ion	process of bending		

Study Tools

Workbook	Activities for Chapter 7
Online Resources	• PowerPoint presentations • Animations

The **Glossary of Terms** provides you with a concise definition for all the *key words* in the textbook. The **Index** serves as an alphabetical listing of topics, terms, concepts, and important names for easy reference. Note that figure page numbers are listed in **boldface** in the index.

DIVERSITY IN EDUCATION

Representing all body types, sexes, and skin tones is an important part of teaching how to care for all patients. No one person is alike, and health care providers will treat a wide range of patients. To accomplish this, health care providers must be aware of how various diseases have unique presentations depending on the patient.

Urticaria, or hives, is a noncontagious skin condition that presents as itchy wheals or welts. On light skin, these welts often appear pink and have a white center. On dark skin, they often appear purple with a light or skin-toned center.

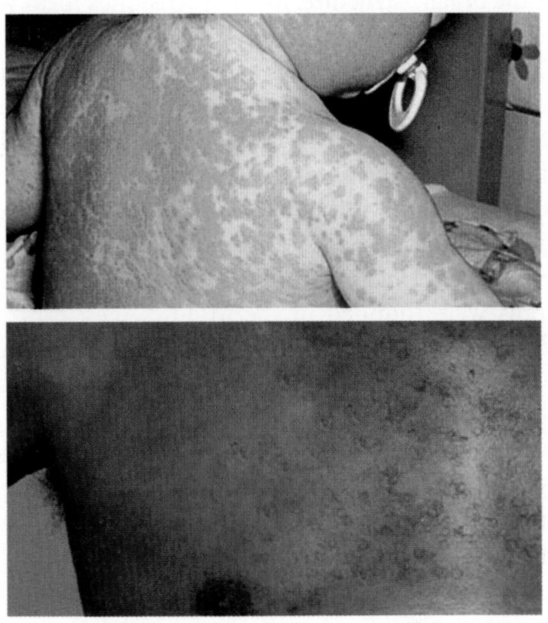

Eczema is an acute, noncontagious inflammatory disease of the skin. The skin becomes, dry, itchy, and scaly. It may appear red on light skin; on dark skin, it may appear dark brown, purple, or gray.

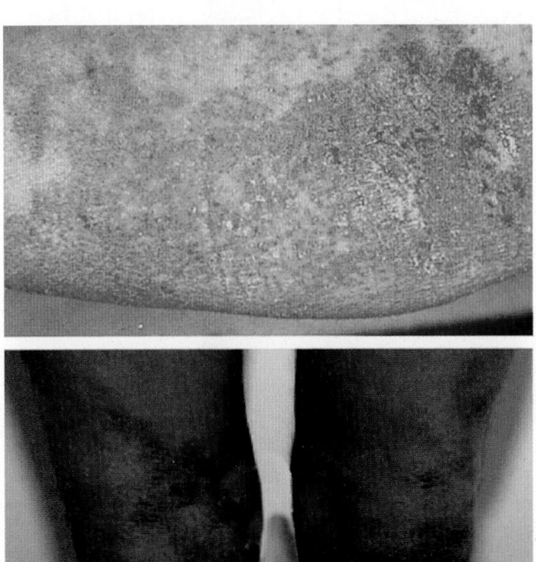

Updated drawings for this edition were designed to reflect gender, age, and race.

  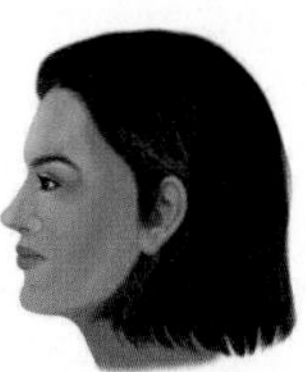   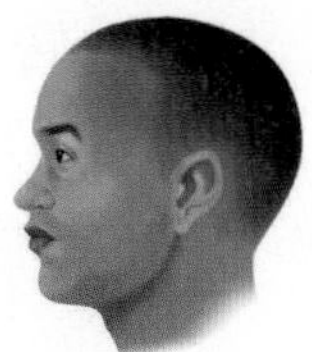

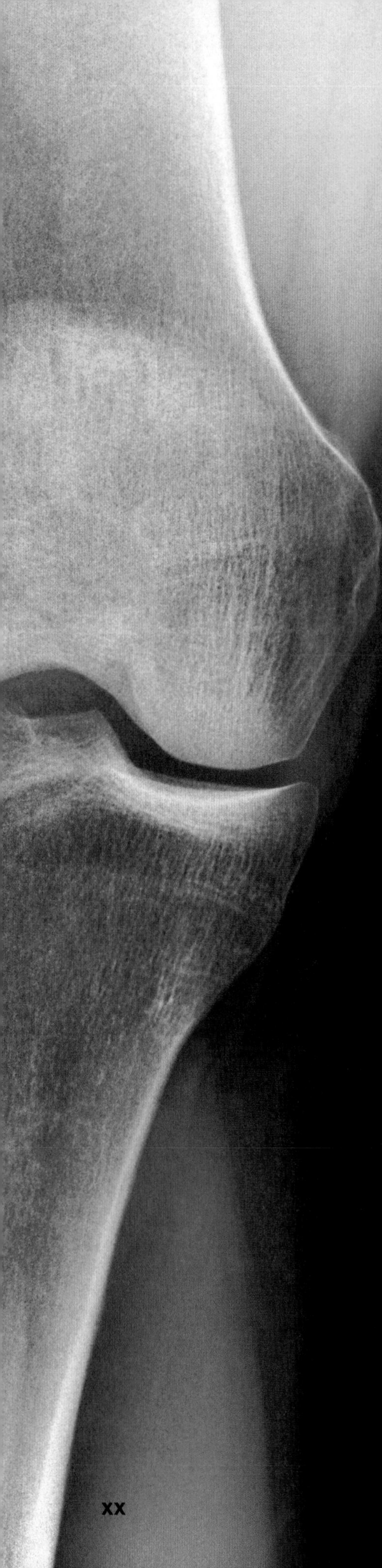

Prologue

THE HISTORY OF ANATOMICAL SCIENCE AND SCIENTISTS

Much of the early study of gross anatomy and physiology comes from Aristotle, a Greek philosopher. Aristotle believed that every organ has a specific function and that function is based on the organ's structure. Most of Aristotle's ideas were based on the dissection of plants and animals. He never dissected a human body.

In the third century BC, Herophilus founded the first school of anatomy and encouraged the dissection of the human body. He is credited with demonstrating that the brain is the center of the nervous system. It was a Greek physician, Galen, however, who is credited with the creation of the first standard medical text expanding on Aristotle's ideas. Galen was the first to discover many muscles and the first to find the value in monitoring an individual's pulse. Galen never performed human dissections, and many of his theories were later proven wrong.

The first medical schools were founded in the Middle Ages; however, instructors at this time were hesitant to question the theories and beliefs founded by the early Greeks such as Aristotle and Galen. As a result, very few ideas or discoveries were made in the medical field in the Middle Ages.

During the Renaissance, however, interest in anatomy was renewed due in part to the work of artist Leonardo da Vinci, who studied the form and function of the human body. It was during this period in history that the first systematic study of the structure of the human body was made. Many of these early scientists were hindered in their pursuit of knowledge of the human body because it was believed by many that human dissections were immoral and illegal. For example, Andreas Vesalius, a founder of modern anatomy, was sentenced to death because of his anatomical dissections of humans.

In the seventeenth century, the invention of the microscope aided in new anatomical discoveries and research. Scientists could now see structures that were invisible to the naked eye. Robert Hooke's investigation of cork under the microscope was the foundation of the theory that the cell is the basic unit of life. This theory was later proved and expanded on by other scientists in the eighteenth century as technological advances continued to improve.

Advances in technology have continued into today, and new anatomical and physiological discoveries are still being made. With the mapping of the human genome, completed in 2003, the complete genetic code has been documented. It is hoped that this knowledge will enable discoveries into disease processes and the development of cures for many of the diseases that continue to plague our society.

The use of new types of medical imaging, such as computerized scanning and digitalized photography, has helped researchers make new discoveries about the body.

Use key words to search the Internet for new discoveries related to a particular body system and the scientists who made those discoveries.

CHAPTER 1

Scientific Foundations and Lab Practices

Objectives

- Research and describe scientific history and the contributions of scientists.
- Evaluate the impact of scientific research on society and the environment.
- Conduct investigations using safe, environmentally appropriate, and ethical practices.
- Apply the scientific method during laboratory and field investigations.
- Choose equipment during laboratory and field investigations.
- Analyze and organize data during laboratory and field investigations.
- Use critical thinking, scientific reasoning, and problem solving to make informed decisions.
- Demonstrate professional standards and employability skills.

Key Words

Age of Enlightenment
comparative investigation
conclusion/results
control group
deductive reasoning
dependent variable
descriptive investigation
empathy
empirical evidence
experiment
experimental group
experimental investigation
feedback
hypothesis
independent variable
inductive reasoning
listening
logical reasoning
nonverbal communication
observation
observational investigation
qualitative data
quantitative data
question
Safety Data Sheets (SDS)
science
scientific method
scientific models
Scientific Revolution
scientific theories
verbal communication
written communication

SCIENCE

Since the first *Homo sapiens* walked the earth, humans have searched for ways to classify and categorize the natural phenomenon of what is observable in nature. Observation, prediction, and explanation are a part of human nature. Humans are naturally inquisitive and want to understand what is happening in the world around them.

Science is the study of what is observable and what can be tested through experimentation. Scientific investigations have allowed discoveries of the natural world and the basic mechanisms of observations, such as the universe and what it contains and how the sun and moon affect seasons, tides, and other natural phenomena.

History of Science

The term "science" is relatively new, being coined by William Whewell in 1834; however, scientific concepts can be traced back millions of years. Science stems from the Latin word *Scientia*, meaning "knowledge." The earliest scientific breakthroughs can be traced back millions of years to the ancestors of *Homo sapiens*. The first tools were made more than three million years ago, and fire was discovered more than one million years ago. *Homo sapiens* discovered the science of agriculture around 15,000 BCE.

THE SCIENTIFIC REVOLUTION

Scientific discoveries can be documented in periods of time. The **Scientific Revolution** began toward the end of the Renaissance period and continued though the late seventeenth century; it saw a major transformation in scientific thought and ideas across multiple disciplines that led to the emergence of modern science (**Table 1-1**). These developments were influenced by early Greek philosophers such as Socrates, Plato, and Aristotle; these philosophers laid the foundation of physics, biology, zoology, ethics, and more. See **Figure 1-1** and **Figure 1-2** for examples of scientific exploration during the Scientific Revolution.

Table 1-1 *Notable Scientists During the Scientific Revolution (1543–1687)*

SCIENTIST	CONTRIBUTION
Francis Bacon	Created the scientific method in which scientists conduct experiments by testing hypotheses
Nicolaus Copernicus	Developed the Copernican heliocentrism, an astronomical model that depicted the sun as the center of the universe and the planets orbiting the sun
Leonardo da Vinci	Made discoveries in anatomy and engineering with foresight to modern engineering designs
Galileo Galilei	Built telescopes to view the sky and was the first to explore the idea of inertia
William Harvey	First to describe the circulatory system and how blood is pumped through the body by the heart
Isaac Newton	Notable mathematics, physics, and astronomy discoveries; built the first reflecting telescope and devised the laws of motion and gravity
Andreas Vesalius	Author and professor who was the first to describe human anatomy by dissection

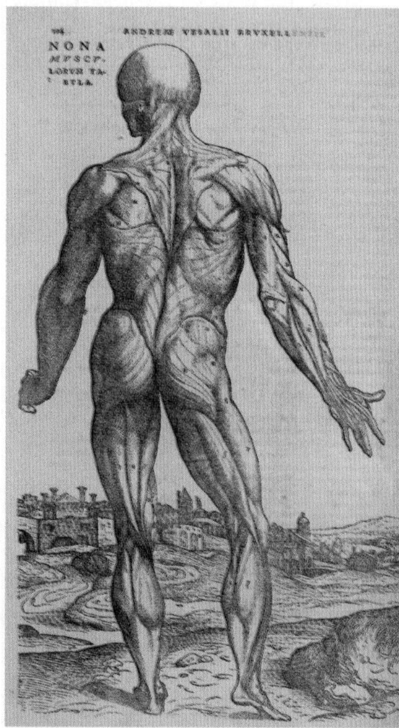

Figure 1-1 *Vesalius woodcut from* On the Structure of the Human Body, *1543*

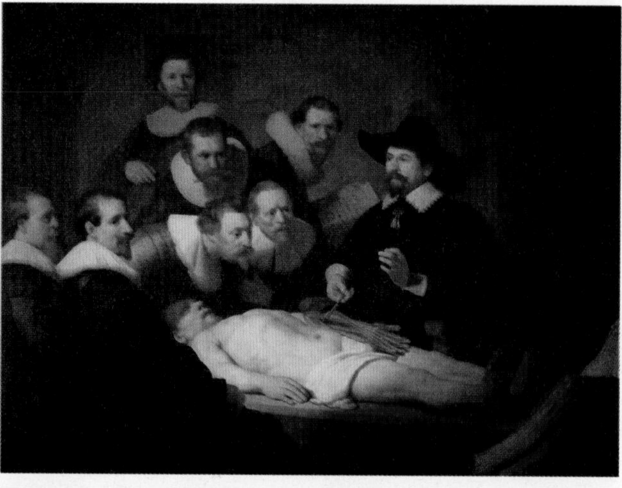

Figure 1-2 *Rembrandt painting illustrating human dissection in 1632*

AGE OF ENLIGHTENMENT

Science in the **Age of Enlightenment**, also known as the Age of Reason, lasted until the late eighteenth century. This period gives rise to scientific societies and academies and saw advancements in medicine, chemistry, taxonomy, mathematics, and physics (**Table 1-2** and **Figure 1-3**).

Figure 1-3 *Nineteenth century engraving of Benjamin Franklin examining the effects of electrical attraction and repulsion*

Table 1-2 *Notable Scientists During the Age of Enlightenment (1715–1789)*

SCIENTIST	CONTRIBUTION
Henry Cavendish	Researched the composition of atmospheric air, the density of gases, and discovered hydrogen; conducted an experiment to measure the density of Earth, known as the Cavendish experiment
Benjamin Franklin	Conducted experiments around electricity and invented the lightning rod, bifocals, and the Franklin stove; researched ocean currents, meteorology, and causes of illness
William Herschel	Discovered Uranus and its moons, nebulae, and infrared radiation in sunlight
James Hutton	Referred to as the father of modern geology; discovered that features of the earth's crust changed by natural processes over a long period of time
Carl Linnaeus	Recognized as the father of modern taxonomy; established the system of naming organisms, known as binomial nomenclature
Joseph Priestley	Contributed to the chemistry of gases; invented carbonated water and discovered oxygen
Daniel Rutherford	Discovered nitrogen gas

NINETEENTH CENTURY

The nineteenth century saw a great amount of scientific progress (**Table 1-3** and **Figure 1-4**). Science became a known profession; previously, it was just considered an interest. There were many influential discoveries that occurred in this century that established current scientific philosophies and reasoning.

Figure 1-4 *Wood engraving of Louis Pasteur experimenting with vaccinations on a rabbit*

Table 1-3 *Notable Scientists During the Nineteenth Century*

SCIENTIST	CONTRIBUTION
Alexander Graham Bell	Experimented with electricity to convey sound; invented the first telephone and founded AT&T (American Telephone and Telegraph Company)
Marie Curie	Pioneered work in radioactivity by isolating isotopes; discovered polonium and radium
John Dalton	Introduced atomic theory into chemistry; researched color blindness
Charles Darwin	Introduced the scientific theory that evolution resulted from natural selection; influential author of *On the Origin of Species*
Gregor Mendel	Established the rules of inheritance through experimentation on pea plants
Dmitri Mendeleev	Created periodic law and the earliest version of the periodic table of elements
Louis Pasteur	Identified causes and prevention of diseases; discovered pasteurization and created the first vaccines for rabies and anthrax
Alessandro Volta	Invented the first electric battery and discovered methane

Table 1-4 Notable Scientific Discoveries of the Twentieth Century

DISCIPLINE	DISCOVERY
Astronomy and Space Exploration	Big Bang theory Hubble Space Telescope Moon landing Space travel
Biology and Medicine	Antibiotics Artificial heart Blood typing, blood banks, and blood transfusions Organ and tissue transplants Psychiatric drugs Structure of DNA Vaccines X-rays
Chemistry	Ammonia Chromatography pH scale Quantum chemistry
Earth Science	Greenhouse effect Paleomagnetism Theory of continental drift
Engineering and Technology	Air travel Assembly line Automobiles Home appliances Internet Jet engines Motorboats New materials: plastic, silicone, Teflon®, PVC Submarines Television Transistor radio
Mathematics	Computers Differential geometry Game theory Mathematical logic Topology
Physics	Atoms Electromagnetism Nuclear energy Nuclear fusion Quantum mechanics Radiocarbon dating Stellar nucleosynthesis

TWENTIETH CENTURY

Science and technology progressed remarkably in the twentieth century. The achievements of this century were more advanced and more pointedly toward specific intentions (**Table 1-4** and **Figure 1-5**). The wars that occurred during this century significantly impacted scientific research and knowledge. This period is easier to characterize by momentous scientific discoveries.

TWENTY-FIRST CENTURY

Twenty-first century scientists made even more discoveries (**Figure 1-6** and **Figure 1-7**). The foundation for these discoveries was provided by scientists in all of the previous centuries and periods. The following is a list of some of these discoveries:

- 3D printing
- Artificial pancreas
- Automated fingerprint identification systems (AFIS)
- Blockchain
- Broadband Internet
- Combined DNA Index System (CODIS)

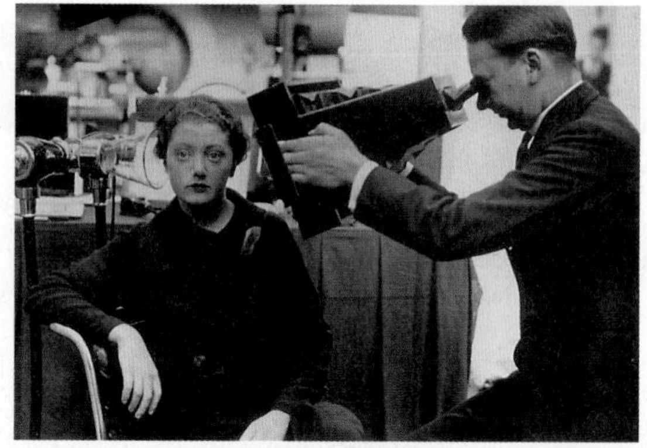

Figure 1-5 *A woman having her head x-rayed in 1934*

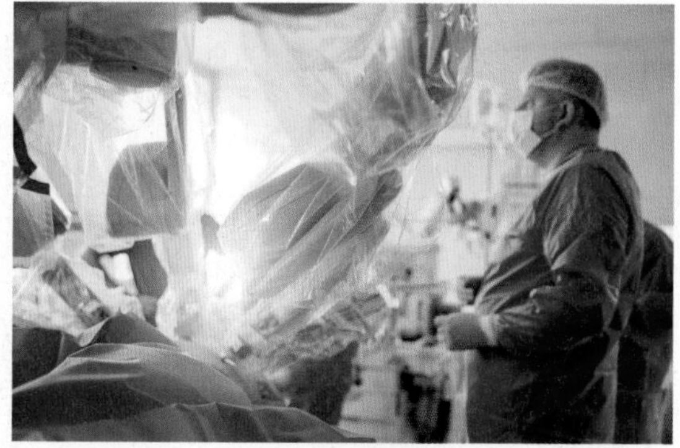

Figure 1-6 *Robotic surgery being performed on a patient*

- Creation of human organs
- Dark matter existence
- Data compression
- Detection of gravitational waves
- Digital assistants
- Digital audio players
- Digital cameras
- Discovery of Eris
- DNA testing kit
- E-readers
- Face transplants
- Gene sequencing
- Human Genome Project (HGP)
- Hybrid cars
- Mars Exploration Rover (MER)
- Mobile operating systems
- Online streaming
- Portable GPS
- Retinal implants
- Radio-frequency identification (RFID)
- Robotic exoskeletons
- Segway
- Smartboards
- Software
- Solid-state lidar
- Teleportation of information to atoms
- Telerobotics
- Tokenization
- Touchscreen glass
- Virtual reality
- Water and ice on the moon
- Water on Mars

Figure 1-7 *Katherine Coleman Goble Johnson received the Presidential Medal of Freedom in 2015.*

Table 1-5 *Influential American Scientists of the Twentieth and Twenty-First Centuries*

SCIENTIST	DISCIPLINE
Alice Augusta Ball	Chemist who developed a successful treatment for Hansen's disease (leprosy)
Charles Drew	Doctor who developed techniques for blood storage and created the first blood bank
Daniel Williams	Cardiologist who performed the first successful open heart surgery
George Washington Carver	Scientist who developed methods to prevent soil depletion
Jonas Salk	Virologist and inventor of polio vaccine
Kizzmekia Corbett	Scientist who pioneered a COVID-19 vaccine
Katherine Coleman Goble Johnson	Mathematician for NASA whose calculations were critical for the success of U.S. crewed spaceflights
Lewis Howard Latimer	Scientist, engineer and inventor of the lightbulb and telephone
Neil DeGrasse Tyson	Astrophysicist, cosmologist, educator, and public figure
Paul Winchell	Ventriloquist, comedian, and trained doctor, built the first mechanical artificial heart that was implantable in the chest cavity
Percy Lavon Julian	Pioneer chemist who produced medicinal chemicals using plants
Roger Wilcott Sperry	Neuropsychologist and neurobiologist who discovered the specializations of the cerebral hemispheres
Sally Ride	Physicist and first female American astronaut in space

SCIENTIFIC INVESTIGATIONS

Science is not pure and has limitations. Only what is observable can be tested and supported or proven false through multiple replications of the same experiment.

A **hypothesis** is an idea that has a foundation in the natural phenomena observed in nature. A hypothesis stems from an observation that can be tested and either proven as "true" or identified as "false." A hypothesis is described as an "educated guess." Hypotheses must be supported or not supported by observational evidence. Hypotheses that have lasting explanatory power and have been tested over a wide variety of conditions are incorporated into theories. Researchers use **empirical evidence** to prove that a hypothesis is true. Empirical evidence remains the same no matter who observes the data. Empirical results support a hypothesis.

Scientific Theories

For a hypothesis to become a theory, it must be tested many times, by multiple independent researchers, and supported by data. **Scientific theories** are based on natural and physical phenomena. Scientific theories are well-established explanations, whereas a hypothesis is an assumption until experimentation has proven it true. Scientific theories provide highly reliable evidence. It is essential for a theory to be accepted.

PLANNING AND IMPLEMENTING SCIENTIFIC INVESTIGATIONS

Scientific research is essential to build scientific knowledge. There are elements that must be systematically planned before an investigation can take place.

Scientific Method

Experimentation involves following specific procedures when carrying out scientific investigations. The **scientific method**, created by Frances Bacon in 1620, is a method of inquiry based on three main concepts: observation, experimentation, and the development of theories or natural laws.

The scientific method consists of six steps (**Figure 1-8**). The first step in the scientific method is **observation**. What is it that piques interest or curiosity?

The second step is to ask a **question** to identify the problem and determine the focus of the investigation. What is the issue or problem that needs to be resolved?

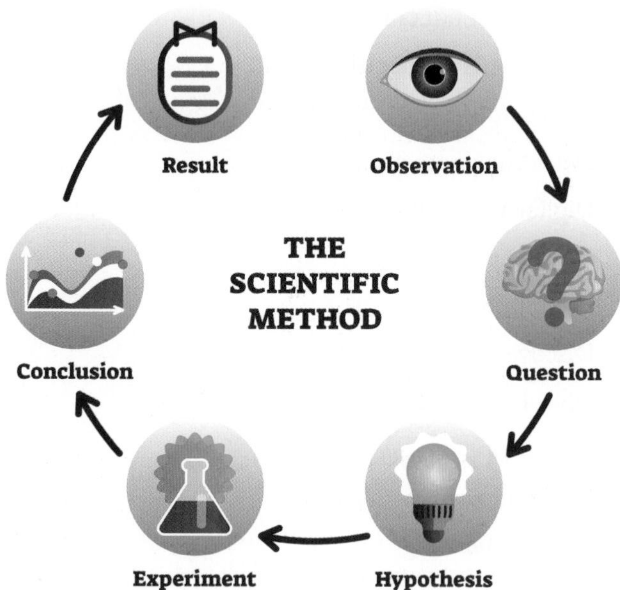

Figure 1-8 *The scientific method*

The third step is to develop a hypothesis that will resolve the question asked. A hypothesis is typically written as an "if…then" statement.

The fourth step is to conduct an **experiment**. An experiment allows scientists to prove or disprove a hypothesis. An important part of conducting an experiment is defining the independent variable and dependent variable. An **independent variable** is controlled and manipulated by the researcher. The **dependent variable** is measured or tested. There must also be a **control group** and an **experimental group**. The control group does not receive treatment. This allows the researcher to compare the results of the experiment to the control group to see if there was actually an effect on the experimental group.

The fifth and sixth steps are the **conclusion/results**. The conclusion discusses the results of the experiment. The following questions should be addressed in the conclusion: Was the hypothesis correct? If the hypothesis was not correct, why? Procedurally, what could change if the experiment were to be conducted again? Are the results easy to understand and the data clearly defined? Were there any errors? What new questions or discoveries developed from the experiment?

Did You Know?

Investigations are imperative to test a hypothesis. All investigations require the researcher to follow specific steps and procedures. Before an investigation, researchers must be able to think critically–they must be able to question, apply, analyze, synthesize, evaluate, make inferences, reflect, and reason.

Types of Data

Qualitative data are data that can be observed or recorded, not through numbers or mathematical equations, but through direct observations, interviews, focus groups, or similar methods. The researcher records the perceptions and feelings of the subject.

Quantitative data are data that are defined by a numerical value, such as information that can be used in a mathematical calculation or statistical analysis. The researcher records numerical data by conducting experiments, controlled observations, studies, and so on.

Types of Reasoning

Logical reasoning is evaluated by deductive and inductive reasoning. **Deductive reasoning** allows one to be able to go from a general idea to focus on a specific outcome. **Inductive reasoning** begins with a specific observation to reach a broad conclusion.

Scientific Investigations

A scientist will conduct many different investigations. Those investigations will fall within four main categories: descriptive investigations, comparative investigations, experimental investigations, and observational investigations (**Table 1-6**). With each type of investigation, data will be collected and analyzed. The data collected will either be qualitative data or quantitative data.

A **descriptive investigation** is one that is used to answer the questions of "what," "how," "where," and "why" and describes and quantifies nature and phenomena using observations and measurements to create data. It is applicable to use a descriptive investigation when the researcher's intent is to identify characteristics, frequencies, trends, and categories, not to make accurate predictions or determine cause and effect. Descriptive research is not necessarily driven by a hypothesis but by exploration.

A **comparative investigation** enlists collecting data on two or more groups for comparison under different conditions with a focus on patterns or trends. Similarities and differences are compared under various conditions over time. Comparative research involves the quantification of the relationship between variables. It is relevant to use comparative investigations to understand the similarities and differences between subjects, when a timeline prevents experimentation, or when there are ethical implications with experimentation.

An **experimental investigation** is conducted to support, refute, or validate a hypothesis. Experiments rely on repeatable procedures so that the experiment can be repeated multiple times. The researcher must identify all variables in an experimental investigation. The dependent variable is reliant upon an independent variable, which is changed and manipulated to see if there is an effect. The relationship between the two variables and the control group will support or disprove the hypothesis.

An **observational investigation**, also called *observational testing*, is used for cause-and-effect relationships. The researcher is not allowed to choose how subjects (independent variables) are assigned to groups or the treatments each group receives due to ethical concerns or logistical constraints.

LABORATORY AND FIELD INVESTIGATION SAFETY

Safety is an important part of conducting laboratory and field investigations. The environments are different; therefore, safety procedures are specific to each area.

Table 1-6 *Investigation Steps of Scientific Investigations*

DESCRIPTIVE	COMPARATIVE	EXPERIMENTAL	OBSERVATIONAL
• observations • hypothesis or exploration • conduct background research • procedures • variables (independent and dependent) • qualitative and/or quantitative data • conclusion	• observations • scientific research question • hypothesis • procedures • variables (independent and dependent) • qualitative and/or quantitative data • conclusion	• observations • scientific research question • hypothesis • procedures • variables (independent and dependent) • control and experimental group • quantitative data • conclusion	• observations • objective • underlying theory research • procedure • observation • qualitative and/or quantitative data • conclusion

Laboratory Safety

Safety is the number one priority for all individuals working in a laboratory setting. Laboratory hazards may be physical, chemical, or biological.

Physical hazards include, but are not limited to, glassware, electricity and electrical equipment, and open flames. Glassware should be free of chips and cracks. If glass is broken, it should be cleaned up and disposed of in a proper receptacle identified within the lab. All electrical equipment and circuits should be in good working order and should not be used if repairs are necessary. Open flames pose a greater risk when wearing loose clothing and for those with long hair—hair should be tied back when in a laboratory. Flammable chemicals should be stored in a flameproof cabinet. A fire extinguisher should be readily available in addition to fire blankets. An escape route should be posted in a visible area and practiced often.

Chemicals may be flammable, toxic (poisonous), caustic (causing severe burns), corrosive, carcinogenic (cancer causing), or mutagenic (causing genetic abnormalities). All chemicals must be labeled clearly with the name of the chemical and hazardous information. **Safety Data Sheets (SDS)** must be available for all chemicals present in a laboratory setting. The SDS provides information on the hazards of the chemical, the personal protective equipment needed when handling the chemical, and the parts of the body that could be affected by exposure. When working with hazardous chemicals, goggles, gloves, an apron, and, if necessary, a face shield should be worn. Jewelry and artificial nails should not be worn when working in a laboratory. Any chemicals that emit fumes should be used only under a fume hood with ventilation. A safety shower and eyewash station should be readily available in the event that chemicals are spilled on the skin or clothes.

Standard precautions are guidelines implemented to protect an individual from biological hazards, such as blood and body fluids. An exposure control plan should be in place for any work area where exposure to biological hazards exists.

Some general rules for working in a laboratory are as follows:

- No horseplay or rowdiness.
- No eating, drinking, or chewing gum in the work area.
- Wear a lab apron or lab coat, long pants, and closed toe shoes.
- Pin up or tie back long hair (chin length or longer).
- Do not wear loose clothing, jewelry, or artificial nails.
- Use gloves when handling chemicals.
- Clean and disinfect the work area before and after use.
- Wash hands before and after any procedure.
- Wear goggles, safety glasses, and/or a face shield when working with chemicals.
- Use a fume hood that ventilates properly when working with chemicals that give off fumes.
- Understand how to transport chemicals and equipment properly.
- Wipe up spills promptly using the appropriate procedure for the type of spill encountered. Use SDS documents as necessary.
- Identify the location of the fire extinguisher, eyewash station, first aid kit, and safety shower.
- Follow the manufacturer's instructions when using equipment.
- Follow the manufacturer's instructions when using chemicals.
- Report broken or frayed electrical cords or damage to equipment.
- Do not use bare hands to pick up broken glass.
- Report accidents to the lab supervisor immediately.

Field Safety

Conducting research in the field provides an entirely different set of hazards. Fieldwork is important to research, but there could be health and safety issues when working in the outdoors (**Figure 1-9**). It is imperative to plan and prepare before going into the field.

Figure 1-9 *Field investigation laboratory settings*

Animals and other wildlife could be a concern in fieldwork. The animals may be the focus of the investigation, or they may be living in the habitat that is being investigated. Additional caution should be paid to ticks, snakes, spiders, bees, wasps, and mosquitoes.

Biological hazards are likely when conducting field research. The air and water could be contaminated by bacteria and viruses. Be aware of insect-borne diseases, take precautions to not get bitten, and check often for ticks.

Chemical safety is important if chemicals are involved in the fieldwork. Gloves, goggles, face shields, and protective clothing should be worn when handling hazardous chemicals. If chemicals are being transported, follow the manufacturer's recommendations on transportation and be sure that SDS documents are available. Chemical waste disposal must be done according to the manufacturer's recommendations.

Electricity may be required if equipment or instruments need power. Equipment should be in good working order without any signs of wear and tear, such as stripped wires.

Environmental hazards might be the most important, yet most neglected, of the hazards. When working outdoors, there is a chance of extreme weather changes. Other concerns include dehydration, sunburn, hypothermia, frostbite, and altitude sickness. Researchers should be prepared for all of the extremes (**Figure 1-10**).

Equipment hazards can pose a threat because some of the equipment used in the field could cause major injuries or even death. Equipment used include chainsaws, drilling equipment, pumps, and so on.

Human hazards are those created and experienced by the researcher and their team, and they may simply be accidents. These include slips, trips, and falls. Also, there could be fragments of sharp metal and glass in the area of work.

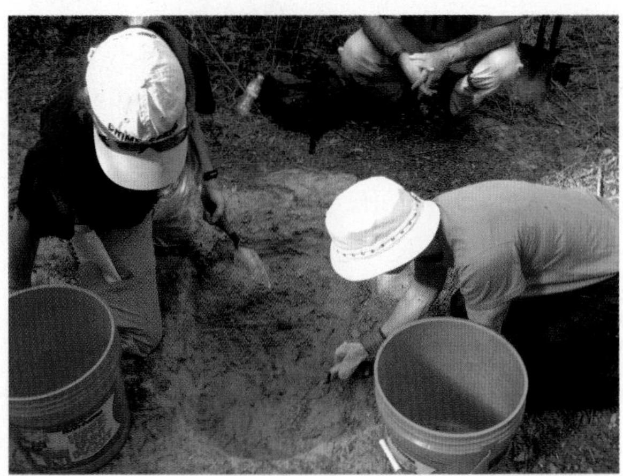

Figure 1-10 *Protective gear being worn during a field investigation*

It is suggested that all researchers assemble a safety and provisions kit. These kits should include, but are not limited to, the following:

- Sunscreen
- Hat
- Flashlight
- Water purification tablets or filter
- First aid kit
- Safety glasses or goggles
- Gloves
- Hard hat
- Work boots
- Food/snacks

Investigative Resources

Resources, such as chemicals, glassware, and thermometers, may be necessary to conduct an investigation. Knowledge of the resources that will be needed may take an investigation in and of itself. Before beginning an investigation, all resources should be gathered and placed in a safe area. This will allow for conservation of resources.

To minimize any adverse effects, all proper safety and disposal measures should be implemented. Chemicals should be stored in proper containers and labeled appropriately. All materials used should be disposed of in proper waste containers and removed as appropriate for the type of material.

Some materials used in an investigation may be recyclable. Glassware can be used over and over again with proper care. If any of the materials used needs to be disposed of, it is important to identify what can be recycled, what should go to the landfill, and what should be disposed of in a special container. Items that can be recycled are newspapers, magazines, paper, cardboard, glass bottles and jars, rigid plastic containers, and certain metals.

Impact of Scientific Research on Society and the Environment

There are many benefits of scientific research on society and the environment. Research creates new knowledge, improves education, discovers new practical applications, and makes life easier in general. Scientific investigations have created a better understanding of the implications of human impact on the earth.

Research can negatively affect society. This is especially true if the research is biased or if an investigator influences the direction of an investigation. This is harmful because knowledge and procedures may influence the way people think about themselves, others, and the environment. If an investigation is skewed and proper procedures are not followed, this could be detrimental to society.

The concern of environmental impact from research and experimentation is legitimate. To minimize any adverse effects, all proper safety and disposal measures should be implemented. Chemicals should be stored in proper containers and labeled appropriately. All materials used should be disposed of in proper waste containers and removed as appropriate for the type of material. It is imperative that resources be conserved so as not to have a negative effect on the environment and the earth.

WORKING ON A TEAM

Communication is fundamental when working as a part of a team (**Figure 1-11**). Dissemination of information comes in different forms, with each being as important as the next.

Communication

Communication amongst team members is essential to reach goals. There are skills that may come naturally for some, whereas other skills must be practiced. See **Figure 1-12** for examples of barriers to effective communication.

Listening is one of the most important components of communication. Communication cannot occur if there is not someone on the receiving end processing the information. Listening instills trust and shows the sender of the message that what they are saying is important.

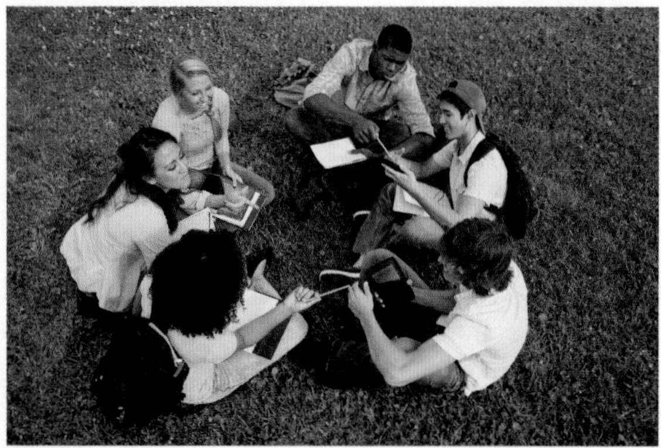

Figure 1-11 *Students working together as a team*

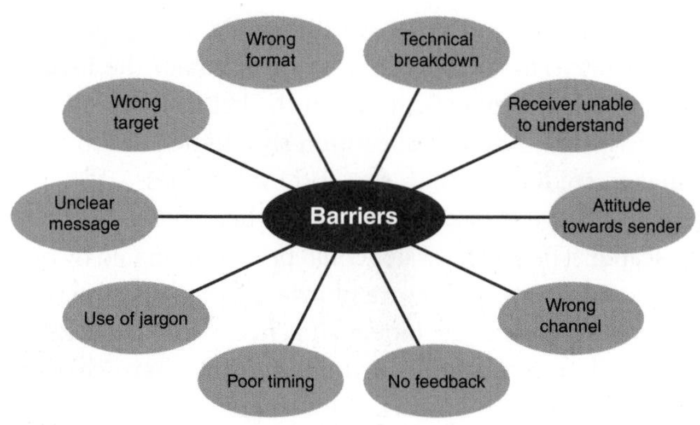

Figure 1-12 *Barriers to communication*

Verbal communication is what is expressed in words. Words can hold a lot of weight when communicating with a team member. Clarity is important when imparting information. Verbal communication is backed up with nonverbal communication.

Nonverbal communication is conveyed though body language or physical behavior. Communication through body language can be conscious or unconscious. When interacting with others, expressions, mannerisms, and general physical behavior are conveyed to the receiver. Nonverbal communication can indicate whether one is interested in the conversation or not.

Empathy is the ability to recognize the emotions and actions of others. When working as a team member, understanding how someone feels in that moment will increase positive communications.

Feedback is the process of receiving communication from another. Feedback can be positive, meaning the receiver is affirming the communication, or negative. There may be times when a team member needs to give or receive negative feedback. Negative feedback should be corrective in nature, not disparaging or demeaning.

Written communication is extremely important when working as a team. Written communication can help support goals, obstacles, clarifications, and solutions. Written communication can take the form of emails, manuals, reports, memos, and so on.

SCIENTIFIC INFORMATION FROM VARIOUS SOURCES

At times, it might be necessary to use information from multiple sources for scientific research. It is imperative to discern valid information from inaccurate or biased information. If it is an online source, the first step is to check the domain name. Those with .edu (educational)

Figure 1-13 *Types of models: (A) physical, (B) conceptual, and (C) mathematical*

and .gov (government) at the end are usually credible. Those with .org, which indicates a nonprofit, could be credible or can be biased in favor of the organization. There are some sites with .com (generic) at the end that could be credible, but additional research would be needed to be certain.

Authors and the date of a publication can be helpful. Many times, the author's credentials will be listed along with their name. The credentials should match the topic or subject of the article or study. Scientific studies can go out of date as technology advances. The newer the article or study, the more reliable it will be.

There is a lot of information that is floating around, whether it is on the Internet, social media, or by word of mouth. At times, it may be difficult to determine if what is being viewed or heard is credible. After valid and reliable evidence is obtained, inferences can be made to reach a conclusion. To make an inference, one must take the valid evidence collected and combine it with something already known to reach a conclusion. For example, data can be used to discern whether the promotional materials truly represent what is being advertised.

Scientific models are used to represent concepts, objects, and phenomena in a way that makes it easier to understand. There are three types of scientific models: physical, conceptual, and mathematical (**Figure 1-13**). Models should be consistent with observations, evidence, phenomenon, inferences, and current interpretations.

While scientific models are useful in helping scientists make predictions and analyze data, they do have limitations. Models are much simpler and not a complete representation of the phenomenon of the actual objects or systems.

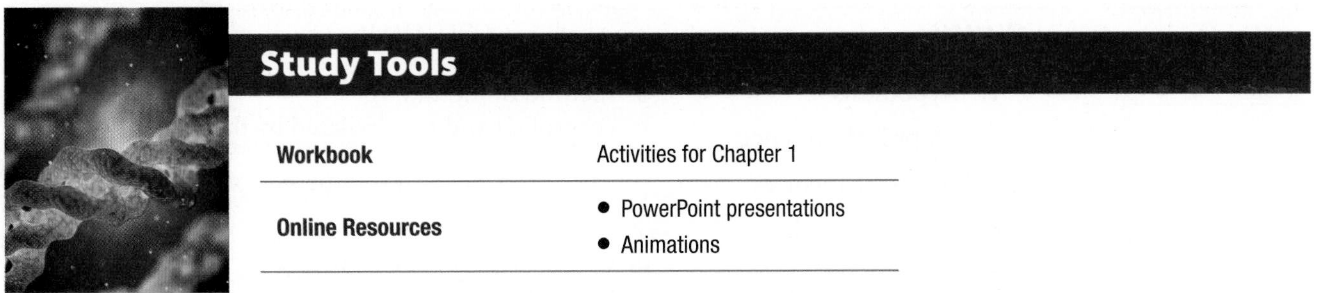

Study Tools

Workbook	Activities for Chapter 1
Online Resources	• PowerPoint presentations • Animations

REVIEW QUESTIONS

Select the letter of the choice that best completes the statement.

1. A hypothesis
a. is testable.
b. can be supported by evidence.
c. can be proven false.
d. all of the above.

2. Data that are observed or recorded are called
a. qualitative data.
b. quantitative data.
c. empirical evidence.
d. experimental evidence.

3. The four types of investigations are
a. observable, quantitative, experimental, and descriptive.
b. qualitative, experimental, variable and quantitative.
c. descriptive, comparative, experimental, and observational.
d. procedural, descriptive, qualitative, and observational.

4. Internet domains that end in .org indicate a
a. nonprofit website.
b. government website.
c. educational website.
d. generic website.

5. The type of investigation that answers the questions what, how, where, and why is the
a. descriptive investigation.
b. comparative investigation.
c. experimental investigation.
d. observational investigation.

6. Observational investigations are used to
a. compare two or more groups.
b. define quantitative data.
c. determine cause-and-effect relationships.
d. define qualitative data.

7. Inductive reasoning begins with
a. a specific observation to reach a broad conclusion.
b. a general idea to a specific outcome.
c. a probable conclusion from what is known.
d. possible alternatives to a problem.

8. Safety Data Sheets (SDS) should be referred to
a. when an accident occurs.
b. when thinking about an investigation.
c. before beginning an investigation.
d. only as necessary.

MATCHING

Match each term in Column I with its correct description in Column II.

	COLUMN I	*COLUMN II*
________	**1.** hypothesis	a. method of inquiry
________	**2.** empirical evidence	b. controlled and manipulated by the researcher
________	**3.** scientific theories	c. tentative and testable statements that can be proven or disproven
________	**4.** scientific method	d. based on natural and physical phenomena
________	**5.** experiment	e. prove or disprove a hypothesis
________	**6.** independent variable	f. an idea
________	**7.** dependent variable	g. does not receive treatment
________	**8.** control group	h. supports a hypothesis

APPLYING THEORY TO PRACTICE

1. Create a timeline of scientific discoveries in a specific discipline of science, such as anatomy, biology, or chemistry. Research the history of that discipline and list the contributions of each notable scientist on the timeline.

2. Investigate concepts that can be proven or disproven through scientific experimentation, and name one concept that science is unable to prove or disprove.

3. Evaluate various promotional materials for goods or services. Draw inferences based on what you already know about the product or service. Conduct research, collect data, and determine if the product or service is legitimate.

4. In a team, plan and conduct a descriptive investigation using the scientific method. Extract and communicate scientific information from various sources. Each team member must exhibit how to cooperate, contribute, and communicate in a verbal and nonverbal manner.

5. Debate the following topic: Should scientific research on society and the environment continue? Prepare by gathering information on ways scientific research has impacted society and the environment negatively or positively. Evaluate your information and use it in the debate.

6. Conduct a simple field investigation in a natural area near your home or school. What will you research? What equipment will you need? What safe practices must you observe? Be sure to demonstrate these safe practices as you conduct your field investigation.

7. Analyze, evaluate, and make inferences from the following graph:

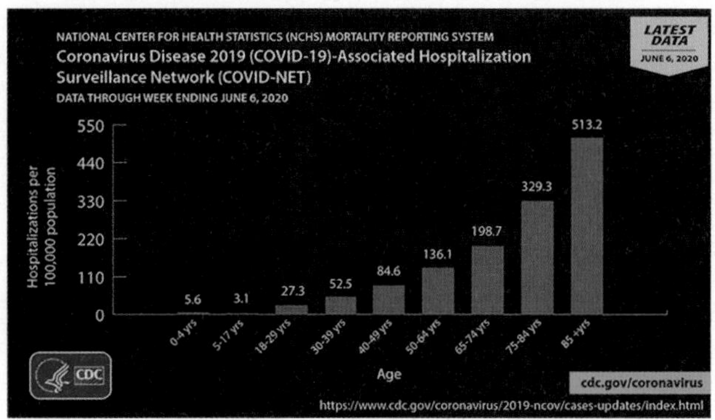

   a. What is the subject of the graph?
   b. What is inferred from the data in this graph?
   c. Analyze the graph, write a paragraph, and justify the results of this study.
   d. How might you organize observational testing to investigate the results of this study? How might you organize experimental testing to investigate the results of the study?
   e. What model is this graph an example of? What are the limitations of this type of graph for the subject?

8. Research a scientific investigation that used observational testing as the investigative method. Using the report from the investigation, analyze the subject using logical reasoning. Evaluate all sides of the scientific evidence. Critique the scientific explanations using empirical evidence.

LAB ACTIVITY 1-1 Comparative Investigation

- ***Objective:*** To plan and implement a comparative investigation following the steps of the scientific method
- ***Materials Needed:*** varies upon the experiment

Step 1: Ask a question about something that piques your interest.

Step 2: Formulate a research question and testable hypothesis.

Step 3: Conduct research using credible sources.

Step 4: Select the equipment needed for the investigation.

Step 5: Select any technology needed for the investigation.

Step 6: Identify variables.

Step 7: Collect data, using accuracy and precision.

Step 8: Analyze the data collected.

Step 9: Evaluate the data collected.

Step 10: Make inferences from the data.

Step 11: Predict trends from the data.

Step 12: Communicate valid conclusions to your class by organizing your data and conclusions into a labeled drawing, oral report, or lab report.

LAB ACTIVITY 1-2 Experimental Investigation

- ***Objective:*** To plan and implement an experimental investigation following the steps of the scientific method
- ***Materials Needed:*** varies upon the experiment

Step 1: Ask a question about something that piques your interest.

Step 2: Formulate a research question and a testable hypothesis.

Step 3: Conduct research using credible sources.

Step 4: Select the equipment and resources needed for the investigation. Determine how you will best conserve your resources.

Step 5: Select any technology needed for the investigation.

Step 6: Determine what safe practices need to be followed to conduct your experiment.

Step 7: Identify variables.

Step 8: Collect data, using accuracy and precision and following safe practices.

Step 9: Identify control and experimental groups.

Step 10: Analyze the data. What qualitative data did you collect? What quantitative data did you collect?

Step 11: Evaluate the data collected.

Step 12: Make inferences from the data.

Step 13: Predict trends from the data.

Step 12: How did you conserve resources? Properly dispose or recycle your conserved materials.

LAB ACTIVITY 1-3 Qualitative and Quantitative Data

- ***Objective:*** To collect and organize qualitative and quantitative data
- ***Materials Needed:*** multiple objects identified by the instructor

Step 1: In teams of 3–4 students, write as many qualitative and quantitative statements as you can about the object given to the group.

Step 2: Rotate objects every 1–2 minutes, and follow the process identified in step 1 for each object.

Step 3: Continue the process until every team has the opportunity to investigate each object.

Step 4: Teams will analyze the data collected and organize the data using logical reasoning.

Step 5: Each team will then present the data collected about each object.

Step 6: Teams will compare and contrast the qualitative and quantitative data they collected versus what was presented by the other teams.

Step 7: Analyze all data. Write a summary of findings.

CHAPTER 2

Introduction to the Structural Units

Objectives

- Identify and discuss the different branches of anatomy.
- Identify terms referring to location, direction, planes, and sections of the body.
- Identify the body cavities and the organs they contain.
- Identify and discuss homeostasis and metabolism.
- Identify the units of measure used in health care.
- Define the key words that relate to this chapter.

Key Words

abdominal cavity
abdominopelvic cavity
anabolism
anatomical position
anatomy
anterior
biology
catabolism
caudal
cell
cephalic
comparative anatomy
coronal (frontal) plane
cranial cavity
cytology
deep
dermatology
developmental anatomy
disease
distal
dorsal
dorsal cavity
embryology
endocrinology
epigastric
external
gross anatomy
histology
homeostasis
hypochondriac
hypogastric
inferior
internal
lateral
life functions
medial
metabolism
metric system
microscopic anatomy
midsagittal plane
nasal cavity
negative feedback loop
neurology
oral cavity
orbital cavity
organ system
organs
pelvic cavity
physiology
planes
positive feedback
posterior
prone
proximal
quadrants
sagittal plane
section
superficial
superior
supine
systemic anatomy
thoracic cavity
tissues
transverse
umbilical
umbilicus
ventral
vertebral (spinal) cavity

ANATOMY AND PHYSIOLOGY

Anatomy and physiology are branches of a much larger science called **biology** (bye-**OL**-oh-jee). Biology is the study of all forms of life. Biology studies microscopic one-celled organisms, multicelled organisms, plants, animals, and humans.

Anatomy (ah-**NAT**-oh-mee) studies the shape and structure of an organism's body and the relationship of one body part to another. The word *anatomy* comes from the Greek *ana*, meaning "apart," and *temuein*, "to cut"; thus, the acquisition of knowledge on human anatomy comes basically from dissection. However, one cannot fully appreciate and understand anatomy without the study of its sister science, **physiology** (fiz-ee-**OL**-oh-jee). Physiology studies the function of each body part and how the functions of the various body parts coordinate to form a complete living organism. Any abnormal change in a structure or function that produces symptoms is considered a **disease** (diz-**EASE**).

Branches of Anatomy

Anatomy is subdivided into many branches based on the investigative techniques used, the type of knowledge desired, or the parts of the body under study.

1. **Gross anatomy** is the study of large and easily observable structures on an organism. This is done through dissection and visible inspection with the naked eye. In gross anatomy, the different body parts and regions are studied with regard to their general shape, external features, and main divisions.
2. **Microscopic anatomy** refers to the use of microscopes to enable one to see the minute details of organ parts. Microscopic anatomy is subdivided into two branches. One branch is **cytology** (sigh-**TOL**-oh-jee), which is the study of the structure, function, and development of cells that comprise the different body parts. The other subdivision is **histology** (hiss-**TOL**-oh-jee), which studies the tissues and organs that make up the entire body of an organism.
3. **Developmental anatomy** studies the growth and development of an organism from fertilization to maturity, also called **embryology** (**em**-bree-**OL**-oh-jee).
4. **Comparative anatomy** is when the different body parts and organs of humans can be studied with regard to similarities to and differences from others in the animal kingdom. Humans are just one of the many animals found in the animal kingdom.
5. **Systematic anatomy** is the study of the structure of various organs or parts that comprise a particular organ system. Depending on the particular organ system under study, a specific term is applied; for example:
   a. **Dermatology** (der-mah-**TOL**-oh-jee)—study of the integumentary system (skin, hair, and nails)
   b. **Endocrinology** (**en**-doh-krin-**OL**-oh-jee)—study of the endocrine or hormonal system
   c. **Neurology** (**new**-**ROL**-oh-jee)—study of the nervous system

ANATOMICAL TERMINOLOGY

In the study of anatomy and physiology, special words are used to describe the specific location of a structure or organ, or the relative position or direction of one body part to another. The initial reference point used is the anatomical position. In the **anatomical position**, a human being is standing erect, with face forward, arms at the side, and palms forward (**Figure 2-1**).

Terms Referring to Location or Position and Direction

See **Figure 2-2** and **Figure 2-3**.

- **Anterior** or **ventral** means "front" or "in front of." For example, the knees are located on the anterior surface of the human body. A ventral hernia may protrude from the front or belly of the abdomen.
- **Posterior** or **dorsal** means "back" or "in back of." For example, human shoulder blades are found on the posterior surface of the body.
- **Supine** (su-**PINE**) refers to lying on one's back, with face and torso facing up.
- **Prone** (pron) refers to lying flat, with face and chest down.
- **Cephalic** (seh-**FAL**-ick) and **caudal** (**KAWD**-al) refer to direction: Cephalic means "skull" or "head end" of the body; caudal means "tail end." For example, a blow to the skull is a cephalic injury that may increase cranial pressure and cause headaches. Caudal anesthesia is injected in the lower spine.
- **Superior** means "upper" or "above another," and **inferior** refers to "lower" or "below another."

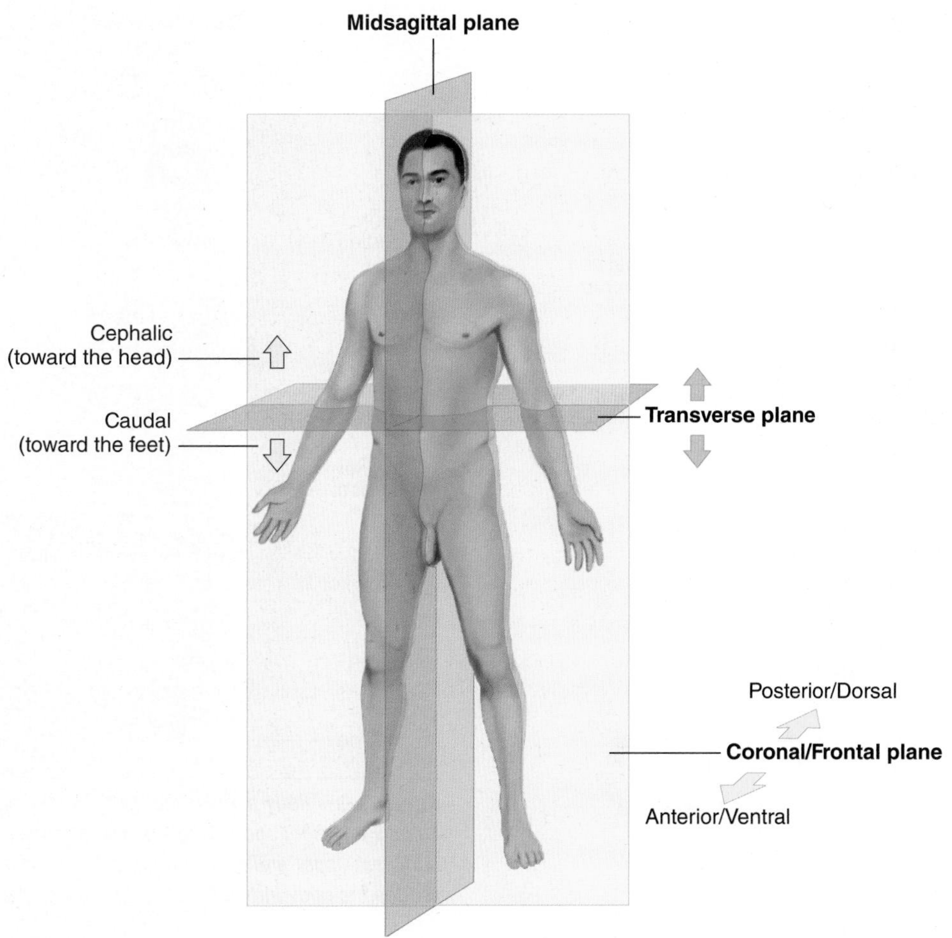

Figure 2-1 *Body directions: Cephalic refers to the skull or head end of the body, and caudal refers to the tail end. Anterior (or ventral) means "front" or "in front of." Posterior (or dorsal) means "back" or "in back of."*

For example, the heart and lungs are situated superior to the diaphragm, whereas the intestines are inferior to the diaphragm.

- **Medial** signifies "toward the midline or median plane of the body"; **lateral** means "away" or "toward the side of the body." For example, the nose is medial to the eyes, and the ears are lateral to the nose.
- **Proximal** means "toward the point of attachment to the body" or "toward the trunk of the body"; **distal** means "away from the point of attachment or origin" or "farthest from the trunk." For example, the wrist is proximal to the hand; the elbow is distal to the shoulder. *Note:* these two words are used primarily to describe the appendages or extremities.
- **Superficial** or **external** implies "on or near the surface of the body." For example, a superficial wound involves an injury to the outer skin. A **deep** or **internal** injury involves damage to an internal organ, such as the stomach. The terms *external* and *internal* are specifically used to refer to body cavities and hollow organs.

Terms Referring to Body Planes and Sections

Planes are imaginary anatomical dividing lines that are useful in separating body structures (**Figure 2-3**). A **section** is a cut made through the body in the direction of a certain plane.

The **sagittal plane** (**SAJ**-ih-tal) divides the body into right and left parts. If the plane started in the middle of the skull and proceeded down, bisecting the sternum and the vertebral column, the body would be divided equally into right and left halves. This would be known as the **midsagittal plane**.

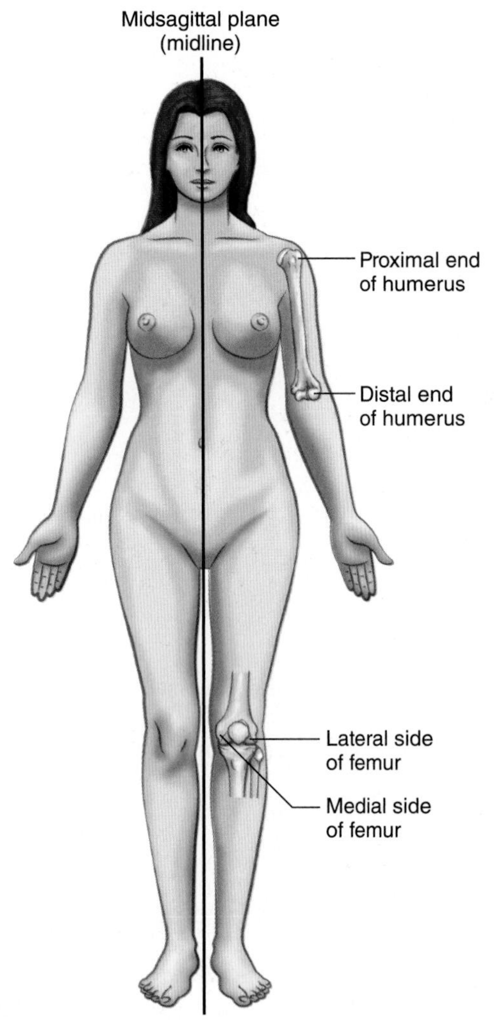

Figure 2-2 *Body directions: Proximal means "toward the point of attachment to the body" or "toward the trunk of the body." Distal means "away from the point of attachment or origin" or "farthest from the trunk." Medial means "toward the midline or median plane of the body," and lateral means "away or toward the side of the body."*

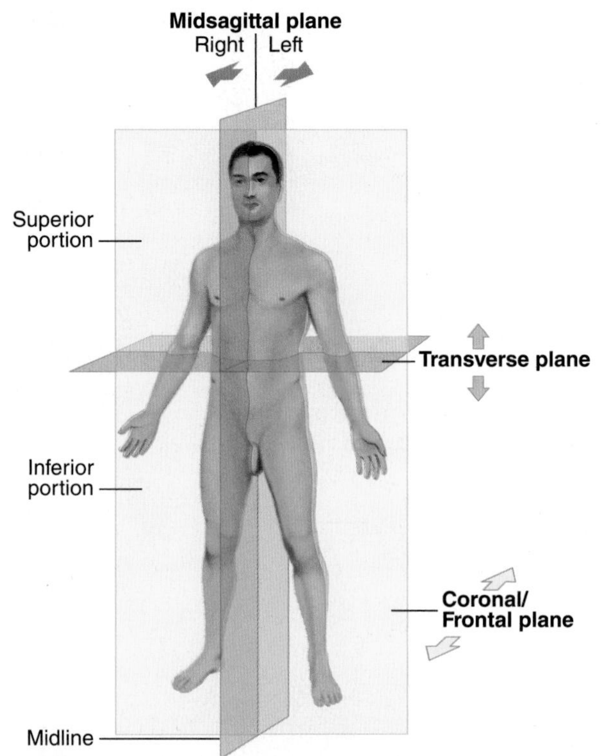

Figure 2-3 *Body planes: The midsagittal plane divides the body equally into right and left halves. The transverse plane divides the body into upper and lower portions. The coronal (or frontal) plane divides the body into anterior and posterior portions.*

A **coronal (frontal) plane** is a vertical cut at right angles to the sagittal plane, dividing the body into anterior and posterior portions. The term *coronal* comes from the coronal suture, which runs perpendicular (at a right angle) to the sagittal suture. A **transverse** or cross section is a horizontal cut that divides the body into upper and lower portions.

Media Link

View the **Body Planes** animation on the Online Resources.

Terms Referring to Cavities of the Body

The organs that comprise most of the body systems are located in four major cavities: cranial, vertebral (spinal), thoracic, and abdominopelvic (**Figure 2-4**). The cranial and spinal cavities are within a larger region known as the posterior (dorsal) cavity. The thoracic and abdominopelvic cavities are found in the anterior (ventral) cavity.

The **dorsal cavity** contains the brain and spinal cord: the brain is in the **cranial cavity**, and the spinal cord is in the **vertebral (spinal) cavity** (**Figure 2-4**). The diaphragm divides the ventral cavity into two parts: the upper thoracic and lower abdominopelvic cavities.

The central area of the **thoracic cavity** (tho-**RASS**-ik) is called the mediastinum. It lies between the lungs and extends from the sternum (breastbone) to the vertebrae of the back. The esophagus, bronchi, lungs, trachea, thymus gland, and heart are located in the thoracic cavity. The heart itself is contained within a smaller cavity called the pericardial cavity.

The thoracic cavity is further subdivided into two pleural cavities. The left lung is in the left cavity; the right lung is in the right cavity. Each lung is covered with a thin membrane called the pleura.

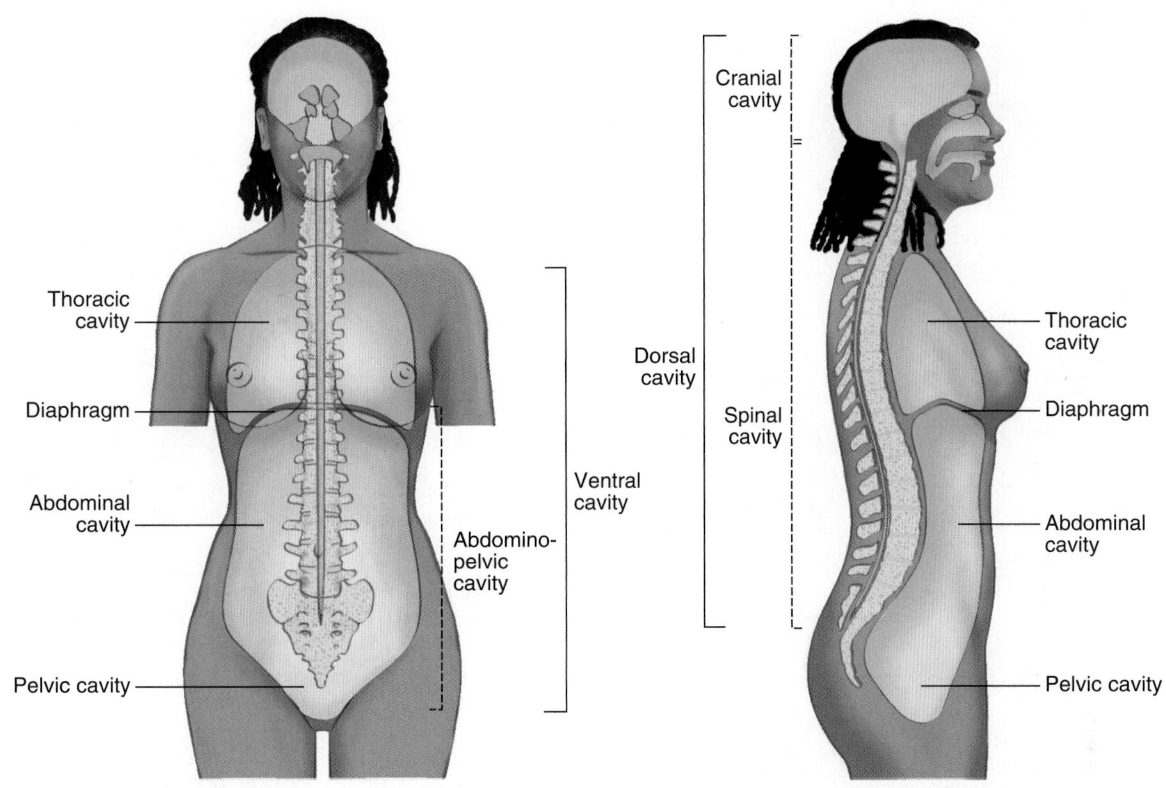

Figure 2-4 *The major body cavities*

The **abdominopelvic cavity** (ab-**dom**-ih-noh-**PEL**-vick) is actually one large cavity with no separation between the abdomen and pelvis. To avoid confusion, this cavity is usually referred to separately as the abdominal cavity and the pelvic cavity. The **abdominal cavity** contains the stomach, liver, gallbladder, pancreas, spleen, small intestine, appendix, and part of the large intestine. The kidneys are close to but behind the abdominal cavity. The urinary bladder, reproductive organs, rectum, and remainder of the large intestine are in the **pelvic cavity**.

Terms Referring to Regions in the Abdominopelvic Cavity

To locate the abdominal and pelvic organs more easily, the abdominopelvic cavity is divided into nine regions (**Figure 2-5**).

The nine regions are located in the upper, middle, and lower parts of the abdomen:

- The upper or **epigastric** (ep-ih-**GAS**-trick) region is located just below the sternum (breastbone). The right and left **hypochondriac** (**high**-poh-**KON**-dree-ack) regions are located below the ribs.
- The middle or **umbilical** area is located around the navel or **umbilicus** (um-**BILL**-ih-kus), and the right and left lumbar regions extend from anterior to posterior. A person will complain of back pain or lumbar pain.
- The lower or **hypogastric** (**high**-poh-**GAS**-trick) region may also be referred to as the pubic area; the left and right iliac may also be called the left and right inguinal areas.

Smaller Cavities

In addition to the cranial cavity, the skull contains several smaller cavities. The eyes, eyeball muscles, optic nerves, and lacrimal (tear) ducts are within the **orbital cavity**. The **nasal cavity** contains the parts that form the nose. The **oral cavity** or *buccal cavity* (**BUCK**-ull) encloses the teeth and tongue.

Terms Referring to Quadrants in the Abdominal Area

Another method for referencing the abdominal area is to divide the area into **quadrants**. This method uses one median sagittal plane and one transverse plane that passes through the umbilicus at right angles. The four resulting quadrants are named according to their positions: right upper quadrant (RUQ), left upper quadrant (LUQ), right lower quadrant (RLQ), and left lower quadrant (LLQ) (**Figure 2-6**).

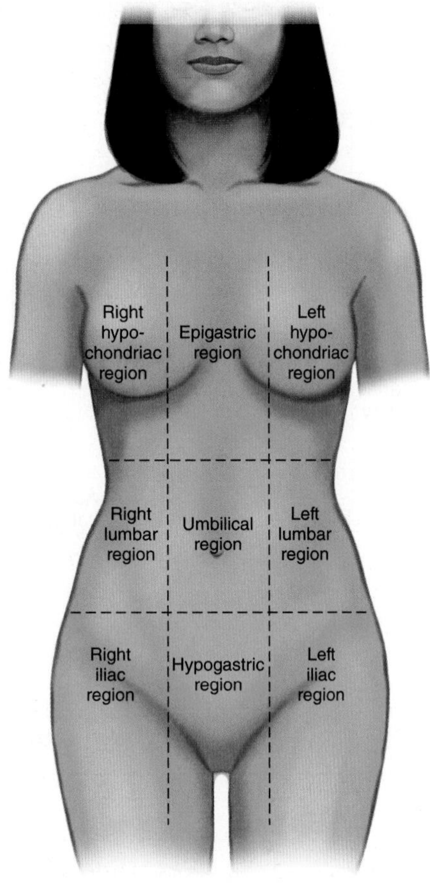

Figure 2-5 *Regions of the thorax and abdomen*

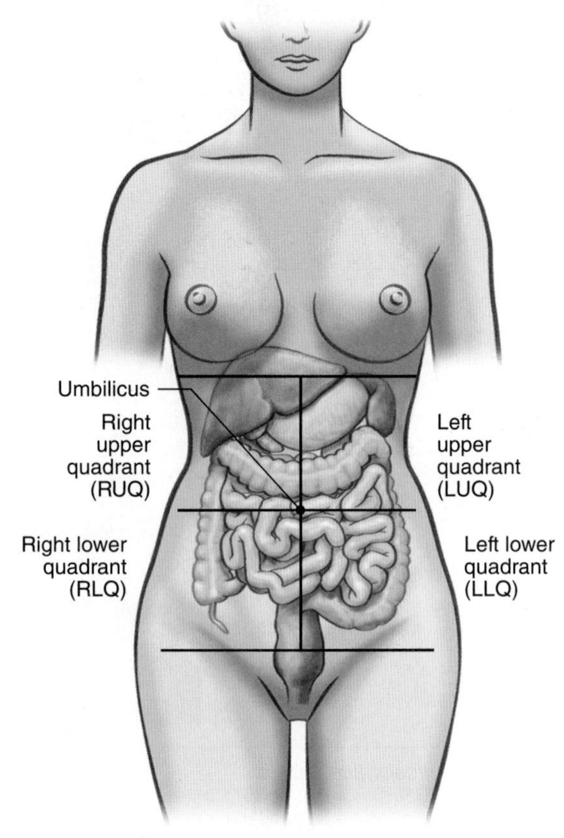

Figure 2-6 *Division of the abdomen into quadrants*

Did You Know?

McBurney's point is not at the top of a mountain but midway between the umbilicus and the iliac crest (the prominent area on the hip bone) in the right lower quadrant, or right inguinal area. This area is painful when a person has appendicitis.

LIFE FUNCTIONS

When humans, plants, one-celled organisms, or multi-celled organisms are examined, it is clear that all of them have one thing in common: they are alive.

All living organisms are capable of carrying on life functions. **Life functions** are a series of highly organized and related activities that allow living organisms to live, grow, and maintain themselves.

These vital life functions include movement, ingestion, digestion, transport, respiration, immunity, protection, growth, secretion, excretion, regulation (sensitivity), and reproduction (**Table 2-1**).

Human Development

The body carries on numerous functions that keep one alive and active. Living depends on the constant release of energy in every cell of the body. Powered by the energy released from food, cells are able to maintain their own living condition and, thus, the life of a human being.

There are six-levels of structural organization, with each level building upon the next. These levels are: chemical, cellular, tissues, organs, systems, and organism.

The first level is the chemical level, with atoms being the smallest unit. Multiple atoms bind together to create molecules, which are the building blocks of all body structures. A complex life-form like a human being consists of more than 50 trillion cells. The **cell** is the basic unit of structure and function of all living things. Early in human development, certain groups of cells become highly specialized for specific functions, such as movement or growth.

Special cells—grouped according to function, shape, size, and structure—are called **tissues**. Tissues, in turn, form larger functional and structural units known as **organs**. For example, human skin is an organ of

Table 2-1 *Review of the Life Functions and Body Systems*

LIFE FUNCTIONS/ BODY SYSTEMS	DEFINITION
Movement / **Muscle System**	The ability of the whole organism—or a part of it—to move
Ingestion	The process by which an organism takes in food
Assimilation	The breakdown of complex food molecules into simpler food molecules
Digestion / **Digestive System**	The transformation of digested food molecules into living tissue for growth and self-repair
Transport / **Circulatory System**	The movement of necessary substances to, into, and around cells, and of cellular products and wastes out of and away from cells
Respiration / **Respiratory System**	The burning or oxidation of food molecules in a cell to release energy, water, and carbon dioxide
Immunity / **Lymphatic System**	The filtering out of harmful bacteria and production of white blood cells (lymphocytes)
Protection / **Integumentary System**	The waterproof covering of the body
Growth / **Skeletal System**	The enlargement of an organism due to synthesis and assimilation, resulting in an increase in the number and size of its cells
Secretion / **Endocrine System**	The formation and release of hormones from a cell or structure
Excretion / **Urinary System**	The removal of metabolic waste products from an organism
Regulation (Sensitivity) / **Nervous System**	The ability of an organism to respond to its environment so as to maintain a balanced state (homeostasis)
Reproduction / **Reproductive System**	The ability of an organism to produce offspring with similar characteristics (This is *essential* for species survival as opposed to individual survival.)

epithelial, connective, muscular, and nervous tissue. In much the same way, kidneys consist of highly specialized connective and epithelial tissue.

Some organs are grouped together because more than one is needed to perform a function. Such a grouping is called an **organ system**. One example is the digestive system, composed of the teeth, esophagus, stomach, small intestine, and large intestine. This textbook will explore the various body systems and the organs that comprise them.

The final level is when all the organ systems are working together, creating a functioning, independent *organism*.

Homeostasis

Homeostasis (**hoe**-mee-oh-**STAY**-sis) is the ability of the body to regulate its internal environment within narrow limits through negative and positive feedback. The body strives to remain in a state of *equilibrium*, where the systems maintain a balance between opposing forces. Maintaining homeostasis is essential to survival; imbalance results in disease. All organ systems contribute to homeostasis. Examples of homeostasis controls are blood sugar levels, body temperature, heart rate, and the fluid environment of cells. Aging cells no longer respond as quickly, which makes it harder to maintain homeostasis.

Most homeostasis control works on a **negative feedback loop**. Feedback responses reverse disturbances to the body's condition. An example of how a negative feedback loop operates is seen in maintaining a body temperature. A normal body temperature is 37°C (98.6°F). Outside, on a hot summer day, a body temperature rises. The hypothalamus in the brain detects this and sends signals to various organs, and one starts to sweat; sweating is a cooling process. As water is excreted by the sweat glands on the skin, it evaporates; evaporation is a cooling mechanism. In addition, blood vessels dilate to bring blood near the skin's surface to dissipate body heat. If one goes outside on a cold day and the body temperature falls below 37°C (98.6°F), the hypothalamus of the brain detects this and sends signals to the muscles, causing them to shiver, which raises the body temperature; increased muscle activity produces heat. In addition, the hypothalamus sends signals to the blood vessels, causing them to constrict, which reduces blood flow near the surface, conserving body heat.

Conversely, if the integumentary, muscular, and cardiovascular systems are not working properly, the negative feedback loop regulating body temperature is interrupted and the entire body could be affected. If homeostasis is not restored, the body is susceptible to diseases, and in severe cases, death may occur.

Positive feedback is the body's ability to increase the level of an event that has already been started. An example of positive feedback is blood clotting. When a person has a cut or damages a blood vessel, platelets in the blood quickly accumulate to clot around the wound and stop the bleeding.

Metabolism

The functional activities of cells that result in growth, repair, energy release, use of food, and secretions are combined under the heading of **metabolism** (meh-**TAB**-oh-lizm). Metabolism consists of two processes that are opposite to each other: anabolism and catabolism. **Anabolism** (ah-**NAB**-oh-lizm) is the building up of complex materials from simpler ones, such as food and oxygen, and it requires energy. **Catabolism** (kah-**TAB**-oh-lizm) is the breaking down and changing of complex substances into simpler ones, with a release of energy and carbon dioxide. The sum of all the chemical reactions within a cell is therefore called metabolism.

METRIC SYSTEM

Knowledge of the metric system is important to the understanding of the language used in *Body Structures and Functions.* The medical community measures length, weight, and volume using this system. The **metric system** is a decimal system based on the power of 10. Just as there are 100 cents in a dollar, there are 100 centimeters in a meter (see Appendix A).

Some of the prefixes used in the metric system are

centi = 1/100 (one/one-hundredth)
milli = 1/1000 (one/one-thousandth)
micro = 1/1,000,000 (one/one-millionth)

Length is measured using meters instead of inches and feet.

1 centimeter (cm) = 0.4 inch
2.5 cm = 1 inch

Weight is measured using grams instead of ounces and pounds.

1 gram (g) = 1 ounce
1 kilogram (kg) = 2.2 pounds
1000 g = 1 kg

In drug dosage, the most familiar unit used is the gram or milligram (mg).

500 mg = 0.5 g

Volume is measured using liters or milliliters instead of quarts, pints, ounces, teaspoons, and tablespoons.

1 liter (L) = 1.06 quarts (a liter is slightly larger than a quart)
1 L = 1000 milliliters (ml)

For liquid drug dosage, milliliters are used.

5 ml = 1 teaspoon
15 ml = 1 tablespoon
30 ml = 1 ounce

Medical Highlights

BIOTECHNOLOGY AND NANOTECHNOLOGY

There have been many advances in the treatment and diagnosis of disease using techniques such as *biotechnology* and *nanotechnology*.

Biotechnology refers to any technological application that uses biological systems, living organisms, or derivatives thereof to make or modify products or processes for specific uses. One field of biotechnology, genetic engineering, has introduced techniques such as gene therapy and recombinant DNA technology. These techniques make use of genes and DNA molecules to diagnose disease and insert new and healthy genes into the body to replace damaged cells. In operating rooms, doctors can operate on patients remotely from their computers, guiding robotic arms with an accuracy of a few nanometers. The use of MRIs and CAT scans allow the doctor to view an entire 3D image of the inside of patients' bodies. Scientists are trying to develop biopharmaceutical drugs to treat diseases such as hepatitis, cancer, and heart disease. Scientists are also trying to develop 3D-printed organs, working on regenerations of nerves, and creating a device that can translate brain signals to audible speech using a voice synthesizer.

Nanotechnology is a science that manipulates atoms and molecules to form new materials. Nanotechnology deals with materials a billion times smaller than a soccer ball. People cannot even visualize such minute dimensions. At this size, matter exhibits unusual properties that can be engineered to perform tasks not otherwise possible.

At present, the signs of disease first appear at a cellular level. To date, instruments used within medicine have only been able to detect abnormalities at the macro level. Being able to diagnose and treat disease at the molecular level will enable physicians to reach the root origins of disease and assist—or even replace—the healing process.

The long-term goals of the National Institutes of Health (NIH) are to be able to use nanoparticles to seek out cancer cells before tumors grow and to remove and/or replace "broken" parts of cells or cell mechanisms with miniature, molecular-sized biological "machines" and use these "machines" as pumps or robots to deliver medicines when and where needed in the body. Pharmaceutical products are reformulated with nano-sized particles to improve their absorption.

ADVANTAGES OF USING MODELS AND COMPUTERS TO REINFORCE LEARNING OBJECTIVES

Students can become living models to achieve an objective, such as learning about body directions, bones, and muscles. When human or plastic models are used, it gives people 3D perception, which encompasses width, length, and depth. Humans are able to perceive the spatial relationship between objects by looking at them because humans have 3D vision, or depth perception. The disadvantage of using models is that they can only be used to meet specific learning situations.

Computerized images of the human body allow the student to get a holistic view of the placement and interaction of structures. Computer technology allows the student to animate an activity, such as seeing the circulation of blood through the circulatory system.

Radiologic images provide students with the opportunity to view accurate images of normal and abnormal structures.

Medical Terminology

ana	apart	**-al**	pertaining to
-tom	cutting	**caud**	tail
-y	process of	**caud/al**	pertaining to the tail
ana/tom/y	process of cutting apart; study of body parts by dissection	**crani**	skull
		crani/al	pertaining to the skull
-ology	study of	**dist**	distant
bio	life	**dist/al**	pertaining to a distant part
bio/logy	study of life	**dors**	back
physio	nature	**dors/al**	pertaining to the back
physi/ology	study of nature or natural function of body	**later**	side
		later/al	pertaining to the side
-ior	compared to something else	**medi**	middle
poster	behind	**medi/al**	pertaining to the middle
poster/ior	in back of	**proxim**	near
super	above		
super/ior	above a part	**proxim/al**	pertaining to nearness or closeness
infer	below	**ventr**	belly, front side
infer/ior	below a part	**ventr/al**	pertaining to the belly or front side

Study Tools

Workbook	Activities for Chapter 2
Online Resources	PowerPoint presentations

REVIEW QUESTIONS

Select the letter of the choice that best completes the statement.

1. The study of the size and shape of the heart is called
   a. physiology.
   b. anatomy.
   c. histology.
   d. embryology.

2. Physiology is the study of
   a. the size of the cell.
   b. the shape of the kidney.
   c. the function of the lungs.
   d. the size and shape of the liver.

3. The anatomical position is described as
   a. body erect, arms at the side, palms forward.
   b. body flat, arms at the side, palms forward.
   c. body erect, arms at the side, palms backward.
   d. body flat, arms at the side, palms backward.

4. A plane that divides the body into right and left parts is a
   a. transverse plane.
   b. coronal plane.
   c. sagittal plane.
   d. frontal plane.

5. If a person is complaining of pain that may indicate appendicitis, the pain would be located in the
   a. left lower quadrant.
   b. right lower quadrant.
   c. right upper quadrant.
   d. left upper quadrant.

6. The heart is described as superior to the diaphragm because it is
   a. in back of the diaphragm.
   b. in front of the diaphragm.
   c. above the diaphragm.
   d. below the diaphragm.

7. The brain and the spinal cord are located in the
   a. ventral cavity.
   b. spinal cavity.
   c. cranial cavity.
   d. dorsal cavity.

8. The epigastric region of the abdominal area is located
   a. just above the sternum.
   b. in the umbilical area.
   c. just below the sternum.
   d. in the pelvic area.

9. Shivering to keep the body warm is an example of
   a. anabolism.
   b. catabolism.
   c. metabolism.
   d. homeostasis.

10. The formation and release of hormones from a cell or structure is called
    a. digestion.
    b. excretion.
    c. synthesis.
    d. secretion.

FILL IN THE BLANKS

1. The standard used for measurement in science is the ________________ system.

2. Danny, age 6, fell off his skateboard and had a 1.5-inch abrasion on his left arm. This is the same as ________________ centimeters.

3. Two teaspoons of cough medicine equal ________________ milliliters of cough medicine.

4. The physician orders 2 grams of penicillin to be divided into 4 doses over 24 hours. This means the average single dose will be ________________ milligrams.

5. A kilogram is equal to ________________ pounds.

MATCHING

Match each term in Column I with its correct description in Column II.

	COLUMN I	COLUMN II
________	**1.** catabolism	a. balanced cellular environment
________	**2.** pelvic cavity	b. constructive chemical processes that use food to build the complex materials of the body
________	**3.** pericardial cavity	c. useful breakdown of food materials, resulting in the release of energy
________	**4.** anabolism	d. contained within the oral cavity
________	**5.** abdominal cavity	e. cavity in which the reproductive organs, urinary bladder, and lower part of the large intestine are located
________	**6.** diaphragm	f. cavity in which the stomach, liver, gallbladder, pancreas, spleen, small intestine, appendix, and part of the large intestine are located
________	**7.** homeostasis	g. cavity containing the heart
________	**8.** tissue	h. a group of cells that together perform a particular job
________	**9.** kidneys	i. portion of the dorsal cavity containing the brain
________	**10.** teeth and tongue	j. divides the ventral cavity into two regions
________	**11.** cranial cavity	k. structure located behind the abdominal cavity
________	**12.** organ system	l. organs grouped together because they have a related function
________	**13.** life function	m. an activity that a living thing performs to help it live and grow

APPLYING THEORY TO PRACTICE

1. In each of the following examples, choose the term that correctly describes the human body according to anatomical position:
   a. The palms are forward or backward.
   b. The liver is superior or inferior to the diaphragm.
   c. The hand is proximal or distal to the elbow.
   d. The shoulder blade is on the anterior or posterior part of the body.
   e. Cranial refers to the head or tail end of the body.
   f. The coronal plane divides the body into front and back or right and left sections.
   g. The arms are located on the medial or lateral side of the body.
   h. The transverse plane divides the body into superior and inferior or anterior and posterior parts.

2. Describe the following to a physician using the correct anatomical term:
   a. The location of an appendectomy scar
   b. A wound that is on the front of the leg
   c. The end of the spine
   d. A pain near the breastbone

3. Think about what your body does within a 24-hour period and name the life functions that take place.

CASE STUDY

An Emergency Medical Technician (EMT) responds to a call about a fall out of a tree. On arrival, the EMT sees a young boy lying at the bottom of the tree; his right arm is visibly deformed. The EMT suspects the arm may be broken.

1. Describe the anatomical terms the EMT will use to describe the injury to the emergency department physician.
2. The boy is right-handed. What life function will be affected by the fall?

LAB ACTIVITY 2-1 Anatomical Directions

- ***Objective:*** To properly use directional terms to reference anatomical regions
- ***Materials needed:*** pencil, paper

Step 1: You may work individually or with a lab partner. Each student will assume the anatomical position. Is it comfortable? Why is the anatomical position used in health care? Record your response.

Step 2: Ask your lab partner if they are comfortable in the anatomical position. Record your partner's response.

Step 3: State the reason why you think this position is comfortable or uncomfortable. Record your response.

Step 4: Locate your own anterior, posterior, lateral, medial, superior, and inferior body surfaces; repeat this step with your partner.

LAB ACTIVITY 2-2 Anatomical Planes

- ***Objective:*** To identify the types of planes used to describe anatomy and what those planes indicate about the anatomical region
- ***Materials needed:*** modeling clay, tongue depressors, pencil, paper

Step 1: Form the clay into a kidney shape.

Step 2: Using the tongue depressor, make a transverse cut of the kidney. What does this type of cut demonstrate? Record your answer.

Step 3: Make a sagittal cut. What does this type of cut demonstrate? Record your answer.

Step 4: Make a coronal cut. What does this type of cut demonstrate? Record your answer.

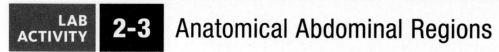

LAB ACTIVITY 2-3 Anatomical Abdominal Regions

- ***Objective:*** To identify each of the cavities of the abdomen and the organs that can be found in those regions
- ***Materials needed:*** anatomical model of a torso; models of a gallbladder, liver, stomach, colon, and pancreas; pencil and paper

Step 1: Place the organs correctly into the anatomical model.

Step 2: Record the name of the abdominal region in which each of the organs is located.

LAB ACTIVITY 2-4 Model of Levels of Organization

- ***Objective:*** To develop and use a model to illustrate the six-levels of hierarchical organization of interacting body systems
- ***Materials needed:*** notepaper, materials for building model (e.g., pencil, paper, clay, putty, styrofoam, wood)

Step 1: Create a model based upon the six levels of structural organization.

Step 2: Label the levels.

Step 3: Identify the functions at each level and write a brief explanation to share with the class during your presentation.

Step 4: Present the model to classmates.

CHAPTER 3

Chemistry of Living Things

Objectives

- Relate the importance of chemistry and biochemistry to health care.
- Define matter and energy.
- Explain the structure of an atom, an element, and a compound.
- Explain the importance of water to the body.
- Describe the four main groups of organic compounds: carbohydrates, fats, proteins, and nucleic acids.
- Explain the difference between the DNA molecule and the RNA molecule.
- Explain the difference between an acid, a base, and salt.
- Explain the acid-base balance.
- Describe why homeostasis is necessary for good health.
- Define the key words that relate to this chapter.

Key Words

acid
acidosis
alkali
alkalosis
amino acids
atom
base
buffer
carbohydrates
chemical bonds
chemistry
cholesterol
compounds
covalent bond
dehydrated
dehydration synthesis
deoxyribonucleic acid (DNA)
disaccharide
electrolytes
electrons
element
energy
enzymes
extracellular fluid
fats
glycogen
hydrogen bond
hydrolysis
hydroxide
intracellular fluid
ion
ionic bond
isotopes
kinetic energy
lipids
matter
molecule
monosaccharides
multicellular
neutralization
neutrons
nucleic acids
organic catalyst
organic compounds
pH scale
phospholipids
polysaccharides
potential energy
proteins
protein synthesis
protons
radioactive
ribonucleic acid (RNA)
steroids
unicellular

To be an effective health care provider, an individual must have a thorough understanding of the normal and abnormal functioning of the human body and knowledge of basic chemistry and biochemistry.

CHEMISTRY

Chemistry is the study of the structure of matter and the composition of substances, their properties, and their chemical reactions. Many chemical reactions occur in the human body. These reactions can range from the digestion of a piece of meat in the stomach to the formation of urine in the kidneys to the manufacture of proteins in a microscopic human cell. The chemical reactions necessary to sustain life occur in the cells. Thus, the study of the chemical reactions of living things is called *biochemistry*.

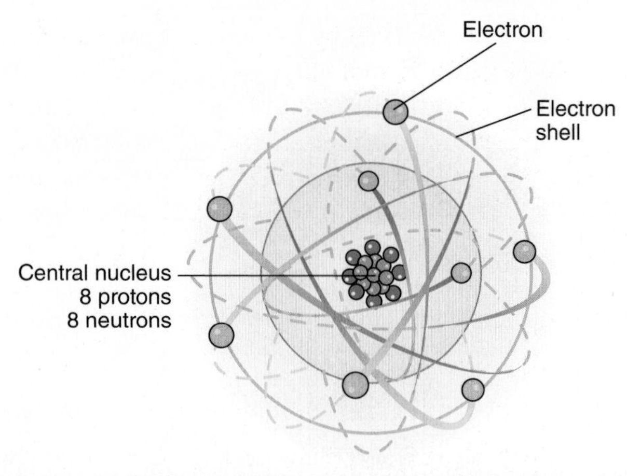

Figure 3-1 *Structure of an atom: Eight protons and eight neutrons are tightly bound in the central nucleus, around which the eight electrons revolve.*

MATTER AND ENERGY

Matter is anything that has weight (mass) and occupies space. Matter exists in solid, liquid, and gas forms. An example of solid matter in a person's body is bone; liquid matter is blood; gas is oxygen.

Matter is neither created nor destroyed, but it can change form through physical or chemical means. A physical change occurs when a person chews a piece of food and breaks it into smaller pieces. A chemical change occurs when the food is acted on by various chemicals in the body to change its composition. For example, imagine a piece of buttered toast that becomes molecules of fat and glucose to be used by the body for energy.

Energy is the ability to do work or to put matter into motion. Energy exists in the body as **potential energy** and **kinetic energy** (kih-**NET**-ik). Potential energy is energy stored in cells waiting to be released, whereas kinetic energy is work resulting in motion. Lying in bed is an example of potential energy; getting out of bed is an example of kinetic energy.

Atoms

An **atom** is the smallest piece of an element. Atoms are invisible to the human eye, yet they surround everything and are part of the human structure. Hydrogen is an example of an atom.

The normal atom is made up of subatomic particles: **protons**, **neutrons**, and **electrons**. Protons have a positive (+) electric charge; neutrons have no electric charge. Protons and neutrons make up the nucleus of an atom, which differs from the nucleus of a cell (**Figure 3-1**). Electrons have a negative (−) electric charge and are arranged around the nucleus in orbital zones, or *electron shells*. Atoms usually have more than one electron shell. The number and arrangement of the subatomic particles dictate how the atoms of one element differ from atoms of another element; for example, the structure of the hydrogen atom is different from the structure of the oxygen atom.

The number of protons of an atom is equal to the number of electrons; atoms are electrically neutral—neither negative nor positive.

Atoms of a specific element that have the same number of protons but a different number of neutrons are called **isotopes** (eye-so-**TOWPS**). All isotopes of a specific element have the same number of electrons. Certain isotopes are called **radioactive** isotopes because they are unstable and decay, meaning they come apart. They emit, or give off, energy in the form of radiation as they decay, which can be picked up by a detector. The detector not only detects the emission from a radioactive isotope but, with the aid of a computer, also forms the image of its distribution within the body. Radioactive isotopes can be used to study the structure and function of particular tissues. Nuclear medicine is a branch of medicine that uses radioactive isotopes to prevent, diagnose (see *Medical Highlights: Medical Imaging*), and treat disease. The most common uses of isotopes are for the treatment of thyroid conditions, prostate cancer, and cancer bone pain. Radioactive isotopes enable the physician to point the selected isotopes directly at the disease and destroy the diseased tissue.

Elements

Atoms that are the same, such as two hydrogen atoms, combine to form the next stage of matter, which is an **element**. An element is a substance that can neither be created nor destroyed by ordinary chemical means. Elements can exist in more than one phase in the body. Bones are solid and contain the element calcium. The air a person takes into their lungs contains the element oxygen, which is a gas. A person's cells are bathed in fluids that contain the elements of hydrogen and oxygen.

Ninety-eight elements are found naturally in the world, some in very small quantities; additional elements have been created by scientists. Each of the elements is represented by a chemical symbol or an abbreviation. **Table 3-1** shows a sampling of elements and their chemical symbols.

Medical Highlights

MEDICAL IMAGING

Medical imaging refers to the noninvasive techniques and processes used to create images of the human body for clinical purposes. Some of these techniques use radioactive isotopes. Many people fear overexposure to radiation from medical imaging processes, but the risk of radiation exposure must be weighed against other risks to one's health. Scientists have not been able to prove satisfactorily that low doses of radiation as used in medical settings increase cancer risks.

X-rays are a type of electromagnetic radiation best known for its ability to see through a person's skin, revealing images of the bones, lungs, or other body parts beneath the skin.

A *computed axial tomography (CAT or CT) scan* is a painless diagnostic X-ray procedure that uses ionizing radiation to produce cross-section images of the body. The computer detects the pattern of the radiation absorption and the variations in tissue density. From the detection of radiation absorption, a series of anatomical pictures is produced. The resulting scan is an analysis of a three-dimensional view of the tissue being evaluated. CT scans are most useful in evaluating the brain, detecting internal injuries or bleeding, and detecting cancer.

Magnetic resonance imaging (MRI) is a scanning procedure that provides visualization of fluids, soft tissue, and bony structures without the use of radiation. The person is placed inside a large, tube-shaped electromagnetic chamber, where specific frequencies of radio signals are generated that change the alignment of hydrogen atoms in the body. The computer analyzes the absorbed radio-frequency energy. Strong magnetic fields are used, causing the radio-frequency waves to produce images, which are projected on a screen. Persons with implanted metal devices, such as pacemakers and prosthetic knees, cannot undergo an MRI because the strong magnetic fields could damage them. An open MRI, which is open on all four sides and does not require placement inside a chamber, can be used for those who are claustrophobic (a pathological fear of confinement).

A *positron emission tomography (PET) scan* is a procedure in which the patient is given a short-lived radioactive isotope, either inhaled or injected, and placed in a scanner. The metabolic activity of the brain and numerous other body structures is shown through computerized color-coded images that indicate the degree and intensity of the metabolic processes. The patient may be asked questions to see how the brain activity changes by reasoning or remembering. PET scans are most useful for diagnosing brain tumors, cerebral palsy, stroke, and heart disease.

Bone scans, liver scans, brain scans, and *spleen scans* are procedures that scan various body parts with a gamma camera. Steps needed to prepare will depend on the type of scan to be done. Fasting may or may not be necessary, but it may be necessary to drink a radionuclide preparation or receive an injection of radionuclide material. The camera's recording of the concentration or collection of the radioactive substance specifically drawn to that area discloses the image of the area.

Ultrasound or *sonography* uses high-frequency sound waves that are directed through a transducer. Those sound waves bounce off the spot just below the transducer and generate an image that can be seen on the ultrasound screen. The screen image changes as the transducer is slowly moved from one location to another. This imaging choice is used for rotator cuff disorders and musculoskeletal disorders, and in obstetrics for visualizing the embryo, fetus, and placenta.

Doppler ultrasound is a variation of sonography in which returning sound waves are transformed into audible sounds that can be detected by earphones. The Doppler method measures blood flow by moving the transducer along the path of a blood vessel.

Mammography (see Chapter 22)

After having a diagnostic test done using radioactive material, patients are advised to drink lots of water to flush out radioactive material.

Table 3-1 *Sample Elements and Their Symbols*

ELEMENT	SYMBOL
Calcium	Ca
Carbon	C
Chlorine	Cl
Hydrogen	H
Iodine	I
Iron	Fe
Magnesium	Mg
Nitrogen	N
Oxygen	O
Phosphorus	P
Potassium	K
Sodium	Na
Zinc	Zn

Compounds

Various elements can combine in a *definite proportion by weight* to form **compounds**. A compound has different characteristics or properties depending on its elements. For example, the compound water (H_2O) is made of two parts hydrogen and one part oxygen. Separately, hydrogen and oxygen are gaseous elements, but when combined, the resulting compound is liquid water. Common table salt is a compound made from the two elements sodium (Na) and chlorine (Cl), chemically called sodium chloride (NaCl). Separately, sodium is a metallic element. It is light, silver-white, and shiny when freshly cut, but rapidly becomes dull and gray when exposed to air. Chlorine, on the other hand, is an irritating, greenish-yellow poisonous gas with a suffocating odor. However, the chemical combination of both sodium and chlorine results in sodium chloride, which is a crystalline powder that can be dissolved in water and has no distinct odor.

Just as elements are represented by symbols, compounds are represented by something called a *formula*. A formula shows the types of elements present and the proportion of each element present by weight. Some common formulas are H_2O (water), NaCl (common table salt), HCl (hydrogen chloride or hydrochloric acid), $NaHCO_3$ (sodium bicarbonate or baking powder), NaOH (sodium hydroxide or lye), $C_6H_{12}O_6$ (glucose or grape sugar), $C_{12}H_{22}O_{11}$ (sucrose or common table sugar), CO_2 (carbon dioxide), and CO (carbon monoxide).

A living organism can be compared with a chemical factory, whether it is a **unicellular** (yoo-nih-**SELL**-yoo-lar), or one-celled, microbe or a **multicellular** (mull-tye-**SELL**-yoo-lar), or many-celled, animal or plant. Most living organisms will take the 20 essential elements and change them into needed compounds for the maintenance of the organism. In many living organisms, the elements carbon, hydrogen, and oxygen are united to form **organic compounds** (or-**GAN**-ik). Compounds found in living things contain the element carbon.

Molecules

The smallest unit of a compound that still has the properties of the compound and the ability to lead its own stable and independent existence is called a **molecule** (**MOL**-eh-kyool). For example, the common compound water can be broken down into smaller and smaller droplets. The absolute smallest unit is a molecule of water (H_2O).

Chemical Bonds

In addition to combining to form elements, atoms can share or combine their electrons with atoms of other elements to form **chemical bonds**. One type of bond is called an **ionic bond** (**Figure 3-2**). If one atom gives up an electron to another atom to form an ionic bond, that atom will now have more protons than electrons and will

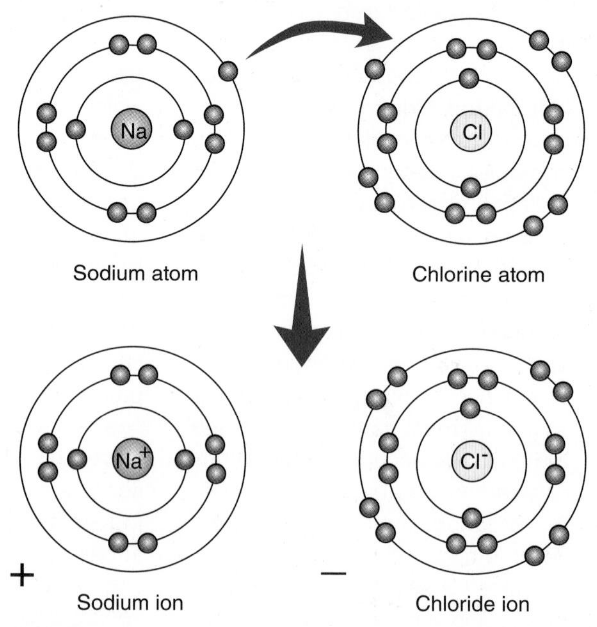

Figure 3-2 *Ionic bond: In this figure, the* Na^+ *atom gives up an electron to the* Cl^- *atom to form an ionic bond.*

have a positive (+) charge. The atom that took the extra electron will now have more electrons than protons and thus have a negative (−) charge. The positively or negatively charged particle is now called an **ion**. Ionically bonded atoms disassociate when immersed in water; an example is sodium chloride (Na^+Cl^-).

A second type of bond is the **covalent bond** (**Figure 3-3**). In this type of bond, the atoms share electrons to fill their outermost levels or shells. Molecules containing covalent bonds do not form ionic bonds and do not disassociate when immersed in water. Four of the most important elements found in cells form this type of bond. They are carbon, oxygen, hydrogen, and nitrogen.

A third type of bond is the **hydrogen bond**. Hydrogen bonds are very weak bonds. They help hold water molecules together by forming a bridge between the negative oxygen atom of one water molecule and the positive hydrogen atoms of another water molecule.

Electrolytes

When compounds are in a solution and act as if they have broken into individual pieces (ions), the elements of the compound are called **electrolytes** (ee-**LEK**-troh-lights). For example, in water, a salt solution consists of sodium (Na^+) ions with a positive charge and chlorine (Cl^-) ions with a negative charge.

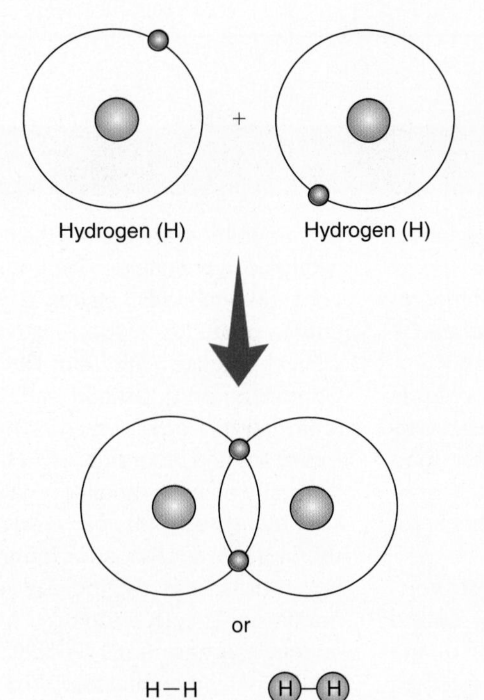

Figure 3-3 *Covalent bond: In this figure, each hydrogen atom will share an electron to form a compound.*

In the cells and tissue fluids of the body, ions make it possible for materials to be altered, broken down, and recombined to form new substances or compounds. Electrolytes are responsible for the acidity or alkalinity of solutions and can conduct an electrical charge. The ability to record electric charges within the tissue is invaluable for diagnostic tools such as *electrocardiograms*, which measure the electrical conduction of the heart.

TYPES OF COMPOUNDS

The various elements can combine to form a great number of compounds. All known compounds, whether natural or synthetic, can be classified into two groups: inorganic compounds and organic compounds.

Inorganic Compounds

Inorganic compounds are made of molecules that do not contain the element carbon (C), such as salt (NaCl). Exceptions include carbon dioxide (CO_2) and calcium carbonate ($CaCO_3$).

WATER

Water is the most important inorganic compound to living organisms. Water makes up 55% to 65% of a person's body weight. It is considered the universal solvent because more substances dissolve in water than in any other fluid. Most of the body's cellular processes take place in the presence of water. In anabolic reactions, water may be removed from the molecule with **dehydration synthesis**. When this occurs, the molecules fuse together and a new substance is formed. In catabolic reactions, water is added to the molecule—**hydrolysis**—to break down larger molecules. Water regulates body temperature, takes nutrients to cells, and takes away the waste products. Water is necessary for homeostasis. Water is essential to life; if a person does not have enough water, their body becomes **dehydrated** (dee-**HYE**-dray-ted), which is a life-threatening condition.

Organic Compounds

Organic compounds are found in living things and the products they make. These compounds always contain the element carbon, combined with hydrogen and other elements. Carbon has the ability to combine with other elements to form a large number of organic compounds. There are more than a million known organic compounds. Their molecules are comparatively large and complex; inorganic molecules are much smaller. The four main groups of organic compounds are carbohydrates, lipids, proteins, and nucleic acids.

CARBOHYDRATES

All **carbohydrates** are compounds of the elements carbon (C), hydrogen (H), and oxygen (O). These compounds have twice as many hydrogen atoms as oxygen and carbon atoms. Carbohydrates are divided into three groups: monosaccharides, disaccharides, and polysaccharides.

MONOSACCHARIDES **Monosaccharides** (mon-oh-**SAK**-ah-rides) are sugars that cannot be broken down any further. They are also called single or simple sugars. The types of monosaccharide sugars are glucose, fructose, galactose, ribose, and deoxyribose.

Glucose is an important sugar. It is the main source of energy in cells. Glucose, sometimes referred to as blood sugar, is carried by the bloodstream to individual cells and is stored in the form of **glycogen** (gly-co-**GEN**) in the liver and muscle cells. Glucose combines with oxygen in a chemical reaction called oxidation, which produces energy.

Fructose is the sweetest of the monosaccharides and is found in fruit and honey. Galactose is found in human breast milk, and nursing infants need it for development. Deoxyribose sugar is found in **deoxyribonucleic acid (DNA)** (dee-ock-see-rye-boh-new-**KLEE**-ik), and ribose sugar is found in **ribonucleic acid (RNA)** (**rye**-boh-new-**KLEE**-ik).

DISACCHARIDES A **disaccharide** (dye-**SAK**-ih-ride) is known as a double sugar because it is formed from two monosaccharide molecules by dehydration synthesis. Refer to **Table 3-2** for an illustration of the process of dehydration synthesis.

Examples of disaccharides are sucrose (table sugar), maltose (malt sugar), and lactose (milk sugar). Disaccharides must first be broken down by the process of digestion (hydrolysis) into monosaccharides to be absorbed and used by the body.

Table 3-2 *The Monosaccharide Composition of Sucrose, Maltose, and Lactose*

MONOSACCHARIDE + MONOSACCHARIDE – H_2O (DEHYDRATION SYNTHESIS)	***FORMS***	***DISACCHARIDE***
Glucose + fructose – H_2O	→	Sucrose
Glucose + glucose – H_2O	→	Maltose
Glucose + galactose – H_2O	→	Lactose

 Career Profile

RADIOLOGIC TECHNOLOGIST

Medical uses of radiation go far beyond the diagnosis of broken bones by X-ray. Radiation is used to produce images of the interior of the body and to treat cancer. The term *diagnostic imaging* does not just apply to X-ray techniques; it also includes ultrasound and MRI scans.

Radiographers produce X-ray films for use in diagnosing disease. They prepare the patient for the procedure by explaining the process, positioning the patient, removing jewelry and other articles through which X-rays cannot penetrate, shielding the patient to prevent unnecessary radiation exposure, and taking the picture. Experienced radiographers may also perform more complex imaging tests, such as mammography and fluoroscopy, or operate CT scanners and MRI machines.

Radiation therapy technologists prepare cancer patients for treatment and administer prescribed doses of ionizing radiation to specific body parts. They check for radiation side effects.

Sonographers project nonionizing, high-frequency sound waves into specific areas of the patient's body; the machine then collects the reflected echoes to form an image.

Education for these positions is offered in hospitals, colleges, and vocational-technical institutes. The course of study includes class and clinical practice. The Joint Review Committee on Education in Radiologic Technology accredits most formal training programs in this field. Specialty areas in radiology include MRI technology, nuclear medicine technology, diagnostic technology, ultrasound technology, and mammography technology. Most specialty areas require additional education and certification. The job outlook in this field is expected to grow faster than average.

POLYSACCHARIDES A large number of carbohydrates found in or made by living organisms are **polysaccharides** (pol-ee-**SAK**-ah-rides). These are large, complex molecules of hundreds to thousands of glucose molecules bonded together in one long, chainlike molecule. Examples of polysaccharides are starch, cellulose, and glycogen. Under the proper conditions, polysaccharides can be broken down into disaccharides and then, finally, into monosaccharides.

Starch is a polysaccharide found in grain products and root vegetables, such as potatoes.

LIPIDS

Lipids are molecules containing the elements carbon, hydrogen, and oxygen. Lipids are different from carbohydrates because they have proportionately much less oxygen in relation to hydrogen. Examples of lipids are fats, phospholipids, and steroids.

Lipids are referred to as "fats." Although "fat-free" foods have become popular, lipids or fats are essential to health. Lipids are an important source of stored energy. They make up the essential steroid hormones and help insulate the body. It is when the intake of lipids in the form of fat becomes excessive that a health problem may occur.

Fats, also called *triglycerides*, consist of glycerol and fatty acids and make up 95% of fats in the human body.

Phospholipids contain carbon, hydrogen, oxygen, and phosphorus. This type of lipid may be found in the cell membranes, the brain, and nervous tissue.

Steroids are lipids that contain **cholesterol** (koh-**LES**-ter-ol). Cholesterol is essential to the structure of the semipermeable membrane of the cell. It is necessary for the manufacture of vitamin D and in the production of male and female hormones. Cholesterol is needed to make the adrenal hormone cortisol. In certain people, however, cholesterol can accumulate in the arteries, becoming problematic.

PROTEINS

Proteins are organic compounds containing the elements carbon, hydrogen, oxygen, nitrogen, and usually phosphorus and sulfur. Proteins are among the most diverse and essential organic compounds found in all living organisms. They are found in every part of a living cell; they are also an important part of the outer protein coat of all viruses. Proteins also serve as binding and structural components of all living things. For example, large amounts of protein are found in fingernails, hair, cartilage, ligaments, tendons, and muscle.

Amino acids are small molecular units that work together to build proteins in the body. They are vital for proper body function. Twenty different amino acids are combined in any number and sequence to make up all the types of protein. They are classified as essential and nonessential. Essential amino acids must be obtained from dietary sources and cannot be made up by the body. Nonessential amino acids are those that the body can manufacture. **Table 3-3** lists the essential and nonessential amino acids.

The DNA creates different types of proteins that are specialized based upon the sequence of the amino acids. Large protein molecules are constructed from any number and sequence of these amino acids. The number of amino acids in any given protein molecule can range from 300 to several thousand. Therefore, the structure of proteins is quite complicated.

ENZYMES **Enzymes** are specialized protein molecules found in all living cells. They help control the various chemical reactions occurring in a cell so that each reaction occurs at just the right moment and at the right speed. Enzymes help provide energy for the cell, assist in the making of new cell parts, and control almost every process in a cell. Because enzymes are capable of such activity, they are a type of **organic catalyst**. An enzyme or organic catalyst affects the rate or speed of a chemical reaction without being changed itself. Enzymes can also be used over and over again. An enzyme molecule is highly specific in its action. Enzymes are either made up of all protein, or they consist of a protein part (apoenzyme) attached to an organic or nonorganic cofactor. The apoenzyme and the cofactor together are called a holoenzyme.

NUCLEIC ACIDS

Nucleic acids (new-**KLEE**-ik) are important organic compounds containing the elements carbon, oxygen, hydrogen, nitrogen, and phosphorus. The two major types of nucleic acids are deoxyribonucleic acid (DNA) and ribonucleic acid (RNA).

Table 3-3 *Essential and Nonessential Amino Acids*

ESSENTIAL AMINO ACIDS	NONESSENTIAL AMINO ACIDS
Arginine	Alanine
Histidine	Asparagine
Isoleucine	Aspartate
Leucine	Cysteine
Lysine	Glutamate
Methionine	Glutamine
Phenylalanine	Glycine
Threonine	Proline
Tryptophan	Serine
Valine	Tyrosine

STRUCTURE OF NUCLEIC ACIDS Nucleic acids are the largest known organic molecules. They are made from thousands of smaller, repeating subunits called *nucleotides*. A nucleotide is a complex molecule composed of three different molecular groups. **Figure 3-4** shows a typical nucleotide. Group 1 is a phosphate or phosphoric acid group (H_3PO_4); Group 2 represents a five-carbon sugar. Depending on the nucleotide, the sugar could be either a ribose or a deoxyribose sugar. Group 3 represents a nitrogenous base. The two groups of nitrogenous bases are the purines and the pyrimidines. The purines are adenine (A) and guanine (G); the pyrimidines are cytosine (C) and thymine (T).

DNA STRUCTURE AND FUNCTION DNA is a double-stranded molecule referred to as a double helix. This structure resembles a twisted ladder. The sides of the ladder are formed by alternating bands of a sugar (deoxyribose) unit and a phosphate unit. The rungs of the ladder are formed by nitrogenous bases that always pair in very specific ways: thymine (T) pairs with adenine (A), and cytosine (C) pairs with guanine (G) (**Figure 3-5**).

DNA is involved in the process of heredity. The nucleus of every human cell contains 46 chromosomes (23 pairs), creating a long, coiled molecule of DNA. These chromosomes contain about 100,000 genes. This genetic information tells a cell what structure it will possess and what function it will have. The DNA molecule passes on the genetic information from one generation to the next.

DNA structures are unique for each person and are used as a means of identification; only a very small amount of DNA is necessary for identification.

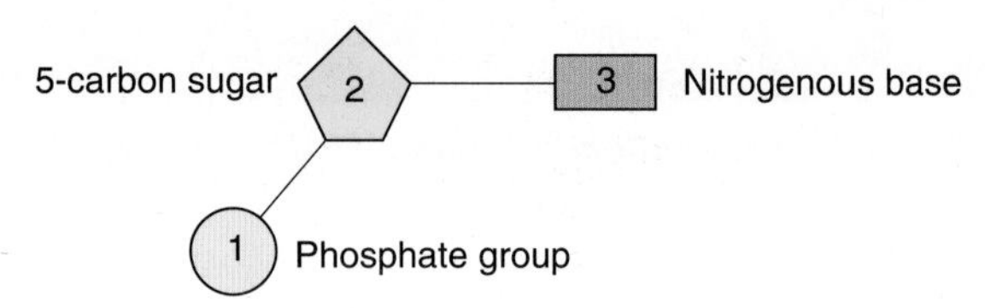

Figure 3-4 *Structure of a typical nucleotide*

Did You Know?

If the strands of DNA from a single cell were stretched out end to end, it would measure 6 feet long but would be so incredibly thin (50 trillionths of an inch wide) that no one could see it.

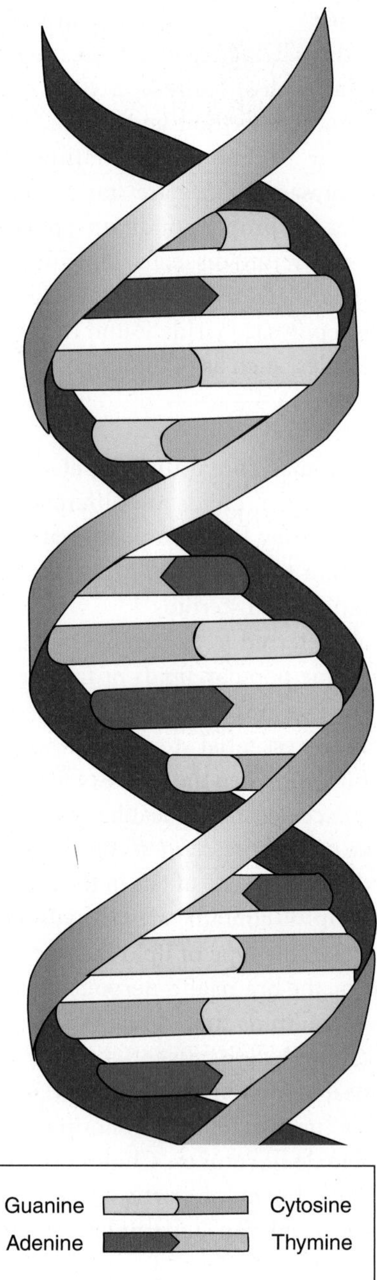

Figure 3-5 *Illustration of DNA*

RNA STRUCTURE AND FUNCTION The RNA nucleotide consists of a phosphate group, the ribose sugar, and any one of the following nitrogenous bases: adenine, cytosine, guanine, and uracil (instead of thymine). The RNA molecule is single-stranded, whereas the DNA molecule is double-stranded.

The three different types of RNA in a cell are messenger RNA (m-RNA), transfer RNA (t-RNA), and ribosomal RNA (r-RNA). Messenger RNA carries the instructions for **protein synthesis** from the DNA molecule located in the nucleus of a cell to the ribosomes in the cytoplasm. The transfer RNA molecule picks up amino

Table 3-4 *Differences between DNA and RNA Molecules*

TYPE OF NUCLEIC ACID	TYPE OF SUGAR PRESENT	TYPES OF BASE PRESENT	PHOSPHATE GROUP	LOCATION	NUMBER OF STRANDS PRESENT
DNA	Deoxyribose	A, T, G, C	Same as RNA	Cell nucleus, chromosomes	2
RNA	Ribose	A, U, G, C	Same as DNA	Cytoplasm, nucleoli, ribosomes	1

acid molecules in the cytoplasm and transfers them to the ribosomes. Ribosomal RNA helps in the attachment of the m-RNA to the ribosome. Protein synthesis is the process by which amino acids are linearly arranged into proteins through the involvement of messenger RNA, transfer RNA, ribosomal RNA, and various enzymes.

Table 3-4 shows the basic differences between the DNA molecule and the RNA molecule.

ACIDS, BASES, AND SALTS

Before ending the discussion of basic chemistry and biochemistry, a brief discussion of acids, bases, salts, and pH is essential.

Many inorganic and organic compounds found in living organisms are ones that people use in their daily lives. They can be classified into one of three groups: acids, bases, and salts. Many are familiar with the sour taste of vinegar and citrus fruits, such as grapefruits, lemons, and limes. The sour taste is due to the presence of compounds called acids. What characteristics do acids have to set them apart from bases and salts?

Acids

An **acid** is a substance that, when dissolved in water, will ionize into positively charged hydrogen ions (H^+) and negatively charged ions of some other element. Basically, an acid is a substance that yields hydrogen ions (H^+) in solution. For example, hydrogen chloride (HCl) in pure form is a gas. But when bubbled into water, it becomes hydrochloric acid. How does this happen? Simply: in a water solution, hydrogen chloride ionizes into one hydrogen ion and one negatively charged chloride ion.

$$HCl + H_2O \longrightarrow H^+ + Cl^-$$

Hydrogen chloride in solution ⟶ Hydrogen ion + Chloride ion

It is the presence of the hydrogen ions that gives hydrochloric acid its acidity and sour taste. However, one should *not* taste any substance to identify it as an acid. There are more reliable and safer methods for identification. A substance can be tested for its acidity through the use of specially treated paper called litmus. In the presence of an acid, blue litmus paper turns red. **Table 3-5** names some common acids, their formulas, and where they are found or how they are used.

Bases

A **base** or **alkali** is a substance that, when dissolved in water, ionizes into negatively charged **hydroxide** (OH^-) ions and positively charged ions of a metal. For example, sodium hydroxide (NaOH) ionizes into one sodium ion (Na^+) and one hydroxide ion (OH^-). The reaction can be shown as follows:

$$NaOH \longrightarrow Na^+ + OH^-$$

Sodium hydroxide in solution ⟶ Sodium ion + Hydroxide ion

Table 3-5 *Names, Formulas, and Locations or Uses of Common Acids*

NAME OF ACID	FORMULA	LOCATION OR USE
Acetic acid	CH_3COOH	Found in vinegar
Boric acid	H_3BO_3	Weak eyewash
Carbonic acid	H_2CO_3	Found in carbonated beverages
Hydrochloric acid	HCl	Found in stomach
Nitric acid	HNO_3	Industrial oxidizing acid
Sulfuric acid	H_2SO_4	Found in batteries and industrial mineral acid

Bases have a bitter taste and feel slippery between the fingers. They turn red litmus paper blue. **Table 3-6** names some common bases, their formulas, and their location or use.

Neutralization and Salts

When an acid and a base are combined, they form a salt. This type of reaction is called a **neutralization** (new-tral-ih-**ZAY**-shun) and is classified as an exchange reaction. In a neutralization reaction, hydrogen ions (H^+) from the acid and hydroxide ions (OH^-) from the base join to form water. At the same time, the negative ions of the acid combine with the positive ions of the base to form the compound salt. For example, hydrochloric acid and sodium hydroxide combine to form sodium chloride and water. The hydrogen ions from the acid unite with the hydroxide ions from the base to form water. The sodium ions (Na^+) combine with the chloride ions (Cl^-) to form sodium chloride (NaCl). When the water evaporates, solid salt remains. The neutralization reaction is shown in **Figure 3-6**.

pH Scale

pH is a measure of the acidity or alkalinity of a solution. Special pH meters determine the hydrogen or hydroxide ion concentration of a solution. The **pH scale** is used to measure the acidity or alkalinity of a solution, and it ranges from 0 to 14. A pH reading of 7 indicates that a particular solution has the same number of hydrogen ions as hydroxide ions. This is a neutral pH; distilled water is neutral with a pH value of 7. Any pH value between 0 and 6.9 indicates an acidic solution. The lower the pH number, the stronger the acid, or the higher the hydrogen ion concentration. A pH value between 7.1 and 14 indicates that a solution is basic, or alkaline. Thus, the greater the number above 7, the stronger the base, or the greater the hydroxide ion concentration. **Figure 3-7** shows the pH values of some common acids, bases, and human body fluids.

Table 3-6 *Names, Formulas, and Locations or Uses of Common Bases*

NAME OF BASE	FORMULA	LOCATION OR USE
Ammonium hydroxide	NH_4OH	Household liquid cleaners
Magnesium hydroxide	$Mg(OH)_2$	Milk of magnesia
Potassium hydroxide	KOH	Caustic potash
Sodium hydroxide	NaOH	Lye

Homeostasis of Acid-Base

Living cells and the fluids they produce are usually nearly neutral, neither strongly acidic nor strongly alkaline. Living cells are very sensitive to even a slight change in the acid-base balance. For instance, human tears have a pH of 7.4, and human blood has a range of 7.35 to 7.45.

A disruption in the acid-base balance can result in disease conditions; if the body becomes too acidic, **acidosis** (ac-i-**DOH**-sis) will occur, and if the body becomes too alkaline, **alkalosis** (al-**KA**-lo-sis) may occur. The lungs and kidneys help to maintain homeostasis; the lungs excrete carbon dioxide, and the kidneys excrete excess acids and bases.

The maintenance of a balanced pH is also achieved through a compound called a **buffer**. Sodium bicarbonate ($NaHCO_3$) acts as a buffer in many living organisms. Buffers help a living organism maintain a constant pH value, which contributes to its homeostasis, or balanced state. A state of homeostasis is required for the body to function at an optimum level of health. If a control system like the acid-base or electrolyte balance is not maintained, cells and tissue will become damaged. A moderate dysfunction causes illness; a severe dysfunction causes death.

Fluids

Optimum cell functioning requires a stable cellular fluid environment. The fluid that bathes the cell and transports nutrients into and out of the cell is known as **extracellular fluid**. This includes the blood, lymph, and fluid between the tissues, known as *interstitial fluid* (in-ter-**STISH**-al). The fluid within the cell is called **intracellular fluid**.

Hydrochloric acid + Sodium hydroxide ⟶ Sodium chloride (salt) + Water

$$HCl + NaOH \longrightarrow NaCl + H_2O$$

Figure 3-6 *Neutralization or exchange reaction*

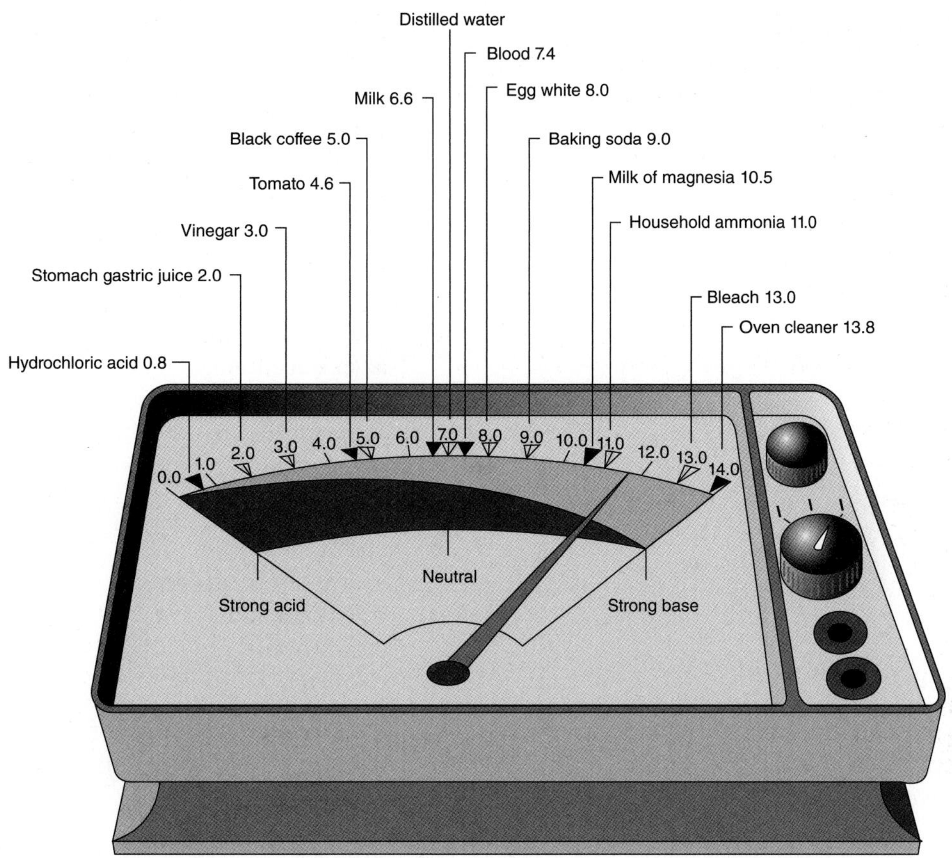

Figure 3-7 *pH values of common acids, bases, and human body fluids*

Medical Terminology

chem	chemical	**intra/cellular**	inside the cell
chemistry	study of chemical composition of matter	**mono**	one
di	two	**mono/saccharide**	contains one sugar
-saccharide	sugar containing carbon, hydrogen, and oxygen	**multi**	many
di/saccharide	contains two sugars	**multi/cellular**	many cells
extra	outside	**poly**	many
-cellular	pertaining to cell(s)	**poly/saccharide**	contains many sugars
extra/cellular	outside the cell	**uni**	one
intra	inside	**uni/cellular**	one-celled

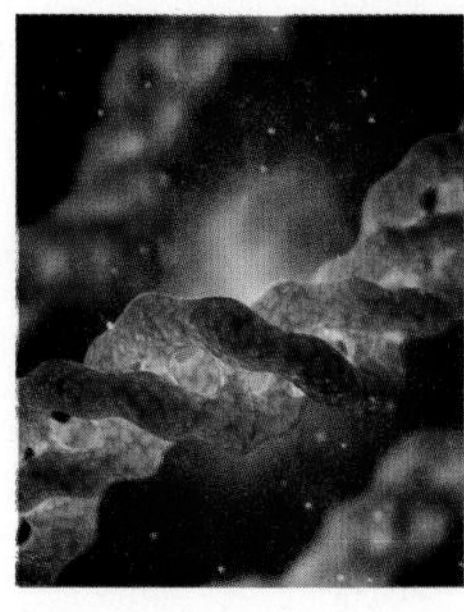

Study Tools

Workbook	Activities for Chapter 3
Online Resources	PowerPoint presentations

REVIEW QUESTIONS

Select the letter of the choice that best completes the statement.

1. A substance that has weight and occupies space is called
a. kinetic energy.
b. a catalyst.
c. matter.
d. potential energy.

2. Walking is an example of
a. a catalyst.
b. kinetic energy.
c. matter.
d. potential energy.

3. Water is classified as a(n)
a. atom.
b. element.
c. mineral.
d. compound.

4. Atoms of a specific element that have the same number of protons but a different number of neutrons are called
a. isotopes.
b. DNA.
c. RNA.
d. compounds.

5. Sugar stored in the liver and muscle cells for energy is called
a. glucose.
b. glycogen.
c. fructose.
d. ribose.

6. A chemical reaction in the cell is affected by
a. enzymes.
b. organic compounds.
c. nucleic acids.
d. energy.

7. Fluid found inside the cell is called
a. extracellular.
b. interstitial.
c. intracellular.
d. intercellular.

8. The compound with a pH of 9 is alkaline and is
a. milk of magnesia.
b. baking soda.
c. ammonia.
d. bleach.

9. When proper amounts of an acid and base are combined, the products formed are salt and
a. gas.
b. water.
c. another base.
d. another acid.

10. The name given to the atomic particle found outside the nucleus of an atom is
a. proton.
b. neutron.
c. electron.
d. ion.

MATCHING

Match each term in Column I with its correct description in Column II.

	COLUMN I	COLUMN II
________	**1.** glucose	a. fluid within the cell
________	**2.** electrolyte	b. double sugar
________	**3.** intracellular	c. triglycerides
________	**4.** disaccharides	d. chromosomes
________	**5.** HCl	e. conducts an electrical charge in a solution
________	**6.** steroid	f. blood sugar
________	**7.** energy	g. positively or negatively charged particle of an atom
________	**8.** ion	h. ability to do work
________	**9.** DNA	i. cholesterol
________	**10.** fats	j. found in the stomach

APPLYING THEORY TO PRACTICE

1. Read the label on a loaf of bread and state why it can say it has "no cholesterol."
2. What diagnostic imaging device would be used for the following conditions?
   a. Brain tumor
   b. Cancer of the stomach
   c. Liver disease
   d. Pregnancy
3. Have a panel discussion on the ethics of DNA testing as part of a pre-employment physical.
4. Explain how the structure of DNA determines the structure of proteins. The explanation should include:
   a. How proteins create specialized cells
   b. How specialized cells are required for the essential functioning of life
5. Explain how carbon, hydrogen, and oxygen from sugar molecules may combine with other elements to form amino acids or other large carbon-based molecules.

CASE STUDY

Isabel Savon is 34 years old. She has come to the clinic because of a general feeling of weakness and some difficulty walking. She also has had problems with her vision. When you bring Isabel to the examining room, she asks you to leave the door open because she is afraid of being shut inside.

The physician does a physical examination on Isabel and orders some diagnostic tests. A possible diagnosis for Isabel is multiple sclerosis.

1. What is the fear that Isabel experiences known as?
2. Understanding Isabel's fears, what type of nuclear imaging test will be ordered for her?
3. Isabel wants to know how nuclear imaging works; she is afraid of radiation. Explain to her how imaging devices work.
4. What additional instructions and information can you give Isabel regarding the test?
5. Are there other imaging tests that could be ordered for Isabel?

3-1 Model of an Atom

- ***Objective:*** To make a model of a helium atom (2 parts neutron, 2 parts proton, 2 parts electron), identifying where the subatomic parts are located
- ***Materials needed:*** colored pieces of red, yellow, and blue clay; 2 strong, 4-inch pieces of wire; toothpicks; scissors; magic marker; pen; notebook
- ***Note:*** Before and after any laboratory experiment, you must wash your hands.

Step 1: Make two red clay balls to represent protons, two balls of yellow clay to represent neutrons, and two smaller balls of blue clay to represent electrons.

Step 2: Attach the red and yellow clay balls to each other using the toothpicks. Cut off any excess toothpick material sticking out.

Step 3: Stick a piece of the wire into each side of the cluster. At the end of each wire, attach the blue balls of clay. This is to illustrate that electrons are outside the nucleus, usually in an orbital ring.

Step 4: Mark the subatomic part that has a positive charge with a plus sign and mark the subatomic part that has a negative sign with a minus sign.

Step 5: Describe the subatomic parts that make up the nucleus of the atom and record your findings.

Step 6: What subatomic part has the positive charge, and what subatomic part has the negative charge? What subatomic part has no charge? Record your findings.

Step 7: Use toothpicks to add another two round balls of yellow into the nucleus. Define the term that is used to describe an atom in which the number of neutrons is more than the number of protons. Record your answer.

Step 8: How are these types of atoms used in medicine?

LAB ACTIVITY 3-2 Test for Carbohydrates–Starch

- ***Objective:*** To test for carbohydrates
- ***Materials needed:*** paring knife, spatula, washed and dried apple, iodine solution, dropper, small glass or plastic dish, paper, pen

Step 1: Use a paring knife (with caution) to peel the apple.

Step 2: Cut the apple into three slices.

Step 3: Use the spatula to place the apple slices on the dish.

Step 4: Add 2–3 drops of iodine solution on the apple slices.

Step 5: Record the changes that occurred on the apple slices.

Step 6: What does this change indicate?

LAB ACTIVITY 3-3 Test for Proteins

- ***Objective:*** To test for proteins
- ***Materials needed:*** ¼ cup of milk, ¼ cup of water, ¼ cup of soup, dropper, 3 test tubes, test tube holder, Biuret solution, marker, pen, paper

Step 1: Place the test tubes in the test tube holder and label the test tubes A, B, and C.

Step 2: Use the dropper to put 40 drops of milk in test tube A.

Step 3: Use the dropper to put 40 drops of water in test tube B.

Step 4: Use the dropper to put 40 drops of soup into test tube C.

Step 5: Add 3 drops of Biuret solution to each test tube.

Step 6: Observe what occurred in test tubes A, B, and C. What does this indicate? Record your observation.

LAB ACTIVITY 3-4 Test for Fats

- ***Objective:*** To test for fats
- ***Materials needed:*** 2 tablespoons of a butter pat, paper towel, pen, paper

Step 1: Place the butter pat on a piece of paper towel.

Step 2: Place the butter pat and paper towel in the sunlight or a warm space.

Step 3: After 15 minutes, observe what has occurred.

Step 4: What does the change in the butter pat indicate? Record your answer.

LAB ACTIVITY 3-5 Acid or Base

- ***Objective:*** To identify the difference between an acidic (containing an acid), a basic (containing a base), and a neutral substance using litmus paper and pH-indicator scale paper
- ***Materials needed:*** paper cups, red or blue litmus paper, pH-indicator scale paper, tap water, vinegar, liquid soap, tomato juice, nail polish remover, baking soda solution, milk, lemon juice, a list of the solutions

Step 1: Place the solutions into separate paper cups and label the contents.

Step 2: Using litmus paper, indicate if the solution is an acid or a base and record your results on the list.

Step 3: Using pH-indicator scale paper, mark the pH of each solution.

Step 4: Which solution is the strongest acid?

Step 5: What is the pH of water?

LAB ACTIVITY 3-6 Effects of Antacid on an Acidic Stomach

- ***Objective:*** To determine the effectiveness of various antacid preparations or household remedies on an acidic stomach; the stomach under normal conditions has a pH of about 2
- ***Materials needed:*** measuring cup, vinegar, water, paper cups, TUMS®, Rolaids®, PEPCID AC®, Alka-Seltzer®, baking soda solution, pH-indicator paper, pencil, paper on which to record your results

Step 1: Mix 1 oz of vinegar with 8 oz of water to make a solution that represents an acidic stomach.

Step 2: Use pH-indicator paper to test the pH of the acidic stomach preparation. Record your result.

Step 3: Place approximately 1.5 oz of the acidic stomach solution into each of five different paper cups.

Step 4: Add one type of antacid preparation or 1 tablespoon of the baking soda solution to separate cups of the acidic stomach solution.

Step 5: After adding antacid preparation, does the solution fizz? What is occurring? Record your results.

Step 6: After the tablets and baking soda solutions have dissolved, retest each of the solutions with pH-indicator paper to measure any changes in the pH of the solution. Record your results.

Step 7: Did the antacid preparation raise the pH of the acidic stomach solution?

Step 8: Which preparation was most effective as an antacid?

Step 9: Obtain the prices of the various antacids. Which preparation is most cost-effective (least expensive to produce the desired result)?

Step 10: Record your results for steps 7, 8, and 9.

LAB ACTIVITY 3-7 Testing Your pH Level

- ***Objective:*** To determine the pH of your body by testing your saliva
- ***Materials needed:*** bottle of pH test strips with color-coded scale, spoon, paper, pencil

Step 1: Recall what you had for breakfast or lunch. If you have not had breakfast or lunch, your instructor may give you a snack and wait until the end of the class session to do the experiment.

Step 2: Assemble pH test strip and spoon.

Step 3: Have test strip ready.

Step 4: Spit on a spoon.

Step 5: Dip the test strip into your saliva.

Step 6: Immediately compare the color on the test strip with the color-coded chart.

Step 7: Record your findings.

Step 8: Wash the spoon and return materials.

LAB ACTIVITY 3-8 Exercise and Homeostasis

- ***Objective:*** Observe how homeostasis returns the pulse rate to normal after exercise
- ***Material needed:*** Student A, Student B, watch, timer, chair, pen, paper

Step 1: Student A will take the pulse of Student B for 30 seconds, multiply it by 2, and record the results.

Step 2: Using the timer, Student A will have Student B run in place for 2 minutes. After two minutes, Student A will take the pulse of Student B for 30 seconds, multiply it by 2, and record the results.

Step 3: Student A will have Student B rest in a chair for 3 minutes. Student A will take the pulse of Student B for 30 seconds, then multiply it by 2. Record the results.

Step 4: Is there a difference in the pulse rates from step 1 and step 2?

Step 5: Is there a difference in the pulse rates recorded in step 2 and step 3? What process has occurred?

Step 6: What will occur if the pulse rate does not return to normal? What happens if homeostasis is not restored?

Step 7: Research biofeedback and explain why biofeedback is an important technique that is used for your body to return to homeostasis.

Step 8: Research the chemical reactions that take place in the body during exercise that cause changes in pulse rate.

CHAPTER 4

Cells

Objectives

- Identify the structure of a typical cell.
- Define the function of each component of a typical cell.
- Relate the functions of cells to the functions of the body.
- Describe the processes that transport materials in and out of a cell.
- Describe a tumor and define cancer.
- Define the key words that relate to this chapter.

Key Words

active transport
adenosine triphosphate (ATP)
anaphase
anoxia
apoptosis
atrophy
benign
biomarkers
cancer
cell membrane
centrioles
centrosomes
chromatid
chromatin
chromosomes
cilia
cytoplasm
cytoskeleton
diffusion
dysplasia
endoplasmic reticulum
equilibrium
eukaryote
filtration
flagella
Golgi apparatus
hyperplasia
hypertonic solution
hypertrophy
hypotonic solution
hypoxia
interphase
isotonic solution
lysosomes
meiosis
metaphase
metastasis
mitochondria
mitosis
necrosis
neoplasia
neoplasms
nuclear membrane
nucleolus
nucleoplasm
nucleus
organelles
osmosis
osmotic pressure
papilloma
passive transport
peroxisomes
phagocytosis
pinocytic vesicles
pinocytosis
prophase
protoplasm
replication
ribosomes
solutes
telophase
tumor
vacuole

The body of a plant or animal seems to be a single entity, but when any portion is examined under a microscope, it is found to be made up of many small, discrete parts. These tiny parts, or units, are called cells. They got their name because of what they resembled when they were first discovered in the 1600s by Robert Hook. When examining a piece of cork under a crude microscope, the units reminded him of a monk's room, which is called a cell. All living things—whether plant or animal, unicellular or multicellular, large or small—are composed of cells. A cell is microscopic in size. A body is made up of trillions of cells that live mostly for a few weeks or months, die, and are replaced by new cells. The major parts of the cell are the protoplasm, the cell/plasma membrane, and the nucleus. Another term for a cell is **eukaryote** (eu-**KARY**-ote), which is defined as any cell that possesses a clearly defined nucleus. *The cell is the basic unit of structure and function of all living things.*

PROTOPLASM

Cells are composed of **protoplasm** (pro-toh-**PLAZM**), an aqueous solution of carbohydrates, proteins, lipids, nucleic acids, and inorganic salts surrounded by a cell membrane. Protoplasm inside the nucleus of a cell is called **nucleoplasm** (**NOO**-klee-oh-plazm); outside the nucleus, it is called **cytoplasm** (**SIGH**-toh-plazm). These components are organized into structures that have a specific function in the cell and are called organelles. **Organelles** (or-guh-**NELZ**) common to human cells include the nucleus, **ribosomes** (**RYE**-boh-sohmz), **centrosomes** (**SEN**-troh-sohmz), **centrioles** (**SEN**-tree-olz), **endoplasmic reticulum** (en-doh-**PLAZ**-mik re-**TICK**-you-lum), **mitochondria** (my-toh-**KON**-dree-ah), **lysosomes** (**LIGH**-soh-sohmz), **peroxisomes** (peh-**ROKS**-ih-sohmz), the **Golgi apparatus** (**GOHL**-jee), and a **cytoskeleton**.

Because cells are microscopic, a special unit of measurement is used to determine their size. This is the micrometer (μm), or micron (μ). It is used to describe both the size of cells and their cellular components. To see or study a cell in fine detail, an electron microscope must be used.

To better understand the structure of a cell, one must compare a living entity—such as a human being—to a house. The many individual cells of this living organism are comparable to the many rooms of a house. Just as each room is bounded by four walls, a floor, and a ceiling, a cell is bounded by a specialized cell membrane with many openings. Cells, like rooms, come in a variety of shapes and sizes. Every kind of room or cell has its own unique function. A house can be made up of a single room or many. In much the same fashion, a living thing can be made up of only one cell—unicellular—or many cells—multicellular. **Figure 4-1** shows the structure of a typical animal cell.

> **Did You Know?**
>
> *Fifty thousand cells in a human body will die and be replaced with new cells in the time it takes to read this sentence.*

Cell/Plasma Membrane

Every cell is surrounded by a **cell membrane**, sometimes called a plasma membrane. The membrane separates the cell from its external environment and from neighboring cells. It also regulates the passage or transport of certain molecules into and out of the cell, while preventing the passage of others. This is why the cell membrane is often called a selective semipermeable membrane. The cell membrane is composed of a double phospholipid layer, with proteins embedded in the layer. The phospholipid looks like a balloon with tails. The round, balloon-like part is hydrophilic, meaning it attracts water, and the double tails are hydrophobic, meaning they repel water. This arrangement allows for the easy passage of water molecules through the cell membrane by osmosis. The proteins embedded in the double phospholipid layer allow for the passage of molecules and ions across the cell membrane (**Figure 4-2**).

Nucleus

The **nucleus** is the most important organelle within a cell. It has two vital functions: to control the activities of the cell and to facilitate cell division. This spherical organelle is usually located in or near the center of the cell. Various dyes or stains, such as iodine, can be used to make the nucleus stand out. The nucleus stains vividly because it contains deoxyribonucleic acid (DNA) and protein. Surrounding the nucleus is a membrane called the nuclear membrane.

When a cell reaches a certain size, it divides to form two new cells. The DNA and protein are arranged in a loose and diffuse state called **chromatin**. The chromatin condenses to form short, rod-like structures called **chromosomes**. Each species has a specific number of chromosomes in the nucleus. The number of chromosomes for the human being is 46.

The nucleus divides first by a process called mitosis. It is only during the process of mitosis that the

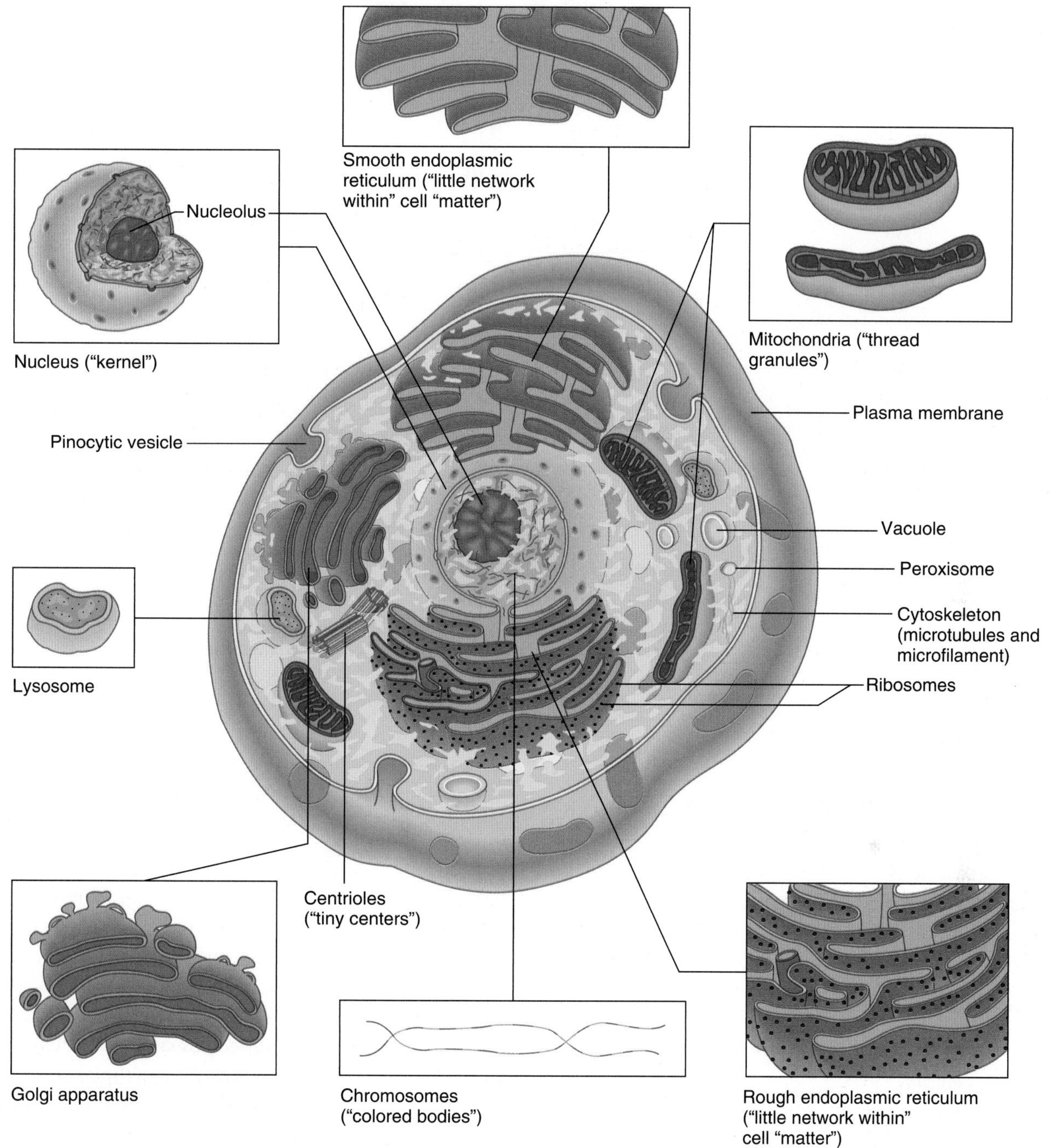

Figure 4-1 *The structure of a typical animal cell*

chromosomes can be seen. Chromosomes store the hereditary material DNA, which is passed on from one generation of cells to the next.

Nuclear Membrane

The nucleus of a cell is contained within a **nuclear membrane**, or nuclear envelope. This membrane is a double-layered structure that has openings, or pores, at regular intervals. Materials can pass through these openings from either the nucleus to the cytoplasm or the cytoplasm to the nucleus. The outer layer of the nuclear membrane is continuous with the endoplasmic reticulum of the cytoplasm and may have small round projections on it, called ribosomes.

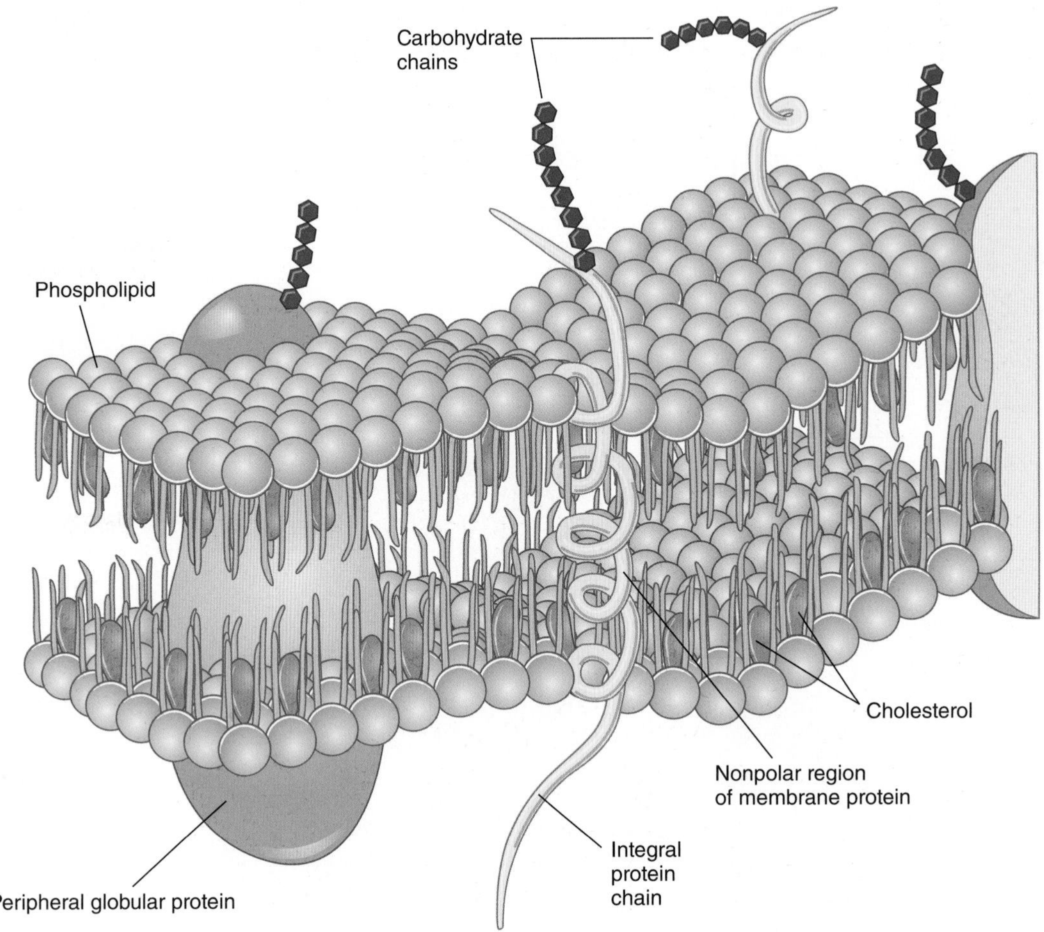

Figure 4-2 *The structure of a plasma (cell) membrane*

Nucleolus and the Ribosomes

Within the nucleus are one or more nucleoli, which is the plural of nucleolus. Each **nucleolus** is a small, round body (refer to **Figure 4-1**). It contains ribosomes composed of ribonucleic acid and protein. The ribosomes can pass from the nucleus through the nuclear pores into the cytoplasm. There, the ribosomes aid in *protein synthesis*. They may exist freely in the cytoplasm, be in clusters called polyribosomes, or be attached to the walls of the endoplasmic reticulum.

Nucleoplasm

Nucleoplasm is a clear, semifluid medium that fills the spaces around the chromatin and the nucleoli within the nucleus.

Cytoplasm

Cytoplasm is a sticky, semifluid material found between the nucleus and the cell membrane. Chemical analysis of cytoplasm shows that it consists of proteins, lipids, carbohydrates, minerals, salts, and water (70% to 90%). Each of these substances, other than water, varies greatly from one cell to the next and from one organism to the next. Cytoplasm is the background for all chemical reactions that take place in a cell, such as protein synthesis and cellular respiration. Molecules are transported about the cell by the circular motion of the cytoplasm.

Table 4-1 summarizes the cell structures and their functions.

Centrosome and Centrioles

The centrioles are two cylindrical organelles found near the nucleus in a tiny, round body called the centrosome. The centrioles are perpendicular to each other; **Figure 4-1** shows two centrioles near the nucleus. During mitosis, or cell division, the two centrioles separate from each other. In the process of separation, thin cytoplasmic spindle fibers form between the two centrioles. This structure is called a spindle-fiber apparatus. The spindle fibers attach themselves to individual chromosomes to help in the equal distribution of these chromosomes to two daughter cells.

Table 4-1 *Cell Structures, Organelles, and Their Functions*

STRUCTURE AND ORGANELLE	FUNCTION
Cell membrane	Regulates transport of substances into and out of the cell
Cytoplasm	Provides an organized, watery environment in which life functions take place by the activities of the organelles contained in the cytoplasm
Nucleus/organelle	Serves as the "brain" for the control of the cell's metabolic activities and cell division
Nuclear membrane	Regulates transport of substances into and out of the nucleus
Nucleoplasm	Fills the spaces around the chromatin and the nucleoli with a clear, semifluid medium
Nucleolus/organelle	Functions as a reservoir for RNA
Ribosomes/organelle	Serve as sites for protein synthesis
Endoplasmic reticulum/organelle	Provides passages through which transport of substances occurs in cytoplasm
Mitochondria/organelle	Serve as sites of cellular respiration and energy production; store ATP
Golgi apparatus/organelle	Manufactures carbohydrates and packages secretions for discharge from the cell
Lysosomes/organelle	Serve as centers for cellular digestion
Peroxisomes/organelle	Use enzymes to oxidize cell substances
Centrosome and centrioles/organelle	Are functional during animal cell division
Cytoskeleton/organelle	Forms the internal framework
Cilia and flagella	Beat and vibrate their hair-like protrusions

Endoplasmic Reticulum

Crisscrossing the cellular cytoplasm is a fine network of tubular structures, or reticulum, called the endoplasmic reticulum. Some of this endoplasmic reticulum connects the nuclear membrane to the cell membrane; thus, it serves as a channel for the transport of materials in and out of the nucleus. Sometimes the endoplasmic reticulum will accumulate large masses of proteins and act as a storage area.

The two types of endoplasmic reticulum are rough and smooth. Rough endoplasmic reticulum has ribosomes studding the outer membrane. The ribosomes are the sites for protein synthesis in the cell. Smooth endoplasmic reticulum has a role in cholesterol synthesis, fat metabolism, and detoxification of drugs.

Mitochondria

Most of a cell's energy comes from spherical or rod-shaped organelles called mitochondria. These mitochondria vary in shape and number. There can be as few as one in each cell or more than a thousand. Cells that need the most energy have the greatest number of mitochondria. Because they supply the cell's energy, mitochondria are also known as the "powerhouses" of the cell.

The mitochondria have a double-membrane structure that contains enzymes. These enzymes help break down carbohydrates, fats, and protein molecules into energy to be stored in the cell as **adenosine triphosphate (ATP)** (ah-**DEN**-oh-seen try-**FOS**-fate). All living cells need ATP for their activities.

Golgi Apparatus

The Golgi apparatus is also referred to as Golgi bodies or the Golgi complex. It is an arrangement of layers of membranes resembling a stack of pancakes. Scientists believe that this organelle synthesizes carbohydrates and combines them with protein molecules as they pass through the Golgi apparatus. In this way, the Golgi apparatus stores and packages secretions for discharge from the cell. These organelles are abundant in the cells of gastric glands, salivary glands, and pancreatic glands.

Lysosomes

Lysosomes are oval or spherical bodies in the cellular cytoplasm. They contain powerful enzymes that digest protein molecules. The lysosome thus helps digest old, worn-out cells, bacteria, and foreign matter. If a lysosome should rupture, as sometimes happens, the lysosome will start digesting the cell's proteins, causing it to die.

Peroxisomes

Membranous sacs that contain oxidase enzymes are called peroxisomes. These enzymes help digest fats and detoxify harmful substances.

Cytoskeleton

The cytoskeleton is the internal framework of a cell. It consists of microtubules, intermediate filaments, and microfilaments. The filaments provide support for the cells; the microtubules are thought to aid in movement of substances through cytoplasm.

Pinocytic Vesicles

Large molecules such as protein and lipids, which cannot pass through the cell membrane, will enter a cell by way of the pinocytic vesicles. The **pinocytic vesicles** form when the cell membrane folds inward to create a pocket. The edges of the pocket then close and pinch away from the cell membrane, forming a bubble or **vacuole** in the cytoplasm. This process, by which a cell forms pinocytic vesicles to take in large molecules, is called **pinocytosis** (pye-noh-sigh-**TOH**-sis), or "cell drinking."

Cilia and Flagella

Cilia and **flagella** are protrusions from the cell membrane. Cilia have short, hair-like protrusions, whereas flagella have a singular tail-like protrusion. They are composed of fibrils that protrude from the cell and beat or vibrate. Cilia move materials across the surface of a cell. An example is the respiratory tract cells, which move the mucous-dust package from the respiratory tree to the throat. The sperm cell of the male has a flagellum that propels the cell to reach the egg in the upper part of the fallopian tube of the uterus of the female.

> **Media Link**
>
> View the **Anatomy of a Typical Cell** animation on the Online Resources.

CELLULAR METABOLISM

For cells to maintain their structure and function, chemical reactions must occur inside the cell. These chemical reactions require energy, most commonly from a molecule called adenosine triphosphate (ATP). The ATP molecule is created from the decomposition of organic molecules from the carbohydrates, proteins, and fats one eats. *Cellular respiration* is the process by which organic compounds are broken apart using oxygen to release the energy that is needed to produce ATP, which is then stored in the mitochondria. The ATP molecule is then available to be used for the maintenance of the cellular structure and function.

CELL DIVISION

Cells divide for two purposes: growth or maintenance of cells in the human body (**mitosis**) and reproduction (**meiosis**). In mitosis, each cell carries a complete set of chromosomes (46); however, in meiosis, each cell carries only half of the chromosomes (23).

Meiosis

Meiosis is the process of cell division of the sex cell, or gamete. During meiosis, the ovum from the female and the spermatozoa from the male *reduce* their respective chromosomes by half, from 46 to 23. When fertilization—the union of the ovum and the spermatozoa—occurs, the two sex cells combine to form a simple cell called the zygote with the full set of 46 chromosomes, 23 from each parent (**Figure 4-3**).

> **Media Link**
>
> View the **Meiosis** animation on the Online Resources.

Mitosis

Cell division is comprised of two distinct processes: the first stage is the division of the nucleus and the second stage is the division of the cytoplasm.

Mitosis essentially is an orderly series of steps by which the DNA in the nucleus of the cell is equally distributed to two daughter, or identical, nuclei. During the process, the nuclear material is distributed to each of the two new nuclei. This is followed by the division of the cytoplasm into two approximately equal parts through the formation of a new membrane between the two nuclei.

All cells do not reproduce at the same rate. Blood-forming cells in the bone marrow, cells of the skin, and cells of the intestinal tract reproduce continuously. Muscle cells only reproduce every few years.

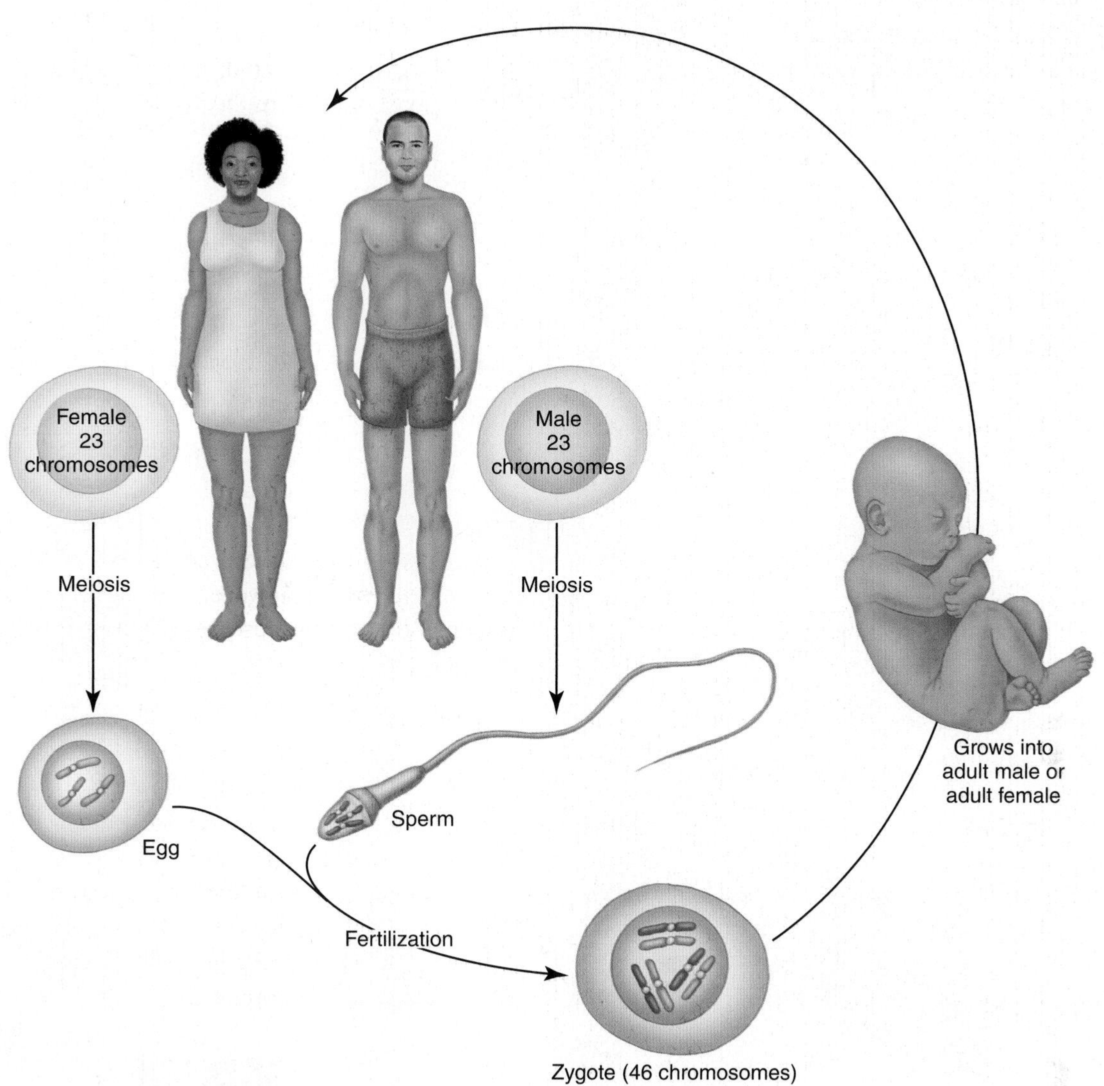

Figure 4-3 *The process of meiosis*

MITOSIS IN A TYPICAL ANIMAL CELL

Mitosis is a smooth, continuous process. For ease and convenience of study, however, five stages, or phases, have been identified by the cell biologist. These five phases are discussed subsequently with accompanying diagrams (**Figure 4-4**). The normal human somatic cell contains 46 chromosomes in the nucleus, which is equal to 23 pairs of chromosomes. This particular chromosome number (46) is called the diploid number of chromosomes. The illustration of a cell in interphase is a representative animal cell with a diploid number of 46 chromosomes. This cell will help illustrate the process of mitosis.

PHASE 1—INTERPHASE (RESTING STAGE)

In the **interphase**, or "resting" stage, an animal cell undergoes *all* metabolic cellular activities to help in the maintenance of cell homeostasis. The term *resting* refers only to the fact that the cell is not undergoing the visible steps of mitosis yet. Interphase occurs between nuclear divisions. During early interphase, an exact duplicate of each nuclear chromosome is made. This process is called **replication**. Replication is the duplication of the molecules of DNA within a chromosome.

At the start of mitosis, each chromosome has already replicated. Each strand of the replicated chromosome is called a **chromatid**. The two chromatid strands are joined by a small structure called the centromere. During interphase, two centrioles located near the periphery of the nucleus are quite visible. The two centrioles are found in an area called the centrosome. They also replicate during interphase in preparation for the next cell division.

PHASE 2—PROPHASE

During **prophase**, the two pairs of centrioles start to separate toward the opposite ends or poles of the cell. As the two pairs of centrioles migrate, an array of cytoplasmic microtubules forms between them.

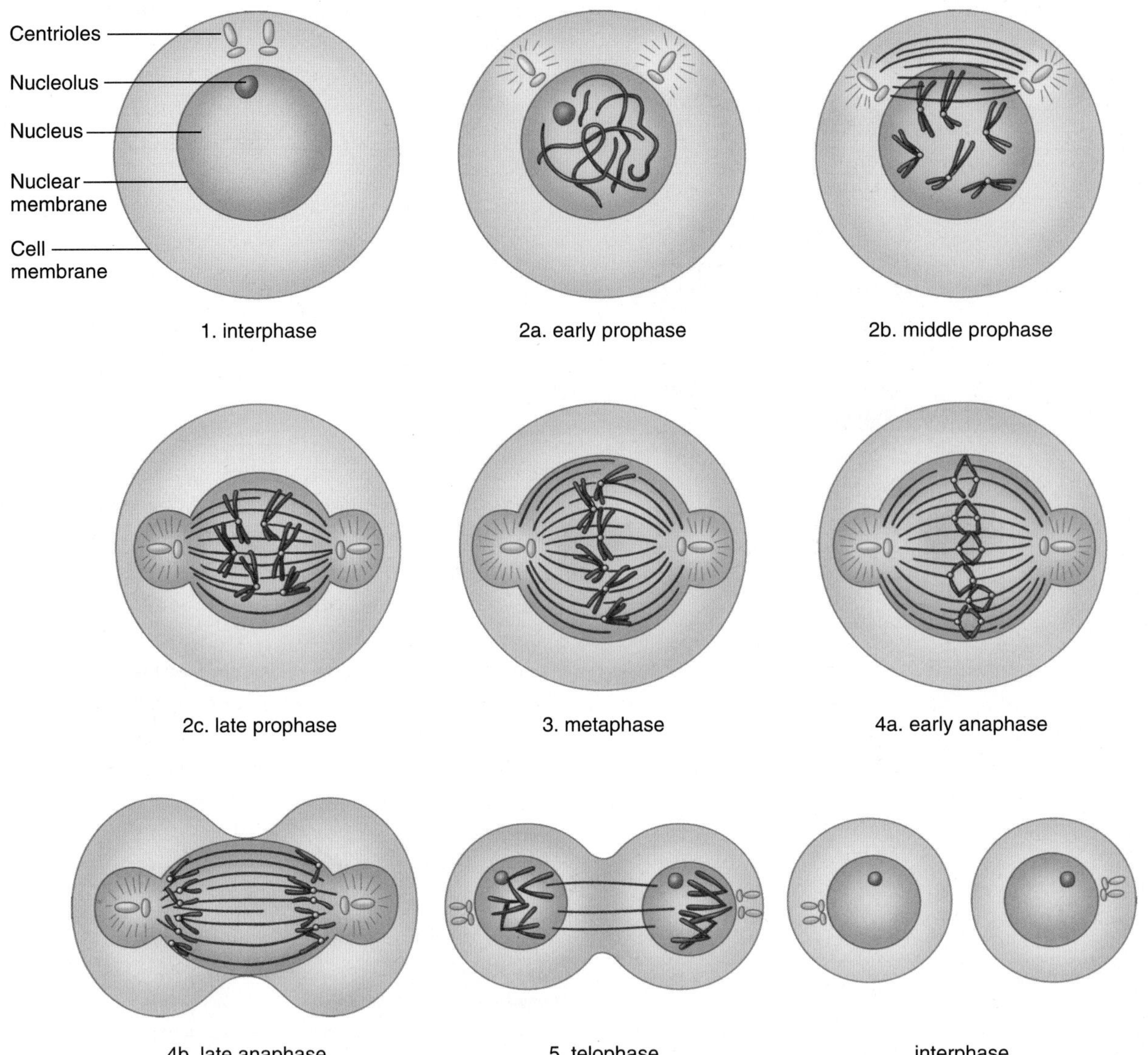

Figure 4-4 *The five phases of mitosis: interphase, prophase, metaphase, anaphase, and telophase*

There are changes in the nucleus as well. The nuclear membrane starts to dissolve and the nucleolus disappears. The DNA in the chromosomes becomes more coiled or condensed and forms very deeply staining, rod-like structures.

PHASE 3—METAPHASE

During **metaphase**, the nuclear membrane has dissolved completely. The chromatid pairs arrange themselves in single file, with one chromatid pair per spindle fiber between the two centrioles. The area where the chromatid pairs align is called the equatorial plate.

PHASE 4—ANAPHASE

During **anaphase**, the chromatid pairs separate and are pulled by the shortening spindle fibers toward the centrioles. The two chromatids of each replicated chromosome are now fully separated.

PHASE 5—TELOPHASE

During **telophase**, the chromosomes migrate to the opposite poles of the cell. There they start to uncoil to become loosely arranged chromatin granules. The nuclear membrane and the nucleolus reappear to help reestablish the nucleus as a definite organelle again.

When cytoplasmic division is finished, two new daughter cells are formed.

Media Link

View the **Mitosis** animation on the Online Resources.

Cell Death

Cells continue to divide until cell death occurs either by **necrosis** (neh-**KROH**-sis) or **apoptosis** (ah-pop-**TOH**-sis). Biologists realize that death is the predestined fate of individual cells or organisms. Necrosis is the name given to the unprogrammed death of cells and living tissue. There are many causes of necrosis, including injury, infection, cancer, infarction (blood clot), toxins, and inflammation. Cells that die from necrosis may release harmful chemicals that damage other cells.

Apoptosis is an orderly process by which cells intentionally die. The cell itself initiates, regulates, and executes its death with an elaborate arsenal of cellular and molecular activity. The term apoptosis is used interchangeably with the term *programmed cell death* (PCD). Apoptosis confers advantages during an organism's life cycle, for example, the differentiation of fingers and toes. In a developing human embryo, the cells in the tissues between the fingers and toes initiate apoptosis so that fingers and toes can separate.

PROTEIN SYNTHESIS

Cells produce proteins that are essential to life through a process called protein synthesis. Within each cell, the DNA determines the kinds of proteins that are produced. The blueprint for each individual kind of protein is contained within a specific gene that resides in the DNA chain. As stated in Chapter 3, messenger RNA carries the instruction for protein synthesis from the DNA to the ribosome in the cytoplasm. The transport RNA molecule picks up the amino acid molecule in the cytoplasm and takes it to the ribosome, where they combine to form a specific protein.

Proteomics is the large-scale study of the proteins that are made as a result of genetic instructions from RNA and ribosomes. Because proteins play a central role in the life of an organism, the study of proteomics will be instrumental in the discovery of biomarkers, the substances found in the blood or tissue in small amounts that may indicate a specific disease.

MOVEMENT OF MATERIALS ACROSS CELL MEMBRANES

The cell plasma membrane controls passage of substances into and out of the cell. The fluid within the cell and the fluid outside the cell must maintain a proper balance to maintain homeostasis. This is important because a cell must be able to acquire materials from its surrounding medium, after which it either secretes synthesized substances or excretes wastes. The physical processes that control the passage of materials through the cell membrane are diffusion, osmosis, filtration, active transport, phagocytosis, and pinocytosis. Diffusion, osmosis, and filtration are examples of **passive transport**, which means they do not need energy to function. Active transport, phagocytosis, and pinocytosis are active processes that require an energy source.

Diffusion

Diffusion is a physical process whereby molecules of gases, liquids, or solid particles spread or scatter themselves evenly through a medium. When solid particles are dissolved within a fluid, they are known as **solutes**. Diffusion also applies to a slightly different process, where solutes and water pass across a membrane to distribute themselves evenly throughout the two fluids, which remain separated by the membrane. Generally, *molecules move from an area where they are greatly concentrated to an area where they are less concentrated.* The molecules will eventually distribute themselves evenly within the space available; when this happens, the molecules are said to be in a state of **equilibrium** (**Figure 4-5**).

The three common states of matter are gases, liquids, and solids. Molecules will diffuse more quickly in gases and more slowly in solids. Diffusion occurs due to the heat energy of molecules. As a result, molecules are in constant motion, except at absolute zero (−273°C; −460°F). In all cases, the movement of molecules increases with an increase in temperature.

A few familiar examples of the rates of diffusion may be helpful. For instance, if a person thoroughly saturates a wad of cotton with ammonia and places it in a far corner of a room, the entire room will soon smell of ammonia. Air currents quickly carry the ammonia fumes throughout the room. Another test for diffusion is to place a pair of dye crystals on the bottom of a water-filled beaker. Eventually the crystals will uniformly permeate and color the water. This diffusion process will take quite a while, especially if no one stirs, shakes, or heats the beaker. In still another test, a dye crystal placed on an ice cube moves even more slowly through the ice. Diffusion of the dye can be accelerated by melting the ice.

The diffusion rate of molecules in gases, liquids, and solids depends on the distances between each molecule and how freely they can move. In a gas, molecules can move more freely and quickly; within a liquid, molecules are more tightly held together. In a solid substance, molecular movement is highly restricted and thus very slow.

Medical Highlights

STEM CELLS

Stem cells are primal cells common to all multicellular organisms. They retain the ability to renew themselves through cell division and to differentiate themselves into a wide range of specialized cell types. Research in the human stem cell field grew out of findings by Canadian scientists Ernest A. McCulloch and James E. Till in the 1960s.

The three broad categories of stem cells are *embryonic stem cells*, derived from the blastocyst stage of development; *adult stem cells*, found in adult tissue; and *cord blood stem cells*, which are found in umbilical cord blood. In adults, stem cells serve as a repair system for the body. Researchers have recently discovered stem cells in the amniotic fluid, which fills the sac that surrounds the fetus in utero.

The types of embryonic stem cells include the *totipotent* stem cells produced from the fusion of an egg and sperm cell. These fertilized eggs have total potential, which means that they give rise to all the different types of cells in the body. *Multipotent* stem cells can give rise to a small number of different cell types; for example, hematopoietic stem cells differentiate into red blood cells, white blood cells, and so forth. *Pluripotent* stem cells can give rise to any type of cell in the body except those needed to develop a fetus.

Embryonic stem cell lines are cultures of cells derived from the inner cell mass of a blastocyst. A blastocyst is an early stage of embryo development—approximately 4 to 5 days old in humans—consisting of 50 to 150 cells. Embryonic stem cells are pluripotent and can develop into each of the more than 200 cell types of the body. Because of their unique combined abilities of unlimited expansion and pluripotency, embryonic stem cells are potential sources for regenerative medicines and tissue replacement after injury or disease.

Adult stem cells are undifferentiated cells found throughout the body that divide to replace dying cells and regenerate damaged tissue. The use of adult stem cells in research and therapy is not as controversial as that of embryonic stem cells because it does not require the destruction of an embryo. Many different kinds of multipotent adult stem cells have been identified, yet adult stem cells that could give rise to all cell and tissue types have not been found. These types of cells are often present in minute quantities and are difficult to isolate and purify. They may not have the same capacity to multiply as embryonic stem cells do.

In 2007, researchers identified another category of stem cells: *induced pluripotent stem cells*. These are adult stem cells that are genetically reprogrammed to act like pluripotent embryonic stem cells. Essentially, one's own normally nonversatile skin cells can be made into a stem cell that can produce any type of tissue. With embryonic stem cells, there is always the issue of potential immune system rejection. Adult stem cells solve the problem of rejection if the stem cells used are one's own. Induced pluripotent stem cells may be more versatile; however, getting reprogrammed stem cells to be predictable could become a challenge (Mayo Clinic Health Letter, October 2011).

Umbilical cord blood is rich in stem cells that can be used to treat disease. The multipotent stem cells are less prone to rejection by the recipient because the cells have not yet developed the features that are recognized and attacked by the recipient's immune system. Cord blood stem cells have been used in treating childhood cancers. Parents are encouraged to bank their child's cord blood to treat diseases that could occur later in the child's life.

The most widely used application of stem cells for therapy is in bone marrow transplants used to treat leukemias and other blood disorders.

Widespread controversy over stem cell research continues, especially over the use of embryonic stem cells. The National Institutes of Health (NIH) has issued guidelines for human stem cell research, stating that embryonic stem cells may only be used from embryos created by in vitro fertilization when the embryo is no longer needed.

Research has been done in creating insulin-producing cells for type 1 diabetes, regenerating cartilage for those with arthritis, creating dopamine neurons to correct Parkinson's disease, repairing spinal cord injury, and restoring sight in those with macular degeneration. Most stem cell applications are still in their infancy but hold great promise for the future.

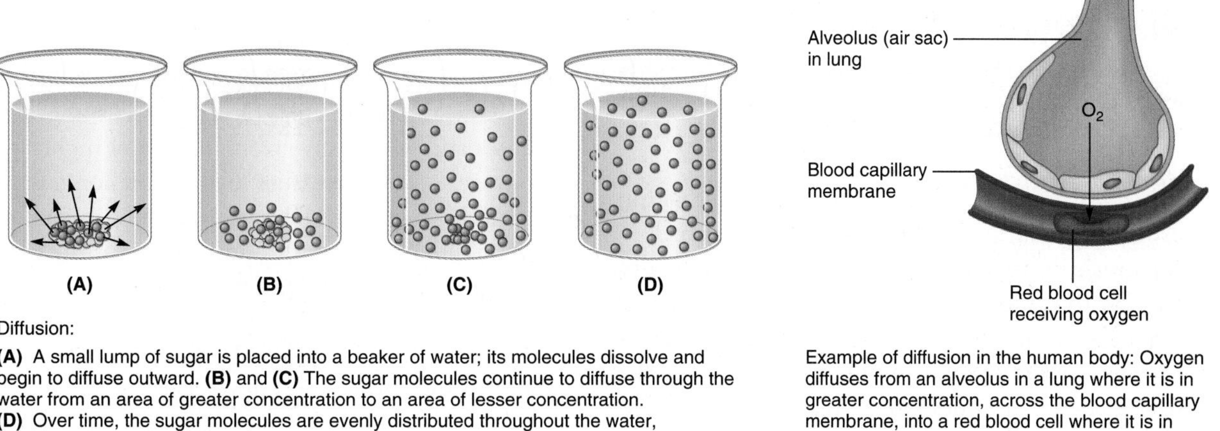

(A)
(B)
(C)
(D)
Diffusion:
(A) A small lump of sugar is placed into a beaker of water; its molecules dissolve and begin to diffuse outward. (B) and (C) The sugar molecules continue to diffuse through the water from an area of greater concentration to an area of lesser concentration.
(D) Over time, the sugar molecules are evenly distributed throughout the water, reaching a state of equilibrium.
Alveolus (air sac) in lung
O2
Blood capillary membrane
Red blood cell receiving oxygen
Example of diffusion in the human body: Oxygen diffuses from an alveolus in a lung where it is in greater concentration, across the blood capillary membrane, into a red blood cell where it is in lesser concentration.

Figure 4-5 *The process of diffusion: The sugar molecules eventually reach a state of equilibrium.*

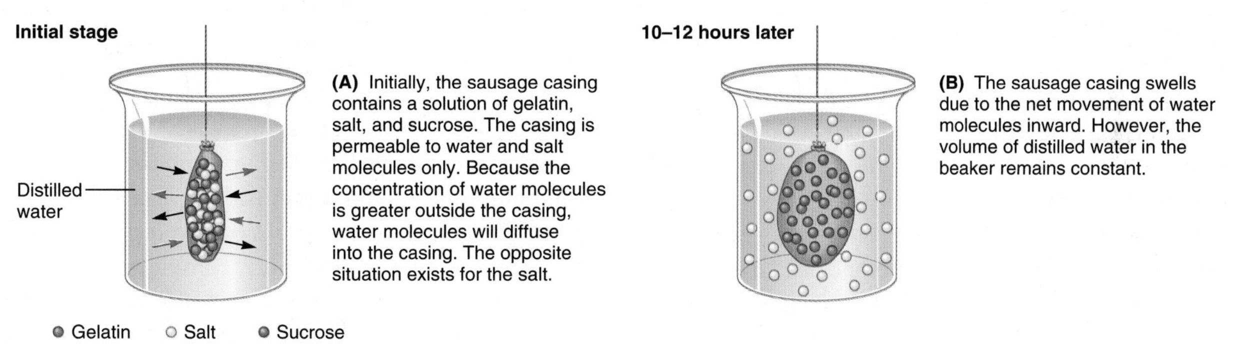

Initial stage
Distilled water
(A) Initially, the sausage casing contains a solution of gelatin, salt, and sucrose. The casing is permeable to water and salt molecules only. Because the concentration of water molecules is greater outside the casing, water molecules will diffuse into the casing. The opposite situation exists for the salt.
10–12 hours later
(B) The sausage casing swells due to the net movement of water molecules inward. However, the volume of distilled water in the beaker remains constant.
Gelatin
Salt
Sucrose

Figure 4-6 *Osmosis is the diffusion of water through a selective permeable membrane, such as a sausage casing; (A) shows the initial stage, and (B) shows the sausage casing 10–12 hours later.*

Diffusion plays a vital role in permitting molecules to enter and leave a cell in maintenance of homeostasis. Oxygen diffuses from the bloodstream, where it dwells in greater concentration. From the bloodstream, the oxygen enters the fluid surrounding a cell and then into the cell itself, where it is far less concentrated (**Figure 4-5**). In this manner, the flow of blood through the lungs and bloodstream provides a continuous supply of oxygen to the cells. Once oxygen has entered a cell, it is utilized in metabolic activities.

Osmosis

Osmosis is the diffusion of water or any other *solvent* molecule through a selective permeable membrane, such as a cell membrane. A selective permeable membrane is any membrane through which some solutes can diffuse, but others cannot.

A sausage casing is a selective permeable membrane that can be used as a substitute for a cell membrane. A solution of salt, sucrose (table sugar), and gelatin is placed into the sausage casing. This mixture is then suspended in a beaker filled with distilled water (**Figure 4-6**). The sausage casing is permeable to water and salt, but not to gelatin and sucrose. Thus, only the water and salt molecules can pass through the casing. Eventually, more salt molecules will move out of the mixture through the casing because the concentration of salt molecules in the distilled water is lower than that in the casing. At the same time, more water molecules move through the casing into the mixture.

The volume of water increases inside the casing, causing it to expand because of the entry of water molecules. When the number of water molecules entering the casing is equal to the number exiting, equilibrium has been achieved and the casing will expand no further.

The pressure exerted by the water molecules within the casing at equilibrium is called **osmotic pressure**.

Osmosis is the movement of water molecules across a semipermeable membrane from an area of higher concentration of a solution to an area of lesser concentration of a solution to maintain homeostasis. The key word is *solute*, the amount of concentration of a dissolved substance.

In the human body, this is well illustrated by a red blood cell in blood plasma (**Figure 4-7**). If a red blood cell is put into blood plasma, which has the same number of sodium particles as a red blood cell, the osmotic pressure of the red blood cell and that of the plasma are the same, representing an **isotonic solution**.

If a red blood cell is put into freshwater, which has fewer sodium particles than the red blood cell does, water will rush into the red blood cell. The freshwater represents a **hypotonic solution**.

If a red blood cell is put into seawater, which has more sodium particles than the red blood cell does, water will leave the red blood cell to dilute the seawater. The seawater represents a **hypertonic solution**.

The health care provider must know which types of solutions are used in various circumstances. When a physician orders intravenous fluids, the patient's condition will determine what type of solution is ordered. Most intravenous fluids are isotonic solutions. Hypertonic solutions are used for patients with edema; hypotonic solutions are used for patients with dehydration.

Filtration

Filtration is the movement of solutes and water across a semipermeable membrane. This results from some mechanical force, such as blood pressure or gravity. The solutes and water move from an area of higher pressure to an area of lower pressure to maintain homeostasis. The size of the membrane pores determines which molecules will be filtered. Thus, filtration allows for the separation of large and small molecules. Such filtration takes place in the kidneys. The process allows larger protein molecules to remain within the body and smaller molecules to be excreted as waste (**Figure 4-8**).

Active Transport

Active transport is a process whereby molecules move across the cell membrane from an area of lower concentration against a concentration gradient to an area of higher concentration. This process requires the high-energy chemical compound ATP. The ATP compound runs the cell's machinery.

How does active transport work? One theory suggests that a molecule is picked up from the outside of the cell membrane and brought inside by a carrier molecule. Both molecule and carrier are bound together, forming a temporary carrier-molecule complex. This carrier-molecule complex shuttles across the cell membrane; the molecule is released at the inner surface of the membrane, where it enters the cytoplasm. At this point, the carrier acquires energy at the inner surface of the cell membrane. Then it returns to the outer surface of the cell membrane to pick up another molecule for transport. Accordingly, the carrier can also convey molecules in the opposite direction, from the inside to the outside (**Figure 4-9**).

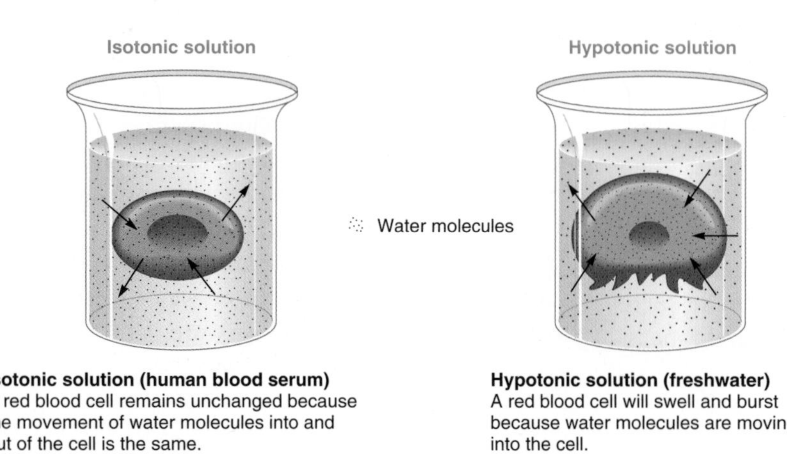
Isotonic solution
Hypotonic solution
Hypertonic solution
Water molecules
Isotonic solution (human blood serum)
A red blood cell remains unchanged because the movement of water molecules into and out of the cell is the same.
Hypotonic solution (freshwater)
A red blood cell will swell and burst because water molecules are moving into the cell.
Hypertonic solution (seawater)
A red blood cell will shrink and wrinkle up because water molecules are moving out of the cell.

Figure 4-7 *Movement of water molecules in solutions of different osmotic pressure*

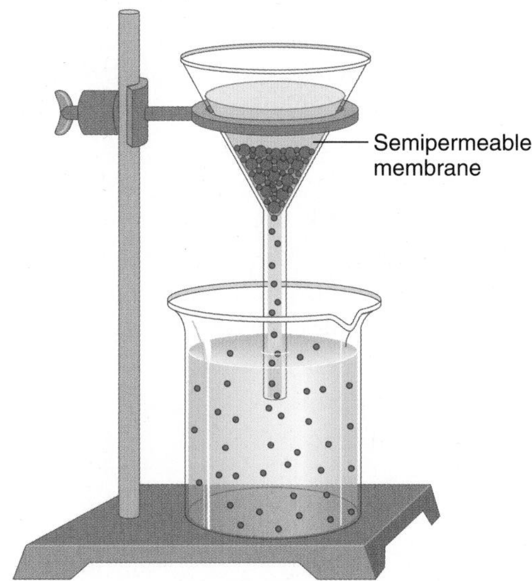

Filtration: Small molecules are filtered through the semipermeable membrane, while the large molecules remain in the funnel.

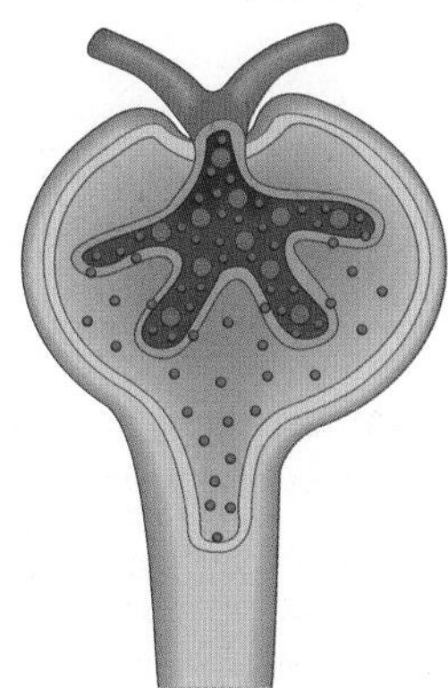

Example of filtration in the human body: Glomerulus of kidney, large particles such as red blood cells and proteins remain in the blood, and small molecules such as urea and water are excreted as a metabolic excretory product: urine.

Figure 4-8 *Filtration: a passive transport process*

Pinocytosis

As stated previously, pinocytosis, or "cell drinking," involves the formation of pinocytic vesicles that engulf large molecules in solution. The cell then ingests the nutrient for its own use.

Phagocytosis

Phagocytosis (fag-oh-sigh-**TOH**-sis), or "cell eating," is quite similar to pinocytosis, with an important difference. In pinocytosis, the substances engulfed by the cell membrane are in solution; however, in phagocytosis, the substances engulfed are within particles. Human white blood cells undergo phagocytosis. The particulate substance is engulfed by an enfolding of the cell membrane to form a vacuole enclosing the material. When the material is completely enclosed within the vacuole, digestive enzymes pour into the vacuole from the cytoplasm to destroy the entrapped substance.

SPECIALIZATION

There are many kinds of cells of different shapes and sizes. Most of them have the characteristics shown in **Figure 4-1**, which is a generalized diagram of a basic cell. Some of the more specialized types, such as nerve cells and red blood cells, look very different (**Figure 4-10**).

Human beings are composed entirely of cells and the nonliving substances that cells build up around themselves. The interaction of the various parts within the cellular structure constitutes the life of the cell. These interactions result in the life activities, life processes, or life functions that were discussed in Chapter 2. In complex organisms, however, groups of cells become specialists and perform particular functions. Nerve cells, for example, have become specialized in response; red blood cells specialize in oxygen transport.

Specialized cells may lose the ability to perform some basic functions, such as reproduction (cell division). Normally, when nerve cells and muscle cells of the heart system are destroyed or damaged, they cannot reproduce. However, recent research has shown that some growth occurs in both of these cells after damage, and scientists are working on ways to encourage these cells to reproduce. Specialization also has resulted in interdependence among cells—certain cellular division cells depend on other kinds of cells to aid them in carrying on the total life activities of the organism. In humans, this specialization and interdependence extends to the organs.

DISORDERS OF CELL STRUCTURE

There are many congenital disorders that may arise during different stages of fetal development.

During blastogenesis, the first 28 days of development, the basic body plan of gene expressions is established. The strongly integrated and interdependent nature of early development defects that occur are usually severe. Severe malformation may include brain and spinal cord injuries, facial clefts, eye defects, and gross heart defects.

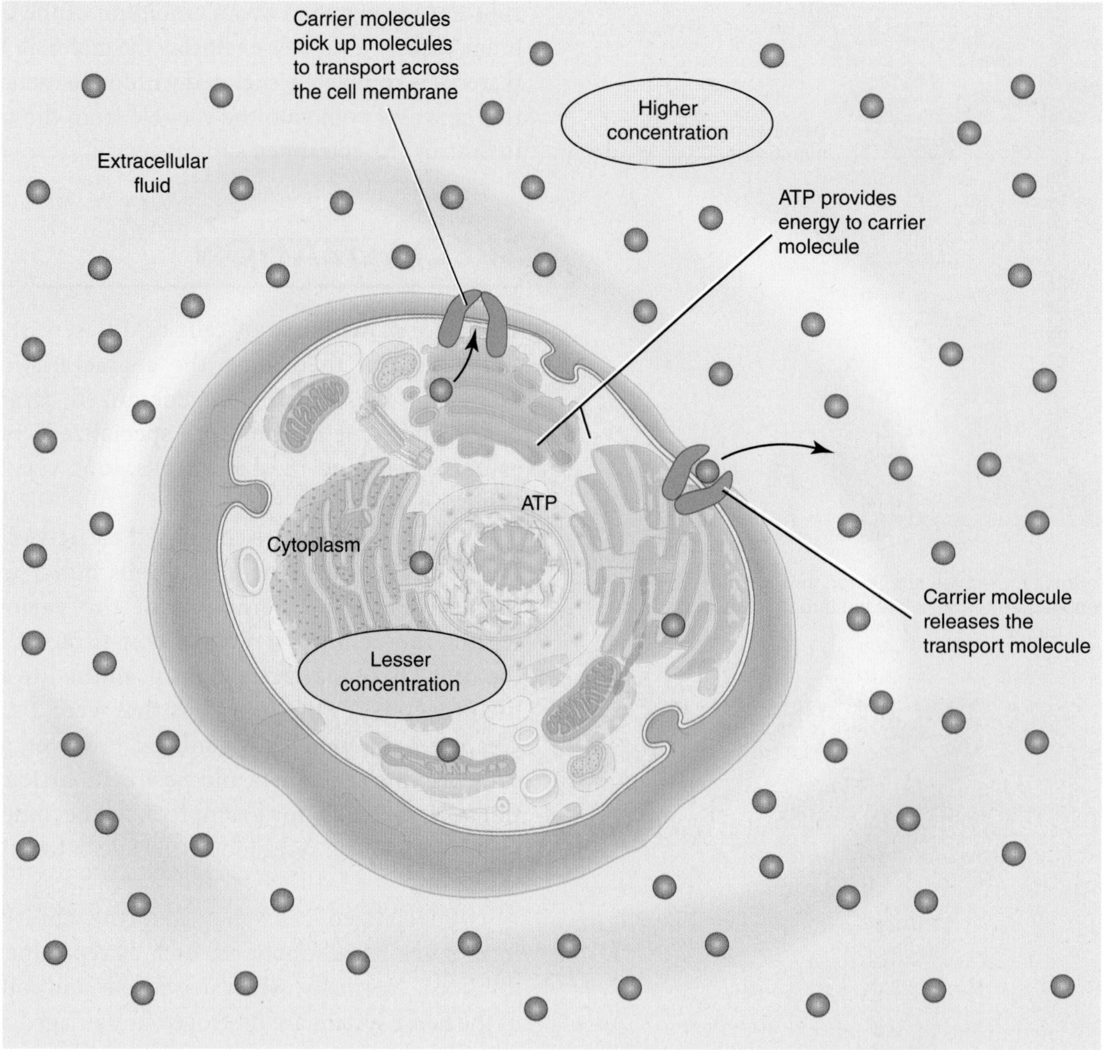

Figure 4-9 *The active transport of molecules from an area of lesser concentration to an area of greater concentration, according to one theoretical model. Active transport is the transportation of materials against a concentration gradient or in opposition to other factors that would normally keep the materials from entering the cell.*

During organogenesis, or days 29–56, organ development occurs. Defects occurring during this stage are usually milder. Defects may include cleft palates; webbed fingers; and hypospadias, the incomplete closure of the urethra in the male.

Between days 57–266, or phenogenesis, the growth and maturation of the fetus occurs. There may be some defects due to genetic factors or environmental factors.

There are several environmental factors that may affect cell development. The mother may have infections that may cross the placenta, such as rubella. Heavy use of alcohol can cause fetal alcohol syndrome; in addition, if there is abuse of controlled substances, the infant may be born with a drug addiction.

Chemicals, such as lead, mercury, thalidomide, and other toxic substances may also alter cell structures.

Atrophy and Hypertrophy

Not all cell growth follows normal physical patterns. Cells may decrease in size, or **atrophy** (**AT**-roh-fee), usually due to aging or disease. Cells may also increase in size, or **hypertrophy** (high-**PER**-troh-fee), usually the result of an increase in workload. Cells can increase in number, or **hyperplasia** (high-per-**PLAY**-zee-ah), which is related to hormonal stimulation. Cells also have the ability to change into another type of cell, called metaplasia; this may be a protective response to a stimulus, such as smoking. **Dysplasia** (dis-**PLAY**-zee-ah) is the change to the size, shape, and organization of cells as a result of a stimulus. This type of cell alteration usually progresses to neoplasia. **Neoplasia** (nee-oh-**PLAY**-zee-ah) is the change in cell structure that occurs in an uncontrolled

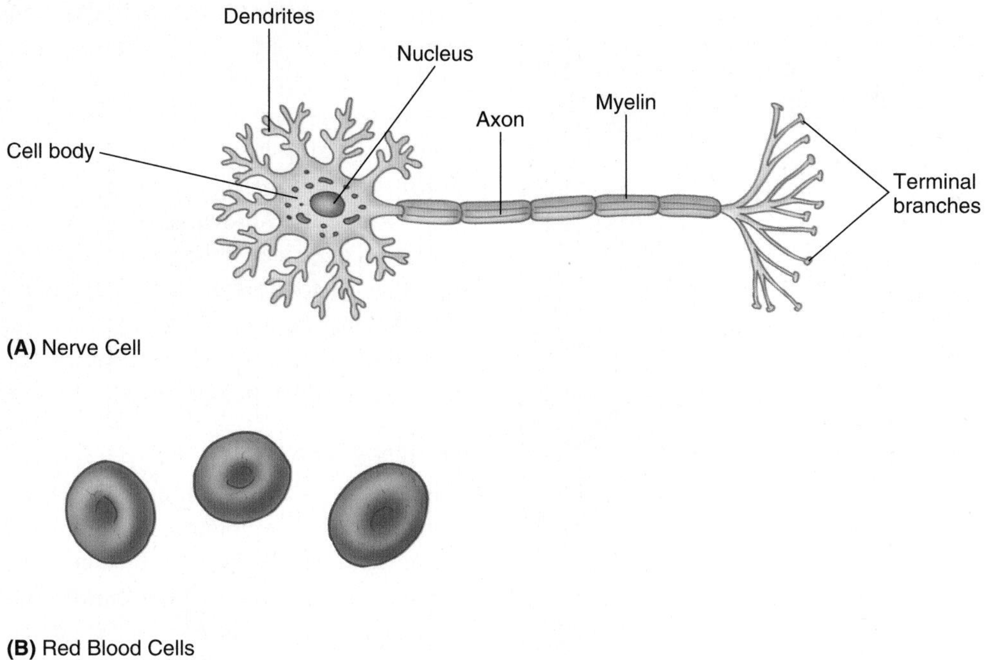

Figure 4-10 *Specialized cells: (A) nerve cells and (B) red blood cells*

growth pattern. Cells can undergo abnormal growth that can range from simple swelling to complex cancerous growths. A variety of abnormalities can occur during cell growth. Some of these disorders are harmless, but others can cause serious illness or death.

Trauma or injury can also affect the structure of a cell. **Hypoxia** (high-**POCKS**-ee-ah), a decrease in the amount of oxygen in the blood flow to cellular structures, and **anoxia** (an-**OCK**-see-ah), a lack of oxygen flow to cellular structures, most commonly cause death in cells. Bacterial toxins or viruses can also result in cell death.

The Effects of Aging **on Cells**

Aging is a phase of normal development. All cells experience changes with aging. They become larger and are less able to divide and multiply. Many cells lose their ability to function, or they function abnormally. Organs have a reserve ability to function beyond the usual needs. After age 30, an average of 1% of this reserve is lost each year. The biggest changes occur in the heart, lungs, and kidneys.

No one really knows how and why people change as they get older; there are many theories. Aging is a complex process that varies in how it affects different people and even different organs (National Institutes of Health).

Tumor

A **tumor** results when cell division does not occur in the usual pattern. If the pattern is interrupted by an abnormal and uncontrolled growth of cells, the result is a tumor (**Figure 4-11**). Tumors are also known as **neoplasms**. Tumors can be divided into two groups: benign and malignant.

A **benign** tumor is composed of cells confined to the local area. Benign tumors are given other names depending on their type or location; for example, a *wart* or **papilloma** (pap-ih-**LOH**-mah) is a type of tumor of the epithelial tissue. Most benign tumors can be surgically removed. A malignant tumor is called **cancer**.

Cancer

Cancerous or malignant tumors continue to grow, crowding out healthy cells, interfering with body functions, and drawing nutrients away from the body tissues. These malignant tumors can spread to other parts of the body through a process called **metastasis** (meh-**TAS**-tah-sis).

Cancer is one of the most common causes of death in the United States, second only to heart disease. There has been a steady decline in cancer rates over the past 25 years (CDC); cancer deaths fell 2.2% from 2016–2017. Improvements in cancer death rates reflect advances in cancer screening and new and improved treatments. Cancer rates have fallen in all four of the most common cancers: lung, colon, breast, and prostate. Cancer rates

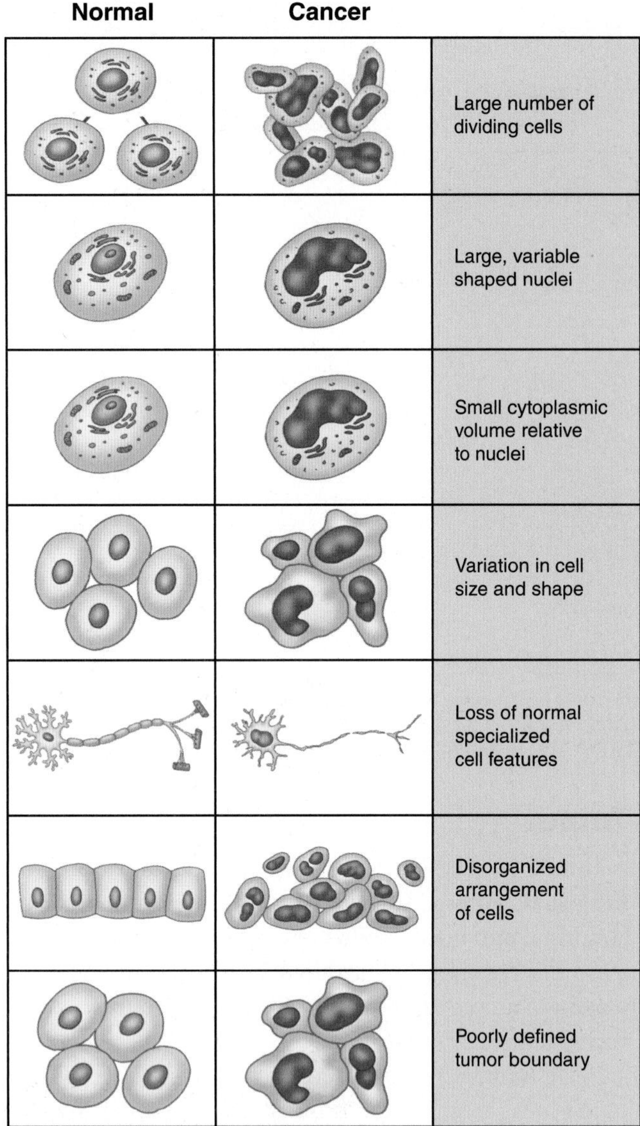

Normal
Cancer
Large number of dividing cells
Large, variable shaped nuclei
Small cytoplasmic volume relative to nuclei
Variation in cell size and shape
Loss of normal specialized cell features
Disorganized arrangement of cells
Poorly defined tumor boundary

Figure 4-11 *Comparison of normal cells to cancerous cells*

Courtesy of the website of the National Cancer Institute (http://www.cancer.gov)

that may show an increase in the coming years include cancers of the pancreas, liver, thyroid, kidney, prostate, and skin (melanoma). While cancer impacts people of all ages, races, ethnicities, and sexes, it does not always affect them equally. Differences in genetics, hormones, environmental exposures, and other factors can lead to differences in risk among different groups of people. For most cancers, though, increasing age is the most important risk factor.

Cancers are grouped into six major categories: carcinoma (kar-sih-**NOH**-mah), sarcoma (sar-**KOH**-mah), myeloma (my-eh-**LOH**-mah), leukemia (loo-**KEE**-mee-ah), lymphoma (lim-**FOH**-mah), and mixed types. Carcinoma refers to a malignant neoplasm of epithelial origin, or cancer of the internal or external lining of the body. Sarcoma refers to cancers that occur in supportive tissue and connective tissue, such as bones, tendons, cartilage, muscle, and fat. Myeloma is cancer that occurs in the plasma cells of bone marrow. Leukemia is cancer of the bone marrow. Lymphoma develops in the glands or nodes of the lymphatic system. Mixed types may include components from more than one of the other five categories.

Any of the following symptoms may be an early indication of cancer: changes in bowel or bladder habits, sores that do not heal, obvious changes in a mole or wart, unusual bleeding or discharge, a new lump or thickening in the breast or elsewhere, difficulty in swallowing or frequent indigestion, a persistent cough, or hoarseness.

CANCER STAGING

Diagnostic tests can detect the early stages of cancer. Tests include X-ray, mammogram, sonogram, and biopsy exams. **Biomarkers** may be used in the diagnosis of cancer. Biomarkers are normal substances found in the blood or tissue in small amounts. Cancer cells can sometimes manufacture these substances. When the amount of a biomarker increases above normal, it may indicate the presence of cancer. Research on markers related to predicting treatment response and studies measuring minimal residual disease and monitoring therapeutic efficacy are also being done.

Cancer staging is a method used to describe the extent or severity of an individual's cancer. Knowing the stage of disease helps the physician plan treatment and the individual's response to treatment. Physical exams, imaging procedures, laboratory tests, pathology reports, and surgical reports provide information to determine the stage of the cancer.

The most commonly used classification systems are the TNM and the Roman numeral, used for most types of cancers except leukemia.

In the acronym TNM, *T* describes the size and extent of the main tumor, *N* describes whether the lymph nodes contain cancer cells and the number of nodes involved, and *M* describes whether the cancer has spread to other parts of the body.

Identifying the stage of cancer can describe location, if it has metastasized, and if it has, what other parts of the body are being affected. Doctors use multiple diagnostic tests to determine the stage of cancer.

- Stage 0: Cancer is in situ (limited to surface cells).
- Stage I: Cancer is limited to the tissue of origin.
- Stage II: There is limited local spread of cancerous cells; it may involve an adjacent lymph node.
- Stage III: There is extensive local and regional spread to lymph nodes.
- Stage IV: Cancers have often metastasized, or spread to other organs or throughout the body.

Treatment of cancer depends on the type of tumor and where it is located. Treatment includes surgery; radiation; and the use of drugs, such as chemotherapy. Other types of treatment include immunotherapy and laser treatment. Disadvantages of cancer treatment include the toxic side effects from drugs and tissue damage caused by radiation. Scientists today are working to develop cancer treatments that are specific to the tumor to help eliminate such side effects.

View the **Cancer Metastasizing** animation on the Online Resources.

Medical Terminology

Term	Meaning	Term	Meaning
chromo	colored	**mei**	lessening or reduction
-some	body	**-osis**	condition of
chromo/some	colored body in the cell; contains the DNA	**mei/osis**	condition of lessening of chromosomes
cyto	cell	**meta**	beyond or after
-skeleton	framework	**-stasis**	controlling or stopping
cyto/skeleton	framework of the cell	**meta/stasis**	beyond control
hyper	excessive	**neo**	new
-tonic	strength, concentration	**-plasm**	growth
hyper/tonic	excessive concentration	**neo/plasm**	new growth
hypo	below normal	**phago**	eat
hypo/tonic	below normal concentration	**-cytosis**	process of
iso	same as	**phago/cytosis**	process of cell eating
iso/tonic	same concentration		

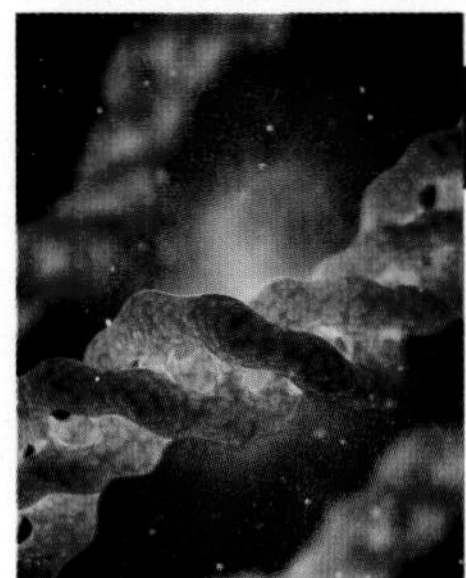

Study Tools

Workbook	Activities for Chapter 4
Online Resources	PowerPoint presentations Animations

REVIEW QUESTIONS

Select the letter of the choice that best completes the statement.

1. Structures found in protoplasm to help cells function are called
   a. the nucleolus.
   b. organelles.
   c. ribosomes.
   d. vacuoles.

2. Regulating transport of substances in and out of the cell is the
   a. cell membrane.
   b. nuclear membrane.
   c. cytoplasm.
   d. nucleus.

3. A structure that digests worn-out cells and bacteria is called a
   a. peroxisome.
   b. ribosome.
   c. lysosome.
   d. mitochondria.

4. The function of the Golgi apparatus of a cell is
   a. protein synthesis.
   b. destroying bacteria.
   c. digesting fats.
   d. storing and packaging secretions.

5. In mitosis, the stage at which the nucleolus disappears is the
   a. prophase.
   b. metaphase.
   c. anaphase.
   d. telophase.

FILL IN THE BLANKS

1. The powerhouse of the cell stores ________________ and is called ________________.
2. The rough endoplasmic reticulum is studded with ________________, which serve as a site for ________________ synthesis.
3. The peroxisomes contain ________________ enzymes, which help digest ________________.
4. During the ________________ stage of mitosis, the two pairs of centrioles start to move toward ________________ ends of the cell.
5. The ________________ for each individual's kind of protein is contained within a specific ________________ in the ________________ chain.

MATCHING

Match each term in Column I with its correct description in Column II.

	COLUMN I	COLUMN II
________	1. solute	a. cells confined to local area
________	2. isotonic solution	b. has a higher concentration of Na than a red blood cell does
________	3. diffusion	c. needs ATP for energy
________	4. phagocytosis	d. malignant tumor
________	5. osmosis	e. solid particles dissolved within a fluid
________	6. benign	f. cell reproduction
________	7. hypertonic solution	g. molecules move from higher concentration to lower
________	8. cancer	h. has the same concentration of Na as a red blood cell does
________	9. mitosis	i. engulfs bacteria
________	10. active transport	j. diffusion of water molecules

LABELING

Study the following diagram of a typical cell and name the labeled structures.

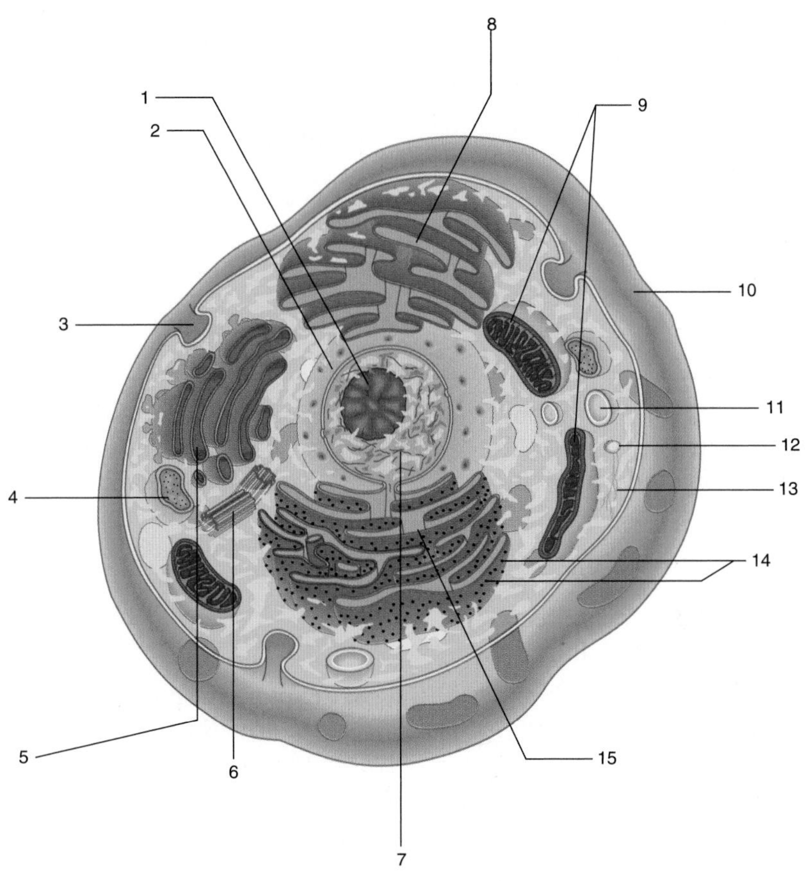

1. ______________________
2. ______________________
3. ______________________
4. ______________________
5. ______________________
6. ______________________
7. ______________________
8. ______________________
9. ______________________
10. ______________________
11. ______________________
12. ______________________
13. ______________________
14. ______________________
15. ______________________

COMPARE AND CONTRAST

Compare and contrast the following terms:

1. Chromatin and chromosomes
2. Cilia and flagella
3. RNA and ribosome
4. Nucleus and nucleolus
5. Anaphase and telophase
6. Hypertonic solution and hypotonic solution
7. Mitosis and meiosis
8. Phagocytosis and pinocytosis
9. Lysosome and peroxisome
10. Osmosis and diffusion

APPLYING THEORY TO PRACTICE

1. The cell is a miniature model of how the body works. Name the cellular structure responsible for each of the following: digestion, respiration, energy, circulation, and the reproductive process.
2. Describe how the cell takes in nutrients.
3. You are working in an emergency care center. A person comes in dehydrated after being out in the sun, and the physician orders a hypertonic solution. Explain why this solution is used instead of an isotonic solution.
4. You are asked to participate in an ethics discussion on stem cell research. Question for debate: "Is it ethical for stem cell research to be done on embryonic stem cells?" Give at least two positive and two negative arguments for this debate.
5. a. If a mother is addicted to alcohol during pregnancy, what will be the effect on the infant?
   b. If a mother has a problem with drug addiction during pregnancy, how will the infant be affected?
   c. Will there be any long-term effects from alcohol or drug use during pregnancy on the infant?
6. Research the ways that chemicals can enter the body. Describe the effect that chemicals can have on the following body systems: reproduction, respiratory, nervous, and endocrine.
7. Investigate how genetics can affect environmental hazards on health.
8. Explore the most common congenital abnormalities and identify how the cells are affected with each.
9. Research preventative measures that can be taken to prevent some congenital abnormalities.
10. Identify causes of hypoxia and anoxia. Describe how trauma affects the function of cells.
11. Investigate how the integration of chemical and physical processes can affect homeostasis within the cells.

CASE STUDY

Jane Fitz, an LPN, admits Mrs. Smith, age 54, to Mercy Hospital. Mrs. Smith has had a persistent cough for 3 months and has lost 10 pounds during the past 6 weeks. Mrs. Smith has been a cigarette smoker for the past 30 years. The physician orders a CT scan to determine if she has cancer of the lung.

1. Describe the CT scan procedure for Mrs. Smith.
2. Name the body system involved in lung cancer and describe the function of the system.

The CT scan reveals a tumor of the lung and a biopsy (removal of tissue for examination) is scheduled.

3. Define cancer and the TNM and Roman Numeral classification system that will be used to describe the tumor.
4. What actions can Jane Fitz take to reduce Mrs. Smith's anxiety?

LAB ACTIVITY 4-1 Parts of a Cell

- ***Objective:*** To identify the parts of a cell and its organelles under a microscope
- ***Materials needed:*** microscope, prepared epithelial cells, prepared muscle cells, textbook, paper, pencil

Step 1: Examine prepared epithelial cells under a microscope.

Step 2: Identify the major parts of the cell and the organelles.

Step 3: Compare your observations with the diagrams in the textbook.

Step 4: Examine prepared muscle cells under a microscope.

Step 5: Describe the parts that are the same as the epithelial cell. Record your observations.

Step 6: Describe the parts that are different from the epithelial cell. Record your observations.

LAB ACTIVITY 4-2 Mitosis

- ***Objective:*** To observe and describe the stages of mitosis
- ***Materials needed:*** prepared slides of mitosis labeled A through E, microscope, textbook, paper, pencil

Step 1: Look at prepared slides of mitosis, labeled A through E.

Step 2: Draw and label what is occurring in each slide.

Step 3: Compare diagrams and answers with a lab partner and the diagrams in the textbook.

Step 4: What additional technologies can you use to learn about and observe mitosis?

LAB ACTIVITY 4-3 Observation of Osmosis

- ***Objective:*** To observe and describe the process of osmosis
- ***Materials needed:*** slices of potato, distilled water and 10% saline solution, saltshaker, small glass containers or test tubes

Step 1: Place some distilled water and saline solution in separate containers and label.

Step 2: Mark the levels of the liquids in each container.

Step 3: Add a potato slice to each container and wait 30 minutes.

Step 4: Remove the potato slices and observe for differences.

Step 5: Observe the levels of liquids in each container.

Step 6: What has happened to the potato slices? To the liquid levels? Record your observations.

Step 7: Make two new containers with solutions and label (same as step 1).

Step 8: Mark the levels of liquid in each container.

Step 9: Take two new potato slices and salt each one with a saltshaker.

Step 10: Place salted potato slices in each container and wait 30 minutes.

Step 11: Remove the potato slices and observe for differences.

Step 12: What has happened to the potato slices? To the liquid levels? Record your observations.

Step 13: Explain what caused the differences between step 6 and step 12.

LAB ACTIVITY 4-4 Diffusion and Active Transport

- ***Objective:*** To observe and describe the effects of diffusion and active transport
- ***Materials needed:*** 2 bags of unpopped popcorn, popcorn maker, bowl, timer, fan (for second part of experiment), classroom setting with 10 or more rows of chairs, 6 students, pen, paper

Step 1: Have two students sit in Row 1, two students in Row 5, and two students in Row 10.

Step 2: Make the popcorn in the popcorn maker.

Step 3: Set timer. When the popcorn is done, put the prepared popcorn in the bowl.

Step 4: Time how long it takes the students in Row 1 to smell the popcorn, the time it takes before the students in Row 5 smell the popcorn, and how long it takes students to smell the popcorn in Row 10. Record the times.

Step 5: Did all the students smell the popcorn at the same time?

Step 6: Describe the process that has occurred and record your response.

Step 7: Share the popcorn.

Step 8: One week later, have two students sit in Row 1, two students sit in Row 5, and two students sit in Row 10.

Step 9: Make the popcorn in the popcorn maker.

Step 10: Set timer and turn on the fan. When the popcorn is done, put the prepared popcorn in the bowl.

Step 11: Time how long it takes for the students in Row 1 to smell the popcorn, the time it takes before students in Row 5 smell the popcorn, and how long it takes students in Row 10 to smell the popcorn. Record the results.

Step 12: What process has occurred? Is it different from the results of the part of the activity without the fan?

Step 13: What does the action of the fan represent?

Step 14: Why did we wait a week between parts of the lab activities?

Step 15: Describe the difference between diffusion and active transport.

Step 16: Enjoy the popcorn.

4-5 Filtration

- ***Objective:*** To observe the process of filtration
- ***Materials needed:*** coffee maker, 4 tablespoons of coffee, 4 cups of hot water, coffee maker basket lined with filter paper, pen, paper

Step 1: Put 4 tablespoons of coffee in the lined basket and put it into the coffee maker.

Step 2: Add 4 cups of water to the coffee maker and turn it on.

Step 3: When coffee is done, note the color of the liquid in the container.

Step 4: Check the coffee basket and describe and record what you see.

Step 5: What process has occurred? Is this an example of active or passive transport?

CHAPTER 5

Tissues and Membranes

Objectives

- List the four main types of tissues.
- Define the function and location of tissues.
- Define the function and location of membranes.
- Define an organ and organ system.
- Relate various organs to their respective systems.
- Describe the processes involved in the two types of tissue repair.
- Define the key words that relate to this chapter.

Key Words

adipose tissue
aponeurosis
areolar tissue
bactericidal
calcified
cardiac muscle
cartilage
Chiari malformation
collagen
connective tissue
elastin
epithelial tissue
fascia
gastric mucosa
grafts
granulation
hyaline
intestinal mucosa
ligaments
Marfan syndrome
membrane
mucosa
mucous membranes
muscle tissue
nervous tissue
osseous (bone) tissue
osteogenesis imperfecta
parietal membrane
pericardial membrane
peritoneal membrane
pleural membrane
primary repair
prune belly disorder
respiratory mucosa
scab
secondary repair
serosa
serous fluid
serous membranes
skeletal muscle
sutures
synovial membrane
tendon
visceral membrane

TISSUES

Multicellular organisms are composed of many different types of cells. Each of these cells performs a special function. These millions of cells are grouped according to their similarity in shape, size, structure, intercellular materials, and function. Cells so grouped are called *tissues.* There are four main types of tissue: (1) **Epithelial tissue** (ep-ih-**THEE**-lee-al) protects the body by covering internal and external surfaces. Epithelial tissue in the lining of the small intestine absorbs nutrients. All glands are made of epithelial tissue. The endocrine glands secrete hormones, mucous glands secrete mucus, and the intestinal glands secrete enzymes. Epithelial tissue also excretes sweat. The epithelial tissue is named according to its structure. (2) **Connective tissue** connects organs and tissues. Connective tissue allows for movement and provides support for other types of tissue. Connective tissue is classified into subgroups: **adipose tissue**, **areolar tissue**, dense fibrous tissue, and supportive tissue. (3) **Muscle tissue** contains cell material that has the ability to contract and move the body. (4) **Nervous tissue** contains cells that react to stimuli and conduct an impulse. Nervous tissue controls and coordinates body activities, controls emotions, and allows people to learn through the memory process. Nervous tissue includes the special senses of sight, taste, touch, smell, and hearing.

Specialization of cells can be seen in a study of the epithelial cells that make up epithelial tissue. Epithelial cells that cover the body's external and internal surfaces have a typical shape: squamous (**SKWAY**-mus), cuboidal, or columnar. This variation is necessary so the epithelial cells can fit together smoothly in order to line and protect the body's surface. Muscle cells that make up muscle tissue are long and spindlelike so they can contract.

Some tissues consist of both living cells and various nonliving substances that the cells build up around themselves. The variations, functions, and locations of each type are described in **Table 5-1**.

Table 5-1 *Different Kinds of Human Tissue*

TYPE OF TISSUE	FUNCTION	CHARACTERISTICS AND LOCATION	MORPHOLOGY
I. Epithelial	Cells form a continuous layer covering internal and external body surfaces; provide protection; produce secretions such as digestive juices, hormones, and perspiration; and regulate the passage of materials across themselves.		
	A. Covering and Lining Tissue These cells can be stratified (layered), ciliated, or keratinized (hard, nonliving substance).	***1. Squamous Epithelial Cells*** These are flat, irregularly shaped cells. They line the heart, blood and lymphatic vessels, body cavities, and alveoli (air sacs) of the lungs. The outer layer of the skin consists of stratified and keratinized squamous epithelial cells. The stratified squamous epithelial cells on the outer skin layer protect the body against microbial invasion.	
		2. Cuboidal Epithelial Cells These cube-shaped cells line the kidney tubules and cover the ovaries and secretory parts of certain glands.	

Table 5-1 *Continued*

TYPE OF TISSUE	FUNCTION	CHARACTERISTICS AND LOCATION	MORPHOLOGY
I. Epithelial (continued)		***3. Columnar Epithelial Cells*** These cells are elongated, with the nucleus generally near the bottom and often ciliated on the outer surface. They line the ducts, digestive tract (especially the intestinal and stomach lining), parts of the respiratory tract, and glands.	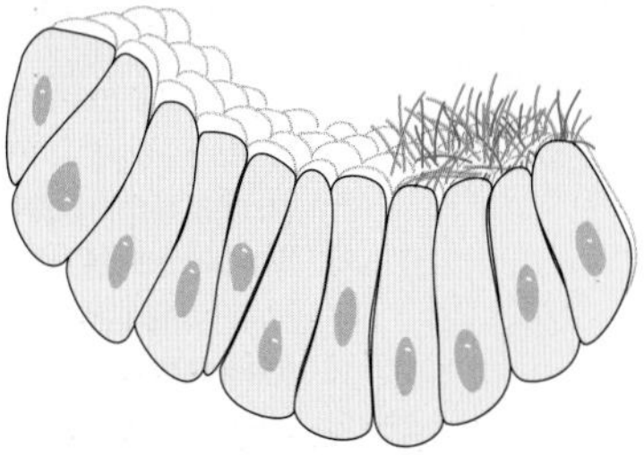
	B. Glandular or Secretory Tissue These cells are specialized to secrete materials such as digestive juices, hormones, milk, perspiration, and wax. They are columnar or cuboidal shaped.	***1. Endocrine Gland Cells*** These cells form ductless glands that secrete their substances (hormones) directly into the bloodstream. For instance, the thyroid gland secretes thyroxine, and adrenal glands secrete adrenaline. ***2. Exocrine Gland Cells*** These cells secrete their substances into ducts. The mammary glands, sweat glands, and salivary glands are examples.	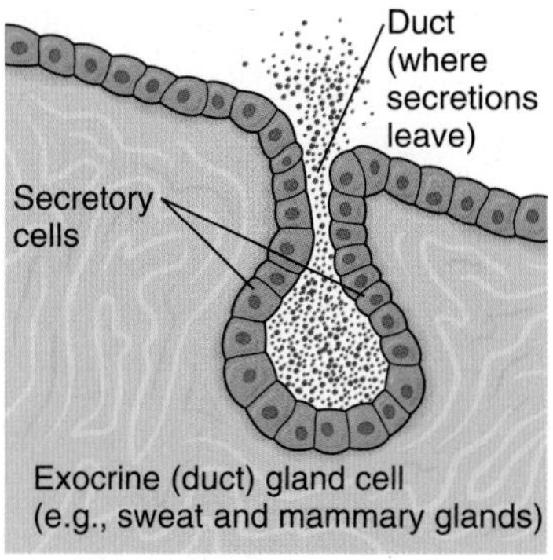
II. Connective	Cells whose intercellular secretions (matrix) support and connect the organs and tissues of the body **A. Adipose Tissue** This tissue stores lipid (fat); acts as filler tissue; and cushions, supports, and insulates the body.	Connective tissue is found almost everywhere in the body: bones, cartilage, mucous membranes, muscles, nerves, skin, and all internal organs. Adipose tissue is a type of loose connective tissue composed of saclike adipose cells; it is specialized for the storage of fat. Adipose cells are found in the subcutaneous skin layer, around the kidneys, within padding around joints, and in the marrow of long bones.	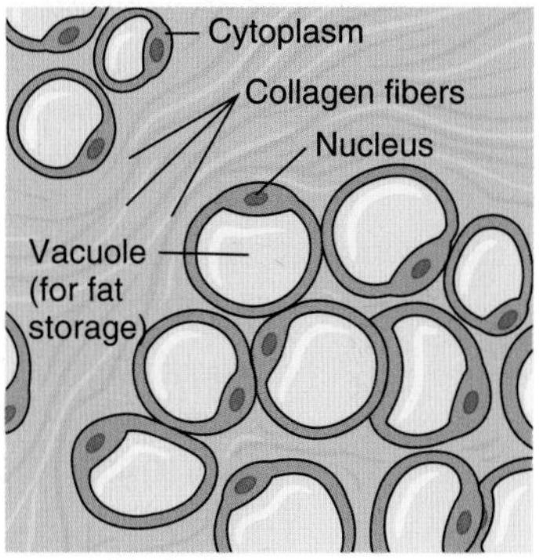
	B. Areolar (Loose Connective) Tissue This tissue surrounds various organs and supports both nerve cells and blood vessels that transport nutrient materials (to cells) and waste (away from cells). Areolar tissue also temporarily stores glucose, salts, and water. Areolar tissue easily stretches and resists tearing.	Areolar tissue is composed of a large, semifluid matrix, with many different types of cells and fibers embedded in it. These include fibroblasts (fibrocytes), plasma cells, macrophages, mast cells, and various white blood cells. The fibers are bundles of strong, flexible, white fibrous protein called **collagen** and elastic single fibers of **elastin**. Areolar tissue is found in the epidermis of the skin and in the subcutaneous layer with adipose cells.	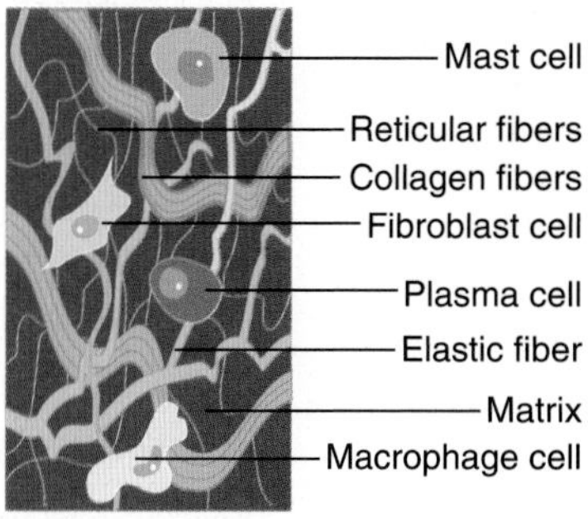

(continues)

Table 5-1 Continued

TYPE OF TISSUE	FUNCTION	CHARACTERISTICS AND LOCATION	MORPHOLOGY
II. Connective (continued)	**C. Dense Fibrous Tissue** This tissue forms ligaments, tendons, aponeuroses, and fasciae. **Ligaments** are strong, flexible bands (or cords) that hold bones firmly together at the joints. **Tendons** are white, glistening bands that attach skeletal muscles to the bones. **Aponeurosis** (**AP**-oh-noo-roh-sis) are flat, wide bands of tissue that hold one muscle to another or to the periosteum (bone covering). **Fascia** (**FASH**-ee-ah) are fibrous connective tissue sheets that wrap around muscle bundles to hold them in place.	Dense fibrous tissue is also called white fibrous tissue because it is made from closely packed white collagen fibers. Fibrous tissue is flexible, but not elastic. This type of tissue has a poor blood supply and heals slowly.	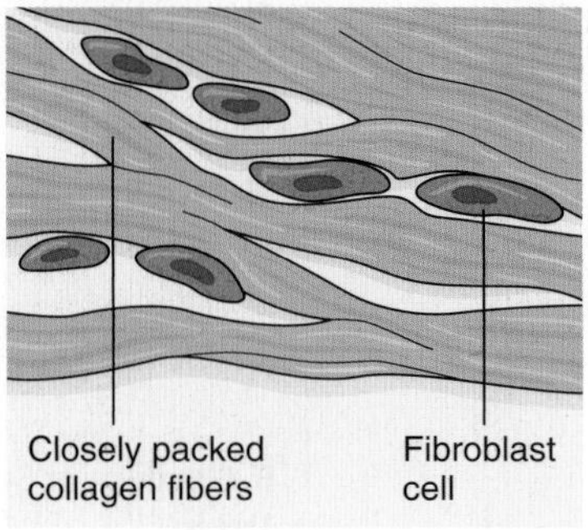
	D. Supportive Tissue ***1. Osseous (bone) tissue*** Comprises the skeleton of the body, which supports and protects underlying soft tissue parts and organs and also serves as attachment points for skeletal muscles	This connective tissue's intercellular matrix is **calcified** by the deposition of mineral salts, such as calcium carbonate and calcium phosphate. Calcification of bone imparts great strength. The entire skeleton is composed of bone tissue.	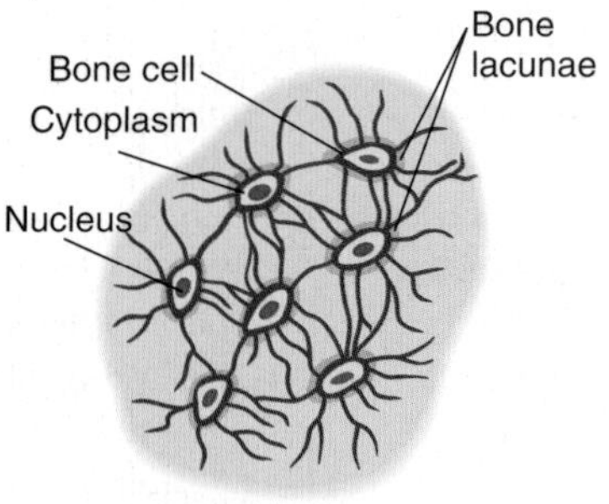
	2. Cartilage Provides firm but flexible support for the embryonic skeleton and part of the adult skeleton ***a. Hyaline*** Forms the skeleton of the embryo	Hyaline cartilage is found on articular bone surfaces and also at the nose tip, bronchi, and bronchial tubes. Ribs are joined to the sternum (breastbone) by the costal cartilage. Costal cartilage is also found in the larynx and in the rings of the trachea.	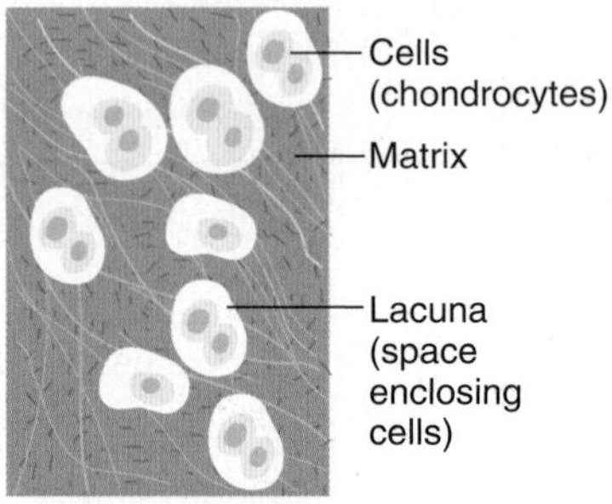
	b. Fibrocartilage A strong, flexible, supportive substance found between bones and wherever great strength (and a degree of rigidity) is needed	Fibrocartilage is located within intervertebral discs and symphysis pubis between the pubic bones.	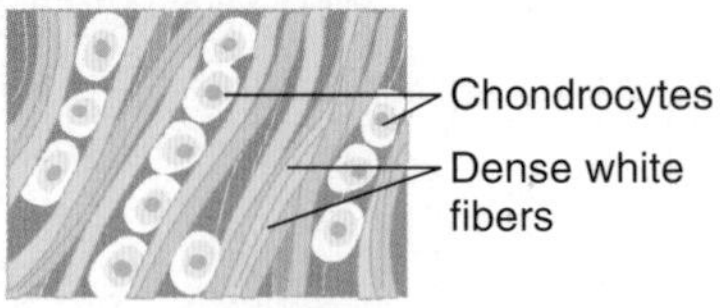
	c. Elastic cartilage The intercellular matrix is embedded with a network of elastic fibers and is firm but flexible.	Elastic cartilage is located inside the auditory ear tube, external ear, epiglottis, and larynx.	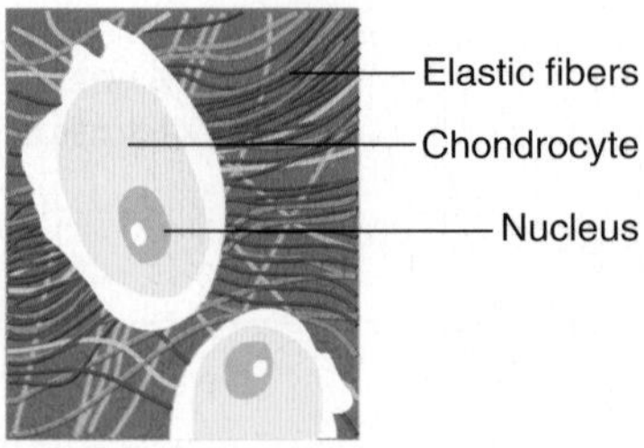

Table 5-1 *Continued*

TYPE OF TISSUE	FUNCTION	CHARACTERISTICS AND LOCATION	MORPHOLOGY
II. Connective (continued)	**E. Vascular (Liquid Blood) Tissue** ***1. Blood*** Blood transports nutrient and oxygen molecules to cells and metabolic wastes away from cells (can be considered a liquid tissue). It contains cells that function in the body's defense and in blood clotting.	Blood consists of two major parts: a liquid called plasma and a solid cellular portion known as blood cells (or corpuscles). The plasma suspends corpuscles, of which there are two major types: red blood cells (erythrocytes) and white blood cells (leukocytes). The types of leukocytes are lymphocytes, monocytes, neutrophils, basophils, and eosinophils. A third cellular component (actually a cell fragment) is platelets (thrombocytes). Blood circulates within the blood vessels (arteries, veins, and capillaries) and through the heart and lungs.	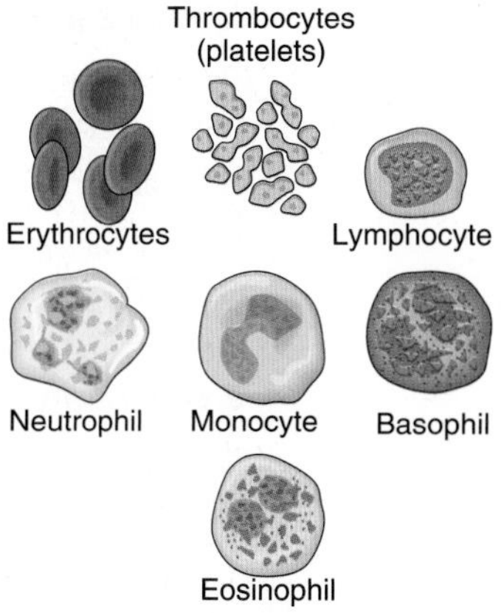
	2. Lymph Lymph transports tissue fluid, proteins, fats, and other materials from the tissues to the circulatory system. This occurs through a series of tubes called the lymphatic vessels.	Lymph fluid consists of water, glucose, protein, fats, and salt. The cellular components are lymphocytes and granulocytes. These components flow in tubes called lymphatic vessels, which closely parallel the veins and bathe the tissue spaces between cells.	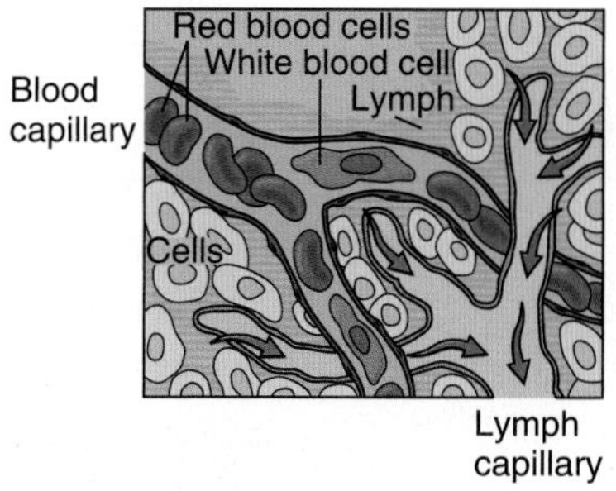
III. Muscle	**A. Cardiac Muscle** These cells have the ability to contract to enable the heart to pump blood throughout and out of the heart.	**Cardiac muscle** is a striated (having transverse bands), involuntary (not under conscious control) muscle. It makes up the walls of the heart.	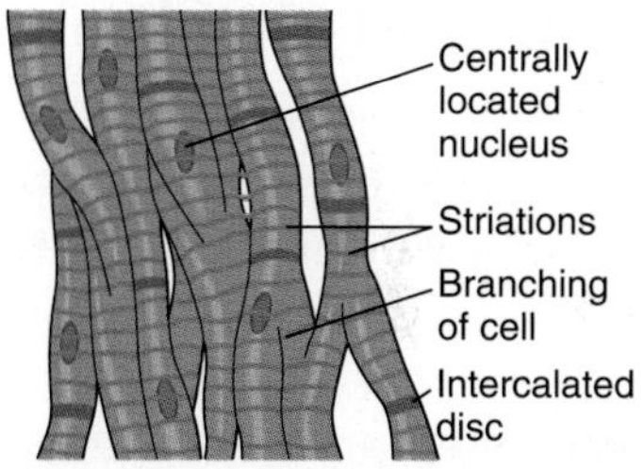
	B. Skeletal (Striated Voluntary) Muscle These muscles are attached to the movable parts of the skeleton. They are capable of rapid, powerful contractions and long states of partially sustained contractions, allowing for voluntary movement.	**Skeletal muscle** is striated (having transverse bands that run down the length of muscle fiber), voluntary because the muscle is under conscious control, and skeletal because these muscles attach to the skeleton (bones, tendons, and other muscles).	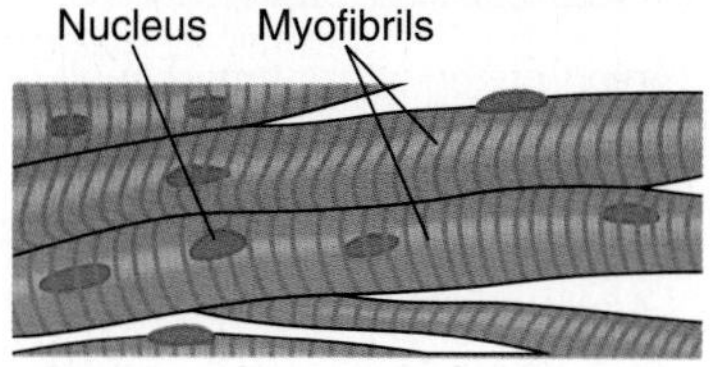
	C. Smooth (Nonstriated Involuntary) Muscle These muscles allow for involuntary movement. Examples include the movement of materials along the digestive tract and control of the diameter of blood vessels and the size of the pupil of the eye.	Smooth muscle is nonstriated because it lacks the striations (bands) of skeletal muscles; its movement is involuntary. It makes up the walls of the digestive tract, genitourinary tract, respiratory tract, blood vessels, and lymphatic vessels.	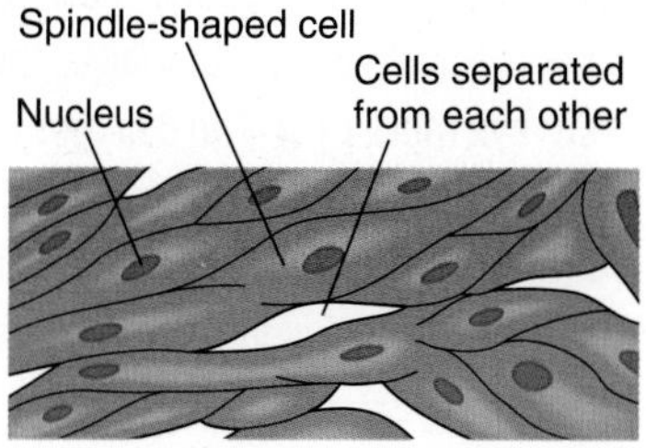

(continues)

Table 5-1 Continued

TYPE OF TISSUE	FUNCTION	CHARACTERISTICS AND LOCATION	MORPHOLOGY
IV. Nervous	**A. Neurons (Nerve Cells)** These cells have the ability to react to stimuli. ***1. Irritability*** Ability of nervous tissue to respond to environmental changes ***2. Conductivity*** Ability to carry a nerve impulse (message)	Nervous tissue consists of neurons (nerve cells). Neurons have branches through which various parts of the body are connected and their activities coordinated. They are found in the brain, spinal cord, and nerves.	

MEMBRANES

A **membrane** is formed by putting two thin layers of tissue together. The cells in the membrane may secrete a fluid. Membranes are classified as epithelial or connective.

Epithelial Membranes

Epithelial membranes are classified as either **mucous membranes** or **serous membranes**, depending on the type of secretions produced (**Figure 5-1**).

The Effects of Aging on Tissue

Aging changes are found in all of the body's cells, tissues, and organs, and they affect the functioning of the body's systems. As an individual ages, the cells become larger and are less able to divide and reproduce. There is an increase in pigments and fatty substances, or lipids, inside the cells. Waste products accumulate in the tissue with aging. A fatty brown pigment called lipofuscin collects in many tissues. Connective tissue becomes progressively stiff. This makes the organs, tissues, and airways more rigid. The cell membranes change, resulting in many tissues having difficulty receiving oxygen and nutrients and getting rid of carbon dioxide and wastes. Many tissues lose mass and atrophy.

MUCOUS MEMBRANES

Mucous membranes line surfaces and spaces that lead to the outside of the body; they line the respiratory, digestive, reproductive, and urinary systems. The mucous membranes produce a substance called mucus, which lubricates and protects the lining. For example, the mucus in the digestive tract protects the lining of the stomach and small intestines from the digestive juices. The term **mucosa** is used for the following specific mucous membranes (**Figure 5-1**):

- **Respiratory mucosa** lines the respiratory passages.
- **Gastric mucosa** lines the stomach.
- **Intestinal mucosa** lines the small and large intestines.

SEROUS MEMBRANES

The serous membrane is a double-walled membrane that produces a watery fluid and lines closed body cavities. The fluid produced is called **serous fluid**. The outer part of the membrane that lines the cavity is known as the **parietal membrane** (pah-**RYE**-eh-tal). The part that covers the organs within is known as the **visceral membrane** (**VIS**-er-al). The fluid produced allows the organs within to move freely and prevents friction. The name **serosa** is given to the specific serous membranes, all beginning with the letter *p*. The serous membranes are as follows (**Figure 5-1**):

- The **pleural membrane** lines the thoracic or chest cavity and protects the lungs. The fluid is called pleural fluid.
- The **pericardial membrane** lines the heart cavity and protects the heart. The fluid is called pericardial fluid.

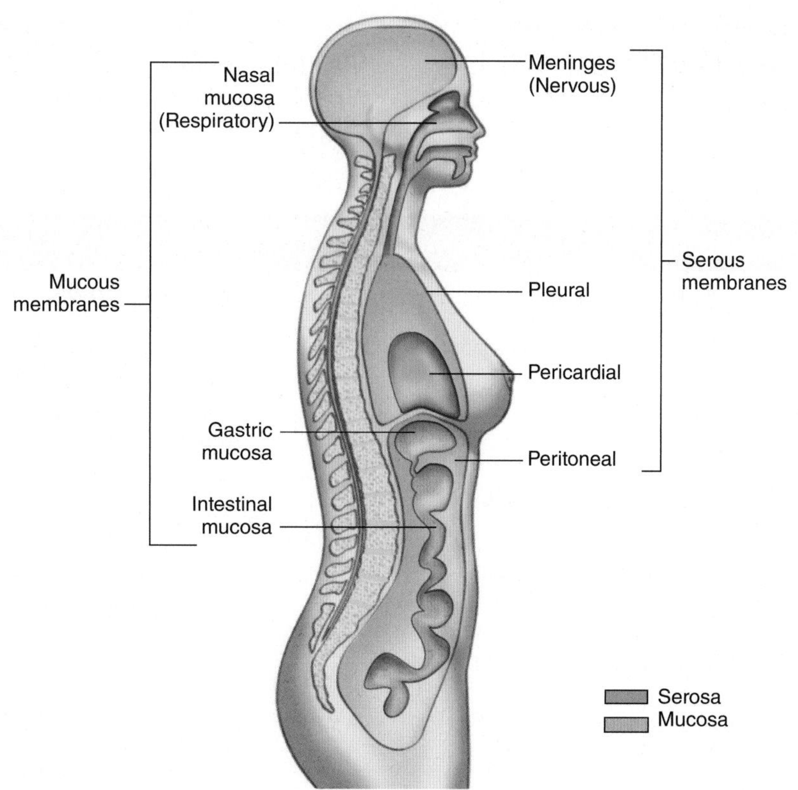

Figure 5-1 *Mucous and serous membranes*

- The **peritoneal membrane** (pehr-ih-toh-**NEE**-al) lines the abdominal cavity and protects the abdominal organs. The fluid is called peritoneal fluid.

CUTANEOUS MEMBRANE (SKIN)

The cutaneous (kyou-**TAY**-nee-ous) membrane, or skin, is a specialized type of epithelial membrane. See Chapter 6 for a complete discussion.

Connective Membranes

Connective membranes consist of two layers of connective tissue. In this classification is the **synovial membrane** (sih-**NOH**-vee-al), which lines joint cavities. Synovial membranes secrete synovial fluid, which prevents friction inside the joint cavity.

ORGANS AND SYSTEMS

An organ is a structure made up of several types of tissues grouped together to perform a single function. For instance, the stomach is an organ consisting of highly specialized vascular, connective, epithelial, muscular, and nervous tissues. These tissues function together to enable the stomach to perform digestion and absorption.

The skin that covers one's body is no mere simple tissue but a complex organ of connective, epithelial, muscular, and nervous tissues. These tissues enable the skin to protect the body and remove its wastes, such as water and inorganic salts.

The various organs of the human body do not function separately. Instead, they coordinate their activities to form a complete, functional organism. A group of organs that acts together to perform a specific, related function is called an *organ system*. For example, the digestive system has the special function of processing solid food into liquid for absorption into the bloodstream. This organ system includes the mouth, salivary glands, esophagus, stomach, small intestine, liver, pancreas, gallbladder, and large intestine. The circulatory system transports materials to and from cells. It consists of the heart, arteries, veins, and capillaries.

Each organ system is highly specialized to perform a specific function; together, they coordinate their functions to form a whole, live, functioning organism.

The systems of the body are the integumentary, skeletal, muscular, nervous, endocrine, circulatory, lymphatic, respiratory, digestive, urinary, and reproductive systems.

The functions and organs of each system are shown in **Table 5-2**, and the progression from simplest organ system to the most complex is shown in (**Figure 5-2**).

Table 5-2 *Body Systems*

BODY SYSTEMS	***MAJOR STRUCTURES***	***MAJOR FUNCTIONS***
Integumentary System Chapter 6	Epidermis, dermis, sweat glands, oil glands, nails, and hair	Helps regulate body temperature; protects the body against invasion by bacteria; synthesizes vitamin D; and has nerve receptors for temperature, pressure, and pain
Skeletal System Chapter 7	Bones, joints, and cartilage	Supports and shapes the body, protects the internal organs, forms some blood cells, and stores minerals
Muscular System Chapter 8	Muscles, fasciae, and tendons	Makes movement possible Moves body fluids and generates heat
Nervous System Chapters 9 and 10	Nerves, brain, and spinal cord	Communicates and coordinates body activities
Special Senses Chapter 11	Eyes, ears, nose, and tongue	Receives visual, auditory, taste, and smell information and transmits it to the brain
Endocrine System Chapter 12	Adrenal glands, gonads, pancreas, parathyroids, pineal, pituitary, thymus, and thyroid	Manufactures hormones to regulate body activities
Circulatory System Chapters 13, 14, and 15	Heart, arteries, veins, capillaries, and blood	Blood circulates to all parts of the body, carrying oxygen and nutrients to the cells and carrying away waste products from the cells
Lymphatic System and Immunity Chapter 16	Lymph, lymphatic vessels, and lymph nodes	Removes and transports waste products from the fluid between the cells Filters fluid through the lymph nodes and removes harmful substances Returns the filtered lymph to the blood stream where it becomes plasma again
Respiratory System Chapter 18	Nose, pharynx, larynx, trachea, bronchi, alveoli, and lungs	Brings oxygen into the body for transportation to the cells Removes carbon dioxide and some water waste from the body
Digestive System Chapter 19	Mouth, salivary glands, teeth, pharynx, esophagus, stomach, intestines, liver, gallbladder, and pancreas	Prepares food for absorption and use by the body through physical and chemical means Eliminates solid waste from the body
Urinary System Chapter 21	Kidneys, ureters, bladder, and urethra	Filters blood to eliminate waste products of metabolism Maintains the fluid and electrolyte balance in the body
Reproductive System Chapter 22	*Male:* testes, epididymis, vas deferens, seminal vesicles, ejaculatory duct, prostate gland, Cowper's glands, penis, and urethra *Female:* ovaries, fallopian tubes, uterus, vagina, external genitalia, and breasts	Reproduces new life Manufactures hormones necessary for development of reproductive organs and secondary sex characteristics

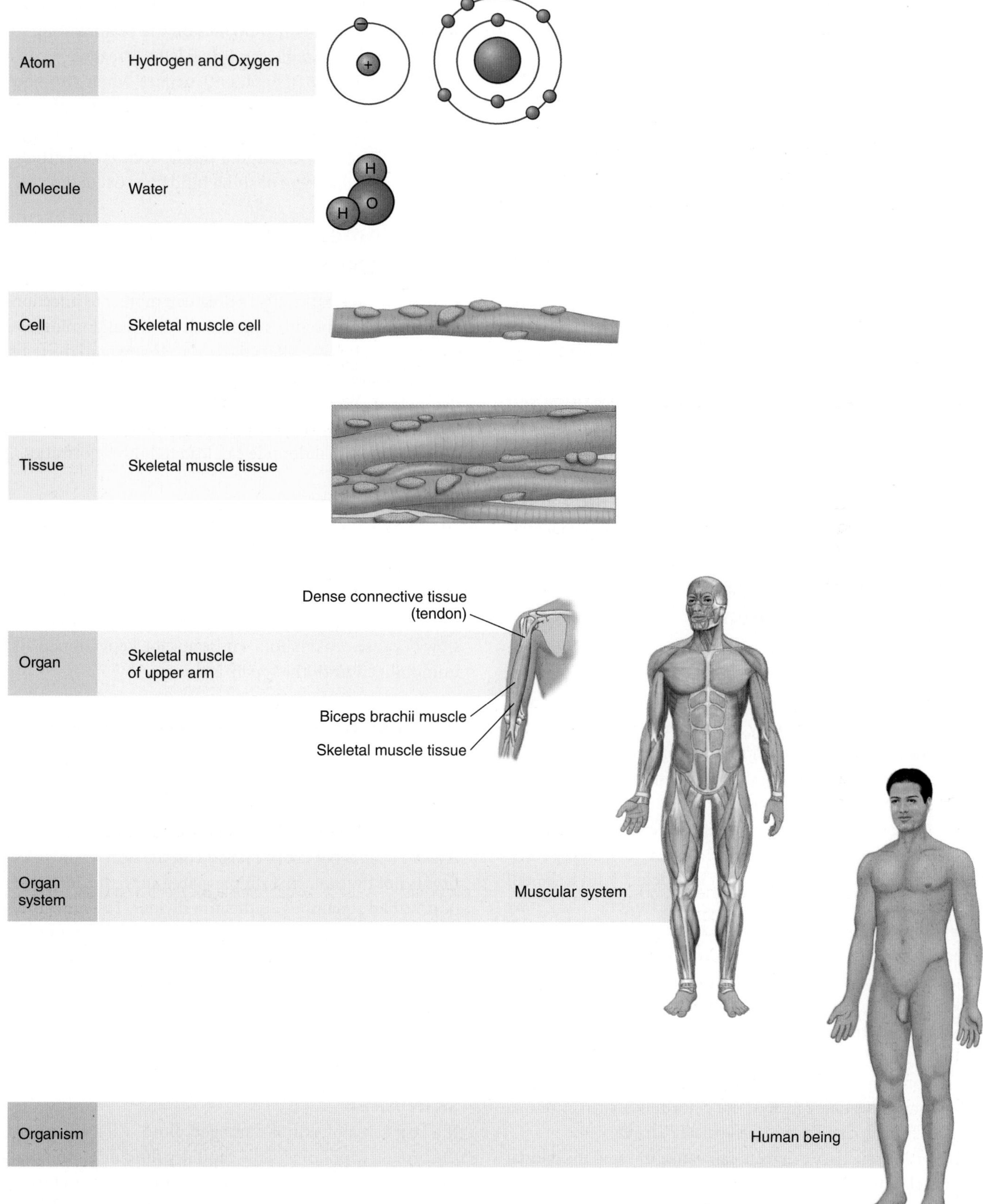

Figure 5-2 *The formation of the human organism progresses from simple to complex*

DISEASE AND TRAUMA TO TISSUE

Tissue can be affected by infection or inflammation. Inflammation is a protective response to an injury or irritant. Inflammation will result in pain, swelling, redness, and loss of motion. Infection refers to the invasion of a microorganism causing disease. Infection usually results in inflammation (see Chapter 13).

Trauma resulting from an external force will cause tissue damage and injury. A wide variety of traumatic events can occur, but the most frequent cause of serious injury is motor vehicle collisions. Emergency management of trauma is necessary to prevent complications, such as shock, hemorrhage, and infection.

Abnormal growth of cells can also alter tissue and cause tissue damage and trauma. These types of growth patterns were discussed in Chapter 4.

Birth defects can also impair tissue. This can result from a change in structure or function of tissue at the chromosomal level or as a result of environmental factors.

Congenital Defects in Tissue Development

Epithelial tissue defects may include changes in the formation of the blood vessels and lead to birthmarks. One example of this is infantile hemangioma. This is sometimes called a "strawberry mark." It can occur anywhere on the body but is seen most often on the face and scalp. Treatment is unnecessary, as most infantile hemangiomas go away on their own.

Connective tissue defects can lead to a condition called **Marfan syndrome**. The degree to which people are affected varies. People with this defect tend to be tall and thin, with long arms, legs, and fingers. They may also have flexible joints and scoliosis (Chapter 7). The most serious implication involves the heart and aorta, which may cause mitral valve prolapse and aortic aneurysms later in life. There is no known cure for Marfan syndrome; treatment is symptomatic as problems arise.

Bone tissue defects can cause **osteogenesis imperfecta**, a disorder in which bones are easily fractured. Other defects of connective tissue include eye deformities, such as retinal detachment; cleft lip; and cleft palate (Chapter 19).

Muscle tissue defects can cause missing individual muscles or groups of muscles, or the muscle may be only partly formed. In **prune belly disorder**, one or more layers of abdominal muscles may be missing at birth, giving the belly a wrinkled appearance.

Nervous tissue defects may cause **Chiari malformation**, a condition in which brain tissue extends into the spinal cord. It occurs when part of the skull is abnormally small, which presses on the brain and forces it downward. In simple cases, it is left untreated.

Spina bifida (Chapter 9) occurs when the neural tube fails to close completely over the spinal cord. There are three forms of spina bifida: myelomeningocele, meningocele, and spina bifida occulta. Treatment depends on what type of spina bifida has occurred.

Environmental Effects on Tissue Development

Environmental exposure of a pregnant mother to infections like rubella or substances such as alcohol, thalidomide, and some antibiotics can affect tissue development in the fetus.

German measles or rubella can cause cataracts, cardiac defects, and fetal growth retardation. Ingestion of alcohol may cause growth retardation and intellectual disability. Drugs such as thalidomide can lead to limb deformities.

Degree of Tissue Repair

Repair of damaged tissues occurs continually during the everyday activities of living. Depending on the type and location of the injury, some tissue is quickly repaired. Muscle tissue heals slowly, and bone tissue repairs are slow because broken bone ends must be kept aligned and immobilized until the repair is done.

Process of Epithelial Tissue Repair

There are two types of epithelial tissue repair: **primary repair** and **secondary repair**.

PRIMARY REPAIR OF A CLEAN WOUND

A *clean wound* is a cut or incision on the skin where infection is not present. In a simple skin injury, the deep layer of stratified squamous epithelium divides. The new stratified squamous epithelial cells "push" themselves upward toward the surface of the skin. The damage or wound is quickly and completely restored to normal. However, if the damage is over a larger area, then the underlying connective tissue cells and fibroblasts are also involved.

PRIMARY REPAIR OVER A LARGE SKIN AREA

If a large area of skin is damaged, fluid will escape from the broken capillaries. This capillary fluid dries and seals the wound, and the typical **scab** forms. The scab formation prevents pathogens from entering the site. Epithelial cells multiply at the edges of the scab and continue to grow over the damaged area until it is covered. If a great or deep area of skin is destroyed, skin **grafts** may be needed to help in wound healing.

Medical Highlights

TISSUE AND ORGAN TRANSPLANT

Organ transplants are proven therapy for a variety of serious diseases. In 2019, more than 39,718 organ transplants were performed, but more than 114,000 people were on waiting lists for organs. On average, 20 people die every day from lack of an available organ.

Thousands of people have received a blood transfusion, which is a tissue transplant. Successful tissue transplants include heart valves, veins to improve circulation, corneas, blood, and bone. Skin tissue is used to promote healing and prevent infection in critically burned individuals. *Apart from corneal transplants, tissue transplants are not subject to rejection and do not require long-term therapy*.

Major organ transplants include bone, kidney, heart, lung, liver, intestines, and pancreas. All organs except the brain have been successfully transplanted. Bone marrow transplants differ from other transplants and are discussed in the chapter on blood (Chapter 13). An organ transplant may be from a deceased donor or a genetically compatible living donor. One question that arises is how long an organ can last outside the body and still be viable for transplant. It depends on the organ; for now, the time window can be between 4–36 hours. Most harvested organs are placed in a cooler filled with ice until the transplant occurs (Journal of Internal Medical Research, 2019).

After an organ has been successfully transplanted, the recipient must continue medical treatment for the rest of their life. This is due to the response of the immune system to a new organ. A transplanted organ is made of cells foreign to the body, which means the body's immune system will attack the organ if left to its own devices. Even with a good match between the donor's and recipient's blood types, the body will see the organ as foreign and act to reject it. There are three types of rejection: *hyperacute rejection, acute rejection,* and *chronic rejection*. *Hyperacute rejection* occurs immediately after transplant to 48 hours after the transplantation of an organ. The host antibodies attack the transplanted organ. The only treatment is removing the donor organ. The most common type of rejection is *acute rejection*, which occurs in the weeks following the transplant. Symptoms of rejection include flu-like symptoms, a temperature of 101°F or higher, decrease in urine output, weight gain, and pain or tenderness over the transplant site. This is the type for which immunosuppressive drugs are given. The drugs used to prevent rejection also suppress parts of the immune system that are necessary to fight infection and disease. Suppressing the immune system is a balancing act; the optimal result weakens the immune system just enough to prevent organ rejection but leaves it strong enough to fight infection and disease. *Chronic rejection* occurs years after the transplant and involves the donor organ becoming fibrotic and no longer able to function.

Living with a transplant is an ongoing challenge; however, in most cases it enables the recipient to live a full and active life. The goal of researchers is to coax the immune system into accepting an organ transplant without the lifelong need for immunosuppressive drugs.

A major issue today is the concern that there has been a decline in the number of living donor organs available for transplants during the past five years. One organ donor can save as many as eight lives.

PRIMARY REPAIR OF DEEP TISSUE

When damage occurs to deep tissues, the edges of the wound must be sewn together with **sutures**. For example, operative incisions or wounds have a tremendous amount of serous fluid that leaks out onto the wound. This helps form a coagulation, or clot, that seals the wound. The coagulum contains tissue fragments and white blood cells. In 24 to 36 hours, the epithelial cells lining the capillaries, or the endothelium, and fibroblasts of connective tissue degenerate rapidly. The newly formed cells remain along the edges of the wound. On the third day following injury, new vascular tissue starts to form. This multiplies across the wound, along with connective tissue formation.

On the fourth or fifth day, fibroblast cells become very active in making new collagen fibers. In addition, capillaries grow and "reach" across the wound, holding the edges firmly together. Toward the end of the healing process, the collagenous fibers shorten, reducing scar tissue to a minimum. Scar tissue is strong but lacks the flexibility and elasticity of most normal tissue. Scar tissue cannot perform the functions of the normal tissue it has replaced.

SECONDARY REPAIR

During secondary repair, a process called **granulation** occurs in a large open wound with small or large tissue loss. The granulation process will form new vertically upstanding blood vessels. These new blood vessels are surrounded by young connective tissue and wandering cells of different types. Granulation causes the surface area to have a pebbly texture. Fibroblasts will be quite active in the production of new collagenous fibers. The activity of this repair causes the large open wound to eventually heal. As granulation occurs, a fluid also is secreted. This fluid has strong **bactericidal,** or bacteria-destructing, properties that help reduce the risk of infection during wound healing.

As in any type of tissue repair, a certain amount of scar tissue will form. The amount of tissue formed depends on the extent of tissue damage. Careful attention must be given to patients whose bodies or body parts are undergoing massive tissue repair, such as victims of burns. These areas *must* be kept immobile and in alignment at the beginning; however, later active movement should be encouraged so that, as new tissue forms, pulling from scar tissue will not occur. It is the role of the health care provider to help prevent or minimize excessive scar tissue formation that can lead to disfigurement.

A health care provider should also be mindful that proper nutrition plays an important part in healing. Newly growing tissues require lots of protein for repair; thus, the need for protein-rich foods is important.

Vitamins also play an essential role in wound repair (Chapter 20). They help the patient develop resistance to infections. **Table 5-3** lists vitamins that are needed in tissue repair.

Media Link

View the **Tissue Repair** animation on the Online Resources.

Table 5-3 *Vitamins Favorable to Tissue Repair*

VITAMIN	*FUNCTION*
Vitamin A	Aids in repair of epithelial tissue, especially the epithelial cells lining the respiratory tract
Vitamin B (thiamine, nicotinic acid, and riboflavin)	Helps promote the general well-being of the individual Specifically helps promote appetite, metabolism, vigor, and pain relief in some cases
Vitamin C	Helps in the normal production and maintenance of collagen fibers and other connective tissue substances
Vitamin D	Is needed for the normal absorption of calcium from the intestine Possibly helps in the repair of bone fractures
Vitamin K	Helps in the process of blood coagulation
Vitamin E	Helps healing of tissues by acting as an antioxidant protector—prevents important molecules and structures in the cell from reacting with oxygen (When delicate components of living protoplasm are attacked by oxygen, they are literally "burned.")

Medical Terminology

adipos	fatty	**muc/ous**	pertaining to a slimy substance
-e	pertaining to	**pariet**	wall
adipos/e	pertaining to fatty	**-al**	pertaining to
bacteria	single-cell microorganisms	**pariet/al**	pertaining to the wall of a body cavity
-cidal	to kill or destroy	**peri**	around
bacteri/cidal	to destroy single-cell microorganisms	**peri/cardi/al**	pertaining to around the heart
cardi	heart	**ser**	watery
-ac	pertaining to	**-ous**	pertaining to
cardi/ac	pertaining to heart	**ser/ous**	pertaining to a watery substance
muc	slime	**viscer**	guts or internal organs
-ous	pertaining to	**viscer/al**	pertaining to the internal organs

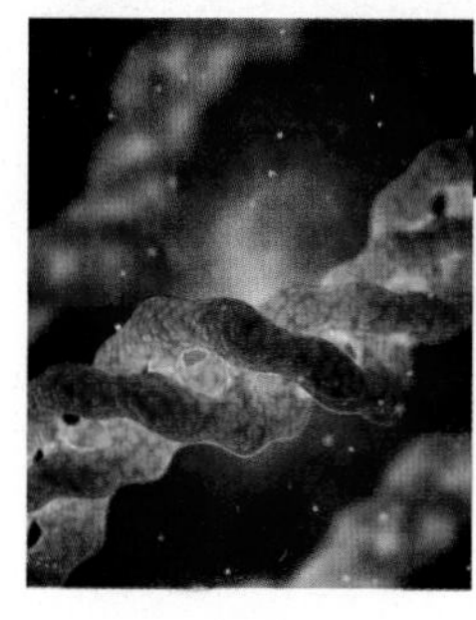

Study Tools

Workbook	Activities for Chapter 5
Online Resources	• PowerPoint presentations • Animation

REVIEW QUESTIONS

Select the letter of the choice that best completes the statement.

1. Cells that are alike in size, shape, and function are called
   a. elements.
   b. tissues.
   c. organs.
   d. systems.

2. The type of tissue found on the outer layer of skin is called
   a. squamous epithelial.
   b. stratified epithelial.
   c. ciliated epithelial.
   d. columnar epithelial.

3. Collagen is a strong, flexible protein found mainly in
   a. adipose tissue.
   b. cartilage tissue.
   c. loose connective tissue.
   d. bone tissue.

4. Connective tissue structures that hold bones firmly together at joints are called
   a. fasciae.
   b. tendons.
   c. aponeuroses.
   d. ligaments.

5. The membrane that lines surfaces with openings to the outside of the body is
   a. cutaneous.
   b. serous.
   c. mucous.
   d. synovial.

6. The membrane that covers the lungs is called
   a. parietal pleura.
   b. visceral pleura.
   c. parietal pericardial.
   d. visceral pericardial.

7. The lining the abdominal cavity is called the
   a. peritoneal membrane.
   b. pleural membrane.
   c. pericardial membrane
   d. mucous membrane.

8. The system that provides for movement of the body is the
   a. skeletal system.
   b. nervous system.
   c. muscular system.
   d. circulatory system.

9. The type of repair that takes place in a clean wound is called
   a. primary repair.
   b. granulation.
   c. secondary repair.
   d. secretion of bactericidal fluid.

10. The vitamin necessary to help as an antioxidant is
   a. A.
   b. D.
   c. K.
   d. E.

FILL IN THE BLANKS

1. The tissue that has the ability to react to stimuli is ________________.
2. The gastric mucosa is the mucous membrane lining of the ________________.
3. The secretion that prevents friction when the bones in a joint rub together is ________________.
4. The lining that protects the lung is the ________________ membrane.
5. In secondary tissue repair when there is a large open wound, the process of tissue repair is called ________________.

APPLYING THEORY TO PRACTICE

1. Feel your skin. What type of tissue is it? Is this tissue the same as the lining in your mouth?
2. Explain how mucus affects the air we breathe or the food we eat.
3. Name the organ systems involved when you eat a slice of pizza.
4. You hear the expression "I sprained a ligament" or "I pulled my tendon." Describe the type of tissue involved with each injury. Describe the function of a ligament and of a tendon.
5. Imagine you have a friend with severe injuries who is hungry. Describe the lunch you would prepare for this person. The menu should include the vitamins necessary for tissue repair and help in pain relief.

CASE STUDY

Anthony, age 8, fell on the school playground and scraped his knee. His mother comes to the school nurse's office. She is worried about infection and whether a big scar will form.

1. Explain to Anthony's mother what a clean wound is.
2. Describe to the mother what takes place during the healing process.
3. Explain to the mother about scarring.
4. Give the mother information on the role of vitamins in healing and developing resistance to other infections.
5. Anthony is worried he cannot run and play. What will you tell Anthony about his mobility?

5-1 Epithelial Tissue

- ***Objective:*** To examine the structure of epithelial tissue and how the structure affects its function
- ***Materials needed:*** slides of epithelial tissue: squamous, cuboidal, columnar, and glandular; microscope; paper; pencil

Step 1: Examine the prepared slides of the different types of epithelial tissue under a microscope. Record the observed differences.

Step 2: Draw a diagram of each type of epithelial tissue. Describe how the structure affects the function.

Step 3: Describe where each type of tissue is located in the body. Record your answers.

Step 4: What are the structural differences between the types of epithelial tissue? Record your conclusions.

LAB ACTIVITY 5-2 Connective Tissue

- ***Objective:*** To examine connective tissue (the most common tissue type) and explain how it serves its general function to fasten, connect, support, and protect
- ***Materials needed:*** labeled slides of connective tissue: adipose, areolar, dense fibrous, bone, cartilage, vascular (blood and lymph); unlabeled slides of connective tissue; microscope; paper; pencil

Step 1: Examine the prepared slides of the different types of connective tissue under the microscope.

Step 2: Describe what you see in each matrix of the tissue. Record your observations.

Step 3: List the categories of connective tissue.

Step 4: What is the difference between adipose tissue and bone tissue? Record your answers.

Step 5: With a lab partner, look at two types of connective tissue that are not labeled. Decide which types of connective tissue are shown on the slides. Record your answers.

CHAPTER 6

Integumentary System

Objectives

- Describe the functions of the skin.
- Describe the structures found in the two skin layers.
- Explain how the skin serves as a channel of excretion.
- Describe the function of the appendages of the skin.
- Describe some common skin, hair, and nail disorders.
- Define the key words that relate to this chapter.

Key Words

abrasion
acne vulgaris
alopecia
arrector pili muscle
athlete's foot
avascular
basal cell carcinoma
boils
cortex
decubitus ulcers
dermatitis
dermis
eczema
epidermis
erythema
eschar
first-degree (superficial) burn
fissure
fungal infections
hair follicle
herpes
herpes simplex
hyperthermia
hypothermia
impetigo
ingrown nails
integumentary system
jaundice
keratin
laceration
lice
malignant melanoma
medulla
melanin
melanocytes
moles
papillae
psoriasis
ringworm
root
rosacea
rule of nines
sebaceous glands
sebum
second-degree (partial-thickness) burn
shaft
shingles (herpes zoster)
skin cancer
squamous cell carcinoma
stratum corneum
stratum germinativum
stratum granulosum
stratum lucidum
stratum spinosum
sweat glands
third-degree (full-thickness) burn
urticaria (hives)
warts

The **integumentary system** (in-teg-you-**MEN**-tah-ree) is made up of the skin and its appendages: hair, nails, sebaceous glands, and sweat glands. The word *integumentary* means "covering." The skin may also be referred to as the cutaneous membrane. It is the largest organ in the body; in the average adult, the integumentary system covers 3000 square inches of surface area.

> **Did You Know?**
>
> *Every minute, 30,000 to 40,000 dead skin cells fall from a person's body—this amounts to about 40 pounds of skin in a lifetime. However, it is not a permanent weight loss because the body makes approximately a whole new layer of skin cells every month.*

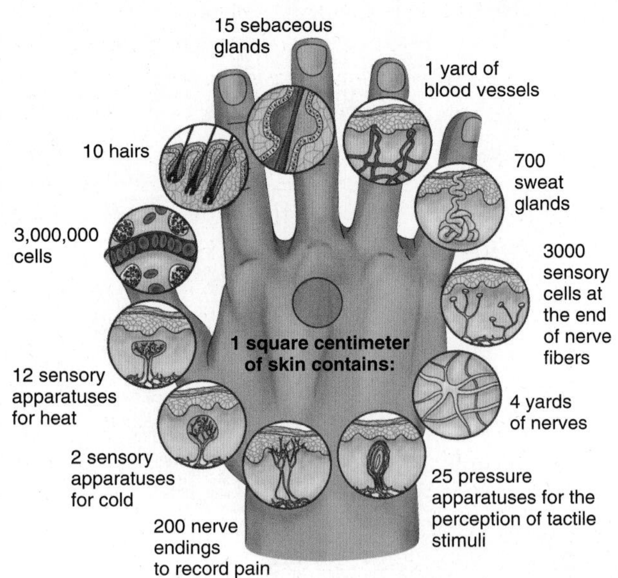

Figure 6-1 *The skin is well supplied with nerves and receptors.*

FUNCTIONS OF THE SKIN

The skin has seven functions:

1. Skin is a covering for the underlying, deeper tissues, protecting them from dehydration, injury, and germ invasion.
2. Skin helps regulate body temperature by controlling the size of the blood vessels in the dermal layer of the skin.
3. Skin helps manufacture vitamin D. The ultraviolet light on the skin is necessary for the first stages of vitamin D formation.
4. Skin is the site of many nerve endings (**Figure 6-1**). A square inch of skin contains about 72 feet of nerves and hundreds of receptors.
5. Skin has tissues for the temporary storage of fat, glucose, water, and salts such as sodium chloride. Most of these substances are later absorbed by the blood and transported to other parts of the body.
6. Skin serves to reduce the harmful ultraviolet radiation contained in sunlight.
7. Skin has special properties that permit it to absorb certain drugs and other chemical substances. Drugs can be applied locally, as in the case of treating rashes, or they can be applied via patches that can be absorbed through the skin and have a general effect in the body.

STRUCTURE OF THE SKIN

The skin consists of two basic layers:

1. The **epidermis** (ep-ih-**DER**-mis), or outermost covering, is made of epithelial cells and contains no blood vessels (**avascular**).
2. The **dermis**, or "true skin," is made of connective tissue and is vascular.

Epidermis

The epidermis consists of four distinct cell types and five layers. The thickness of the epidermis varies: It is thinnest on the eyelids and thickest on the palms of the hands and the soles of the feet. The surface layer, or stratum corneum, consists of dead cells rich in keratin. **Keratin** (**KER**-ah-tin) is a protein that renders the skin dry and provides a waterproof covering, thus resisting evaporation and preventing excessive water loss. It also serves as a barrier against ultraviolet light, bacteria, abrasions, and some chemicals.

The *epidermal cells* are as follows:

- Keratinocytes comprise most of the epidermis and produce the protein keratin.
- Merkel cells are the sensory receptors for touch.
- **Melanocytes** (**MEL**-ah-noh-sightz) make the protein **melanin** (**MEL**-ah-nin), which protects the skin against the ultraviolet rays of the sun.
- Langerhans cells—not the same as the islets of Langerhans in the pancreas—are macrophages that are effective in the defense of the skin against microorganisms.

The following are the epidermal layers, from the deepest to the most superficial:

1. The **stratum germinativum** (**STRAT**-um jer-mih-**NAY**-tih-vum), or stratum basale, undergoes continuous cell division; it is the deepest

epidermal layer. It consists of a layer of cells that are mostly keratinocytes. They grow upward and become part of the more superficial layers, the stratum spinosum. Melanocytes and Merkel cells are also found in the germinativum layer.

As seen in **Figure 6-2**, the lower edge of the stratum germinativum is thrown into ridges. These ridges are known as the **papillae** (pah-**PILL**-ee) of the skin. The papillae actually arise from the dermal layer of the skin and push into the stratum germinativum of the epidermis. In the skin of the fingers, soles of the feet, and the palms of the hands, these papillae are quite pronounced; they raise the skin into permanent ridges. These ridges are arranged so that they provide maximum resistance to slipping when grasping and holding objects; thus, they are also referred to as *friction ridges*. The ridges form the fingerprint patterns used in identification. Feet contain the same structures, and newborn infants are footprinted as a means of identification.

2. The **stratum spinosum** (spye-**NOH**-sum) is 8–10 cell layers thick. Contained in it are melanocytes, keratinocytes, and Langerhans cells. When seen under a microscope, the cells in this layer look prickly, thus the name *spinosum*, meaning "little spine."
3. The **stratum granulosum** (gran-yoo-**LOH**-sum) is where the keratinization process begins and the cells begin to die. Keratinization is the process whereby the keratinocyte cells change their shape, lose their nucleus and most of their water, and become mainly hard protein or keratin.
4. The **stratum lucidum** (**LOO**-sid-um) is found only on the palms of the hand and the soles of the feet. The cells in this layer appear clear.
5. The **stratum corneum** (**COR**-nee-um) is composed of dead, flat, scalelike, keratinized cells that slough off daily. Complete cell turnover occurs every 28 to 30 days in young adults, while the same process takes 45 to 50 days in older adults. This layer is also slightly acidic, which helps in the defense against harmful microorganisms.

Skin Color

Three pigments contribute to skin color: melanin, carotene, and hemoglobin.

Melanocytes produce two distinct classes of melanin: pheomelanin (fee-oh-**MEL**-ah-nun), which is red to yellow in color, and eumelanin (you-**MEL**-ah-nun), which is dark brown to black. People who have light skin generally have a greater proportion of pheomelanin in their skin than do those who have dark skin. Both classes of melanin bind to a wide variety of compounds, including some drugs. Because of their affinity for melanin, some drugs make people photosensitive, meaning their skin burns more easily. In older adults, melanin collects in spots, often called "aging" or "liver" spots.

The environment can modify skin coloring. Exposure to sunlight may result in a temporary increase of eumelanin, causing a darkened or tanned effect. Prolonged exposure to the ultraviolet rays of sunlight is dangerous because it may lead to skin cancers.

Carotene (**KAR**-oh-teen) is a yellow to orange pigment found in the stratum corneum and deeper layers of the skin. The yellowish tinge of the skin of some people of Asian descent is due to variations in both melanin and carotene. The pinkish color of some fair-skinned people is due to the presence of oxygen in the hemoglobin of the red blood cells circulating through the dermal capillaries.

Alterations in skin color may indicate disease conditions or emotional states, as listed in **Table 6-1**. The degree to which alterations in skin color occur depends on the normal coloring of the skin. It is important to ask the patient what their normal skin color is as a baseline.

Dermis

The dermis, or corium, is the thicker, inner layer of the skin that lies directly below the epidermis. It is composed of dense connective tissue; collagen tissue bands; elastic fibers, through which pass numerous blood vessels; muscle fibers; some mast cells and white blood cells; oil and fat glands; and fat cells.

Table 6-1 *Skin Color as Indicator of a Disease Condition*

SKIN COLOR	*CAUSE*	*CONDITION*
Redness—**erythema** (er-ih-**THEE**-mah)	Dilation of capillary network	Fever, allergic reaction, inflammation, or embarrassment
Bluish tint—*cyanosis* (sigh-ah-**NOH**-sis) Grayish tint may be present in darker-skinned people	Decrease in oxygen in capillary network	Heart or respiratory disease
Yellow—the term **jaundice** is also used to denote yellow coloring	Accumulation of bile in capillary network	Gallbladder or liver disease
Pallor	Constriction of capillary network or decrease in red blood cells	Emotional stress or anemia

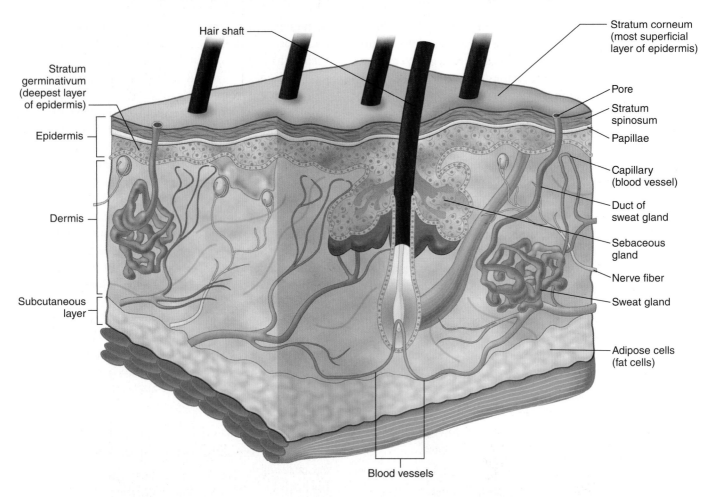

Hair shaft
Stratum corneum (most superficial layer of epidermis)
Stratum germinativum (deepest layer of epidermis)
Pore
Stratum spinosum
Epidermis
Papillae
Capillary (blood vessel)
Duct of sweat gland
Dermis
Sebaceous gland
Nerve fiber
Sweat gland
Subcutaneous layer
Adipose cells (fat cells)
Blood vessels

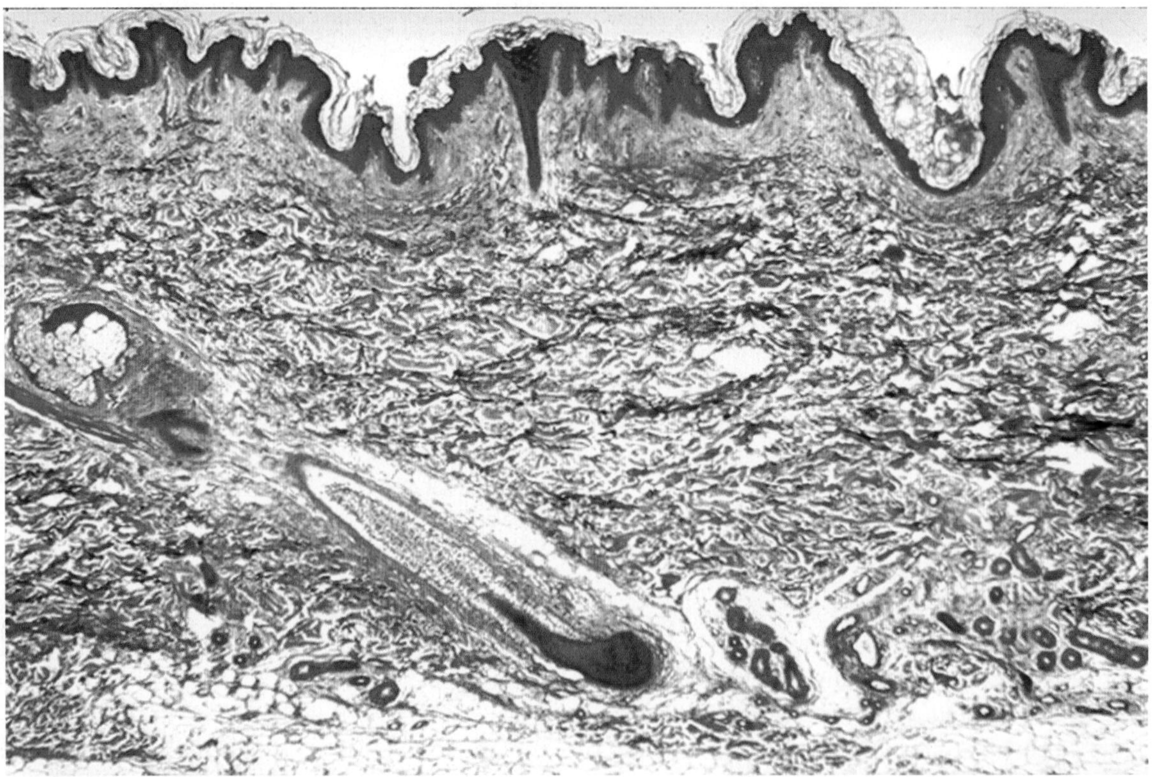

Figure 6-2 *A cross section of the skin*

B, Courtesy of University of Wisconsin Medical School, Madison, WI.

The mast cells in the dermis respond to injury, infection, or allergy and produce histamine and heparin. Histamine is released in response to allergens, causing the signs of an allergic reaction: itching and increased mucus secretions. Heparin is released in response to injury and prevents blood clotting (anticoagulant).

The thickness of the dermis varies over different parts of the body. It is, for instance, thicker over the soles of the feet and the palms of the hands. The skin covering the shoulders and back is thinner than that over the palms, but thicker than the skin over the abdomen and thorax.

The dermal layer contains many nerve receptors of different types. The sensory nerves end in nerve receptors, which are sensitive to heat, cold, touch, pain, and pressure. The locations of nerve endings vary. The receptors for touch are closer to the epidermis so a person can feel someone's touch. However, the pressure receptors are deeper in the dermal layer. This explains why a person can sit for a long period before they feel uncomfortable. Nerve endings that sense pain are located under the epidermis and around the hair follicles. These pain receptors are especially numerous on the lower arm, breast, and forehead.

Blood vessels in the dermis aid in the regulation of body temperature to maintain homeostasis. When external temperatures increase, blood vessels in the dermis dilate to bring more warmed blood flow to the surface of the body from deeper tissues. On a hot day, the heat brought to the skin's surface can be lost through the process of radiation, the transfer of heat from a warm body to a cooler environment; convection, air currents that pick up and transfer heat away from a warm surface; conduction, transfer of heat from a warm object to a cooler object with which it is in contact; or evaporation, the transfer of heat into body fluids, which are then evaporated from the body surface to the air. Heat loss through these means will cool the body. If the body is exposed to cold for an extended period of time, the blood vessels will constrict to bring warmed blood closer to vital organs to warm and preserve them. This process cannot be maintained for long periods of time.

Subcutaneous or Hypodermal Layer

The subcutaneous or hypodermal layer is not a true part of the integumentary system. It lies under the dermis and sometimes is called superficial fascia. It consists of loose connective tissue and contains about one-half of the body's stored fat. The hypodermal layer attaches the integumentary system to the surface muscles underneath. Injections given in this area are called hypodermic or subcutaneous.

APPENDAGES OF THE SKIN

The appendages of the skin include the hair, nails, sudoriferous (sweat) glands, and sebaceous (oil) glands and their ducts.

Hair

Hairs are distributed over most of the surface area of the body. They are missing from the palms of hands, soles of feet, glans penis, and inner surfaces of the vaginal labia.

The length, thickness, type, and color of hair vary with the different body parts and different ethnicities. The hairs of the eyelids, for example, are extremely short, whereas hair from the scalp can grow to a considerable length. Facial and pubic hair is quite thick.

A hair is composed of a root shaft, the outer *cuticle* layer, the **cortex**, and the inner **medulla**. The cuticle consists of a single layer of flat, scalelike, keratinized cells that overlap each other and protect the inside of the hair shaft. The cortex consists of elongated, keratinized, nonliving cells. Hair pigment is located in the cortex. In dark hair, the cortex contains pigment granules; as one ages, pigment granules are replaced with air, which gives a gray or white appearance to hair. The medulla, also called the pith or marrow, is the innermost layer. This layer is only present in thick and coarse hair types and is usually lacking in individuals who are naturally blonde.

The **root** is the part of the hair that is implanted in the skin. The **shaft** projects from the skin's surface. The root is embedded in an inpocketing of the epidermis called the **hair follicle** (**Figure 6-3**). Hair varies from straight to curly. The shape of the hair follicle determines the curl of the hair. A round follicle makes straight hair, an oval follicle makes wavy hair, and a flat follicle makes curly hair. Toward the lower end of the hair follicle is a tuft of tissue called the papilla, which extends upward into the hair root. The papilla contains capillaries that nourish the hair follicle cells. This is important because the division of cells in the hair follicle gives rise to a new hair.

There is a genetic predisposition in some people to a condition known as **alopecia**, (al-oh-**PEE**-shee-ah) or baldness, which is a permanent hair loss. The normal hair is replaced by a very short hair that is transparent and—for practical purposes—invisible. Males typically experience more hair loss than women do and at a younger age. Treatment for baldness includes medications, both topical and oral, and hair transplants.

Attached to each hair follicle on the side toward which it slopes is a smooth muscle called the **arrector pili muscle** (ah-**RECK**-tor **PYE**-lye). It is stimulated by the autonomic nervous system; for example, when the

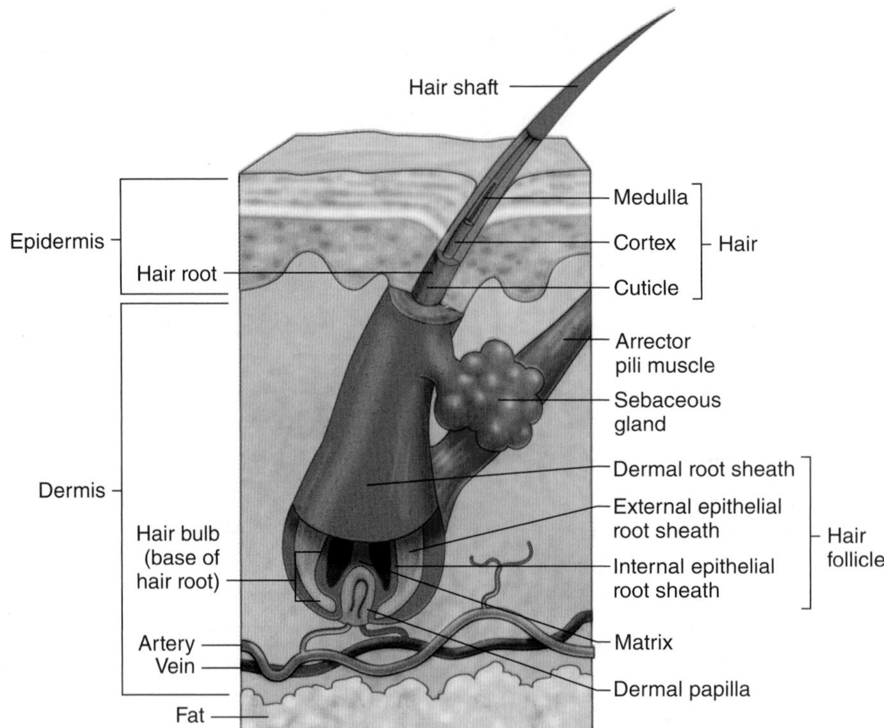

Figure 6-3 *The anatomy of an individual hair*

body feels a sudden chill, it contracts and causes the skin to pucker around the hair, generating heat. This reaction may be called "goosebumps" or "gooseflesh."

Nails

The nails are hard structures at the ends of the fingers and toes. They are slightly convex on their upper surfaces and concave on their lower surfaces. A nail is formed in the nail bed or matrix (**Figure 6-4**). Here, the epidermal cells first appear as elongated cells. These then fuse together to form hard, keratinized plates. Air mixed in the keratin matrix forms the white crescent at the proximal end of each nail called the lunula (**LOO**-noo-lah) and the white at the free edge of the nail. As long as a nail bed remains intact, a nail will always be formed. Healthy nails are usually pink in color and grow 1 millimeter (mm) per week. Fingernails grow faster than toenails; as people age, their nails grow more slowly.

Some disease conditions may be revealed by the color of a person's nails, as discussed in **Table 6-2**.

Sweat Glands

While actual excretion is a minor function of the skin, certain wastes dissolved in perspiration are removed. Perspiration is 99% water, with only small quantities of salt and organic materials, such as waste products. **Sweat glands**, also called sudoriferous (**soo**-doh-**RIF**-er-us) glands, are distributed over the entire skin surface. They are present in large numbers under the arms, and on the palms of the hands, soles of the feet, and forehead.

Sweat glands are tubular, with a coiled base and a tubelike duct that extends to form a pore in the skin (**Figure 6-2**). Perspiration is excreted through the pores. Under the control of the nervous system, these glands may be activated by several factors including heat, pain, fever, and nervousness.

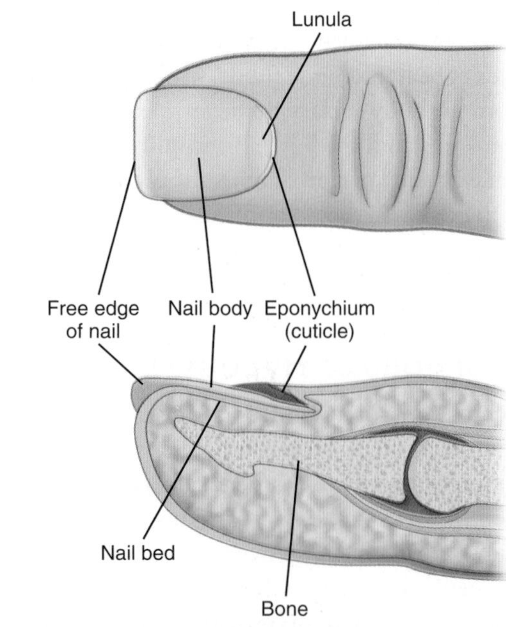

Figure 6-4 *Diagram of the fingernail bed*

Table 6-2 Disease Conditions and Nail Color

CONDITION	NAIL COLOR
Liver disease	White nails
Kidney disease	Half of nail is pink or reddish brown; half is white
Heart condition	Nail bed is red
Lung disease	Yellow and thickening nails
Anemia	Pale nail bed
Diabetes	Yellowish with a slight blush at the base
Hypoxia	Bluish nails

The amount of water lost through the skin is almost 500 milliliters (mL) a day, although this varies according to the type of exercise and the environmental temperature. With profuse sweating, a great deal of water may be lost; it is vital to replace the lost water as soon as possible.

Ceruminous glands (seh-**ROO**-mih-nus), or wax glands, are modifications of the sweat glands. These are found in the ear canals and produce earwax.

Sebaceous Glands

The skin is protected by a thick, oily substance known as **sebum** (**SEE**-bum), which is secreted by the **sebaceous glands** (see-**BAY**-shus). Sebum contains amino acids, lactic acids, lipids, salt, and urea. Sebum lubricates the skin, keeping it soft and pliable.

> **Media Link**
>
> View the **Skin** animation on the Online Resources.

INTEGUMENT AND ITS RELATIONSHIP TO MICROORGANISMS

An intact skin surface is the best way the body can defend itself against pathogens (disease-producing toxins) and water loss. If skin is especially dry, lotions or creams may prevent cracking.

Most of the skin surface is not a favorable place for microbial growth because it is too dry. Microbes live only on moist skin areas where they adhere to and grow on the surfaces of dead cells that compose the outer epidermal layer. The types of microbes found are of the *Staphylococcus* or *Corynebacterium* bacterial species. The other common types found on the skin are *fungi* and *yeasts*. (See Chapter 17.)

Most skin bacteria are associated with the hair follicles or sweat glands, where nutrients are present and moisture content is high. Underarm perspiration odor is caused by the interaction of bacteria on perspiration. This odor can be minimized or prevented either by decreasing perspiration with antiperspirants or killing the bacteria with deodorant soaps.

Handwashing

The number one way to prevent the spread of disease is by handwashing. Wash hands in running water, using soap and friction. Be sure to scrub the backs of hands, between the fingers, and under the nails. Rub hands for at least 20 seconds; singing the tune "Happy Birthday" twice is a good way to determine if it has been long enough. Rinse and dry hands using a paper towel or air dryer. If a person comes in contact with infectious material, washing time should be between 2–4 minutes. If a person comes in contact with blood, infectious material, or body secretions, wash the hands and apply gloves before exposure. After exposure, remove the gloves and wash the hands again.

If soap and water are not available, use an alcohol-based hand sanitizer that contains at least 60% ethanol or 70% isopropanol. Sanitizers quickly reduce the number of germs on the hands in some situations but do not eliminate all types of germs.

> **Media Link**
>
> View the **Handwashing** video on the Online Resources.

Environmental Factors That Affect the Skin

Many skin disorders are caused by certain allergens in the air, such as dust pollen and pollution, or by sensitivity to certain foods, materials, or chemicals. The result may be a skin rash or hives.

Heat rash, or "prickly heat," resulting from the plugging of the sweat glands gives rise to tiny, watery blisters. Treatment includes keeping the body cool or the application of creams or lotions to relieve discomfort.

REPRESENTATIVE DISORDERS OF THE SKIN

There are many skin disorders, ranging from minor to severe. They can be temporary, intermittent, or permanent.

Congenital Skin Disorders

Congenital deformities of the skin are rare and vary in severity; they may be present at birth or may appear in the first months of life. Common locations are the head, nose, and neck. Some types of congenital deformities include the following:

- *Albinism* is a partial or complete loss of pigmentation caused by a genetic mutation that reduces the production of melanin. It causes an extreme sensitivity to sunlight, especially in the eyes.

The Effects of Aging on the Integumentary System

The skin presents the most visible signs of aging. Adult stem cell production declines with aging, causing the epidermal cells to slow down their reproduction process. This causes thinner, more translucent skin, leading to more frequent skin injuries, tearing, and infections. The rate of healing takes twice as long in older adults as in much younger people.

With aging, the sebaceous glands secrete fewer lubricants, and the outer layer of the skin becomes drier and more fragile. The elastin fibers shrink, becoming more rigid and leading to a loss of elasticity in the skin. Loss of subcutaneous fat results in lines, wrinkles, and sagging. The dermal vascular network decreases in its ability to respond to heat and cold. This predisposes an older person to **hypothermia**, a condition in which body temperature drops below normal, and **hyperthermia**, a condition in which body temperature rises above normal.

The number of melanocytes decreases, making the skin more sensitive to the ultraviolet rays of the sun. Selected melanocytes increase their production in areas exposed to the sun, resulting in brown spots on the skin (liver spots). Small, cherry-red bumps (cherry angiomas) may appear, which are benign skin tumors.

Vascular supply to the nail bed decreases, resulting in dull, brittle, hard, and tough nails.

The physiological changes in the skin may affect a person's feelings of self-worth. This can be the most difficult part of aging in a society where youthful appearance translates into beauty.

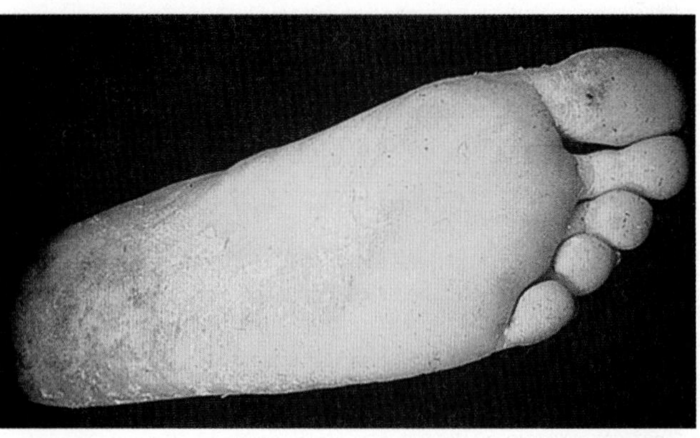

Figure 6-5 *Athlete's foot*

Courtesy of the Centers for Disease Control and Prevention

- *Birthmarks* include two types: *vascular,* where too many blood vessels cluster in one area, such as port wine stain and hemangioma, and *pigmented,* where more pigment occurs in one part of the skin, such as moles.

General Skin Disorders

Acne vulgaris (**ACK**-nee vul-**GAY**-ris), usually referred to as acne, is a common, noncontagious, and chronic disorder of the sebaceous glands. The sebaceous glands secrete excessive oil, which is deposited at the openings of the glands. Eventually this oily deposit becomes hard, or keratinized, plugging up the opening. This prevents the escape of the oily secretions, and the area becomes filled with leukocytes, which are white blood cells. The leukocytes cause an accumulation of pus. Acne occurs most often during adolescence and is marked by blackheads, cysts, pimples, and scarring. Treatment may be topical medications that dry up oil and promote skin peeling. The physician may also order antibiotics if the skin becomes infected.

Athlete's foot is a contagious fungal infection that infects the superficial skin layer and leads to skin eruptions (**Figure 6-5**). These eruptions are characterized by the formation of small blisters between the fingers and most often the toes. Accompanied by cracking and scaling, this condition is usually contracted in public baths or showers or by wearing tight, closed-toe shoes. Treatment involves thorough cleansing and drying of the affected area. In addition, special antifungal agents are administered, and antifungal powders are applied liberally.

Dermatitis (der-mah-**TYE**-tis) is a noncontagious inflammation of the skin that may be nonspecific (**Figure 6-6**). For example, some people may use a particular soap and develop a rash, or they may develop a rash as a result of emotional stress. To treat contact dermatitis, remove the irritant that is causing the problem. Wash the area and apply topical ointments to reduce inflammation and itching.

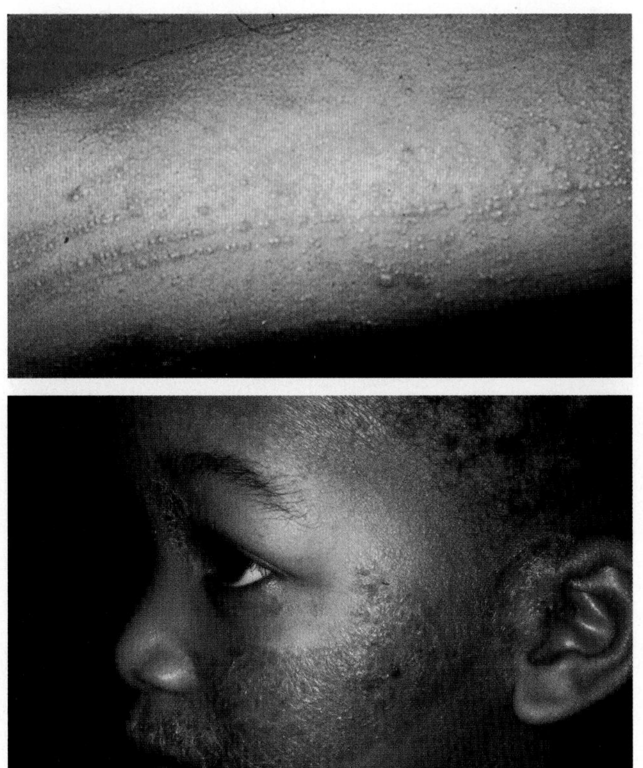

Figure 6-6 *Allergic contact dermatitis*

Courtesy of the Centers for Disease Control and Prevention

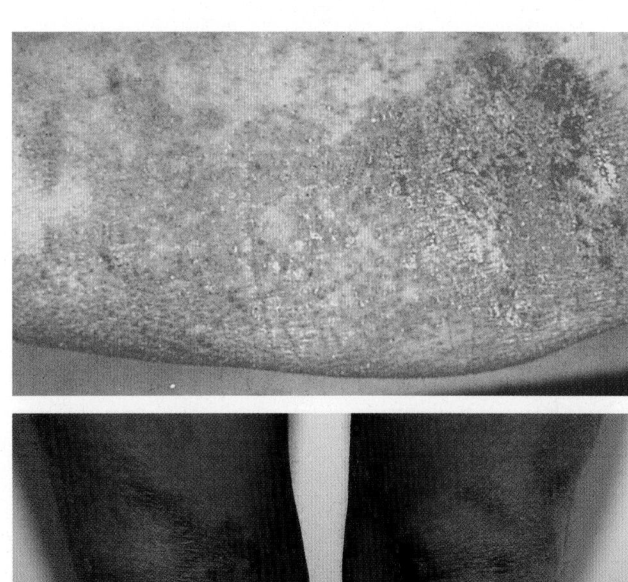

Figure 6-7 *Eczema*

Courtesy of the Centers for Disease Control and Prevention

Eczema (**ECK**-zeh-mah) is an acute, or chronic, noncontagious inflammatory skin disease. The skin becomes dry, itchy, and scaly . It may appear red on light skin; on dark skin, it may appear darker brown, purple, or gray. (**Figure 6-7**). Various factors can lead to eczema, such as soaps, heat, pollen, or pollution. The most common type is atopic eczema, an allergic reaction that usually occurs in the first year of life. Treatment consists of removal or avoidance of the causative agent, as well as topical medications to help alleviate the symptoms.

Impetigo (im-peh-**TYE**-goh) is an acute, inflammatory, and contagious skin disease seen in babies and young children. It is caused by *Staphylococcus* or *Streptococcus* organisms. This disorder is characterized by the appearance of vesicles that rupture and develop distinct yellow crusts (**Figure 6-8**). Treatment is with a topical antibacterial cream and oral antibiotics. Impetigo is not contagious after the medication has been taken for 24 hours.

Psoriasis (soh-**RYE**-uh-sis) is a chronic, noncontagious, inflammatory autoimmune skin disease characterized by the development of dry, reddish patches covered with silvery-white scales. It affects the skin surface over the elbows, knees, shins, scalp, and lower back (**Figure 6-9**). Onset may be triggered by stress, trauma, or infection. Psoriasis has no definitive treatment at present. The best treatment is topical corticosteroids, but methotrexate may

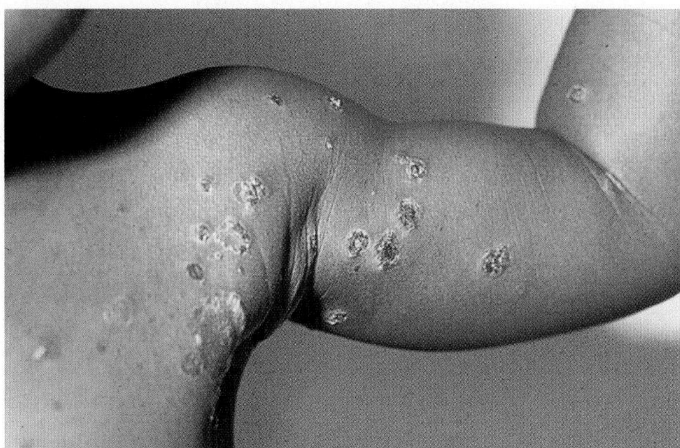

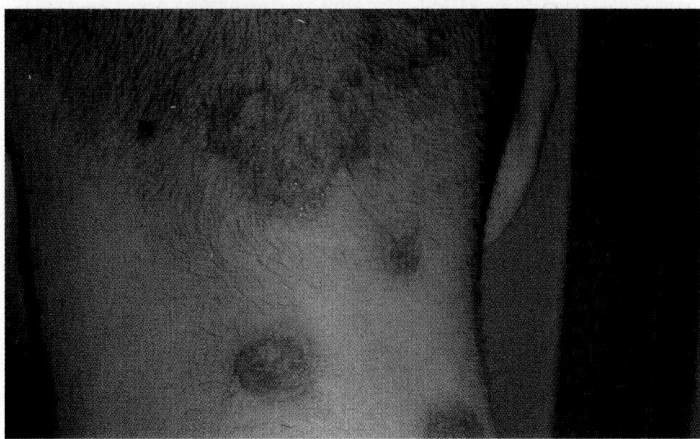

Figure 6-8 *Impetigo*

Courtesy of Robert A. Silverman, M.D., Clinical Associate Professor, Department of Pediatrics, Georgetown University

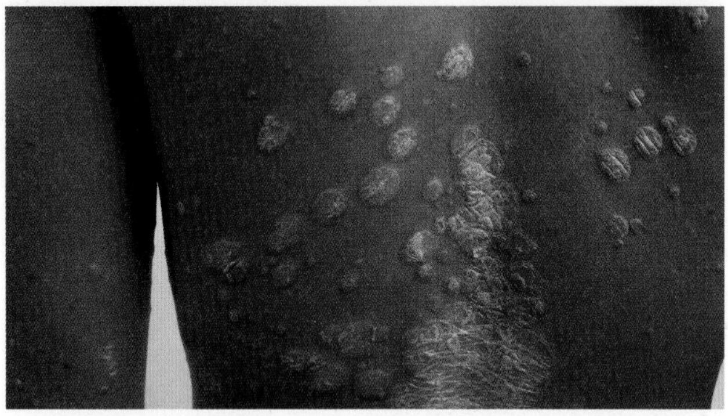

Figure 6-9 *Psoriasis*

Courtesy of Robert A. Silverman, M.D., Clinical Associate Professor, Department of Pediatrics, Georgetown University

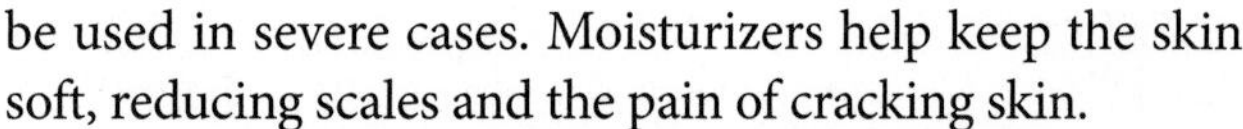

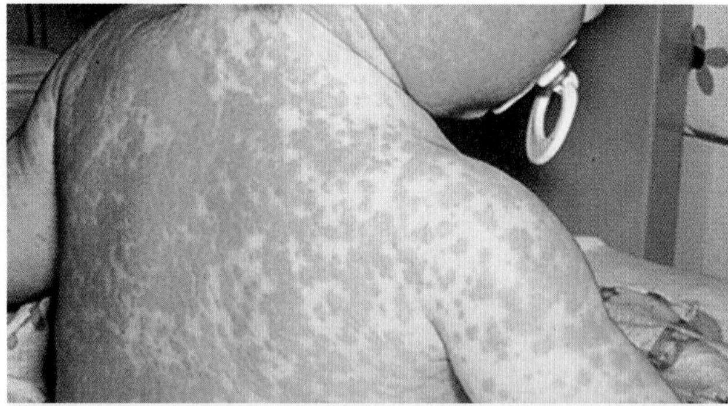

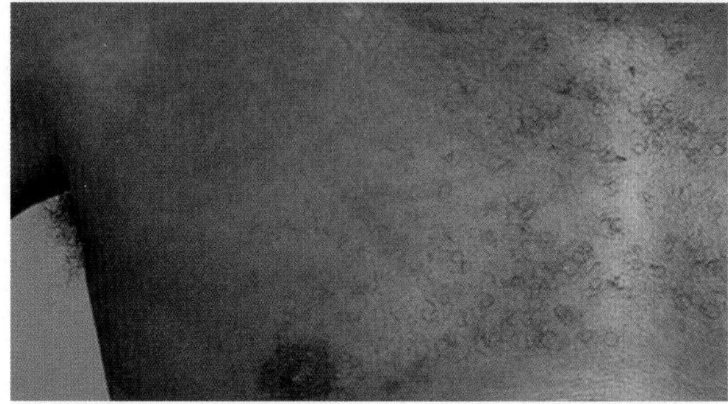

Figure 6-10 *Urticaria (hives)*

Courtesy of Robert A. Silverman, M.D., Clinical Associate Professor, Department of Pediatrics, Georgetown University

be used in severe cases. Moisturizers help keep the skin soft, reducing scales and the pain of cracking skin.

Ringworm is a highly contagious fungal infection marked by raised, itchy, circular patches with crusts. It may occur on the skin, scalp, and underneath the nails. Ringworm can be effectively treated with antifungal drugs.

Urticaria (hives) (ur-tih-**KAY**-ree-ah) is a noncontagious skin condition recognized by the appearance of intensely itching wheals or welts. These welts may last from 1–2 days and have an elevated, usually white or skin-toned center with a surrounding pink or purple area. They appear in clusters distributed over the entire body surface (**Figure 6-10**). Urticaria is generally a response to an allergen, such as an ingested drug or food. Elimination of the causative factor(s) and medications, such as antihistamines, alleviate the problem.

Boils, or carbuncles, are noncontagious and very painful. A boil is a bacterial infection of the hair follicles or sebaceous glands usually caused by a *Staphylococcus* organism. If the boil becomes more extensive and is deeply embedded, it is called a *carbuncle*. The treatment for boils is the application of warm compresses to relieve pain and promote drainage. The treatment for carbuncles consists of antibiotics and the excision and drainage of the affected area.

Moles are benign growths that occur when melanocytes grow in a cluster with tissue surrounding them. Moles darken after exposure to the sun and may be present at birth, or they may develop during teen years or during pregnancy. If a change is noted, it should be evaluated by a physician.

Rosacea (roh-**ZAY**-shee-ah) is a noncontagious, common, inflammatory disorder characterized by chronic redness and irritation to the face. It most often affects fair-skinned adults. It may begin as a simple tendency to flush or blush easily, then progress to a persistent redness in the central portion of the face. Small red bumps or pustules may appear and spread across the face.

Each person has their own trigger factor for the condition; these may include hot foods or beverages, spicy foods, alcoholic beverages, temperature extremes, strenuous exercise, stress, or menopausal hot flashes. Treatment includes avoiding triggers that aggravate the condition and the use of a topical or oral antibiotic medication prescribed by a physician.

Herpes (**HER**-peez) is a viral infection that is usually seen as a blister. Herpes viruses cycle between periods of active disease that last for 2–21 days followed by remission. After initial infection, the virus resides in sensory nerve cell bodies, where it becomes latent and resides there lifelong, meaning that it is subject to recurrence. The most common types are herpes simplex, genital herpes, and herpes zoster (shingles).

Herpes simplex occurs around the face and mouth; the symptoms are called cold sores or fever blisters (**Figure 6-11**). It may be spread through oral

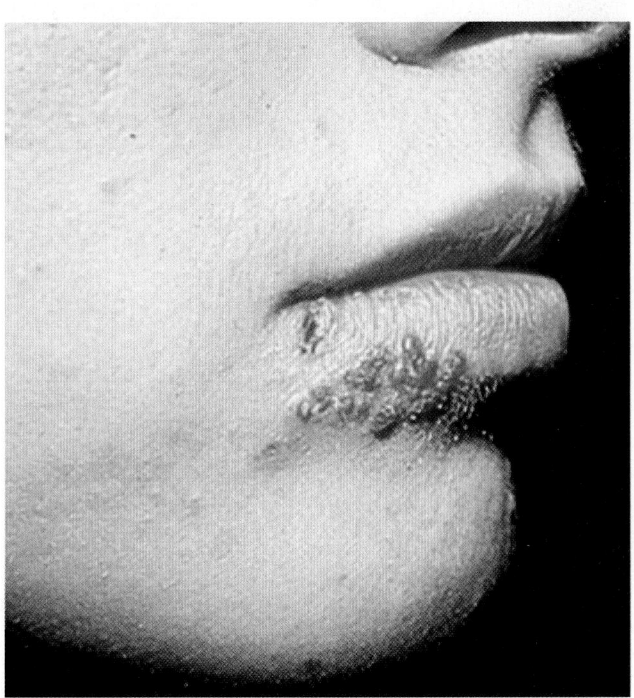

Figure 6-11 *Herpes simplex*

Courtesy of Robert A. Silverman, M.D., Clinical Associate Professor, Department of Pediatrics, Georgetown University

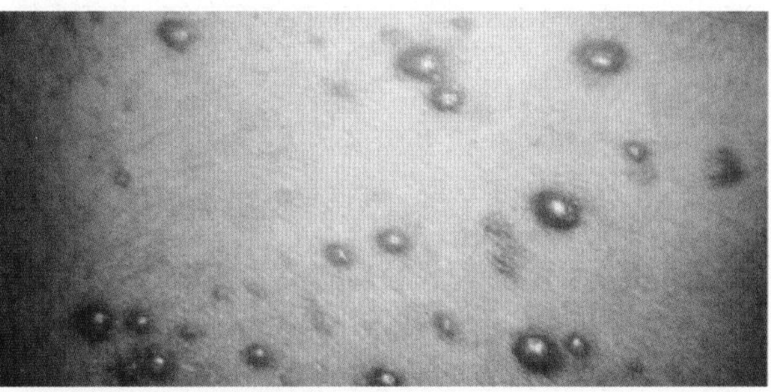

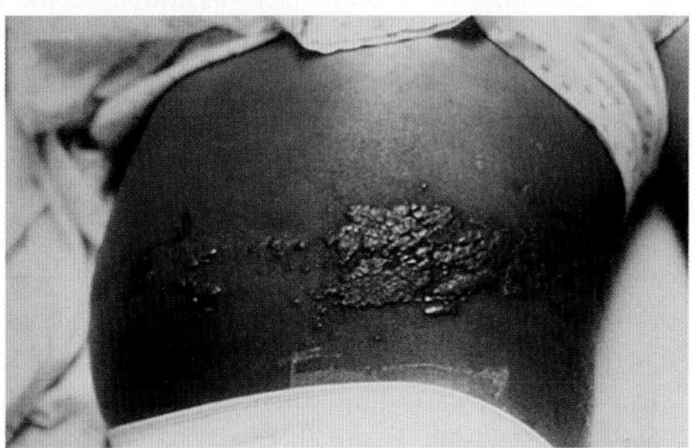

Figure 6-12 *Shingles*

Courtesy of the Centers for Disease Control and Prevention

contact. Antiviral medications can reduce the frequency and duration of the outbreak.

Genital herpes is another form of the virus, which may appear as a blister in the genital area. Genital herpes is discussed in more detail in Chapter 22 under sexually transmitted diseases.

Shingles (herpes zoster) is a skin eruption due to a virus infection of the nerve endings. The virus is the same one that causes chickenpox in children. It is commonly seen on the chest or abdomen, accompanied by severe pain known as herpetic neuralgia (**Figure 6-12**). The condition is especially serious in older adults or people with disabilities. Treatment consists of medication for pain and itching and for protecting the area. A vaccine is available for people over age 60 to prevent shingles.

Disorders of the Hair and Nails

Head **lice** are parasitic insects found on the heads of people. The condition is contagious and affects millions of people each year. It is found most often in preschool and school-age children. Symptoms are a feeling of something moving on the head, intense itching, and sores on the head caused by scratching. Treatment includes using lotions or shampoos designed to kill adult lice and the use of a fine-tooth comb every 2 to 3 days to comb the hair and to remove any nits, or immature lice. Treat the head again with the shampoo or lotion in 7 to 10 days. Check the hair again in 2 to 3 weeks to be sure the treatment was effective.

Ingrown nails are a common nail problem. The great toenail is the one most often affected because the nail may curve downward into the skin. This is painful, and the toe may become infected. Ingrown toenails are often caused by improper nail trimming or tight shoes. Have the ingrown nail treated by a physician.

Fungal infections make up approximately 50% of nail disorders. Conditions for fungus to grow include an enclosed, warm, moist environment, which occurs when wearing closed types of shoes. Fungal infections often cause the nail to separate from the nail bed. Additionally, debris may build up under the nail and discolor the nail bed. Fungal infections are difficult to eradicate and may cause lifelong symptoms.

Warts are human papillomavirus (HPV) infections that affect the skin around or underneath the nail. They are painful and sometimes cause limited use of the affected finger or toe. Treatment usually involves freezing or the application of a chemical to dissolve the wart blister. Other types of warts include plantar warts, which show up on the soles of the feet, and genital warts, a sexually transmitted disease.

Skin Cancer

Skin cancer has been associated with exposure to ultraviolet light, so scientists and health care providers are cautioning people to limit their exposure to direct sunlight. Skin cancer is the most common type of cancer in people.

Basal cell carcinoma is the most common and least malignant type of skin cancer, usually occurring on the face. The abnormal cells start in the epidermis and extend to the dermis or subcutaneous layer. This cancer may be treated by surgical removal, radiation, or cryosurgery. *Cryosurgery* is the destruction of tissue by freezing, using liquid nitrogen. Full recovery occurs in 99% of cases.

Squamous cell carcinoma arises from the epidermis and occurs most often on the scalp and lower lip. This type grows rapidly and metastasizes to the lymph nodes. This cancer may be treated by surgical removal or radiation. Chances for recovery are good if found early (**Figure 6-13**).

Malignant melanoma (mel-ah-**NOH**-mah) occurs in pigmented cells of the skin called melanocytes. The cancer cells metastasize to other areas quickly. This type of tumor may appear as a brown or black irregular patch that occurs suddenly (**Figure 6-14**). A color or size change in a preexisting wart or mole may also indicate the presence of a melanoma. Treatment is surgical removal of the melanoma and the surrounding area and chemotherapy. The overall 6-year survival rate for melanoma is 92%.

Burns

Burns are a traumatic injury that result from exposure to radiation from the sun, or sunburn; a heat lamp; or contact with boiling water, steam, fire, chemicals, or electricity. When the skin is burned, dehydration and infection may occur—either condition can be life threatening.

Burns are usually referred to as first, second, or third degree, depending on the skin layers affected and the symptoms (**Figure 6-15**). It is important to note that it can be more difficult to detect burns on darker skin, as the redness may be less obvious. However, the patient may experience a feeling of tightening of the skin, and a darkening or shininess may be detectable.

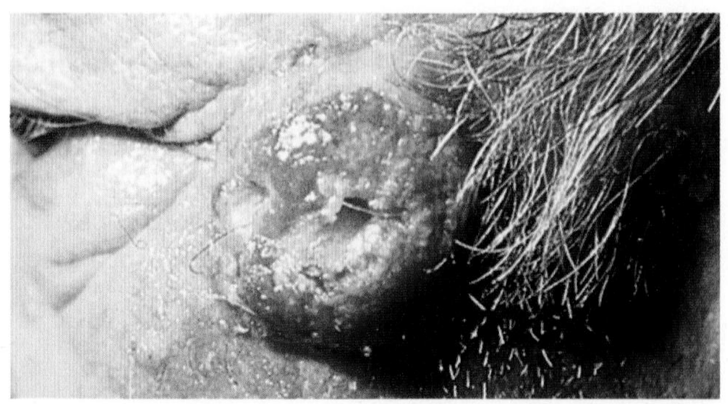

Figure 6-13 *Squamous cell carcinoma of the face*

Courtesy of Robert A. Silverman, M.D., Clinical Associate Professor, Department of Pediatrics, Georgetown University.

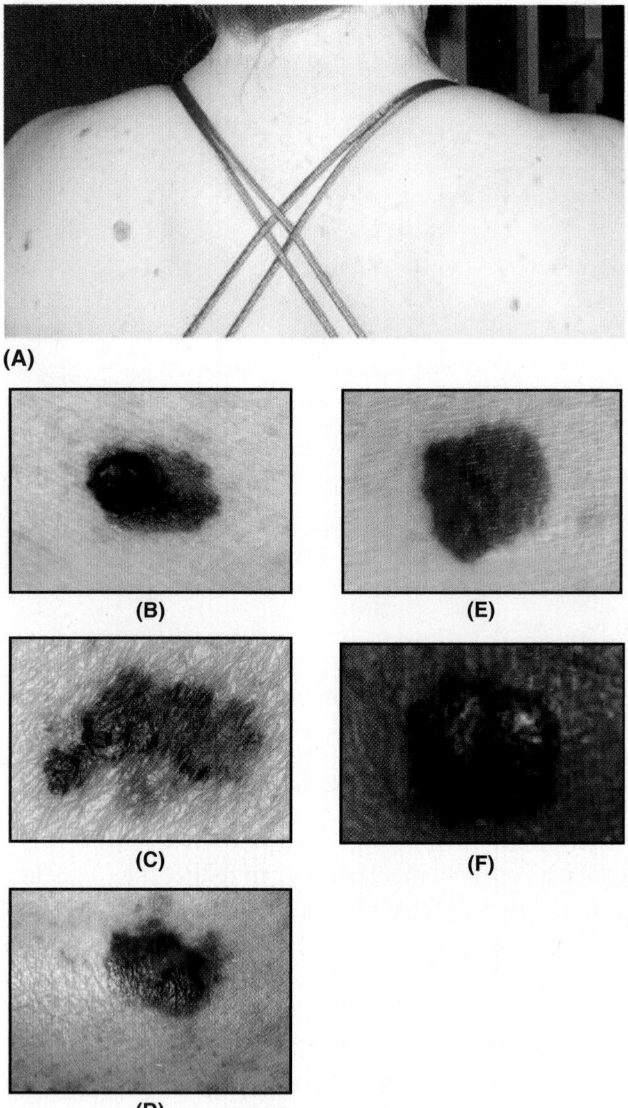

Figure 6-14 *(A) Malignant melanoma on the left shoulder blade. The signs of melanoma are (B) asymmetry, (C) border irregularity, (D) color variation, (E) diameter larger than a pencil eraser, and (F) evolving, or changing in size, shape, or shade of color.*

A, Sherry Morris; B-E, The Skin Cancer Foundation (http://www.skincancer.org).

Researchers are developing methods of treating burns that take advantage of stem cell therapy.

A **first-degree (superficial) burn** involves only the epidermis. Sunburns are often considered first-degree burns. Symptoms are redness, swelling, and pain. Hold burned area under cool, not cold, running water for 10–15 minutes or apply cool compresses to burned area. Healing usually occurs within one week.

A **second-degree (partial-thickness) burn** may involve the epidermis and dermis. Symptoms include pain, swelling, redness, and blistering. The skin may also be exposed to infection. Treatment begins with immersion in cool water for 10–15 minutes and may include pain medication and dry sterile dressings applied to open skin areas. Topical antibiotic cream is usually given. Healing generally occurs within two weeks.

Third-degree (full-thickness) burn involves complete destruction of the epidermis, dermis, and subcutaneous layers. Symptoms include loss of skin and **eschar**, or blackened skin, yet possibly no pain because nerve endings may have also been affected by the burns. This may be a life-threatening situation depending on the amount of skin damaged and fluid and blood plasma lost. The **rule of nines** measures the percent of the body burned: The body is divided into 11 areas, and each area accounts for 9% of the total body surface. For example, each arm is 4.5%; the perineal area accounts for 1% (**Figure 6-16**) The person requires immediate hospitalization. Treatment consists of prevention of infection; contracture, a condition of shortening and hardening of tissue often leading to deformity and rigidity of joints; and fluid replacement. Skin grafting is done as soon as possible.

Media Link

View the **Burns** animation on the Online Resources.

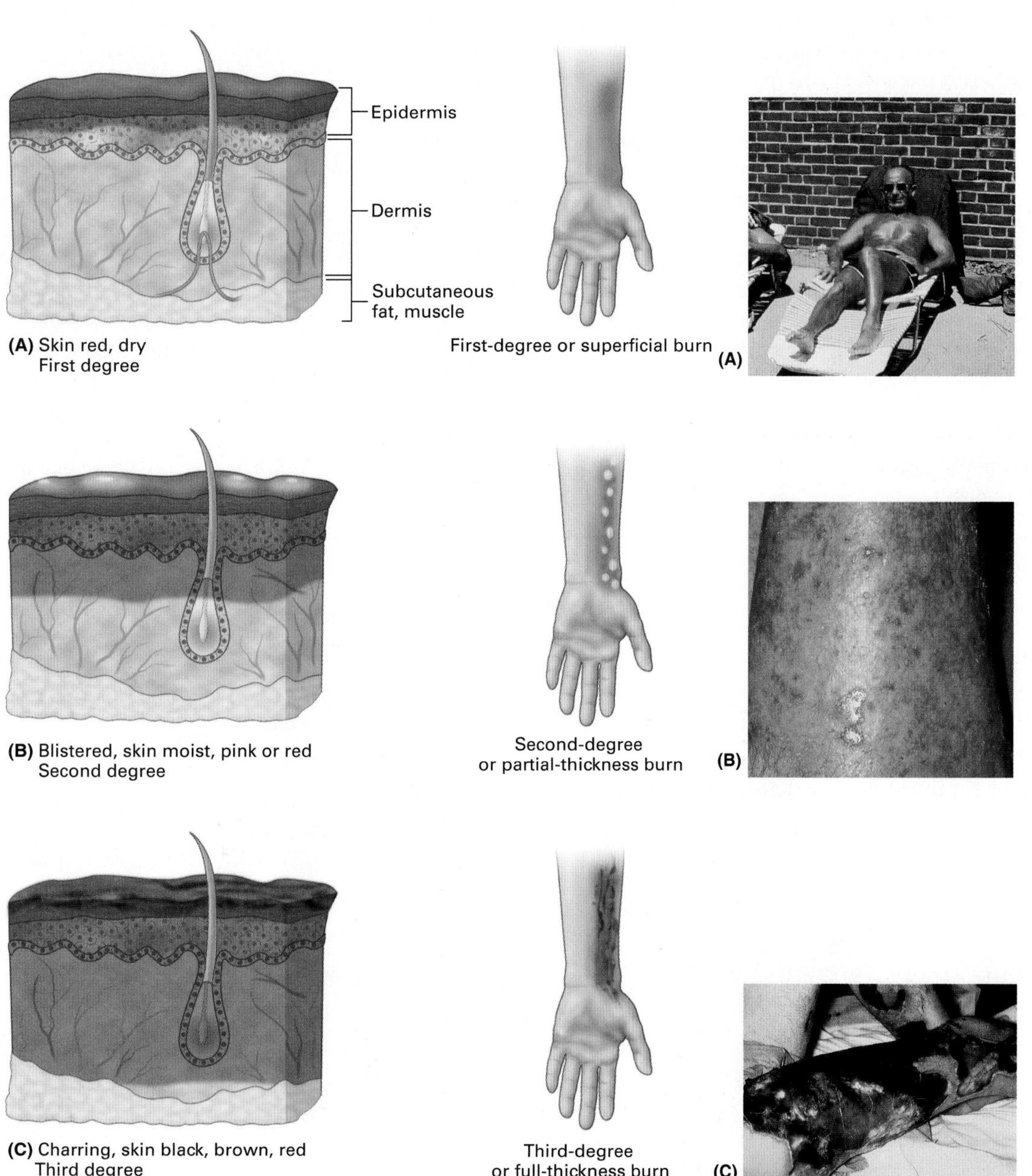

Figure 6-15 *Burns are usually referred to as (A) first degree (superficial), (B) second degree (partial thickness), or (C) third degree (full thickness)*

A, B, and C, Photo courtesy of The Phoenix Society for Burn Survivors, Inc.

Medical Highlights

HAZARDS OF THE SUN

The skin's number one enemy is the sun. It is estimated that more than 5 million cases of basal cell and squamous cell cancer are found in the United States each year (American Cancer Society). Basal cell cancer typically does not metastasize and is not usually fatal. However, those who develop basal cell cancers or squamous cell cancers have a higher risk of developing melanoma, the most serious of the skin cancers. The reasons are clear, according to physicians: people become overexposed to the sun or to ultraviolet (UV) rays from tanning beds, which have an effect similar to that of the sun's rays.

Ultraviolet radiation from the sun is a principal cause of skin cancer. There are three types of UV rays. Ultraviolet A (UVA) is the most abundant source of solar radiation at the earth's surface and penetrates beyond the top layer of skin. Ultraviolet B (UVB) rays are less abundant at the earth's surface than UVA rays because a significant portion of the UVB rays is absorbed by the ozone layer. Ultraviolet B rays penetrate less deeply than UVA rays do but can also be damaging. Ultraviolet C (UVC) rays are completely absorbed by the stratosphere's ozone layer and do not reach the surface of the earth.

Some limited exposure to the sun—5 to 10 minutes about two to three times a week—is necessary for skin to manufacture vitamin D. Most people like to feel the warmth of the sun on their bodies and find it very relaxing. As a result, it is easy to exceed safe limits for sun exposure.

The Skin Cancer Foundation recommends that people who go out in the sun use a broad-spectrum sunscreen with a sun protection factor (SPF) of at least 15. For extended outdoor activity, sunscreen should have a SPF of 30 or higher. This will block the UVA and UVB rays, which are primarily responsible for skin cancer. Sunscreen should be applied every two hours and right after swimming. Wear a wide-brimmed hat that shades the face and use sunglasses with 100% UV protection. Wear a long-sleeved shirt and long pants. Stay out of the sun between the hours of 10 a.m. and 4 p.m., when the sun is the most intense. Light reflecting from snow, sand, water, and shiny surfaces can burn as easily as direct sunlight.

Many drugs are photosensitive, which means they increase a person's sensitivity to sunlight and the risk of getting sunburned. The prescription label for these types of drugs usually states that users should avoid prolonged or excessive exposure to direct or artificial sunlight.

Remember, a person can have fun in the sun if they use preventive measures to avoid overexposure.

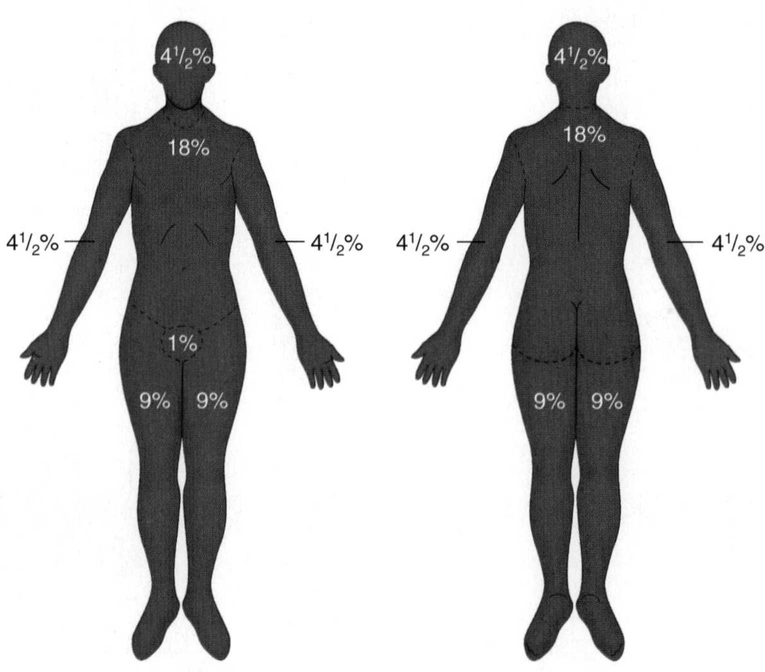

Figure 6-16 *The rule of nines is used to calculate the percentage of body surface burned.*

Skin Lesions and Trauma to the Skin

The health care provider should be familiar with the different types of skin disorders or lesions. Sometimes skin lesions indicate only an outer skin disorder. An **abrasion** (ah-**BRAY**-zhun) is an injury in which superficial layers of the skin are scraped or rubbed away. A **fissure** is a groove or cracklike break in the skin. A **laceration** (lass-er-**AY**-shun) is a torn or jagged wound. Treatment depends on the type of skin lesion.

Pressure/Decubitus Ulcer

Pressure ulcers, also known as **decubitus ulcers** (de-**KEW**-bih-tus) or bedsores, are preventable and are a primary concern of health care providers. Decubitus ulcers occur when a person is constantly sitting or lying in the same position without shifting their weight. Any area of tissue that lies over a bone is much more likely to develop a decubitus ulcer. These areas include the spine, coccyx, hips, elbows, and heels. The constant pressure against the area causes a decrease in the blood supply, and thus, the tissue begins to decay. These ulcers are classified in stages according to their severity:

- *Stage I*—involves surface reddening in light skin; in dark skin, hyper- or hypo-pigmentation can occur, as well as a dark blue-purple tint. The skin is unbroken. Treatment is to alleviate the pressure.
- *Stage II*—characterized by blistered areas that are either broken or unbroken; the surrounding area is red and irritated. Treatment is to protect and clean the area and alleviate the pressure.
- *Stage III*—presents with skin breaks through all layers of skin. It becomes a primary site for infection. Medical treatment is necessary to treat and prevent infection and promote healing.
- *Stage IV*—characterized by ulcers that have an ulcerated area that extends through skin and involves underlying muscles, tendons, and bones. This can produce a life-threatening situation. Treatment is the surgical removal of necrotic (dead) or decayed areas and antibiotics.

The best treatment for decubitus ulcers is prevention. Frequent turning and relief of pressure on bony prominences is essential. If the person is at home, family members must be educated on how to prevent the disorder.

Table 6-3 and **Figure 6-17** describe different types of skin lesions.

Table 6-3 *Different Types of Skin Lesions, Their Characteristics, Sizes, and Examples of Each*

TYPE OF SKIN LESION	CHARACTERISTICS	SIZE	EXAMPLE(S)
Bulla (large blister)	Fluid-filled area	Greater than 10 mm across	Bullous impetigo
Macule	Flat area usually distinguished from surrounding skin by its change in color	Less than 1 cm in diameter	• Freckle • Petechia
Nodule	Elevated, solid area, deeper and firmer than a papule	Greater than 10 mm across	Wart
Papule	Elevated, solid area	Less than 1 cm in diameter	Elevated nevus
Pustule	Discrete, pus-filled raised area	Less than 0.5 cm in diameter	Acne
Ulcer	A deep loss of skin surface that may extend into the dermis, which can bleed periodically and scar	Varying size	• Venous stasis ulcer • Decubitus ulcer
Tumor	Solid abnormal mass of cells that may extend deep through cutaneous tissue	Larger than 2 cm	• Benign (harmless) epidermal tumor • Basal cell carcinoma (rarely metastasizing)
Vesicle (small blister)	Fluid-filled raised area	Less than 10 mm	• Chickenpox • Herpes simplex
Urticaria (hives)	Itchy, temporarily elevated area with an irregular shape formed as a result of localized skin edema	Varying size	• Wheal • Insect bite

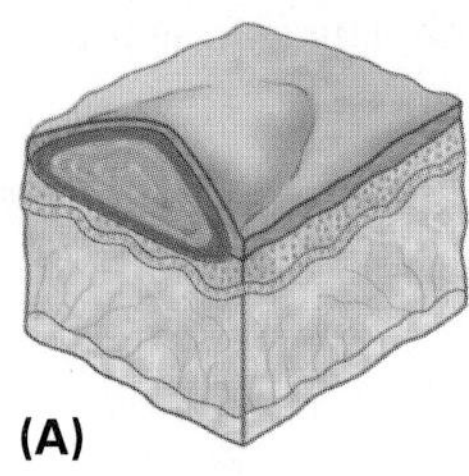
(A)

Bulla: (Large blister)
Same as a vesicle only greater than 10 mm
Example:
Contact dermatitis, large second-degree burns, bullous impetigo, pemphigus

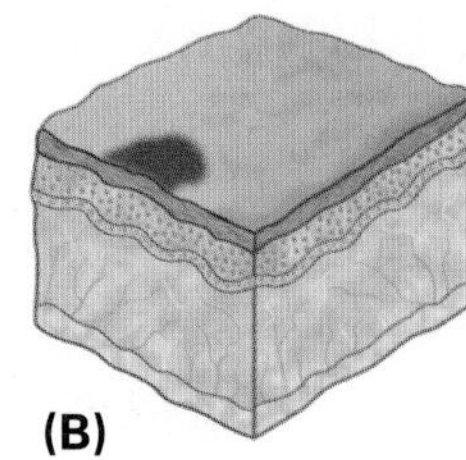
(B)

Macule:
Localized changes in skin color of less than 1 cm in diameter
Example:
Freckle

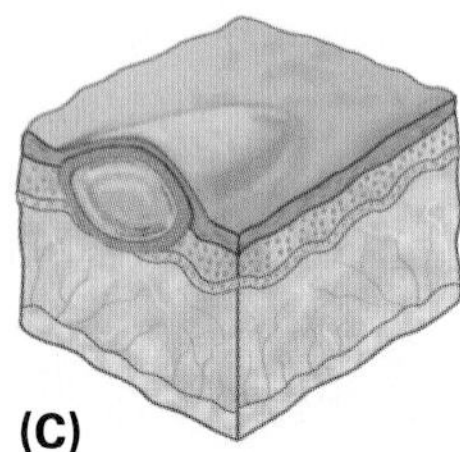
(C)

Nodules:
Solid and elevated; however, they extend deeper than papules into the dermis or subcutaneous tissues, greater than 10 mm
Example:
Lipoma, erythema, cyst, wart

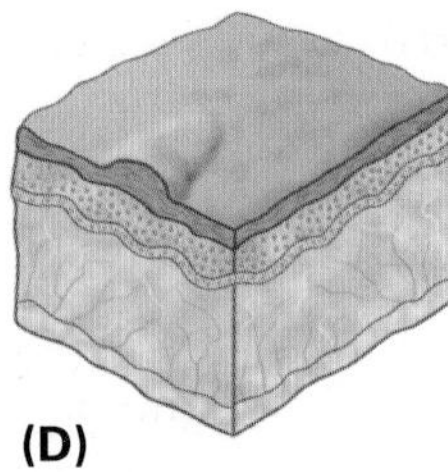
(D)

Papule:
Solid, elevated lesion less than 1 cm in diameter
Example:
Elevated nevus

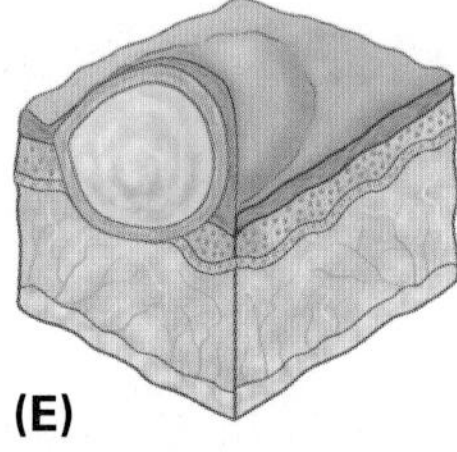
(E)

Pustule:
Vesicles or bullae that become filled with pus, usually described as less than 0.5 cm in diameter
Example:
Acne, impetigo, furuncles, carbuncles

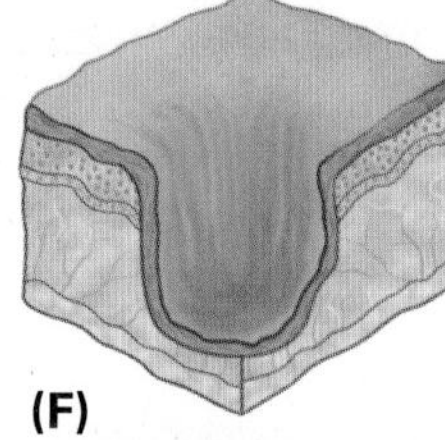
(F)

Ulcer:
A depressed lesion of the epidermis and upper papillary layer of the dermis
Example:
Stage 2 pressure ulcer

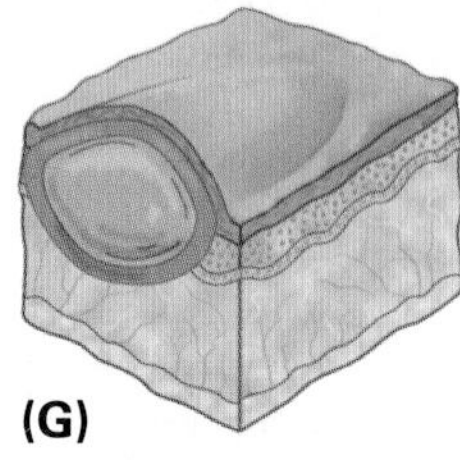
(G)

Tumor:
The same as a nodule only greater than 2 cm
Example:
Benign epidermal tumor basal cell carcinoma

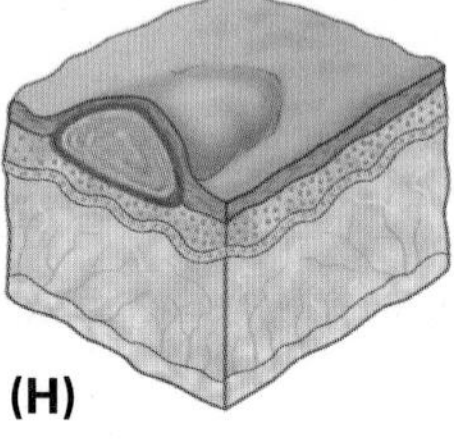
(H)

Vesicle: (Small blister)
Accumulation of fluid between the upper layers of the skin; elevated mass containing serous fluid; less than 10 mm
Example:
Herpes simplex, herpes zoster, chickenpox

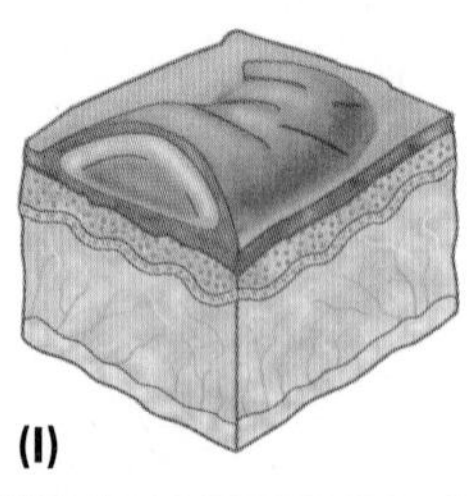
(I)

Urticaria, Hives:
Localized edema in the epidermis causing irregular elevation that may be red or pale, may be itchy
Example:
Insect bite, wheal

Figure 6-17 *Different types of skin lesions*

Career Profile

PHYSICIAN

Physicians diagnose illnesses and prescribe and administer treatments for people suffering from illness and disease. Physicians examine patients; obtain medical histories; and order, perform, and interpret diagnostic tests. They counsel patients on hygiene, diet, and preventive health care.

Two types of physicians are the doctor of medicine (M.D.) and the doctor of osteopathy (D.O.). Both physicians may use all methods of treatment. Doctors of osteopathy place special emphasis on the body's musculoskeletal system and preventive and holistic medicine.

Some physicians are primary care physicians who practice general and family medicine. Some are specialists who are experts in their medical field, such as dermatology, cardiology, or pediatrics.

Most physicians work long, irregular hours. Increasingly, they practice in groups or health care organizations. Becoming a physician requires four years of undergraduate study, four years of medical school, and 3–8 years of internship and residency, sometimes referred to as a fellowship depending on the specialization. Physicians must pass their medical boards to obtain a license to practice. Formal education and training requirements are among the longest of any occupation, but the earnings are among the highest.

One BODY — How the Integumentary System Interacts with Other Body Systems

SKELETAL SYSTEM

- Acts as a protective covering for bones.
- The ultraviolet rays of sun and skin are precursors to vitamin D, which is needed for absorption of calcium and phosphorus for bones.

MUSCULAR SYSTEM

- Provides protective covering for muscles.
- Provides the vitamin D necessary for muscle contraction.

NERVOUS SYSTEM

- Provides a protective covering for the nervous system.
- The skin has sensory receptors for pain, touch, pressure, and temperature.

ENDOCRINE SYSTEM

- Protects the glands.
- Stores fat necessary for production of hormones.

CIRCULATORY SYSTEM

- Provides a protective covering for the capillary network.
- Dilation and constriction of the capillary network regulates body temperature.
- Mast cells help in production of heparin.

LYMPHATIC SYSTEM

- Provides a protective covering for lymph vessels.
- Acts as a waterproof, intact covering to protect against infection.
- Macrophages help activate the immune system.

RESPIRATORY SYSTEM

- Provides a protective covering for organs of respiration.
- Hair on skin guards the entrance to the nasal cavity.

DIGESTIVE SYSTEM

- Acts as a protective covering for the organs of digestion.
- Provides the vitamin D necessary for absorption of calcium and phosphorus.

URINARY SYSTEM

- Provides a protective covering for the organs of the urinary system.
- Helps excrete waste products through sweat glands.

REPRODUCTIVE SYSTEM

- Provides a protective covering for the organs of reproduction.
- Sensory receptors in genitalia stimulate sexual interest.

Medical Terminology

alopec	baldness	**-thermia**	heat
-ia	abnormal condition	**hyper/thermia**	above normal heat
alopec/ia	abnormal condition of baldness	**hypo-**	below
a-	without	**hypo/thermia**	below normal heat
vascul	blood vessels	**melan**	black
-ar	pertaining to	**-oma**	tumor
a/vascul/ar	being without blood vessels	**melan/oma**	tumor of blackness, usually malignant
decubit	bedsore	**papill**	pimple
-us	presence of	**-a**	presence of
decubit/us	presence of bedsore, pressure sore	**papill/a**	presence of pimple
dermat	skin	**sebac**	grease or oil
-itis	inflammation	**-ous**	pertaining to
dermat/itis	inflammation of the skin	**sebac/e/ous**	pertaining to oil glands
epi-	upon	**stratum**	layer
epi/dermis	upon the skin; top layer of skin	**corneum**	horny
hyper-	above normal	**stratum corneum**	horny layer of skin

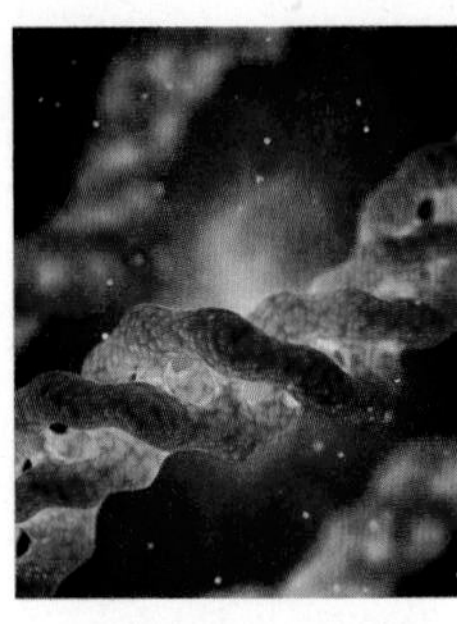

Study Tools

Workbook	Activities for Chapter 6
Online Resources	• PowerPoint presentations • Animations

REVIEW QUESTIONS

Select the letter of the choice that best completes the statement.

1. The outermost layer of the skin is the
   a. epidermis.
   b. dermis.
   c. hypodermis.

2. The substance that best serves to keep our skin smooth and protected is
   a. melanin.
   b. keratin.
   c. cortex.

3. Nerve receptors are found in the
   a. epidermis.
   b. dermis.
   c. hypodermis.

4. Hair contains keratinized cells, which are found in the
   a. cuticle layer.
   b. cortex.
   c. medulla.

5. The glands that secrete 99% water, small amounts of salt, and organic matter are called
   a. endocrine glands.
   b. sudoriferous glands.
   c. sebaceous glands.

MATCHING

Match each item in Column I with its correct description in Column II.

	COLUMN I	*COLUMN II*
______	**1.** eschar	a. gooseflesh
______	**2.** papillae	b. chickenpox
______	**3.** head lice	c. white nails
______	**4.** stratum germinativum	d. sebum
______	**5.** hypoxia	e. chronic redness in the face
______	**6.** herpes zoster	f. ridges in the epidermis
______	**7.** sebaceous gland	g. parasitic insects
______	**8.** arrector pili muscle	h. melanocytes
______	**9.** rosacea	i. blackened skin
______	**10.** liver	j. bluish nails

FILL IN THE BLANKS

1. A bluish tint to the skin of a fair-skinned person is called ________________ and is caused by ________________. On darker skin, this may appear as a gray tint.
2. A common and chronic disorder that occurs in the teen years is called ________________.
3. Inflammation of the skin is called ________________.
4. Urticaria, or hives, is usually a reaction to a(n) ________________.
5. A chronic inflammatory disease characterized by silvery patches is known as ________________.
6. A cold sore or fever blister is known as ________________.
7. Painful viral infections of the nerve endings are called ________________.
8. The most common type of cancer is ________________.
9. A skin cancer that occurs as a large brown or black patch is ________________.
10. To determine the percentage of the body that is burned, a health care provider may use a formula called the ________________.
11. A skin condition that is present at birth is known as ________________.
12. Abrasion injuries involve the ________________ layer of skin.

APPLYING THEORY TO PRACTICE

1. If you get a cut on your skin, what may be the result?
2. The skin helps regulate body temperature by evaporation of water from the skin. Why do you feel uncomfortable on a hot, humid day?
3. A person is brought to the emergency room with third-degree burns but is not complaining of pain. How is this possible?
4. The cosmetic industry sells many products that remove or prevent wrinkles. What happens to the skin during the aging process that causes wrinkles?
5. Name at least three representative contagious skin disorders. Research the technology used to diagnose these disorders and describe the treatment and the best method to prevent their spread.

CASE STUDY

Meghan is a 22-year-old female home from college on summer break. She goes to her physician's office for a checkup. She tells the physician assistant, Nichole, all about her year at school and states, "I can't wait to get to the beach and soak up the sun." Meghan tells Nichole, "I hear so much about the sun and the danger of skin cancer, but the sun on my body makes me feel so relaxed." Meghan asks Nichole some questions about sun exposure. Pretend you are Nichole and give Meghan information about the following:

1. What causes sunburn?
2. What diagnostic technologies are used to investigate the following types of skin cancer: basal cell carcinoma, squamous cell carcinoma, and melanoma? What types of therapeutic technologies are used to treat the following types of skin cancer: basal cell carcinoma, squamous cell carcinoma, and melanoma?
3. What does "SPF" mean? Will a sunscreen protect from the sun's rays?
4. How is sunscreen properly applied?
5. What does photosensitive mean? Should people who take medication that increases photosensitivity stay out of the sun?

6-1 Gross Examination of the Skin

- ***Objective:*** To describe and examine the anatomical structures visible on the surface of the skin with the naked eye
- ***Materials needed:*** high-power magnifying glass, latex glove, paper, pencil

Step 1: Examine the back of your left hand visually and then under the magnifier. Locate hair follicles. Record the features you observe.

Step 2: Examine the palm of your left hand. List the differences between the back and the palm of your hand.

Step 3: Put the latex glove on your left hand. Remove the glove after five minutes.

Step 4: Visually examine the back of your left hand again. Is there a difference between what you see now and what you saw in step 1?

Step 5: Reexamine the back of your left hand under the magnifying glass. Record the differences you observe between now and step 1.

Step 6: Compare your results with those of a lab partner.

6-2 Microscopic Examination of the Skin

- ***Objective:*** To observe and examine the anatomical structure of the skin as visible with a microscope
- ***Materials needed:*** prepared slides of skin, microscope, textbook, paper, pencil

Step 1: Examine the slides of skin under the microscope.

Step 2: Examine the layers of the skin and compare with photos/drawings of skin in the textbook. Observe and record any differences you see.

Step 3: Examine the structures of the skin and compare with photos/drawings of skin in the textbook.

Step 4: Draw a diagram of what you see under the microscope. Label your diagram to identify the layers and structures you observed.

LAB ACTIVITY 6-3 The Sweat Glands

- ***Objective:*** Observe the number and activity of the sweat glands with a lab partner using bond paper and iodine (If you are allergic to iodine or shellfish, do not attempt this lab.)
- ***Materials needed:*** bond paper cut in 1-inch squares, tincture of iodine solution, adhesive tape, cotton-tipped applicators (Q-tips), paper, pencil
- ***Note:*** Tincture of iodine is a poison. It may cause burns to the skin, and it permanently stains clothing.

Step 1: Using the applicator, paint an area on your left forearm with iodine solution. Allow it to dry thoroughly.

Step 2: Have your lab partner securely tape a square of the bond paper over the iodine area. Leave in place for 20 minutes.

Step 3: Have your lab partner do step 1, and you perform step 2 on your lab partner. Apply to the same area of skin.

Step 4: After 20 minutes, remove the paper and tape; count the number of blue-black dots in the square of bond paper.

Step 5: Have your lab partner repeat step 4.

Step 6: Compare your results. Are the number of dots the same? What is the reason for the difference, if any? Record your observations.

CHAPTER 7

Skeletal System

Objectives

- List the main functions of the skeletal system.
- Explain the process of bone formation.
- Name and locate the bones of the skeleton.
- Name and define the main types of joint movement.
- Identify common bone and joint disorders.
- Define the key words that relate to this chapter.

Key Words

abduction
acetabulum
adduction
amphiarthroses
appendicular skeleton
arthritis
articular cartilage
atlas
axial skeleton
axis
ball-and-socket joints
bursae
bursitis
calcaneus
cancellous bone
carpals
cervical vertebrae
circumduction
clavicles
coccyx
compact bone
diaphysis
diarthroses
dislocation
endosteum
epiphysis
ethmoid
extension
femur
fibula
flatfeet
flexion
fontanel
fracture
frontal
gliding joints
gout
hammertoe
hinge joints
humerus
hyoid
ilium
inferior concha
ischium
joints
kyphosis
lacrimals
lordosis
lumbar vertebrae
mandible
manubrium
maxillae
medullary canal
menisci
metacarpal
metatarsal
nasal
obturator foramen
occipital
ossification
osteoarthritis
osteoblasts
osteoclasts
osteocytes
osteomyelitis
osteoporosis
osteosarcoma
palatines
parietals
patella
pelvis
periosteum
phalanges
pivot joints
pronation
pubis
radius
rheumatoid arthritis
rickets
rotation

(*continues*)

Key Words *continued*

sacrum	sternum	talus	ulna
scapulae	supination	tarsal	vomer
scoliosis	suture	temporals	whiplash injury
sphenoid	synarthroses	thoracic vertebrae	xiphoid process
spongy bone	synovial fluid	tibia	zygomatics

Sometimes during a visit to a beach, it is possible to spot a jellyfish floating lightly near the surface. The organs of the jellyfish are buoyed up by the water. If a wave should deposit the jellyfish on the beach, however, it would collapse into a disorganized mass of tissue because the jellyfish has no supportive framework or skeleton. Fortunately, humans do not suffer such a fate because they have a solid, bony skeleton to support body structures.

The skeletal system comprises the bony framework of the body. It consists of 206 individual bones in the adult. Some bones are hinged; others are fused to one another.

FUNCTIONS

The skeletal system has five specific functions:

1. It *supports* body structures and provides shape to the body.
2. It *protects* the soft and delicate internal organs. For example, the cranium protects the brain, the inner ear, and parts of the eye. The ribs and breastbone protect the heart and lungs; the vertebral column encases and protects the spinal cord.
3. It *allows movement and anchorage* of muscles. Muscles that are attached to the skeleton are called skeletal muscles. Upon contraction, these muscles exert a pull on a bone and thus move it. In this manner, bones play a vital part in body movement, serving as passively operated levers. Ligaments are fibrous bands that connect bones to bones and cartilage and serve as support for muscles. Joints are also bound together by ligaments. Tendons are fibrous cords that connect muscles to bone.
4. It *provides mineral storage*. Bones are a storage depot for minerals such as calcium and phosphorus. In case of inadequate nutrition, the body is able to draw upon these reserves. For example, if the body's blood calcium level dips below normal, the bone releases the necessary amount of stored calcium into the bloodstream. When calcium levels exceed normal, calcium release from the skeletal system is inhibited. In this way, the skeletal system helps maintain blood calcium homeostasis.
5. It *is the site for hematopoiesis* (**hem**-ah-toe-poy-**EE**-sis), that is, the formation of blood cells. Stem cells in the red marrow tissue of the bone differentiate into red blood cells, white blood cells, and platelets. In an adult, the ribs, sternum, and bones of the **pelvis** contain red marrow. Red bone marrow at the ends of the humerus and femur are plentiful at birth but gradually decrease as people age.

STRUCTURE AND FORMATION OF BONE

Bones are formed of cells called **osteocytes** (**os**-tee-uh-sahyts); this word comes from the Greek word *osteon*, meaning "bone." Bone is made up of 35% organic material, 65% inorganic mineral salts, and water.

The organic part derives from a protein called bone collagen, a fibrous material. Between these collagenous fibers is a jellylike material. The organic substances of bone give it a certain degree of flexibility. The inorganic portion of bone is made from mineral salts, such as calcium phosphate, calcium carbonate, calcium fluoride, magnesium phosphate, sodium oxide, and sodium chloride. These minerals give bone its hardness and durability.

A bony skeleton can be compared with steel-reinforced concrete. The collagenous fibers may be compared with flexible steel supports, and mineral salts with concrete. When pressure is applied to a bone, the flexible organic material prevents bone damage, whereas the mineral elements resist crushing or bending under pressure.

Bone Formation

The embryonic skeleton initially consists of collagenous protein fibers secreted by the osteoblasts, which are primitive embryonic cells. Later, during embryonic development, cartilage is deposited between the fibers. Cartilage, a connective tissue, is also found at the ends of certain bones in adults, providing a smooth surface for adjacent bones to move against each other. During the eighth week of embryonic development, **ossification** (oss-sih-fih-**KAY**-shun) begins; that is, calcium-based mineral matter starts to replace the formed cartilage, creating bone. Infant bones are very soft and pliable because of incomplete ossification at birth. An example is the soft spot on a baby's head, the **fontanel** (fon-tah-**NEL**). The bone has not yet been formed there, although it will harden later. Ossification due to mineral deposits continues through childhood. As bones ossify, they become hard and more capable of bearing weight.

Structure of Long Bone

A typical long bone contains a shaft, or **diaphysis** (dye-**AF**-ih-sis). This is a hollow cylinder of hard, **compact bone**. It is what makes a long bone strong and hard yet light enough for movement. At each extreme of the diaphysis is an **epiphysis** (eh-**PIF**-ih-sis) (**Figure 7-1A**).

In the center of the shaft is the broad **medullary canal** (**MED**-you-lehr-ee). This is filled with yellow bone marrow, mostly made of fat cells (**Figure 7-1B**). The marrow also contains many blood vessels and some cells that form white blood cells, called leukocytes. The yellow marrow functions as a fat storage center. The **endosteum** (en-**DOS**-tee-um) is the lining of the marrow canal that keeps the cavity intact.

The medullary canal is surrounded by compact or hard bone. Haversian (ha-**VER**-shun) canals branch into the compact bone. They carry blood vessels that nourish the osteocytes, or bone cells. Where less strength is needed in the bone, some of the hard bone is dissolved away, leaving **spongy bone**, or **cancellous bone** (**KAN**-suh-luh-s). Spongy bone is located at the ends of long bones and forms the center of all other bones. It consists of a meshwork of interconnecting sections of bone called trabeculae (trah-**BEK**-you-lee), creating the sponge-like appearance of cancellous bone.

The ends of the long bones contain the red marrow where some red blood cells, called erythrocytes, and some white blood cells are made. The outside of the bone is covered with the **periosteum** (pehr-ee-**OSS**-tee-um) (**Figure 7-1C**), a tough, fibrous tissue that contains blood vessels, lymph vessels, and nerves. The periosteum is necessary for bone growth, repair, and nutrition.

Covering the epiphysis is a thin layer of cartilage known as the **articular cartilage** (ar-**TICK**-you-lar **KAR**-tih-lidj). This cartilage acts as a shock absorber between two bones that meet to form a joint.

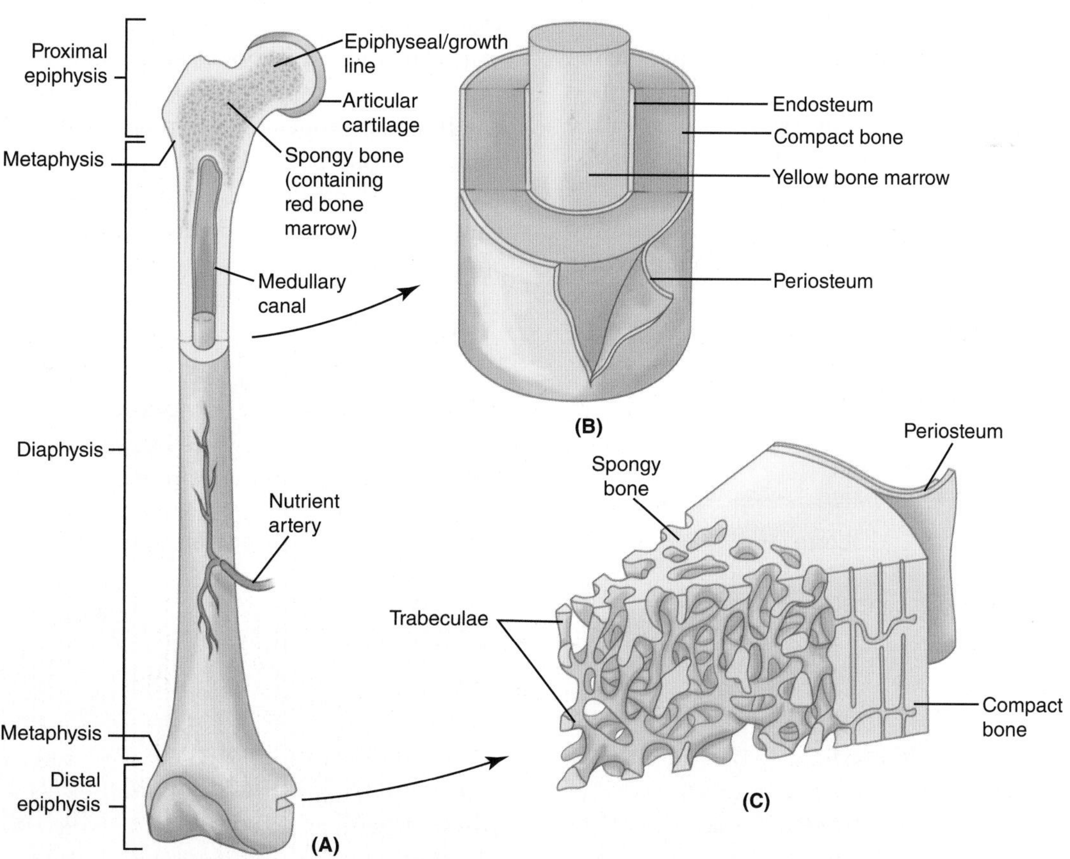

Figure 7-1 *Structure of a typical long bone: (A) Diaphysis, epiphysis, and medullary canal; (B) Compact bone surrounding yellow bone marrow in the medullary canal; (C) Spongy bone and compact bone in the epiphysis*

Growth

Bones grow in length and ossify from the center of the diaphysis toward the epiphyseal extremities. Using a long bone by way of example, it will grow lengthwise in an area called the growth zone. Ossification occurs here, causing the bone to lengthen; this causes the epiphyses to grow away from the middle of the diaphysis. It is a sensible growth process because it does not interfere with the articulation between two bones.

A bone increases its circumference by the addition of more bone to the outer surface of the diaphysis by osteoblasts. **Osteoblasts** are bone cells that deposit new bone. As girth increases, bone material dissolves from the central part of the diaphysis. This forms an internal cavity called the marrow cavity, or medullary canal. The medullary canal gets larger as the diameter of the bone increases.

The dissolution of bone from the medullary canal results from the action of cells called osteoclasts. **Osteoclasts** are immense bone cells that secrete enzymes. These enzymes digest the bony material, splitting the bone minerals, calcium, and phosphorus and enabling them to be absorbed by the surrounding fluid. The medullary canal eventually fills with yellow marrow.

The length of a bone shaft continues to grow until all the epiphyseal cartilage is ossified. At this point, bone growth stops. This fact is helpful in determining further growth in a child. First, an X-ray of the child's wrists is taken. If some epiphyseal cartilage remains, there will be further growth. If no epiphyseal cartilage is left, the child has reached their full stature, or height.

Throughout life, bone is constantly renewed through a two-step process called remodeling, which consists of resorption and formation. During resorption, old bone tissue is broken down and removed by the osteoclasts. During bone formation, new bone tissue is laid down by the osteoblasts, replacing the old bone tissue.

The average bone growth in females continues to about 18 years of age; males grow for approximately 20 or 21 years. However, new bone growth can occur in a broken bone at any time. The process of bone repair includes bleeding at the site of the injury with clot and granulation tissue formation, proliferation of cells at the site to form a soft bone deposit over the site of injury or fracture, cells becoming osteoblasts or cartilage at the site, the bone calcifying, and the remodeling of the bone to the shape necessary to complete the repair. The process of bone healing proceeding efficiently depends on the age and health of the individual.

Bone Types

Bones are classified as one of four types based on their shape (**Figure 7-2**). *Long* bones are found in both upper and lower arms and legs. The bones of the skull are examples of *flat* bones, as are the ribs. *Irregular* bones are represented by bones of the spinal column. The wrist and ankle bones are examples of *short* bones, which appear cube-like in shape.

The individual bones in the hand are short, making flexible movement possible. The same is true of the irregular bones of the spinal column. The thighbone is a long bone needed for support of the strong leg muscles and the weight of the body. The degree of movement at a joint is determined by bone shape and joint structure.

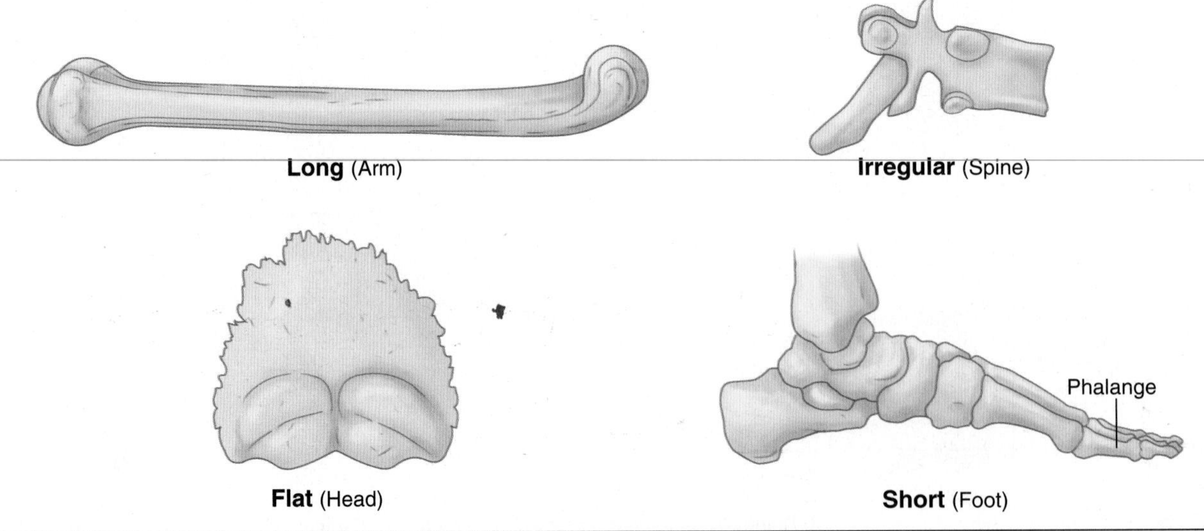

Figure 7-2 *Bone shapes*

PARTS OF THE SKELETAL SYSTEM

The skeletal system consists of two main parts: the axial skeleton and the appendicular skeleton.

Axial Skeleton

The **axial skeleton** consists of the skull, spinal column, ribs, sternum (breastbone), and **hyoid** bone (**Figure 7-3**). The hyoid bone is a U-shaped bone in the neck to which the tongue is attached (not seen in **Figure 7-3**).

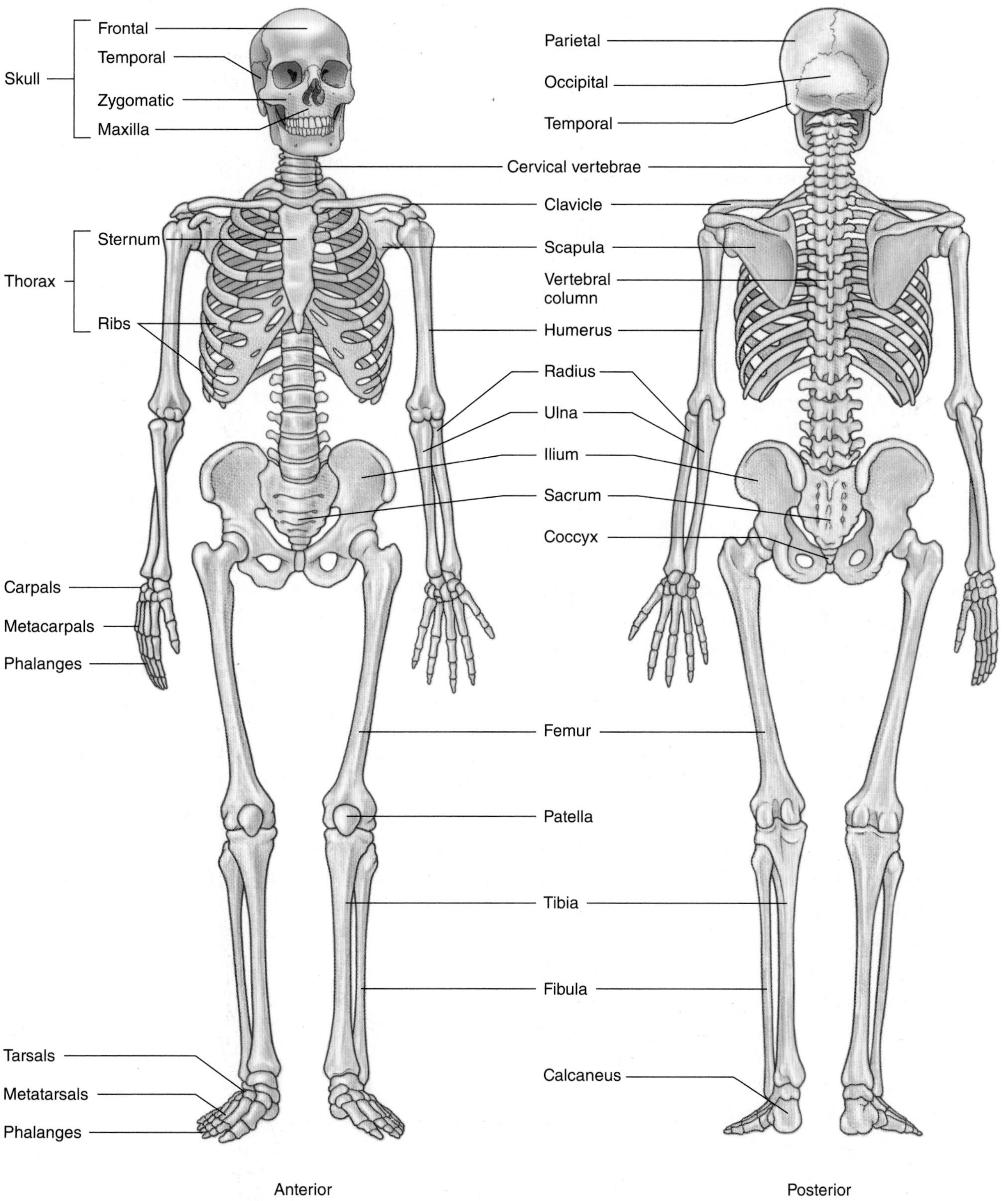

Figure 7-3 *The axial skeleton (blue) and the appendicular skeleton*

The six tiny bones of the ear are discussed in Chapter 11.

SKULL

The skull is made up of the cranium and facial bones. The cranium houses and protects the delicate brain, whereas the facial bones guard and support the eyes, ears, nose, and mouth. Some of the facial bones, such as the nasal bones, are made of bone and cartilage. For example, the upper part of the nose, the bridge, is bone, whereas the lower part of the nose is cartilage.

Cranial bones are thin and slightly curved. During infancy, these bones are held snugly together by an irregular band of connective tissue called a **suture**. As the child grows, this connective tissue ossifies and turns into hard bone. Thus, the cranium becomes a highly efficient, domed shield for the brain. The dome shape affords better protection than a flat surface, deflecting blows directed toward the head.

Collectively, the skull has 22 bones (**Figure 7-4**). The following eight bones are in the cranium:

1. The **frontal** (1) forms the forehead, the roof of the nasal cavity, and the eye orbits of the cranium.
2. The **parietals** (2) (pah-**RYE**-eh-tals) form the roof and sides of the cranium.
3. The **temporals** (2) form the sides of the cranium and house the ears.
4. The **occipital** (1) (ock-**SIP**-eh-tal) forms the posterior, or base of the cranium, and contains the foramen (faw-**RAY**-men) magnum. The foramen magnum is the large opening on the inferior portion of the occipital bone where the spinal cord passes through to connect to the brain.
5. The **ethmoid** (1) (**ETH**-moid) is located between the eyes and forms the principal part of the nasal cavity and helps form part of the eye orbit.
6. The **sphenoid** (1) (**SFEE**-noid) resembles a bat and is considered the key bone of the cranium; all other bones of the cranium connect to it.

Following are the 14 facial bones:

7. The **nasal** bones (5): two are nasal bones that form the bridge of the nose where glasses sit; one is the **vomer** (**VOH**-mer) bone, which forms the lower part, or midline, of the nasal septum; and two are **inferior concha** (**KONG**-kee) bones, which make up the side walls of the nasal cavity.
8. The **maxillae** (2) (**MAK**-si-lee) make up the upper jaw.
9. The **lacrimals** (2) (**LACK**-rih-mals) make up part of the eye orbit at the inner angle of the eye; they contain the tear ducts.
10. The **zygomatics** (2) (zye-goh-**MAT**-icks) form the prominence of the cheek.
11. The **palatines** (2) (**PAL**-ah-tines) form the hard palate of the mouth.
12. The **mandible** (1) (**MAN**-dih-bull) is the lower jaw, the only movable bone in the face.

The skull contains large spaces within the facial bones, referred to as paranasal sinuses. The sinuses are filled with air to help lighten the bones of the skull. These sinuses are lined with mucous membranes. When a person suffers from a cold, flu, or hay fever, the membranes become inflamed and swollen, producing a copious amount of mucus. This may lead to sinus pain and a "stuffy" nasal sensation.

SPINAL COLUMN/VERTEBRA

The spine, or vertebral column, is strong and flexible. It supports the head and provides for the attachment of the ribs. The spine also encloses the spinal cord of the nervous system.

The spine consists of small bones called vertebrae (**VER**-teh-bray). They are separated from each other by pads of cartilage tissue called intervertebral discs (**Figure 7-5**). These discs serve as cushions between the vertebrae and act as shock absorbers. These discs become thinner with age, which accounts for a loss in height.

The vertebral column is divided into the following five sections, named according to the area of the body in which they are located (**Figure 7-5**):

1. **Cervical vertebrae** (7) (**SER**-vih-kal) are located in the neck area. The **atlas** (**Figure 7-6A**) is the first cervical vertebra. It articulates, or is jointed, with the occipital bone of the skull. This permits heads to nod. On the **axis** (**Figure 7-6B**), the second cervical vertebra is the odontoid process, which forms a pivot on which the atlas rotates; this permits heads to turn. The cervical vertebrae are also known as C1 through C7.
2. **Thoracic vertebrae** (12) (thoh-**RASS**-ick) are located in the chest area. They articulate with the ribs. The thoracic vertebrae are also known as T1 through T12.
3. **Lumbar vertebrae** (5) are located in the back. They have large bodies that bear most of the body's weight. They are also known as L1 through L5.

Did You Know?

The giraffe, with its long neck, has the same number of cervical vertebrae as a human. The giraffe's cervical vertebrae, however, are much longer.

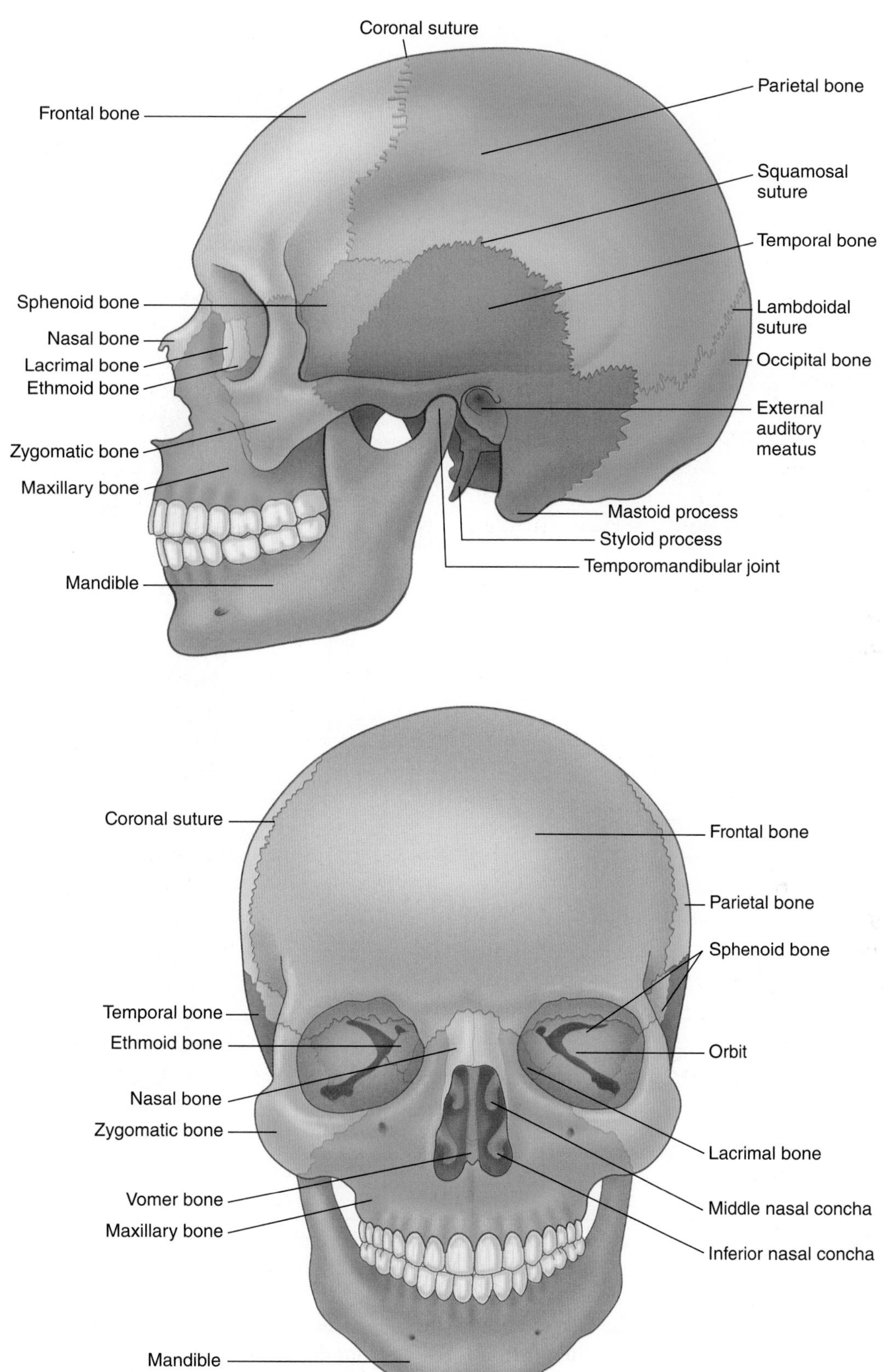

Figure 7-4 *Bones and sutures of the skull and facial bones*

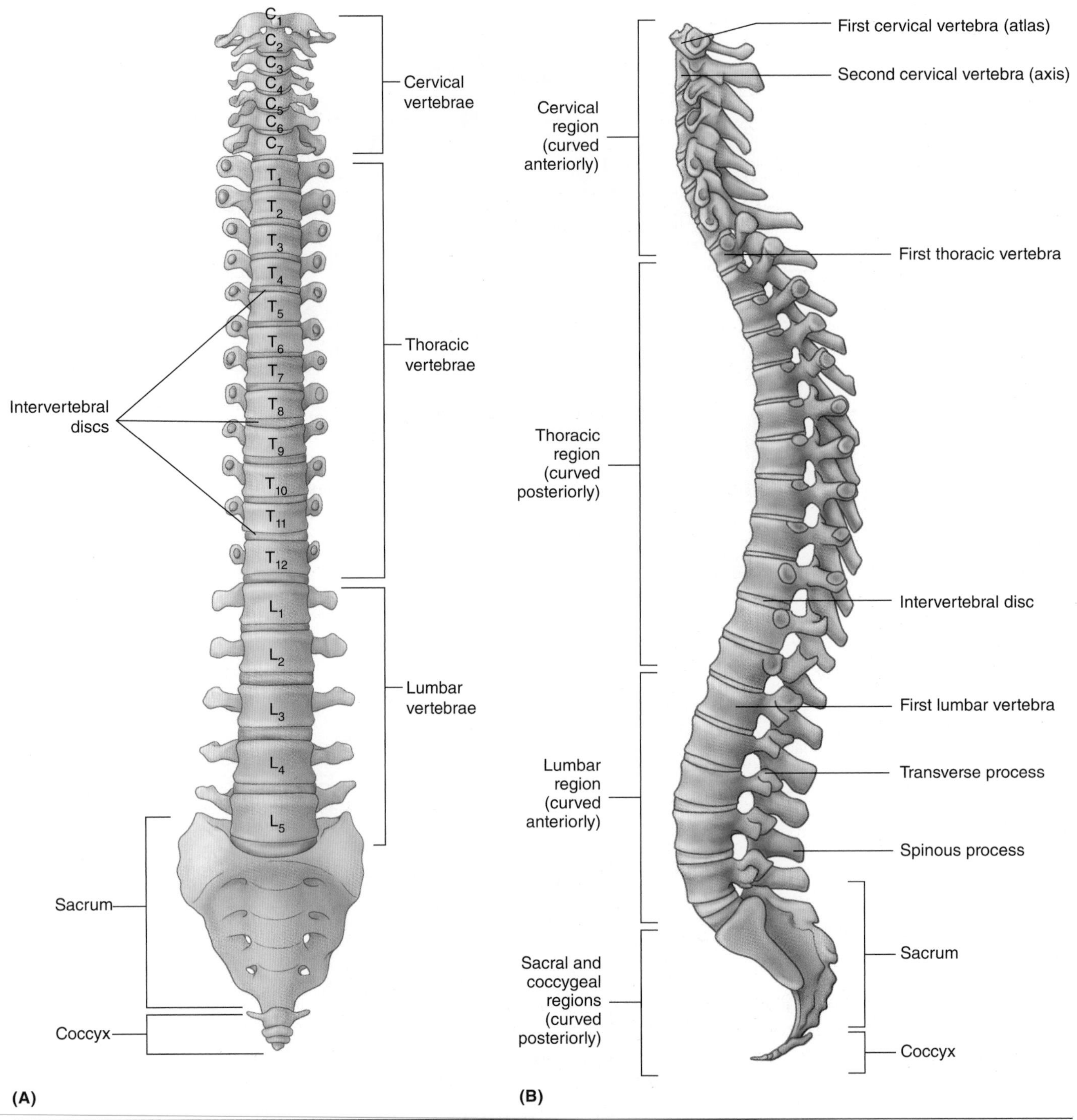

Figure 7-5 *(A) Anterior view of the vertebral column; (B) lateral view of the vertebral column*

4. The **sacrum** (**SAY**-krum) is a wedge-shaped bone formed by five fused bones. It forms the posterior pelvic girdle and serves as an articulation point for the hips.
5. The **coccyx** (**KOCK**-sicks) is also known as the tailbone. It is formed by four fused bones.

The spinal nerves enter and leave the spinal cord through openings between the vertebrae called foramens.

The shape of the spine changes through growth. In a model of the human skeleton, note that the spine is curved instead of straight. A curved spine has more strength than a straight one would have. Before birth, the thoracic and sacral regions are convex curves. As the infant learns to hold up its head, the cervical region becomes concave. When the child learns to stand, the lumbar region also becomes concave. This completes the four curves of a normal, adult human spine.

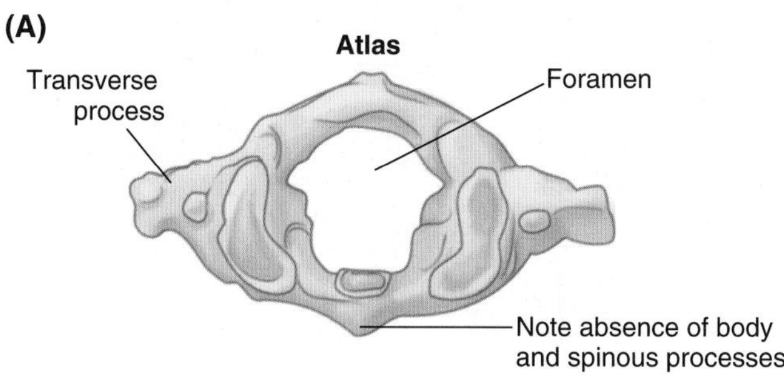

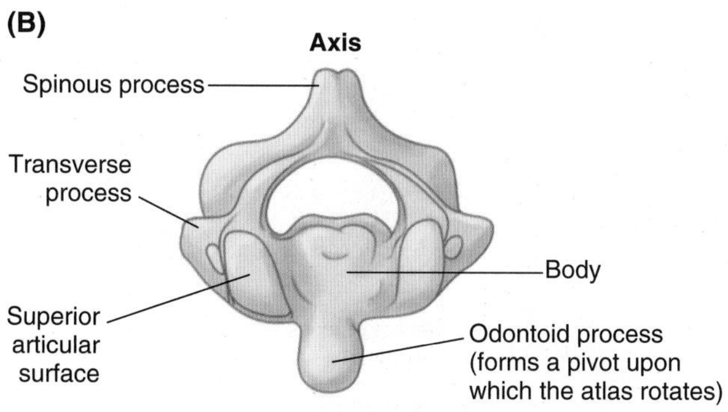

Figure 7-6 *(A) View of the atlas. (B) View of the axis.*

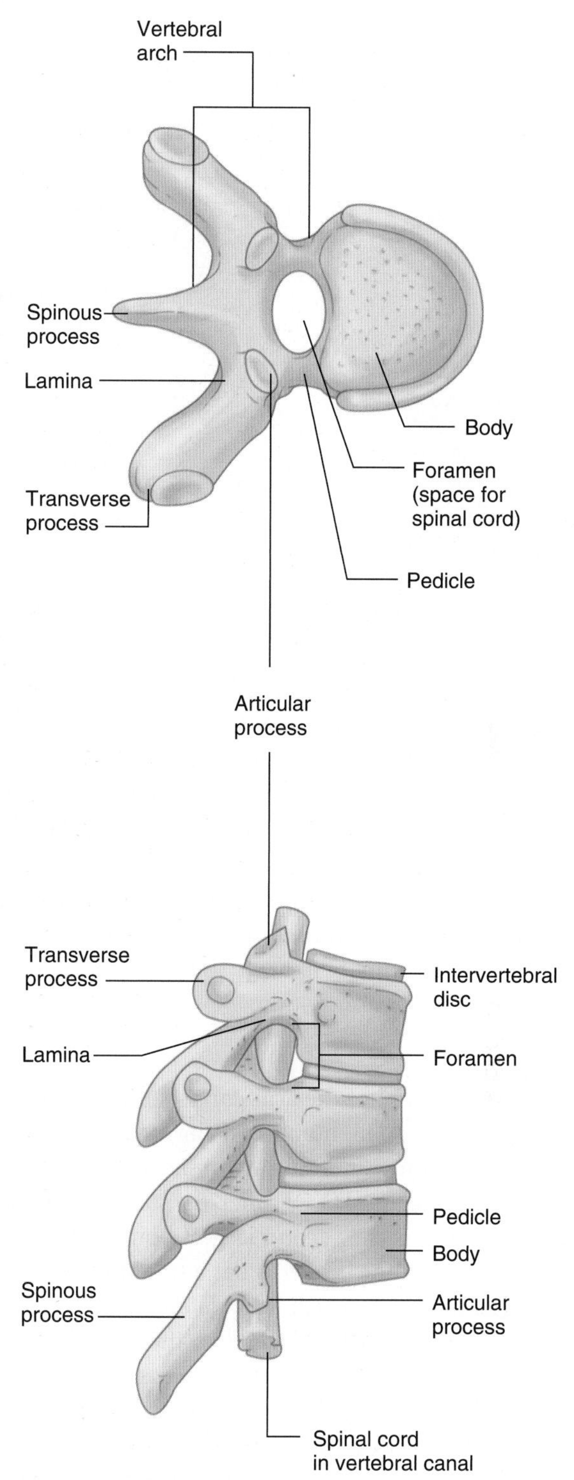

Figure 7-7 *A typical vertebra*

A typical vertebra, as seen in **Figure 7-7**, contains three basic parts: body, foramen, and several processes. The large, solid part of the vertebra is known as the body; the central opening for the spinal cord is called the foramen. Above the foramen protrude two winglike, bony structures called transverse processes. The roof of the foramen contains the spinous process (spine) and the articular processes.

RIBS AND STERNUM

The thoracic area of the body is protected and supported by the thoracic vertebrae, ribs, and the sternum.

The **sternum** (**STER**-num), or breastbone, is divided into three parts: the upper region (**manubrium** [mah-**NEW**-bree-um]), the body, and a lower cartilaginous part called the **xiphoid process** (**ZIF**-oid). Attached to each side of the upper region of the sternum, by means of ligaments, are the two clavicles, or collarbones.

Seven pairs of costal cartilages join seven pairs of ribs directly to the sternum (**Figure 7-8**). The human body contains 12 pairs of ribs. The first seven pairs are called true ribs because they join directly to the sternum. The next three pairs are called false ribs because their costal cartilages are attached to the seventh rib instead of directly to the sternum. Finally, the last two pairs of ribs, connected neither to the costal cartilages nor the sternum, are called floating ribs.

Appendicular Skeleton

The **appendicular skeleton** includes the upper extremities: shoulder girdles, arms, wrists, and hands; and the lower extremities: hip girdle, legs, ankles, and feet. There are 126 bones in the appendicular skeleton. Refer to (**Figure 7-3**).

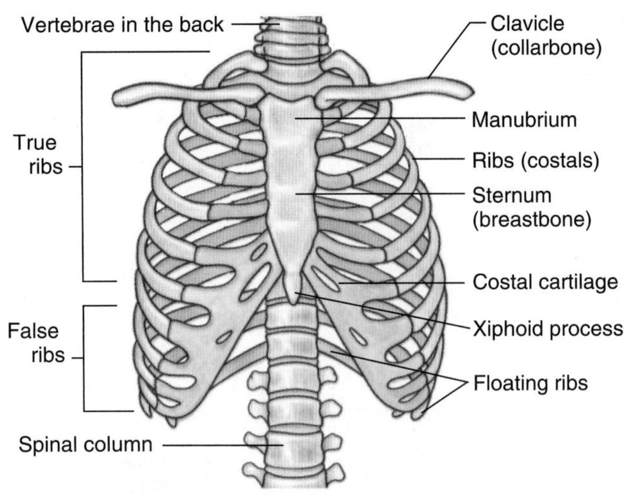

Figure 7-8 *Ribs and sternum*

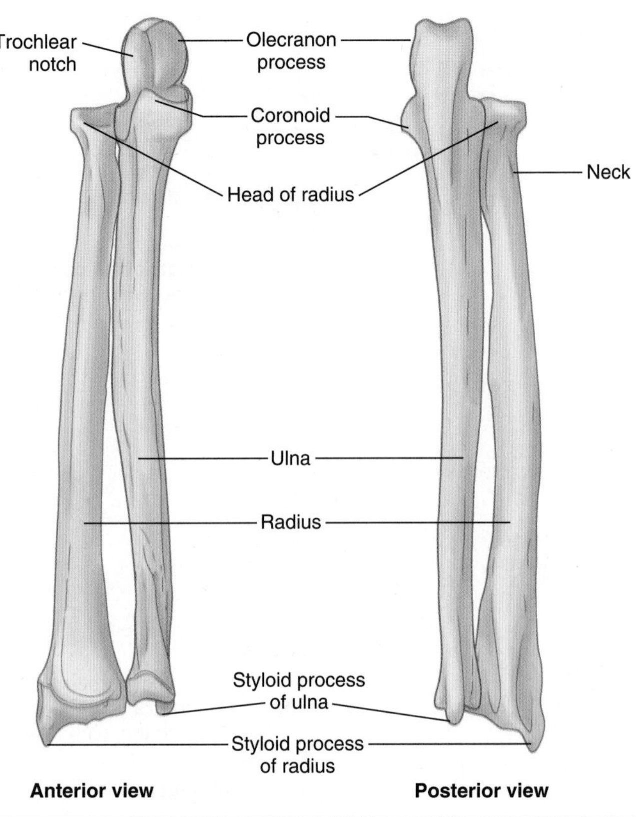

Figure 7-9 *Radius and ulna*

SHOULDER GIRDLE

The shoulder girdle, also called the pectoral girdle, consists of four bones: two curved **clavicles** (**KLAV**-ih-kuls), or collarbones, and two triangular **scapulae** (**SKAP**-you-lee), or shoulder bones. On the skeleton, the two scapulae are on the upper posterior surface. They permit the attachment of muscles that assist in arm movement and serve as a place of attachment for the arms. The two clavicles, attached at one end to the scapulae and at the other to the sternum, help brace the shoulders and prevent excessive forward motion.

ARM

The bone structure of the arm consists of the humerus, radius, and ulna. The humerus is located in the upper arm, and the radius and ulna are in the forearm.

The **humerus** (**HYOU**-mer-us), the only bone in the upper arm, is the second largest bone in the body. The upper end of the humerus has a smooth, round surface called the head, which articulates with the scapula. The upper humerus is attached to the scapula socket (glenoid fossa) by muscles and ligaments.

The forearm consists of two bones: the radius and the ulna. The **radius** (**RAY**-dee-us) is the bone running up the thumb side of the forearm. Its name derives from the fact that it can rotate around the ulna. This is an important characteristic, permitting the hand to rotate freely and with great flexibility. The **ulna** (**ULL-nah**), by contrast, is far more limited. It is the largest bone in the forearm: at its upper end, it produces a projection called the olecranon process, forming the elbow (**Figure 7-9**). When a person bangs their elbow, the ulnar nerve gets compressed against the ulna, referred to as "hitting the funny bone." The olecranon process articulates with the *humerus*.

HAND

The human hand is a remarkable piece of skeletal engineering; the design provides for great dexterity. It contains more bones for its size than any other part of the body. Collectively, the hand has 27 bones (**Figure 7-10**).

The wrist bone, or **carpals** (**KAR**-palz), consists of eight small bones arranged in two rows. These are held together by ligaments that permit sufficient movement to allow the wrist a great deal of mobility and flexion. There is only slight lateral (side) movement of these carpal bones, however. There are several short muscles attached on the palm side of the hand that supply mobility to the little finger and thumb.

The hand consists of two parts: the palmar surface with five **metacarpal** (met-ah-**KAR**-pal) bones, and five fingers with 14 **phalanges** (fah-**LAN**-jeez) (singular, phalanx). Each finger, except for the thumb, has three phalanges, whereas the thumb has two. There are hinge joints between each phalanx, allowing the fingers to bend easily. The thumb is the most flexible finger because the end of the metacarpal bone is more rounded, and there are muscles attached to it from the hand itself. Thus, the thumb can be extended across the palm of the hand. Only humans and other primates possess such a digit, known as an opposable thumb. This gives the hand the ability to grip.

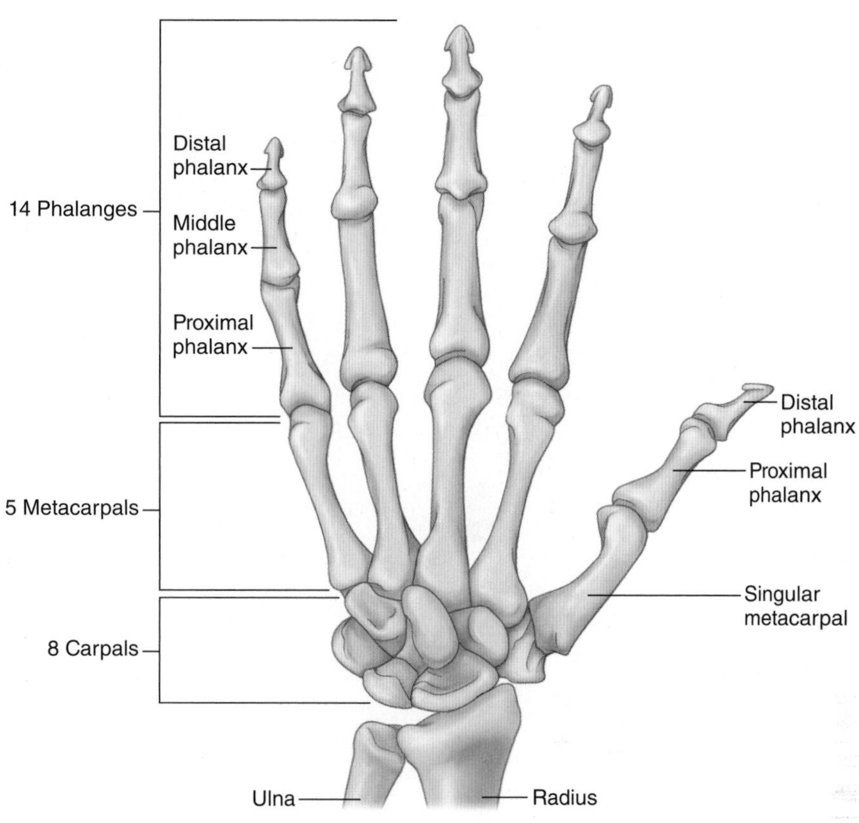

Figure 7-10 *The 27 bones of the left hand*

PELVIC GIRDLE

The pelvic girdle consists of paired hip bones. In youth, the pelvic girdle consists of six bones: three on either side of the midline of the body. These bones are the ilium, ischium, and pubis, which fuse as the body grows. The **ilium** (**ILL**-ee-um) is the broad, blade-shaped bone that forms the back and sides of the hip bone. The **ischium** (**ISS**-kee-um) is the strongest portion of the hip bone. It has the rounded and thick tuberosities that a person sits on and bears the weight of the body. The **pubis** (**PEW**-bis) forms the anterior portion of the hip bone. As the body grows, these bones articulate with the sacrum to form a bowl-shaped structure called the pelvic girdle (**Figure 7-11A**). These two sets of hip bones form a joint with the bones in front, called the symphysis pubis, and with the sacrum in back, at the sacroiliac (SI) joint. The SI joint is held together by strong ligaments, distributes the shock of motion across the pelvis, and allows for an upright position while standing.

Between the pubis and the ischium is the **obturator foramen** (**OB**-tuh-ray-tohr), a large opening that allows for the passage of nerves, blood vessels, and tendons. On the lateral side of the hip bone located just above the obturator foramen is the deep socket called the **acetabulum** (ass-eh-**TAB**-you-lum). All three parts of the hip bone meet and unite at this socket. This structure receives the head of the femur to form the hip joint (**Figure 7-11B**).

The pelvic girdle serves as an area of attachment for the bones and muscles of the leg. It also provides support for the soft organs of the lower abdominal region. The female pelvis is different from the male in that it is oval-shaped and much wider, with a bigger pubic arch and shorter iliac crests. This shape is necessary for pregnancy and childbirth.

UPPER LEG

The upper leg contains the longest and strongest bone in the body: the thighbone, or **femur** (**FEE**-mur). The upper part of the femur has a smooth, rounded head (**Figure 7-12**). It fits into the acetabulum of the pelvic girdle, forming a ball-and-socket joint.

LOWER LEG

The lower leg consists of two bones: the **tibia** (**TIB**-ee-ah) and the **fibula** (**FIB**-you-lah). The tibia is the largest of the two lower leg bones. The tibia is also known as the shinbone. It is the larger, weight-bearing bone in the anterior of the lower leg. The fibula is the smaller bone of the lower leg (**Figure 7-13**).

The **patella** (pah-**TELL**-ah), or kneecap, is found in front of the knee joint. It is a flat, triangular, sesamoid

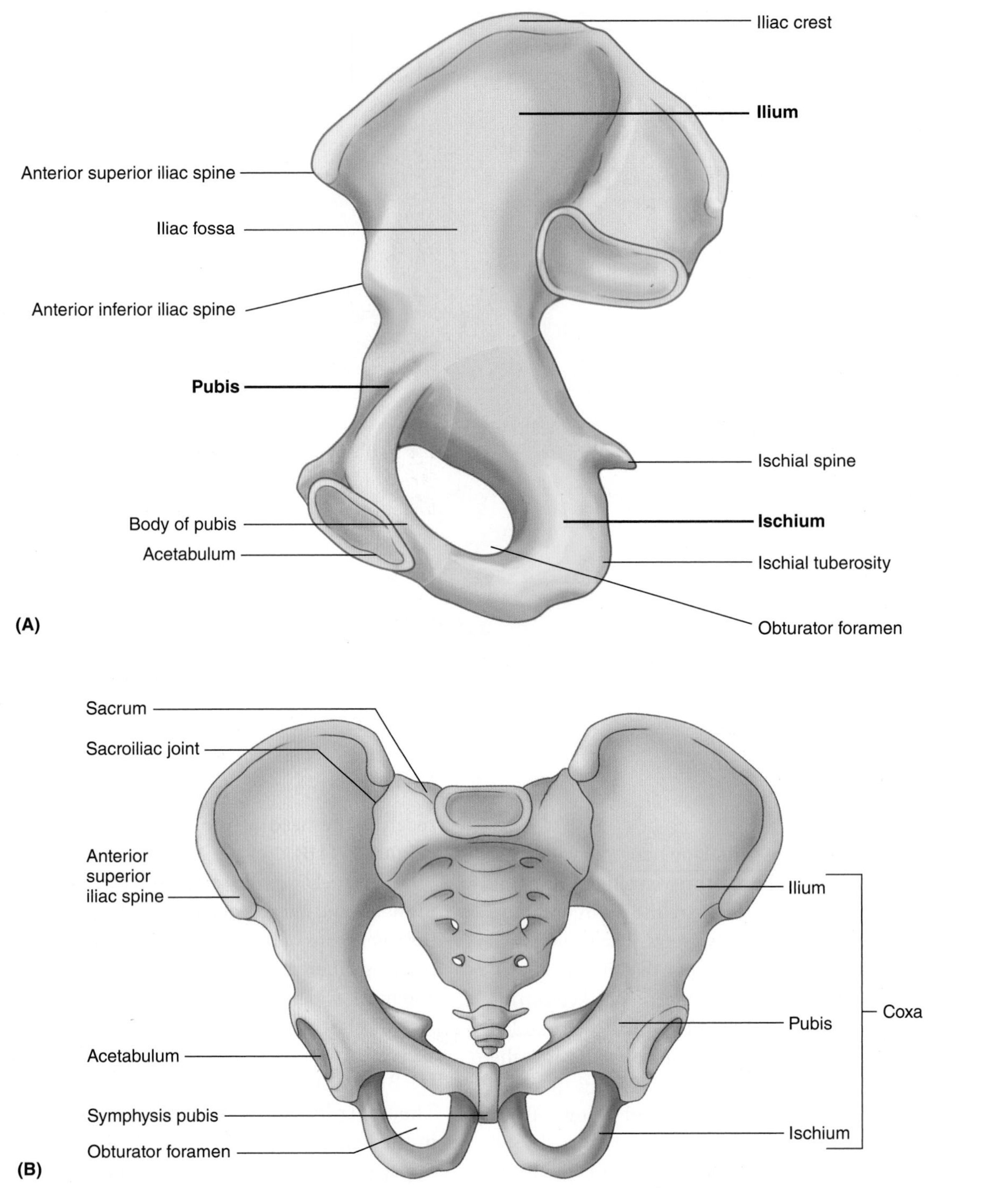

Figure 7-11 *(A) Right hipbone medial view; (B) anterior view of the pelvis*

bone (**Figure 7-14**); a sesamoid bone is a bone embedded within a tendon or muscle. The patella is found in the tendons of the large muscle in front of the femur: the quadriceps femoris. The **menisci** (me-NIS-ci) of the knee are the medial meniscus and the lateral meniscus; these are crescent-shaped bands of thick, rubbery cartilage attached to the tibia. They act as shock absorbers and stabilize the knee. The patella is also attached to the tibia by a series of ligaments that allow movement. Surrounding the patella are four bursae, which serve to cushion the knee joint.

ANKLE

The ankle, or tarsus, contains seven **tarsal** (**TAHR**-sal) bones. These bones provide a connection between the foot and leg bones. The largest anklebone is the heel bone, or **calcaneus** (kal-**KAY**-nee-us). The tibia and fibula articulate with a broad tarsal bone called the

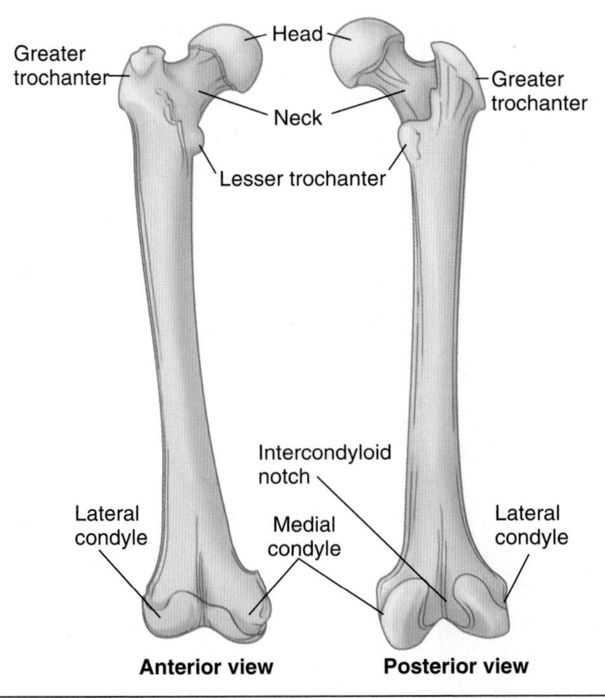

Figure 7-12 *Anterior and posterior views of the femur*

talus (**TAY**-luss). Ankle movement is a sliding motion, allowing the foot to extend and flex when walking.

FOOT

Feet are designed to take a great deal of punishment from the body. With every mile a person walks, 200,000 to 300,000 pounds of stress bears down on their feet. By the time a person is 50, they may have walked 75,000 miles. The foot has five **metatarsal** (met-ah-**TAHR**-sal) bones, which are somewhat comparable to the metacarpals of the hand, but the metatarsal bones are arranged to form two distinct arches, which are not found in the palm of the hand. One arch runs longitudinally from the calcaneus to the heads of the metatarsals; it is called the *longitudinal arch*. The other, which lies perpendicular to the longitudinal arch in the metatarsal region, is known as the *transverse arch*. Strong ligaments and leg muscle tendons help hold the foot bones in place to form those two arches. In turn, arches strengthen the foot and provide flexibility and springiness to the stride. In certain cases, these arches

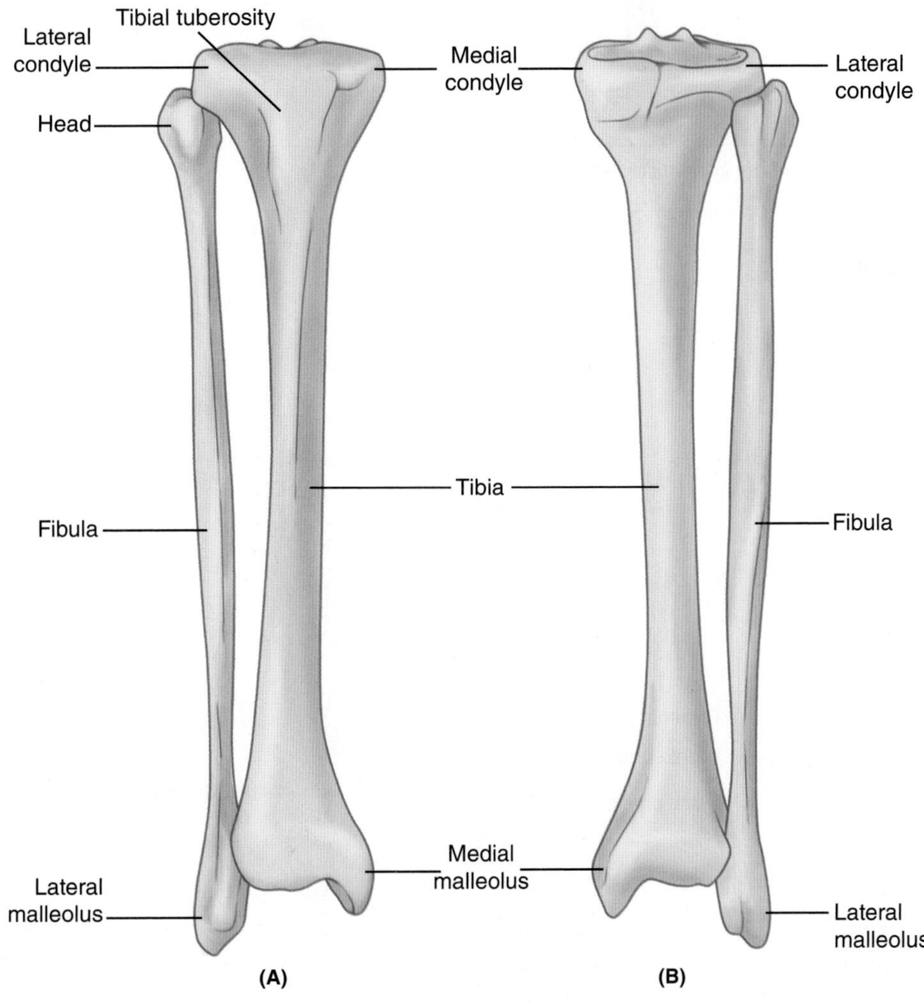

Figure 7-13 *(A) Anterior view of the tibia and fibula. (B) Posterior view of the tibia and fibula.*

may "fall" due to weak foot ligaments and tendons. Then downward pressure caused by the weight of the body slowly flattens them, resulting in fallen arches, or **flatfeet.** Flatfeet cause a good deal of stress and strain on the foot muscles, leading to pain and fatigue.

The toes are similar in composition to the fingers. There are three phalanges in each, with the exception of the big toe, which has only two. Because the big toe is not opposable like the thumb, it cannot be brought across the sole. Each foot has 14 phalanges (**Figure 7-15**).

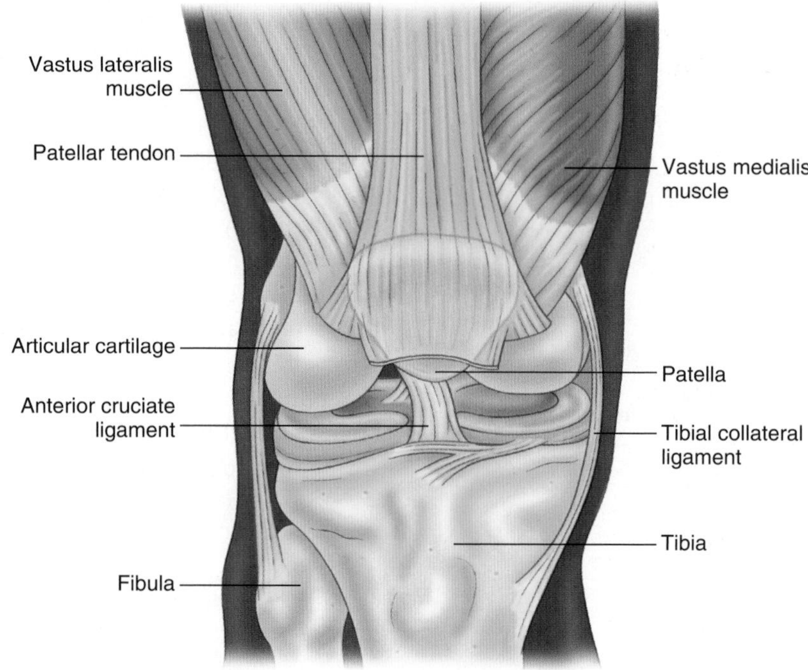

Figure 7-14 *The major ligaments of the knee make joint movement possible.*

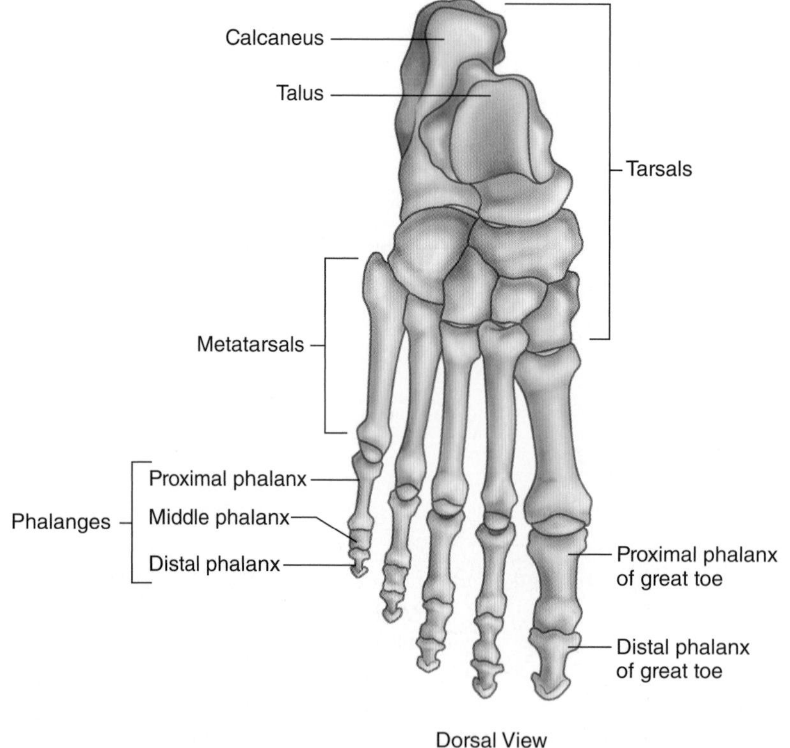

Figure 7-15 *Dorsal view of the foot*

JOINTS AND RELATED STRUCTURES

Joints, or articulations, are points of contact between two bones. They are classified into three main types according to their degree of movement: diarthrosis (movable) joints, amphiarthrosis (partially movable) joints, and synarthrosis (immovable) joints (**Figure 7-16**).

Most of the joints in the body are **diarthroses** (dye-ahr-**THROH**-seez). They tend to have the same structure. These movable joints consist of three main parts: an articular cartilage; a bursa, or joint capsule; and a synovial cavity, or joint cavity.

When two movable bones meet at a joint, their surfaces do not touch one another. The two articular (joint) surfaces are covered with a smooth, slippery cap of cartilage known as articular cartilage. As mentioned, this cartilage helps absorb shock and prevent friction between parts.

Enclosing two articular surfaces of the bone is a tough, fibrous connective tissue capsule called an articular capsule. Lining the articular capsule is a synovial membrane, which secretes **synovial fluid** (sih-**NOH**-vee-al), a lubricating substance, into the synovial cavity, an area between the two articular cartilages. The synovial fluid reduces the friction of joint movement.

Bursae (**BURR**-see) are closed sacs with a synovial membrane lining found in spaces of connective tissue between muscles, tendons, ligaments,

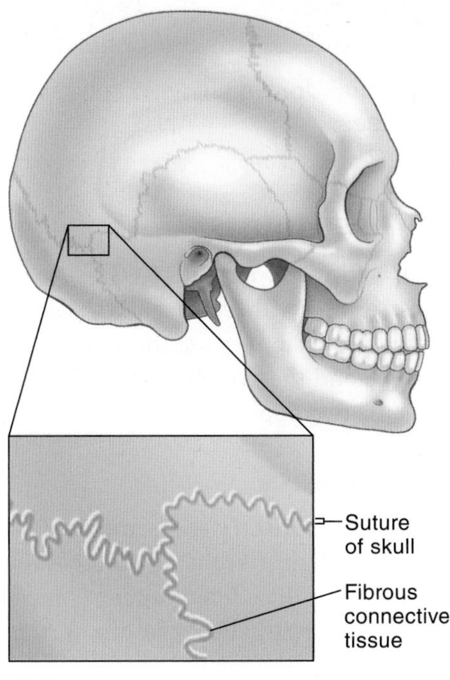

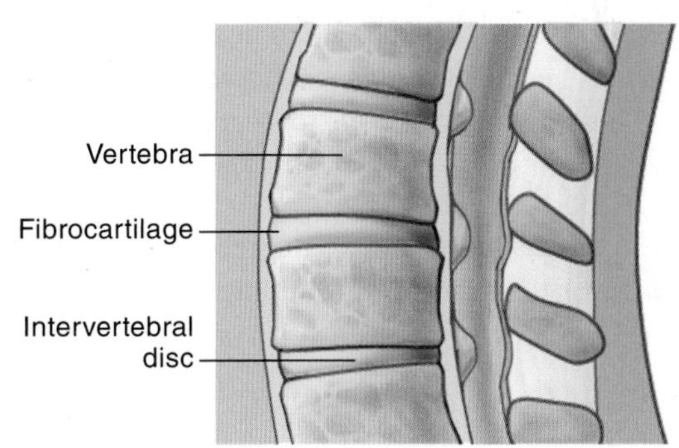

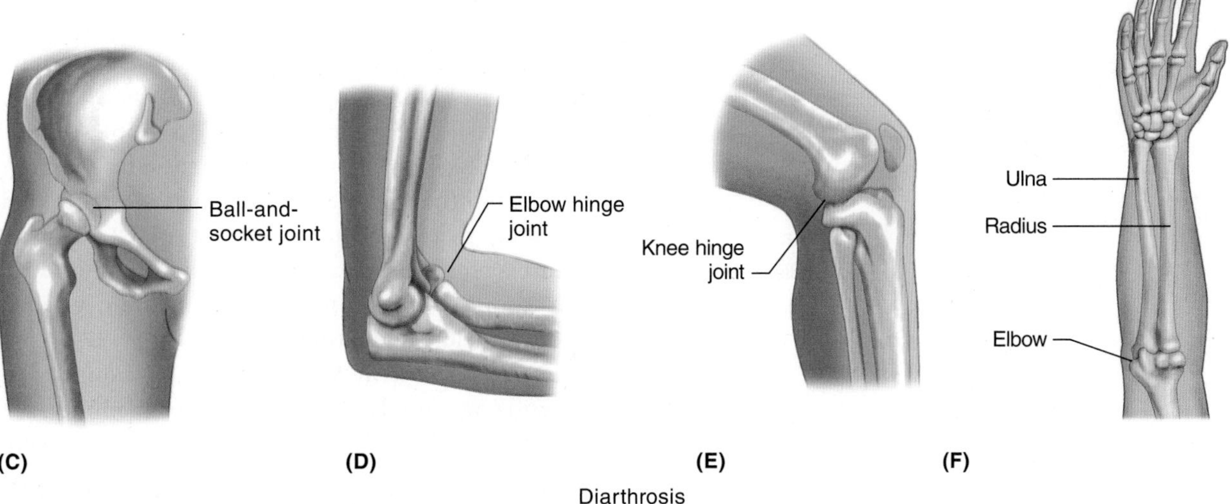

Figure 7-16 *Types of joints: (A) a synarthrosis, an immovable fibrous joint (cranial bones); (B) an amphiarthrosis, a slightly movable cartilaginous joint (ribs or vertebra); (C–F) diarthroses, freely movable hinge, or ball-and-socket joints*

and bones. The secreted synovial fluid serves as a lubricant to prevent friction between a tendon and a bone. If this sac becomes irritated, injured, or inflamed, **bursitis** (bur-**SIGH**-tis) develops. The synovial fluid can be aspirated, or withdrawn, from the bursa sacs to be examined for diagnostic purposes (**Figure 7-17**).

Diarthroses Joints

There are six types of diarthroses, or movable joints:

1. **Ball-and-socket joints** allow the greatest freedom of movement. One bone has a ball-shaped head that nestles into a concave socket of the second bone. The shoulders and hips have ball-and-socket joints.
2. **Hinge joints** move in one direction or plane, as in the knees, elbows, and outer joints of the fingers.
3. **Pivot joints** are those with an extension rotating in a second, arch-shaped bone. The radius and ulna are pivot joints. Another example is the joint between the atlas that supports the head, and the axis that allows the head to rotate.
4. **Gliding joints** are those in which nearly flat surfaces glide across each other, as in the vertebrae of the spine. These joints enable the torso to bend forward, backward, and sideways, as well as rotate.
5. *Condyloid joints* allow for two degrees of movement, such as flexion/extension, adduction/abduction, and circumduction. Some examples are the fingers and the wrist.

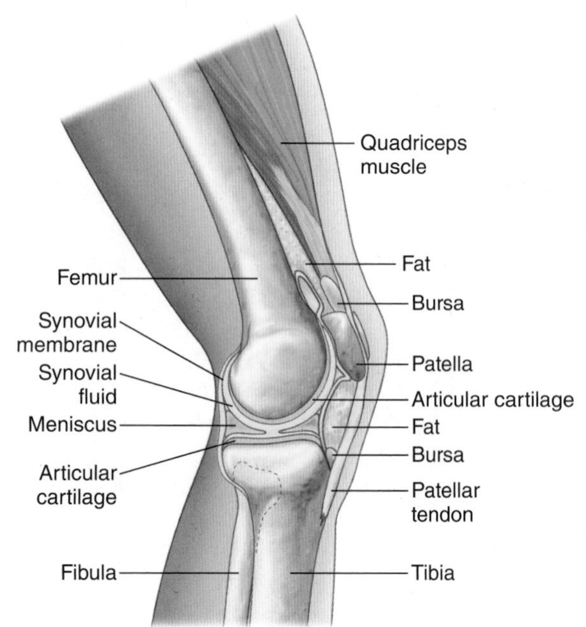

Figure 7-17 *In this lateral view of the knee, note the structure of the synovial membrane and bursa.*

6. *Saddle joints* allow for movement back and forth and side to side. Some examples are the joint at the base of the thumb and the sternoclavicular joint of the chest.

Between each body of the vertebrae are cartilage discs. These discs serve as a buffer between the vertebrae to decrease the forces of weight and shock resulting from running, walking, or jumping. Discs can be compressed by sudden, forceful jolts to the spine, causing a portion of the disc to protrude from the vertebrae and impinge on the spinal nerves, resulting in extreme pain. This is known as a *herniated* or *slipped disc*.

Media Link

View the **Synovial Joints** animation on the Online Resources.

Amphiarthroses Joints

Amphiarthroses (am-fee-ahr-**THROH**-seez) are partially movable joints, with cartilage between their articular surfaces. Two examples are the attachment of the ribs to the spine and the symphysis pubis, which is the joint between the two pubic bones.

Synarthroses Joints

Synarthroses (sin-ahr-**THROH**-seez) are immovable joints connected by tough, fibrous connective tissue. These joints are found in the adult cranium. The bones are fused together in a joint called a suture, forming a heavy, protective cover for the brain.

Types of Motion

Joints can move in many directions (**Figure 7-18**). **Flexion** is the act of bringing two bones closer together, which decreases the angle between the two bones. **Extension** is the act of increasing the angle between two bones, which results in a straightening motion. **Abduction** (ab-**DUCK**-shun) is the movement of an extremity away from the midline, an imaginary line that divides the body from head to toe. **Adduction** (add-**DUCK**-shun) is movement toward the midline. **Circumduction** (sir-kum-**DUCK**-shun) includes flexion, extension, abduction, and adduction.

A **rotation** movement allows a bone to move around one central axis. This type of pivot motion occurs when the head turns from side to side or movement within a ball-and-socket joint. In **pronation** (proh-**NAY**-shun), the forearm turns the hand so the palm is downward or backward. In **supination** (soo-pih-**NAY**-shun), the palm is forward or upward.

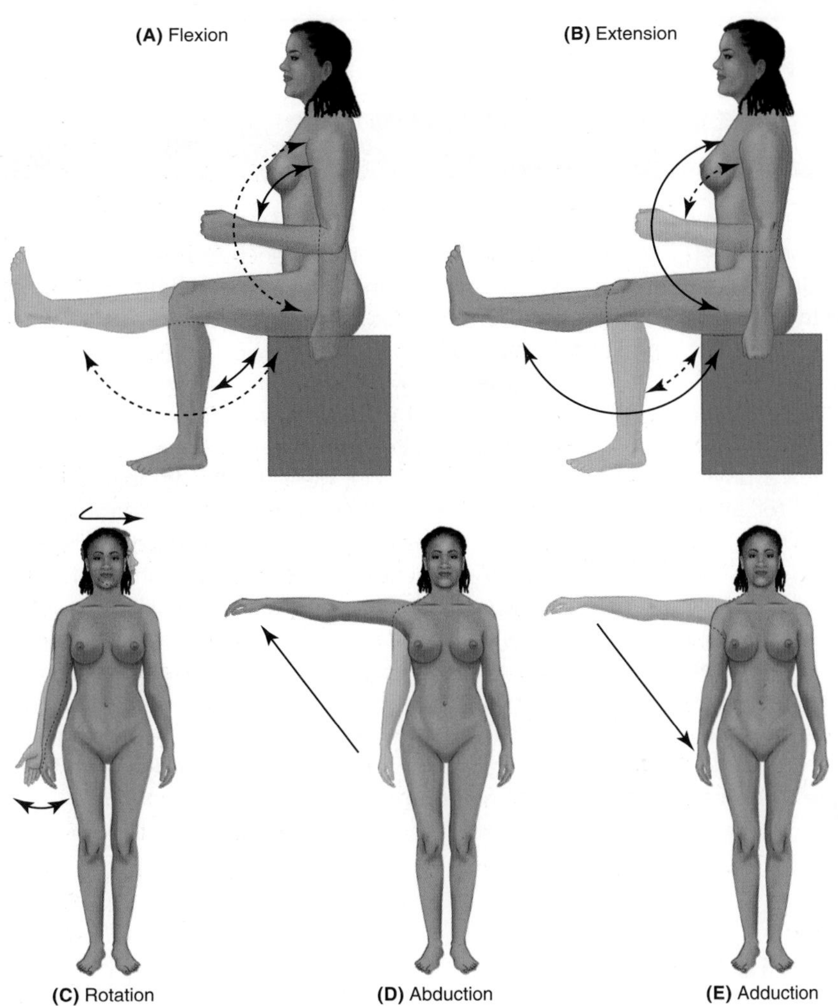

Figure 7-18 *Joint movements*

Career Profile

PHYSICAL THERAPIST AND PHYSICAL THERAPIST ASSISTANT

PHYSICAL THERAPIST (PT)

Physical therapists improve mobility, relieve pain, and prevent or limit permanent disability of patients suffering from injuries or disease. Therapists evaluate patient histories, test and measure patients' strength and range of motion, and develop a treatment plan. Treatment often includes exercises to increase flexibility and range of motion. Physical therapists must have moderate strength because the job can be physically demanding. Job prospects are excellent. The education required is preparation in a bachelor's or master's degree program in physical therapy. Entry is highly competitive; some schools require volunteer activity in therapy departments in a hospital or a clinic prior to admission. All states require physical therapists to pass a licensure examination.

PHYSICAL THERAPIST ASSISTANT (PTA)

Physical therapist assistants work under the direction and supervision of a physical therapist. The PTA assists patients who are recovering from illnesses, injuries, and surgeries to regain movement and manage pain. This job requires a moderate degree of strength because of its physical demands. The education requirement for the PTA is an associate's degree from an accredited program for physical therapist assistants. Education requires a clinical component. At the present time, most states require licensure and/or certification. Job prospects are excellent because of the aging population.

Medical Highlights

ORTHOPEDIC DIAGNOSTIC TECHNOLOGIES

There are a variety of diagnostic technologies used to identify skeletal injuries and disorders. These technologies are used to determine the course of treatment.

A *physical examination* is usually the first diagnostic test used to identify injuries or disorders. A medical provider uses a physical examination to look for signs of swelling, bruising, growths, and range of motion.

Laboratory tests are usually ordered after the initial physical examination. Tests involving the blood, urine, and the synovial (joint) fluid can be used to diagnose certain conditions; for example, a high level of uric acid in the blood can be an indication of gout, high white blood cell count in synovial fluid can point to inflammation, and calcium levels can be measured to determine the risk of osteoporosis.

Imaging tests may be done to give additional insight into an injury or disorder. *X-rays* are the most common diagnostic test used for skeletal injuries and disorders. An X-ray beam passes through the part of the body that has been positioned between the machine and the film. The film captures the image of dense matter. X-rays can be used to diagnose fractures, joint dislocations, arthritis, scoliosis, bone tumors, and other ailments.

An *arthrography* is used to diagnose conditions surrounding the joint and the joint structures. A contrast iodine solution is injected into the joint and several X-rays are taken, using a specialized X-ray devise called a fluoroscope. A *CT scan* combines X-rays with computer technology to produce a more detailed, cross-sectional image of the body. CT scans are used to diagnose fractures, joint damage, and other disorders when X-rays and physical examinations are not definitive. An *MRI* uses magnets, radio waves, and computer technology to take high-resolution pictures of bones and soft tissues. MRIs are used to diagnose joint abnormalities, torn ligaments, and bone cancer.

DISORDERS OF THE BONES AND JOINTS

Fractures

The most common traumatic injury to a bone is a **fracture**, or break. When this occurs, there is swelling due to injury and bleeding tissues. The common types of fractures are as follows (**Figure 7-19**):

- A *greenstick* fracture is the simplest type of fracture. The bone is partly bent, but it never completely separates. The break is similar to that of a young, sap-filled woodstick, where the fibers separate lengthwise when bent. Such fractures are common among children because their bones contain flexible cartilage (**Figure 7-19A**).
- A *closed/simple* fracture is one in which the bone is broken, but the broken ends do not pierce through the skin to form an external wound (**Figure 7-19B**).

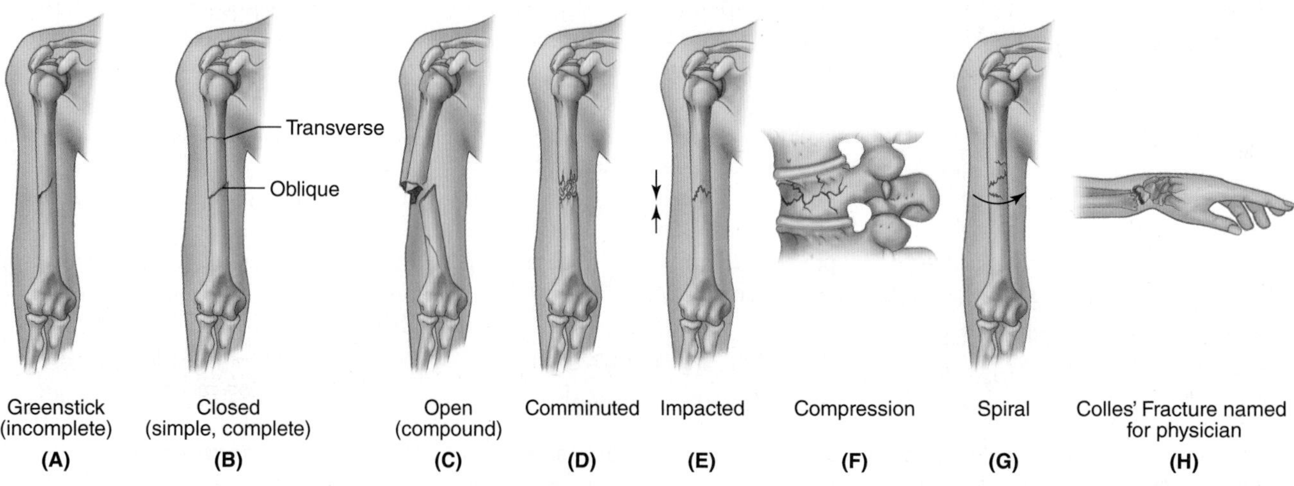

Figure 7-19 *Types of bone fractures: (A) greenstick, (B) closed, (C) open, (D) comminuted, (E) impacted, (F) compression, (G) spiral, (H) Colles'*

- An *open/compound* fracture is the most serious type of fracture, where the broken bone ends pierce and protrude through the skin. This can provide a site for infection of the bone and neighboring tissues (**Figure 7-19C**).
- A *comminuted* fracture occurs when the bone is splintered or broken into many pieces that can become embedded in the surrounding tissue (**Figure 7-19D**).

The Effects of Aging on the Skeletal System

Around the age of 40, bone mass and density begin to decline. The body starts to lose calcium and other minerals. Women are more vulnerable to bone loss (osteoporosis) than men are. Bone loss in women occurs especially in the decade following menopause.

The change in bones is gradual and is due to reabsorption of the interior matrix of the long and flat bones. The external surfaces of the bones begin to thicken. These changes are not directly observable, but they are evident by alterations in position and stature. The intervertebral cartilage discs shrink, narrowing the space between discs and resulting in a loss of height. Posture is also affected; the center of balance is altered due to the shortening of the spinal column. The long bones of the arms and legs become more brittle but do not change in length. As a result, the arms and legs appear longer compared with the trunk of the body. Foot arches become less prominent, and the foot lengthens. Joints, by the age of 70, reflect a lifetime of wear and tear. The joints become less mobile because the cartilage shrinks (loses water) and the joints fuse at the cartilage surface.

Protein synthesis slows as one ages; as a result, there is no new formation of collagen fibers. These fibers give bones strength and flexibility; without them, bones become brittle and easily fracture. Hardening of ligaments, tendons, and joints leads to an increase in rigidity and a decrease in flexibility. Stiff, painful joints are due to the general wear and tear on the ligaments and synovial membrane. The discomfort and physically limiting changes decrease the range of motion of the joints. The psychological fear of falling due to physical changes further adds to potential for inactivity and injury.

- A *stress* or *hairline* fracture is a tiny crack in the bone that typically occurs from overuse. This fracture can be quite painful but usually heals itself.
- An *impacted* fracture (**Figure 7-19E**) or *compression* fracture (**Figure 7-19F**) is a fracture in which the bone breaks into multiple fragments that are driven into each other.
- A *spiral* fracture or *torsion* fracture occurs when torque, a rotating force, is applied only on the axis of the bone, and the body is in motion while one extremity is firmly planted. An example of this is when a skier locks their feet into ski boots; if the skier loses control and the ski rotates, the leg may be twisted in one direction (**Figure 7-19G**).
- A *Colles'* fracture is a fracture of the distal radius in the forearm with dorsal (posterior) and radial displacement of the wrist and hand; this is often called a broken wrist (**Figure 7-19H**).

Media Link

Watch the **Types of Fractures** animation on the Online Resources.

The process of restoring bone occurs through the following three main methods:

1. *Closed reduction*: The bony fragments are brought into alignment by manipulation, and a cast or splint is applied.
2. *Open reduction*: Through surgical intervention, devices such as wires, metal plates, or screws are used to hold the bone in alignment; a cast or splint may be applied.
3. *Traction*: A pulling force is used to hold the bones in place. This is used for fractures of the long bone.

Other Bone and Joint Injuries

Dislocation occurs when a bone is displaced from its proper position in a joint. This may result in the tearing and stretching of the ligaments. Reduction or return of the bone to its proper position is necessary, along with rest to allow the ligaments to heal.

Media Link

View the **Ankle Sprain** and **Knee and ACL Tear** animations on the Online Resources.

A **hammertoe** is a toe that is curled, or flexed, due to a bend in the middle joint of one or more toes. It may be caused by shoes that are too tight or heels that are too high. The longest of the four smaller toes may be forced against the front of the shoe, resulting in an unnatural bending of the toe with pain and pressure in the affected area. Proper footwear with a deep toe box and flexible material covering the toes may be helpful, as can using a special device ordered by a physician to wear in the shoe to help position the toe.

A **whiplash injury** is trauma to the cervical vertebrae, usually the result of an automobile accident. The force generated by a car's abrupt change in speed or direction whips the head, putting tremendous strain on the cervical spine and neck muscles. Treatment includes bracing, heat or cold, pain medication, and physical therapy. The goal of treatment is to alleviate pain and restore a normal range of motion; recovery time may be up to 6 weeks.

DISEASES OF THE BONES

Diseases of the Bones can be congenital, ranging from innocuous to severe, or they can evolve as a person begins to age.

Congenital Bone Disorders

The following are the common congenital disorders of the skeletal system:

- *Limb abnormalities:* This could be the absence of fingers or toes, or the partial or complete absence of an arm or leg.
- *Club foot:* The front part of the foot turns toward the inside of the heel. Treatment is the use of a cast or boot and physical therapy.
- *Congenital hip dislocation:* The hip and femur are underdeveloped, which leads to a dislocation of the hip. Treatment is the use of splints and sometimes surgery to allow the hip to develop properly.

Medical Highlights

RICE TREATMENT

RICE is the acronym for rest, ice, compression, and elevation, the recommended immediate treatment for bone, joint, and muscle injuries. Treatment that occurs in the first 24 to 72 hours after an injury can do a lot to relieve or even prevent aches and pains.

R = REST

Injuries heal faster if rested. Rest means staying off the injured body part. Using any part of the body increases the blood circulating to that area, which can cause more swelling of an injured part. In the case of an ankle sprain, there should be no weight bearing for at least the first 24 hours.

I = ICE

An ice pack should be applied to the injured area as soon as possible after the injury. Apply for 20 minutes every 4 hours during the first 48 hours. Skin treated with cold passes through four stages: cold, burning, aching, and numbness. Do not ice for more than 20 minutes at a time.

C = COMPRESSION

Compressing the injured area may squeeze some fluid and debris out of the injured area. Compression limits the ability of the skin and other tissues to expand and reduces internal bleeding. Apply an elastic bandage to the injured area, especially the foot, ankle, knee, thigh, hand, or elbow. Fill in the hollow areas with padding such as a washcloth or sock before applying the elastic bandage.

Caution: DO NOT apply an elastic bandage too tightly because this may restrict circulation. Leave fingers or toes exposed so possible skin color change due to restricted circulation can be observed. Compare the injured side with the uninjured side. Pale skin, numbness, pain, and tingling are signs of impaired circulation. Remove the elastic bandage immediately if any of these signs appear.

E = ELEVATION

Gravity slows the return of blood to the heart from the lower parts of the body. Once fluid gets to the hands or feet, the fluid has nowhere to go, and those parts of the body swell. Elevating the injured part, in combination with ice and compression, limits circulation to that area, which in turn helps limit internal bleeding and minimize swelling.

Source: http://athletics.mckenna.edu/sportsmedicine/rice.html; © Cengage Learning 2014

- *Polydactyl:* This is the presence of extra fingers or toes. Treatment is surgery to remove the extra digit.
- *Syndactyl:* This is the fusion of one or more fingers or toes. Treatment is surgery to separate the fingers or toes.
- *Anencephaly:* This is the absence of a major part of the skull or brain.

General Bone Disorders

Many of the diseases that affect a person's bones are more likely to develop as they age.

Arthritis is one of the most common and debilitating health problems in the world. In the United States alone, about 54 million adults have some type of arthritis, according to the Centers for Disease Control and Prevention. Arthritis is an inflammation of one or more joints, accompanied by pain, stiffness, swelling, and other problems that limit normal activities of daily living. The joint pain and stiffness associated with arthritis is most noticeable in the morning, after a period of rest. There are at least 20 different types, the most common being rheumatoid arthritis, osteoarthritis, and gout. Some autoimmune diseases, such as psoriasis, lupus, scleroderma, and Sjögren's syndrome, often present with arthritis (these will be covered in Chapter 16).

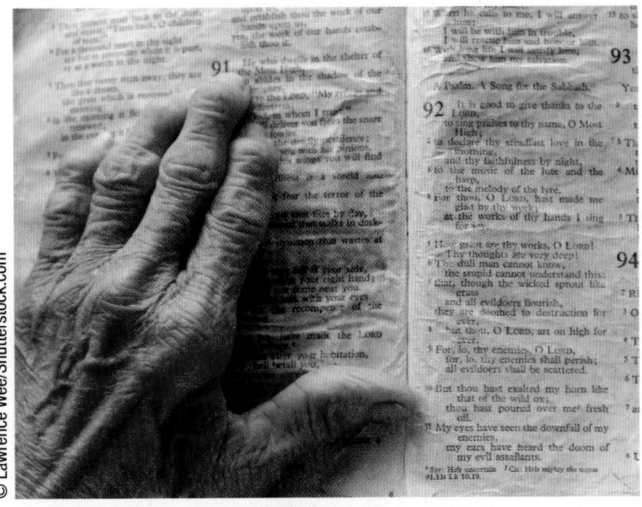

Figure 7-20A *Arthritic nodes*

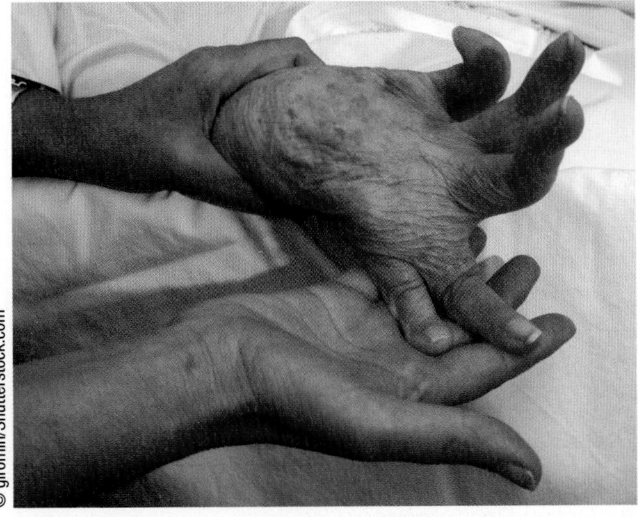

Figure 7-20B *Ulnar deviation*

Rheumatoid arthritis is a chronic autoimmune disease—a disease in which the body's immune system attacks the tissue—that affects the connective tissue and joints. There is acute inflammation of the connective tissue; thickening of the synovial membrane; and ankyloses, or fusing, of joints. The joints are badly swollen and painful.

The pain, in turn, causes muscle spasms that can lead to deformities in the joints. In addition, the cartilage that separates the joints degenerates, and hard calcium fills the spaces. When the joints become stiff and immobile, muscles attached to these joints slowly atrophy, or shrink in size (**Figure 7-20A** and **Figure 7-20B**). The signs and symptoms of rheumatoid arthritis may vary in severity and may even go into periods of remission. It typically begins in middle age and occurs more frequently in older adults. This disease affects approximately three times more women than men.

Osteoarthritis is known as degenerative joint disease. It occurs with aging. In this disease, the articular cartilage degenerates and a bony spur formation occurs at the joint. The joints may enlarge and there is pain and swelling, especially after activity.

Gout is characterized by an acute inflammation commonly affecting the big toe, although it may affect other joints as well. One of the main causes of gout is an excess of protein in the diet. The pain and swelling are the body's responses to the accumulation of uric acid crystals in the affected joint. Uric acid is formed by the breakdown of molecules called purines. Treatment is with nonsteroidal anti-inflammatory drugs, gout medications such as colchicine, and diet.

Learning to live with a chronic disease such as arthritis is the biggest challenge of all. At the present time, there is no cure for arthritis, although there are many treatments to relieve pain and increase mobility.

Medication and Other Therapies

Medications include painkillers, nonsteroidal anti-inflammatory drugs (ibuprofen), anti-rheumatic drugs that suppress the immune system (methotrexate),

> **Media Link**
>
> View the **Osteoarthritis and Rheumatoid Arthritis Compared** animation on the Online Resources.

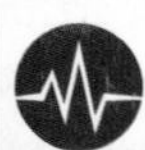

Medical Highlights

PROCEDURES FOR DAMAGED HIPS AND KNEES

Two terms frequently heard when discussing treatment of a damaged hip or knee are arthroscopy and arthroplasty.

Arthroscopy (ar-**THROS**-koh-pee) is the visual examination of the internal structure of a joint using an arthroscope. The arthroscope is a small fiber-optic viewing instrument used to view the interior of the joint. Through an incision of about ¼ inch, the physician can diagnose and treat injuries of the joint area. This procedure is frequently used to remove loose cartilage in the knee.

Microfracture is a surgical option using arthroscopic technique. It creates small holes in the hard surface of the bone to allow the deeper, more vascular bone to access the surface layer. The cells can then get to the surface layer and stimulate cartilage growth. This procedure is frequently done for patients with damaged cartilage.

Arthroplasty (**AR**-throh-plas-tee) means the surgical repair of a damaged joint but has come to mean the surgical placement of an artificial joint.

Total hip replacement is the replacement of a damaged hip joint. A plastic lining is fitted into the acetabulum to restore a smooth surface. The head of the femur is removed and replaced with a metal ball attached to a metal shaft that is fitted into the femur. The smooth surfaces restore the function of the joint.

Total knee replacement involves replacement of all parts of the knee. In a partial knee replacement, only the damaged parts are replaced.

Resurfacing is a procedure that matches the shape and contour of the affected joint. It is a patch for an area of damaged cartilage and is done to prevent further damage to the area. It results in minimal bone loss.

Hip resurfacing involves the placement of a metal cap over the head of the femur to allow it to move smoothly over a metal lining in the acetabulum.

Knee resurfacing involves the use of an implant only on the part of the knee surface that is damaged. By retaining the undamaged part of the knee, the joint may bend better and function more normally.

The type of joint surgery is determined by the physician, the patient, and the patient's age and health.

Regenerative medicine attempts to restore function and form by relying on the body's ability to heal itself. With one procedure known as bone marrow aspirate concentrate (BMAC), bone marrow is withdrawn, concentrated, and reinjected in a stronger form; this includes stem cells and is used to treat damaged tissue and improve function. A study at the Mayo Clinic is using BMAC therapy to treat osteoarthritis of the knee.

biologic response modifiers designed to selectively block parts of the immune response (Humira®, Remicade®, Enbrel®), and corticosteroids to reduce inflammation and suppress the immune system.

LIFESTYLE AND HOME REMEDIES

Weight loss, physical activity, rest, and heat and cold packs can ease pain. Assistive devices such as a cane may help.

COMPLEMENTARY AND ALTERNATIVE THERAPIES

Herbal and dietary supplements, such as fish oils or turmeric, may alleviate pain in some individuals.

Yoga and tai chi movements may help by improving strength and flexibility. Acupuncture and massage might also be useful.

SURGICAL PROCEDURES

Joint replacement surgery involves removing all or part of an affected joint and replacing it with an artificial one. The knee and hip are the most commonly replaced joints (**Figure 7-21**). Also see *Medical Highlights: Procedures for Damaged Hips and Knees.*

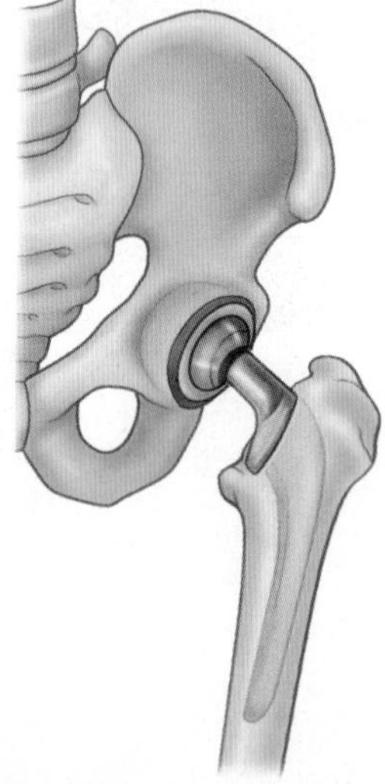

Figure 7-21 *Arthroplasty: total hip replacement*

Media Link

View the **Arthroscopy** animation on the Online Resources.

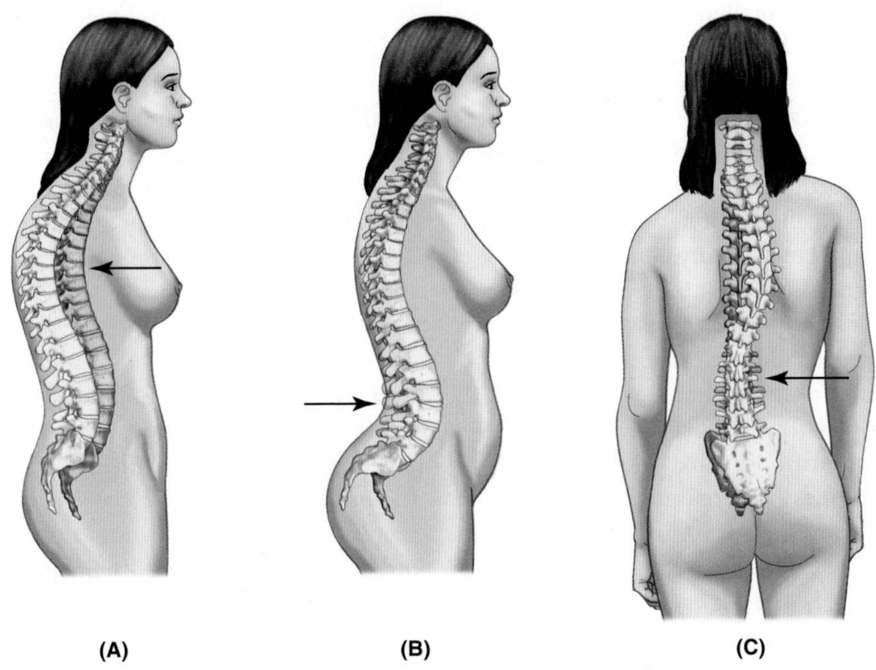

Figure 7-22 *Abnormal curvatures of the spine: (A) kyphosis, (B) lordosis, (C) scoliosis (Note the normal curvature is shown in shadow.)*

ABNORMAL CURVATURES OF THE SPINE

Kyphosis (kye-**FOH**-sis), or hunchback, is a humped curvature in the thoracic area of the spine (**Figure 7-22A**).

Lordosis (lor-**DOH**-sis), or swayback, is an exaggerated inward curvature in the lumbar region of the spine just above the sacrum (**Figure 7-22B**).

Scoliosis (skoh-lee-**OH**-sis) is a side-to-side or lateral curvature of the spine (**Figure 7-22C**).

Media Link

View the **Curvatures of the Spine** animation on the Online Resources.

Other Medically Related Disorders

Osteoporosis (oss-tee-oh-poh-**ROH**-sis), or porous bone disease, is characterized by low bone mass and structural deterioration of bone tissue. The National Osteoporosis Foundation estimated in July of 2019 that 10.2 million adults in the U.S. have osteoporosis, and another 43.4 million more have low bone mass; this means that 54 million American adults, or half of the population over 50, are at risk of a fracture due to osteoporosis. Women are at greater risk because they start with 30% less bone mass than men do, and they experience estrogen depletion after menopause, which increases bone loss. In osteoporosis, the mineral density of the bone is reduced from 65% to 35%. This loss of bone mass leaves the bone thinner, more porous, and more susceptible to fracture (**Figure 7-23**).

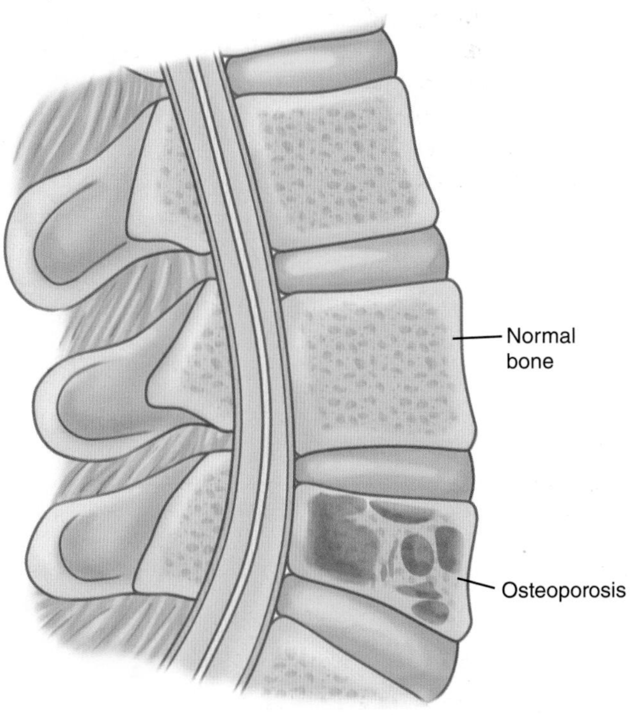

Figure 7-23 *Comparison of normal bone tissue to that of osteoporosis*

Osteoporosis is often called a "silent disease" because bone loss occurs without symptoms. Symptoms may not be evident until a sudden strain, bump, or fall causes a fracture or vertebrae to collapse. Collapsed vertebrae may be seen in the form of loss of height, severe back pain, or spinal deformities (**Figure 7-24**).

A specialized test, called a bone mineral density (BMD) test, can measure bone density in various sites of the body and show signs of early bone loss. Treatment is aimed at preventing or slowing the process.

A person may take calcium and vitamin D supplements. Drugs such as bisphosphonates (Fosamax®) help maintain bone density. The drug denosumab is a human monoclonal antibody that interferes with the body's normal process of breaking down bone, thereby preventing bone loss. In April of 2019, the FDA approved another monoclonal antibody drug, Evenity®, which reduces the risk of bone fracture and helps to regrow bone. Weight-bearing exercise is also helpful in increasing bone mass. Postmenopausal women may take estrogen to help maintain bone mass.

Osteomyelitis (oss-tee-oh-my-eh-**LYE**-tis) is an infection that may involve all parts of the bone. It may result from injury or systemic infection and most commonly occurs in children between the ages of 5 and 14 years. Treatment is antibiotic therapy.

Osteosarcoma, or bone cancer, may occur in younger people, typically affecting the long bone of a limb. The most common site of affliction is just above the knee. Treatment is usually amputation of the affected part and chemotherapy.

Rickets is usually found in children and caused by a lack of vitamin D. Bones become soft due to lack of calcification, causing such deformities as bowlegs and pigeon breast. Presently, there seems to be an increase in the number of cases of rickets; physicians think that an increase in breastfeeding or the overuse of sunscreens may be the cause in young adults. Breast milk does not contain a lot of vitamin D, and sunscreens block out the ultraviolet rays of the sun. The disease may be prevented with sufficient quantities of calcium, vitamin D, and moderate exposure to sunshine.

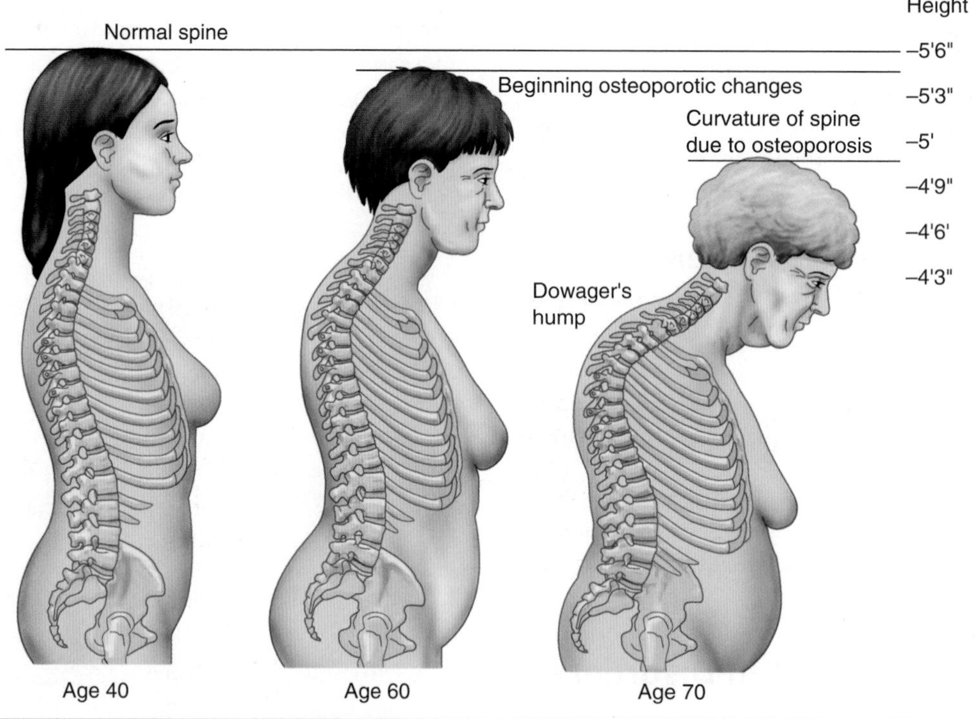

Figure 7-24 *Osteoporosis: loss in height and the Dowager's hump*

Career Profile

ORTHOTIST AND PROSTHETIST

Orthotists design, manufacture, and fit braces and other orthopedic devices such as artificial limbs, or prostheses.

Prosthetists measure, design, and fit or service prosthetic devices as prescribed by a physician.

Both of these fields require a master's degree in orthotics and prosthetics and a one-year residency program. After completion of the residency program, the candidate must pass a certification exam given by the American Board for Certification in Orthotics, Prosthetics, and Pedorthics. Employment is in hospitals, medical supply houses, and physicians' offices. The job outlook is expected to rise about 12% annually.

One BODY How the Skeletal System Interacts with Other Body Systems

INTEGUMENTARY SYSTEM

- Helps give shape to the body.
- Skin produces the precursor to vitamin D, which is necessary for the absorption of the calcium stored in the bones.

MUSCULAR SYSTEM

- Bones serve as attachments for tendons of skeletal muscle, which move the body.
- Provides the calcium necessary for muscle contraction.

NERVOUS SYSTEM

- The cranium protects the brain. The vertebrae protect the spinal cord.
- Sensory receptors in joints send signals to the brain about the body's position.

ENDOCRINE SYSTEM

- Protects some endocrine glands.
- Interacts with calcitonin of thyroid and parathormone of parathyroid glands to act on the bone marrow to regulate blood calcium levels.

CIRCULATORY SYSTEM

- Ribs protect the heart and major blood vessels.
- Bone marrow cavities are the site for the manufacture of red blood cells to carry oxygen, white blood cells for defense, and platelets for blood clotting.

LYMPHATIC SYSTEM

- Provides some protection of lymph glands.
- Bone marrow is the site of the manufacture of some lymphocytes necessary for the immune system.

RESPIRATORY SYSTEM

- The rib cage protects the organs of respiration.
- Ribs and intercostal muscles assist with the mechanics of breathing.

DIGESTIVE SYSTEM

- Provides protection for teeth, esophagus, stomach, liver, and gallbladder.
- Absorbs calcium in the intestines; blood vessels take calcium to be stored in the bones.

URINARY SYSTEM

- Bones protect the kidneys and bladder.

REPRODUCTIVE SYSTEM

- The pelvic girdle protects the organs of reproduction.

Medical Terminology

ab-	away from
duc	move
-tion	process
ab/duc/tion	process of moving away from
ad-	to or toward
ad/duc/tion	process of moving toward
arthr	joint
-itis	inflammation
arthr/itis	inflammation of a joint
burs	small purselike sac
burs/itis	inflammation of a small sac
carp	wrist
-al	pertaining to
carp/al	pertaining to the wrist
circum	around
circum/duc/tion	process of moving around
end-	within
oste	bone
-um	presence of
end/oste/um	presence of lining within the bone
extens	straightening
-ion	process of
extens/ion	process of straightening
flex	bend
flex/ion	process of bending
kyph	humpback or hunchback
-osis	process of
kyph/osis	process of being hunchbacked
lord	bending backward, swayback
lord/osis	process of bending backward, inward curvature of the spine
meta	beyond
meta/carp/al	pertaining to beyond the wrist, bones of the palm of the hand
tars	ankle
meta/tars/al	pertaining to beyond the ankle bones of the sole of the foot
osteo/arthr/itis	inflammation of the joint
poro	pores in the bone
-sis	abnormal condition
osteo/poro/sis	abnormal condition of pores in the bone
peri-	around
peri/oste/um	presence of lining around the bone
pronat	placing face down
pronat/ion	process of being face down
rheumat	painful changes in the joints
-oid	resembling
rheumat/oid	resembling painful changes in the joints
supinat	placing on the back
supinat/ion	process of placing on the back

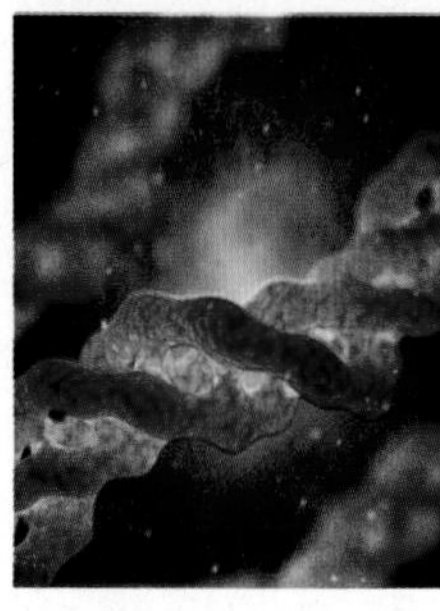

Study Tools

Workbook	Activities for Chapter 7
Online Resources	• PowerPoint presentations • Animations

REVIEW QUESTIONS

Select the letter of the choice that best completes the statement.

1. Supination is one type of
   a. extension.
   b. abduction.
   c. adduction.
   d. rotation.
2. The bones found in the skull are
   a. irregular bones.
   b. flat bones.
   c. short bones.
   d. long bones.
3. The cranium protects the
   a. lungs.
   b. brain.
   c. heart.
   d. stomach.
4. Pivot joints may be found in the
   a. vertebral column.
   b. skull.
   c. wrist.
   d. shoulder.
5. Bones are a storage place for minerals such as
   a. calcium and sodium.
   b. calcium and potassium.
   c. sodium and potassium.
   d. calcium and phosphorus.
6. The site of blood cell formation is
   a. yellow marrow.
   b. periosteum.
   c. articular cartilage.
   d. red marrow.
7. Immovable joints are found in the
   a. infant's skull.
   b. adult cranium.
   c. adult spinal column.
   d. child's spinal column.
8. Flexion means
   a. bending.
   b. rotating.
   c. extending.
   d. abduction.
9. The degree of motion at a joint is determined by
   a. the amount of synovial fluid.
   b. the number of bursae.
   c. the amount of exercise.
   d. bone shape and joint structure.
10. The bone that forms the base of the skull is the
    a. parietal.
    b. temporal.
    c. occipital.
    d. frontal.
11. The key bone of the skull is the
    a. ethmoid.
    b. frontal.
    c. parietal.
    d. sphenoid.
12. The only movable bone of the face is the
    a. lacrimatic.
    b. mandible.
    c. maxilla.
    d. palatine.
13. The central opening on the vertebrae for passage of the spinal cord is the
    a. transverse process.
    b. intervertebral disc.
    c. foramen.
    d. spinous process.
14. The shoulder girdle consists of two bones:
    a. radius and ulna.
    b. clavicle and scapula.
    c. tibia and fibula.
    d. metatarsal and tarsal.
15. The ribs attached directly to the sternum are called
    a. floating.
    b. true.
    c. false.
    d. humerus.
16. The bone of the arm located on the thumb side is called the
    a. ulna.
    b. radius.
    c. humerus.
    d. carpal.
17. The bones of the wrist are called
    a. tarsals.
    b. metatarsals.
    c. carpals.
    d. metacarpals.

18. The longest, strongest bone in the body is the
    a. humerus.
    b. tibia.
    c. femur.
    d. fibula.

19. The heel bone is known as the
    a. calcaneus.
    b. patella.
    c. fibula.
    d. talus.

20. An inflammation of the bone is known as
    a. arthritis.
    b. bursitis.
    c. osteomyelitis.
    d. osteoarthritis.

LABELING

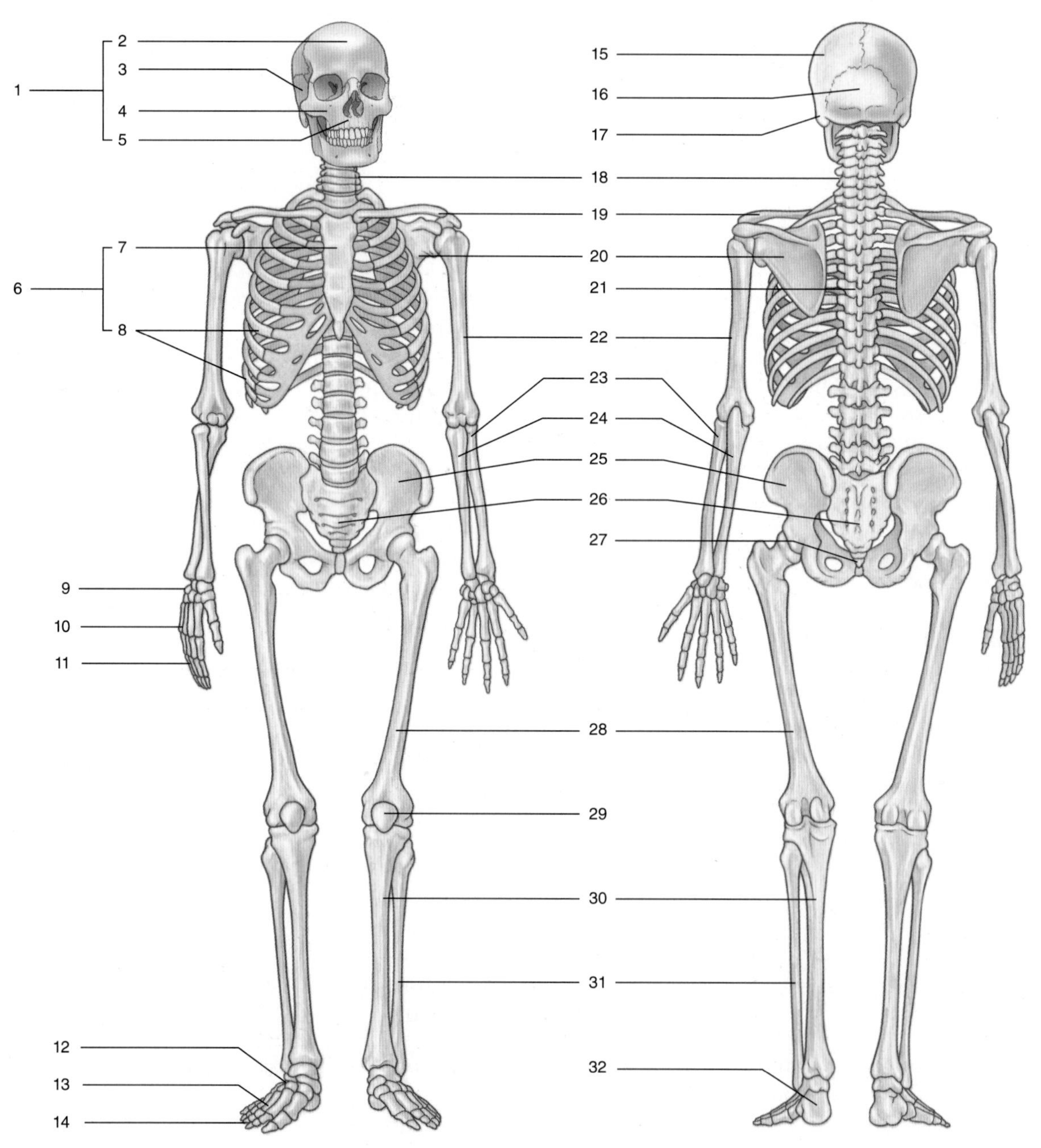

1. Label the parts of the skeleton.

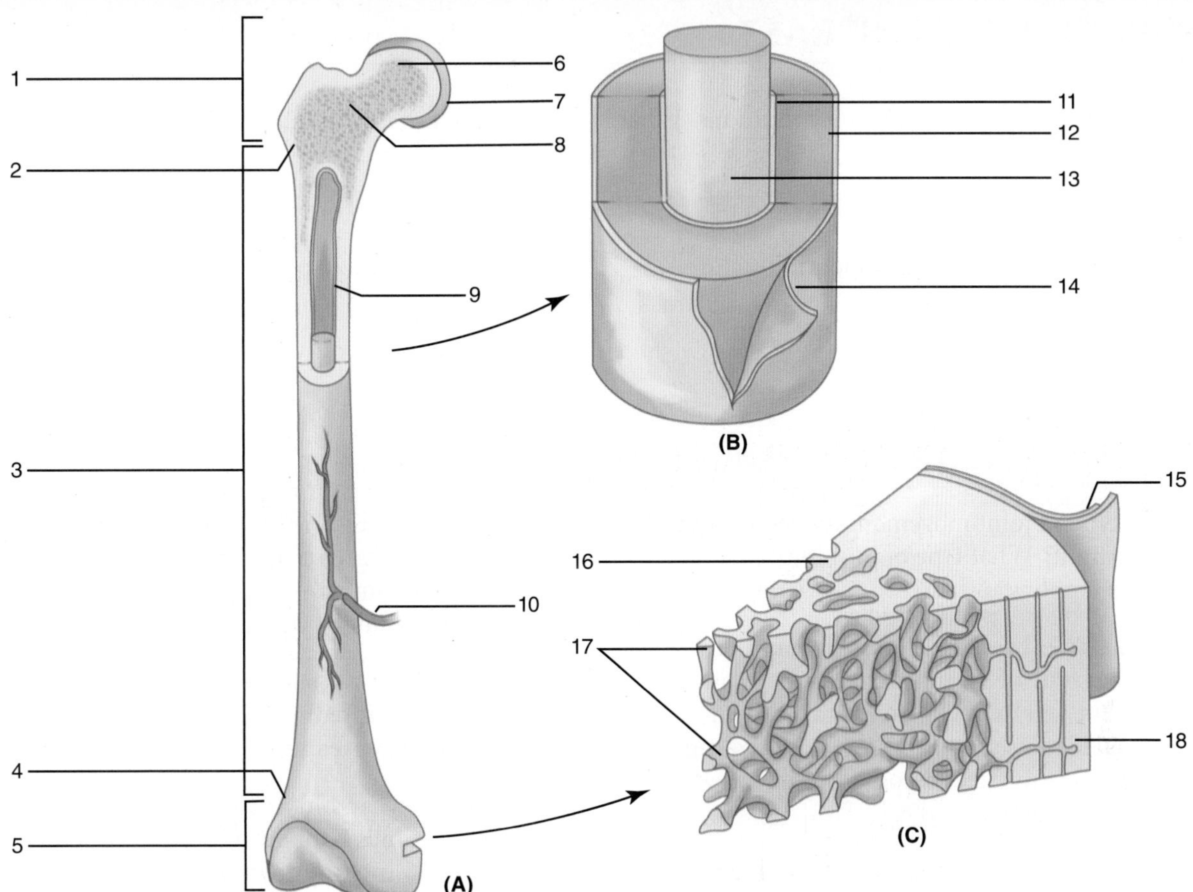

2. Label the parts of the long bone.

MATCHING

Match each term in Column I with its correct description in Column II.

	COLUMN I	COLUMN II
________	**1.** osteoarthritis	a. first cervical vertebra
________	**2.** closed fracture	b. shock absorbers
________	**3.** tendon	c. movable joint
________	**4.** endosteum	d. degeneration of articular cartilage
________	**5.** bursa	e. bone broken, skin intact
________	**6.** epiphysis	f. joint capsule
________	**7.** periosteum	g. fibrous cords that connect muscles to bone
________	**8.** atlas	h. lining of the marrow cavity
________	**9.** intervertebral disc	i. calcium and phosphorus
________	**10.** diarthrosis joint	j. end structure of long bone
		k. bone cells or osteocytes
		l. bone covering that contains blood vessels

COMPARE AND CONTRAST

Compare and contrast the following conditions. What is similar about these conditions? What is different about these conditions?

1. Osteoarthritis and osteoporosis
2. Greenstick fracture and compound fracture
3. Flatfeet and hammer toe
4. Rheumatoid arthritis and gout
5. Kyphosis and scoliosis

APPLYING THEORY TO PRACTICE

1. What type of joint movement is used to shut off a lamp light? What type of joint movement is used to comb your hair?
2. When a skier breaks the long bone of their leg, what type of diagnostic technology and treatment will be used? Investigate the initial diagnostic tools that could be used. Investigate what therapeutic technology could be used after initial treatment.
3. Your sister was in a car accident and hurt her neck; it was described as a whiplash injury. Describe a whiplash injury and the usual treatment prescribed. Investigate the initial diagnostic tools that could be used. Investigate whether using a therapeutic technology would be beneficial for whiplash.
4. Your friend has sudden, severe pain, along with swelling and redness in their big toe. What is it? Research causes and treatments.
5. Your 70-year-old uncle, Mike, says, "I don't know what's happening to me. I used to be 5 feet 10 inches; now I'm only 5 feet 8 inches." Explain to your uncle why he is becoming shorter.

CASE STUDY

Your 80-year-old grandmother, Tess, fell off a step stool while putting dishes on a shelf and was unable to get up. She activated her medical lifeline, and the emergency medical team arrived at the scene. They noticed her right leg was abducted, and she was complaining of pain in her right leg and hip. Tess was taken to the emergency department, where an X-ray revealed that the neck of her right femur was fractured. Further X-rays revealed a reduced bone mass in her right hip, femur, and vertebrae. Surgery was done to repair the hip. Your grandmother is now recuperating and having physical therapy treatment daily.

1. What organ and body system were affected by the injury?
2. Name the type of tissue involved and the cells responsible for healing.
3. The physician says she will do an open reduction to repair Tess's hip. Explain the process of an open reduction.
4. What disease condition did the X-ray of the vertebrae reveal?
5. What is the significance of your grandmother's age and sex?
6. What other body systems may be affected by the fall?
7. What will the role of the physical therapist be in your grandmother's rehabilitation?
8. Name the test that might have revealed the disease condition.
9. What measures can be taken to prevent osteoporosis?
10. What limitations will your grandmother have after her rehabilitation?

LAB ACTIVITY 7-1 Long Bones

- ***Objective:*** To describe and examine the structures that make up a long bone
- ***Materials needed:*** long bone cut in half longitudinally from butcher or prepared lab specimen, disposable gloves, textbook, paper, pencil
- ***Note:*** Your observations of bones should include location, size, shape, and any special bone markings you observe.

Step 1 Examine the long bone (if using a fresh bone, use disposable gloves).

Step 2 Identify the shaft. Describe and record its appearance.

Step 3 If using a fresh bone, peel away the periosteum.

Step 4 Locate and describe the compact bone. Record your observations.

Step 5 Locate and describe the epiphysis. Record your observations.

Step 6 Locate and describe the marrow cavity. Record your observations.

Step 7 Locate and describe the red marrow. In what part of the long bone is it located? Record your observations.

Step 8 If using a fresh bone, dispose of it in the designated container. Remove your gloves and wash your hands.

LAB ACTIVITY 7-2 Axial and Appendicular Skeleton

- ***Objective:*** To examine the size, shape, and location of the bones in the human skeleton
- ***Materials needed:*** articulated skeleton, textbook, paper, pencil
- ***Note:*** Your observations of bones should include location, size, shape, and any special bone markings you observe.

Step 1 Locate and describe the bones of the cranium and the facial bones. Record your observations.

Step 2 Locate and describe the bones of the rib cage. Note any differences. Record your observations.

Step 3 Locate the xiphoid process.

Step 4 Locate and describe the vertebrae. Compare your observations with the illustration in the textbook. Describe the types of vertebrae. Record your descriptions.

Step 5 Locate and describe the bones of the pectoral girdle. Record your observations.

Step 6 Locate the olecranon process. Record its location.

Step 7 Locate the radius and ulna bones. Which is the longer of the two bones? Record your answer.

Step 8 Count the bones located in a hand. Record your answer.

Step 9 Locate and describe the pelvic girdle. Record your observations.

Step 10 Find the acetabulum. What bone fits into this structure? Record your answer.

Step 11 Is the tibia longer or shorter than the fibula? What bone is called the shinbone? Record your answers.

Step 12 Locate the ankle bones. How many are in each foot? Record your answer.

Step 13 Locate and describe the structure of bones of the foot. Record your descriptions.

Step 14 Name three hinge joints you can locate on the articulated skeleton. Record their names and locations.

Step 15 Locate and describe two amphiarthroses joints. Record the location and features of this type of joint.

LAB ACTIVITY 7-3 Joint Motion

- ***Objective:*** Demonstrate and describe joint motion
- ***Materials needed:*** Student A, Student B, pen, paper

Step 1 Student A should observe as Student B demonstrates flexion and extension. Describe and record the action that has occurred, and name two places this occurs in the body.

Step 2 Student B should observe as Student A demonstrates abduction and adduction. Describe and record this action. Name an activity that uses these actions.

Step 3 Student A observes as Student B demonstrates circumduction. Describe and record this action. Where in the body does this type of action occur?

Step 4 Student B observes as Student A demonstrate rotation. Describe and record this action. Where in the body does this activity occur?

CHAPTER 8

Muscular System

Objectives

- Describe the function of muscle.
- Describe each of the muscle groups.
- List the characteristics of muscle.
- Describe how pairs of muscles work together.
- Explain origin and insertion of muscle.
- Locate the important skeletal body muscles.
- Describe the function of these skeletal muscles.
- Analyze the effects of torque on the human body.
- Discuss how sports training affects muscles.
- Identify some common muscle disorders.
- Define the key words that relate to this chapter.

Key Words

acetylcholine
action potential
all or none law
antagonist
belly
biceps
biomechanics
blepharospasm
contractility
dilator muscles
dystonia
elasticity
excitability
extensibility
fibromyalgia
force
heel spur
hernia
insertion
intramuscular
irritability
isometric
isotonic
motor unit
muscle fatigue
muscle spasm
muscle strain
muscle tone
muscular dystrophy
myalgia
myasthenia gravis
neuromuscular junction
origin
physiotherapy
plantar fasciitis
prime mover
rehabilitation
remission
rotator cuff injury
sarcolemma
sarcoplasm
shin splints
smooth muscle
sphincter muscles
sprain
strain
strength
synergists
tennis elbow
tetanus
torque
torticollis
triceps
vastus lateralis

The ability to move is an essential activity of the living human body that is made possible by the unique function of contractility in muscles. Muscles comprise a large part of the human body. Nearly half of body weight comes from muscle tissue. If a person weighs 140 pounds, about 60 pounds of it comes from the muscles attached to their bones. Collectively, there are more than 650 different muscles in the human body. Muscles are responsible for all body movement. They allow people to move from place to place and to perform involuntary functions, such as the heart beating and breathing. Muscles give the human body form and shape.

Major functions of the muscular system include the following:

1. Body movement: gross and fine motor movement, such as walking and talking
2. Maintenance of posture
3. Protection of the internal organs
4. Circulation: cardiac muscle and smooth muscle help the heart beat and blood flow through the body
5. Involuntary movement: smooth muscle aids in movements such as digestion
6. Maintenance of body temperature

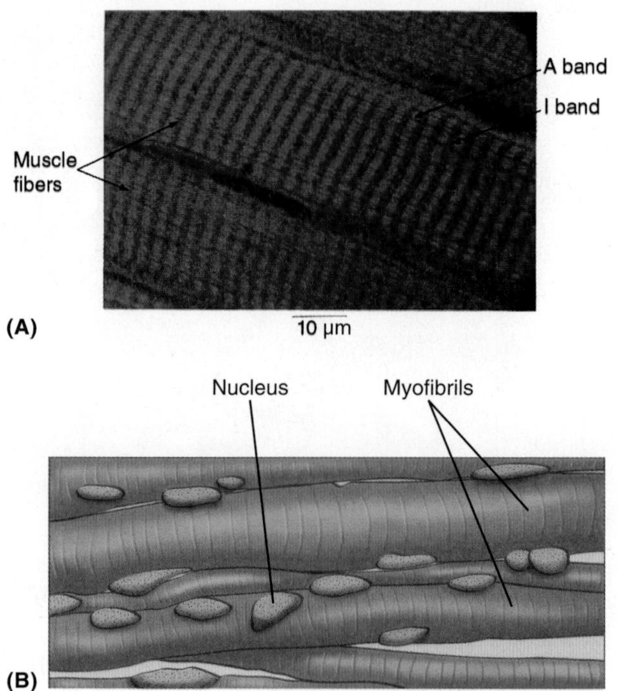

Figure 8-1 *Striated (voluntary) muscle cells: (A) microscopic view, (B) illustration*

Photo is from Atlas of Microscopic Anatomy: A Functional Approach: Companion to Histology and Neuroanatomy, by R. Bergman, A. Afifi, P. Heidger, 1999, www.vh.org/Providers/Textbooks/MicroscopicAnatomy.html. Reprinted with permission.

TYPES OF MUSCLES

Body movements are determined by one or more of the three principal types of muscles: skeletal, smooth, and cardiac muscle. These muscles are also described as striated, spindle shaped, and nonstriated (involuntary) because of the way their cells look under a microscope.

Skeletal muscles are attached to the bones of the skeleton. They are called striped, or striated, because under a microscope they show crossbandings (striations) of alternating light and dark bands running perpendicular to the length of the muscle (**Figure 8-1**). Skeletal muscle is also called voluntary muscle because it contains nerves and is under voluntary control. Each cell is multinucleate, or contains many nuclei. The cell membrane is **sarcolemma** (sahr-koh-**LEM**-ah), and the cytoplasm is **sarcoplasm** (**SAHR**-koh-plazm). Each muscle cell is known as a muscle fiber. Skeletal muscle consists of groups of these fibers bound together by connective tissue. Fascia (**FASH**-ee-ah) is the name given to a band of connective tissue that envelops, separates, or binds muscles or groups of muscle. Fascia is flexible to allow muscle movements.

The fleshy body parts are made of skeletal muscles. They provide movement to the limbs but contract quickly, fatigue easily, and lack the ability to remain contracted for prolonged periods. Blinking the eye, talking, breathing, dancing, eating, and writing are all produced by the motion of these muscles. This chapter focuses on skeletal muscle.

Smooth muscle (visceral muscle) cells are small and spindle shaped. There is only one nucleus, located at the center of the cell. They are called smooth muscles because they are unmarked by any distinctive striations. Unattached to bones, they act slowly, do not tire easily, and can remain contracted for a long time (**Figure 8-2**).

Smooth muscles are not under conscious control; for this reason, they are also called involuntary muscles. Their actions are controlled by the autonomic (automatic) nervous system. Smooth muscles are found in the walls of the internal organs, including the stomach, intestines, uterus, and blood vessels. They help push food along the length of the alimentary canal, contract the uterus during labor and childbirth, and control the diameter of the blood vessels as the blood circulates throughout the body.

Cardiac muscle is found only in the heart. Cardiac muscle cells are striated and branched, and they are involuntary (**Figure 8-3**). Cardiac cells are joined in a continuous network without a sheath separation. The membranes of adjacent cells are fused at places

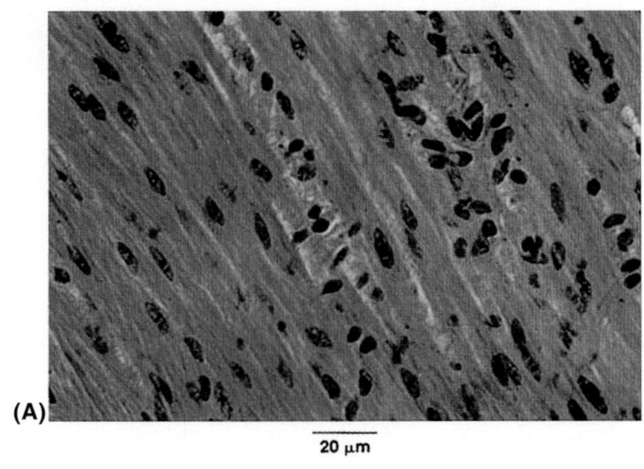

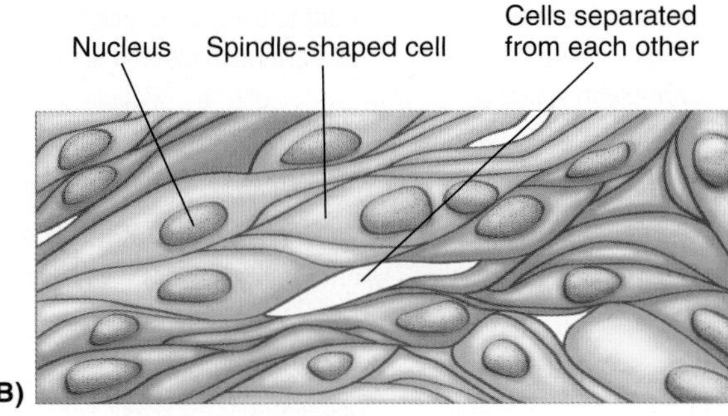

Figure 8-2 *Smooth (involuntary) muscle cells: (A) microscopic view, (B) illustration*

Photo is from Atlas of Microscopic Anatomy: A Functional Approach: Companion to Histology and Neuroanatomy, by R. Bergman, A. Afifi, P. Heidger, 1999, www.vh.org/Providers/MicroscopicAnatomy.html. Reprinted with permission.

called intercalated (in-**TER**-kah-lay-ted) discs. A communication system at the fused area will not permit independent cell contraction. When one cell receives a signal to contract, all neighboring cells are stimulated and contract together to produce the heartbeat. When the heart beats normally, it holds a rhythm of about 72 beats per minute; however, the activity of various nerves leading to the heart can increase or decrease its rate. Cardiac muscle requires a continuous supply of oxygen to function. Should its oxygen supply be cut off for as few as 30 seconds, the cardiac muscle cells start to die.

Sphincter muscles (**SFINK**-ter), or **dilator muscles**, are special circular muscles in the openings between the esophagus and stomach, and the stomach and small intestine. They are also found in the walls of the anus, the urethra, and the mouth. They open and close to control the passage of substances.

Table 8-1 summarizes the characteristics of the three major muscle types.

Media Link

View the **Types of Muscle Tissue** animation on the Online Resources.

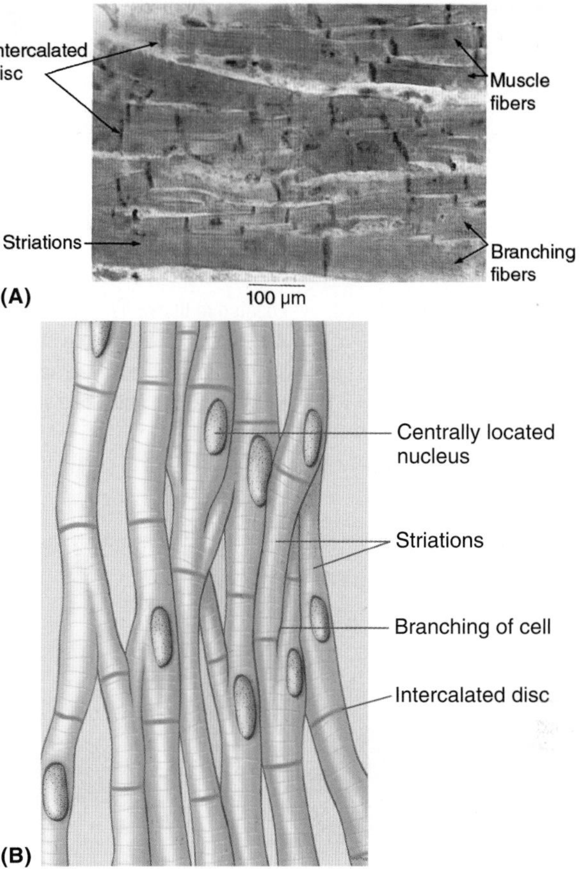

Figure 8-3 *Cardiac muscle cells: (A) microscopic view, (B) illustration*

Photo is from Atlas of Microscopic Anatomy: A Functional Approach: Companion to Histology and Neuroanatomy, by R. Bergman, A. Afifi, P. Heidger, 1999, www.vh.org/Providers/MicroscopicAnatomy.html. Reprinted with permission.

CHARACTERISTICS OF MUSCLES

All muscles, whether they are skeletal, smooth, or cardiac, have four common characteristics. Collectively, these four characteristics of muscles—contractility, excitability, extensibility, and elasticity—produce a veritable mechanical device capable of complex, intricate movements. **Contractility** is a quality possessed by no other body tissue. Contractility is a capacity for the muscle to shorten in response to a suitable stimulus. The contraction of skeletal muscles that connect a pair of bones brings the attachment points closer together, thus causing the bone to move. When cardiac muscles contract, they reduce the area in the heart chambers, pumping blood from the heart into the blood vessels. Likewise, smooth muscles surround blood vessels and the intestines, causing the diameter of these tubes to decrease upon contraction.

Excitability, or **irritability**, is a characteristic of both muscle and nervous cells (neurons). It is the capability

Table 8-1 Characteristics of Major Muscle Types

MUSCLE TYPE	LOCATION	STRUCTURE	FUNCTION
Skeletal muscle (striated, voluntary)	Attached to the skeleton and also located in the wall of the pharynx and esophagus	A skeletal muscle fiber is long, cylindrical, and multinucleated, and contains alternating light and dark striations. Nuclei are located at the edge of fiber.	Contractions occur voluntarily and may be rapid and forceful. Contractions stabilize the joints.
Smooth muscle (nonstriated, involuntary)	Located in the walls of tubular structures and hollow organs, such as in the digestive tract, urinary bladder, and blood vessels	A smooth muscle fiber is long and spindle shaped, with no striations.	Contractions occur involuntarily and are rhythmic and slow.
Cardiac (heart) muscle	Located in the heart	These are short, branching fibers with a centrally located nucleus; striations are not distinct.	Contractions occur involuntarily and are rhythmic and automatic.

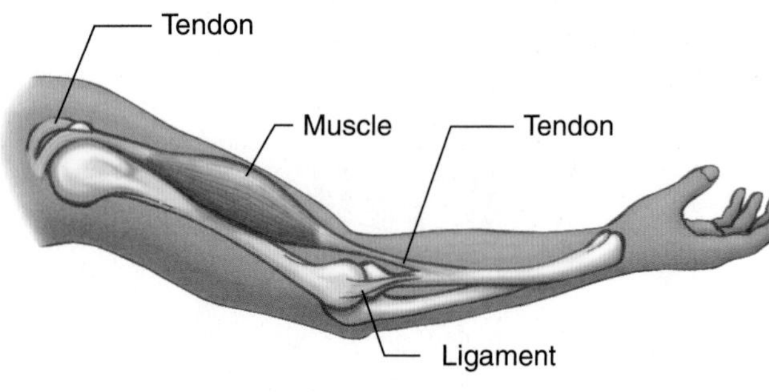

Figure 8-4 *Tendons attach skeletal muscle to bone. Ligaments join bone to bone firmly at the joint.*

of generating an **action potential**, which is the ability to respond to certain stimuli with electrical impulses.

Extensibility is the ability to be stretched. When a forearm bends, the muscles on the back of it are extended or stretched.

Muscles also exhibit **elasticity**, or the ability to return to original length when relaxing.

MUSCLE ATTACHMENTS AND FUNCTIONS

There are more than 650 different muscles in the body. For any of these muscles to produce movement in any part of the body, they must be able to exert force on a movable object. Muscles must be attached to bones for leverage (see Muscle Movement section) in order to have something to pull against. Muscles only pull, never push.

Muscles are attached to the bones of the skeleton by nonelastic, dense, fibrous connective tissue called *tendons*. The connection of one bone to another is by bands of fibrous connective tissue called *ligaments*

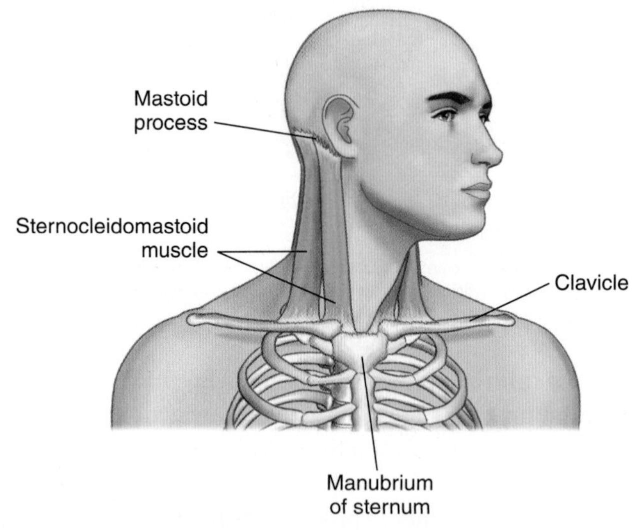

Figure 8-5 *The sternocleidomastoid muscle is named for its point of origin and insertion.*

(**Figure 8-4**). Bones are connected at joints. Skeletal muscles are attached to bones in such a way as to bridge these joints. When a skeletal muscle contracts, the bone to which it is attached will move.

Skeletal muscles are attached at both ends. They may attach to bones, cartilage, ligaments, tendons, skin, or even each other. The **origin** is the part of a skeletal muscle that is attached to a fixed structure or bone; it moves least during muscle contraction. The **insertion** is the other end, attached to a movable part; it is the part that moves most during a muscle contraction.

The sternocleidomastoid muscle is named for its origins and insertion (**Figure 8-5**). For example, the origin of the sternocleidomastoid muscle is the sternum and the clavicle; the insertion is the mastoid process of the temporal bone.

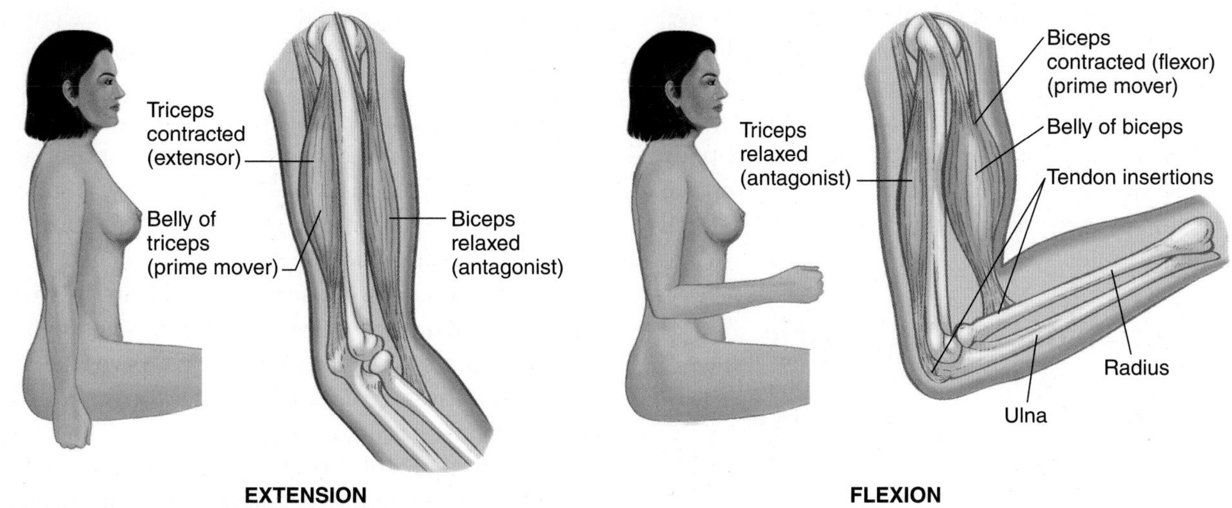

Figure 8-6 *Coordination of prime mover and antagonistic muscles*

The **belly** is the central body of the muscle (**Figure 8-6**). The muscles of the body are arranged in pairs. One, called the **prime mover**, produces movement in a single direction; the other, called the **antagonist**, pulls from the opposite direction. This arrangement of muscles with opposite actions is known as an antagonist pair.

For example, upper arm muscles are arranged in antagonist pairs (**Figure 8-6**). The muscle located on the front part of the upper arm is the **biceps**. One end of the biceps is attached to the scapula and humerus (its origin). When the biceps contracts, the scapula and humerus remain stationary. The opposite end of the biceps is attached to the radius of the lower arm (its insertion); the radius moves upon contraction of the biceps.

The muscle on the back of the upper arm is the **triceps**. Try this simple demonstration: Bend the elbow. With the other hand, feel the contraction of the belly of the biceps. At the same time, stretch the fingers out (around the arm) to touch the triceps; it will be in a relaxed state. Then, extend the forearm; feel the simultaneous contraction of the triceps and relaxation of the biceps. Now, bend the forearm halfway and contract the biceps and triceps. They cannot move because both sets of muscles are contracting at the same time. In some muscle activity, the role of prime mover and antagonist may be reversed. When a person flexes, the biceps is the prime mover, and the triceps is the antagonist. When a person extends their arm, the triceps is the prime mover, and the biceps is the antagonist.

Another group of muscles, called the **synergists**, helps steady a movement or stabilize joint activity.

Muscle Movement

The word **torque** is commonly used in conjunction with **biomechanics** (study of structure, function, and motion) when discussing movement of the human body. Torque is the application of **force** on a lever to cause rotation on an axis. In the human body, the force is the group of muscles that are pulling, the lever is the skeletal bone, and the axis is the joint that is being pulled upon.

Effects of Pressure on Muscle

Intramuscular pressure (IMP) measures the mechanical forces produced by the muscle. Intramuscular pressure can be used to determine muscular changes that can occur due to age, diseases, or other conditions. If intramuscular pressure is too high, it can be an indication of compartment syndrome, which is dangerously high pressure in a muscle; it can decrease blood flow to that muscle. An external pressure that can affect the body is barometric pressure; barometric pressure can cause muscles and tendons to expand and contract. Low pressure can cause these tissues to expand, which in turn puts excess pressure on joints and can cause pain.

SOURCES OF ENERGY AND HEAT

When muscles do their work, they not only move the body but also produce the heat the body needs. For example, to get warm on a cold day, a person can use their muscles to jump up and down and generate heat. This protects the body and keeps the body temperature from dropping too low. Human beings usually maintain their body temperatures within a narrow range (37°C to 37.7°C [98.6°F to 99.8°F]). For muscles to contract and do their work, they need energy. The major source of this energy is adenosine triphosphate (ATP) (ah-**DEN**-oh-seen tri-**FOS**-fate), a compound found in the muscle cell. To make ATP, the cell requires oxygen, glucose, and other material that is brought to the cell by the circulating blood. Extra glucose can be stored in the cell in the form of glycogen. When a muscle is stimulated, the ATP is released, thus producing the heat the body needs and the energy the muscle needs to contract. During this process, lactic acid, a by-product of cell metabolism, builds up.

Contraction of Skeletal Muscle

Movement of muscles occurs as a result of two major events: myoneural stimulation and contraction of muscle proteins. Skeletal muscles must be stimulated by nerve impulses to contract. A single muscle contraction is called a *muscle twitch*. A motor neuron (nerve cell) stimulates all of the skeletal muscles within a **motor unit**. A motor unit is a motor neuron plus all the muscle fibers it stimulates. The junction between the motor neuron's fiber (axon), which transmits the impulse, and the muscle cell's sarcolemma (muscle cell membrane) is the **neuromuscular junction**. The gap between the axon and the end of the muscle cell is known as the synaptic cleft.

When the nerve impulses reach the end of the axon, the chemical neurotransmitter **acetylcholine** (ah-**see**-till-**KOH**-leen) is released. Acetylcholine diffuses across the synaptic cleft and attaches to receptors on the sarcolemma. The sarcolemma then becomes temporarily permeable to sodium ions (Na^+), which go rushing into the muscle cell. This gives the muscle cell excessive positive ions, which upset and change the electrical condition of the sarcolemma. This electrical upset causes an action potential, or an electric current. See the discussion in Chapter 9 on the function of the nerve cell.

Skeletal muscle contraction begins with the action potential, which travels along the muscle fiber length. The basic source of energy is glucose, and the energy derived is stored in the form of ATP and phosphocreatine. The latter serves as a trigger mechanism by allowing energy transfer to the protein molecules, actin and myosin, within the muscle fibers. Once begun, the action potential travels over the entire surface of the sarcolemma, conducting the electric impulse from one end of the cell to the other. This results in the contraction of the muscle cell. The movement of electrical current along the sarcolemma causes calcium ions (Ca^{2+}) to be released from storage areas inside the muscle cell. When calcium ions attach to the action myofilaments (contractile elements of skeletal muscle), the sliding of the myofilaments is triggered and the whole cell shortens. The sliding of the myofilaments is energized by ATP.

The events that return the cell to a resting phase include the diffusion of potassium and sodium ions back to their initial positions outside the cell. When the action potential ends, calcium ions are reabsorbed into their storage areas and the muscle cell relaxes and returns to its original length. The amazing part is that this entire activity takes place in just a few thousandths of a second.

While the action potential is occurring, acetylcholine, which began the process, is broken down by enzymes on the sarcolemma. For this reason, a single nerve impulse produces only one contraction at a time. The strength of the contraction depends on a number of factors:

- The strength of the stimulus—a weak stimulus will not bring about a contraction.
- The duration of the stimulus—even if the stimulus is strong when only applied for a millisecond, it may not be long enough to be effective.
- The speed of the application—a strong stimulus applied quickly and then quickly pulled away may not have enough time to take effect.
- The weight of the load—one may be able to pick up a basket with one hand, but not a table.
- The temperature—muscles operate best at normal body temperature.

A muscle cell, when stimulated properly, contracts all the way. This is known as the **all or none law**. The muscle cell relaxes until it is stimulated by the next release of acetylcholine (**Figure 8-7**).

The Effects of Aging on the Muscular System

As an individual ages, their muscles undergo a great amount of atrophy, and there is a gradual decrease in both the number of muscle fibers and their individual bulk. Fibrous tissue replaces the muscle tissue. There is a decrease in muscular strength and endurance associated with a decrease in muscle fibers. A diminished storage of muscular glycogen may cause a loss of energy reserve, which contributes to a rapid onset of fatigue. Simple everyday activities, such as getting up from a chair, walking, or climbing stairs, can become more difficult.

Regular exercise improves strength and stamina. Generally, a man of 70 years has 50% of the strength of a man of 30 years.

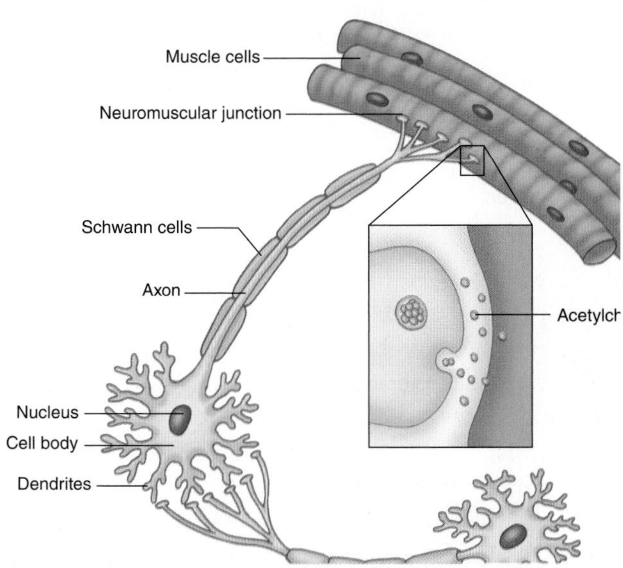

Figure 8-7 *A neuron stimulating muscle cells*

Muscle Fatigue

Muscle fatigue is caused by an accumulation of lactic acid in the muscles. During periods of vigorous exercise, the blood is unable to transport enough oxygen for the complete oxidation of glucose in the muscles. This causes the muscles to contract anaerobically, or without oxygen.

The lactic acid normally leaves the muscle, passing into the bloodstream; but if vigorous exercise continues, the lactic acid level in the blood rises sharply. In such cases, lactic acid accumulates within the muscle. This impedes muscular contraction, causing muscle fatigue and cramps. After exercise, a person must stop, rest, and take in enough oxygen through respirations to change the lactic acid back to glucose and other substances to be used by the muscle cells. The amount of oxygen needed is called the oxygen debt. When the debt is paid, respirations resume a normal rate.

Muscle Tone

To function, muscles should always be slightly contracted and ready to pull. This is **muscle tone**, which maintains body posture. Muscle tone can be achieved through proper nutrition and regular exercise. Muscle contractions may be **isotonic** or **isometric**. When muscles contract and shorten, it is called an isotonic contraction. This occurs when people walk, talk, and so on. When the tension in a muscle increases but the muscle does not shorten, it is called an isometric contraction. This occurs with exercises such as tensing the abdominal muscles. If people fail to exercise, their muscles become weak and flaccid and may atrophy (**AT**-roh-fee), or shrink from disuse. If people overexercise, their muscles will hypertrophy (high-**PER**-troh-fee), or become enlarged.

PRINCIPAL SKELETAL MUSCLES

The skeletal or voluntary muscles are the muscles that are attached to, and help move, the skeleton.

Naming of Skeletal Muscles

Muscles are named by location, size, direction, number of origins, location of origin and insertion, and action; however, not *all* muscles are named in this manner. **Table 8-2** lists examples of how some muscles are named.

Table 8-2 *Naming of Skeletal Muscles*

• Location	Frontalis—forehead
• Size	Gluteus maximus—largest muscle in buttock
• Direction of fibers	External abdominal oblique—edge of the lower rib cage
• Number of origins	Biceps—two-headed muscle in humerus
• Location of origin and insertion	Sternocleidomastoid—origin in sternum and clavicle; insertion is the mastoid process of the temporal bone
• Action flexor	Flexor carpi ulnaris—flexes the wrist
• Extensor	Extensor carpi ulnaris—extends the wrist
• Depressor	Depressor anguli oris—depresses the corner of the mouth; raises or lowers body parts

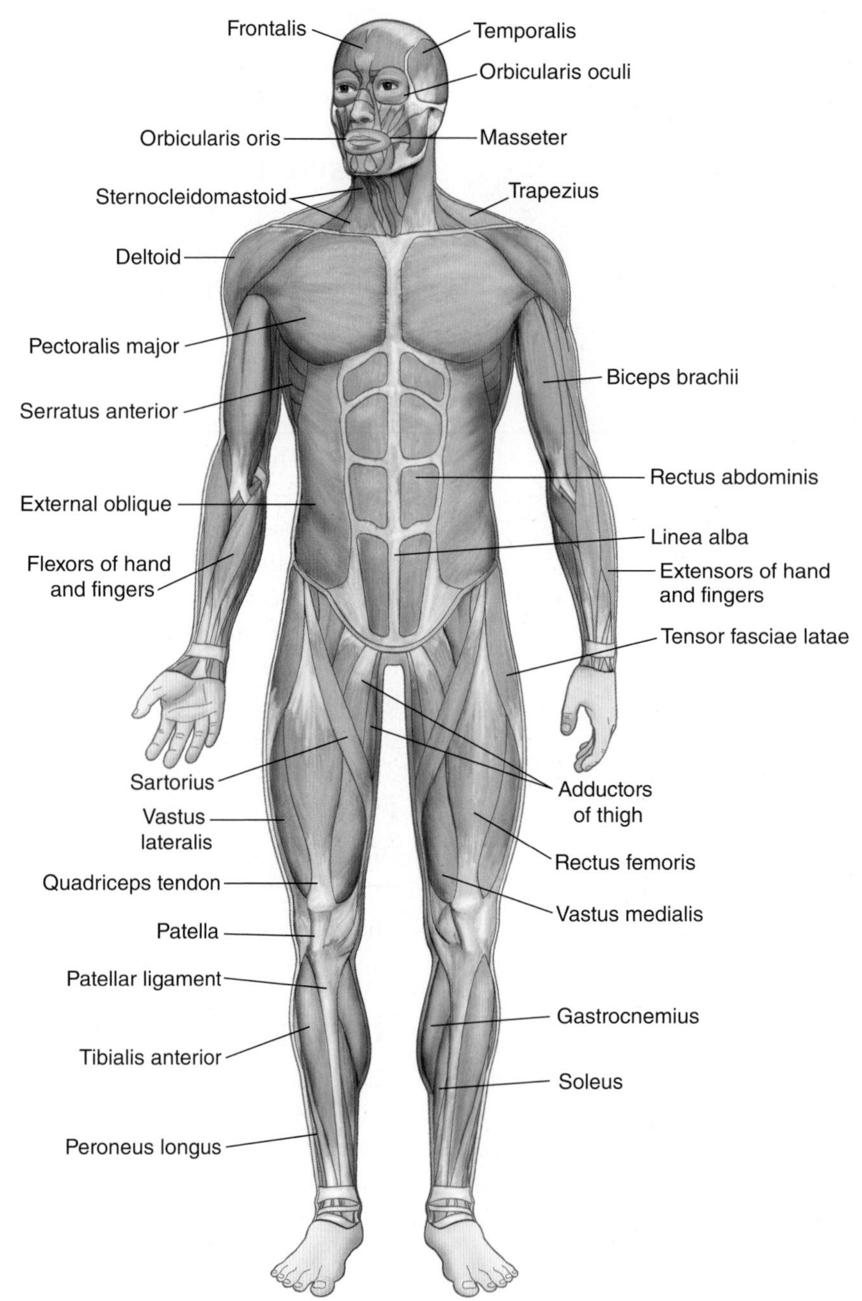

Figure 8-8 *Principal skeletal muscles of the body—anterior view*

Look at **Figure 8-8** and **Figure 8-9** and find other muscles named by location, size, direction, number of origins, and action.

There are 656 skeletal muscles in the human body. This breaks down to 327 antagonistic muscle pairs and two unpaired muscles. These two unpaired muscles are the orbicularis oris and the diaphragm. The 656 skeletal muscles can be divided and subdivided into the following muscle regions:

A. *Head muscles*

1. Muscles of expression
2. Muscles of mastication (chewing)
3. Muscles of the tongue
4. Muscles of the pharynx
5. Muscles of the soft palate

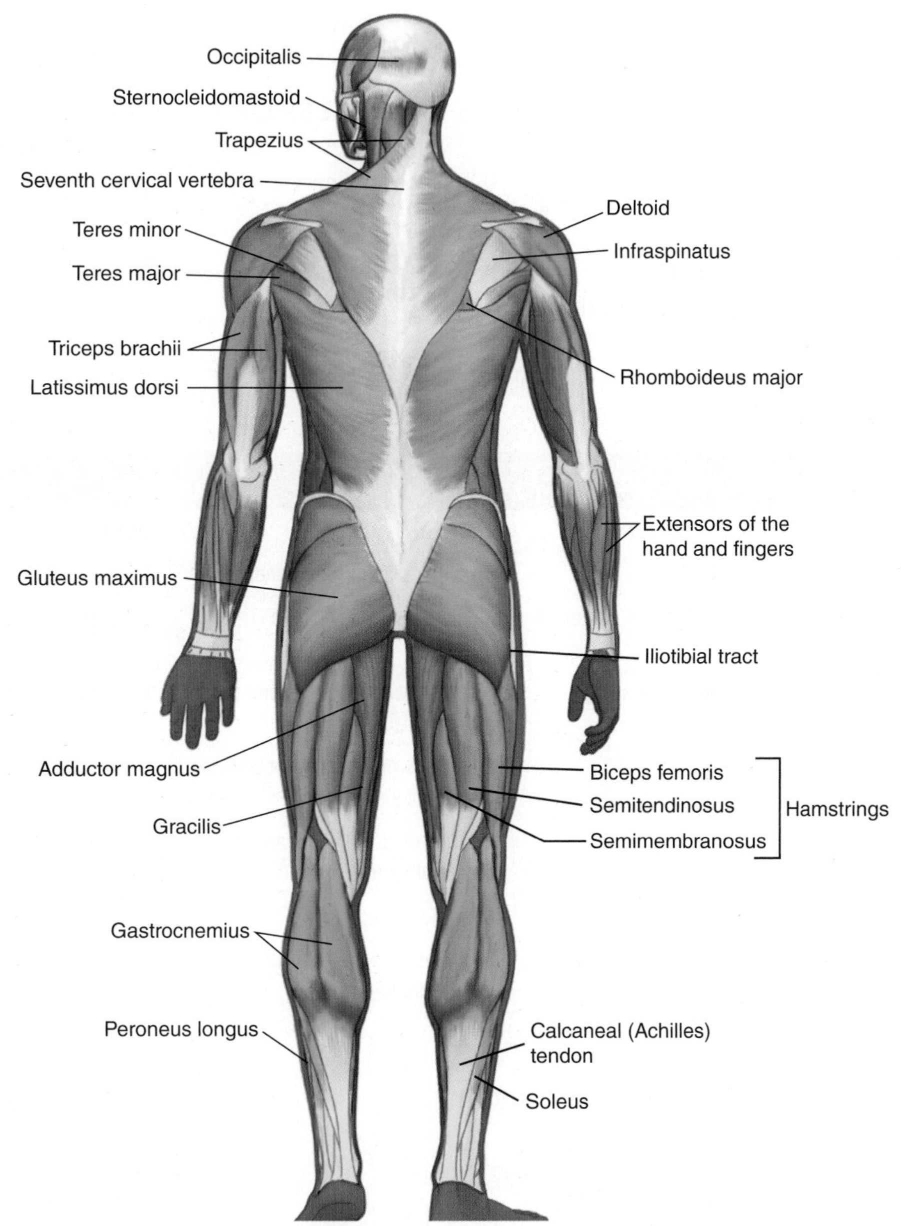

Figure 8-9 *Principal skeletal muscles of the body—posterior view*

B. *Neck muscles*

1. Muscles moving the head
2. Muscles moving the hyoid bone and the larynx
3. Muscles moving the upper ribs

C. *Trunk and extremity muscles*

1. Muscles that move the vertebral column
2. Muscles that move the scapula
3. Muscles of breathing
4. Muscles that move the humerus
5. Muscles that move the forearm
6. Muscles that move the wrist, hand, and finger digits
7. Muscles that act on the pelvis
8. Muscles that move the femur
9. Muscles that move the leg
10. Muscles that move the ankles, feet, and toe digits

Table 8-3 through **Table 8-8** list some representative skeletal muscles involved in various types of bodily movements.

MUSCLES OF THE HEAD AND NECK

Muscles of the head and neck control human facial expressions, such as anger, fear, grief, joy, pleasure, and pain. Refer to **Table 8-3** and **Figure 8-10**.

Muscles of mastication control the mandible, or lower jaw, raising it to close the jaw and lowering it to open the jaw. Refer to **Table 8-4** and **Figure 8-10**.

Muscles that move the head cause extension, flexion, and rotation. Refer to **Table 8-5** and **Figure 8-10**.

MUSCLES OF THE UPPER EXTREMITIES

Muscles of the upper extremities help move the shoulder (scapula) and arm (humerus), as well as

Did You Know?

Only 17 muscles are needed to smile, but 43 muscles are needed to frown. Every 2000 frowns create one wrinkle.

Table 8-3 *Muscles of Facial Expression*

MUSCLE	***EXPRESSION***	***LOCATION***	***FUNCTION***
Frontalis (frohn-**TAL**-is)	Surprise	On either side of the forehead	Raises eyebrow and wrinkles forehead
Depressor anguli oris (de-**PRESS**-or **ANG**-you-lye **OR**-is)	Doubt, disdain, contempt	Along the side of the chin	Depresses corner of mouth
Orbicularis oris (or-**BICK**-you-lah-ris **OR**-is)	pursed lips (kissing)	Ring-shaped muscle around the mouth	Compresses and closes the lips
Platysma (plah-**TIZ**-mah) broad sheet muscle	Horror	Broad, thin muscular sheet covering the side of the neck and lower jaw	Draws corners of mouth downward and backward
Zygomaticus major (**zye**-goh-**MAT**-ick-is)	Laughing or smiling	Extends diagonally upward from corner of mouth	Raises corner of mouth
Nasalis	Anger	Over the nasal bones	Closes and opens the nasal openings
Orbicularis oculi (**OCK**-you-lye)	Sadness	Surrounds the eye orbit underlying the eyebrows	Closes the eyelid and tightens the skin on the forehead

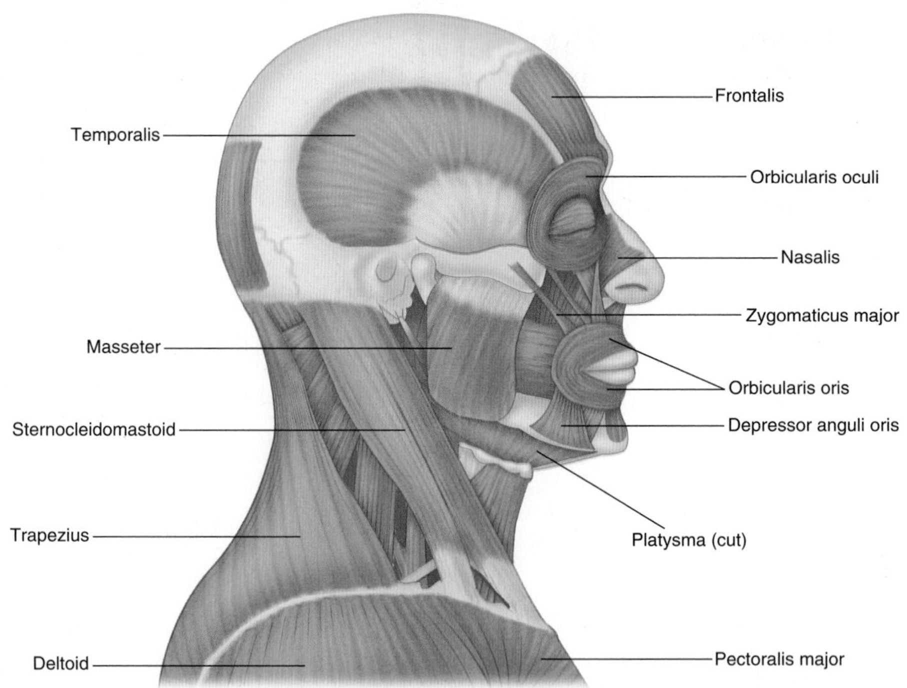

Figure 8-10 *Head and neck muscle arrangement*

Table 8-4 *Muscles of Mastication*		
MUSCLE	***LOCATION***	***FUNCTION***
Masseter (mass-**SUH**-ter)	Covers the lateral surface of the ramus (angle) of the mandible	Closes the jaw
Temporalis (tem-poh-**RAL**-is)	On the temporal fossa of the skull	Raises the jaw, closes the mouth, and draws the jaw backward

Table 8-5 *Muscles of the Neck*		
MUSCLE	***LOCATION***	***FUNCTION***
Sternocleidomastoid (two heads) (**stir**-noh-**klye**-doh-**MASS**-toyd)	Large muscles extending diagonally down sides of neck	Flexes head; rotates the head toward opposite side from muscle

the forearm, wrist, hand, and fingers. Refer to **Table 8-6** and **Figure 8-11**.

MUSCLES OF THE TRUNK

The trunk muscles control breathing and the movements of the abdomen and the pelvis. Refer to **Table 8-7** and **Figure 8-12**.

MUSCLES OF THE LOWER EXTREMITIES

Muscles of the lower extremities (**Figure 8-13**) assist in the movement of the thigh (femur), leg, ankle, foot, and toes (**Table 8-8**). The group of muscles that comprise the semitendinosus, biceps femoris, and semimembranosus muscles are called the *hamstrings.* The reason this group of muscles is so named is that these are the muscles by which a butcher hangs a slaughtered pig. The tendons of these muscles attach posteriorly to the tibia and fibula. They can be felt behind the knee. The hamstring muscle group is responsible for flexing the knee.

Media Link

View the **Hamstring Muscles** animation on the Online Resources.

Table 8-6 *Muscles of the Upper Extremities*		
MUSCLE	***LOCATION***	***FUNCTION***
*Trapezius (trah-**PEE**-zee-us)	A large, triangular muscle located on the upper surface of the back	Moves the scapula and contributes to shoulder movement; extends the head
*Deltoid (**DEL**-toyd)	A thick, triangular muscle that covers the shoulder joint	Abducts, flexes, and extends the shoulder/upper arm
*Pectoralis major (peck-toh-**RAL**-is)	Anterior part of the chest	Flexes the upper arm and helps adduct the upper arm
Serratus (sir-**AYE**-tis)	Anterior chest	Moves scapula forward and helps raise the arm
*Biceps brachii (**BYE**-seps **BRAY**-kee-eye)	Upper arm to radius	Flexes the lower arm and turns and holds the forearm
*Triceps brachii (**TRI**-seps)	Posterior arm to ulna	Extends the lower arm
Extensor and flexor carpi (**KAHR**-pye) muscle groups	Extends from the anterior and posterior forearm to the hand	Moves the hand and extends the wrist
Extensor and flexor digitorum (dij-ih-**TOHR**-um) muscle groups	Extends from the anterior and posterior forearm to the fingers	Moves and extends the fingers

*Major prime movers

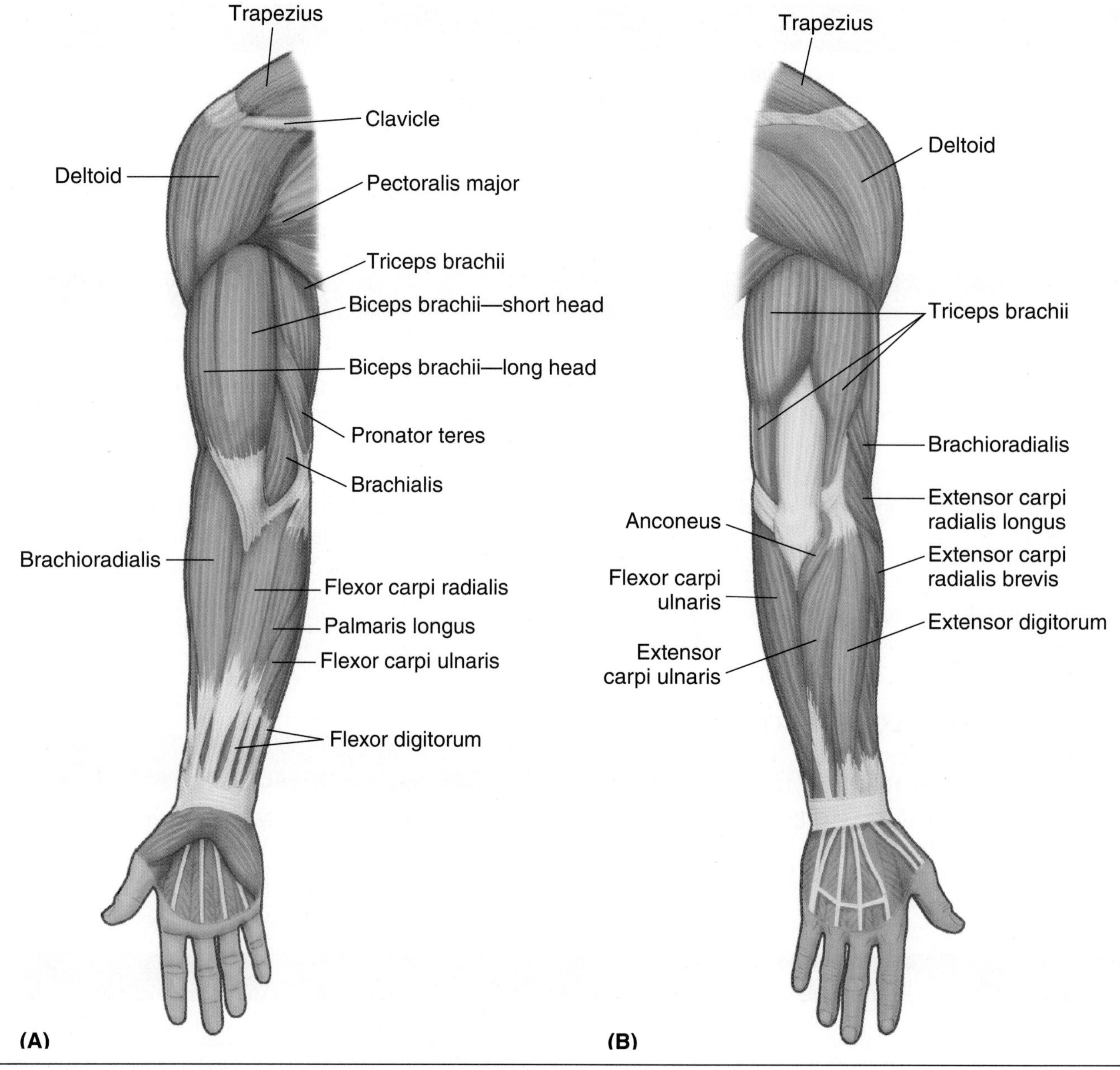

Figure 8-11 *Muscles of the upper extremity: (A) anterior view, (B) posterior view*

Table 8-7 *Muscles of the Trunk*

MUSCLE	***LOCATION***	***FUNCTION***
External intercostals (in-ter-**KOS**-talz)	Between the ribs	Raises the ribs to help in breathing
Diaphragm (**DYE**-ah-fram)	A dome-shaped muscle separating the thoracic and abdominal cavities	Contracts automatically to expand the lungs when breathing
Rectus abdominis (**REK**-tus ab-**DOM**-ih-nus)	Extends from the ribs to the pelvis	Compresses the abdomen
External oblique (oh-**BLEEK**)	Anterior inferior edge of the last eight ribs	Depresses ribs, flexes the spinal column, and compresses the abdominal cavity
Internal oblique	Directly beneath the external oblique; its fibers run in the opposite direction	Same as above

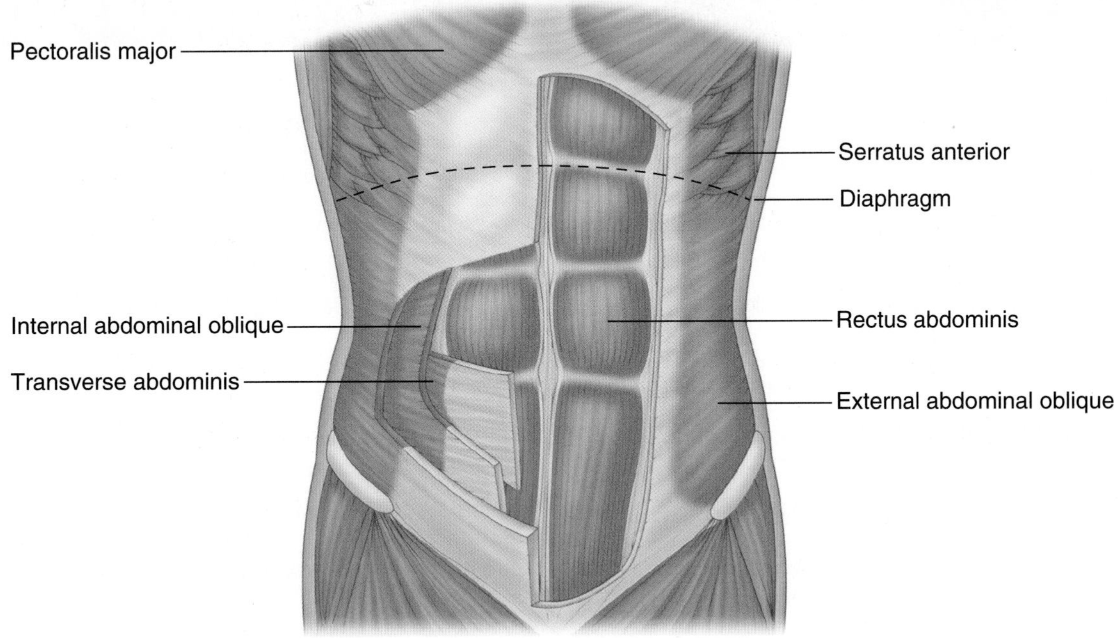

Figure 8-12 *Muscles of the trunk*

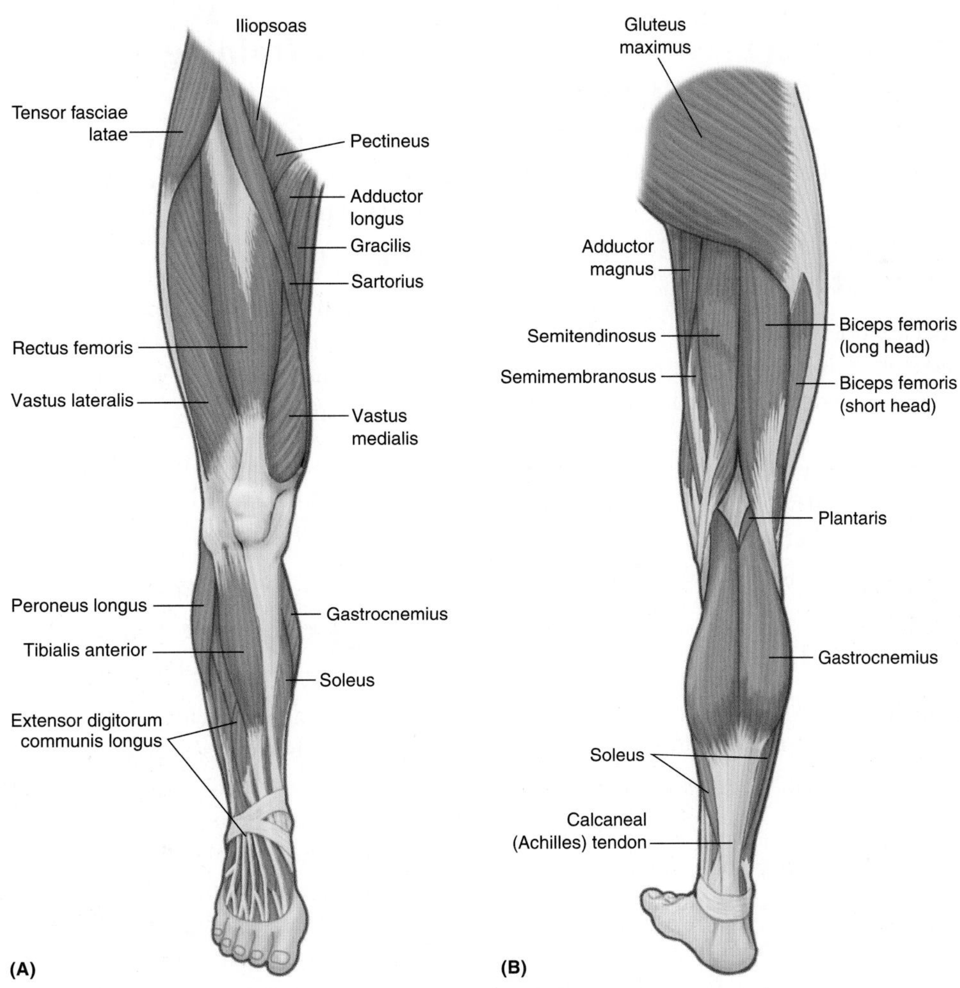

Figure 8-13 *Muscles of the lower extremity: (A) anterior view, (B) posterior view*

Table 8-8 Muscles of the Lower Extremities

MUSCLE	*LOCATION*	*FUNCTION*
*Gluteus (**GLOO**-tee-us) maximus	Muscle forms the buttocks	Extends femur and rotates it outward
Gluteus medius	Extends from the deep femur to the buttocks	Abducts and rotates the thigh
Tensor fasciae latae (**TEN**-sohr **FASH**-ee-ee LAY-tay)	A flat muscle found along the upper lateral surface of the thigh	Flexes, abducts, and medially rotates the thigh
*Rectus femoris (**REK**-tus **FEM**-oh-ris)	Anterior thigh	Flexes thigh and extends the lower leg
*Sartorius (tailor's muscle) (sar-**TOR**-ee-us)	A long, straplike muscle that runs diagonally across the anterior and medial surface of the thigh	Flexes and rotates the thigh and leg
*Tibialis anterior (**TIB**-ee-ah-lis)	In front of the tibia bone	Dorsiflexes the foot; permits walking on the heels
*Gastrocnemius (gas-trok-**NEE**-mee-us)	Calf muscle	Points toes and flexes the lower leg
*Soleus (**SO**-lee-us)	A broad, flat muscle found beneath the gastrocnemius	Extends foot
Peroneus (payr-oh-**NEE**-us) longus	A superficial muscle found on the lateral side of the leg	Extends and everts the foot and supports arches

*Major prime movers

HOW EXERCISE AND TRAINING CHANGE MUSCLES

Exercise and training will alter the size, structure, and strength of a muscle.

Size and Muscle Structure

Skeletal muscles that are not used will atrophy, and those that are used excessively will hypertrophy. Muscles that have been injured can regenerate only to a limited degree. If the muscle damage is extensive, then the muscle tissue is replaced by connective (scar) tissue. Muscles that are overexercised or overworked will have a tremendous increase of connective tissue between the muscle fibers. This causes the skeletal muscle to become tougher.

Effect of Training on Muscle Efficiency

The following will occur as a result of training:

- Improved coordination of all muscles involved in a particular activity
- Improvement of the respiratory and circulatory system to supply the needs of an active muscular system
- Elimination or reduction of excess fat
- Improved joint movement involved with that particular muscle activity

Effect of Training on Muscle Strength

Strength, or the capacity to do work, is increased by proper training. Training can have the following effects on skeletal muscles:

- Increased muscle size
- Improved antagonistic muscle coordination, or when antagonistic muscles are relaxed at the right moment and do not interfere with the functioning of the working muscle
- Improved functioning in the cortical brain region, where the nerve impulses that start muscular contraction originate

MUSCLE MASSAGE

Occasionally, a health care provider must give a patient either a total body massage or a massage to a specific body area. The correct type of massage is essential in either providing the proper **physiotherapy** or a general sense of comfort and well-being to a patient.

The health care provider must be aware of the specific skeletal muscles involved in therapeutic massage. The importance of these skeletal muscles comes from their proximity to the body's surface and their relatively large size. **Table 8-9** gives the names of these superficial skeletal muscles and their general locations. It is essential for the health care provider to be able to locate these skeletal muscles not only on the muscle diagrams, but also on the living bodies of patients with different physiques: muscular, male or female, or weigh less or more than average.

Medical Highlights

MASSAGE THERAPY AND HEALTH

Massage, or hands-on therapy, is one of the oldest forms of therapy. Ancient cultures used this type of therapy for many years. As medicine and technology advanced, this therapy seemed unscientific. Today, however, there is growing interest in the benefits of massage when it is used in addition to conventional medicine. Massage can also be used as a preventive measure to avoid injury or illness.

Massage uses positioning, hands-on pressure, and movement to promote relaxation and to loosen and increase motion in muscles.

Potential health benefits of massage include the following:

- Improvement in circulation, which aids in wound healing after surgery, improves blood pressure, and relieves edema in arms or legs
- The release of stress-reducing hormones, such as endorphins, to increase energy and reduce the risk of illness caused by chronic stress
- Fostering faster healing of strained muscles and sprained ligaments, reducing pain and swelling, and reducing formation of excessive scar tissue
- The relief of symptoms of some conditions, such as arthritis, asthma, fibromyalgia, gastrointestinal disorders, reduced range of motion, headaches, and eye strain

Terms used in massage therapy are the following:

- *Swedish massage* uses five methods to give a massage: effleurage; kneading; tapping; friction, which is the use of circular pressure with the palms of the hand; and vibration, a movement that shakes or vibrates the body.
- *Deep tissue massage* releases the chronic patterns of tension in the body through slow strokes and deep finger pressure on the contracted areas, focusing on the deeper layers of muscle tissue.
- *Effleurage* is a gliding stroke used to relax soft tissue that is applied using both hands.
- *Reflexology* is massage based on a system of points in the hands or feet that correspond, or "reflex," to all areas of the body.
- *Acupressure* and *shiatsu* use finger pressure to treat special points along the acupuncture meridians, the invisible channels of energy found in the body.
- *Sports massage* is a therapy that focuses on the muscular system relevant to a particular sport.

Some facts to consider when using massage therapy include a person's health history (massage therapy may be contraindicated in some conditions), the type of massage preferred, and the services of a trained provider. Any massage causing pain should be immediately discontinued.

Career Profile

MASSAGE THERAPIST

Massage therapists treat clients by using touch to manipulate the soft tissues of the body. This helps relieve pain, reduce stress, increase relaxation, and aid in the general well-being of an individual. Most states require a specific number of hours of in-class theory and practice, which must include anatomy and physiology, theory and practice of massage therapy, and elective subjects. This requirement varies from 500–1000 hours for each state. The National Certification Board does certification for therapeutic massage and bodywork. This group administers the national certification examination and certifies therapists who pass the exam and maintain their status through continuing education. Most states and the District of Columbia have passed laws to regulate massage therapists. The job outlook in the area is expected to increase by about 22%.

Sources: www.amtamassage.org/about/terms.html; *United States Occupational Outlook Handbook.*

Table 8-9 Skeletal Muscles Involved in Massage

NAME OF SKELETAL MUSCLE	LOCATION
Sternocleidomastoid	Side of neck
Trapezius	Back of the neck and upper back
Latissimus dorsi	Lower back
External oblique	Anterior and lateral abdomen
Deltoid	Shoulder
Biceps brachii	Anterior aspect of arm
Triceps brachii	Posterior aspect of arm
Brachioradialis	Anterior and proximal forearm
Gluteus maximus	Buttock
Tensor fasciae latae	Lateral and proximal thigh
Sartorius	Anterior thigh
Quadriceps femoris group (rectus femoris, vastus lateralis, vastus medialis, vastus intermedius)	Anterior thigh
Hamstring group (biceps femoris, semitendinosus, semimembranosus)	Posterior thigh
Gracilis	Medial thigh
Tibialis anterior	Anterior leg
Gastrocnemius	Posterior leg
Soleus	Posterior (deep) leg
Peroneus longus	Lateral leg

ELECTRICAL STIMULATION

The passage of electrical currents through skin into the body for therapeutic uses has been used for a number of years. Electrical modalities achieve their effect by stimulating nerve tissue and do not produce heat or cold.

Electrical stimulation is a commonly used modality in physical therapy and has proven to be effective for many purposes, including increasing range of motion (ROM), increasing muscle strength, reeducating muscles, improving muscle tone, enhancing function, controlling pain, accelerating wound healing, decreasing muscular atrophy, and reducing muscle spasms.

INTRAMUSCULAR INJECTIONS

Health care providers often have to administer prescribed **intramuscular** (into the muscle) medications to patients. Therefore, a working knowledge of the major skeletal muscles and the underlying anatomy of the area to be injected is needed. The most common sites for an intramuscular injection are the deltoid muscle of the upper arm, **vastus lateralis** (anterior thigh), and the dorsal gluteal area or ventral gluteal area of the buttocks.

MUSCULOSKELETAL DISORDERS

Muscular and skeletal systems work as a team to move the body. Muscular coordination is vital if a person is to perform daily functions efficiently. Injuries and diseases, which may affect the musculoskeletal system, sometimes interfere with these functions. The retraining of injured or unused muscles is a type of **rehabilitation** called therapeutic exercise.

Muscle atrophy can occur to muscles used infrequently. They shrink in size and lose muscle strength; an example is the effect of a stroke (cerebrovascular accident). The muscles are understimulated and gradually waste away. Muscle atrophy due to nerve paralysis may reduce a muscle by up to 25% of its normal size. Muscle atrophy can also be caused by prolonged bed rest or the immobilization of a limb in a cast. Muscle atrophy can be minimized by massage, electrical stimulation, or special exercise.

Muscle strain is the overstretching or tearing of a muscle. Individuals frequently develop muscle strain from lifting too much weight, lifting improperly, or using a muscle excessively. Symptoms include soreness, pain, and tenderness. In some cases, bleeding may occur. Treatment includes using the RICE method discussed in Chapter 7. A **strain** is less serious than a sprain.

A **muscle spasm**, or cramp, is a sustained contraction of the muscle. These contractions may occur because of overuse of the muscle. Night leg cramps occur during the night in the calf of the leg and sometimes in the thigh or foot. Their frequency increases with age, and their cause is unknown. This type of cramp is sometimes called a charley horse because it feels like one has been kicked by a horse. Relief may be obtained by stretching the affected leg and pointing the toes toward the knees; do this until the cramp subsides.

Myalgia (my-**AL**-jee-ah) is a term used to describe muscle pain.

Fibromyalgia (figh-broh-my-**AL**-jee-ah) is a collection of symptoms, or a syndrome. In fibromyalgia, the most definite symptom is chronic pain lasting three or more months in specific muscle points. Other symptoms may include morning stiffness, fatigue, numbness, tingling, and joint pain. Treatment is directed at pain relief and instructions to get enough sleep, exercise regularly, and utilize massage therapy. Medications to treat symptoms may include analgesics, anti-depressants, and anti-seizure drugs.

Dystonia (dis-**TOH**-nee-ah) is a condition characterized by involuntary muscle contractions that cause repetitive movements or abnormal postures. The cause is unknown. There are several different types; one is **torticollis** (tor-tih-**KOL**-is), or wryneck. This is a condition in which the muscles in the neck are affected, causing the head to turn to one side or be pulled forward or back. Another type of dystonia is **blepharospasm** (**BLEF**-ah-roh-spazm), which affects the muscle controlling the eye blink, causing increased blinking. Both eyes are usually affected. *Craniofacial dystonia* affects the muscles of the head, face, and neck; the jaws, lips, and tongue may also be affected.

Treatments with BOTOX® (botulinum toxin) for dystonia are effective. Injections of small amounts into the affected muscle can provide temporary improvement by blocking the release of acetylcholine, which normally causes muscle to contract. This results in a decrease in the muscle contractions. Several classes of drugs that affect different neurotransmitters may also be effective.

A **hernia** occurs when an organ protrudes through a weak muscle. An abdominal hernia occurs when organs protrude through the abdominal wall. Inguinal hernia occurs in the right or left iliac (inguinal) areas (see **Figure 2-5** in Chapter 2), and hiatal hernia occurs when the stomach pushes through the diaphragm. While most hernias do not affect the body in the long term, serious cases can change the structure and function of the organs; for example, if a portion of the intestine is trapped outside of the abdominal wall, oxygen may be cut off to the organ, causing the organ to die. These cases need immediate medical attention.

Obesity (oh-**BEE**-set-ee) is excessive fat accumulation. Obesity may be due to excess energy (calorie) consumption or genetic predisposition. Being obese presents a health risk to multiple body systems. It can affect the muscular system by decreasing mobility and strength and cause problems with posture and balance.

Career Profile

CHIROPRACTOR

Chiropractors, also known as chiropractic doctors, diagnose and treat patients whose health problems are associated with the body's muscular, nervous, or skeletal systems. The chiropractic approach to health care is holistic, stressing the patient's overall well-being. Chiropractors use natural, nonsurgical health treatments, such as water, heat, light, and massage. For difficulties involving the muscular system, the chiropractor manually manipulates or adjusts the spinal column.

The education required is a bachelor's degree or at least two years of college in addition to the completion of a 4-year course of study at a chiropractic college. All states require licensure. To qualify, a candidate must meet educational requirements and pass the state boards. Job prospects are excellent, and employment is expected to grow faster than the average for all other careers.

Individuals who are obese may become bedridden due to lack of muscle mass, painful joints, and overall body pain and discomfort. Treatment for obesity can include a calorie-controlled diet, exercise, and surgery.

Tetanus (**TET**-an-us), or lockjaw, is an infectious disease characterized by continuous spasms of the voluntary muscles. It is caused by a toxin from the bacillus *Clostridium tetani*, a bacterium that can enter the body through a puncture wound. This disease can be prevented by a tetanus anti-toxoid vaccine.

Muscular dystrophy (**MUS**-kyou-lar **DIS**-troh-fee) is a group of diseases in which the muscle cells deteriorate. The most common type is Duchenne's muscular dystrophy, which is caused by a genetic defect. At birth, the child appears to have normal muscle mass; as growth occurs and muscle cells die, the child becomes weak. The child loses the ability to walk between the ages of 9 and 11. Treatment includes physical therapy, respiratory therapy, orthopedic appliances, and drug therapy. The prognosis varies depending on the progression of the disease.

Myasthenia gravis (my-as-**THEE**-nee-ah **GRAH**-vis), or grave muscle weakness, occurs when the connection between the nerves and muscle is lost. It is considered an autoimmune disease; see Chapter 16 for more on autoimmune diseases. This chronic disease results in muscle weakness of the voluntary muscles that is more pronounced with activity and improves with rest. Treatment includes rest, use of cholinesterase inhibitors, and, if necessary, removal of the thymus gland. The patient may experience a **remission**, or a long-term disappearance of symptoms, but most people need to be on medication indefinitely.

A **heel spur** is a calcium deposit in the plantar fascia near its attachment to the calcaneus bone; it can be one cause of plantar fasciitis.

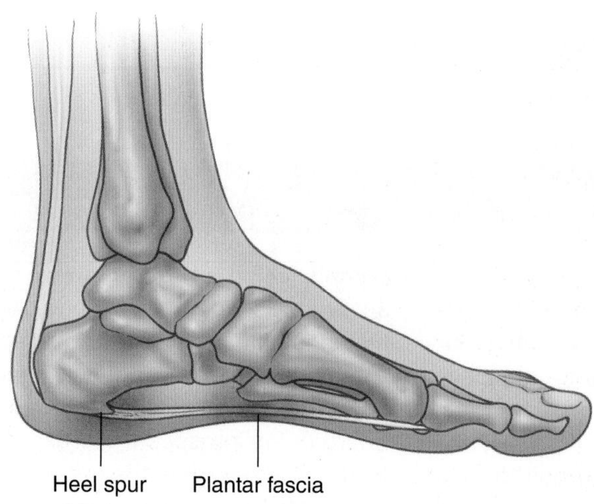

Figure 8-14 *A heel spur and plantar fasciitis, conditions that cause heel and foot pain*

Plantar fasciitis (**PLAN**-tar fas-ee-**EYE**-tis) is an inflammation of the plantar fascia on the sole of the foot. This condition causes foot or heel pain when walking or running. It is a very common condition. Treatment options include rest, massage therapy, stretching, anti-inflammatory medications, and orthotics (foot supports) (**Figure 8-14**).

Injuries and Trauma

The need to exercise can sometimes lead to excessive stress on the tendons. The tendons are cords of connective tissue that attach the muscles to bone. They become stretched and are not able to contract and return to their original place; therefore, they are more susceptible to straining and tearing. For example, the sudden, severe muscle contractions used when playing tennis can cause the tendons to tear.

A **sprain** is an injury or trauma to a joint caused by any sudden or unusual motion, such as "turning the ankle." The ligaments are either torn from their attachment to the bones or torn across, but the joint is not dislocated. A common athletic injury is to the ligaments of the knee, sometimes called an anterior cruciate ligament (ACL) injury. The sprain is accompanied by swelling and acute pain in the area and is treated with nonsteroidal anti-inflammatory drugs and the RICE treatment.

Tennis elbow, or lateral epicondylitis, occurs at the bony prominence (lateral epicondyle) on the sides of the elbow. The tendon that connects the arm muscle to the elbow becomes inflamed because of the repetitive use of the arm and underconditioning (**Figure 8-15**). Carrying luggage, playing tennis, swinging a golf club, or pounding a hammer can cause tennis elbow. Treatment consists of relief of pain and ice packs to reduce the inflammation. Sleeping on the affected arm should be avoided. Surgery is used as a last resort.

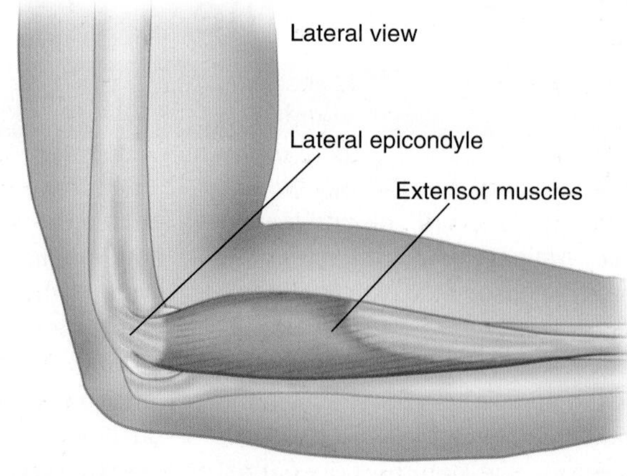

Figure 8-15 *Tennis elbow*

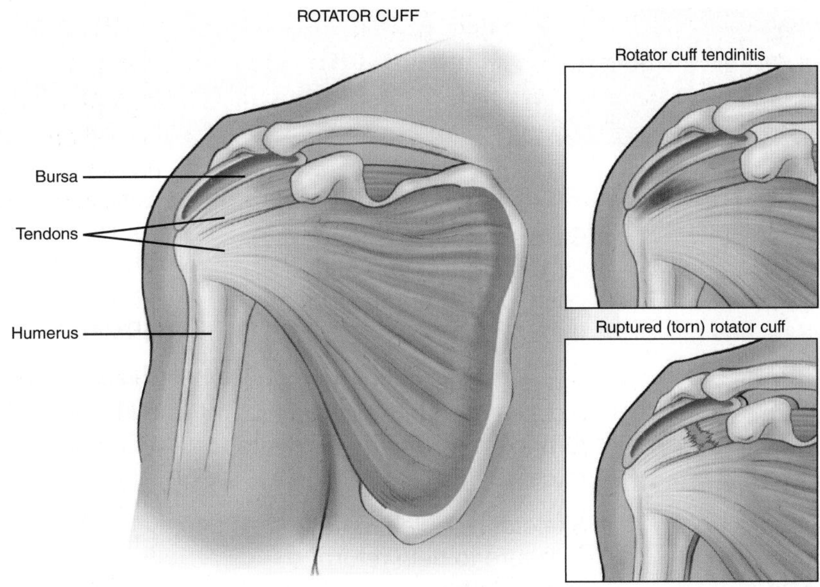

Figure 8-16 *Views of a healthy rotator cuff and rotator cuff injuries showing inflammation and tear*

Shin splints (medial tibial stress syndrome) refer to pain along the tibia due to damage of the tibialis posterior muscle and tendon. It can be caused by increased activity (muscle overuse), change in activity, flat feet, and improper footwear. Rest, ice, and anti-inflammatory drugs are the treatments for shin splints.

Rotator cuff injury is an inflammation of a group of tendons that act together to stabilize the shoulder joint. This injury can occur because of repetitive overhead swinging, such as swinging a tennis racquet or pitching a ball. The most common complaint is aching in the top and front of the shoulder. Pain increases when the arm is lifted overhead. Treatment includes RICE, physical therapy, and a steroid injection to reduce pain and inflammation. If the rotator cuff has sustained a complete tear, arthroscopic surgery may be necessary to repair the injury (**Figure 8-16**).

Media Link

View the **Shoulder Injury** animation on the Online Resources.

Career Profile

SPORTS MEDICINE/ATHLETIC TRAINING

Sports medicine refers to many different areas of exercise and sports science that relate to performance as well as care of injury. Within sports medicine are areas of specialization, such as clinical medicine, orthopedics, exercise physiology, biomechanics, physical therapy, athletic training, sports nutrition, and sports psychology. Studying sports medicine involves one of these fields, which combines medical principles and science with sports and physical performance. Many careers in sports medicine require certification. Before beginning study in these areas, find out what the requirements are for entering the profession and what type of license or certification is required in the state or workplace.

One field in sports medicine is athletic training. An athletic trainer provides a variety of services, including injury prevention; injury recognition; and immediate care, treatment, and rehabilitation after athletic trauma. Certification in this field is by the National Athletic Trainers' Association (NATA) Board of Certification. This certification identifies for the public all quality health care providers through a system of certification, adjudication, standards of practice, and continuing competency programs. Most states and places of employment require athletic trainers to be certified.

One BODY How the Muscular System Interacts With Other Body Systems

INTEGUMENTARY SYSTEM

- Muscles in skin constrict and expand the blood vessels according to the external temperature; this helps to protect the body by controlling the body's temperature.

SKELETAL SYSTEM

- Muscles are attached to bones by tendons, which then create body movement.
- Muscles store calcium, which is necessary for muscle contraction.

NERVOUS SYSTEM

- Muscles of the face show emotion.
- Muscles receive stimuli from the nervous system, which causes muscles to contract.

ENDOCRINE SYSTEM

- Growth hormone of the pituitary gland influences skeletal muscle development.

CIRCULATORY SYSTEM

- Cardiac muscle is responsible for the contraction of the heart, which pumps blood through the body.
- Smooth muscle in arteries moves blood into capillaries to bring oxygen and nutrients to cells.
- Skeletal muscle creates pressure to help return venous blood from cells with waste products for excretion.

LYMPHATIC SYSTEM

- Skeletal muscle creates pressure on lymph vessels to return lymph fluid to the heart.

RESPIRATORY SYSTEM

- The diaphragm and the external intercostal muscles assist in breathing.

DIGESTIVE SYSTEM

- Skeletal facial muscles are responsible for taking in, chewing food, and pushing food back to the esophagus.
- Smooth muscle is responsible for moving food through the digestive tracts to allow the processes of digestion and absorption to occur.
- Skeletal muscle in the anus eliminates the waste products of digestion.

URINARY SYSTEM

- Smooth muscle moves urine from the kidney to the bladder.
- Skeletal muscle of the urethra forms the voluntary sphincter muscle to eliminate urine.

REPRODUCTIVE SYSTEM

- Smooth muscle moves sperm along the reproductive tract of the male for ejaculation.
- Smooth muscle moves eggs from fallopian tubes to the uterus in females.
- Smooth muscle contractions in the uterus bring about delivery.

Medical Terminology

a-	without	**muscul**	muscle
troph	nourishment	**-ar**	pertaining to
-y	process of	**intra/muscul/ar**	pertaining to inside the muscle
a/troph/y	muscle without nourishment; muscles shrink	**my/algia**	muscle pain
bi-	two	**my**	muscle
-ceps	head	**-asthenia**	weakness
bi/ceps	two-headed muscle	**gravis**	heavy, grave
fibro	fiber	**my/asthenia gravis**	grave muscle weakness
my	muscle	**neuro**	nerve
-algia	pain	**neuro/muscul/ar**	pertaining to the nerve and muscle
fibro/my/algia	pain in the muscle fiber	**physio**	nature
gastrocnemi	calf or belly of the leg	**physio/therapy**	treatment with natural means
-us	pertaining to	**sarco**	flesh
gastrocnemi/us	pertaining to the calf of the leg	**lemma**	husk or covering
hyper-	excessive	**sarco/lemma**	covering around muscle flesh
hyper/troph/y	pertaining to excessive nourishment; causes enlargement	**plasm**	tumor
intra-	into	**sarco/plasm**	tumor of the flesh

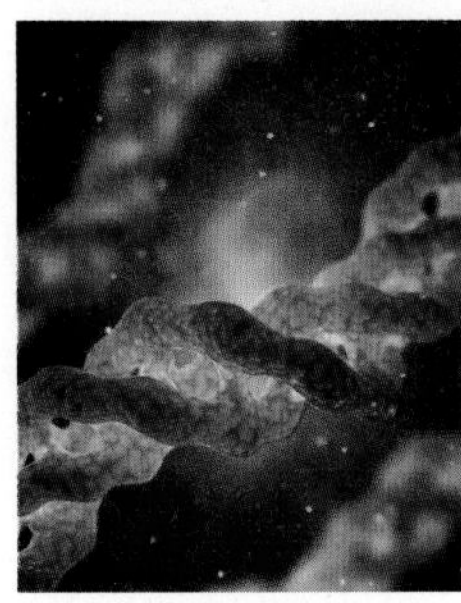

Study Tools

Workbook	Activities for Chapter 8
Online Resources	• PowerPoint presentations • Animations

REVIEW QUESTIONS

Select the letter of the choice that best completes the statement.

1. The muscular system is responsible for
   a. producing red blood cells.
   b. providing a framework.
   c. moving the body.
   d. conducting impulses.
2. Skeletal muscle is also known as
   a. involuntary.
   b. voluntary.
   c. cardiac.
   d. smooth.
3. The muscle responsible for action in a single direction is called
   a. prime mover.
   b. antagonist.
   c. synergistic.
   d. adduction.
4. The constant state of partial contraction of muscles is called
   a. muscle atrophy.
   b. muscle tone.
   c. tetanus.
   d. muscle hypertrophy.
5. The muscle used to turn the head is the
   a. trapezius.
   b. sternocleidomastoid.
   c. orbicularis.
   d. temporalis.
6. The muscle in the upper arm that is used as an injection site is the
   a. triceps.
   b. biceps.
   c. trapezius.
   d. deltoid.
7. The muscle used in breathing is the
   a. oblique.
   b. diaphragm.
   c. rectus abdominis.
   d. smooth.
8. A muscle located on the chest wall is the
   a. trapezius.
   b. frontalis.
   c. pectoralis major.
   d. rectus abdominis.
9. Muscle fatigue is caused by a buildup of
   a. glycogen.
   b. oxygen.
   c. lactic acid.
   d. ATP.
10. The muscle on the calf portion of the leg is the
    a. gastrocnemius.
    b. sartorius.
    c. rectus femoris.
    d. tibialis anterior.

COMPARE AND CONTRAST

Compare and contrast the following terms:

1. Strain and sprain
2. Biceps and triceps
3. Isotonic and isometric
4. Contractility and elasticity
5. Muscular dystrophy and myasthenia gravis

APPLYING THEORY TO PRACTICE

1. Your body feels very warm after exercising. What has happened?
2. After running up a hill, you are out of breath and have a cramp in your leg. What caused the cramp? How can you relieve it? When will your breathing return to normal?
3. You want to get a massage. What are the benefits of getting a massage?
4. Name the leg muscles you would use to kick a soccer ball or ride a skateboard.
5. A friend who was involved in an accident is wearing a leg cast. Describe how wearing the cast will temporarily affect their muscular system. How can you prevent this condition?
6. A patient comes to the office and explains that she is on the school all-star tennis team, but her entire right shoulder and arm are hurting. The physician states that her condition is known as rotator cuff injury. Explain recreational injuries to the patient and the details of this disease.
7. A baseball player "winds up" before a pitch. Describe how torque is working in the player's body. What muscles do you think are involved in winding up for a baseball pitch?
8. A colleague has recently been diagnosed as obese. Research how storing excess energy (calories), either because of lifestyle or a genetic predisposition, can affect the body.
9. Describe diseases and disorders that can be diagnosed by measuring intramuscular pressure (IMP).

CASE STUDY

Carolyn is a 36-year-old female who had an accident while skiing. After 4 months, she is still experiencing pain in her right knee and is walking with a limp. Carolyn visits the orthopedic physician and is told she needs an arthroscopic examination. While doing the arthroscopy, the physician also removes scar tissue in her knee joint. The follow-up care requires intensive physical therapy.

1. What types of treatments should Carolyn have done immediately after her injury?
2. Explain what occurs during an arthroscopic examination.
3. Identify the types of physical therapy exercises that Carolyn will need to do in order to facilitate recovery.
4. How can regular exercise help prevent future injuries?

LAB ACTIVITY 8-1 Types of Muscle Tissue

- ***Objective:*** To compare and contrast the different types of muscle tissue found in the human body
- ***Materials needed:*** prepared slides of skeletal, smooth, and cardiac muscle; microscope; textbook; paper; pencil

Step 1: Examine skeletal muscle under a microscope. How many nuclei do you see? Compare with the textbook. Record your observations, including any differences between the slide and the textbook illustration.

Step 2: Examine smooth muscle under the microscope and identify the nucleus. Compare with the textbook. Record your observations per instructions for step 1.

Step 3: Examine cardiac muscle under a microscope. What is the purpose of the intercalated discs? Compare with the textbook. Record your observations.

Step 4: List the similarities among the various types of muscle tissue.

Step 5: List the differences among the various types of muscle tissue.

8-2 Muscles of Facial Expression and of the Upper Extremity

- ***Objective:*** To observe and examine the location and function of the muscles of facial expression and of the upper extremity
- ***Materials needed:*** anatomical model, textbook, paper, pencil

Step 1: Review the illustration and table of the muscles of facial expression in the textbook.

Step 2: Locate these muscles on the anatomical model.

Step 3: Record the name and function of the muscles and where they are located using the anatomical description terminology for locations.

Step 4: With a partner, do the following activities and name the muscles used. While you are doing the activities, see how the muscles contract.

1. Show sadness.
2. Show surprise.
3. Show doubt.
4. Laugh.

Step 5: Review the illustration and table of the muscles of the upper extremity.

Step 6: With a partner, abduct the arm, flex the lower arm, extend the lower arm, move the hand, and name the muscles used. While you are doing the activity, feel each muscle contract and relax.

Step 7: Record the name and function of the muscles and where they are located using the anatomical description terminology for location.

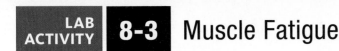

8-3 Muscle Fatigue

- ***Objective:*** To examine the function of the muscle and the effect work has on the muscle
- ***Materials needed:*** textbook, stopwatch, blood pressure cuff, paper, pencil

Step 1: Perform this activity with a lab partner. Rest your elbow on the lab table or desk, with your hand facing you. Open and close your hand, making a fist, as many times as you can in 30 seconds. Have your lab partner count and record the number of times you open and close your hand.

Step 2: Repeat this activity three more times. Have your lab partner record the number of times you can make the fist in each cycle.

Step 3: Has the number of fists you made in the 30-second period changed?

Step 4: Do you have any sign of muscle aches? Record your answer.

Switch places with your lab partner and repeat steps 1 through 4. Are there any differences in the numbers between you and your lab partner? Record your answer.

Step 5: Stand up and hold the textbook in your left arm, with the arm hanging straight down. Keep your arm straight and raise your arm with the book to shoulder level, lower it, and then count the number of times you can raise and lower the book in 30 seconds. Have your lab partner record the number of times you lifted the textbook.

Step 6: Repeat this activity three more times.

Step 7: Has the number of times you raised and lowered the textbook changed? Record your results.

Step 8: Do you have any sign of muscle aches? Record your answer.

Switch places with your lab partner and repeat steps 5 through 8. Are there any differences in the numbers between you and your lab partner? Record your answer.

Step 9: Apply a blood pressure cuff to your left arm. Inflate the blood pressure cuff to place tension on the muscle. Repeat steps 5 through 8.

Step 10: What differences occur when tension is applied to the muscle?

Step 11: What has occurred in your muscles during these activities that may have caused muscle fatigue? Write a brief paragraph to describe the events, including how the muscle returns to its normal state after muscle activity.

This lab activity can be done at home with different family members or friends. Are there differences that might be related to age, gender, or physical fitness of the individual?

LAB ACTIVITY 8-4 Force Variances of the Muscles

- ***Objective:*** To investigate the causes and effects of force variance on the muscular system
- ***Materials needed:*** textbook, stopwatch, various weights between 3–10 pounds, table, paper, pencil

Step 1: Perform this activity with a partner. Sit comfortably upright in a chair with a supportive back. With a light weight in your dominant hand (between 3–5 pounds), laterally raise your arm, palm down, until it is level with your shoulder. Hold this position until your shoulder begins to feel fatigued. Have your lab partner time you. Record your findings.

Step 2: Repeat this activity three more times, with the same arm, increasing the weight by at least one pound each time. Have your lab partner time you each round. Record your findings after each round.

Step 3: Create a line graph to communicate your results from each round.

Step 4: With your nondominant hand, use the same weight sequence as you did in steps 1 and 2. This time, rest your extended elbow on the corner of a table. Keep the palm down and be sure the weight is level with your shoulder. Hold this position until your shoulder begins to feel fatigued. Have your lab partner time you. Record your findings.

Step 5: Repeat this activity three more times, with the same arm, increasing the weight by at least one pound each time. Have your lab partner time you each round. Record your findings after each round.

Step 6: Create a line graph to communicate your results from each round.

Step 7: Compare and contrast the results of the two line graphs.

Step 8: Answer the following questions:

a. Did the time you were able to hold the weight, before fatiguing, increase or decrease with each round? Why?

b. How did the time differ between your dominant hand and your nondominant hand? What are the causes of the time difference?

CHAPTER 9

Central Nervous System

Objectives

- Describe the functions of the central nervous system.
- List the main divisions of the central nervous system.
- Describe the neuron.
- Describe the structure of the brain and spinal cord.
- Describe the functions of the parts of the brain.
- Describe the functions of the spinal cord.
- Describe disorders of the brain and spinal cord.
- Define the key words that relate to this chapter.

Key Words

afferent neurons
Alzheimer's disease
amyotrophic lateral sclerosis (ALS)
arachnid cysts
arachnoid mater
associative neurons
autonomic nervous system
axons
blood-brain barrier
brain stem
brain tumors
central nervous system
cerebellum
cerebral aqueduct
cerebral cortex
cerebral palsy (CP)
cerebral ventricles
cerebrospinal fluid
cerebrum
choroid plexus
concussion
contrecoup
coup
dementia
dendrites
diencephalon
dura mater
efferent neurons
encephalitis
encephalocele
epilepsy
essential tremor
fissures
fourth ventricle
frontal lobe
glial cells
gyri
hematoma
hydrocephalus
hypothalamus
interneurons
interventricular foramen
lateral ventricles
limbic lobe
lumbar puncture
medulla oblongata
membrane excitability
memory
meninges
meningitis
motor neuron diseases (MND)
motor neurons
multiple sclerosis (MS)
myelin sheath
neuroglia
neuron
neurotransmitters
nystagmus
occipital lobe
paraplegia
parietal lobe
Parkinson's disease
peripheral nervous system

(*continues*)

Key Words *continued*

pia mater	sensory neurons	sulci	tethered cord syndrome
poliomyelitis	spastic quadriplegia	synapse	thalamus
pons	spina bifida	synaptic cleft	third ventricle
quadriplegia	spinal cord	temporal lobe	West Nile virus (WNV)

THE NERVOUS SYSTEM

The study of body functions reveals that the body consists of millions of small structures that perform many different activities. These separate structures and functions are coordinated by the central nervous system, which helps them work together in all the complicated functions of the human body. The two main communications systems are the endocrine system and the nervous system. They send chemical messengers and nerve impulses to all structures. The endocrine system and hormonal regulation are discussed in Chapter 12. Hormonal regulations are slow, whereas neural regulation is comparatively rapid.

DIVISIONS OF THE NERVOUS SYSTEM

The nervous system can be divided into three divisions: the central, peripheral, and autonomic.

1. The **central nervous system** consists of the brain and spinal cord.
2. The **peripheral nervous system** consists of nerves of the body: 12 pairs of cranial nerves extending out from the brain and 31 pairs of spinal nerves extending out from the spinal cord.
3. The **autonomic nervous system** is part of the peripheral nervous system. It includes peripheral nerves and ganglia; ganglia are a group of cell bodies outside the central nervous system that carry impulses to involuntary muscles and glands.

When a decision is called for and action must be considered, the central and peripheral nervous systems are involved. They carry information to the brain, where it is interpreted, organized, and stored. An appropriate command is then sent to organs or muscles. The autonomic nervous system supplies heart muscle, smooth muscle, and secretory glands with nervous impulses as needed. It is usually involuntary in action.

Central Nervous System

The central nervous system (CNS) consists of the brain and spinal cord. The CNS is the most highly organized system of the body. Functions of the CNS include the following:

1. It is the communication and coordination system in the body.
   - It receives messages from stimuli all over the body.
   - The brain interprets the messages.
   - The brain responds to the messages and carries out activities.
2. The brain is also the seat of intellect and reasoning.

Neuron

A **neuron** (**NEW**-ron), or nerve cell, is specially constructed to carry out its function: transmitting a message from one cell to the next. In addition to the nucleus, cytoplasm, and cell membrane, the neuron has extensions of cytoplasm from the cell body. These extensions, or processes, are called **dendrites** (**DEN**-drytes) and **axons** (**ACK**-sonz). Each cell may have several dendrites, but only one axon. These processes, or fibers, are paths along which nerve impulses travel (**Figure 9-1**). The axon has a specialized covering called the **myelin sheath** (**MY**-eh-lin) (**Figure 9-1**). This covering speeds the nerve impulse as it travels along the axon. The myelin sheath produces a fatty substance called myelin, which protects the axon. This substance is also called white matter. The nodes of Ranvier (rahn-vee-**AY**) is an area where no myelin is present. This area is important in the conduction of a nerve impulse. Axons carry messages away from the cell body. Dendrites carry messages to the cell body.

Nervous Tissue

Nerve tissue consists of two major types of nerve cells: **neuroglia** (noo-**ROG**-lee-ah) or **glial cells** (**GLEE**-al), and neurons. Glial cells insulate, support, and protect the neurons. They are sometimes referred to as "nerve glue." Glial cells do not carry nerve impulses. Types of glial cells include the following:

1. *Astrocytes* transport nutrients to neurons and restrict what substances can enter the brain, helping to form the *blood-brain barrier.*
2. *Microglia* remove cellular debris and toxic agents.
3. *Oligodendrocytes* are CNS structures that wrap around neuronal axons, forming the myelin sheath.

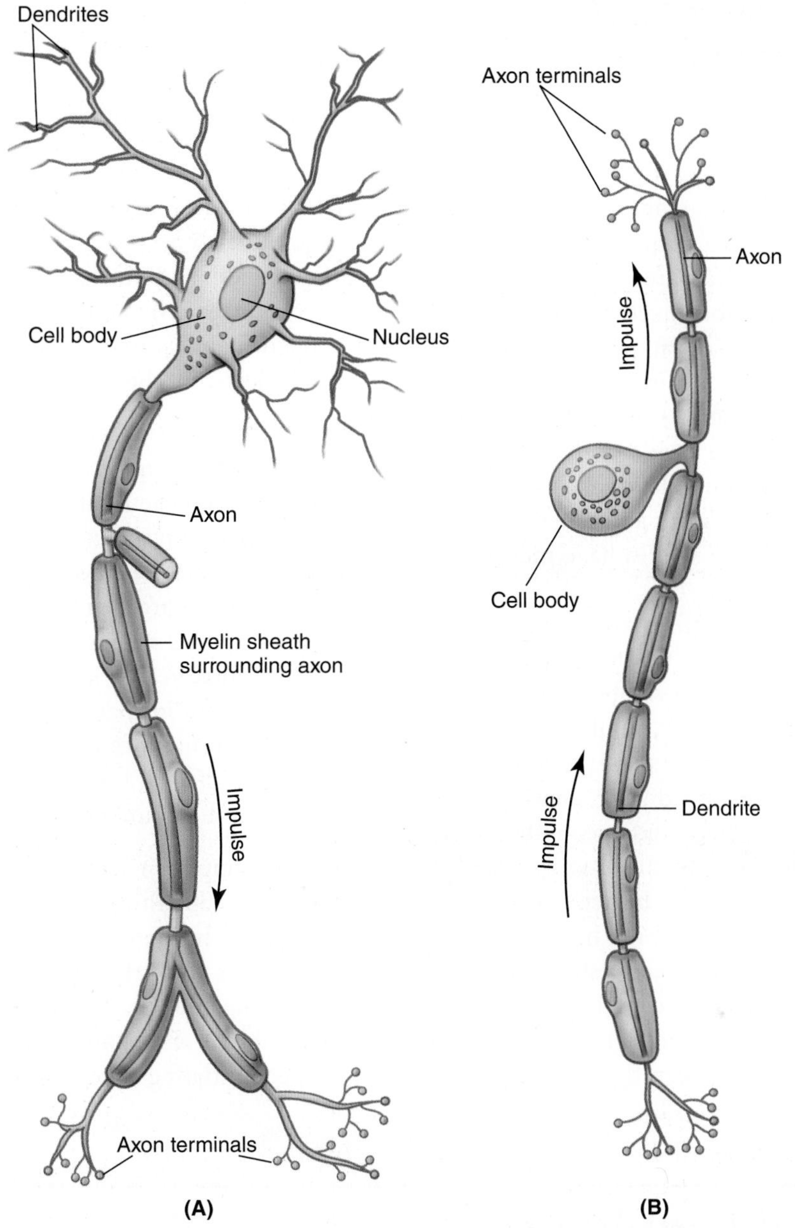

Figure 9-1 *Two types of neurons showing cell body, axon, and dendrite: (A) efferent (motor) neuron, (B) afferent (sensory) neuron*

4. *Schwann cells* are peripheral nervous system structures that wrap some neuronal axons to form the myelin sheath.

All neurons possess the characteristics of being able to react when stimulated and being able to pass the nerve impulse generated on to other neurons. These characteristics are irritability, or the ability to react when stimulated, and conductivity, which is the ability to transmit a disturbance to distant points. The dendrites receive the impulse and transmit it to the cell body, and then to the axon, where it is passed to another neuron or to a muscle or gland.

There are three types of neurons:

1. **Efferent neurons**, or **motor neurons**, carry messages from the brain and spinal cord to the muscles and glands (**Figure 9-1A**).
2. **Associative neurons**, or **interneurons**, carry impulses from the sensory neurons to the motor neurons.
3. **Afferent neurons**, or **sensory neurons**, receive stimuli from receptor sites in the sensory organs (skin, eyes, ears, nose, and taste buds) and carry messages or impulses toward the spinal cord and brain (**Figure 9-1B**).

Function of the Nerve Cell/ Membrane Excitability

Nerves carry impulses by creating electric charges in a process known as **membrane excitability**. Neurons have a membrane that separates the cytoplasm inside from the extracellular fluids outside the cell, thereby creating two chemically different areas. Each area has differing amounts of potassium and sodium ions and some other charged substances, with the inside part of the cell being more negatively charged than the outside. When a neuron is stimulated, ions move across the membrane, creating a current that, if large enough, will briefly cause the inside of the neuron to be more positive than the outside area. This state is known as *action potential*. Neurons and other cells that produce action potentials are said to have membrane excitability.

To understand how impulses are carried along nerves or throughout a muscle when it contracts, it is important to learn a little more about membrane excitability. During membrane excitability, ions cross a membrane through channels, some of which are open and allow ions to "leak," or diffuse, continuously. Other channels are called "gated" and open only during action potential. Another membrane opening is called a sodium-potassium pump, which, by active transport, maintains the flow of ions from lower to higher concentration levels across the membrane and restores the cytoplasm and extracellular fluid to their original electrical state after an action potential occurs. This action is in response to the imbalance between the cytoplasm and the extracellular fluid. When diffusion takes place, ions move from an area of greater concentration to an area of lesser concentration.

The following simplified description explains how this process works:

1. A neuron membrane is "at rest." There are large amounts of potassium (K^+) ions inside the cells but not many sodium (Na^+) ions. The reverse is true outside the cell in the extracellular fluid. Most of the open channels are for potassium to pass through, so it leaks out of the cell.

2. As the K^+ ions leave, the inside becomes relatively more negative until some K^+ ions are attracted back in and the electrical force balances the diffusion force and movement stops. The inside is still more negative, and the amount of energy between the two differently charged areas is ready to work, or carry an impulse. This state is called *resting membrane potential* (**Figure 9-2A**). The membrane is now polarized. The sodium ions are not able to move "in" because their channels are closed during the resting state; however, if a few leak in, the membrane pump sends out an equal number.

3. Now, suppose a sensory neuron receptor is stimulated by something such as a sound. This will cause a change in the membrane potential. The stimulus energy is converted to an electrical signal, and if it is strong enough, it will depolarize a portion of the membrane and allow the gated Na^+ ion channels to open, initiating an action potential (**Figure 9-2B**).

Medical Highlights

SPECIALIZED BRAIN CELLS: MIRROR NEURONS

In the early 1990s, Italian researchers studying brain activity in monkeys discovered that the neuron activity that fired up when the monkey reached for its own food also fired up when the monkey watched another monkey perform the same action. This discovery was named *mirror neurons*.

Mirror neurons are a special class of brain cells that fire not only when an individual performs an action, but also when an individual observes someone else performing the same action. Mirror neurons appear to lead a person to simulate not just another person's actions but also the intention and emotions behind those actions. The brain's mirror neuron system plays a significant role in the ability to empathize and socialize with others. Major neuron research is helping scientists reinterpret the neurological foundation of social interactions. These studies may lead to new insights into autism, schizophrenia, and other brain disorders characterized by socialization difficulties.

Implications of the brain's mirror neuron system include learning more about speech and language development, which may be useful in treating patients who have had strokes.

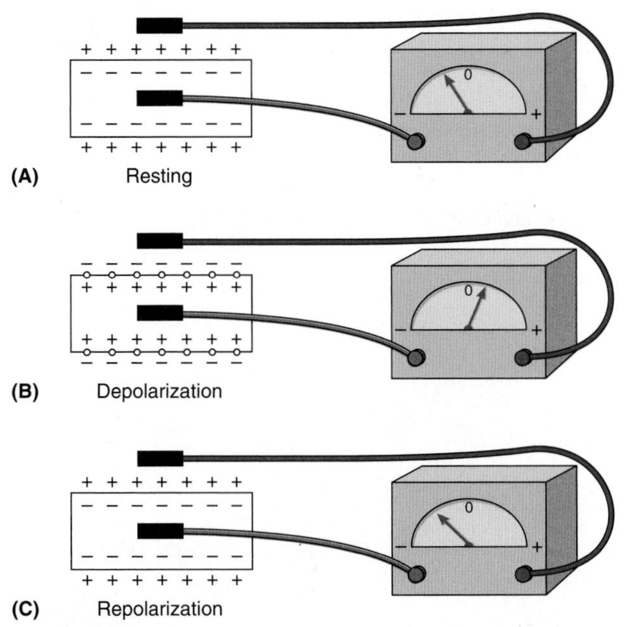

Figure 9-2 *Sequence of events in membrane potential and relative positive and negative states: (A) normal resting potential (negative inside/positive outside); (B) depolarization (positive inside/negative outside); (C) repolarization (negative inside/positive outside)*

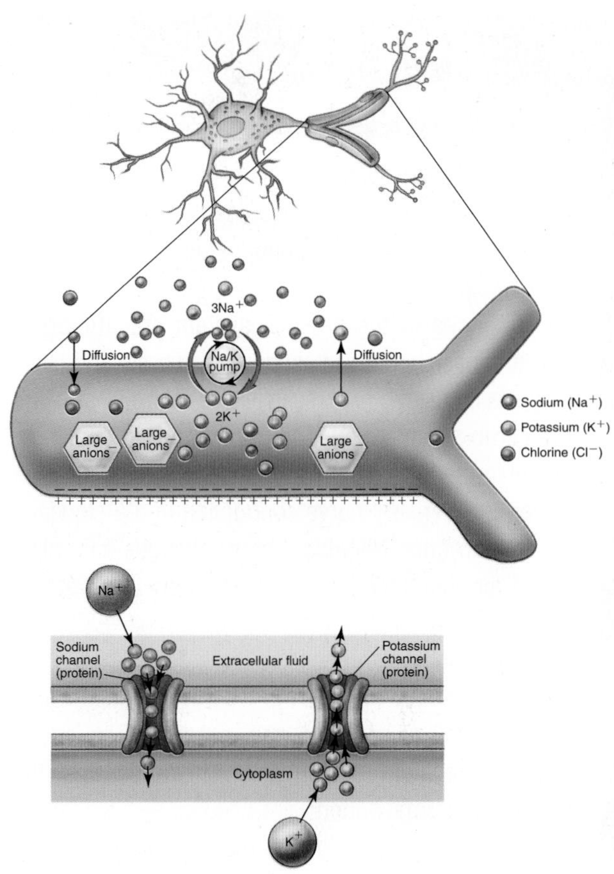

Figure 9-3 *The sodium-potassium pump of a nerve cell's membrane*

4. The Na^+ ions move through the gated channels into the cytoplasm, and the inside becomes more positive until the membrane potential is reversed and the gates close to Na^+ ions.
5. Next, the K^+ gates open and large amounts of potassium leave the cytoplasm, resulting in the repolarization of the membrane (**Figure 9-2C**). After repolarization, the sodium-potassium pump restores the initial concentrations of Na^+ and K^+ ions inside and outside the neuron (**Figure 9-3**).

This entire process occurs in a few milliseconds. When this action occurs in one part of the cell membrane, it spreads to adjacent membrane regions, continuing away from the original site of stimulation and sending "messages" over the nerve. This cycle is completed millions of times a minute throughout the body, day after day, year after year.

Synapse

A **synapse** (sin-**APPS**) is where the messages go from one cell to the next cell. The nerve cell has both an axon and a dendrite. Messages go from the axon of one cell to the dendrite of another; they never actually touch. The space between them is known as the **synaptic cleft**. The conduction is accomplished through **neurotransmitters** (new-roh-trans-**MIT**-erz). Neurotransmitters are chemical substances that make it possible for messages to cross the synapse of a neuron to a target receptor. There are between 200 and 300 neurotransmitters with a specialized function. Some neurons produce only one type of neurotransmitter, whereas others may produce two or three. The best known are acetylcholine and norepinephrine (nor-ep-ih-**NEH**-frin) (**Figure 9-4**).

An impulse travels along the axon to the end, where the neurotransmitter is released. This helps the impulse "jump," or move across, the space between, and the impulse is sent to the dendrite of the next nerve cell. The neurotransmitter between muscle cells and the nervous system cells is acetylcholine. The autonomic nervous system uses the transmitters norepinephrine and acetylcholine.

Media Link

View the **Firing of Neurotransmitters** animation on the Online Resources.

The Effects of Aging on the Nervous System

As an individual ages, there is a general slowing of nerve conduction due to a decrease in the number of functioning neurons and a degeneration of the existing nerves.

Changes in the nervous system are primarily due to diminished blood supply to the brain and loss of neurons. Slow, progressive loss in brain size in the cerebral cortex leads to impairment in thinking, reasoning, and remembering. There is a decrease in motor and sensory nerve conduction and a slowing of reaction time. Nervous system changes basically affect all voluntary and automatic nervous system functions.

Alterations also occur in the sleep patterns of aging individuals. They are more easily wakened, take longer to fall asleep, and awaken more frequently through the night. Napping, which increases with aging, is a normal pattern.

There is some evidence that continued physical and mental activity helps keep cognitive abilities sharp, such as reasoning and thinking. Disrupting, an established routine can stimulate nerve cells, enhance blood flow, and increase chemicals called neurotrophins that protect the brain cells. Activities may include using the left hand instead of the right, changing the furniture in a room, learning sign language, starting a new hobby, taking classes, or learning a new skill.

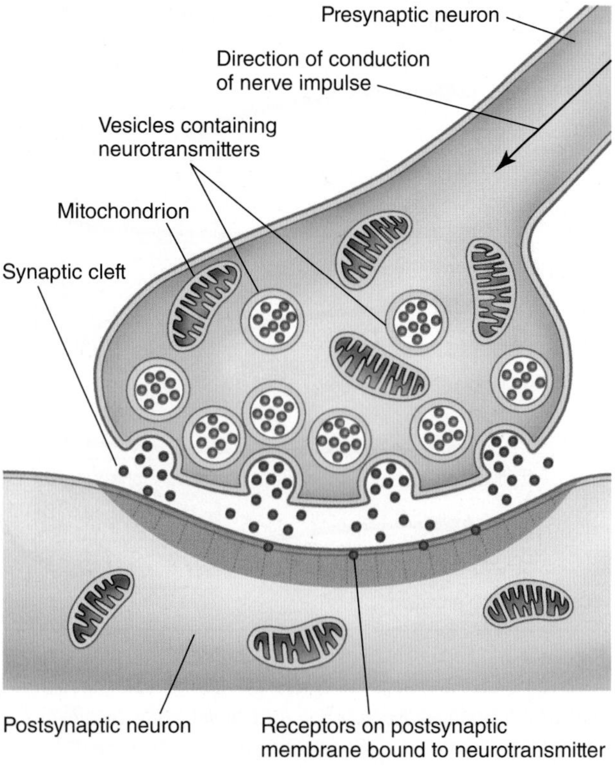

Figure 9-4 *The release of neurotransmitter molecules by a presynaptic neuron into the synaptic cleft, transmitting the nerve impulse to the postsynaptic neuron*

THE BRAIN

The adult human brain is a highly developed, complex, and intricate mass of soft nervous tissue. It weighs about 1400 grams (3 pounds) and consists of 100 billion neurons. The brain is protected by the bony cranial cavity. Further protection is afforded by three membranous coverings called **meninges** (meh-**NIN**-jeez) and the cerebrospinal fluid. The brain contains both white and gray matter. The outer cortex, known as the **cerebral cortex**, is gray. This is the highest center of reasoning and intellect; this is why some people say, "I need to use my gray matter" when trying to solve a problem. The deeper part of the cerebral cortex, which consists of myelinated nerve tracts, is called the white matter. An adequate oxygenated blood supply to the brain is critical. Without oxygen, brain damage will occur within 4–8 minutes. The brain is divided into four major parts: the cerebrum, diencephalon, cerebellum, and brain stem (**Figure 9-5**).

Did You Know?

One human brain generates more electrical impulses in one day than all the world's telephones put together.

Memory

The brain is the warehouse that stores "old" information that has been learned and packages and stores new information. This process is called **memory**. To create a memory, nerve cells are thought to form new interconnections. No one area of the brain stores all

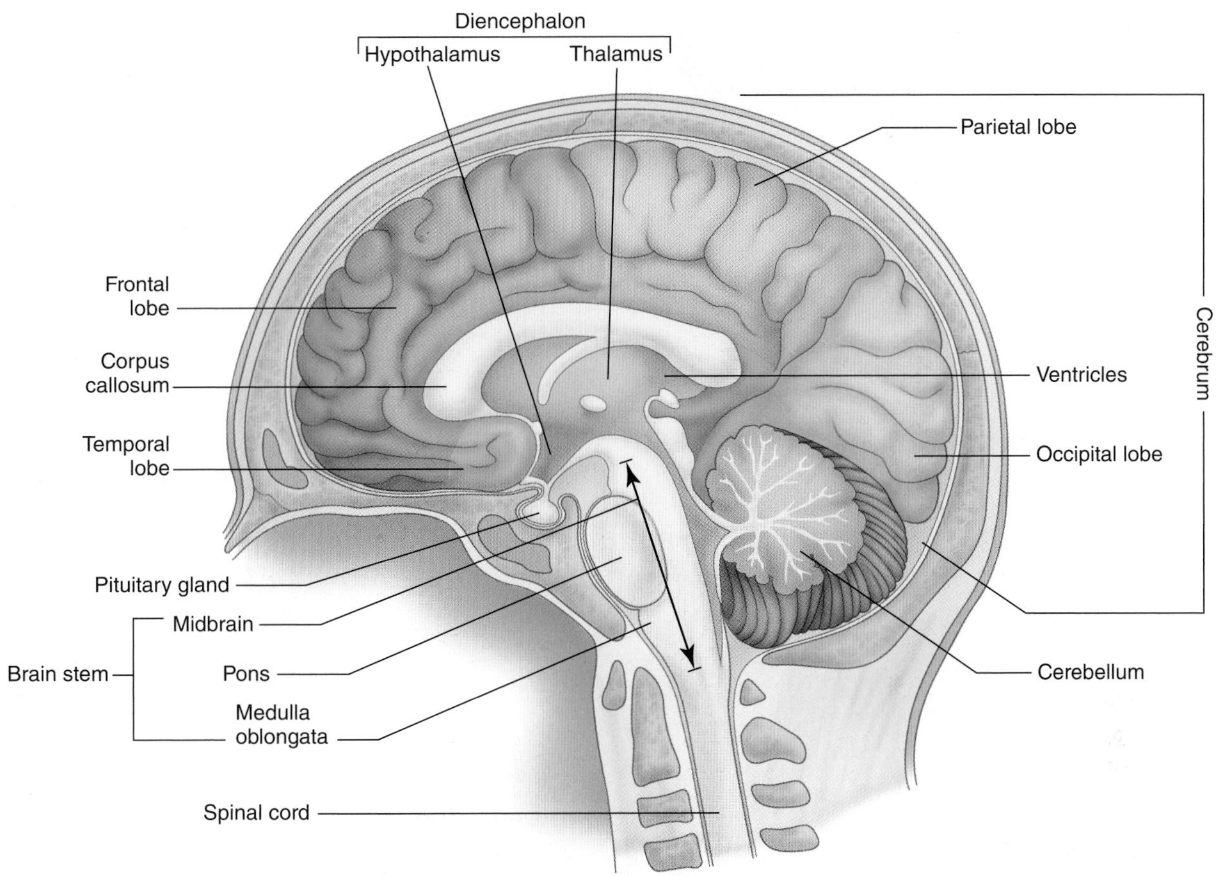

Figure 9-5 *A cross section showing the major parts of the brain: cerebrum, cerebellum, diencephalon, and brain stem*

memories because the storage site depends on the type of memory. For example, information on how to swim would be held in the motor area of the brain, whereas visual memories would be stored in the visual area of the brain. Scientists believe that the hippocampus of the limbic system acts like a front desk manager, deciding on the significance of the event and determining where in the brain the information should be stored.

Memory may be short term or long term, depending on how much attention a person gives an event, how many times they repeat an activity, and the kinds of memory associations. People frequently recall what took place during a significant event, such as the first day at school. However, a commercial may have to be seen many times before it is committed to memory.

Coverings of the Brain

The three meninges are the dura mater, arachnoid mater, and pia mater (**Figure 9-6**). The **dura mater** (**DOO**-rah **MAH**-ter) is the outer brain covering that lines the inside of the skull. This is a tough, dense membrane of fibrous connective tissue containing an abundance of blood vessels. The subdural space is between the dura mater and the arachnoid mater. The **arachnoid mater** (ah-**RACK**-noid) is the middle layer. It resembles a fine cobweb with fluid-filled spaces. Covering the brain surface itself is the **pia mater** (**PEE**-ah), consisting of blood vessels held together by fine areolar connective tissue. The subarachnoid space between the arachnoid and pia mater is filled with cerebrospinal fluid, produced within the ventricles of the brain. This fluid acts as a shock absorber as well as a source of nutrients for the brain.

Ventricles of the Brain

The brain contains four lined cavities filled with cerebrospinal fluid. These cavities are called **cerebral ventricles** (**Figure 9-7**). The ventricles lie deep within the brain. The two largest, located within the cerebral hemispheres, are known as the right and left **lateral ventricles**.

The **third ventricle** is placed behind and below the lateral ventricles. It is connected to the two lateral ventricles via the **interventricular foramen**. The **fourth**

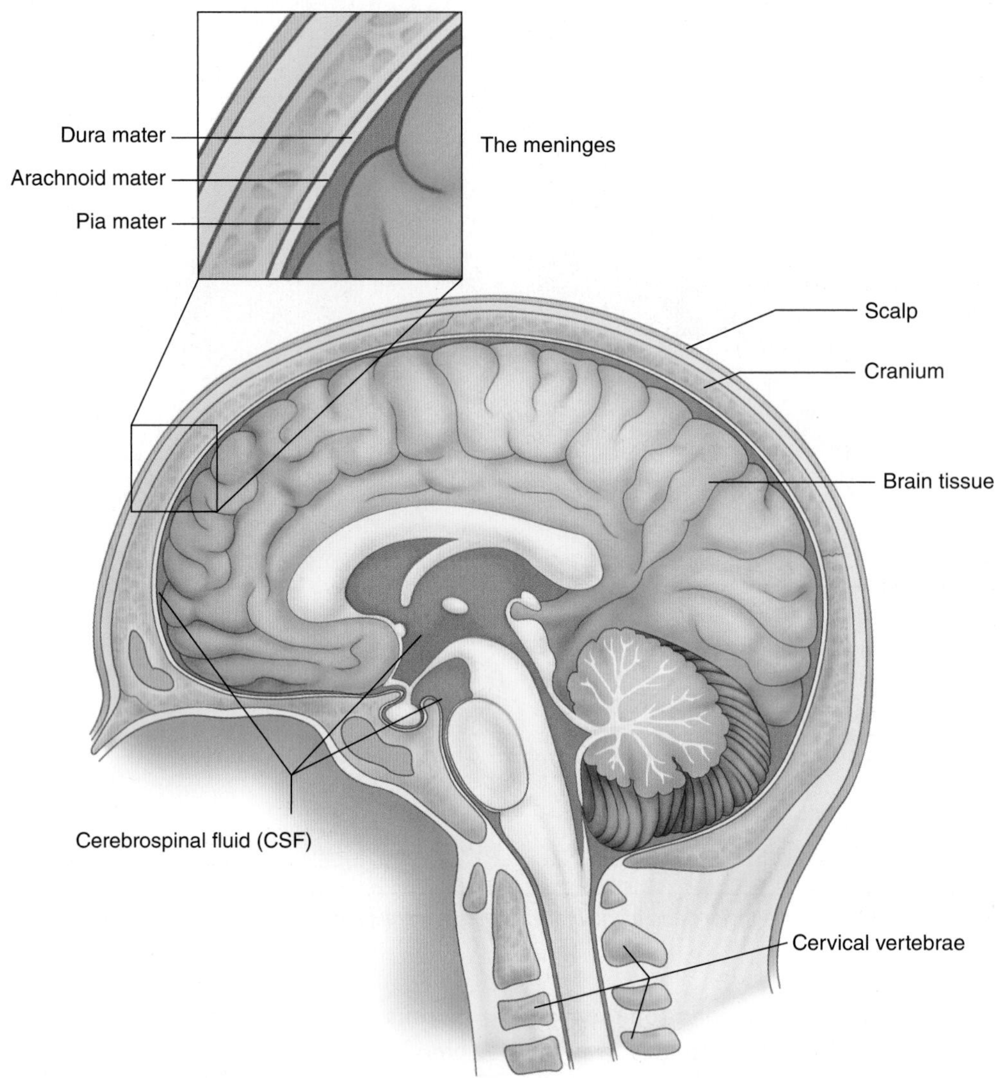

Figure 9-6 *A cross section of the brain showing the meninges and the protective coverings. The cerebrospinal fluid is shown in purple.*

ventricle is situated below the third ventricle; in front of the cerebellum; and behind the pons and the medulla oblongata, or the brain stem. The third and fourth ventricles are interconnected via a narrow canal called the **cerebral aqueduct**.

Each of the four ventricles contains a rich network of blood vessels of the pia mater, referred to as the **choroid plexus** (**KOH**-roid **PLEX**-us). The choroid plexus is in contact with the cells lining the ventricles, which helps in the formation of cerebrospinal fluid.

CEREBROSPINAL FLUID AND ITS CIRCULATION

Cerebrospinal fluid (**ser**-eh-broh-**SPY**-nal) is a substance that forms inside the four brain ventricles from the blood vessels of the choroid plexuses. This fluid serves as a liquid shock absorber, protecting the delicate brain and spinal cord. It is formed by filtration from the intricate capillary network of the choroid plexuses. The fluid transports nutrients to, and removes metabolic waste products from, the brain cells.

Choroid plexus capillaries differ significantly in their selective permeability from capillaries in other areas of the body. As a result, drugs carried in the bloodstream may not effectively penetrate brain tissue, rendering brain infections difficult to cure. This phenomenon is commonly referred to as the **blood-brain barrier**.

After filling the two lateral ventricles of the cerebral hemispheres, the cerebrospinal fluid seeps into the third ventricle through the interventricular foramen, the foramen of Monro. It flows through the cerebral aqueduct from the third ventricle to the fourth ventricle. The roof of the fourth ventricle has three openings, through which it connects with the subarachnoid spaces of the brain

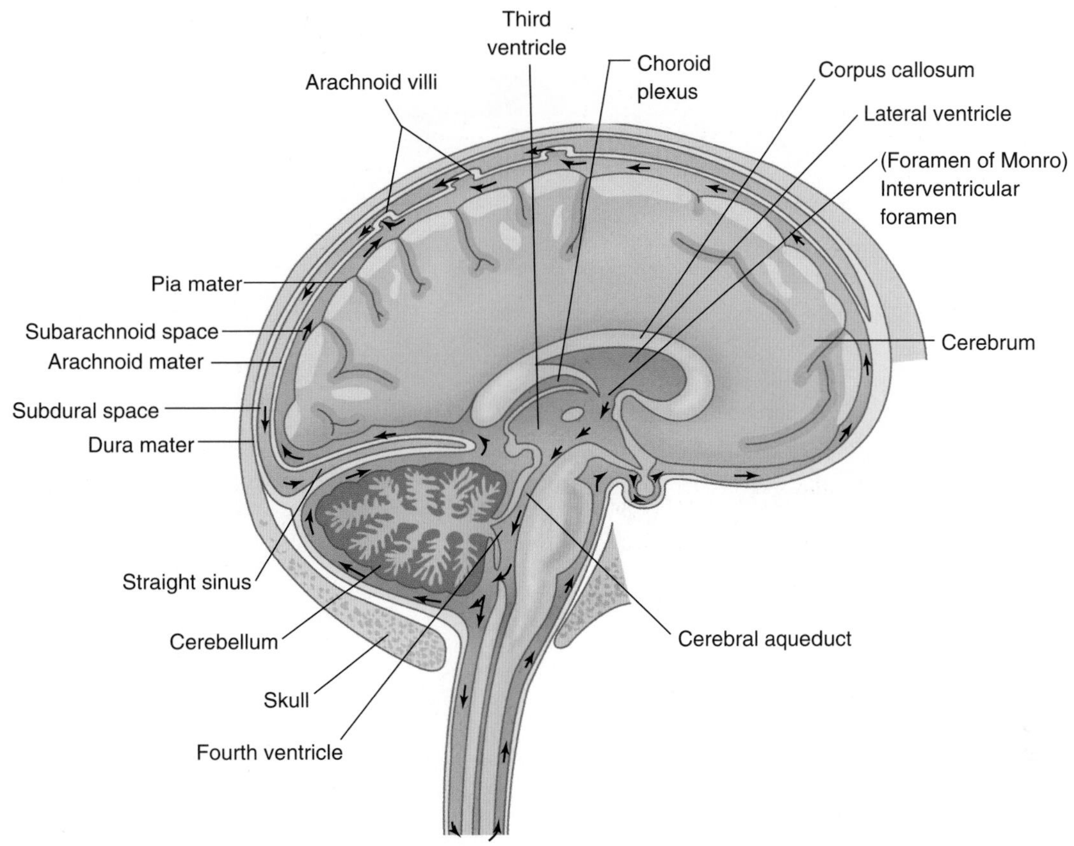

Figure 9-7 *Cerebral ventricles and circulation of the cerebrospinal fluid*

and spinal meninges (**Figure 9-7**). The subarachnoid spaces are thus filled with cerebrospinal fluid that bathes the brain, spinal cord, and meninges. Ultimately, the cerebrospinal fluid returns to the bloodstream via the venous structures in the brain, called arachnoid villi.

Cerebrospinal fluid is used by members of the health team to detect some defects and diseases of the brain. For example, inflammation of the cranial meninges quickly spreads to the meninges of the spinal cord. This leads to an increased secretion of cerebrospinal fluid, which collects in the confined bony cavity of the brain and spinal column. The accumulation of excess fluid causes headaches, reduced pulse rate, slow breathing, and partial or total unconsciousness.

Removal of cerebrospinal fluid for diagnostic purposes is accomplished with a **lumbar puncture**, also known as a spinal tap. The needle used to withdraw the cerebrospinal fluid is inserted between the third and fourth lumbar vertebrae; the spinal cord ends at the second lumbar vertebrae. Changes in the composition of the fluid withdrawn can be an indication of injury, infection, or disease. A spinal tap also serves to alleviate the pressure caused by meningitis and/or hydrocephalus.

Cerebrum

The **cerebrum** (seh-**REE**-brum) is the largest part of the brain. It occupies the whole upper part of the skull. Covering the upper and lower surface of the cerebrum is a layer of gray matter called the cerebral cortex.

The cerebrum is divided into two hemispheres—right and left—by a deep groove known as the longitudinal fissure. The cerebral surface is completely covered with furrows and ridges. The deeper furrows, or grooves, are referred to as **fissures**; the shallower ones are called **sulci** (**SULL**-sigh).

The elevated ridges between the sulci are the **gyri**, or convolutions (**Figure 9-8**). These convolutions serve to increase the surface area of the brain, resulting in a proportionately larger amount of gray matter. The arrangement of the gyri and sulci on the brain's surface varies from one brain to another. Certain fissures, however, are constant and represent important demarcations. They help localize specific functional areas of the cerebrum and divide each hemisphere into four lobes.

Each cerebral hemisphere is divided into a frontal, parietal, occipital, and temporal lobe. The middle region of

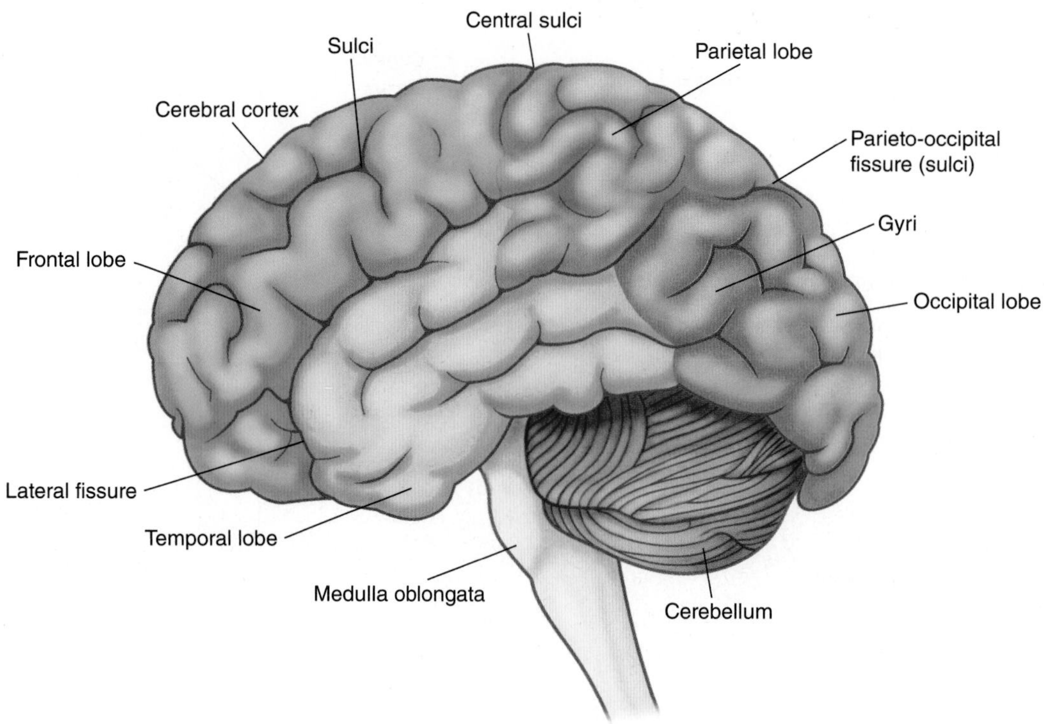

Figure 9-8 *Lateral view of the brain*

the two hemispheres is held together by a wide band of axonal fibers called the *corpus callosum*. The lobes correspond to the cranial bones by which they are overlaid (**Figure 9-8**).

The major fissures or sulci dividing the cerebral hemispheres include the following:

1. *Longitudinal fissure*—a deep groove that divides the cerebrum into two hemispheres
2. *Transverse fissure*—divides the cerebrum from the cerebellum
3. *Central sulci*—divides the frontal lobe from the parietal lobe
4. *Lateral fissure*—divides the frontal lobe and temporal lobe
5. *Parieto-occipital fissure (sulci)*—serves to separate the occipital lobe from the parietal and temporal lobe, although no definite demarcation between these two lobes exists

CEREBRAL FUNCTIONS

Each lobe of the cerebral hemisphere controls specific functions (**Figure 9-9**).

1. The **frontal lobe** forms the anterior portion of each hemisphere. It controls voluntary muscle movement. Cells in the right hemisphere activate movements that occur on the left side of the body, whereas the left hemisphere controls movements on the right side. The frontal lobe includes the area that makes speech possible. This is usually located in the left hemisphere, also called the *Broca area*. When right-sided paralysis occurs in someone who has suffered a stroke, their speech is usually affected. Damage to this area means that the person may know what to say, but they cannot vocalize the words.
2. The **parietal lobe** is located behind the frontal lobe. It receives and interprets nerve impulses from the sensory receptors for pain, touch, heat, cold, and balance. It also helps determine distance, sizes, and shapes.
3. The **occipital lobe** is located over the cerebellum; it houses the visual area controlling eyesight.
4. The **temporal lobe** is located beneath the frontal and parietal lobes. The anterior portion is occupied by the olfactory (smell) area, and the temporal lobe also contains the auditory area. The *Wernicke area* of the temporal lobe is the central language area for speech understanding and comprehension.

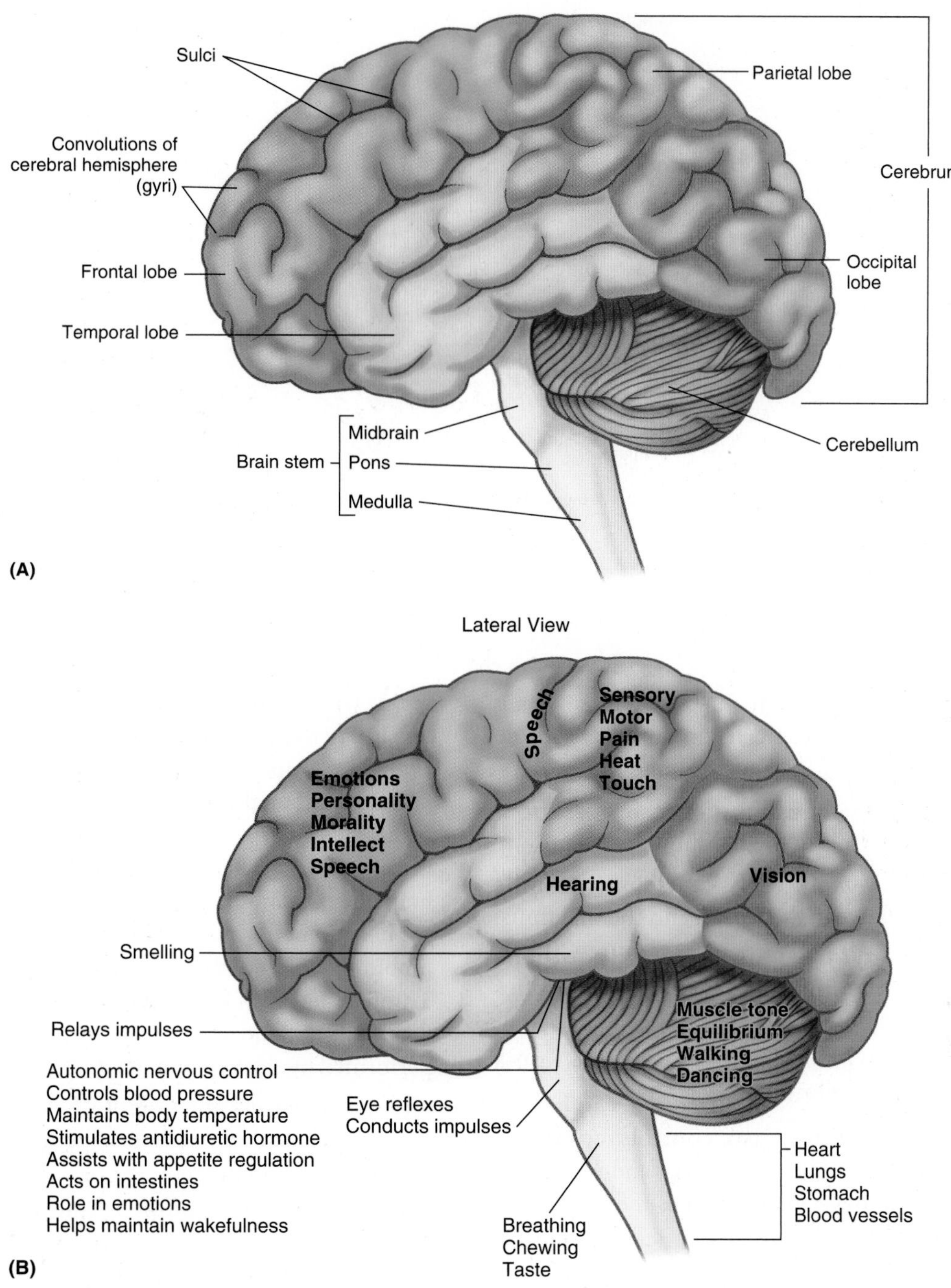

Figure 9-9 *(A) The parts of the brain; (B) areas of brain function*

5. The **limbic lobe** or system—sometimes called a lobe, other times a system—is located at the center of the brain beneath the other four lobes; it encircles the top of the brain stem (**Figure 9-10**). The limbic system influences unconscious and instinctive behaviors that relate to survival. The behavior is modified by the action of the cerebral cortex. Parts of the limbic system include the following:
   - *Olfactory bulb*—This connection explains why the sense of smell is associated with emotions, such as when smells recall happy memories.
   - *Amygdala*—This influences behavior appropriate to meet the body's needs and is associated with emotional reactions, especially fear, anxiety, and aggression.
   - *Hippocampus*—This involves memory and learning, recognizes new information, and recalls spatial relationships.

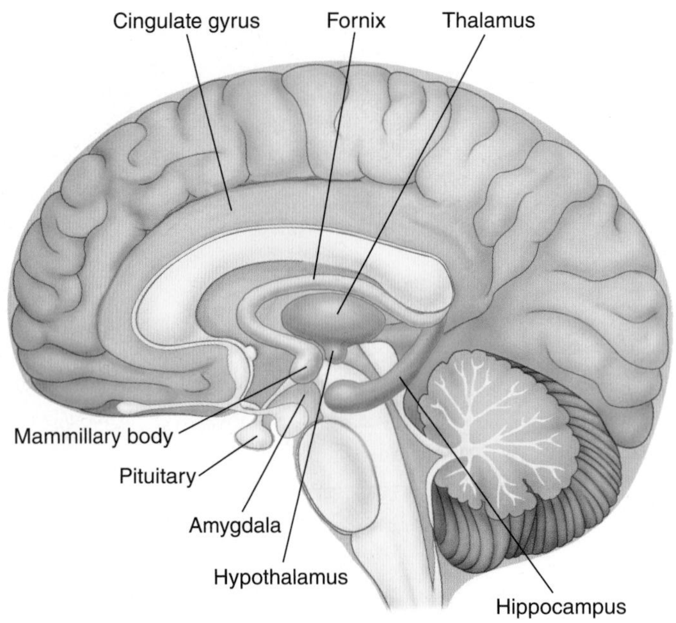

Figure 9-10 *The limbic system*

- *Parahippocampus*—This helps monitor strong emotions, like rage and fright, and plays a role in memory.
- *Fornix*—This is the pathway of nerve fibers from the hippocampus to the mammillary body.
- *Mammillary body*—This nucleus transmits messages between the fornix and the thalamus.
- *Cingulate gyrus*—This area, with others, comprises the limbic cortex, which modifies behavior and emotion.
- *Septum pellucidum*—This connects the fornix to the corpus callosum.

The cerebral cortex also controls conscious thought, judgment, memory, reasoning, and willpower. This high degree of development makes the human the most intelligent of all animals.

Diencephalon

The **diencephalon** (dye-en-**SEFF**-ah-lon) is located between the cerebrum and the midbrain (**Figure 9-11**). It contains two major structures: the **thalamus** (**THAL**-ah-mus) and the **hypothalamus** (**Figure 9-10**). The thalamus is a spherical mass of gray matter. It is found deep inside each of the cerebral hemispheres, lateral to the third ventricle. The thalamus acts as a relay station for incoming and outgoing nerve impulses. It receives direct or indirect nerve impulses from the various sense organs of the body, with the exception of olfactory sensations. These nerve impulses are then relayed to the cerebral cortex. The thalamus also receives nerve impulses from the cerebral cortex, cerebellum, and other areas of the brain. Damage to the thalamus may result in increased sensitivity to pain or total loss of consciousness.

The hypothalamus lies below the thalamus. It forms part of the lateral walls and floor of the third ventricle. A bundle of nerve fibers connects the hypothalamus to the posterior pituitary gland, the thalamus, and the midbrain. The main function of the hypothalamus is to keep the body in homeostasis. The hypothalamus is part of the limbic system and is considered the "brain" of the brain. Through the use of feedback, the hypothalamus stimulates the pituitary gland to release its hormones. Vital functions performed by the hypothalamus are the following:

1. *Autonomic nervous control*—Regulates the parasympathetic and sympathetic systems of the autonomic nervous system
2. *Cardiovascular control*—Controls blood pressure, regulating the constriction and dilation of blood vessels and the beating of the heart
3. *Temperature control*—Helps in the maintenance of normal body temperature (37°C or 98.6°F)
4. *Appetite control*—Assists in regulating the amount of food ingested. The "feeding center," found in the lateral hypothalamus, is stimulated by a feeling of hunger that prompts a person to eat. The "satiety center" in the medial hypothalamus becomes stimulated when a person has eaten enough.
5. *Water balance*—Within the hypothalamus, certain cells respond to the osmotic pressure of the blood. When osmotic pressure is high, due to water deficiency, the antidiuretic hormone (ADH) is secreted. A "thirst area" is found near the satiety area. It becomes stimulated when the blood's osmolality, or

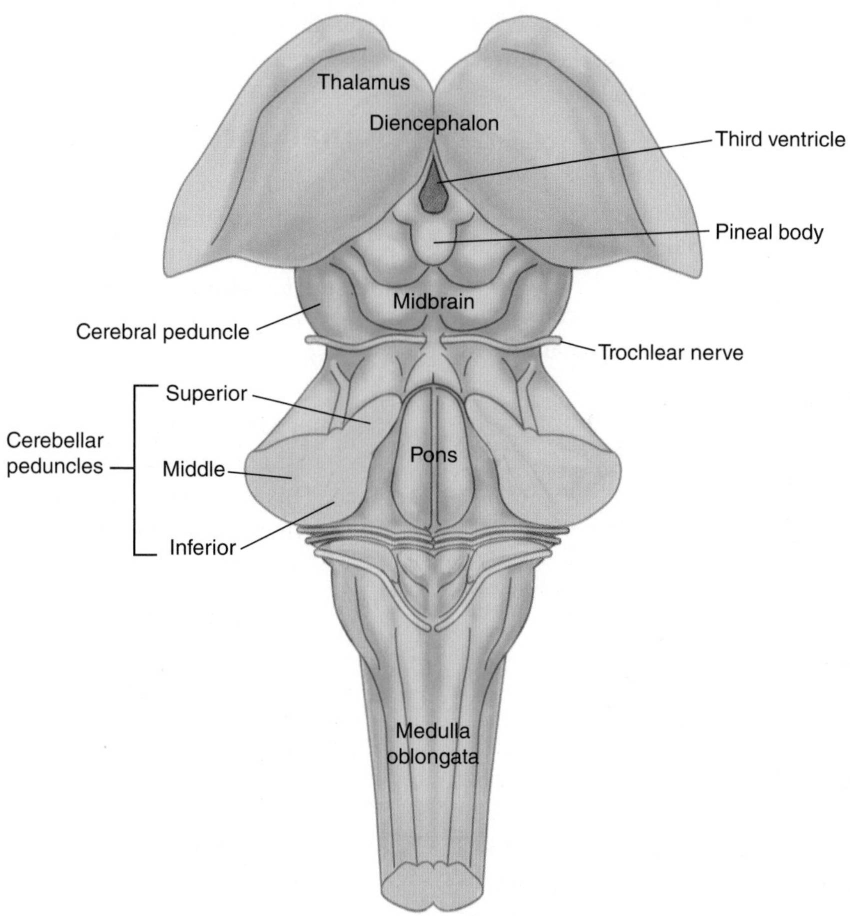

Figure 9-11 *The brain stem*

the measure of the blood's concentration, is high. This causes people to consume more liquids.

6. *Manufacture of oxytocin*—Contracts the uterus during labor
7. *Gastrointestinal control*—Increases intestinal peristalsis and secretion from the intestinal glands
8. *Emotional state*—Plays a role in the display of emotions, such as fear and pleasure
9. *Sleep control*—Helps keep a person awake when necessary
10. *Mind-over-body experiences*—The hypothalamus may be involved in cases of patients who, when diagnosed with terminal illness, refuse to accept the diagnosis and experience an unexplainable cure.

Cerebellum

The **cerebellum** (ser-eh-**BELL**-um) is located behind the pons and below the cerebrum (**Figure 9-9**). It consists of two hemispheres, or wings: the right cerebellar hemisphere and the left cerebellar hemisphere. These two hemispheres are connected to a central portion called the vermis. The cerebellum consists of gray matter on the outside and white matter on the inside. The white matter is marked with a treelike pattern called arbor vitae, meaning "tree of life."

The cerebellum communicates with the rest of the CNS by three pairs of tracts called peduncles (puh-**DUNK**-els). These three peduncles are composed of "incoming" axons that carry nerve messages into the cerebellum, and "outgoing" axons that transmit messages out of the cerebellum. The incoming axons carry messages to the cerebellum regarding movement within joints, muscle tone, position of the body, and tightness of ligaments and tendons. Any and all information relating to skeletal muscle activity is carried to the cerebellum. This information reaches the cerebellum directly from sensory receptors, including the inner ear, the eye, and the proprioceptors in skeletal muscle. The "outgoing" axons carry nerve messages to the different parts of the brain that control skeletal muscles.

CEREBELLAR FUNCTION

The cerebellum controls all body functions that have to do with skeletal muscles, such as the following:

- Maintenance of balance: If the body is imbalanced, sensory receptors in the inner ear send nerve messages to the cerebellum. There, the cerebellum carries impulses to the motor-controlling areas of the brain. These brain areas, in turn, stimulate muscle contraction, which move the body to restore balance.

- Coordination of muscle movement: Any voluntary movement is initiated in the cerebral cortex. However, once the movement is started, its smooth execution is the role of the cerebellum. The cerebellum allows each muscle to contract at the right time, with the right strength, and for the right amount of time so that the overall movement is smooth and flowing. This is important when doing complex or skilled movements, such as speaking, walking, or writing. Even simple movements need the coordinating abilities of the cerebellum. A simple action, such as raising the hand to the face, requires the synchronized action of 50 or more muscles. These muscles then act on 30 separate bones of the arm and hand.

The removal of or injury to the cerebellum results in motor impairment.

Brain Stem

The **brain stem** is a tube-shaped mass of nervous tissue located at the base of the brain, superior to the spinal cord and inferior to the cerebrum. The brain stem has three parts: the midbrain, pons, and the medulla oblongata (**Figure 9-11**). The brain stem provides a pathway for the ascending and descending tracts (messages going to and coming from the cerebrum). Extending the length of the brain stem is the gray matter of the reticular formation system. These are neurons that are involved in the sleep-wake cycle. A coma will result if there is damage to this area. The **pons** (**PONZ**), or bridge, is located in front of the cerebellum, between the midbrain and the medulla oblongata. It contains interlaced transverse and longitudinal myelinated, white nerve fibers mixed with gray matter. The pons serves as a two-way conductive pathway for nerve impulses between the cerebrum, cerebellum, and other areas of the nervous system; is the site for the emergence of four pairs of cranial nerves; and contains a center that controls respiration.

The midbrain extends from the mammillary bodies to the pons. The cerebral aqueduct travels through the midbrain. It contains the nuclei for reflex centers involved with vision and hearing.

The **medulla oblongata** (meh-**DULL**-ah ob-long-**GAH**-tah) is a bulb-shaped structure found between the pons and the spinal cord. It lies inside the cranium and above the foramen magnum of the occipital bone. The medulla is white on the outside because of the myelinated nerve fibers that serve as a passageway for nerve impulses between the brain and spinal cord. It contains the nuclei for vital functions such as heart rate and the rate and depth of respiration; the vasoconstrictor center, which affects blood pressure; and the center for swallowing and vomiting.

Spinal Cord

The **spinal cord** continues down from the brain. It begins at the foramen magnum of the occipital bone and continues to the second lumbar vertebrae. It is white and soft and lies within the vertebrae of the spinal column. It is made up of a series of 31 segments, each giving rise to a pair of spinal nerves. The spinal cord is also protected by the three layers of meninges. The cerebrospinal fluid circulates to bathe the spinal cord. The meninges do not attach to the vertebrae; they are separated by the epidural space. This space contains loose connective tissue and adipose tissue that further protects the spinal cord.

The gray matter in the spinal cord is located in the internal section; the white matter composes the outer part (**Figure 9-12**). In the gray matter of the cord, connections can be made between incoming and outgoing nerve fibers, which provide the basis for reflex action. A major function of the spinal cord is to carry messages from the sensory neurons to the brain

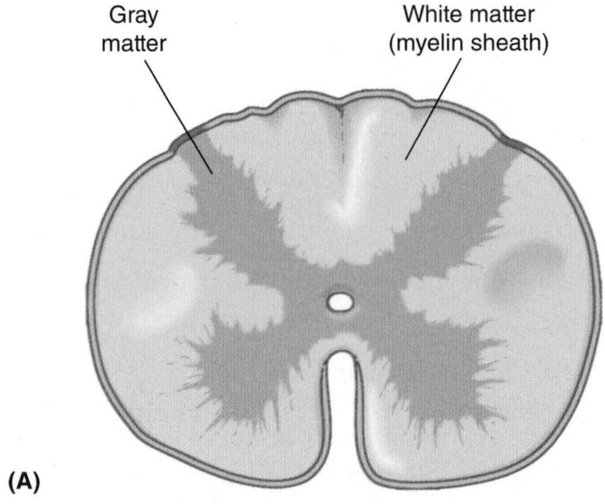

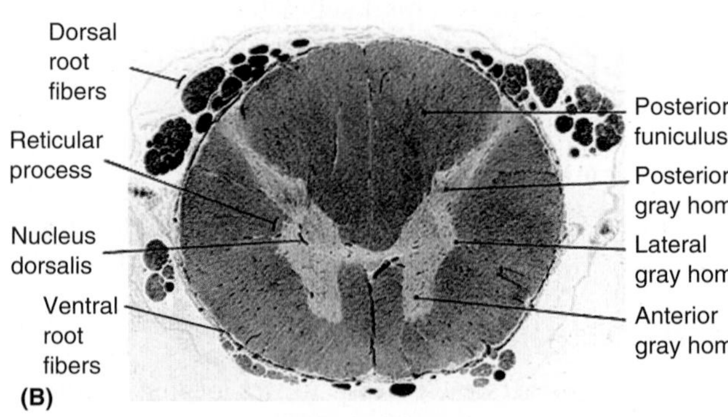

Figure 9-12 *Cross sections of the spinal cord*

B is from *Atlas of Microscopic Anatomy: A Functional Approach: Companion to Histology and Neuroanatomy*, by R. Bergman, A. Afifi, and P. Heidger, 1999, www.vh.org/Providers/Textbooks/MicroscopicAnatomy.html. Reprinted with permission.

for interpretation; the response is carried back from the brain through the motor neurons to the muscles and glands. The second major function is to serve as the reflex center for the body.

DISORDERS OF THE CENTRAL NERVOUS SYSTEM

Disorders of the central nervous system are caused by birth defects or problems with the brain or spinal cord.

Congenital Nervous System Disorders

The most common congenital malformation is neural tube defects. In the case of neural tube defects, the neural tube fails to close completely during embryonic development. During the second trimester of pregnancy, the mother may choose to have an alpha-fetoprotein (AFP) test done. The fetus produces AFP, and this mixes with the mother's blood. An elevated AFP may indicate a neural tube defect or Down's syndrome. Some neural tube defects are spina bifida, tethered cord syndrome, Chiari malformation, encephalocele, and arachnid cysts. Epilepsy and cerebral palsy may also occur as a result of a genetic disorder. The following list contains details about neural tube defects:

- **Spina bifida** is a malformation of the vertebral bones and skin surrounding the spine that leads to serious infections, bladder and bowel dysfunction, hydrocephalus, and paralysis. In *spina bifida occulta,* there may be a tuft of hair at the site of the defect; in this case, no treatment is necessary. In *meningocele,* there may be a sac protruding through the infant's back. In most cases, surgical correction of the neural tube is done sometime after birth, and the prognosis is good. *Myelomeningocele* occurs when the bones of the spine do not completely form, causing the spinal cord, nerves, spinal fluid, and meninges to protrude from the back within a sac. Prenatal surgery or surgery within a few days of birth is required to prevent further damage or infection. This type of spina bifida causes moderate to severe disabilities.
- **Tethered cord syndrome** occurs when the spinal cord is abnormally attached to the surrounding tissue. Failure to detect this condition can lead to a catastrophic injury during childhood. This condition can be diagnosed through the detection of skin lesions along the midline of the back. Surgery may be done to prevent further neurological damage.
- *Chiari malformation* occurs in as many as 1 out of every 1000 births. It is when portions of the brain protrude through the opening of the skull into the upper spine, which puts pressure on the brain and spinal cord. This may block the flow of cerebral spinal fluid. Treatment focuses on relieving the pressure.
- **Encephalocele** is a condition in which the brain is exposed to the outside instead of being covered by the skull and skin. It can lead to infection and hydrocephalus.
- **Arachnid cysts** are benign lesions that occur as a result of the splitting of the arachnoid layer. Treatment involves shunting or draining the cyst by making small openings in the cyst to open the natural fluid pathways in the brain.

General Nervous System Disorders

Although the following disorders are not caused by birth defects, they may start from birth and continue throughout life. However, disorders of the central nervous system can also present later in life or even be temporary and treatable.

Meningitis (men-in-**JIGH**-tis) is inflammation of the linings of the brain and spinal cord meninges. The cause may be bacterial or viral. Symptoms include headache, fever, and stiff neck. In its severe form, it may lead to paralysis, coma, and death. If the cause is bacterial, it may be treated with antibiotics. There is a vaccine available for some types, and the Centers for Disease Control (CDC) recommends the vaccine for pre-teens and teens.

Encephalitis (en-sef-ah-**LYE**-tis) is an inflammation of the brain. The disease may be caused by a virus. The symptoms of this disorder are usually fever, lethargy, extreme weakness, and visual disturbances. Strong antiviral medications may be ordered as well as treatment of the symptoms.

Epilepsy (**EP**-ih-lep-see) is a seizure disorder of the brain characterized by a recurring and excessive discharge from neurons. According to the Epilepsy Foundation, 1 in 60 people in the United States will develop some form of epilepsy. The cause is uncertain. One portion of the brain stimulates another, setting off a cycle of activity that accelerates and runs its course until the neurons become fatigued. The subject may suffer hallucinations; a seizure, or convulsion; and loss of consciousness. Grand mal, or severe seizures, are less frequent than petit mal seizures, which are milder. In petit mal seizures, some individuals seem to be staring or daydreaming. Medications used to control seizures are referred to as anticonvulsants.

First aid for epilepsy includes easing the person to the floor and turning them on their side, if possible. Make sure the person is safe and stay with them until the seizure is over; they usually last between a few seconds and a few minutes. Loosen tight clothing, remove glasses, and time seizures. *Do not put anything in the person's mouth.*

Cerebral palsy (CP) (seh-**REE**-bral **PAWL**-zee) is a disturbance in voluntary muscular action due to brain damage. It is caused by abnormalities in the parts of the brain that cause movement. The most pronounced characteristic is **spastic quadriplegia** (kwad-rih-**PLEE**-jee-ah), which involves spastic paralysis of all four limbs. The person with cerebral palsy frequently exhibits head rolling, grimacing, and difficulty in speech and swallowing. In cerebral palsy, there is usually no impairment of the intellect; the person frequently has normal or above normal intelligence. There is no cure for CP. The disease usually occurs in infancy or early childhood. The therapy regimen for a patient with cerebral palsy is tailored to the individual needs of the patient.

Poliomyelitis (poh-lee-oh-my-eh-**LYE**-tis), also known as polio, is a contagious viral disease of the nerve pathways of the spinal cord that causes paralysis. This disease has been 99% eradicated, and the United States has been polio-free for more than 30 years because of vaccines. However, the disease may still occur in other countries.

Hydrocephalus (high-droh-**SEF**-ah-lus) is a condition that involves an increased volume of cerebrospinal fluid within the ventricles of the brain due to a blockage. Enlargement of the head occurs. A bypass or shunt operation is performed to divert the cerebrospinal fluid around the blocked area. This operation prevents a buildup of pressure on brain tissue.

Parkinson's disease is characterized by tremors; slowed movement, or bradykinesia; a shuffling gait; pill-rolling, the movement of the thumb and index finger; balance problems; and muscular rigidity. Diagnosis is based on symptoms. The patient with Parkinson's has difficulty initiating movement. The cause may be a decrease of the neurotransmitter dopamine. Persons with Parkinson's disease are treated with the drug L-dopa and other drugs that help control the symptoms of the disease. A person may also have a deep brain stimulation procedure done (see *Medical Highlights: Parkinson's Disease and Deep Brain Stimulation*).

Essential tremor is a disorder that causes rhythmic shaking. It is a progressive, often inherited disorder that occurs in later adulthood. Tremors occur when the arms are held up or when the hands are being used for activities like eating, drinking, or writing. The tremors also may affect the head, voice, tongue, and legs; it may worsen with fatigue, stress, and stimulant medications. This is a benign condition affecting movement or voice quality. If the tremors interfere with the ability to perform activities of daily living, medication may be given or deep brain stimulation may be done to reduce these tremors.

Motor neuron diseases (MND) are conditions that cause the nervous system to lose function over time. One type, **amyotrophic lateral sclerosis (ALS)**, causes muscle weakness in a limb or the muscles of the mouth and throat. The cause is unknown. Gradually all the muscles under voluntary control are affected. Most people with ALS die from respiratory failure within 3–5 years of the onset of symptoms.

Multiple sclerosis (MS) (skleh-**ROH**-sis) is an autoimmune, chronic inflammatory disease of the CNS in which the immune cells attack the myelin sheath of nerve cell axons. The myelin sheaths are destroyed, leaving scar tissue on the nerve cells. This destruction delays or completely blocks the transmission of nerve impulses in the affected areas. The cause is unknown. There is no definitive test for MS. The diagnosis is based on signs of impairment to more than one area of the CNS, occurring at more than one time; neurological examination; and an MRI showing two or more abnormal areas or lesions in the brain.

Medical Highlights

PARKINSON'S DISEASE AND DEEP BRAIN STIMULATION

Deep brain stimulation (DBS) is a surgical procedure used to treat the debilitating tremors, rigidity, stiffness, slowed movement, and walking problems associated with Parkinson's disease. At present, the procedure is only used for patients whose symptoms cannot be adequately controlled with medications. DBS uses a surgically implanted, battery-operated medical device called a neurostimulator to deliver artificial stimulation to targeted areas in the brain that control movement, blocking the abnormal nerve signals that cause tremors and Parkinson's symptoms.

The neurostimulator is usually implanted under the skin near the clavicle. Impulses are sent from the neurostimulator; these impulses interfere or block the electrical signals that cause Parkinson's symptoms. DBS does not destroy healthy nerve cells. Although most patients still need to continue with their medications, many patients state that with the DBS implant, their Parkinson's symptoms are greatly reduced.

Symptoms include weakness of extremities, numbness, double vision, **nystagmus** (nis-**TAG**-mus) (involuntary movement of the eyes), speech problems, loss of coordination, and possible paralysis. It typically strikes young adults between the ages of 20 and 40; about two-thirds are women. With MS, there are outbreaks of the symptoms, and then the disease may go into remission for a long period of time. Drugs such as interferon and Avonex® are used. These can slow progression of the disease and decrease the number of flare-ups. Adequate rest, exercise, and minimal stress may also lessen the effects of MS.

West Nile virus (WNV) is a mosquito-borne virus. Most people infected either have no symptoms or experience mild flulike symptoms. In older adults, the virus may cause encephalitis or meningitis. Treatment is symptomatic. To prevent infection, wear protective clothing, use insect repellent, screen windows and doors to keep mosquitoes out, and get rid of mosquito breeding sites by emptying standing water.

Dementia (dih-**MEN**-shah) is a general term that includes specific disorders such as Alzheimer's disease, vascular dementia, and others. Dementia is defined as a loss in at least two areas of complex behavior, such as language, memory, visual and spatial abilities, or judgment that significantly interferes with a person's daily life. *Note:* Everyone has weak areas, and people are frequently forgetful. This does not necessarily mean that the person is experiencing dementia.

Alzheimer's disease (**ALTZ**-high-merz) is a progressive degenerative disease in which the initial symptom is usually a problem with remembering recently learned information. With Alzheimer's disease, the nerve endings in the cortex of the brain degenerate and block the signals that pass between nerve cells. Two types of abnormal lesions clog the brains of people with the disease: beta-amyloid plaques (**Figure 9-13B**), which are sticky clumps of protein fragments and cellular material that form outside and around the neurons; and neurofibrillary tangles, which are insoluble twisted fibers composed largely of the protein *tau* that build up inside nerve cells. Although these structures are hallmarks of the disease, scientists are not sure whether they cause the disease or are a by-product of it.

The specific cause of Alzheimer's is unknown. Scientists believe that for most people, Alzheimer's disease results from a combination of genetic mutation, lifestyle, and environmental factors that affect the brain over time.

The National Institute of Aging's new guidelines on Alzheimer's disease propose that the disease progresses in the following three stages:

- Mild Alzheimer's Disease: As the disease progresses, people experience greater memory loss and other cognitive difficulties. Problems include wandering, managing money, repeating questions, taking longer to complete activities of daily living (ADL), and personality changes. People are often diagnosed at this stage.
- Moderate Alzheimer 's Disease: In this stage, memory loss and confusion grow worse. Damage occurs in parts of the brain that control language, reasoning, sensory perception, and conscious thought. People may have trouble recognizing family and friends, and they can't carry out ADL. At this stage, people may have delusions, hallucinations, and paranoia.

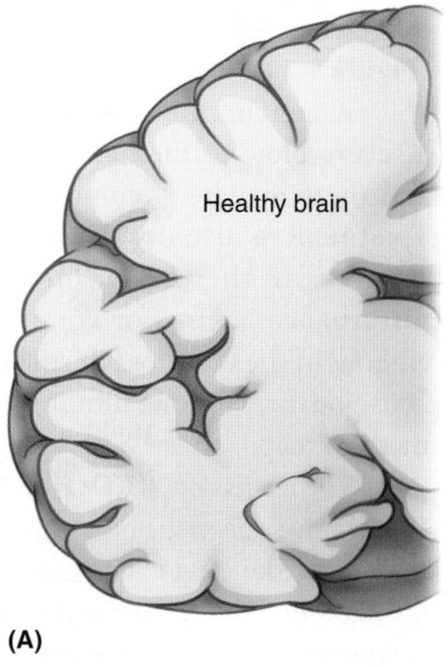

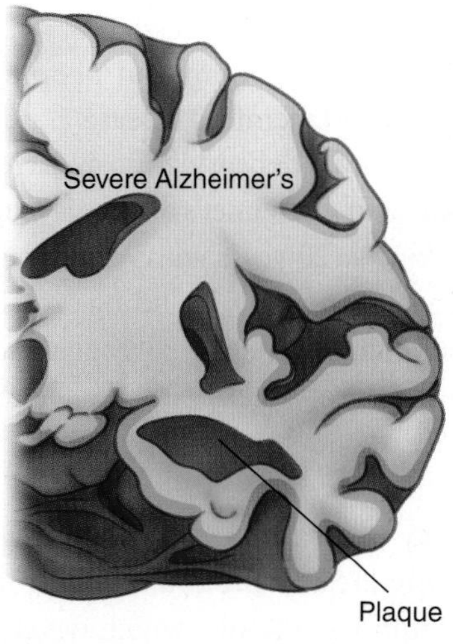

Figure 9-13 *(A) A healthy brain and (B) the brain of a patient with Alzheimer's disease; note the areas of plaque formation*

Medical Highlights

HEADACHES

Many people get headaches. The three most common types of headaches are tension, migraine, and cluster headaches. All cause different types of pain. In most cases, headache pain is not related to a separate underlying disease.

A tension headache is usually a dull, squeezing pain that builds slowly and may involve the forehead, scalp, back of the neck, and both sides of the head. Researchers believe the cause is related to levels of the chemical serotonin and endorphins in the brain. Triggers may be stress, poor posture, depression, and anxiety. Treatment involves pain relievers, rest, ice packs, warm compresses, and relaxation techniques.

Migraine headaches affect 28 million people in the United States; women are three times more likely than men are to be affected. Migraines often run in families. Pain may last from a few hours to days. The throbbing pain occurs on one side of the head and gradually spreads. Migraines may be accompanied by nausea and vomiting. Lights, sounds, and odors may aggravate the migraine. In some people, a migraine is preceded by a visual distortion or aura. The cause is not fully understood. Trigger mechanisms for women may include a change in hormonal levels, dietary factors, lifestyle factors, and certain medications. Treatment includes pain medications to prevent or stop the pain, such as the prescribed medications Imitrex®, Lasmiditian®, and Ubrelvy®. Exercise and rest in a darkened, quiet room are also recommended. Botox® injections every 12 weeks may help prevent migraines in some adults.

Cluster headache pain is worse than migraine pain. This type of headache occurs more frequently in men. Cluster headaches can occur one or more times daily for weeks—often at the same time each day—and then will disappear for months. Treatment includes breathing 100% oxygen and prescription medications for pain.

Source: Mayo Health Clinic Letter, September 2001.

- Severe Alzheimer's Disease: In this stage, people cannot communicate and are completely dependent on others. Gradually the body shuts down.

Medications approved by the FDA to treat general symptoms of Alzheimer's include Aricept®, Exelon®, and Razadyne®. They may treat mild to moderate Alzheimer's. They work by regulating the neurotransmitters and may reduce symptoms and behavior problems. However, they don't change the underlying disease process.

There are now more than five million people in the United States living with Alzheimer's disease. Studies indicate that people ages 65 and older survive an average of 3 to 11 years after a diagnosis of Alzheimer's disease is made, yet some people live as long as 20 years with Alzheimer's. Some researchers say a healthy lifestyle, including exercise, a low-fat diet, and foods rich in omega-3 fatty acids and antioxidants can delay the onset of Alzheimer's. There is some limited evidence that staying mentally active can lower the risk for developing Alzheimer's.

Brain tumors may be benign or cancerous, and they may develop in any area of the brain. The most common type of brain tumor is glioma, which develops from glial cells. The symptoms depend on which area of the brain is involved. Early detection, surgery, and chemotherapy may cure some cases of brain tumors.

Brain Injuries

A **hematoma** (hee-mah-**TOH**-mah) is a localized mass of blood collection. Hematomas may occur in the spaces between the meninges. The cause may be a traumatic blow to the head; the person may have an epidural hematoma, which is located above the dura mater, or a subdural hematoma. In some cases, surgery may be needed to remove the blood and reduce swelling (**Figure 9-14**).

A **concussion** (kon-**KUSH**-un) is the result of a severe blow to the head. It may be mild or severe, and there may be a temporary loss of consciousness. Treatment is usually rest, fluids, and a mild pain reliever. If there is prolonged dizziness, vision disturbance, nausea or vomiting, impaired memory, ringing in the ears, loss of smell or taste, or any loss of consciousness for one minute or more, seek immediate medical help.

The terms **coup** (**KOO**) and **contrecoup** (kon-trah-**KOO**) injury describe head injuries. Coup injuries occur within the skull near the point of impact, such as when the skull hits the windshield in an automobile accident. A contrecoup injury, also described as a counterblow, is one that occurs beneath the skull opposite to the area of impact (**Figure 9-15**).

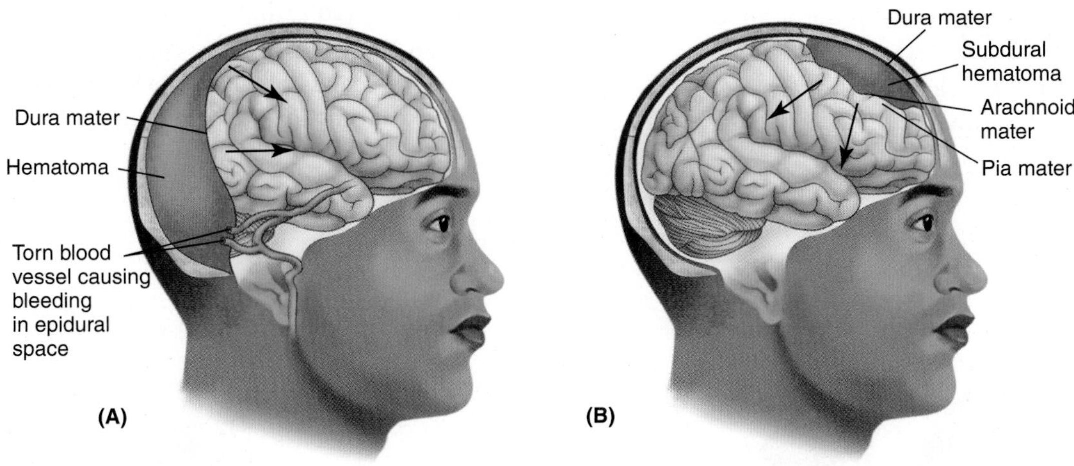

Figure 9-14 *Cranial hematomas: (A) epidural hematoma located above the dura mater, (B) subdural hematoma located below the dura mater*

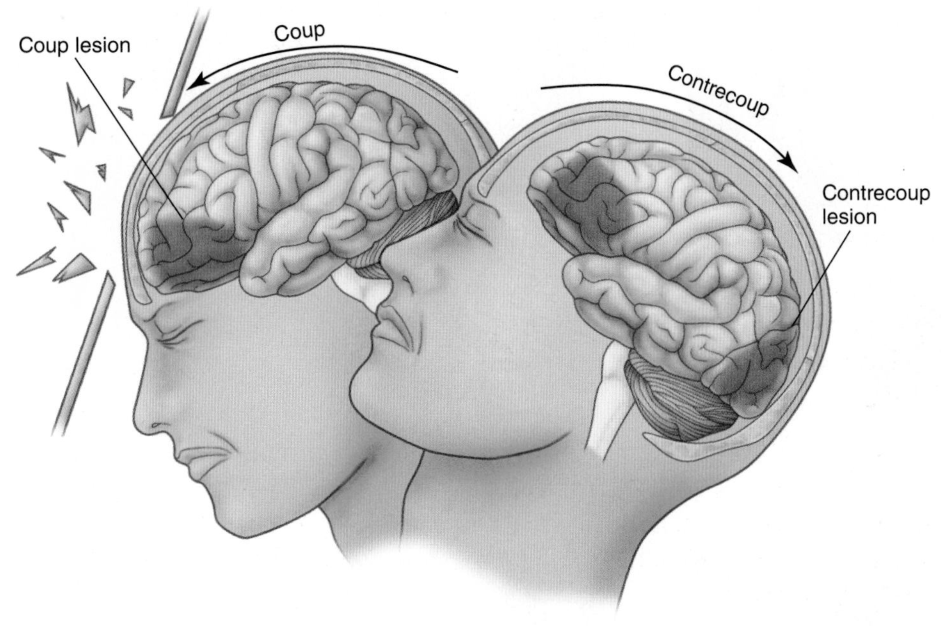

Figure 9-15 *In coup and contrecoup injuries, the brain hits the skull with force.*

View the **Coup and Contrecoup** animation on the Online Resources.

Spinal Cord Injury

The spinal cord may be injured at any level, but the mobility of the neck causes this area to be most vulnerable. The site of the injury, the type of trauma, and the degree of injury all play a role in determining whether paralysis will occur and whether it will be temporary or permanent. **Figure 9-16** shows the most common types of injury. The areas that are affected correspond to the vertebrae involved.

- C1–C3 is the highest level of the cervical spine. Injury there is usually fatal.
- Injuries at C1–C4 may lead to **quadriplegia**. Quadriplegia is the loss of movement and feeling in the trunk

and all four extremities, with the accompanying loss of bowel, bladder, and sexual function.

- Injuries at C5–C7 may lead to varying degrees of paralysis of the arms and shoulders.
- Injuries at T1–T12 and L1–L5 may lead to **paraplegia** (par-ah-**PLEE**-jee-ah), a loss of movement and feeling in the trunk and both legs. The loss of bladder, bowel, and sexual function are common.

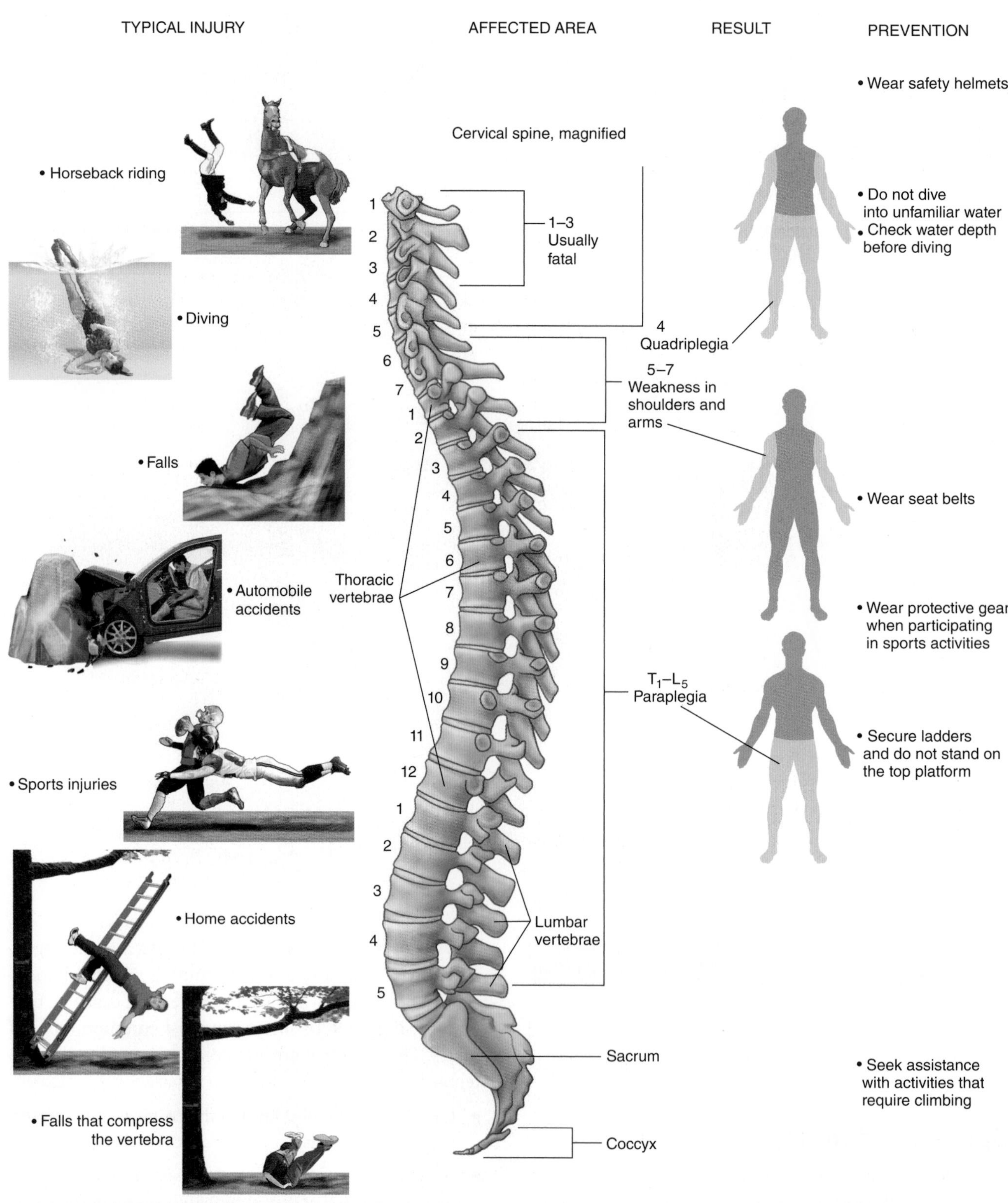

Figure 9-16 *Spinal cord injuries and suggestions for preventing them*

Suspected spinal injuries need immediate emergency medical treatment; *do not* move the victim unless the surroundings are life threatening. Emergency medical treatment is aimed at maintaining the position of the spine by limiting movement using special collars and boards. Treatment includes realignment, stabilization, and release of pressure on the spinal cord; surgery, if necessary; and special medications. Much of the early treatment is aimed at preventing further injury. Intensive rehabilitation is necessary for the best prognosis. See **Figure 9-16** for ways to prevent spinal cord injuries.

Career Profile

ELECTRONEURODIAGNOSTIC TECHNICIAN/EEG TECH

An electroneurodiagnostic technician works in a medical laboratory performing neurological tests on patients. One type of test is electroencephalogram (EEG). An electroencephalogram is a test to check for abnormalities in the brain and to detect and record the brain's electrical activity patterns. The length of study for an EEG tech is usually 2 years, leading to an associate's degree, which includes a clinical component. The Commission on Accreditation of Allied Health Education Programs accredits EEG programs. Personal qualities for this career include good interpersonal skills.

One BODY — How the Central Nervous System Interacts with Other Body Systems

INTEGUMENTARY SYSTEM

- Sensory receptors send messages to the brain regarding temperature, pain, and body position.

SKELETAL SYSTEM

- Skull protects the brain, and vertebrae protect the spinal cord.
- Bones store the calcium necessary for nerve transmission.

MUSCULAR SYSTEM

- Cerebrum controls voluntary muscle movements.
- Cerebellum controls fine motor movement.

ENDOCRINE SYSTEM

- Hypothalamus controls the action of the pituitary gland.

CIRCULATORY SYSTEM

- Hypothalamus controls blood pressure, regulating the constriction and dilation of blood vessels of the heart.
- Medulla oblongata helps regulate the heart rate.

LYMPHATIC SYSTEM

- The hypothalamus is involved with mind-over-body experiences, which helps the immune system of the body fight diseases.

RESPIRATORY SYSTEM

- Medulla oblongata is the vital control center for rate and depth of respiration.

DIGESTIVE SYSTEM

- Hypothalamus regulates the amount of food a person consumes and lets them know when they are full.
- Brain controls muscles necessary for taking in food and eliminating the waste products of digestion.

URINARY SYSTEM

- Hypothalamus responds to the concentration of the blood as it is filtered through the kidney, and releases antidiuretic hormone (ADH) when necessary to enable the kidneys to absorb more water.
- Receptors let a person know when the bladder is full and to eliminate the urine.

REPRODUCTIVE SYSTEM

- Hypothalamus releases oxytocin, which contracts the uterus during labor.
- Hypothalamus plays a role in experiencing pleasure during sexual activity.

Medical Terminology

arach	spider's web	**-itis**	inflammation
-oid	resembling	**en/cephal/itis**	presence of inflammation within the head
arach/noid	structure resembling a spider's web	**hemat**	blood
cerebell	little brain	**-oma**	tumor
-um	presence of	**hemat/oma**	blood tumor
cerebell/um	presence of little brain	**hydro-**	water
cerebr	brain	**-us**	presence of
cerebr/um	presence of brain	**hydro/cephal/us**	presence of water in the head
-al	pertaining to	**mening**	membrane
aqua	water	**mening/itis**	inflammation of the membranes
duct	channel	**neuro**	nerve
cerebr/al aque/duct	channel pertaining to brain fluid	**-glia**	glue
en-	within	**neuro/glia**	nerve glue
cephal	head		

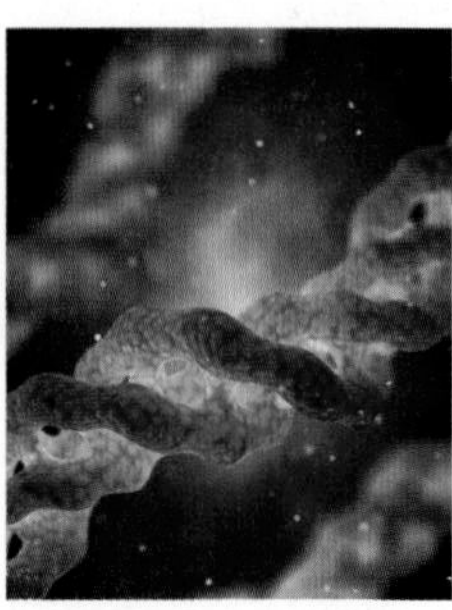

Study Tools

Workbook	Activities for Chapter 9
Online Resources	• PowerPoint presentations • Animations

REVIEW QUESTIONS

Select the letter of the choice that best completes the statement.

1. Each nerve cell has only one
   a. axon.
   b. nodes of Ranvier.
   c. dendrite.
   d. myelin.

2. The fatty substance that helps protect the axon is called
   a. neurotransmitter.
   b. myelin.
   c. dendrite.
   d. nodes of Ranvier.

3. The junction between the axon of one cell and the dendrite of another is called
   a. neurilemma.
   b. myelin.
   c. synaptic cleft.
   d. nodes of Ranvier.

4. The neurons that carry messages to the brain are called
   a. motor.
   b. associative.
   c. connective.
   d. sensory.

5. The nervous system that consists of 12 pairs of cranial nerves and 31 pairs of spinal nerves is called
   a. central.
   b. peripheral.
   c. sympathetic.
   d. parasympathetic.

6. The outermost covering of the meninges is
   a. arachnoid mater.
   b. arachnoid villi.
   c. dura mater.
   d. pia mater.

7. The lumbar puncture must be done below the
   a. first lumbar vertebrae.
   b. second lumbar vertebrae.
   c. third lumbar vertebrae.
   d. sacrum.

8. The frontal, parietal, temporal, and occipital lobes make up the
   a. cerebrum.
   b. cerebellum.
   c. midbrain.
   d. brain stem.

9. The part of the brain associated with muscle movement is
   a. midbrain.
   b. thalamus.
   c. cerebrum.
   d. medulla.

10. The thalamus and hypothalamus are parts of the
    a. cerebrum.
    b. cerebellum.
    c. diencephalon.
    d. brain stem.

MATCHING

Match each term in Column I with its correct description in Column II.

	COLUMN I	*COLUMN II*
______	**1.** frontal lobe	a. auditory
______	**2.** occipital lobe	b. receptor for pain, touch, and so on
______	**3.** hypothalamus	c. reflex center
______	**4.** temporal lobe	d. speech area
______	**5.** parietal lobe	e. maintain balance
______	**6.** cerebellum	f. eyesight
______	**7.** thalamus	g. respiratory center
______	**8.** spinal cord	h. appetite control
______	**9.** medulla	i. site for four pairs of cranial nerves
______	**10.** pons	j. relay station for nerve impulses

COMPARE AND CONTRAST

List the similarities and differences between the following terms:

1. Meningitis and encephalitis
2. Cerebral palsy and multiple sclerosis
3. Parkinson's and essential tremors
4. Dementia disease and Alzheimer's disease

APPLYING THEORY TO PRACTICE

1. The central nervous system serves as the communication center of the body. Explain how your hand touches something cold and you know it; refer to a sensory neuron and the correct lobe of the cerebrum.
2. A blow to the head can cause a loss of consciousness. What centers in the brain are associated with alertness?
3. Depending on the type of injury and where the injury occurred, trauma to the spinal cord can have varied effects. Define the following: paralysis, monoplegia, hemiplegia, paraplegia, and quadriplegia. For each type of paralysis, determine the main causes, describe how the trauma may affect the structure and function of the body system involved, and identify the potential outcomes and treatments for the injured individuals.
4. What are some of the actions you can take in your home to prevent the West Nile virus?

CASE STUDY

Mr. Anwari, age 73, is brought to the physician's office by his daughter, Lucy, who is a licensed practical nurse (LPN). She states her concerns about her father: During the past 2 months, he has been found wandering in the neighborhood because he forgets where he lives. Neighbors see him, note that he appears confused, and bring him home. Lucy is worried that her father is showing early signs of Alzheimer's disease.

1. Describe the physical changes that occur in the cortex of the brain.
2. Describe the stages of Alzheimer's disease.
3. Describe the physiological and psychological changes that occur during Alzheimer's disease.
4. What are the functions of the frontal lobe of the cerebral cortex?
5. What parts of the limbic system may be affected in Alzheimer's disease?
6. What diagnostic technologies are used to diagnose Alzheimer's disease and eliminate diseases with similar symptoms? What would be the concerns of the family when a person is diagnosed with this disease?

9-1 Peripheral Nerves

- ***Objective:*** To examine and describe the function of a neuron
- ***Materials needed:*** slide of a neuron, textbook, microscope, paper, pencil

Step 1: Examine the neuron; identify and describe the nerve fiber. Record your observations.

Step 2: Locate and identify the myelin sheath. What is its function? Record your answer.

Step 3: Compare what you see with the diagram shown in your textbook.

LAB ACTIVITY 9-2 Sheep Brain

- ***Objective:*** To compare and contrast the sheep brain with the human brain and to identify the structures of the brain
- ***Materials needed:*** anatomical human brain model, preserved sheep brain, dissecting tray and instruments, disposable gloves, paper, pencil

Step 1: Put on disposable gloves.

Step 2: Examine the structures of the sheep brain. Locate and describe the cerebral cortex, cerebellum, and brain stem. Record your observations of the location and appearance of these structures.

Step 3: Is there a difference in the structure and size of the cerebral cortex, cerebellum, and brain stem between the human brain and the sheep brain? Record your answer.

Step 4: Locate the dura mater on the sheep brain. Describe how it looks and feels. Record your observations.

Step 5: Place the sheep brain ventral side down on the dissecting pan.

Step 6: Using your dissecting knife, carefully cut along the longitudinal fissure of the sheep brain (it separates the two cerebral hemispheres); separate the sheep brain into right and left hemispheres.

Step 7: Examine the right portion of the sheep brain.

Step 8: Locate the arachnoid mater and pia mater in the right hemisphere. Describe the differences between these two meninges layers. Record your observations.

Step 9: Locate and describe the sizes and structures of the following: lateral ventricle, corpus callosum, midbrain, medulla, pons, and pituitary gland. Record your observations and describe the functions of these structures.

Step 10: Observe the anatomical model of the human brain. Compare the size and structure of the lateral ventricle, corpus callosum, midbrain, medulla, pons, and pituitary gland. Record your observations of the similarities and differences.

Step 11: Dispose of the sheep brain in the appropriate disposal containers.

Step 12: Clean all equipment.

Step 13: Remove your gloves and wash your hands.

Step 14: Compare your observations with those of your lab partner.

CHAPTER 10

Peripheral and Autonomic Nervous System

Objectives

- Describe a mixed nerve.
- Describe the functions of the cranial and spinal nerves.
- Relate the functions of the sympathetic and parasympathetic nervous systems.
- Explain the simple reflex arc pattern.
- Describe common disorders of the peripheral nervous system.
- Define the key words that relate to this chapter.

Key Words

analgesics
Bell's palsy
biofeedback
carpal tunnel syndrome
Charcot-Marie-Tooth (CMT) disease
congenital insensitivity to pain
cranial nerves
effectors
electromyography (EMG)
femoral nerve
ganglia
mixed nerve
motor nerve (efferent nerve)
neuralgia
neuritis
parasympathetic system
paresthesia
peripheral neuropathy
phrenic nerve
plexus
radial nerve
receptors
reflex
sciatic nerve
sciatica
sensory nerve (afferent nerve)
somatic nervous system
spinal nerves
stimulus
sympathetic system
trigeminal neuralgia

PERIPHERAL NERVOUS SYSTEM

The *peripheral nervous system (PNS)* is subdivided into several smaller units (**Figure 10-1**). The PNS consists of all the nerves that connect the brain and spinal cord with sensory receptors, muscles, and glands.

The PNS can be divided into two subcategories. The afferent peripheral system consists of afferent or sensory neurons that convey information *from* receptors in the periphery of the body *to the brain and spinal cord*. The efferent peripheral system consists of efferent neurons, or motor neurons, that convey information *from* the brain and spinal cord *to the muscles and glands*.

The efferent peripheral system can be further subdivided into two subcategories. The first is the **somatic nervous system**, which conducts impulses from the brain and spinal cord to skeletal muscles, thereby causing the body to respond to changes in the external environment. The second is the autonomic nervous system, which is further subdivided into the sympathetic and parasympathetic divisions (**Figure 10-2**). The autonomic nervous system is involuntary (think automatic) and conducts impulses from the brain and spinal cord to smooth muscle tissue. Some examples of autonomic nervous system processes include how food is pushed along the digestive tract and how the glands of the endocrine system are stimulated. The sympathetic division stimulates (speeds up) activity, requires energy expenditure, and uses norepinephrine as a neurotransmitter. The parasympathetic system speeds up the body's vegetative activities, such as urination and digestion, and restores or slows down other activities. It uses acetylcholine as a neurotransmitter.

Nerves

A nerve is made up of bundles of nerve fibers enclosed by connective tissue. If the nerve's fibers carry impulses from the sense organs to the brain or spinal cord, it is called a **sensory nerve (afferent nerve)**; if its fibers carry impulses from the brain or spinal cord to muscles or glands, it is known as a **motor nerve (efferent nerve)**; and if it contains both sensory and motor fibers, it is called a **mixed nerve**.

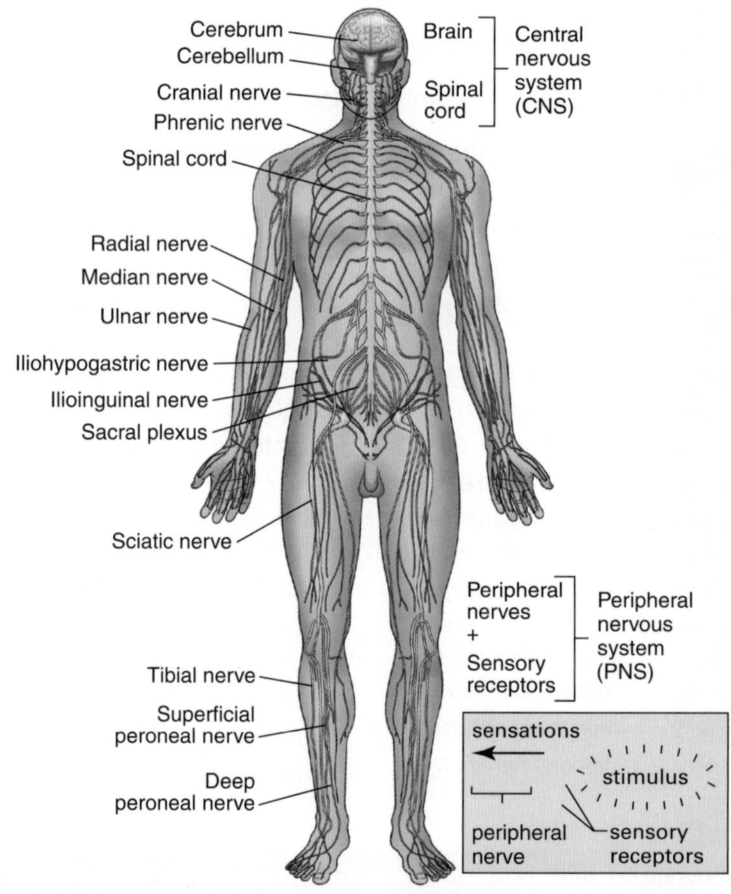

Figure 10-1 *The peripheral nervous system connects the central nervous system to structures of the body.*

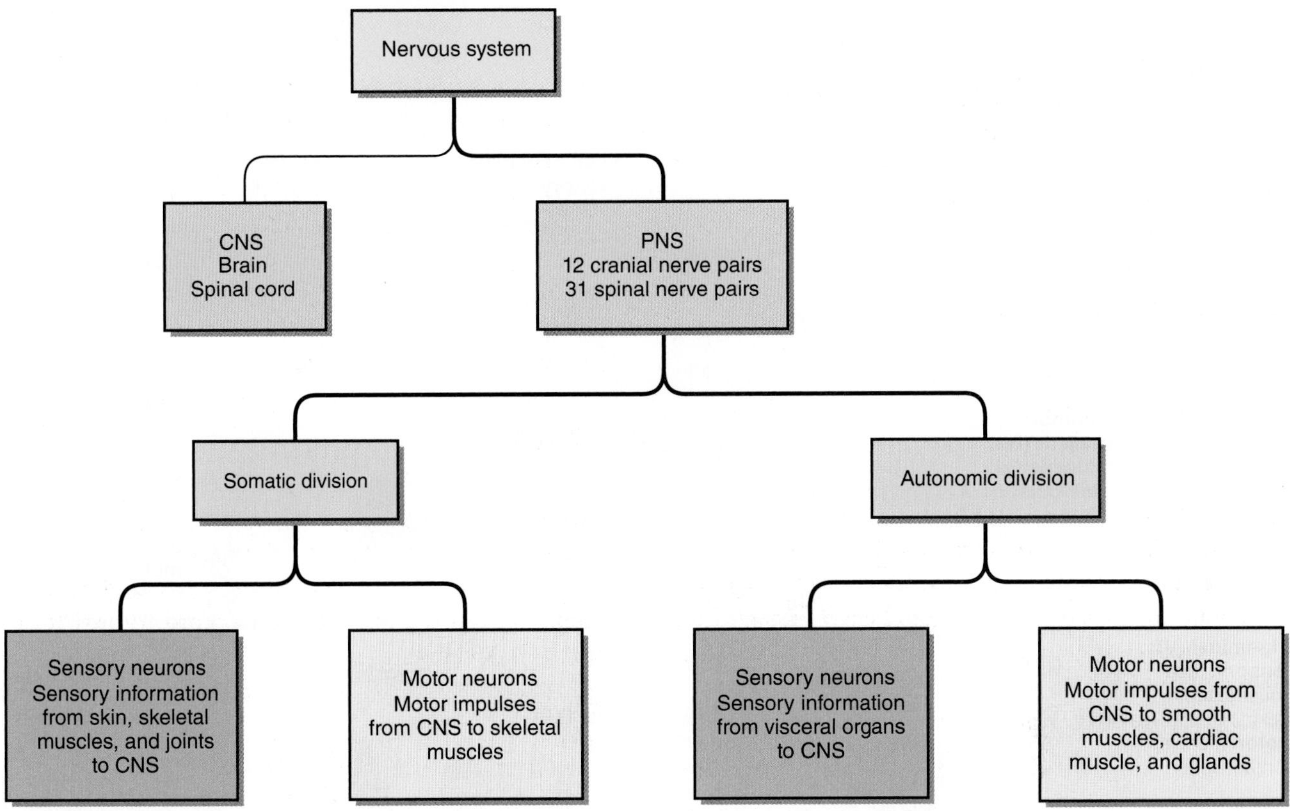

Figure 10-2 *Divisions of the nervous system*

CRANIAL NERVES

Cranial nerves are part of the peripheral nervous system. The 12 pairs of **cranial nerves** originate from the brain and brain stem. The cranial nerves are designated by number and name; the name may give a clue as to its function (**Figure 10-3**). Cranial nerves are identified with Roman numerals and are named for the area or function they serve (**Table 10-1**). For example, the olfactory nerve, cranial nerve I, is responsible for the sense of smell. The optic nerve, cranial nerve II, is responsible for vision. The functions of the cranial nerves are concerned mainly with the activities of the head and neck, with the exception of the vagus nerve. The vagus nerve, cranial nerve X, is responsible for activities involving the throat as well as regulating the heart rate; it also affects the smooth muscle of the digestive tract. Most of the cranial nerves are mixed nerves: they carry both sensory and motor fibers. The olfactory, optic, and vestibulocochlear nerves, however, carry only the sensory fibers, meaning they only respond to stimuli.

SPINAL NERVES

The **spinal nerves** originate at the spinal cord and are each connected to a specific segment of the spinal cord. They exit through the openings in the vertebrae. Each pair of spinal nerves is connected to that segment of the cord by two pairs of attachments called roots (**Figure 10-4**). The posterior, or dorsal, root is the sensory root and contains only sensory nerves. It conducts impulses from the periphery, such as the skin, to the spinal cord. The other point of attachment is the anterior, or ventral, root; this is the motor root. It contains motor nerve fibers only and conducts impulses from the spinal cord to the periphery, such as muscles. It connects with the ventral gray horn of the spinal cord.

There are 31 pairs of spinal nerves, and all are mixed nerves because they contain both sensory and motor fibers. They are named according to the region and level from which they emerge (**Figure 10-5**). There are 8 pairs of cervical nerves, 12 pairs of thoracic nerves, 5 pairs of lumbar nerves, 5 pairs of sacral nerves, and 1 coccygeal nerve.

Each of these spinal nerves divides and branches. They either go directly to a particular body segment, or they form a network with other spinal nerves and veins, bunching together to form a **plexus** (**Figure 10-5** and **Table 10-2**). The largest spinal nerve is the **sciatic nerve** (sigh-**AT**-ic), which is part of the sacral plexus. The sciatic nerve leaves the dorsal part of the pelvis, passes beneath the gluteus maximus muscle, and extends down the back of the thigh. It branches to the thigh muscle near the knee and forms two subdivisions that supply the leg and the foot.

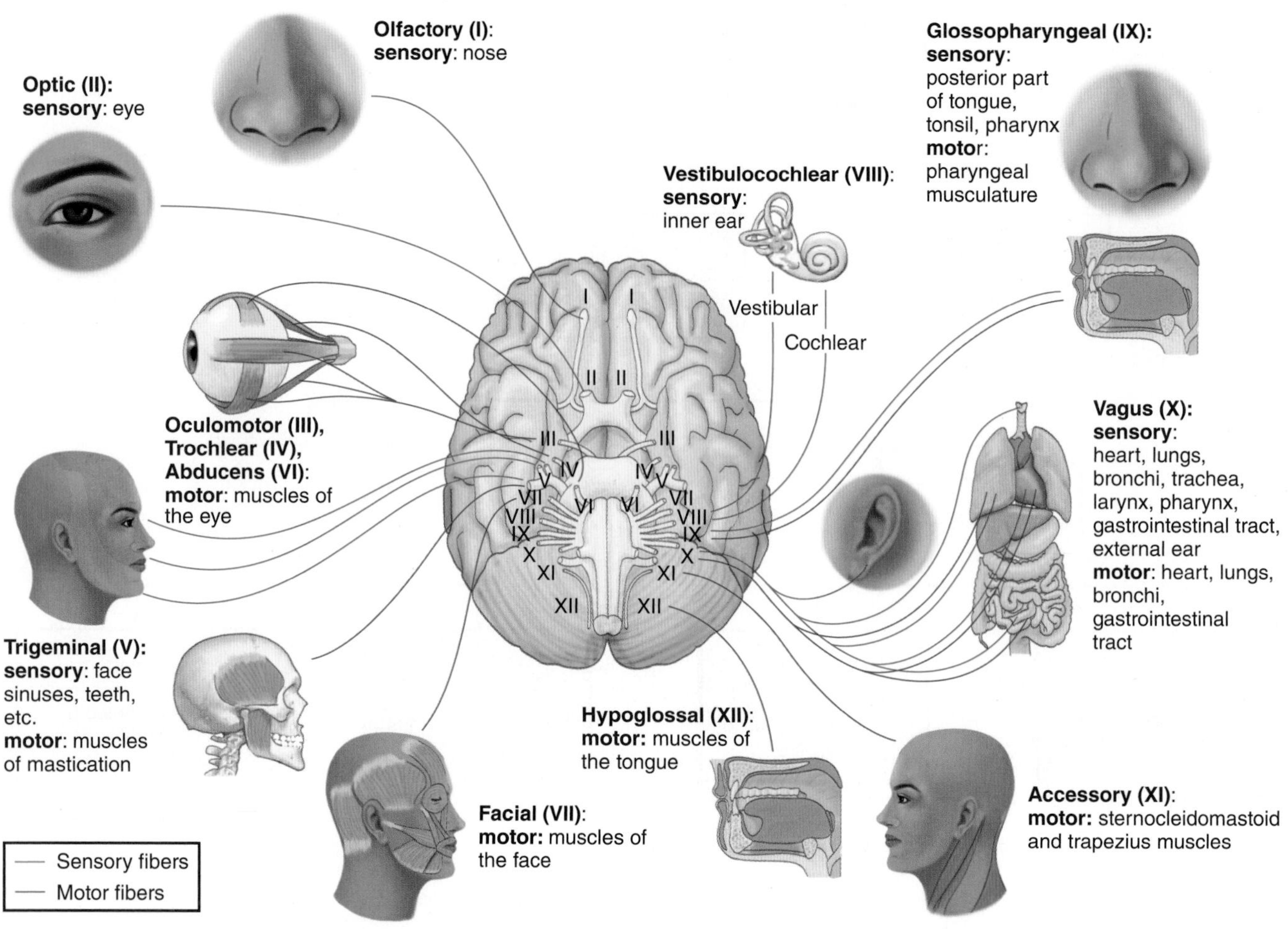

Figure 10-3 *Cranial nerves are named for the area or function they serve.*

Table 10-1 ***Cranial Nerves***

NUMBER	NAME	FUNCTION
I	Olfactory	Sensory: smell
II	Optic	Sensory: vision
III	Oculomotor	Motor: eyelid and eyeball movement
IV	Trochlear	Motor: innervates superior oblique muscle, turns eye downward and laterally
V	Trigeminal	Sensory: sensation of face and mouth Motor: chewing
VI	Abducens	Motor: turns eye laterally
VII	Facial	Sensory: taste Motor: controls most facial expressions, secretion of tears and saliva
VIII	Vestibulocochlear (ves-tib-yoo-loh-**KOK**-lee-ar) (auditory)	Sensory: hearing and equilibrium
IX	Glossopharyngeal (**gloss**-oh-fair-in-**GEE**-al)	Sensory: taste Motor: controls swallowing and secretion of saliva
X	Vagus	Sensory: sensation from the larynx, pharynx, liver, and stomach Motor: movement within the organs of the thoracic and abdominal area Vagus nerve supplies heart, lungs, respiratory tract, gastrointestinal tract, and the back of the external ear
XI	Accessory	Motor: controls trapezius and sternocleidomastoid muscles
XII	Hypoglossal	Motor: movement of tongue muscles

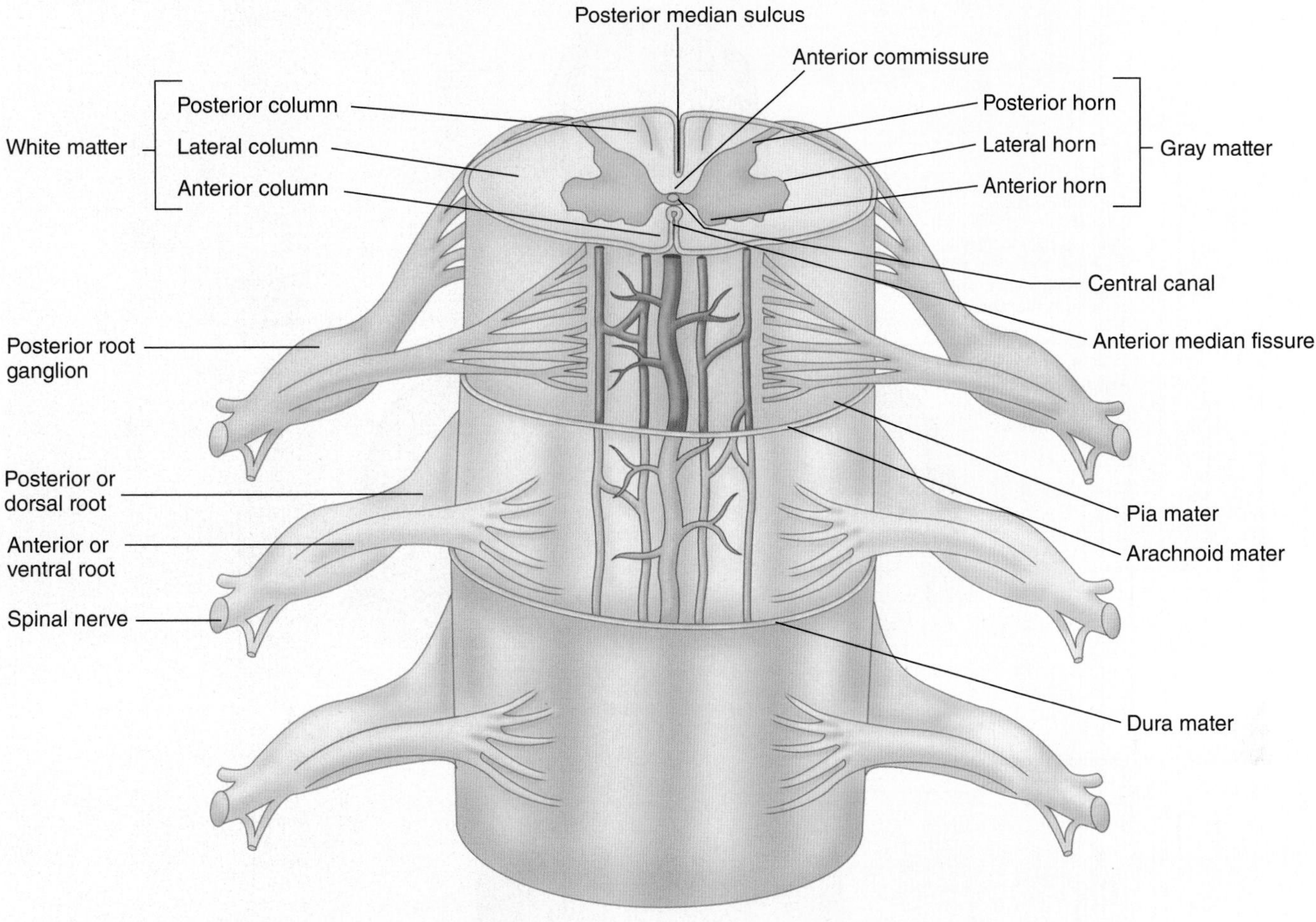

Figure 10-4 *The anatomy of the spinal cord and spinal nerve*

Table 10-2 ***Spinal Nerve Plexus***

NAME	*LOCATION*	*FUNCTION*
Cervical plexus	C1–C4	Supplies motor movement to muscles of neck and shoulders and receives messages from these areas. **Phrenic nerve** is part of this group and stimulates the diaphragm.
Brachial plexus	C5–C8, T1	Supplies motor movement to shoulder, wrist, and hand and receives messages from these areas. **Radial nerve** is part of this group and stimulates the wrist and hand. Impairment to radial nerve results in wrist drop.
Lumbar plexus	T12, L1–L4	Supplies motor movement to buttocks, anterior leg, and thighs and receives messages from these areas. **Femoral nerve** is part of this group and stimulates the hip and leg.
Sacral plexus	L4–L5, S1–S2	Supplies motor movement to posterior of leg and thighs and receives messages from these areas. Sciatic nerve is the largest nerve in the body and is part of this group. It passes through the gluteus maximus and down the back of the thigh and leg. It extends the hip and flexes the knee; avoid this nerve when giving an intramuscular injection.

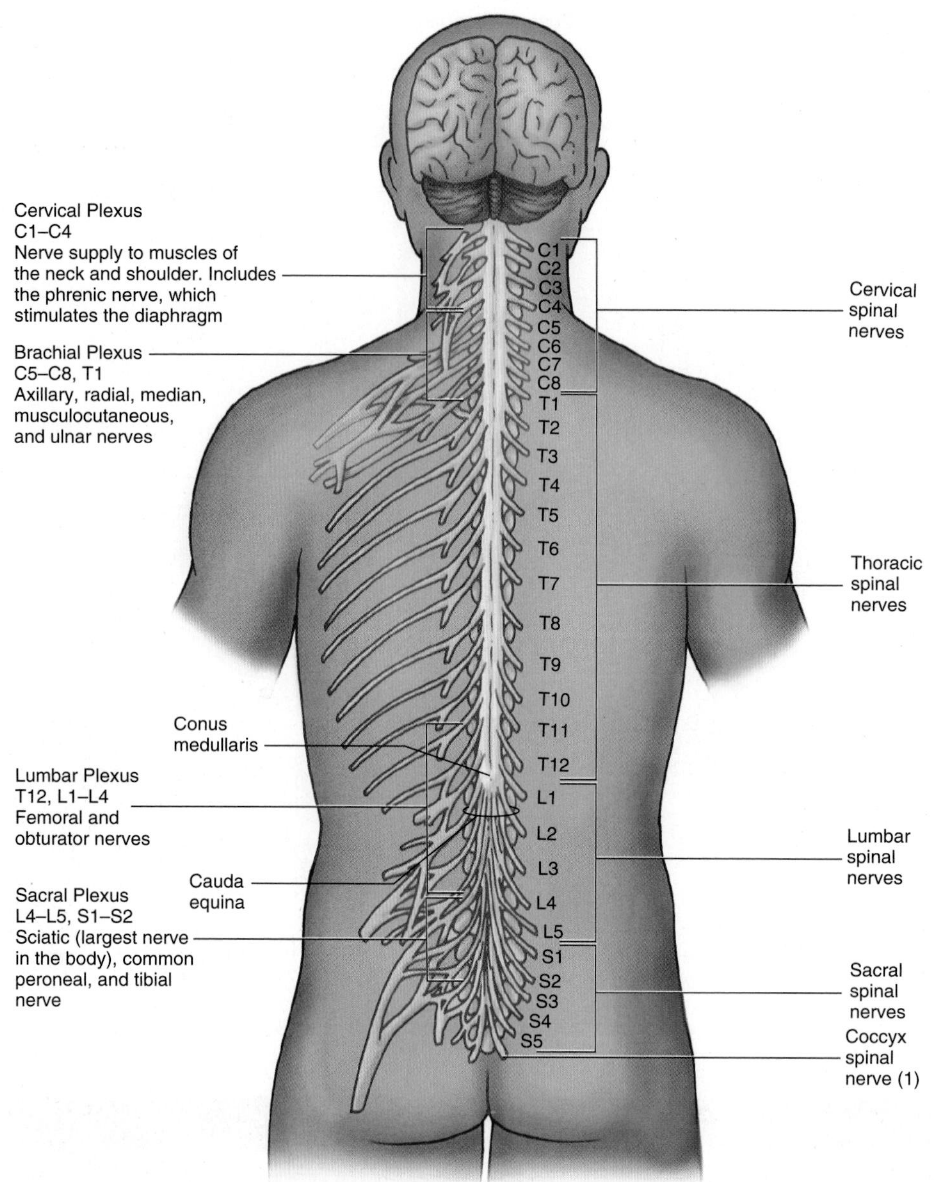

Figure 10-5 *Spinal nerve plexus and important nerves*

AUTONOMIC NERVOUS SYSTEM

The autonomic nervous system includes nerves, **ganglia** (mass of nerve cell bodies), and plexuses, which carry impulses to all smooth muscle, secretory glands, and heart muscle. It regulates the activities of the visceral organs (heart and blood vessels, respiratory organs, alimentary canal, kidneys, urinary bladder, and reproductive organs). The activities of these organs are usually automatic, meaning they are not subject to conscious control.

The autonomic system has two divisions: the sympathetic system and the parasympathetic system (**Figure 10-6**). These two divisions may be antagonistic in their action. The sympathetic system may accelerate the heartbeat in response to fear, whereas the parasympathetic system slows it down. Normally, the two divisions are in balance; the activity of one or the other becomes dominant as dictated by the needs of the organism.

The **sympathetic system** consists primarily of two cords, beginning at the base of the brain and proceeding down both sides of the spinal column, just lateral to the vertebrae. These consist of nerve fibers and ganglia of nerve cell bodies. The cord between the ganglia is a cable of nerve fibers, closely associated with the spinal cord. Sympathetic nerves extend to all the vital internal organs, including the liver, pancreas, heart, stomach, intestines, blood vessels, the iris of the eye, sweat glands, and the bladder (**Figure 10-6A**). The sympathetic nervous system is often referred to as the "fight-or-flight system." When the body perceives that it is in danger or under stress, it prepares to either run away or stand and

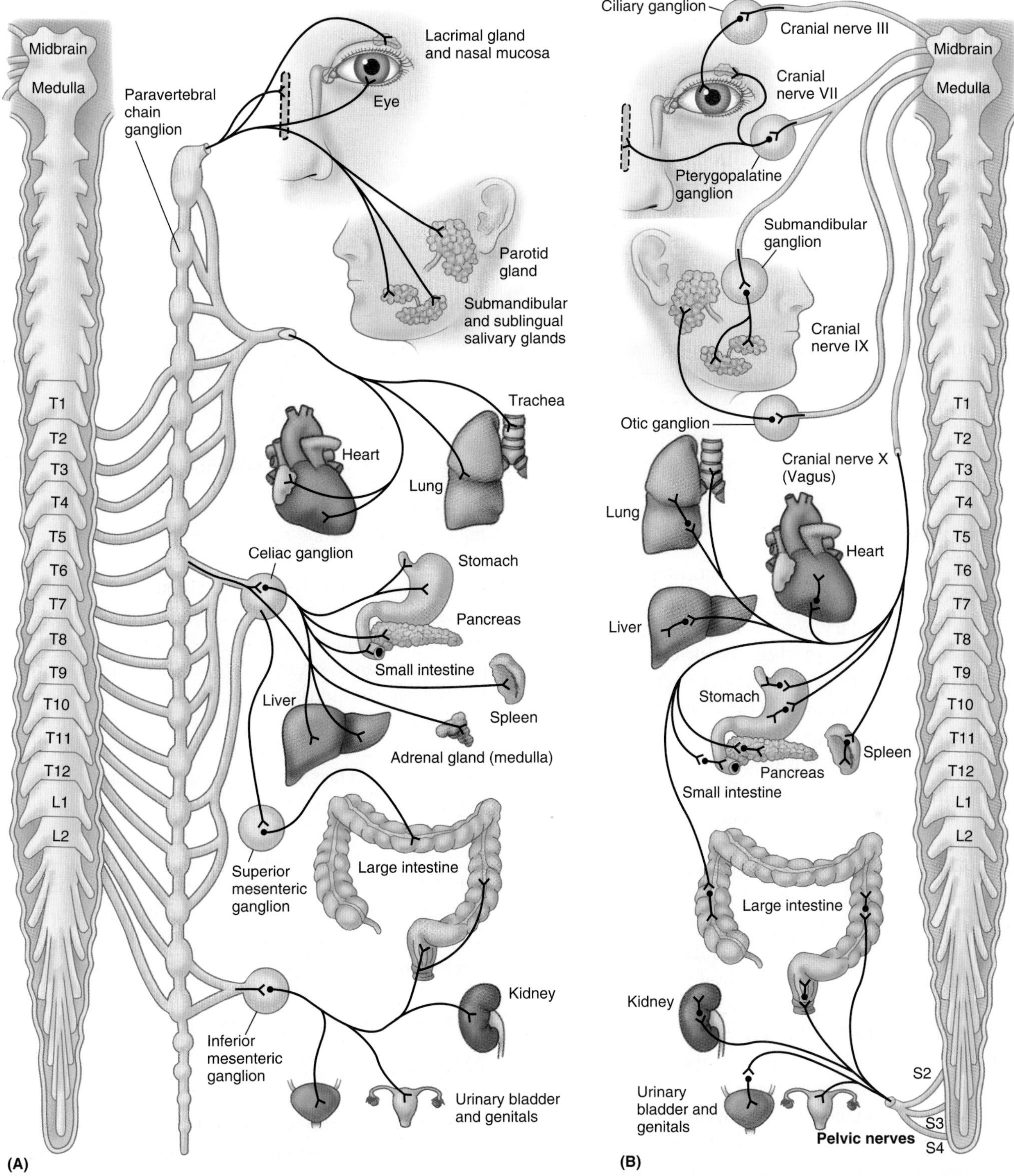

Figure 10-6 *(A) The sympathetic division of the autonomic nervous system, (B) the parasympathetic division of the autonomic nervous system*

fight. The sympathetic nervous system sends the message to the adrenal medulla, which secretes its hormones to prepare the body for this action. Some examples of situations that might trigger this response are an upcoming major test or waiting for medical test results. The heart beats faster and the mouth may go dry—all results of the automatic response to danger. When the danger passes, the parasympathetic nervous system will help restore balance to the body system. If the system gets too much of the "stress hormones," health problems may result. Learning to live with stress is the key to a healthier body.

The **parasympathetic system** has two important active nerves: the vagus and the pelvic nerves. The vagus nerve, which extends from the medulla and proceeds

> **Did You Know?**
>
> *Nerve impulses travel to and from the brain at a rate of 170 miles per hour.*

down the neck, sends branches to the chest and neck. The pelvic nerve, emerging from the spinal cord around the hip region, sends branches to the organs in the lower part of the body (**Figure 10-6B**).

Both the sympathetic and parasympathetic nerves are strongly influenced by emotion. During periods of fear, anger, or stress, the sympathetic division acts to prepare the body for action. The effects of the parasympathetic are generally to counteract the effects of the sympathetic. For example, the sympathetic nervous system increases the rate of heart muscle contraction, and the parasympathetic decreases the rate. The two systems operate as a pair, striking a nearly perfect balance, or homeostasis, when the body is functioning properly.

Reflex Arc

The reflex arc is a basic pathway that results in a **reflex**, the simplest type of nervous response, which is unconscious and involuntary. Some examples of reflex actions are the blinking of an eye when a particle of dust touches it; the removal of a finger from a hot object; the secretion of saliva at the sight or smell of food; and the movements of the heart, stomach, and intestines.

Every reflex action is preceded by a change in the environment, called a **stimulus**. Examples of stimuli are sound waves, light waves, heat energy, and odors. Special structures called **receptors** pick up these stimuli. For example, the retina of the eye is the receptor for light; special cells in the inner ear are the receptors for sound waves; and special structures in the skin are the receptors for heat and cold.

A simple reflex is one in which only a sensory nerve and a motor nerve are involved. The classic example is the knee-jerk reflex. The knee is tapped, and the leg extends (**Figure 10-7**). This test is used by physicians to test both the muscular and nervous systems.

A reaction to a stimulus is called a response. The response may be in the form of movement, in which case the muscles are the **effectors**, or responding organs. If the response is in the form of a secretion, the glands are the effectors. Reflex actions, or autonomic reflexes, that involve the skeletal muscles are controlled by the spinal cord. They also may be called somatic reflexes.

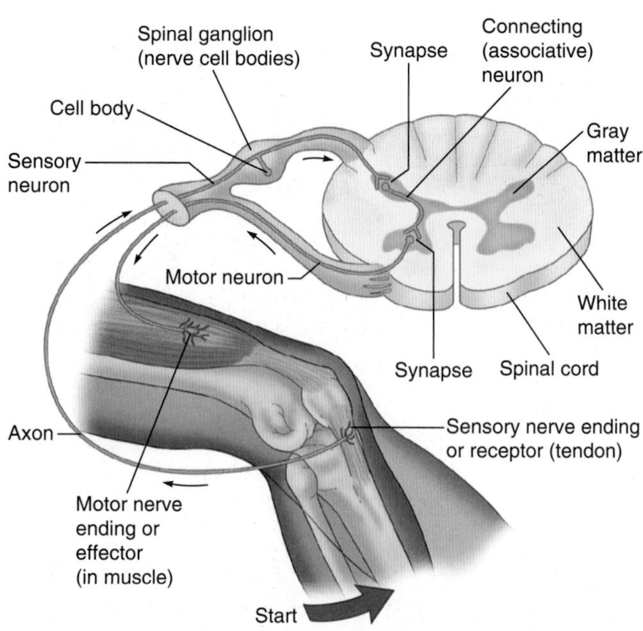

Figure 10-7 *In this example, tapping the knee (patellar tendon) results in extension of the leg, producing the knee-jerk reflex.*

Biofeedback

Biofeedback is the process of gaining greater awareness of the many physiological functions of one's own body. Biofeedback yields information about the different relationships between the mind and body and helps people learn how to manipulate their bodily responses through mental activity. While attached to sensors, which are sensitive devices that measure such bodily responses as skin temperature, blood pressure, galvanic skin resistance, and electrical activity in the muscles, the individual imagines stressful experiences. The person's physiological responses are then measured and recorded. The individual receives an interpretation of these responses and is taught methods for practicing relaxation to aid in the maintenance of homeostasis.

Biofeedback and the biofeedback loop enhance relaxation in tense muscles, relieve tension headaches, reduce bruxism (grinding of the teeth), reduce pain perception, and may even lower blood pressure. Some biofeedback techniques include deep breathing exercises, creative imagery, and progressive muscle relaxation.

DISORDERS OF THE PERIPHERAL NERVOUS SYSTEM

Charcot-Marie-Tooth (CMT) disease, also called hereditary motor and sensory neuropathy, is named for the three physicians who identified the disease in 1866. Charcot-Marie-Tooth disease is the most common inherited neurological condition and is caused by a gene defect. Symptoms appear in adolescents and young adults. These symptoms include hammertoes; high foot arch; and foot drop, which is when the foot cannot be picked up and held in a horizontal position. There is weakness in the lower legs and hands. Nerve degeneration leads to a reduced ability to sense heat, cold, or pain. Treatment includes physical therapy, occupational therapy, and assistive devices; however, CMT does not affect life span.

Congenital insensitivity to pain with anhidrosis is a condition in which there is insensitivity to pain; anhidrosis, or the inability to sweat; and intellectual disability. Treatment is focused on management of symptoms and prevention of injury or infection.

Neuritis (noo-**RIGH**-tis) is an inflammation of a nerve. Symptoms may be severe pain, hypersensitivity, loss of sensation, muscular atrophy, weakness; **paresthesia** (pair-es-**THEE**-zee-ah), or tingling, burning, and crawling of the skin, is also a symptom. The causes of neuritis may be infection; injury; chemical; or due to other conditions, such as chronic alcoholism. In the patient who is diagnosed with alcoholism, neuritis usually occurs because of a lack of vitamin B or an improper diet.

In the treatment of neuritis, it is necessary to determine the cause to eliminate the symptoms. The pain of neuritis may be relieved with **analgesics** (an-al-**JEE**-zicks), or painkillers; rest or reduced activity; and ice or heat therapy.

Sciatica is a form of neuritis that affects the sciatic nerve. The cause may be the rupture of a lumbar disc or arthritic changes. The most common symptom is pain that radiates through the buttock and behind the knee down to the foot; typically, only one side of the body is affected. The pain is usually worse after prolonged sitting. The person may have difficulty walking. Treatment includes hot or cold packs, anti-inflammatory medications, muscle relaxants, physiotherapy, exercises, and possible surgery to alleviate the symptoms. Alternative therapy for sciatica includes acupuncture or spinal adjustment by a chiropractor. Bed rest is not recommended.

Peripheral neuropathy (new-**ROP**-ah-thee) is the term used to describe damage or injury to the peripheral nerves. It is a frustrating and painful nerve condition that affects millions of Americans. The most common form involves damage to multiple nerves, or polyneuropathy, and is frequently caused by diabetes. Sensorimotor neuropathy usually starts with numbness or tingling in the toes or the soles of the feet, and symptoms slowly spread upward. It may also begin in the hands and extend up the arms. It may feel as if there is a sock or glove on the appendage. There may be muscle weakness, extreme sensitivity to touch, loss of balance and coordination, burning or freezing pain, and possible skin injury because of reduced pain perception. Diagnosis is through **electromyography (EMG)** (ee-leck-troh-my-**OG**-rah-fee). An electrograph is an instrument used to determine the electrical activity of muscles, and measurements can be taken of muscle strength. Another type of diagnostic technology that may be used is a nerve conduction study.

Treatment of peripheral neuropathy is through managing the underlying condition, repairing damage to the nerve if possible, and providing relief for the pain. Alternative therapies include the use of a transcutaneous electrical nerve stimulator (TENS) to deliver tiny electrical impulses to specific nerve pathways through small electrodes placed on the skin near the site of the pain. These impulses stimulate the nerve fibers and reduce pain signals to the brain. A TENS machine can also generate the release of hormones called endorphins, which are natural pain blockers.

Health care providers should instruct the patient on proper foot care: soak hands or feet in cool water for 15 minutes twice a day and, after soaking, apply a light coat of lotion to retain moisture in the skin. Massage the hands or feet to improve circulation.

Neuralgia (noo-**RAL**-jee-ah) is a sudden, severe, sharp, stabbing pain along the pathway of a nerve. The pain is often brief, and it may be a symptom of a disease. The various forms of neuralgia are named according to the nerve they affect. Treatment for neuralgia may include surgery to relieve pressure on the nerve; it can also include a nerve block, or an injection directed at a particular nerve or nerve group that blocks the pain signals and reduces inflammation. Medications prescribed include antidepressants, antiseizure medications, analgesics, and topical cream with capsaicin.

Shingles (herpes zoster) is an acute viral nerve infection. It is characterized by a unilateral (one-sided) inflammation of a cutaneous nerve (**Figure 10-8**). The intercostal nerves are the ones most commonly affected. The course of nerve inflammation can spread to any

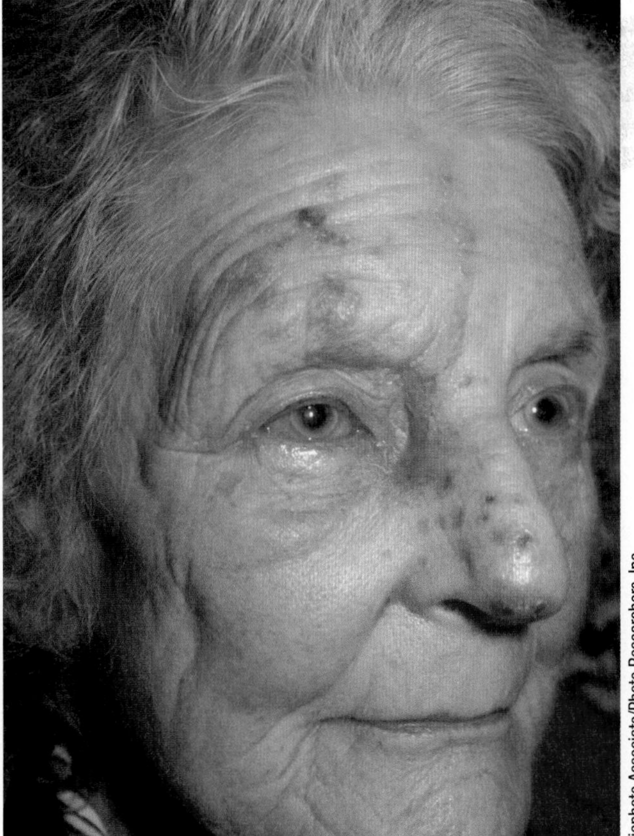

Figure 10-8 *Shingles are caused by the herpes zoster virus.*

nerve. For more discussion on shingles (herpes zoster), see Chapter 6. A vaccine for shingles has been approved by the FDA and is recommended for people over the age of 50.

Trigeminal neuralgia is a condition that involves cranial nerve V (trigeminal); refer to **Table 10-2**. The cause is unknown, the onset is rapid, and the pain is severe. The spasm of pain can be brought on by so slight a stimulus as a breeze, a piece of food in the mouth, or even a change in temperature. The term *tic douloureux* (tic doo-luh-**ROO**) is sometimes applied to this condition because the pain lasts only 2–5 seconds. Treatments include anticonvulsant drugs that block nerve firing, analgesics, or partial removal of cranial nerve V.

Bell's palsy is a condition that involves cranial nerve VII (facial); refer to **Table 10-2**. The patient with Bell's palsy appears as if they have had a stroke on one side of the face because Bell's palsy affects only one side. The eye does not close properly, the mouth droops, and there is numbness on the affected side. The cause is unknown.

There is no standard treatment for Bell's palsy. Some studies show that steroid and antiviral drugs may be effective by limiting or reducing damage to the nerve. Facial massage, physical therapy, moist heat, and analgesics may help control the pain. Another important treatment is eye protection. The natural blinking ability of the eyelid is interrupted with Bell's palsy, leaving the eye exposed to irritation and drying. Keeping the eye moist and protected from debris and injury is important. Lubricating eye drops and eye patches are effective. The patient must do exercises, such as whistling, to prevent atrophy of the cheek muscles. The symptoms usually disappear within a few weeks with no residual effects.

Carpal tunnel syndrome is a condition that affects the median nerve and the flexor tendons that attach to the bones of the carpal, or wrist. At the base of the palm, there is

Medical Highlights

TYPES OF ANESTHESIA

A variety of anesthetic techniques are available to ensure surgery without pain. The type of anesthesia used depends on the type of procedure being performed; other treatments the patient may be receiving; the medical condition of the patient; and the personal preference of the patient, if that is an option.

For major surgery, options include regional anesthesia or general anesthesia. In some conditions, the anesthetist may combine regional and general anesthesia.

REGIONAL

Regional anesthesia types include the following:

- Spinal anesthesia—Temporarily blocks nerve signals to and from the lower part of the body. The spinal anesthetic is delivered into the subarachnoid space. A local injection is given to numb the injection site. The patient remains conscious; however, the patient may receive a sedative for relaxation. Once the drug has taken effect, numbness occurs in the lower body. This type of anesthesia is most often used in orthopedic procedures.
- Epidural anesthesia—Similar to spinal anesthesia. Medications are injected into the epidural space, which is outside the subarachnoid space. A catheter is inserted to allow for repeated injections if necessary. This method is used frequently in childbirth.

Side effects of spinal and epidural anesthesia may include headache, itching, backache, and difficulty in urination. Feeling will generally return within 1–4 hours.

GENERAL

General anesthesia is appropriate for extensive surgery that requires the patient to be unconscious, or when regional anesthesia is not an option. Medications are delivered intravenously or are inhaled. All vital signs are closely monitored during surgery. A breathing tube may also be inserted. General anesthesia induces a deep sleep, blocks the memory of the surgery, and keeps the brain from perceiving pain. Because medications are delivered to the patient's general circulation, they may produce side effects, such as nausea and vomiting, muscle aches, dry mouth, sore throat, shivering, sleepiness, and inhibited bowel function. These side effects are generally temporary.

LOCAL

For local anesthesia, an anesthetic agent is used to temporarily stop the sense of pain in a particular area; the patient is conscious.

IV/MONITORED SEDATION

IV/Monitored sedation is an intravenous type of anesthesia that entails the injection of a pain-killing drug into the site of a specific body part followed by IV drugs to keep the patient sedated. This type of procedure is used for less complex surgeries, such as biopsies or foot surgery. The patient is in a very relaxed state with this type of anesthesia and may be able to communicate.

Source: Anesthesia Option, Mayo Clinic Health Letter, Volume 23, Number 8, August 2005. Mayo Foundation for Medical Education and Research, Rochester MN.

a tight canal or "tunnel" through which tendons and nerves pass on their way from the forearm to the hand and fingers. The median nerve and flexor tendon pass through this tunnel (**Figure 10-9**). The syndrome develops because of repetitive movement of the wrist, while the hands are held in an unusual position. Swelling (edema) develops around the carpal tunnel, causing pressure on the nerve that results in pain, muscle weakness, and tingling sensations of the hand. The diagnostic test for carpal tunnel syndrome is electromyography. Treatment consists of immobilizing the wrist joint. If this treatment is not effective, surgery may be done.

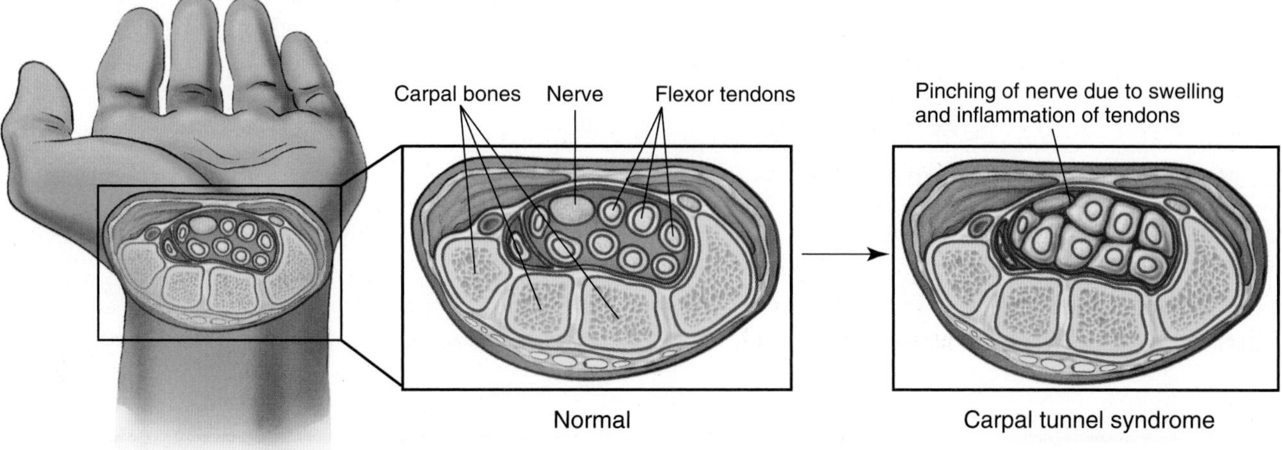

Figure 10-9 *Carpal tunnel syndrome occurs with inflammation of the carpal tunnel, resulting in pressure on the median nerve.*

One BODY — How the Peripheral and Autonomic Nervous System Interacts with Other Body Systems

INTEGUMENTARY SYSTEM

- Sympathetic nervous system reacts to heat and cold and influences the sweat glands, blood vessels, and muscles to regulate body temperature.

SKELETAL SYSTEM

- Spinal radial nerve influences the action of the wrist and hand.
- Spinal femoral nerve influences the action of the hip and leg bones.
- Spinal sciatic nerve influences the action of the knee and hip.

MUSCULAR SYSTEM

- Cranial nerves III, IV, and VI act on eye muscles.
- Cranial nerve VII influences facial muscles.
- Cranial nerve XI influences swallowing muscles.
- Autonomic nervous system (ANS) influences smooth muscle action.

ENDOCRINE SYSTEM

- Sympathetic nervous system stimulates the adrenal medulla to produce epinephrine, or adrenaline.

CIRCULATORY SYSTEM

- Cranial nerve X slows the heart rate.
- The ANS influences heart rate, blood pressure, and blood flow.

LYMPHATIC SYSTEM

- Anxiety and stress decrease the immune response as a reaction to epinephrine production.

RESPIRATORY SYSTEM

- Spinal phrenic nerve stimulates the diaphragm for regular breathing.
- The ANS regulates the rate and depth of respiration.

DIGESTIVE SYSTEM

- Cranial nerve X stimulates digestion.
- The ANS influences the smooth muscle of the digestive tract.

URINARY SYSTEM

- The ANS influences the smooth muscle of the urinary tract and aids in the elimination of urine.

REPRODUCTIVE SYSTEM

- The ANS regulates sexual erection and ejaculation in males and erection of the clitoris in females.
- Smooth muscle contraction initiates childbirth and delivery.

Medical Terminology

anal	without	**electro/myo/graphy**	process of recording electrical activity in the muscle
-gesic	sensitivity to pain	**neuro**	nerve
anal/gesic	without sensitivity to pain	**-algia**	pain
crani	skull	**neur/algia**	nerve pain
-al	pertaining to	**-itis**	inflammation
crani/al nerve	pertaining to a nerve in the skull	**neur/itis**	inflammation of a nerve
electro	electrical activity	**par-**	near, beyond, beside, around
myo	muscle	**-esthesia**	abnormal condition of feeling sensation
-graphy	process of recording	**par/esthesia**	an abnormal condition of feeling sensation; tingling sensation

Study Tools

Workbook	Activities for Chapter 10
Online Resources	• PowerPoint presentations • Animation

REVIEW QUESTIONS

Select the letter of the choice that best completes the statement.

1. A nerve that contains fibers that send as well as receive messages is called a
a. sensory nerve.
b. afferent nerve.
c. efferent nerve.
d. mixed nerve.

2. The cranial nerve responsible for chewing is
a. trochlear.
b. facial.
c. glossopharyngeal.
d. trigeminal.

3. The cranial nerves responsible for eye muscle movement are the oculomotor, trochlear, and
a. abducens.
b. vestibulocochlear.
c. accessory.
d. hypoglossal.

4. A network of spinal nerves is called
a. mixed.
b. efferent.
c. plexus.
d. afferent.

5. The autonomic nervous system is also called
a. voluntary.
b. involuntary.
c. neuralgic.
d. carpal.

6. The autonomic nervous system is part of the
a. central nervous system.
b. peripheral nervous system.
c. sympathetic nervous system.
d. parasympathetic nervous system.

7. The sympathetic nervous system is influenced by the action of the
a. adrenal cortex.
b. pancreas.
c. thyroid.
d. adrenal medulla.

8. The nerve that activates the diaphragm is called
a. sciatic.
b. phrenic.
c. radial.
d. femoral.

9. The simplest type of nervous system response is called
a. stimulus.
b. effector action.
c. reflex.
d. affector action.

10. The acute viral infection that usually affects the intercostal nerves is called
a. Bell's palsy.
b. neuralgia.
c. sciatica.
d. shingles.

FILL IN THE BLANKS

1. The ________________ nervous system conducts impulses from the brain and spinal cord to the skeletal muscles.
2. A nerve composed of fibers carrying impulses from sense organs to the brain or spinal cord is called a ________________ or ________________ nerve.
3. A nerve composed of fibers carrying impulses from the brain or spinal cord to muscles or glands is called a ________________ or ________________ nerve.
4. A mixed nerve contains both ________________ and ________________ fibers.
5. The autonomic nervous system has two parts that counterbalance each other; these are the ________________ and ________________ systems.

TRUE OR FALSE

Mark your answers true or false. Correct the false statements.

T F **1.** There are 24 pairs of cranial nerves that begin in areas of the brain.

T F **2.** The phrenic nerve is located in the lumbar plexus.

T F **3.** An inflammation of a nerve is called neuritis.

T F **4.** Paresthesia is tingling, burning, and crawling of the skin.

T F **5.** Trigeminal neuralgia is a condition that affects cranial nerve III.

T F **6.** Bell's palsy is a condition that affects only one side of the face.

T F **7.** The sciatic nerve is located in the brachial plexus.

T F **8.** The diagnostic test for carpal tunnel syndrome is an electrocardiogram (EKG).

T F **9.** Peripheral neuropathy may occur as a result of diabetes.

T F **10.** Treatment for peripheral neuropathy includes soaking hands or feet for 30 minutes at least three times a day.

LABELING

Study the following diagram and name the numbered structures.

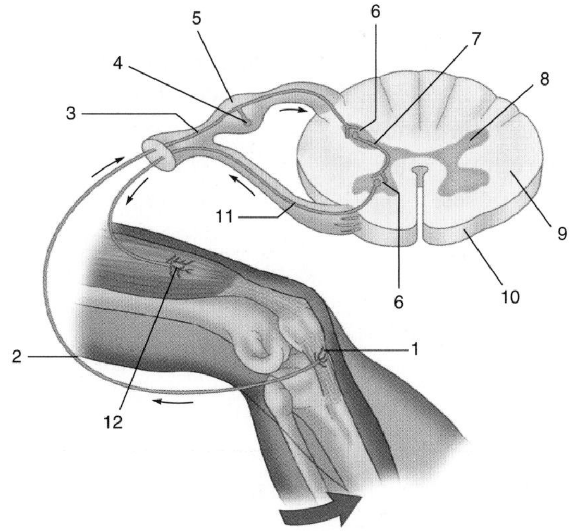

1. ______________
2. ______________
3. ______________
4. ______________
5. ______________
6. ______________
7. ______________
8. ______________
9. ______________
10. ______________
11. ______________
12. ______________

APPLYING THEORY TO PRACTICE

1. You are passing by a pizzeria and smell the pizza cooking. Describe what happens to your salivary glands. What other feelings do you notice? Relate these reactions to your peripheral nervous system.
2. The knee jerk is the most common reflex known in health care. Name at least five reflexes you were born with. What are their purposes?
3. A physician has a patient who is experiencing facial and cheek pain, sometimes called trigeminal neuralgia. Describe this condition and the appropriate treatment.
4. After a lengthy car ride, your elderly uncle gets out of the car and complains, "I can hardly walk. It must be sciatica." Explain what this means.
5. Peripheral neuropathy is a common complication of diabetes. As a health care provider, what are some therapeutic technologies you could advise a person with this condition to try? What effects should a person expect to see with these therapies?

CASE STUDY

Paula is a 38-year-old administrative assistant. She visits the medical assistant at ABC Company's health office. During the interview, Paula explains that she has been waking up at night with pain in both wrists. Paula also states that the wrist pain becomes worse after she has been working on the computer. Paula says she has been using wrist supports, but they do not seem to help. The medical assistant refers Paula to the physician.

1. The diagnosis is carpal tunnel syndrome. Name the nerves and bones involved in this disorder.
2. Explain the test that will be done to confirm the diagnosis.
3. Describe the symptoms that occur in carpal tunnel syndrome.
4. What is the treatment for this disorder?

LAB ACTIVITY 10-1 Simple Reflex

- ***Objective:*** To observe the response of the simple knee-jerk reflex
- ***Materials needed:*** reflex hammer, stopwatch, textbook, paper, pencil

Step 1: Work in groups of three; the third person is needed to do the timing. Have your lab partner sit on the lab bench or chair and cross their right knee over their left knee.

Step 2: Tap the right knee with the reflex hammer (see Figure 10-7). Time the response. Observe the action that occurred. What leg muscle and nerves were involved? Record the timing, your observations, and answers.

Step 3: Reverse the process and have your lab partner cross their left knee over their right knee.

Step 4: Tap the left knee with the reflex hammer. Distract your partner by reciting the multiplication table by 5s while you are doing the experiment. Time the response. What action occurred? Was there a difference between the response times in the left and right knees? Record your answer.

Step 5: Switch places with your lab partner and have your lab partner conduct steps 1–4 on you.

Step 6: Was there any difference in the timing of the responses between you and your partner? Explain the differences, if any. Record your answer.

LAB ACTIVITY 10-2 Salivary Reflex Response

- ***Objective:*** To observe the response of the salivary reflexes
- ***Materials needed:*** lemon juice, measuring cup, paper cup, pH paper, stopwatch, paper, pencil

Step 1: Have your lab partner refrain from swallowing for two minutes.

Step 2: After two minutes, have the partner spit saliva into a paper cup.

Step 3: Measure the amount of saliva and use the pH paper to determine the pH of the saliva.

Step 4: Place two drops of lemon juice on your lab partner's tongue.

Step 5: Allow the lemon juice to mix with the saliva for 5–10 seconds.

Step 6: After 5–10 seconds, touch a piece of the pH paper to your lab partner's tongue. Record the results.

Step 7: Switch roles with your partner. Refrain from swallowing for two minutes.

Step 8: After two minutes, spit saliva into a paper cup. Have your partner measure the amount of saliva, use pH paper to determine the pH of the saliva, and record their findings.

Step 9: Does the amount and pH of the saliva secretions differ between the ordinary saliva and the saliva that was mixed with lemon juice? Record the differences, if any.

LAB ACTIVITY 10-3 Normal Function of Cranial Nerve XII

- ***Objective:*** To test the function of cranial nerve XII, the hypoglossal nerve
- ***Materials needed:*** textbook, paper, pencil

Step 1: Work in pairs.

Step 2: Stand in front of your lab partner and ask them to read one or two sentences from the textbook.

Step 3: Record if speech is normal; note any deficiencies.

Step 4: Ask your lab partner to protrude their tongue; check for any deviation to the right or left.

Step 5: Record whether the tip of the tongue was in the midline or deviated to the right or left.

Step 6: Record any further observations.

Step 7: Change places with your lab partner and repeat steps 2–6.

Step 8: Compare your results. Explain the normal function of the hypoglossal nerve.

LAB ACTIVITY 10-4 The Function of Cranial Nerve XII

- ***Objective:*** To demonstrate how deep breathing affects relaxation
- ***Materials needed:*** chairs, pen, pencil

Step 1: Sit in a relaxed position and write down how you are feeling.

Step 2: Inhale through your nose, expanding the diaphragm. To check whether you are breathing using your diaphragm, place your hand on your stomach to check that it rises with each inhale.

Step 3: Hold your breath for five seconds.

Step 4: Slowly exhale through the mouth as you say, "My body is relaxed and calm."

Step 5: Repeat activity 10 times.

Step 6: Record how you feel after this exercise. Is how you feel different than before this exercise?

CHAPTER 11

Special Senses

Objectives

- Describe the function of the sensory receptors in the body.
- Identify the parts of the eye and describe their functions.
- Trace the pathway of light from outside to the occipital lobe.
- Identify the parts of the ear and describe their functions.
- Trace the pathway of sound from pinna to temporal lobe.
- Describe the process involved with the sense of smell.
- Describe common disorders of the eye, ear, nose, and tongue.
- Define the key words that relate to this chapter.

Key Words

accommodation
amblyopia
anterior chamber
anvil (incus)
aqueous humor
astigmatism
auricle
cataract
cerumen
choroid coat
ciliary body
cochlea
cochlear duct
color blindness
cones
congenital hearing loss
conjunctivitis
cornea
detached retina
deviated nasal septum
diabetic retinopathy
diplopia
dry eye
epistaxis
eustachian tube
extrinsic muscles
eyestrain
fovea centralis
glaucoma
hammer (malleus)
hyperopia (farsightedness)
intrinsic muscles
iris
lens
macula lutea
macular degeneration
Ménière's disease
miotic
myopia (nearsightedness)
myringotomy
nasal polyps
night blindness
optic disc (blind spot)
organ of Corti
otitis media
otosclerosis
oval window
posterior chamber
presbycusis
presbyopia
pupil
retina
rhinitis
rods
sclera
semicircular canals
stirrup (stapes)
strabismus (crossed eyes)
sty (hordeolum)
suspensory ligaments
tinnitus
tympanic membrane
umami
vertigo
vitreous humor

GENERAL SENSORY RECEPTORS

Sensory receptors are structures that are stimulated by changes in the environment. Sensory receptors for touch, pain, temperature, and pressure are found all over the body in the skin, connective tissue, and muscle. There are special sensory receptors, which include the taste buds of the tongue, cells in the nose, the retina of the eye, and the special cells in the inner ear. When a sense organ is stimulated, the impulse travels along nerve pathways to the brain, where the sensation is registered in a certain area. Sensation actually takes place in the brain, but it is mentally referred back to the sense organ. This is called projection of the sensation. The sensory receptors become less sensitive with age. A decrease in the number of receptors makes it more difficult for older adults to feel pain or cope with changes in temperature.

This chapter will focus on the special senses.

SPECIAL SENSES

The special senses are those organs and receptors that are associated with touch (sensory receptors), vision, hearing, smell, and taste. Sight, hearing, and smell are distance senses; they bring information from far away. Touch and taste can only reveal information about things with which a person actually comes in direct contact. Functions of the special senses are to receive stimuli from the sensory receptors—the eye (sight), the ear (hearing), the nose (smell), and the tongue (taste)—and to transmit these impulses to the brain for interpretation.

THE EYE

The human eye is a tender sphere about 1 inch (2.5 cm) in diameter. It is protected by the orbital socket of the skull, the eyebrows, eyelids, and eyelashes. When people blink, the eyes are continuously bathed in fluid by tears secreted from lacrimal (**LACK**-rih-mal) glands, which are located on the underside of the upper lid of each eye. Tears flow across and downward to the lacrimal canal, which consists of a duct at the inner corner of each eye. Ducts collect the tears and empty them into the nasal lacrimal duct that drains the excess tears into the nose (**Figure 11-1**). This explains why, when a person cries, they may also need to blow their nose. Lacrimal secretions contain lysozymes, which help combat bacterial infections. Tears cleanse and moisten the eyes on a continuous basis.

Along the border of each eyelid are glands that secrete an oily substance that lubricates the eye. The canthus (**KAN**-thus) is the angle where the upper and lower eyelids meet.

The conjunctiva (**con**-junk-**TYE**-vah) is the thin membrane that lines the eyelids and covers part of the eye. The conjunctiva secretes mucus that helps lubricate the eye.

The location of the eyes in the front of the head allows for the superimposition of images from each eye. This enables a person to see stereoscopically in three dimensions: length, width, and depth.

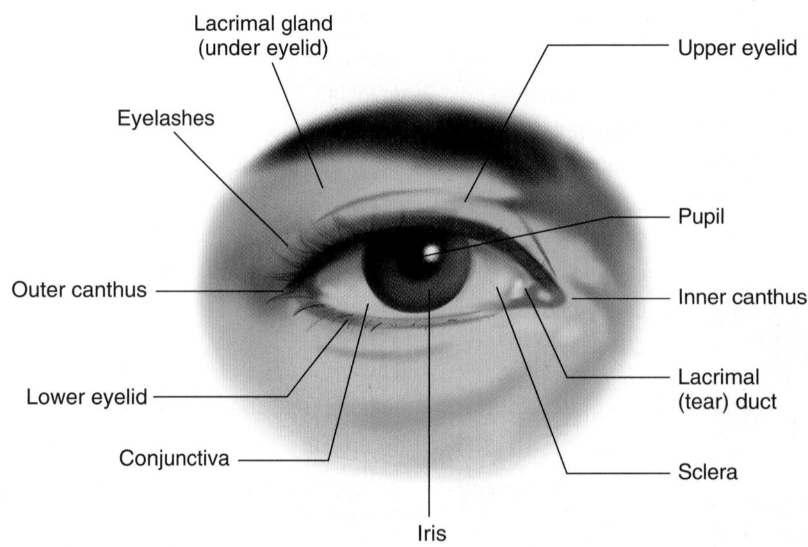

Figure 11-1 *External view of the eye*

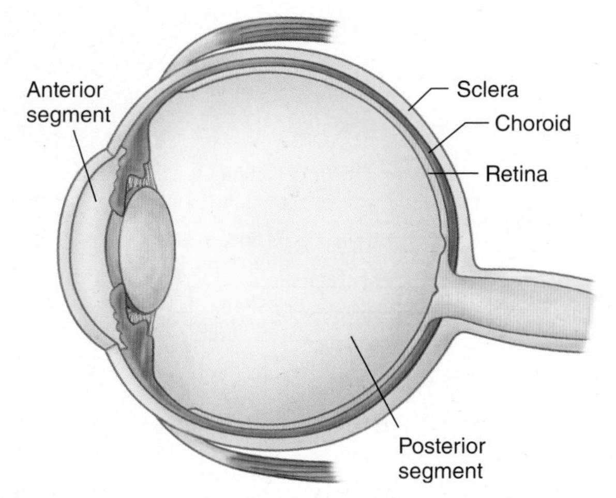

Figure 11-2 *The layers of the eyeball are made up of the sclera, choroid, and retina.*

The wall of the eye is made up of three concentric layers, or coats, each with its specific function. These three layers are the sclera, choroid, and retina (**Figure 11-2**).

Sclera

The outer layer is called the **sclera** (**SKLEHR**-ah), or white of the eye. It is a tough, unyielding, fibrous capsule that maintains the shape of the eye and protects the delicate structures within. Muscles responsible for moving the eye within the orbital socket are attached to the outside of the sclera. These muscles are referred to as the **extrinsic muscles** (**Figure 11-3**). They include the superior, inferior, lateral, and medial rectus, as well as the superior and inferior oblique. See **Table 11-1** for a listing of the extrinsic eye muscles and their functions.

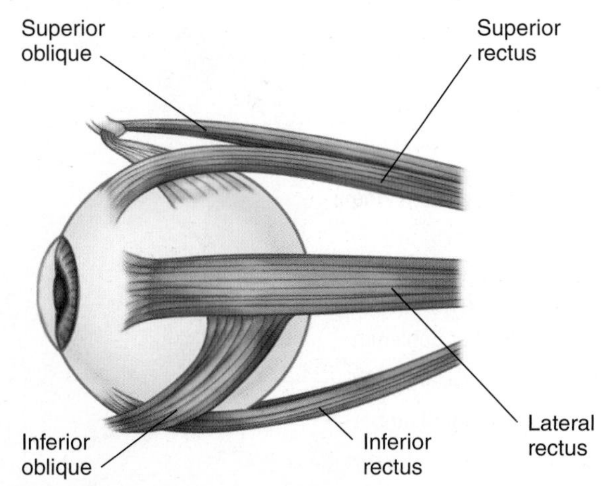

Figure 11-3 *The extrinsic eye muscles—six muscles arranged as three pairs—make eye movement possible. The medial rectus muscle is not visible here.*

Table 11-1 *Extrinsic and Intrinsic Eye Muscles*

EYE MUSCLE	***FUNCTION***
A. Extrinsic	
1. Superior rectus	Rolls eyeball upward
2. Inferior rectus	Rolls eyeball downward
3. Lateral rectus	Rolls eyeball laterally, away from the nose
4. Medial rectus	Rolls eyeball medially, inward toward the nose
5. Superior oblique	Rolls eyeball on its axis; moves cornea downward and laterally
6. Inferior oblique	Rolls eyeball on its axis; moves cornea upward and laterally
B. Intrinsic	
1. Sphincter pupillae	Constricts pupil
2. Dilator pupillae	Dilates pupil

CORNEA

In the very front center of the sclerotic coat lies a circular, clear area called the **cornea** (**KOR**-nee-ah). The cornea is sometimes referred to as the "window" of the eye. It is transparent to permit the passage of light rays. The cornea consists of five layers of flat cells. Possessing pain and touch receptors, it is sensitive to any foreign particles that come in contact with its surface. An injury to the cornea may cause scarring and impaired vision.

Choroid Coat and the Iris

The middle layer of the eye is the **choroid coat** (**KOH**-roid). It contains blood vessels to nourish the eye and a nonreflective pigment that renders it dark and opaque. The pigment provides the choroid coat with a deep, red-purple color; this darkens the eye chamber, preventing light reflection within the eye. In front, the choroid coat has a circular opening called the **pupil**. Light passes into the eye through the pupil. A colored, muscular layer surrounds the pupil; this is the **iris**, or colored part of the eye. The iris may be blue, green, gray, brown, or black. Eye color is related to the number and size of melanin pigment cells in the iris. If little melanin is present, the eye is blue because light is scattered to a greater extent. With increasing quantities of melanin, eye color ranges from green to black. The total absence of melanin results in a pink eye color, a characteristic of albinism. Such

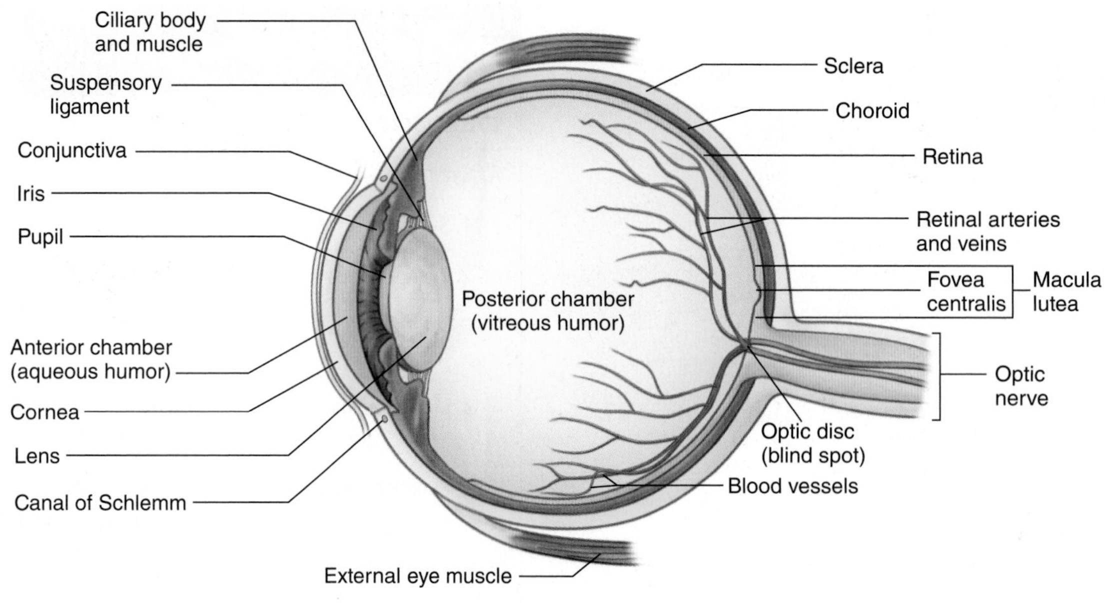

Figure 11-4 *Internal view of the eye*

irises are pink because the blood inside the choroid blood vessels shows through the iris (**Figure 11-4**).

There are two sets of antagonistic smooth muscles within the iris: the sphincter pupillae and the dilator pupillae. These **intrinsic muscles** help the iris control amounts of light entering the pupil. When the eye is focused on a close object or stimulated by bright light, the sphincter pupillae muscle contracts, rendering the pupil smaller. Conversely, when the eye is focused on a distant object or stimulated by dim light, the dilator pupillae muscle contracts. This causes the pupil to grow larger, permitting as much light as possible to enter the eye (refer to **Table 11-1**).

THE LENS AND RELATED STRUCTURES

The **lens** is a crystalline structure located behind the iris and pupil. The purpose of the lens is to focus images on the retina. It has concentric layers of fibers and crystal-clear proteins in solution. It is an elastic, disc-shaped structure with anterior and posterior convex surfaces that form a biconvex lens.

The curvature of each surface changes with age. The capsule surrounding the lens loses its elasticity over time. The lens is held in place behind the pupil by **suspensory ligaments** from the **ciliary body** (**SIL**-ee-ehr-ee) of the choroid layer. The ciliary body consists of smooth muscle. The muscle controls the shape of the lens for vision at near and far distances; this is the process of **accommodation** (ah-**KOM**-oh-day-shun).

The lens is situated between the **anterior chamber** and **posterior chamber**. The anterior chamber is filled with a watery fluid called **aqueous humor** (**AH**-kwee-uhs). It is constantly replenished by blood vessels behind the iris (**Figure 11-5**). Aqueous humor fluid is constantly filtered and drained through the trabecular network and canal of Schlemm. **Vitreous humor** (**VIT**-ree-us), a transparent, jellylike substance, fills the posterior chamber. Both of these substances help maintain the eyeball's

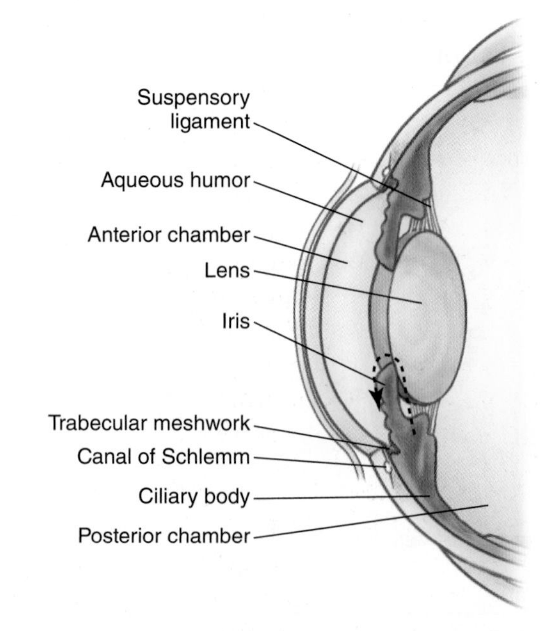

Figure 11-5 *Aqueous humor fluid in the anterior chamber of the eye*

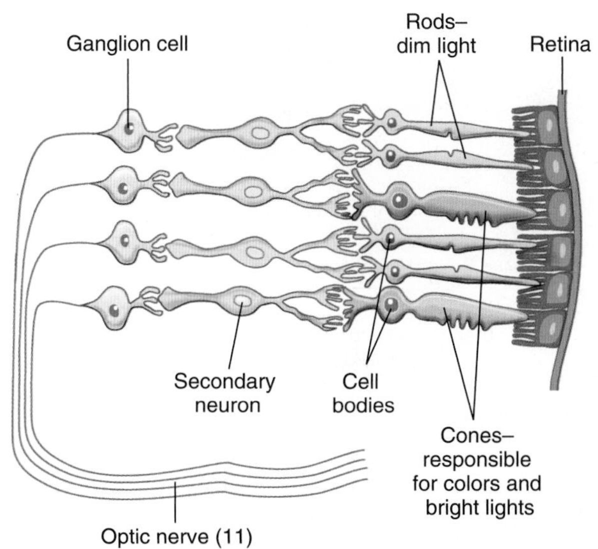

Figure 11-6 *Diagram of visual neurons showing rods and cones*

spherical shape. Intraocular pressure is a measurement of the fluid pressure inside the eye. This pressure is regulated by the rate at which aqueous humor enters and leaves the eye.

Retina

The **retina** (**RET**-ih-nah) of the eye is the innermost, or third, coat of the eye. It is located between the posterior chamber and the choroid coat. The retina does not extend around the front portion of the eye. It is on this light-sensitive layer that light rays from an object form an image. After the image is focused on the retina, it travels via the optic nerve to the visual part of the cerebral cortex (occipital lobe). If light rays do not focus correctly on the retina, the image is not sharp. This condition may be corrected with properly fitted contact lenses or eyeglasses, which bend the light rays as required.

The retina contains pigment and specialized cells known as **rods** and **cones** (**Figure 11-6**), which are sensitive to light. The rod cells are sensitive to dim light, and the cones are sensitive to bright light. The cones are also responsible for color vision. There are three varieties of cone cells. Each type is sensitive to a special color. The part of the retina where the nerve fibers enter the optic nerve to go to the brain is called the optic disc.

THE OPTIC DISC AND THE FOVEA

Viewing the retina through an ophthalmoscope, one can observe a yellow disc called the **macula lutea** (**MACK**-you-lah **LOO**-tee-ah). Within this disc is the **fovea centralis** (**FOH**-vee-ah sen-**TRAH**-lis), which contains the cones for color vision (**Figure 11-4**). The area around the fovea centralis is the extrafoveal, or peripheral, region. This is where the rods for dim and peripheral vision can be found.

Slightly to the side of the fovea lies a pale disc called the **optic disc (blind spot)**. Nerve fibers from the retina gather here to form the nerve. The optic disc contains no rods or cones; therefore, it is unable to convert images into nerve impulses.

See **Figure 11-7** for how to locate the blind spot.

1. Close your left eye and focus your right eye on the cross.
2. Move the page slowly away from your eye and then slowly toward your eye.
3. At a distance of about 6–8 inches the black circle "disappears."

Figure 11-7 *Testing for the blind spot*

Pathway of Vision

Images in the light → cornea → pupil → lens → where the light rays are bent or refracted → retina → rods and cones (nerve cells) pick up the stimulus → optic nerve → optic chiasma (kye-**AZ**-mah) (where the two optic nerves cross) → optic tracts → occipital lobe of the brain for interpretation (**Figure 11-8**).

View the **Vision** animation on the Online Resources.

Eye Disorders

It is important to have an annual eye examination. Early detection of eye problems can save one's vision.

CONGENITAL EYE DISORDERS

Newborns are frequently given antibiotic eye drops at birth to prevent bacterial infections after birth, such as gonococchal opthalmia neonaturum (GON), an infection in the eye that can occur in babies born to a female with gonorrhea.

It may be difficult at first to see if a baby has a vision problem; focusing usually does not occur until about 12 weeks of age. Six to 12 weeks after birth, some signs a

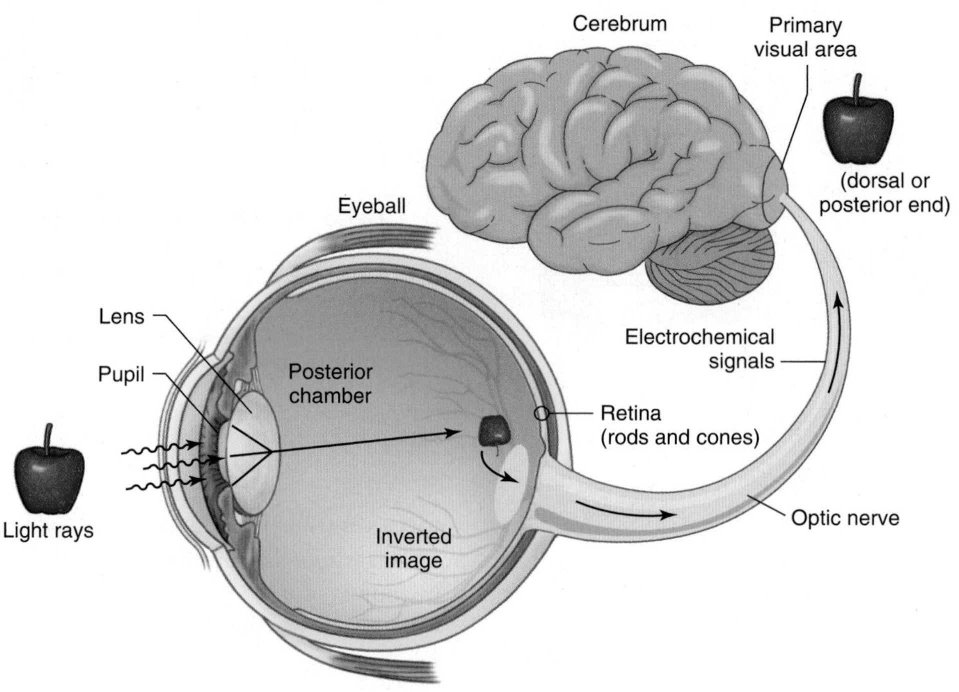

Figure 11-8 *Pathway of vision: Images in light → cornea → pupil → lens → retina → optic nerve → occipital lobe (cerebrum)*

baby may have a vision problem include not reacting to mobiles, lights, and other distractions, or if the baby does not respond to their mother's face. The baby may squint or rub their eyes even when not sleepy, never open an eye, or an eye may look cloudy.

Babies may be born with an astigmatism, cataracts, glaucoma, or myopia. Treatment is surgery and/or corrective lenses. Congenital ptosis is drooping of the eyelids, which does not affect vision.

GENERAL EYE DISORDERS

Conjunctivitis (kon-junk-tih-**VYE**-tis), or pink eye, is an inflammation of the conjunctival membranes. Redness, pain, swelling, and discharge of mucus occur. Conjunctivitis usually begins in one eye and spreads rapidly to the other by a washcloth or hands. Because it is highly contagious, other family members should not share washcloths or towels with the infected person. Good handwashing is important. Treatment includes eyewashes or eye irrigations, which will cleanse the conjunctiva and relieve the inflammation and pain. Bacterial conjunctivitis responds to antibiotic drug therapy.

Glaucoma (glaw-**KOH**-mah) is a condition of excessive intraocular pressure resulting in the destruction of the retina and atrophy of the optic nerve. The condition results from overproduction of aqueous humor or the obstruction of its outflow through the canal of Schlemm for absorption into the venous circulation. Symptoms are gradual. They include mild aching in the eye, a loss of peripheral vision (**Figure 11-9C**), and a halo around lights.

Glaucoma may occur with aging and has no initial symptoms. It is important for people to be tested for glaucoma annually after age 40. Some tests that are used for the diagnosis and continued evaluation of glaucoma are tonometry, which measures the pressure in the eye; ophthalmoscopy, with visualization of the optic nerve; perimetry, or visual field testing, in which the person being tested focuses on a single object straight ahead but must also say what can be seen above, below, or to the side of the object; gonioscopy, which evaluates the internal drainage system of the eye; and pachymetry, which measures the thickness of the corneas.

Treatment involves **miotic** drugs, which constrict the pupil and thus increase the outflow of aqueous humor, or drugs that reduce the amount of aqueous fluid produced by the eye. Today, laser surgery or incisional surgery helps increase the flow of aqueous humor. All treatments are focused on lowering the intraocular pressure.

A **cataract** is a condition in which the lens of the eye gradually becomes cloudy (**Figure 11-9B** and **Figure 11-10**). This frequently occurs in people over 70 years of age. The condition causes a painless, gradual blurring and loss of vision. The pupil appears to change color from black to milky white. People with cataracts may complain of seeing halos around lights or being blinded at night by oncoming headlights.

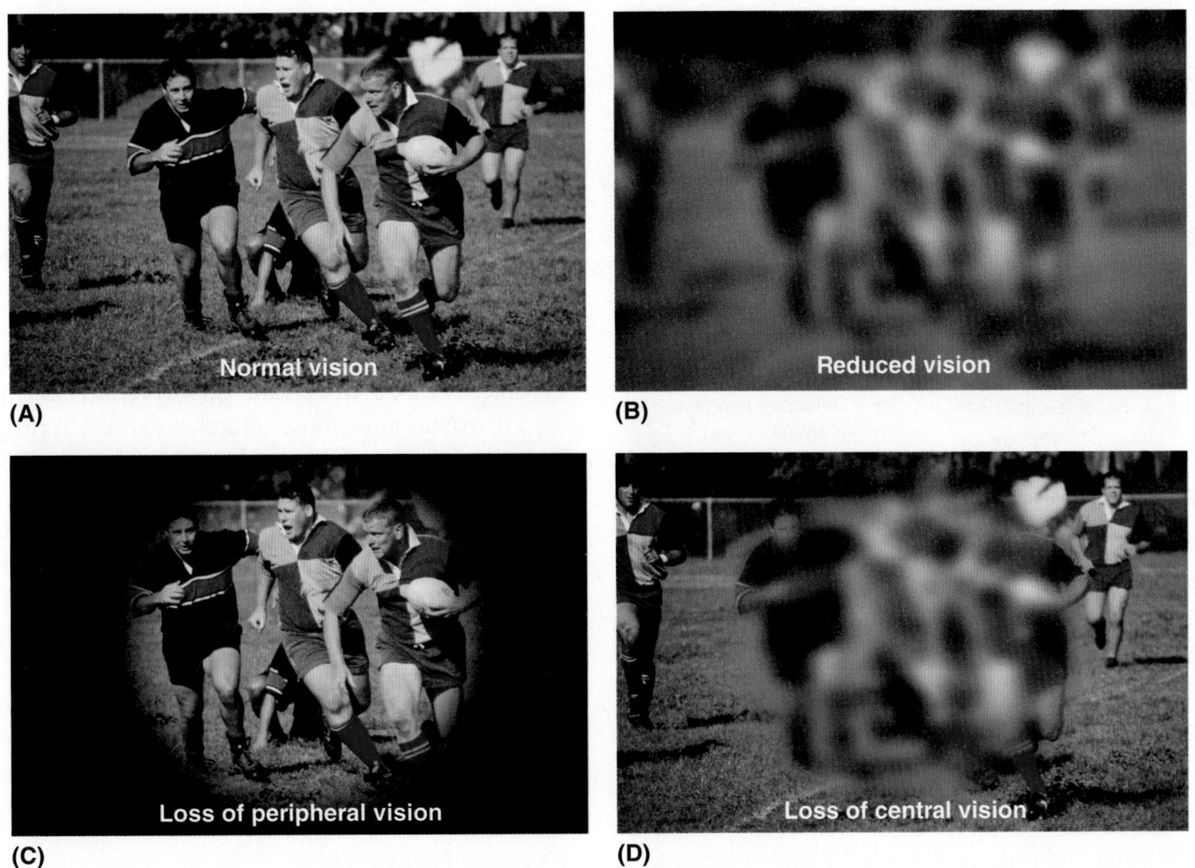

Figure 11-9 *Normal vision and pathologic changes: (A) normal vision, (B) vision reduced by cataracts, (C) the loss of peripheral vision caused by untreated glaucoma, (D) the loss of central vision caused by macular degeneration*

Cataracts are either treated by laser surgery or by the surgical removal of the lens and the subsequent implantation of an intraocular lens directly behind the cornea.

Macular degeneration is another eye disorder that occurs as a person ages. In the central part of the retina is the macula, which is responsible for sharp central vision. Symptoms include a dimming or distortion of vision that is most obvious when reading. In one form of the disease, straight lines look wavy, and blind spots may develop in the visual field. The two types of macular degeneration are dry—the most common type—and wet. In the dry type, the main defect is a gradual thinning of the retina. This is slowly progressive, and there is no known treatment. Central vision will be greatly reduced, but there is not usually total blindness (**Figure 11-9D**).

In wet macular degeneration, new blood vessels grow behind the macula, leaking, bleeding, and distorting its shape. Drug treatment with Avastin® inhibits the protein that prompts destructive new blood vessels to form. The injectable drug treatment is given every few weeks, may be needed indefinitely, and is not successful for everyone with the disease. Laser treatment may also be used with this type of macular degeneration. The good news about macular degeneration is that the majority of people who develop it will be able to maintain their independence of movement with low-vision aids.

A **detached retina** is another problem that can occur with aging. It can also occur as the result of trauma to the eye at a younger age. The vitreous fluid contracts as it ages and pulls on the retina, causing a tear (**Figure 11-11**). Symptoms include the loss of peripheral

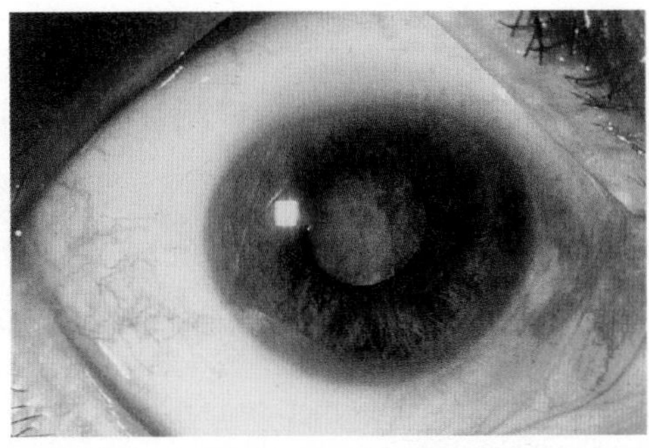

Figure 11-10 *Cataract*

Courtesy of the National Eye Institute, NIH.

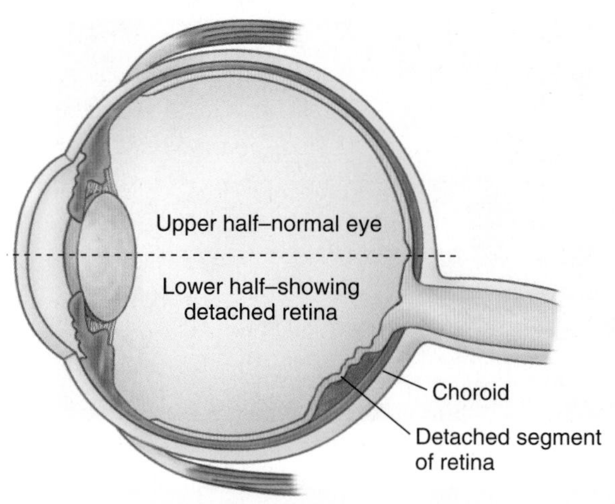

Figure 11-11 *Retinal detachment*

vision and then the loss of central vision. A detached retina is a medical emergency; with early detection, it can be repaired with pneumatic retinopathy, vitrectomy, or scleral buckling. If a detached retina goes without treatment, blindness in the eye may occur.

Diabetic retinopathy (ret-ih-**NOP**-ah-thee) is the leading cause of blindness in American adults. It is caused by changes in the blood vessels in the retina. Blood vessels may swell and leak, or abnormal blood vessels may grow on the retina. Because there are no symptoms in the early stages, people with diabetes must have regular eye examinations. With advanced diabetic retinopathy, people can see red spots if bleeding occurs. Laser surgery for diabetic retinopathy is usually effective and reduces the risk of blindness by 90%.

A **sty (hordeolum)** (hor-**DEE**-oh-lum) is a tiny abscess at the base of an eyelash caused by inflammation of a tiny sebaceous gland of the eyelid. The eye is red, painful, and swollen. Treatment consists of warm, wet compresses to relieve pain and promote drainage.

Dry eye is a scratchy feeling in the eyes that occurs when the tears fail to keep the eye surface adequately lubricated. It is important to avoid pollutants, wear protective eyewear to block dry air, humidify dry air indoors, and use artificial tears to help this condition. Dry eye should be evaluated by a physician to determine if there is an underlying disease causing the eyes to be dry.

EYE INJURIES

In most cases of simple eye irritation, the natural flow of tears will help cleanse the eye. Eye irritations can be caused by chemicals or fragments that get into the eye. If this occurs, rinse the eyes with water for at least 15 minutes and seek medical treatment. In situations where a piece of glass or other fragment gets into the eye, do not attempt to remove the object. Patch both eyes and get medical treatment.

Corneal abrasions, scratches, and scarring may occur as a result of an accident or irritation. Treat by blinking a few times, leaving contact lenses out, and wearing sunglasses to avoid the glare of light. In a severe case, a corneal transplant may be done. The cornea is avascular; that is, it does not have any blood vessels. Since there are no vessels present, tissue rejection is not an issue for corneal transplants.

Flash burns are burns to the eyes from sunlight, arc welding, or sun lamps. Treatment includes padding the eyes; using sunglasses even indoors; avoiding television, tablets, cell phones, computers, and so on; and using artificial tears.

VISION DEFECTS

Eyestrain is experienced as burning; tightness; sharp or dull pain; watery, blurry vision; and headaches. If a person has any discomfort when viewing something, it can be called eyestrain. The most common cause is too much computer usage. Too much light may be coming from the computer screen, or a strong light behind the user may be causing a glare on the computer monitor. These problems can be resolved by repositioning the monitor, using a glare screen, or wearing dark clothing to reduce glare caused by reflection.

If glasses are necessary, the user should wear glasses that allow for near; intermediate, such as when using the computer; and far vision. *Most important—Give the eyes a break. Remember the 20/20 rule—every 20 minutes, look 20 feet away for 20 seconds.*

Medical Highlights

LASERS

Laser, short for light amplification by stimulated emission of radiation, is based on the principle that certain atoms, molecules, or ions can be excited by absorption of thermal, electrical, or light energy. After such energy absorptions, the atoms, molecules, or ions give off a beam of synchronized light waves. The laser beam is a narrow, intense, and monochromatic (single-color) light beam that can be used for a variety of purposes. For example, it can stop bleeding, make incisions, or remove tissue.

Night blindness is a condition that makes it difficult to see at night. The rod cells in the retina are affected in this condition.

Color blindness is the inability to distinguish colors. The retina has three specific types of cone cells that are related to the primary colors: blue, red, and green. The cone cells are affected in color blindness. Color blindness is identified as a hereditary characteristic and is more common in males than females.

Presbyopia (pres-bee-**OH**-pee-ah) is a condition in which the lenses lose their elasticity, resulting in a decrease in the ability to focus on close objects. It usually occurs after age 40. This condition can be corrected by glasses or contact lenses.

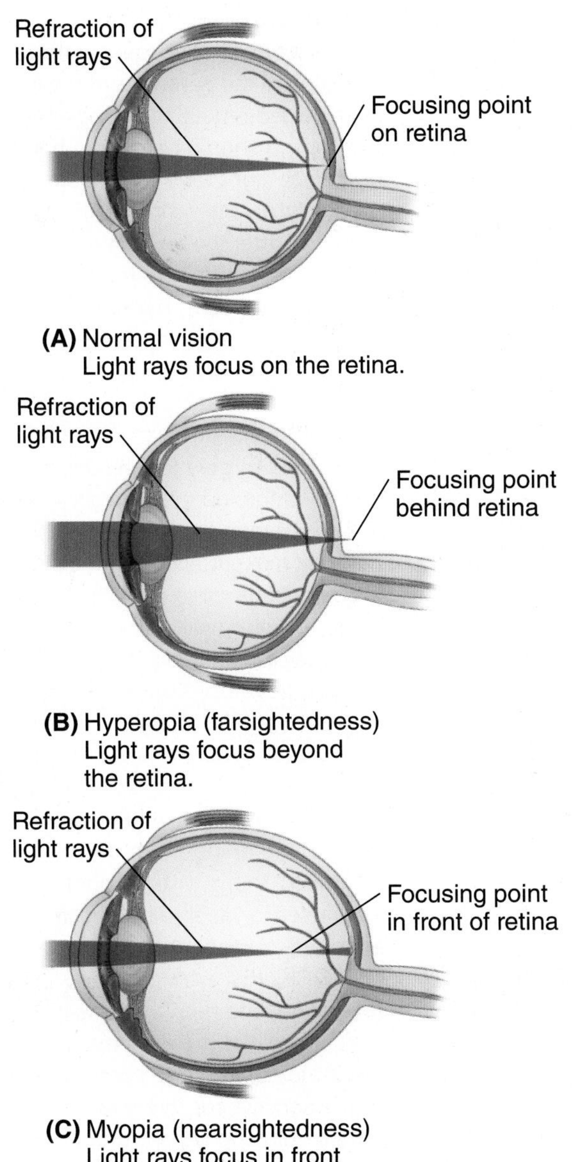

(A) Normal vision
Light rays focus on the retina.

(B) Hyperopia (farsightedness)
Light rays focus beyond the retina.

(C) Myopia (nearsightedness)
Light rays focus in front of the retina.

Figure 11-12 *Refraction of light rays*

Hyperopia (farsightedness) (high-per-**OH**-pee-ah) is a condition in which the focal point is beyond the retina because the eyeball is shorter than normal (**Figure 11-12B**). Objects must be moved farther away from the eye to be seen clearly. Convex lenses help correct this situation.

Myopia (nearsightedness) (my-**OH**-pee-ah) is a condition in which the focal point is in front of the retina because the eyeball is elongated (**Figure 11-12C**). Objects must be brought close to the eye to be seen clearly. Concave lenses help correct this condition. Various surgical techniques, such as photorefractive keratectomy (see *Medical Highlights: Eye Surgery*), can be used to correct refraction errors, particularly myopia.

Amblyopia (**am**-blee-**OH**-pee-ah), or "lazy eye," is a reduction, or dimness, of vision. It is the most common cause of decrease in vision in children. Early detection improves the success of treatment. Treatment is with glasses, eye exercises, the use of a patch over the stronger eye, or surgery.

The Effects of Aging **on the Eye**

Tear production and quality decrease with age, causing dry eyes. Dry eyes make the eyes feel hot and gritty and appear irritated but seldom cause eye damage. Artificial tears are helpful with this condition.

The loss of elasticity, opacity of the lens, and atrophy of the ciliary muscle decrease a person's ability to focus on fine details, or presbyopia. This change compromises the accommodation of the lenses. Older adults need more time for the eyes to adjust from light to dark; thus, they have a loss of night vision.

With age, the fluid in the eyes becomes cloudy, reducing light sensitivity. The cornea becomes thick and less transparent; the thickened cornea scatters light inside the retina, making glare more of a problem. The changes in the eye also affect color perception, making it more difficult to distinguish between blues, greens, and violets.

Peripheral vision and depth perception decline with age. An adequate visual field is necessary for driving and walking in crowded places. The depth perception loss leads to falls and mobility problems because of miscalculations about the distance and height of objects.

Loss of visual acuity occurs from changes in the lenses. Cataract formation, glaucoma, and macular degeneration occur more often as one ages.

Astigmatism (ah-**STIG**-mah-tizm) is a condition in which there is an irregular curvature of the cornea or lens, which causes blurred vision and possible eyestrain. A special prescription eyeglass helps this condition.

Diplopia (dih-**PLOH**-pee-ah) is also known as double vision; it is the perception of two images in a single object. Treatment includes eyeglasses, wearing an opaque contact lens, eye exercises, or Botox® injections.

Strabismus (crossed eyes) (strah-**BIZ**-mus) is a condition in which the muscles of the eyeball do not coordinate their action. This condition is usually seen at an early age in children and can be corrected by eye exercises or surgery.

Medical Highlights

EYE SURGERY

CATARACT

Phacoemulsification is the preferred technique for cataract extraction through small incisions. An ultrasound or laser probe is used to break the lens apart without harming the lens capsule. These fragments are then aspirated out of the eye. A foldable intraocular lens is then introduced through the 3-mm incision; the lens unfolds to take position inside the capsule. No sutures are needed because the incision is self-sealing. The most common type of lens used in cataract surgery is a monofocal lens; glasses may still be needed for near or intermediate distances. A trifocal lens implant was approved in 2019, which can be used for both distance and near focus. Visual rehabilitation after cataract surgery is extremely fast; two weeks after surgery, the patient will be able to perform any activity without risk.

Extracapsular extraction is an older technique in which a 12-mm incision is made in the eye to extract the lens as a whole. This method is used when the cataract is in a very advanced stage. The lens capsule is left in place to hold an intraocular lens. Multiple sutures are required to seal the eye after surgery. These sutures must be carefully tightened so as not to produce astigmatism.

DETACHED RETINA

Most retinal tears are treated with laser surgery or cryotherapy (freezing), which seals the retina back to the back wall of the eye.

Pneumatic retinopexy—A gas bubble is injected into the vitreous space inside the eye. The gas bubble pushes the retinal tear close against the back wall of the eye to reattach. The head must be held in a certain position for several days. The gas bubble will gradually deflate and disappear, and the retina will reattach.

Vitrectomy—The vitreous humor is removed from the eye and usually replaced with a gas bubble. The body's own fluid will gradually replace the gas bubble.

Scleral buckle—A flexible band, called a scleral buckle, is placed around the eyeball. This band compresses or buckles the eye inward, reducing the pulling of the retina and allowing the retina to reattach to the interior wall of the eye.

TREATMENT FOR VISION DEFECTS

LASIK (laser-assisted in situ keratomileusis)—The refractive surgery LASIK permanently changes the shape of the cornea. A special knife is used to cut a flap in the cornea. A hinge is left at one end of this flap. The flap is folded back, revealing the middle section of the cornea. Pulses from a computer-controlled laser vaporize a portion of the cornea, and the flap is replaced. After the procedure, the cornea should be able to bend, or refract, light rays to focus more precisely on the retina rather than before or after the retina. LASIK eye surgery is an option for patients who have myopia, hyperopia, and astigmatism. Most eye physicians still recommend using glasses and contact lenses as the first option to treat these conditions.

LASIK surgery for presbyopia is not recommended; the surgery may give the person clear distance vision but may make it even more difficult to see objects close-up. With LASIK surgery, vision will not necessarily be better immediately, but patients should develop 20/25 or better vision over the following 2–3 months.

Photorefractive keratectomy (PRK)—The surgeon removes the top layer of cells of the cornea, often by scraping them away after loosening them with alcohol, and then uses a laser directly on the exposed surface of the cornea. With PRK, the eyes generally take several days to heal, vision is blurry for a week, and then clears. *Dry eye is the number one complication after laser eye surgery and may be helped with special eye drops.*

Did You Know?

Eyes can be used to help diagnose a variety of nonvisual disturbances. A neurological assessment called PERLA, which stands for pupils equal and reactive to light and accommodation, can be used to assess brain damage. The REM, or rapid eye movement, stage of sleep is measured during sleep studies to help diagnose sleep disorders.

THE EAR

The ear is a special sense organ that is especially adapted to pick up sound waves and send these impulses to the auditory center of the brain. The auditory center is located in the temporal area just above the ears. The receptor for hearing is the delicate **organ of Corti**, which is located within the cochlea of the inner ear.

The ear is also involved with equilibrium. The receptors in the inner ear send a message to the cerebellum in the brain about head position to help maintain balance. Other receptors include proprioceptors in the eyes and receptors located around the joints. The information picked up by these receptors is processed by the cerebellum and cerebral cortex to enable the body to cope with changes in equilibrium.

The ear has three parts: the outer or external ear, the middle ear, and the inner ear (**Figure 11-13**).

The Outer Ear

The **auricle**, or outer ear, collects sound waves and directs them into the external auditory canal. The external auditory canal is lined with sebaceous or ceruminous

Medical Highlights

HEARING AIDS

As many as 38 million Americans have hearing loss. Hearing aids make sounds louder so the user can hear them better. A hearing aid collects sounds from the environment and converts them into digital code. Then it analyzes and adjusts the sound based on the user's hearing loss. The signals are then converted into sound waves and delivered to the ears through speakers. Some problems associated with hearing aids are that they are too costly, fall short of solving the problem, and demand upkeep and adjustment. Hearing aids can also be susceptible to earwax clogging.

The different types of hearing aids are as follows:

1. *Behind-the-ear (BTE)* types are crescent shaped and worn behind the ear, with a wire leading from it to a speaker within a visible molded piece in the ear canal. The microphone and amplifier in the crescent pick up the sound, process it into digital codes, and transmit it through the wire to the molded piece. Mini-BTEs do not have a visible earpiece. Both of these types can be used by people with mild to profound hearing loss.
2. *In-the-ear (ITE)* types are molded from plastic to the shape of the outer ear and are used for mild to severe hearing loss. ITE aids can accommodate added technical mechanisms, such as a small magnetic coil that improves sound transmission during a phone call. ITE aids can be damaged by earwax, and their small size can cause adjustment problems. They may pick up more wind noise than smaller devices.
3. *In-the-canal (ITC)* hearing aids are customized to fit the size and shape of the ear canal and are used for mild or moderate hearing loss. Because of their small size, the ITC aids may be difficult for the user to adjust and remove.
4. *Open-fit* hearing aids are a variation of the BTE hearing aid. This style keeps the ear canal open, allowing low-frequency sounds to enter the ear naturally and high-frequency sounds to be amplified through the hearing aid.

Experts stress that it takes time to get used to a hearing aid. Hearing aids do not restore the user's hearing to what it was before or eliminate background noise. The person should start using the hearing aid for an hour a day and lengthen the period gradually. More suggestions are as follows:

1. Ask a hard-to-follow speaker to talk more slowly and at a lower pitch. Hearing loss is most pronounced at higher frequencies.
2. Many people with hearing impairments unconsciously try to lip-read. Good lighting and a full-face view of the speaker may help.
3. Plan for social situations. If dining out with a group, try to be seated against a wall to reduce sound coming from behind the listener.

Reference: Mayo Foundation for Medical Education and Research, http://nided.nih

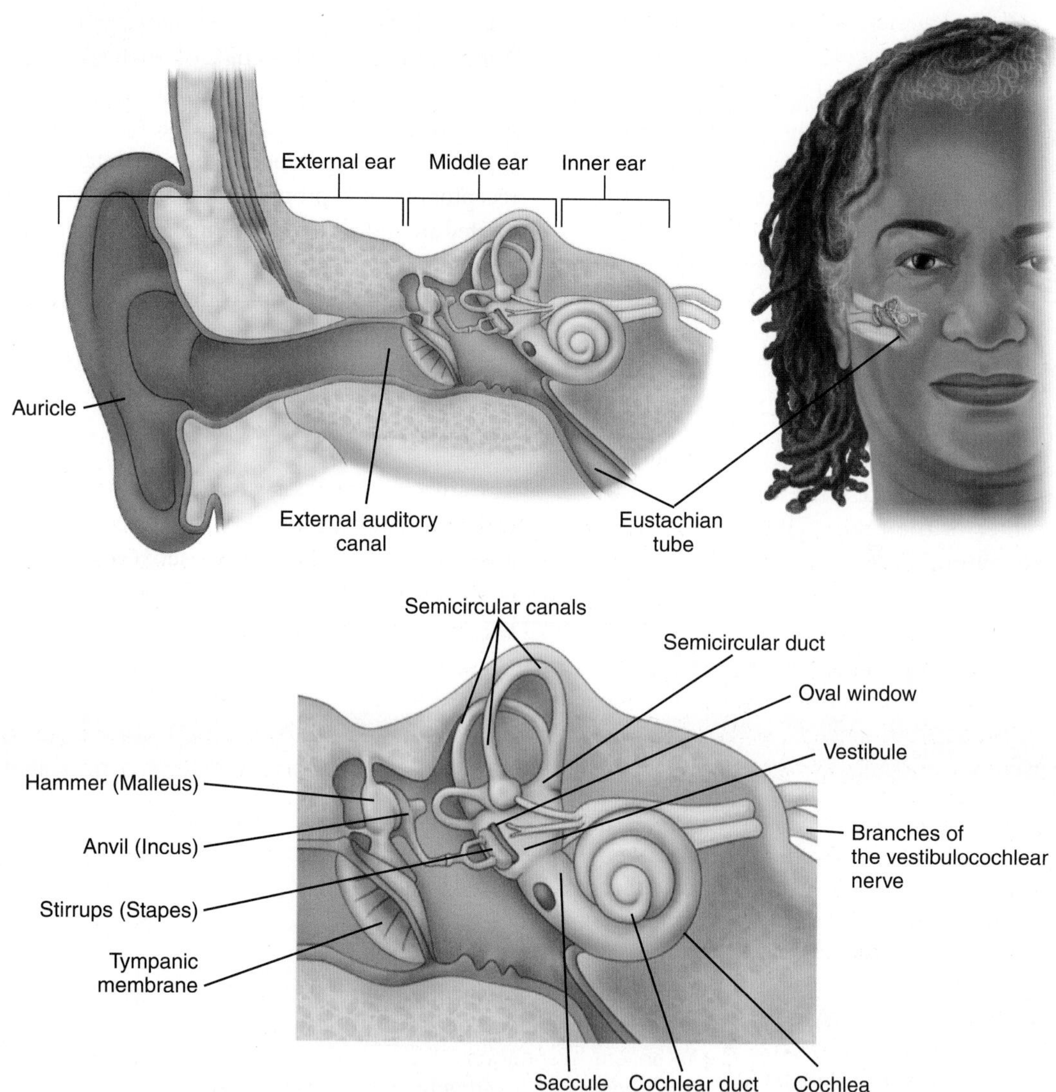

Figure 11-13 *The ear and its structures*

glands that secrete a waxlike or oily substance called **cerumen** (seh-**ROO**-men). This substance protects the ear. The auditory canal leads to the eardrum, or **tympanic membrane** (tim-**PAN**-ick), which separates the outer and middle ear. When sound waves reach the eardrum, the membrane transmits the sound by vibrating.

The mastoid process is the temporal bone containing hollow air space that surrounds the middle ear.

The Middle Ear

The middle ear is really the cavity in the temporal bone. It connects with the pharynx, or throat, by means of a tube called the **eustachian tube** (you-**STAY**-shun). This tube serves to equalize the air pressure in the middle ear with that of the outside atmosphere. A chain of three tiny bones is found in the middle ear: the **hammer (malleus)** (**MAL**-ee-us), the **anvil (incus)** (**ING**-kus), and the **stirrup (stapes)** (**STAY**-peez). These bones transmit sound waves from the eardrum to the inner ear by vibration.

The Inner Ear

The inner ear contains the sensory receptors for hearing and balance. The structure of the inner ear is known as the labyrinth (**LAB**-ih-rinth). The vestibule is the central egg-shaped cavity of the labyrinth. The **oval window**, which is located just under the base of the stapes, is the membrane that separates the middle ear from the inner ear. Vibrations reach the inner ear through this structure.

The inner ear contains several membrane-lined channels that lie deep within the temporal bone. The **cochlea** (**KOCK**-lee-ah) is the snail-shaped structure where sound vibrations are converted into nerve impulses. Located within the cochlea is a membranous tube called the **cochlear duct**, the organ of Corti, the semicircular canals, and the auditory nerves. These delicate cells of the organ of Corti pick up the nerve impulses and transmit them through the vestibulocochlear (auditory) nerve to the hearing center of the cerebrum of the brain.

Three **semicircular canals** also lie within the inner ear. They contain a liquid and delicate, hairlike cells that bend when the liquid is set in motion by head and body movements. These impulses are sent to the cerebellum, helping to maintain body balance, or equilibrium. They have nothing to do with the sense of hearing.

Pathway of Hearing

Sound waves → *auricle*, or outer ear → external *auditory canal* → *tympanic membrane* → *ear ossicles* (hammer, anvil, and stirrup) → stimulate the receptors in the *cochlea* → *auditory nerve* (part of the vestibulocochlear nerve) → *temporal lobe* of the brain for interpretation (**Figure 11-14**)

Media Link

View the **Hearing** animation on the Online Resources.

Pathway of Equilibrium

Movement of head → stimulates equilibrium receptors in the semicircular and vestibule areas of the inner ear → vestibular nerve (part of the vestibulocochlear nerve) → cerebellum of the brain for interpretation

Loud Noise and Hearing Loss

Hearing is both sensitive and fragile. Loud noise heard for too long will damage hearing. If the delicate hair cells in the organ of Corti become overstimulated, they will become damaged. Repeated exposure to the loud noise causes the loss to become permanent, as more cells and their nerve receptors are destroyed. When the same sound keeps reaching the ears, the auditory receptors adapt to the sound, and the sound is not heard.

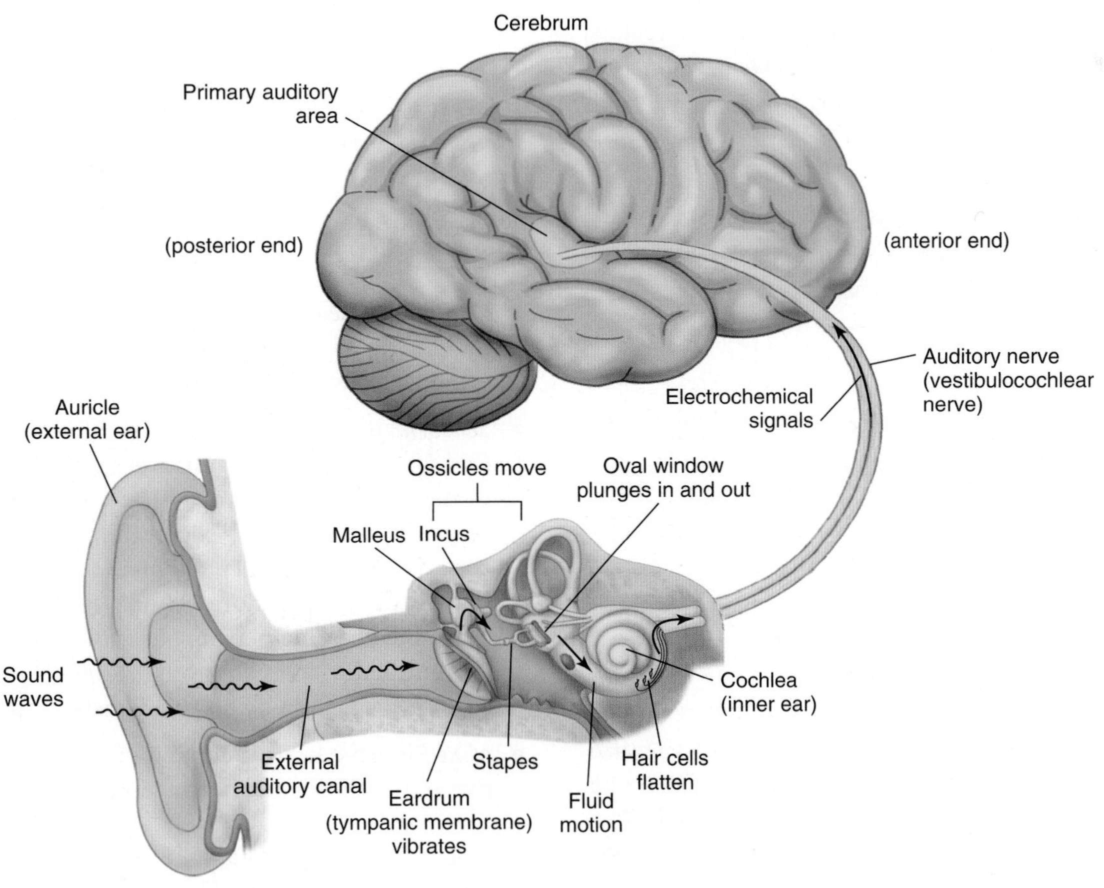

Figure 11-14 *Pathway of hearing: sound waves → external ear → external auditory canal → eardrum → ear ossicles → cochlea → auditory nerve → temporal lobe (cerebrum)*

The Effects of Aging on the Ears

The physiologic changes of aging result in three major types of hearing deficit: conductive, sensorineural, and mixed. Conductive hearing loss occurs when there is an interference with conduction of sound waves. In the outer ear, cerumen, or earwax, becomes embedded and drier because of a decrease in the number and activity of ceruminous glands. The tympanic membrane becomes fibrotic, reducing the transmission of sound. There is a degeneration of the ear bones, vestibular structure, cochlea, and organ of Corti, which affects sensitivity to sound, understanding of speech, and maintenance of equilibrium.

Sensorineural hearing loss, or presbycusis, involves changes in the neural, sensory, and mechanical structures of the inner ear. It is characterized by a loss of hearing of high-pitched frequencies and diminished ability to hear consonants such as *f, g, s, t, z, sh,* and *ch*. These changes result in impaired recognition of words as opposed to volume. The speech of other people sounds garbled and a normal conversation is difficult to follow.

Mixed hearing loss is a combination of both types.

Many older people experience changes in balance due to changes in the inner ear. Changes can cause postural hypotension, or lowered blood pressure on standing, due to the inability to quickly respond to changes in position. This often causes dizziness or light-headedness when one moves quickly from a lying or sitting position to standing.

The alarming increase of hearing loss in young people is most likely caused by loud music heard through headphones or earbuds. The symptoms of hearing loss may be tinnitus (tih-**NIGH**-tus) (ringing in the ears) or difficulty in understanding what people are saying (they seem to be mumbling). Words with high-frequency sounds, such as *pill, hill,* and *fill*, may sound alike.

Sound is measured in decibels (dB). The scale runs from a whisper—the faintest sound the human ear can hear (10 dB)—to a shotgun blast (more than 165 dB). Exposure to more than 90 dB for eight hours (busy city traffic) may be dangerous to one's hearing (**Figure 11-15**).

NOTE: Noise heard long and loud enough over time will cause damage. To protect their hearing, a person should turn down the volume on music devices and wear earplugs or noise-cancelling headphones.

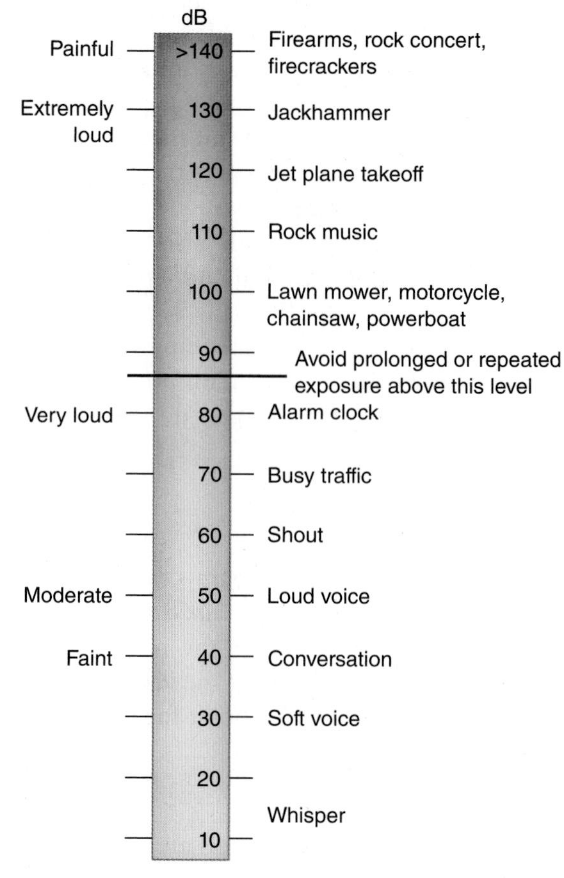

Figure 11-15 *A decibel scale of frequently heard sounds from lowest to highest*

TYPES OF HEARING LOSS

The following are the types of hearing loss:

- *Conductive hearing loss* occurs when sounds to the inner ear are blocked by earwax, fluid in the middle ear, or abnormal bone growth.
- *Sensorineural hearing loss,* or **presbycusis**, is when damage to parts of the inner ear or auditory nerve results in a partial or complete deafness. In cases of profound deafness, cochlear implants improve communication ability (**Figure 11-16**). Cochlear implants cannot restore sounds as they would be heard by a hearing person, but with communication training, children as young as 12 months and adults can use cochlear implants to understand speech.
- *Mixed hearing loss* is a combination of both conductive and sensorineural hearing loss.

Ear Disorders

Some general ear disorders have external causes, but it can also be inherited.

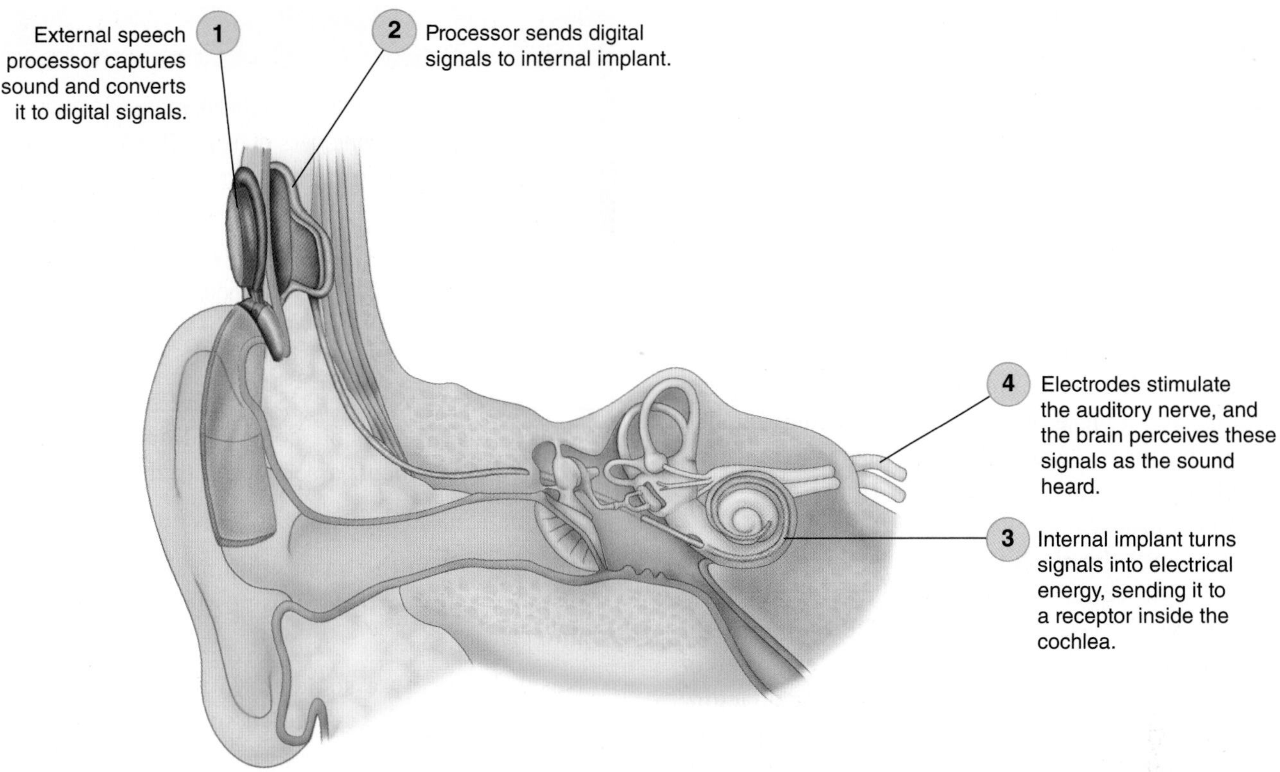
External speech processor captures sound and converts it to digital signals.
1
2
Processor sends digital signals to internal implant.
4
Electrodes stimulate the auditory nerve, and the brain perceives these signals as the sound heard.
3
Internal implant turns signals into electrical energy, sending it to a receptor inside the cochlea.

Figure 11-16 *A cochlear implant provides limited hearing for an individual who has been deaf since birth or for an adult who has a profound hearing loss.*

CONGENITAL HEARING LOSS

More than 12,000 babies are born each year with some type of **congenital hearing loss** (CDC Hearing and Screening Follow-Up Data, 2017). Signs of hearing loss may include a baby not reacting to loud noise, not turning to the source of sound after six months, and turning their head when they see someone but not when they hear a voice. Causes of hearing loss include prematurity; cytomegalovirus, which a mother can pass through the placenta; congenital rubella; infection; neurodegenerative disorder; or family history of deafness. Treatment may include antiviral medication, hearing aids, cochlear implant, lip speech, finger speech, or a combination of treatments. Hearing loss will affect speech and language development.

GENERAL EAR DISORDERS

Otitis media (oh-**TYE**-tis **MEE**-dee-ah) is an infection of the middle ear. It usually causes earache. This disorder is often a complication of the common cold in children. Treatment with antibiotics will cure the infection. In some cases, there may be a buildup of fluid or pus; this can be relieved by a **myringotomy** (mir-in-**GOT**-oh-mee), which is an opening made in the tympanic membrane. Tubes may be placed in the ear to allow fluids to drain off, especially in cases of chronic otitis media.

Otosclerosis (oh-toh-skleh-**ROH**-sis) is a condition in which the stapes bone of the middle ear first becomes spongelike and then hardens. This causes the stapes to become fixed or immovable, altering the structure of the ear. Otosclerosis is a common cause of deafness in young adults. Stapedectomy (stay-peh-**DECK**-toh-mee), or total replacement of the stapes, is the treatment of choice.

Tinnitus, or ringing in the ears, may affect 40 to 50 million Americans. The microscopic hairs in the inner ear move in relation to the pressure of sound waves, which then trigger the cell to discharge impulses through the auditory nerve to the brain. If the hairs are bent or broken, the hairs move randomly in a constant state of irritation. The auditory cells then send random electrical impulses to the brain as noise that is sometimes perceived as a ringing in the ears. Damage to these cells is commonly caused by loud noise. Other causes may be impacted wax; otitis media; otosclerosis; blockage of normal blood supply to the cochlea; or the effects of various drugs, such as salicylates, which are a type of painkiller. Treat the underlying cause if possible.

Ménière's disease (men-**YEHRS**) is a rare condition that affects the semicircular canals of the inner ear and causes marked **vertigo**, or dizziness. Vertigo can occur at any time and without warning. In addition, vertigo is sometimes accompanied by nausea, vomiting, and tinnitus. Bed rest is sometimes necessary during an acute attack. Medication may be given to relieve vertigo and nausea and to alleviate stress associated with repeated attacks.

The patient should avoid salt, caffeine, and nicotine and should eat a balanced diet with plenty of water. The cause is unknown; the symptoms subside after the episode has run its course and then return again without warning.

THE NOSE

The human nose can detect about 10,000 different smells. Smell accounts for about 90% of what is perceived as taste. For example, if a person holds their nose, they may not be able to taste the difference between a piece of orange and a piece of pear. Odor molecules inhaled through the nose get warmed and moistened as they pass through the nasal cavity (see Chapter 18).

In the nasal cavity (**Figure 11-17**), there is a patch of tissue about the size of a postage stamp called the olfactory epithelium, which has a plentiful supply of nerve cells with specialized receptors. The receptors send signals to the adjoining olfactory bulbs, an extension of the brain. The stimulus is transmitted by the olfactory nerve to the limbic system, thalamus, and frontal cortex. The limbic system generates basic emotions, such as affection, aggression, and fear. This relationship may explain why odors are tied to feelings. For example, a person may associate the smell of something cooking with a good experience. Scientists are starting to do research on how smells may affect learning, weight loss, aggression levels, and behavior.

Nose Disorders

Nose disorders may occur during during fetal development, or they may come up later in life.

> **The Effects of Aging on Smelling**
>
> A decrease in the number of olfactory neurons during aging reduces the awareness of odors. This decrease in the sense of smell can affect appetite, social relationships, and detection of warning smells such as gas leaks. Senile rhinitis is a clear, continuous, watery discharge from the nose that is not associated with underlying disease.

CONGENITAL NOSE DISORDERS

The following are possible congenital disorders of the nose and the treatments associated with them:

- Traumatic nasal deformity: there is no treatment.
- Abnormality of the nasal septum: since the nose at birth is mainly cartilage, it is possible to twist it back into place.
- Nasal deformity associated with cleft lip or cleft palate: this can be corrected surgically with good results.

GENERAL NOSE DISORDERS

Rhinitis (rye-**NYE**-tis) is an inflammation of the lining of the nose that may cause nasal congestion, nasal drainage, sneezing, or itching. The cause may be allergies, infection, or other factors, such as fumes, odors, emotional changes, or drugs. Treatment includes eliminating the allergens or use of a steroid nasal spray to reduce the inflammation. Some antihistamines are effective for short periods of time.

Nasal polyps are soft, noncancerous growths in the lining of the nasal passages or sinuses. They may

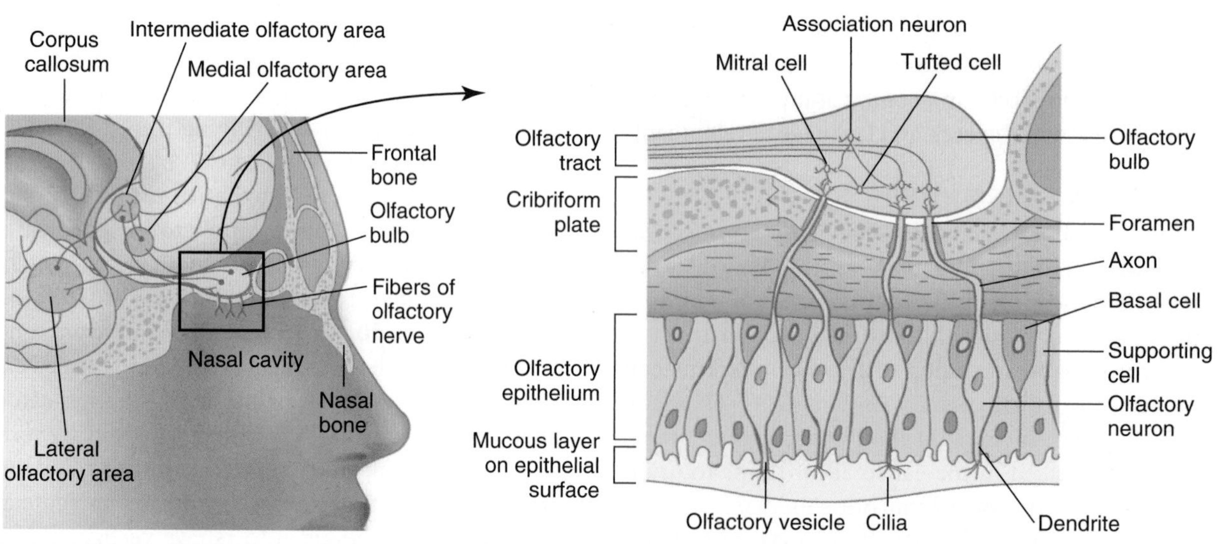

Figure 11-17 *The nasal cavity*

result from chronic inflammation of the nasal cavity. Medication may shrink the polyps. In severe cases, surgery may be necessary to remove the polyps.

A **deviated nasal septum** is a condition in which there is a bend in the cartilage structure of the septum. Symptoms that result are a blockage in the airflow through one nostril, headaches, loud breathing or snoring, dry nose, and nosebleeds. Treatment involves surgical correction. For temporary relief of breathing problems, an external adhesive strip placed across the nose is used to pull apart the nostrils, allowing for better airflow.

INJURY OR TRAUMA TO THE NOSE

A broken nose may be caused by falls, sports injuries, or a physical assault. To reduce swelling, apply cold compresses; if there is no improvement in 3–5 days, seek medical attention.

Epistaxis, or nosebleeds, can often be treated without medical attention. Treat minor nosebleeds by sitting upright and leaning forward to reduce the blood pressure to the nose. Pinch both nostrils shut at the soft portion of the nose for 5–15 minutes. Refrain from blowing the nose. Nosebleeds lasting more than 20 minutes need medical attention.

Foreign objects in the nose may be treated by gently blowing the object out of the nose by pinching the unaffected nostril. If the object can be seen, gently remove with tweezers or seek medical attention for removal.

THE TONGUE

The tongue is a mass of muscle tissue with structures called papillae (**Figure 11-18**). The taste buds are located on the papillae. Taste buds are stimulated by the flavors of foods. A single taste bud contains 50 to 199 taste cells. Each taste cell has receptors that connect to a sensory neuron leading from the three cranial nerves to the cerebral cortex for interpretation. The sensation of taste resides in the brain.

Before food can be tasted, it must first be dissolved in a fluid. The saliva produced by the salivary gland provides the fluid medium.

All taste buds can detect all five sensations: **umami** (oo-**MAH**-mee) (a savory taste), sweet, salty, sour, and bitter. Four of these sensations help meet nutritional needs. Umami guides the intake of meat and cheese to meet protein needs; sweet guides the intake of sugars to meet carbohydrate needs; salty guides the need for necessary minerals; and sour guides the intake of certain fruits to meet the need for vitamin C. Bitter taste protects a person by helping them detect spoiled food and poisons.

Tongue Disorders

Tongue disorders can make speech difficult, cause pain, or can be a sign of another underlying issue.

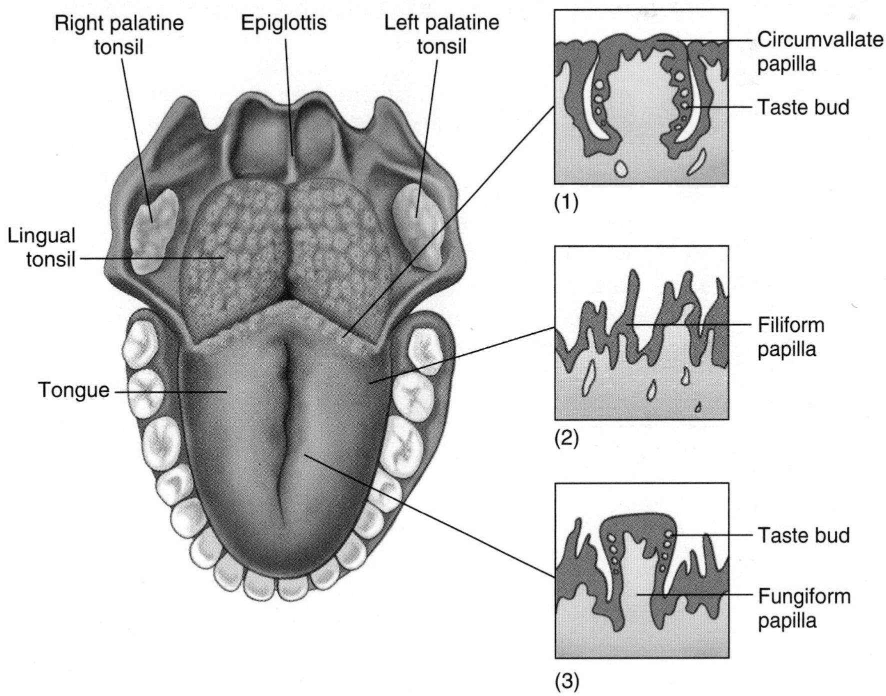

Figure 11-18 *The three types of papillae on the tongue*

CONGENITAL TONGUE DISORDERS

The following are congenital disorders of the tongue and their associated treatments:

- Microglossia is an abnormally small tongue that causes difficulty in speaking; there is no treatment.
- Macroglossia is an abnormally large tongue; it may be treated with surgical stripping of the tongue.
- Ankyloglossia, or a tongue tie, is the fusion of the lingual frenulum (see Chapter 19, **Figure 19-2**) to the floor of the mouth, which may lead to a speech problem. Surgical removal is done between the frenulum and the floor of the mouth.

GENERAL TONGUE DISORDERS

Injury—traumatic injury; the tongue may be bitten accidentally but heals quickly.

Discoloration—the tongue may appear black if the person takes bismuth preparations for an upset stomach. Iron-deficiency anemia may make the tongue look pale. White patches may accompany fever, dehydration, or mouth breathing.

Infection—may be the result of tongue piercing. Regularly clean area 2–3 times a day with saline solution. Use clean cloth; do not use cotton balls or tissue, which can get caught in the piercing. Suck on ice to alleviate symptoms.

Cancer—signs of cancer of the tongue include any unexplained red or white areas; sores; or lumps, particularly when hard, and especially if they are painless. Most oral cancers grow on the sides of the tongue or on the floor of the mouth. These should be examined by a physician or dentist immediately.

The Effects of Aging on Tasting

Older adults experience a reduction in the number of taste buds. The loss of taste receptors means that food is not as appetizing to the older person, and it may also mean that they are unable to detect when food has spoiled. Wearing full upper dentures diminishes taste sensation because they cover the taste buds in the upper palate.

TASTE: UMAMI

Taste occurs when chemicals from food reach the taste receptors in the taste buds. The taste receptors send messages to the brain about the flavor. The five primary taste types are sweet, salty, sour, bitter, and umami. The human tongue has receptors for the protein L-glutamate, which is the source of umami flavor. As such, scientists consider umami distinct from the other taste types.

Umami is a Japanese word meaning "savory." Umami has a mild but lasting aftertaste that is difficult to describe. It induces salivation and a sense of furriness on the tongue while stimulating the roof and the back of the mouth. By itself it is not palatable, but it makes a great variety of foods pleasant, especially in the presence of a matching aroma.

Umami, an amino acid, is found in the commonly used food additive monosodium glutamate (MSG) and in various other foods such as aged cheese, tomatoes, corn, meat (it gives beef its steak flavor), carrots, and fish. Older adults may benefit from umami taste because their taste and smell sensitivity are impaired by age and multiple medications.

Career Profile

AUDIOLOGIST

Audiologists assess and treat patients with hearing and hearing-related disorders. They use audiometers and other testing devices to measure the loudness at which a person begins to hear sounds, the ability to distinguish between sounds, and the extent of any hearing loss. Audiologists coordinate the results with medical and educational information to make a diagnosis and determine a course of treatment. Treatment may consist of cleaning the ear canal, fitting a hearing aid, auditory training, and instruction in speech or lipreading.

A master's degree is the standard credential in this field. Patience and compassion are critical traits because the patient's progress may be very slow. The job outlook is higher than average because hearing loss is associated with the aging process.

Career Profile

OPTOMETRIST

More than half of the people in the United States wear glasses. Optometrists, or doctors of optometry, provide most of the primary vision care people need.

Optometrists examine eyes to diagnose vision problems and eye disease. Optometrists use instruments and observations to examine eye health and to test patients' visual acuity, depth and color perception, and ability to focus and coordinate the eyes. They analyze test results and develop a treatment plan. Optometrists prescribe eyeglasses, contact lenses, and vision therapy. They prescribe drugs for other eye problems such as conjunctivitis, glaucoma, and corneal infection.

Optometrists differ from ophthalmologists. Ophthalmologists diagnose and treat eye diseases, perform surgery, and prescribe drugs. All states require optometrists to be licensed. Applicants must have a doctor of optometry degree from an accredited school and pass a licensing examination.

Career Profile

DISPENSING OPTICIAN

Dispensing opticians fit eyeglasses and contact lenses. Dispensing opticians help customers select appropriate frames, order the necessary ophthalmic laboratory work, and adjust the finished eyeglasses. They examine written prescriptions to determine lens specification and measure the patient's eyes. They prepare work orders that give laboratory technicians the information needed to grind and insert lenses.

Dispensing opticians keep records, work orders, and payments, as well as track inventory and perform other administrative duties.

Employers generally hire individuals with no background in opticianry and then provide the required training. Mechanical drawing is particularly useful because training in this field usually includes instruction in optical mathematics, optical physics, and the use of precision measuring instruments and other machinery and tools. Formal training may be offered in community colleges. The job outlook is greater than average in response to rising demand for corrective lenses. Fashion also influences demand, encouraging people to buy more than one pair of eyeglasses.

Medical Terminology

ambly	dull or dim
op	eyes
-ia	condition of
ambly/op/ia	condition of dim eyes
cochle	snail shell
-a	relating to, pertaining to
cochle/a	spiral; snail shell
conjunctiv	eyelid lining
-itis	inflammation of
conjunctiv/itis	inflammation of eyelid lining
corne	tough, hornlike
corne/a	horny web
tympan	eardrum
-ic	relating to
tympan/ic	relating to eardrum
dipl	double
dipl/op/ia	condition of seeing double
hyper-	over, excessive
hyper/op/ia	condition of excessive eye vision; farsightedness
lacrim	tears
-al	pertaining to
lacrim/al gland	tear gland
my	squinting
my/op/ia	condition of squinting; nearsightedness
myring	eardrum
-otomy	opening into
myring/otomy	opening into eardrum
ot	ear
media	middle
ot/itis media	inflammation of middle ear
oto	ear
-sclerosis	hardening
oto/sclerosis	hardening of the ear
strabism	distorted squinting, cross-eyed
-us	presence of
strabism/us	presence of crossed eyes

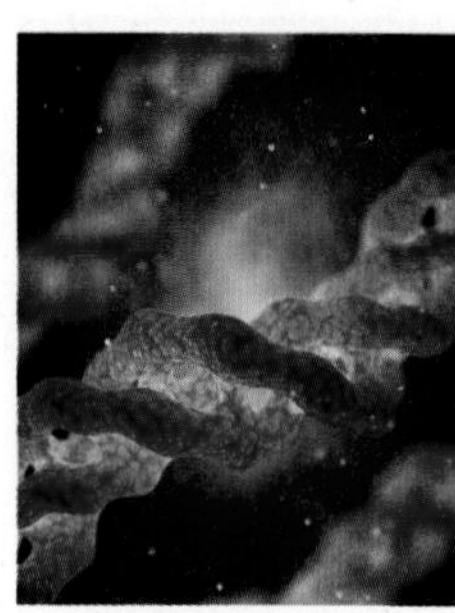

Study Tools

Workbook	Activities for Chapter 11
Online Resources	• PowerPoint presentations • Animations

REVIEW QUESTIONS

Select the letter of the choice that best completes the statement.

1. The outer tough coat of the eye is the
a. retina.
b. sclera.
c. choroid.
d. lens.

2. The clear anterior portion of the sclera is called the
a. cornea.
b. lens.
c. pupil.
d. iris.

3. The muscle that regulates how much light enters the eye is called the
a. conjunction.
b. iris.
c. cornea.
d. lens.

4. The posterior chamber of the eye is filled with fluid called
a. tears.
b. ciliary body.
c. vitreous humor.
d. aqueous humor.

5. The area of the eye that contains the rods and cones is called the
a. retina.
b. choroid.
c. sclera.
d. cornea.

6. The tube that connects the throat to the ear is called the
a. pinna tube.
b. eustachian tube.
c. cochlear tube.
d. auditory tube.

7. Hardening of the bones of the middle ear is called
a. otitis media.
b. presbycusis.
c. otosclerosis.
d. presbyopia.

8. Nearsightedness is also known as
a. myopia.
b. hyperopia.
c. presbyopia.
d. strabismus.

9. A clouding of the lens is called
a. myopia.
b. glaucoma.
c. hyperopia.
d. cataract.

10. An infectious disease known as pink eye is also called
a. kernicterus.
b. otitis.
c. conjunctivitis.
d. strabismus.

LABELING

Study the following diagram of the eye and name the numbered structures.

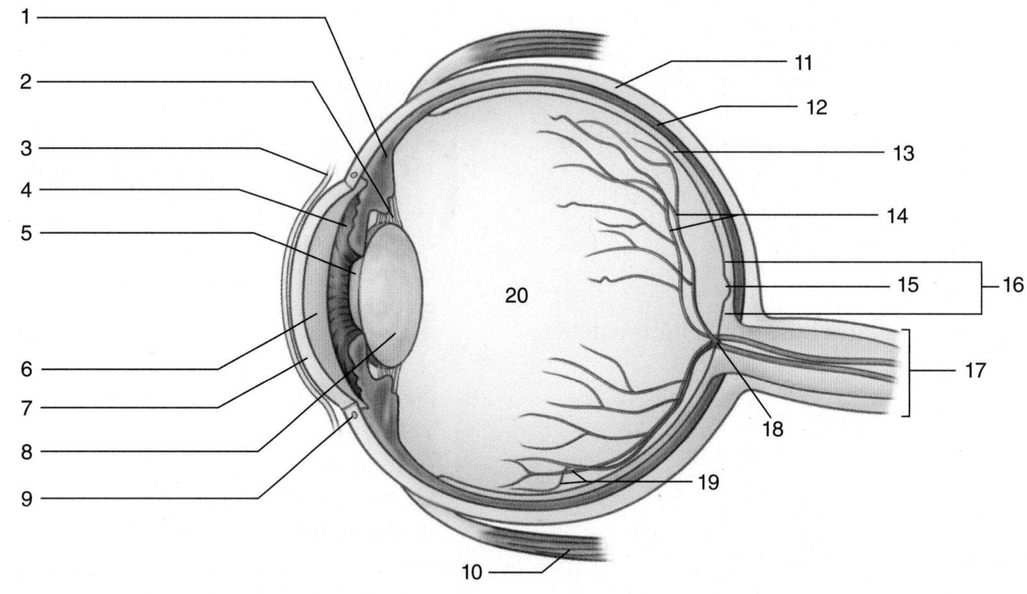

1.	______	**6.**	______	**11.**	______	**16.**	______
2.	______	**7.**	______	**12.**	______	**17.**	______
3.	______	**8.**	______	**13.**	______	**18.**	______
4.	______	**9.**	______	**14.**	______	**19.**	______
5.	______	**10.**	______	**15.**	______	**20.**	______

APPLYING THEORY TO PRACTICE

1. Explain how you see and track the pathway of light from the cornea to the occipital lobe of the brain.

2. Explain to a friend how your outer ear catches a sound and where in the brain it is interpreted.

3. Since the age of 10, you have used glasses for myopia. You have heard about the new types of surgery to correct this condition so that you would no longer need glasses. Investigate and report on the types of surgery you can get to correct this condition.

4. In what areas of the brain are the following senses interpreted?
   a. Sight
   b. Hearing
   c. Balance
   d. Smell
   e. Taste

5. You observe your grandmother not finishing a meal she normally enjoys. You ask her why. Her response is that it doesn't taste good. Explain to your grandmother the reason for this change.

CASE STUDY

Wayne is a 75-year-old retired teacher who comes to the physician's office feeling quite upset. He tells Rebecca, the licensed practical nurse, that he does not know what is happening to him. When people speak to him their speech seems garbled, and lately he is having difficulty reading. After an examination, the physician tells him he has a conductive hearing loss because of accumulated cerumen.

1. What is a conductive hearing loss?

2. What is cerumen?

3. Describe other causes of conductive hearing loss.

4. What is the role of an audiologist?

5. Explain the pathway of sound.

The physician tells Rebecca to make an appointment for Wayne to be seen by an optometrist for his vision problem. The optometrist tells Wayne he has the beginning of macular degeneration.

6. Describe the duties of the optometrist.

7. Explain and describe the symptoms of macular degeneration.

8. What is the treatment for macular degeneration?

9. What reassurance can the optometrist give Wayne regarding macular degeneration?

LAB ACTIVITY 11-1 Anatomy of the Eye

- ***Objective:*** To observe the anatomy of a cow eye and compare it with the human eye
- ***Materials needed:*** preserved cow eye, anatomical model of the human eye, dissecting kit and tray, disposable gloves, paper, pencil

Step 1: Put on disposable gloves.

Step 2: Examine the structure of the model of the human eye; identify the conjunctiva, sclera, cornea, and optic nerve. Describe the structure and function of these parts.

Step 3: Compare the cow eye with the anatomical model. Record the differences, if any.

Step 4: Hold the cow eye and make an incision with the scalpel into the sclera just above the cornea. (Use caution when making the incision because the sclera may be hard to cut.) Cut around the cornea.

Step 5: Lift the anterior portion (cornea section) of the eyeball away from the posterior part. The vitreous humor should still be in the posterior part.

Step 6: Locate and identify the lens, iris, and cornea. Record your observations.

Step 7: Examine the posterior portion of the eye; remove the vitreous humor with dissecting kit tweezers.

Step 8: Identify the retina and the choroid coat. Describe these structures and record their function.

Step 9: Dispose of the cow eye in the appropriate disposal container.

Step 10: Clean all equipment.

Step 11: Remove your gloves and wash your hands.

LAB ACTIVITY 11-2 Test for Visual Acuity

- ***Objective:*** To observe the function of the eye
- ***Materials needed:*** Snellen eye chart, measuring device card, paper, pencil, contact lens cases, contact lens solution

Step 1: Find a lab partner. Measure 20 feet from the eye chart, which is where your partner will stand.

Step 2: Have your partner cover their left eye with one hand or a card.

Step 3: Have your partner read each line with their right eye; check for accuracy.

Step 4: Record the line with the smallest number read for the right eye.

Step 5: Repeat the process to record the visual acuity for the left eye. Record the number. If your lab partner wears glasses or contact lenses, have the person do the test first with the corrective lenses and then without them.

Step 6: Switch places and have your lab partner repeat steps 1 to 5 with you as the subject.

Step 7: Is there a difference between your test results and your lab partner's results? Record your answers. What use do you think this has as a diagnostic technique?

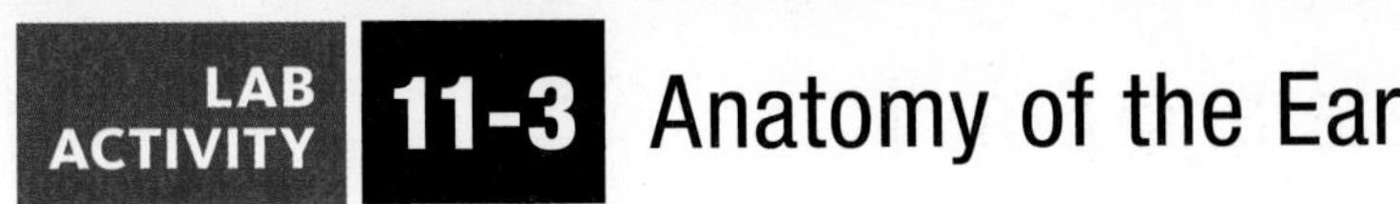

Anatomy of the Ear

- ***Objective:*** To observe the anatomical structure of the ear
- ***Materials needed:*** anatomical model of the ear, paper, pencil

Step 1: Using the anatomical model, locate and identify the structures of the outer ear. List them, describe them, and state their functions.

Step 2: Locate, identify, and list the structures of the middle ear. Record their descriptions and functions as in step 1.

Step 3: Locate, identify, and list the structures of the inner ear. Again, record their descriptions and functions as in step 1. What fluid fills the inner ear, and what is the function of this fluid?

11-4 Test for Hearing Acuity

- ***Objective:*** To observe the function of the ear
- ***Materials needed:*** ticking clock, cotton, tape measure, paper, pencil

Step 1: Find a lab partner and move to a quiet area. Carefully pack your lab partner's left ear with cotton.

Step 2: Hold the ticking clock close to the right ear, then slowly walk away until your lab partner can no longer hear the ticking.

Step 3: Measure the distance and record. Remove the cotton from the left ear and dispose of it in an appropriate container.

Step 4: Repeat steps 1 to 3 to determine the acuity in the left ear, placing the cotton in the right ear.

Step 5: Switch places with your lab partner and have your lab partner repeat steps 1 to 4 to determine your auditory acuity.

Step 6: Is there a difference between your auditory acuity and your lab partner's?

You may repeat this test at home with members of your family. Note the difference, if any, between persons of different ages or sexes. What use do you think this has as a diagnostic technique?

LAB ACTIVITY 11-5 Sense of Taste and Smell

- ***Objective:*** To observe the function of the nose and mouth
- ***Materials needed:*** cubes of apple, pear, orange, cheese; blindfold; spoon; paper towels; paper; pencil

Step 1: Blindfold your lab partner.

Step 2: Have your partner pinch their nostrils together.

Step 3: Using the spoon, place one of the four foods in your partner's mouth.

Step 4: Have your partner chew the food and then spit it out into the paper towel.

Step 5: Have your partner identify the food. Record the information.

Step 6: Repeat steps 3 to 5 for the remaining three foods.

Step 7: Leave the blindfold in place, but do not ask your partner to pinch their nostrils this time. Repeat steps 3 to 5.

Step 8: Is there a difference in the identification of food? Can food be identified by taste alone?

CHAPTER 12

Endocrine System

Objectives

- List the glands that make up the endocrine system.
- Describe hormones and their classification.
- Describe negative feedback hormonal control.
- Name the hormones of the endocrine system and their function.
- Describe the role of the other hormones produced in the body.
- Describe some disorders of the endocrine system.
- Define key words that relate to this chapter.

Key Words

acromegaly
Addison's disease
adrenal glands
adrenaline
adrenocorticotropic hormone (ACTH)
androgens
anterior pituitary lobe
antidiuretic hormone (ADH)
calcitonin
cretinism
Cushing's syndrome
diabetes insipidus
diabetes mellitus
endocrine glands
epinephrine
estrogen
exocrine gland
exophthalmos
follicle-stimulating hormone (FSH)
ghrelin
gigantism
glucagon
glucocorticoids
goiter
gonads
growth hormone (GH)
hormones
hyperglycemia
hyperthyroidism
hypoglycemia
hypothyroidism
insulin
interstitial cell-stimulating hormone (ICSH)
islets of Langerhans
leptin
luteinizing hormone (LH)
melanocyte-stimulating hormone (MSH)
melatonin
mineralocorticoids
myxedema
negative feedback
neurohormones
norepinephrine
oxytocin
pancreas
parathormone
parathyroid glands
pheochromocytoma
pineal gland
pituitary gland
pituitary short stature
polydipsia
polyphagia
polyuria
posterior pituitary lobe
progesterone
prolactin hormone (PR)
prostaglandins
somatotropin
testosterone
tetany
thymus
thyroid gland
thyroid-stimulating hormone (TSH)
thyroxine (T_4)
triiodothyronine (T_3)
vasopressin

A gland is any organ that produces a secretion. **Endocrine glands** (**Figure 12-1**) are organized groups of tissues that use materials from the blood or lymph to make new compounds called **hormones**. Endocrine glands are also called ductless glands and glands of internal secretion; the hormones are secreted directly into the bloodstream as the blood circulates through the gland. The secretions are transported to all areas of the body. There is another type of gland called an **exocrine gland**, in which the secretions from the gland must go through a duct. This duct then carries the secretion to a body surface or organ. Exocrine glands include sweat, salivary, lacrimal, and the pancreas. Their functions are included in chapters on the relevant body systems. See **Figure 12-2**.

One of the endocrine glands, the pancreas, performs as both an exocrine gland and an endocrine gland. The pancreas produces pancreatic juices, which go through a duct into the small intestines. The pancreas also has a special group of cells known as **islets of Langerhans**, which secrete the hormone insulin directly into the bloodstream.

Media Link

View the **Exocrine and Endocrine Glands** animation on the Online Resources.

HORMONES

The term *hormone* comes from the Greek word *hormao*, meaning "I excite." Hormones are chemical messengers secreted by endocrine glands with specialized functions in regulating the activities of specific cells, organs, or both.

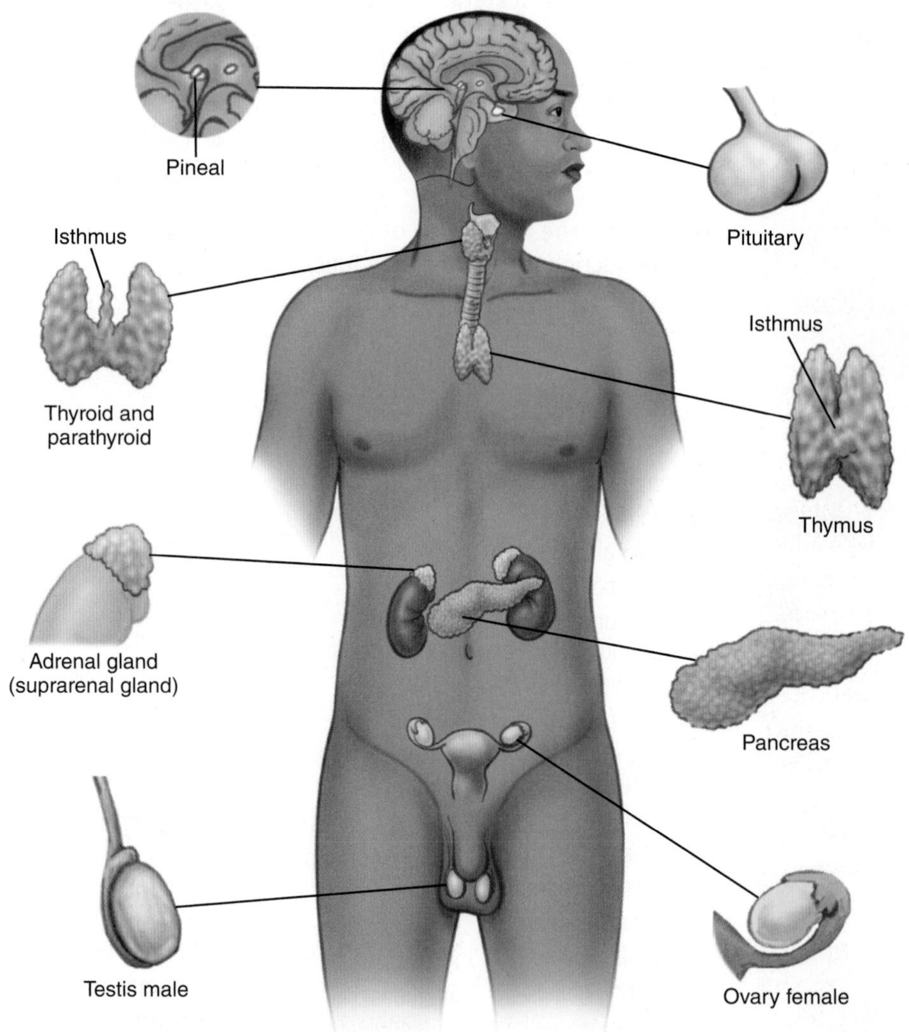

Figure 12-1 *Locations of the endocrine glands*

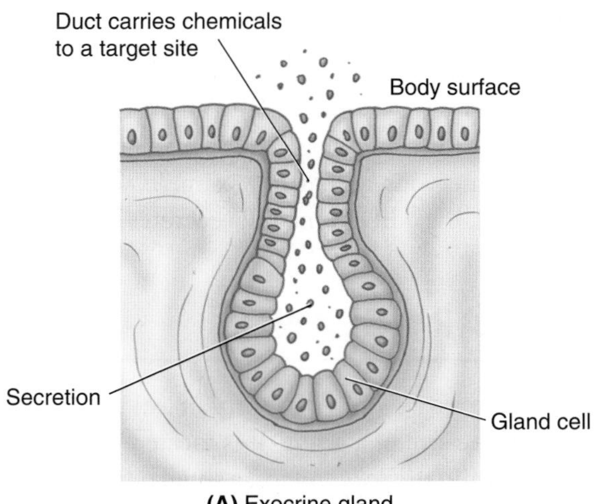

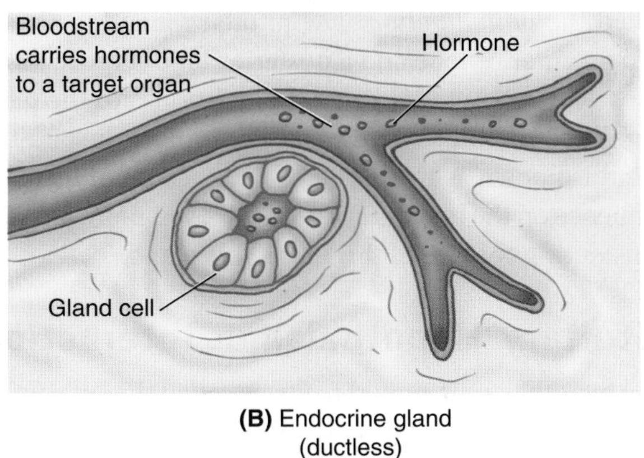

Figure 12-2 *(A) Exocrine gland; (B) endocrine gland*

The following are three different classes of hormones based on their chemical composition:

1. *Amines*, such as norepinephrine, epinephrine, and dopamine, are derived from single amino acids. Thyroid hormones make up a subset of this class because they are derived from a combination of two iodinated tyrosine amino acids.
2. *Peptides* and *protein* hormones consist of three to more than 200 amino acid residues. All hormones secreted by the pituitary gland are peptide hormones, as is the leptin found in adipose cells, ghrelin found in the stomach and pancreas, and insulin found in the pancreas.
3. *Steroid* hormones are converted from their parent compound, cholesterol. Steroid hormones can be grouped into five groups by the receptors to which they bind: glucocorticoids, mineralocorticoids, androgens, estrogens, and progestogens.

Other Hormones Produced in the Body

There are additional hormones produced in the body: prostaglandins, neurohormones, leptin, and ghrelin.

PROSTAGLANDINS

In various tissues throughout the body, hormonelike chemicals called **prostaglandins** are secreted. On discovery, it was believed they came from the prostate gland and thus were called prostaglandins. Their activity depends on which tissue secretes them. Some prostaglandins cause constriction of the blood vessels; others may cause dilation. Prostaglandins can be used to induce labor and cause severe muscular contractions of the uterus. The exact nature and function of the prostaglandins are being extensively studied by scientists.

A host of hormones are produced throughout the body. They can originate from many different glands or other organs. A complete description of all the hormones in the body is beyond the intent of this anatomy textbook.

NEUROHORMONES

Neurohormones (new-roh-**HOR**-mohnz) are produced and released by the neurons in the hypothalamic area of the brain rather than the endocrine glands and are delivered to organs and tissue through the bloodstream for systemic effect. An example is neurohormones secreted by the hypothalamus that influence the secretion of the pituitary gland.

LEPTIN

Leptin (**LEP**-tin) is secreted by the fat cells in adipose tissue. Leptin travels via the bloodstream to the hypothalamus appetite center where it acts on the hypothalamus to suppress appetite and burn fat stored in adipose tissue.

GHRELIN

Ghrelin (**GRE**-lin) is produced by the stomach and is an appetite stimulant. It appears to be the wake-up call to eat. Its level peaks just before mealtime and diminishes after meals.

FUNCTION OF THE ENDOCRINE SYSTEM

The human body must coordinate all of its functions to reach states of equilibrium, or *homeostasis*. The maintenance of homeostasis involves growth, maturation, reproduction, and metabolism. Human behavior is shaped by the endocrine system and the nervous system working in a unique partnership. The hypothalamus of the brain (nervous system) sends directions via chemical signals (neurohormones) to the pituitary (endocrine system). The secretions of the pituitary gland then send chemical signals or messengers in the form of hormones, stimulating other endocrine glands to secrete their unique hormones.

The hormones thus produced coordinate and direct the activities of target cells and target organs.

The major glands of the endocrine system include the pituitary, pineal, thyroid, parathyroid, thymus, adrenal, the pancreas, and the gonads (ovaries in the female and testes in the male).

Refer to **Figure 12-1** for the locations of the endocrine glands in the body. Each has specific functions to perform. Any disturbance in the functioning of these glands may cause changes in the appearance or functioning of the body.

Hormonal Control

The secretion of the hormones operates on a negative feedback system or under the control of the nervous system.

NEGATIVE FEEDBACK

Negative feedback occurs when there is a drop in the blood level of a specific hormone. This drop triggers a chain reaction of responses to increase the amount of hormone in the blood. Negative feedback operates like air conditioning. The thermostat for the air conditioner is set at a certain temperature, and when the temperature rises above the set temperature, the thermostat sends a signal to turn on the air conditioner. Once the set temperature is reached, the thermostat sends another signal to turn it off. A description follows of how the negative feedback system functions as it relates to the thyroid gland:

1. The blood level of thyroxine, the main thyroid hormone, falls.
2. The hypothalamus in the brain gets the message.
3. The hypothalamus responds by sending a releasing hormone for thyroid-stimulating hormone (TSH).
4. This goes to the anterior pituitary gland, which responds by releasing TSH.
5. The TSH stimulates the thyroid gland to produce thyroxine.
6. The thyroxine blood level rises, which in turn causes the hypothalamus to shut off the releasing hormone for TSH.

Nervous Control

The nervous system controls the glands that are stimulated by nervous stimuli, as in the adrenal medulla. Stimulation of the adrenal medulla is controlled by the sympathetic nervous system. For example, when a person is frightened, the adrenal medulla secretes adrenaline.

PITUITARY GLAND

The **pituitary gland** (pih-**TOO**-ih-tair-ee) is a tiny structure about the size of a grape. It is located at the base of the brain within the *sella turcica*, a small bony depression in the sphenoid bone of the skull (**Figure 12-3**). The pituitary gland is connected to the hypothalamus by a stalk called the *infundibulum*. The pituitary gland is divided into an anterior lobe and a posterior lobe (**Figure 12-4**).

Figure 12-4 highlights the hormones of the pituitary gland and the structures they act upon. The pituitary gland is known as the *master gland* because of its major influence on the body's activities. It is even more amazing considering the size of this incredible gland.

Pituitary-Hypothalamus Interaction

The hypothalamus should really be called the "master" of the master gland because the hormones of the pituitary are controlled by the releasing chemicals of the hypothalamus. As the pituitary hormones are needed by the body, the hypothalamus releases chemical signals called releasing hormones and releasing inhibitory hormones (see **Figure 12-3** and **Figure 12-4**). For example, the thyroid-stimulating hormone (TSH) of the pituitary gland has a TSH-releasing hormone in the hypothalamus that allows for TSH's release to the body. In addition, when a sufficient amount of the hormone is released, a different releasing hormone in the hypothalamus will inhibit the anterior pituitary from secreting TSH.

The hypothalamus is considered part of the nervous system. However, it produces two hormones: vasopressin, which converts to **antidiuretic hormone (ADH)**, and oxytocin. These are stored in the posterior lobe of the pituitary and release into the bloodstream in response to nerve impulses from the hypothalamus.

Hormones of the Pituitary Gland

The pituitary gland is divided into two lobes. The larger anterior pituitary lobe produces six hormones. The smaller **posterior pituitary lobe** consists primarily of nerve fibers and neuroglial cells that support the nerve fibers. Neurons in the hypothalamus produce hormones secreted by the posterior pituitary lobe.

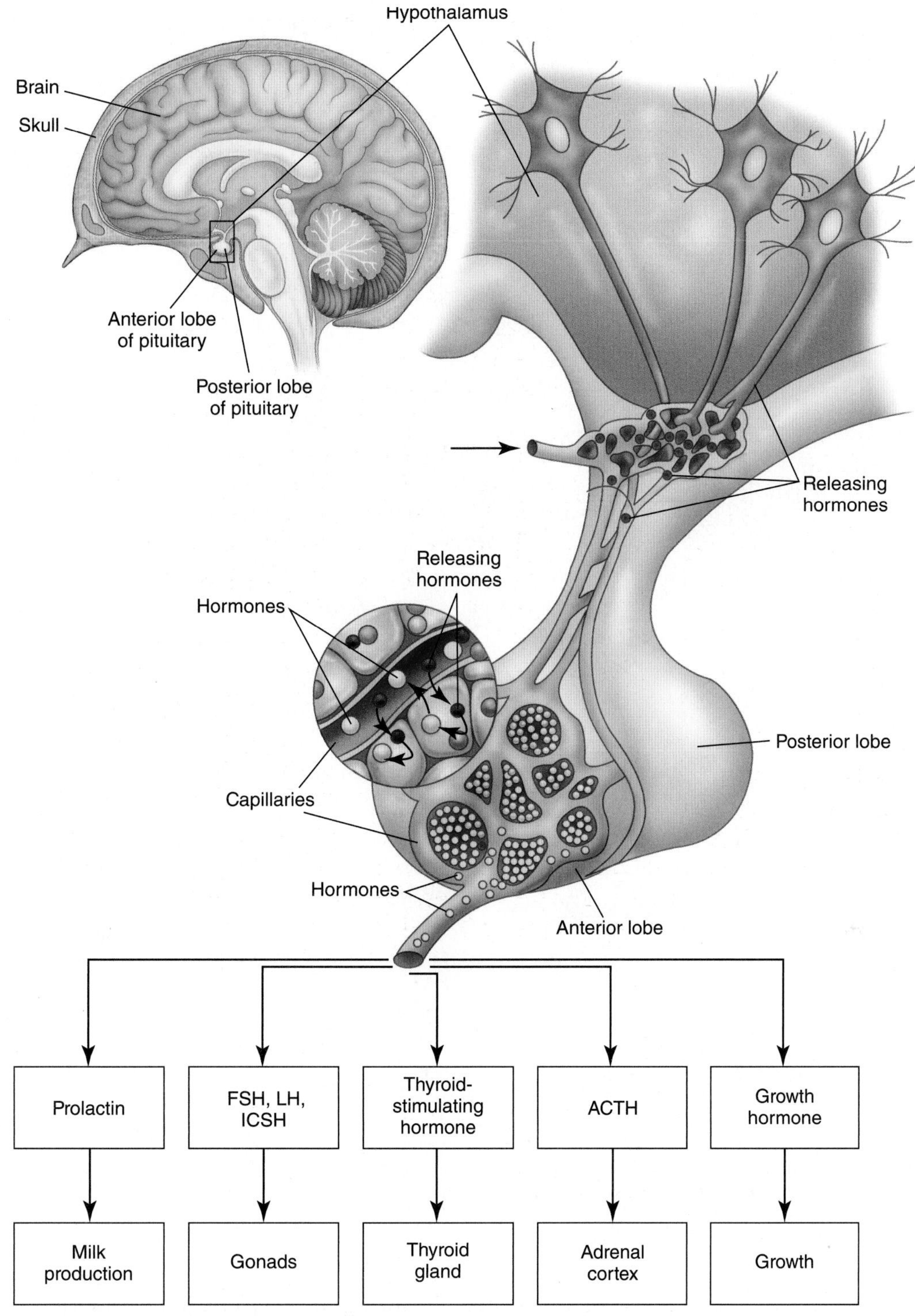

Figure 12-3 *The relationship of the hypothalamus of the brain with the anterior lobe of the pituitary gland*

ANTERIOR PITUITARY LOBE

The **anterior pituitary lobe** secretes the following hormones (see **Figure 12-4**):

1. **Growth hormone (GH)**, or **somatotropin**, is responsible for growth and development. GH decreases the utilization of glucose, which in turn affects blood sugar levels. GH also influences the metabolism of fats, proteins, and carbohydrates.
2. **Prolactin hormone (PR)**, also called lactogenic hormone (LTH), develops breast tissue and stimulates the production of milk after childbirth. The function in males is unknown.

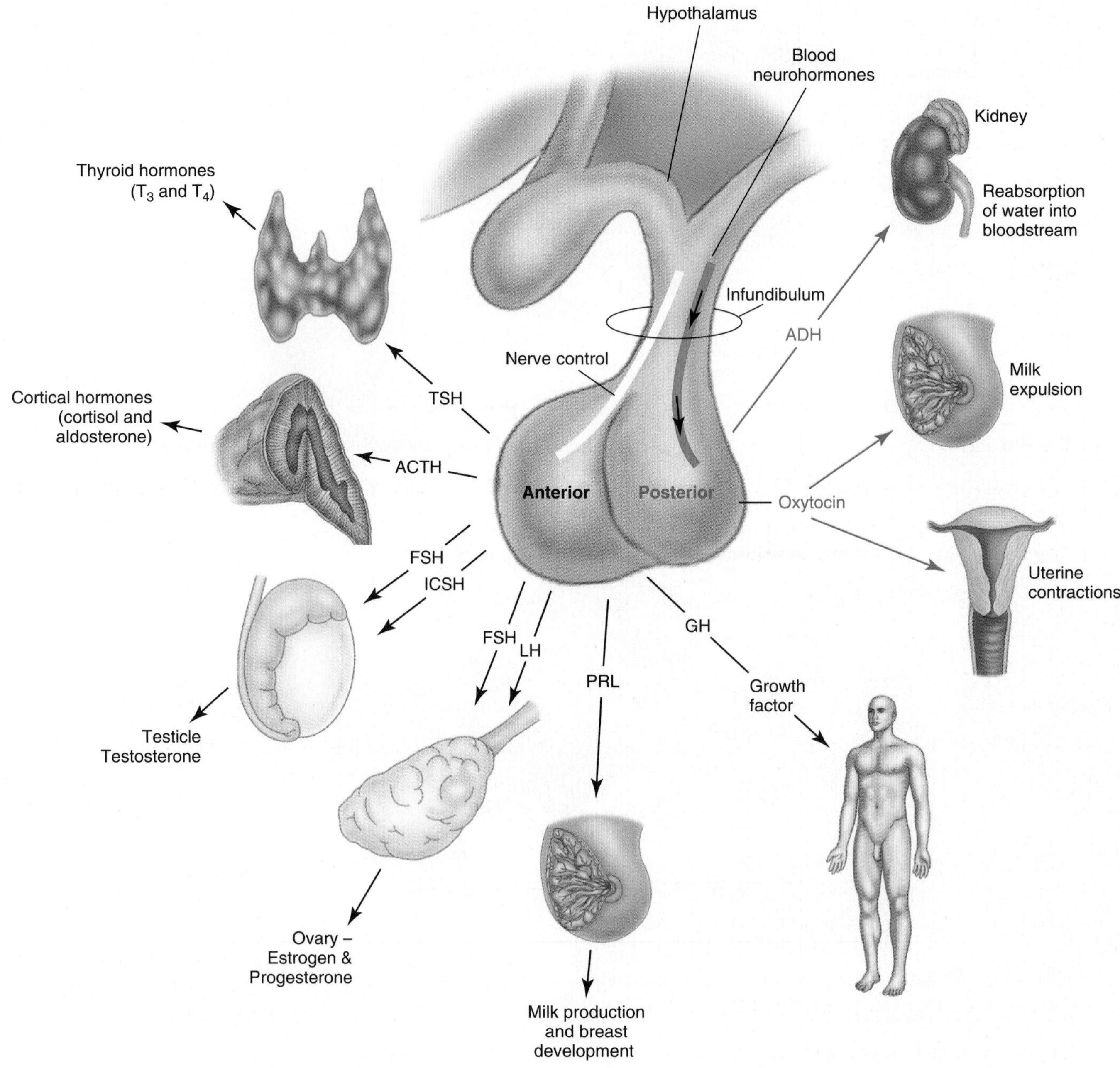

Figure 12-4 *The pituitary gland and its hormonal secretions*

3. **Thyroid-stimulating hormone (TSH)** stimulates the growth and secretion of the thyroid gland.

4. **Adrenocorticotropic hormone (ACTH)** (ad-ree-noh-kor-tih-coh-**TROP**-ik) stimulates the growth and secretion of the adrenal cortex.

5. **Follicle-stimulating hormone (FSH)** stimulates the growth of the graafian follicle and the production of estrogen in females, and it stimulates the production of sperm in males.

6. **Luteinizing hormone (LH)** stimulates the growth of the graafian follicle; the production of estrogen; and the formation of the corpus luteum after ovulation, which produces progesterone in females. LH in males may also be called the **interstitial cell-stimulating hormone (ICSH)** and is necessary for testosterone production by the interstitial cells of the testes.

Table 12-1 *Pituitary Hormones and Their Known Functions*

PITUITARY HORMONE	***KNOWN FUNCTION***
Anterior Lobe	
TSH—thyroid-stimulating hormone (thyrotropin)	Stimulates the growth and secretion of the thyroid gland
ACTH—adrenocorticotropic hormone	Stimulates the growth and secretion of the adrenal cortex
FSH—follicle-stimulating hormone	Stimulates the growth of new graafian (ovarian) follicle and secretion of estrogen by follicle cells in the female and production of sperm in the male
LH—luteinizing hormone (female)	Stimulates ovulation and formation of the corpus luteum; the corpus luteum secretes progesterone
LH/ICSH—interstitial cell-stimulating hormone (male)	Stimulates testosterone secretion by the interstitial cells of the testes
PRL—prolactin	Stimulates secretion of milk in females; function in males is unknown
GH—growth hormone (somatotropin, STH)	Accelerates body growth and causes fat to be used for energy; this helps maintain blood sugar
Intermediate Lobe—Cells Along the Border of the Anterior and Posterior Lobes	
MSH—melanocyte-stimulating hormone	Stimulates the melanocytes to produce melanin in the skin
Posterior Lobe—Hormones Produced by the Hypothalamus	
ADH—antidiuretic hormone (vasopressin)	Maintains water balance by reducing urinary output; acts on kidney tubules to reabsorb water into the blood more quickly. In large amounts, it causes constriction of arteries.
Oxytocin	Promotes milk ejection and causes contraction of the smooth muscles of the uterus

INTERMEDIATE PITUITARY LOBE

The intermediate pituitary lobe is not a full-fledged lobe; it consists of only a few cells dispersed along the border of the posterior and anterior lobes. These cells produce **melanocyte-stimulating hormone (MSH)**, which stimulates the melanin cells in the skin.

POSTERIOR PITUITARY LOBE

The following hormones produced by the hypothalamus are stored in the posterior pituitary lobe:

1. **Vasopressin** converts to ADH in the bloodstream. The name *vasopressin* may cause confusion because it causes little or no vasoconstriction. ADH maintains the water balance by increasing the absorption of water in the kidney tubules. Sometimes drugs called diuretics are used to inhibit the action of ADH. The result is an increase in urinary output and a decrease in blood volume, thus decreasing blood pressure.
2. **Oxytocin** (ock-see-**TOH**-sin) is released during childbirth, causing strong contractions of the uterus. It also causes strong contractions when a mother is breast-feeding. A synthetic form of oxytocin called Pitocin® is given to help start labor or make uterine contractions stronger.

For a list summarizing the pituitary hormones and their functions, see **Table 12-1**. To view the posterior pituitary lobe, its relationship to the hypothalamus, and hormone production, see **Figure 12-5**.

THYROID AND PARATHYROID GLANDS

The thyroid and parathyroid glands are located in the neck close to the cricoid cartilage, or the "Adam's apple." The thyroid regulates body metabolism. The parathyroid maintains the calcium-phosphorus balance.

Thyroid Gland

The **thyroid gland** is a butterfly-shaped mass of tissue located in the anterior part of the neck. It lies on either side of the larynx, over the trachea. It is about 2 inches long, with two lobes joined by strands of thyroid tissue called the isthmus. Coming from the isthmus is a

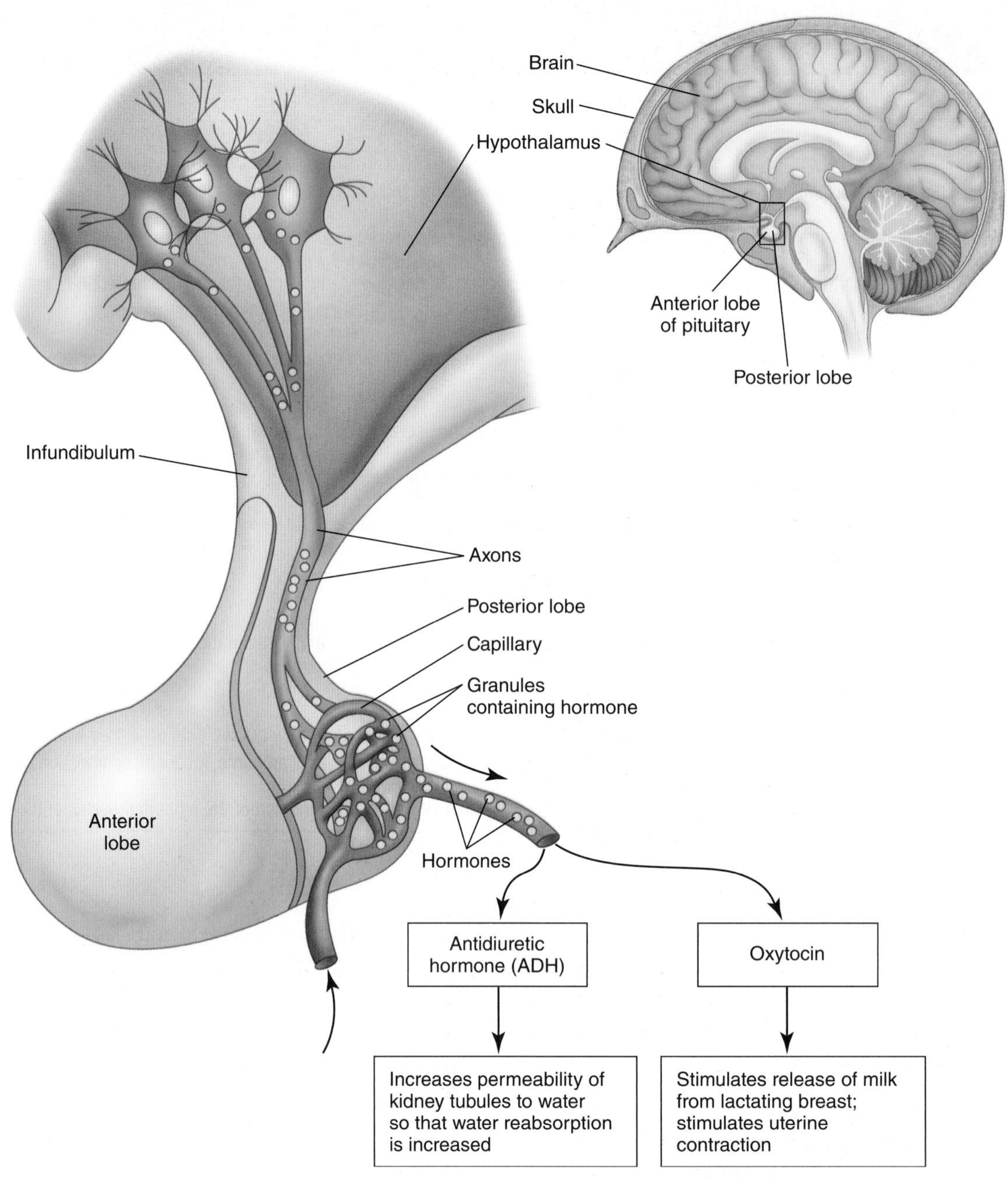

Figure 12-5 *The relationship of the hypothalamus of the brain with the posterior lobe of the pituitary gland*

fingerlike lobe of tissue known as the intermediate lobe. This intermediate lobe projects upward toward the floor of the mouth, as far up as the hyoid bone. See **Figure 12-6** for an illustration of the thyroid gland. The thyroid gland has a rich blood supply. In fact, it has been estimated that about 4 to 5 liters (some 8.5 to 10.5 pints) of blood pass through the gland every hour.

The thyroid gland secretes three hormones: **thyroxine (T_4)** (thigh-**ROCK**-seen), **triiodothyronine (T_3)** (try-eye-oh-doh-**THIGH**-roh-neen), and **calcitonin** (kal-sih-**TOH**-nin). The first two, T_3 and T_4, are iodine-bearing derivatives of the amino acid tyrosine. T_3 is 5–10 times more active than T_4, but its activity is less prolonged. However, the two have the same effect. Both hormones are produced in the follicle cells of the thyroid gland. The TSH of the pituitary controls the production and secretion of the thyroid hormones from the thyroid gland. The thyroid hormones contain iodine. Most of the iodine needed for their synthesis comes from the diet. Iodides are circulated to the thyroid gland, where they are "trapped." Here the iodides combine with the amino acid tyrosine to form the hormones T_3 and T_4.

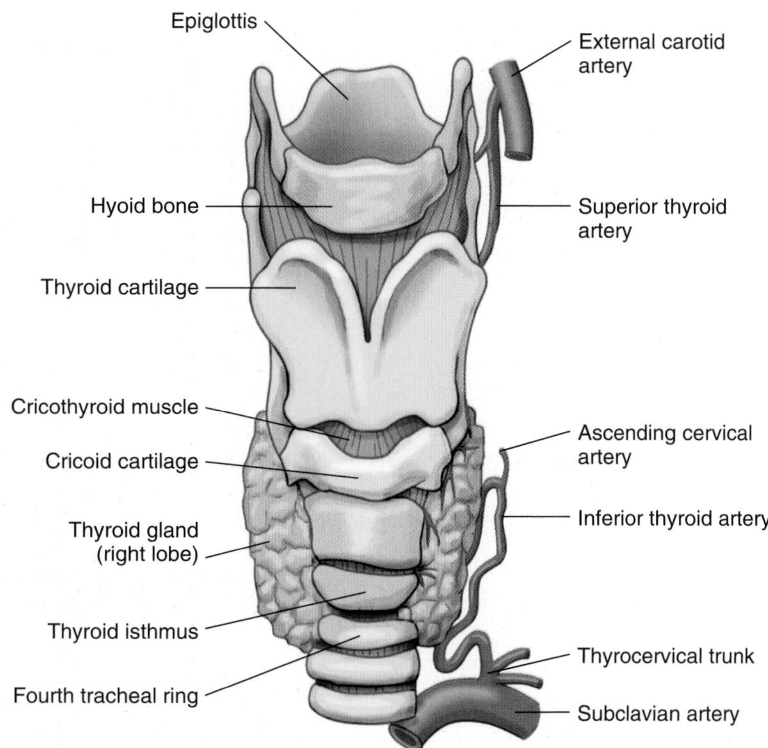

Figure 12-6 *Thyroid gland*

The concentration of these two hormones in the bloodstream is controlled by the negative feedback system previously discussed. The consequences of hyposecretion and hypersecretion of the thyroid hormones is discussed later in this chapter.

Thyroxine controls the rate of metabolism, heat production, and oxidation of all cells, with the possible exception of the brain and spleen cells. The functions of T_4 and T_3 are as follows:

1. Control the rate of metabolism in the body: how cells use glucose and oxygen to produce heat and energy
2. Stimulate protein synthesis and thus help in tissue growth
3. Stimulate the breakdown of liver glycogen to glucose

CALCITONIN

Another hormone produced and secreted by the thyroid gland is calcitonin. It controls the calcium ion concentration in the body by maintaining a proper calcium level in the bloodstream.

Calcium is an essential body mineral. Approximately 99% of the calcium in the body is stored in the bones. The rest is located in the blood and tissue fluids. Calcium is necessary for blood clotting, holding cells together, and neuromuscular functions. The constant level of calcium in the blood and tissues is maintained by the action of calcitonin and parathormone, which is produced by the parathyroid gland.

When blood calcium levels are higher than normal, calcitonin secretion is increased. Calcitonin lowers the calcium concentration in the blood and body fluids by decreasing the rate of bone resorption—*osteoclastic* activity—and by increasing the calcium absorption by bones—*osteoblastic* activity. Proper secretion of calcitonin into the bloodstream prevents hypercalcemia, a harmful rise in the blood calcium level.

Parathyroid Glands

The **parathyroid glands**, usually four in number, are tiny glands the size of grains of rice. These are attached to the posterior surface of the thyroid gland and secrete the hormone **parathormone**. Parathormone, like calcitonin, also controls the concentration of calcium in the bloodstream. When the blood calcium level is lower than normal, parathormone secretion is increased.

Parathormone stimulates an increase in the number and size of specialized bone cells referred to as osteoclasts. Osteoclasts quickly invade hard bone tissue, digesting large amounts of the bony material containing calcium. As this process continues, calcium leaves the bone and is released into the bloodstream, increasing the calcium blood level.

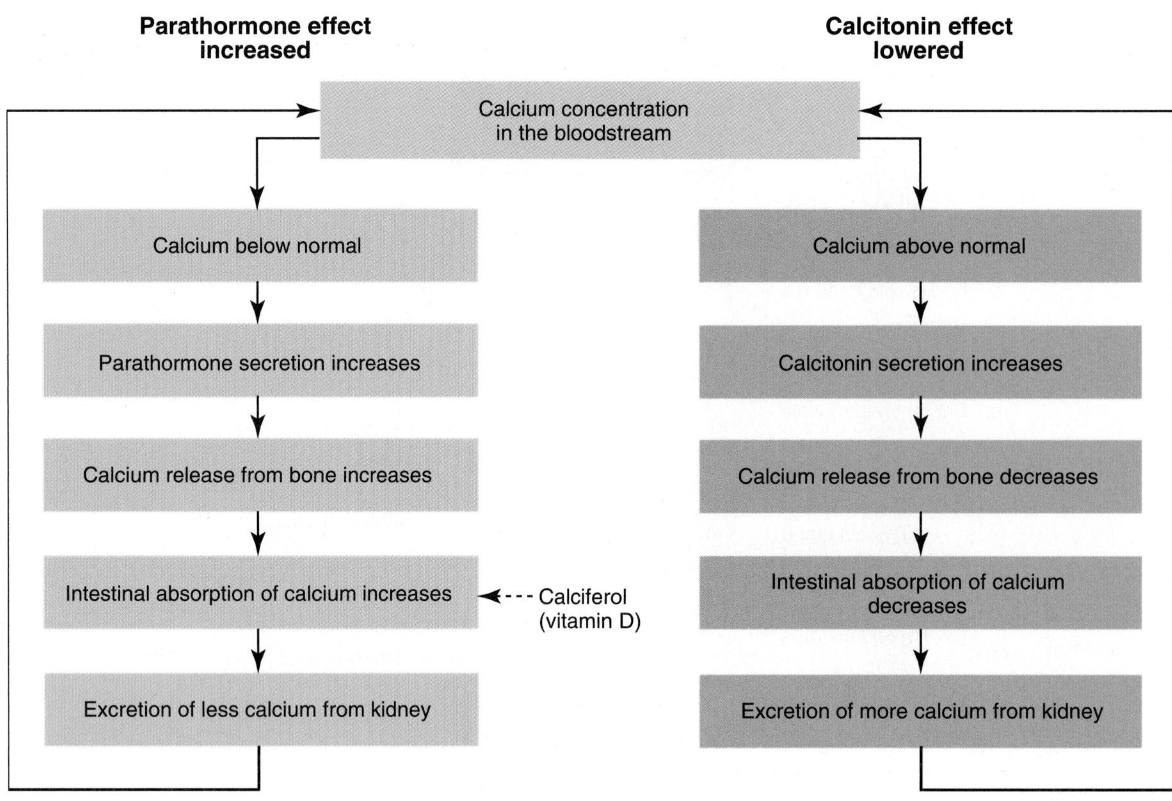

Figure 12-7 *Effects of parathormone and calcitonin on the level of calcium in the blood*

Bone calcium is bonded to phosphorus in a compound called calcium phosphate ($CaPO_4$). When calcium is released into the bloodstream, phosphorus is released along with it. Parathormone stimulates the kidneys to excrete any excess phosphorus from the blood; at the same time, it inhibits calcium excretion from the kidneys. Consequently, the concentration of blood calcium rises.

Thus, parathormone and calcitonin of the thyroid have opposite, or antagonistic, effects to one another (see **Figure 12-7** for a summary of their actions). Parathormone, however, acts much more slowly than calcitonin does. It may be hours before the effects of parathormone become apparent. In this manner, the secretions of parathormone and calcitonin serve as complementary processes controlling the level of calcium in the bloodstream.

THYMUS GLAND

The **thymus** gland is both an endocrine gland and a lymphatic organ. It is located under the sternum, anterior and superior to the heart. Fairly large during childhood, it begins to disappear at puberty. The thymus gland secretes a large number of hormones. The major hormone is thymosin (**THIGH**-moh-sin), which helps stimulate the lymphoid cells that are responsible for the production of T cells, which fight certain diseases. The thymus gland is critical to the development of the immune system. *Myasthenia gravis* occurs when the thymus is abnormally large and produces antibodies that block or destroy the muscle cells.

ADRENAL GLANDS

The two **adrenal glands** are located on top of each kidney (**Figure 12-8**). Each gland has two parts: the cortex and the medulla. ACTH from the pituitary glands stimulates the activity of the cortex of the adrenal gland. The hormones secreted by the adrenal cortex are known as corticoids (**KOR**-tih-koidz). The corticoids are also very effective as anti-inflammatory drugs.

The cortex secretes the following three groups of corticoids, each of which is of great importance:

1. **Mineralocorticoids**, mainly aldosterone (al-**DOSS**-ter-ohn), affect the kidney tubules by speeding up the reabsorption of sodium into the blood circulation and increasing the excretion of potassium from the blood. They also speed up the reabsorption of water by the kidneys. This means that aldosterone helps to regulate the blood pH by controlling the levels of electrolytes in the blood. Aldosterone is used in the treatment of Addison's disease to replace deficient secretion of mineralocorticoids.

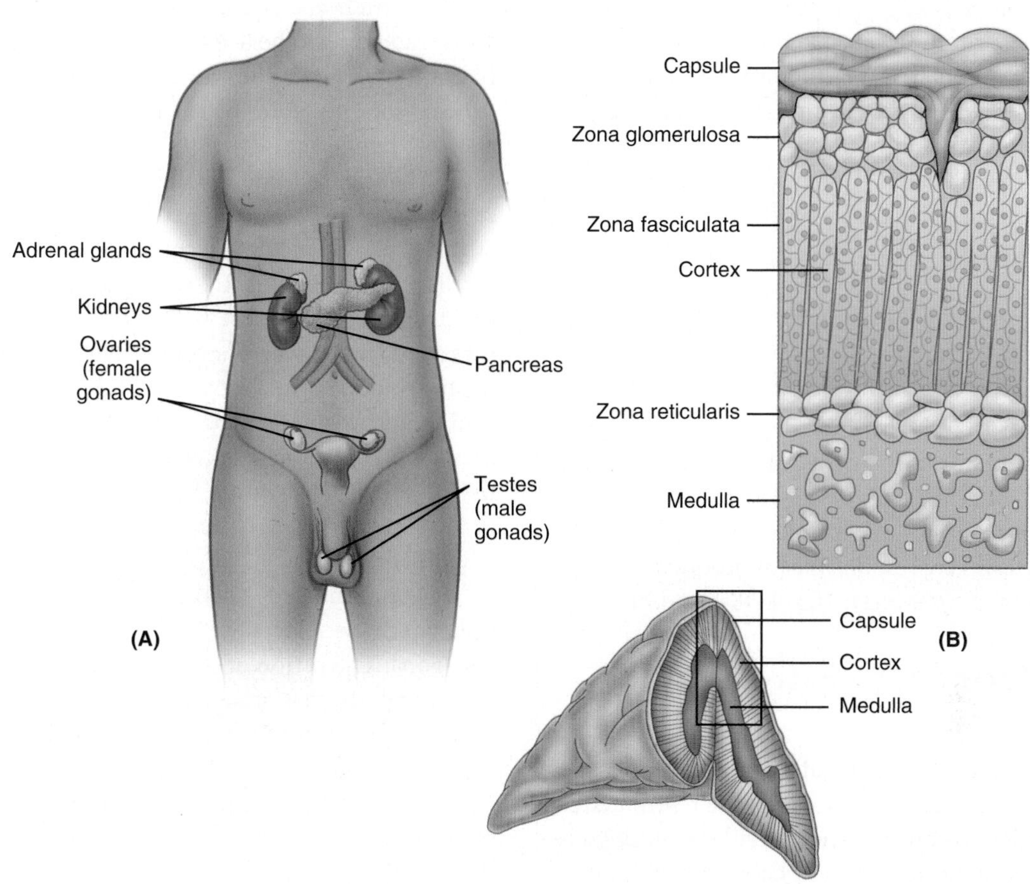

Figure 12-8 *(A) Location of adrenal glands and gonads; (B) adrenal capsule, cortex, and medulla*

2. **Glucocorticoids**, namely cortisone and cortisol, increase the amount of glucose in the blood. This is done by (1) the conversion of proteins and fats to glycogen in the liver, followed by (2) the breakdown of the glycogen into glucose. These glucocorticoids also help the body resist the aggravations caused by various everyday stresses. In addition, these hormones seem to decrease edema in inflammation, regulate the blood pressure, and reduce pain by inhibiting pain-causing prostaglandin.
3. **Androgens** and DHEA are weak male hormones that are found in both males and females. They are precursor hormones that are converted into estrogen in the ovaries of females and into male hormones in the testes of males. Much larger amounts of estrogen and androgenic hormones are then produced by the ovaries and the testes.

Medulla of the Adrenal Gland

The adrenal medulla responds to the sympathetic nervous system and secretes **epinephrine**, also called **adrenaline**, and **norepinephrine** (**Table 12-2**). Epinephrine is a

Table 12-2 *Comparison of the Effects of Epinephrine and Norepinephrine*

EPINEPHRINE	*NOREPINEPHRINE*
1. Bronchial relaxation	No effect
2. Dilation of iris	No effect
3. Excitation of central nervous system	No effect
4. Increased conversion of stored glycogen to glucose	Much less effect
5. Increased heart rate	Little effect
6. Increased cardiac output and venous return	Slight effect
7. Increased blood flow to muscles	Vasoconstriction in muscles
8. Increased myocardial strength	About the same
9. Increased basal metabolic rate (BMR)	Much less effect
10. Increased systolic blood pressure	Raises both systolic and diastolic blood pressure

powerful cardiac stimulant. It functions by bringing about a release of more glucose from stored glycogen for muscle activity and increasing the force and rate of the heartbeat. This chemical activity increases cardiac output and venous return and raises the systolic blood pressure. The hormones produced are referred to as the fight-or-flight hormones because they prepare the body for an emergency situation.

GONADS

The **gonads**, or sex glands, include the ovaries in the female and the testes in the male. The ovary is responsible for producing the ova, or egg, and the hormones **estrogen** (**ESS**-troh-jen) and **progesterone** (proh-**JES**-ter-ohn). The testes are responsible for producing sperm and the hormone **testosterone** (tes-**TOS**-teh-rohn).

Female Hormones—Estrogen and Progesterone

Estrogen is produced by the graafian follicle cells of the ovary. It stimulates the development of the reproductive organs, including the breasts, and secondary sex characteristics, such as pubic and axillary hair.

Progesterone is produced by the cells of the corpus luteum of the ovary. Progesterone works with estrogen to build up the lining of the uterus for the fertilized egg. If no fertilization occurs, menstruation takes place. This cycle depends on the secretion of the anterior pituitary gland (see Chapter 22).

Male Hormone—Testosterone

Testosterone is produced by the interstitial cells of the testes and is responsible for the development of the male reproductive organs and secondary sex characteristics. Testosterone influences the growth of a beard and other body hair, deepening of the voice, increases in musculature, and the production of sperm. The secretion of the hormone depends on the pituitary gland (see Chapter 22).

PANCREAS

The **pancreas** is located behind the stomach and functions as both an exocrine and an endocrine gland. The exocrine portion secretes pancreatic juices, which are excreted through a duct into the small intestines. There they become part of the digestive juices. The endocrine portion is involved in the production of insulin by the beta cells (B cells) of the islets of Langerhans.

The islet cells are distributed throughout the pancreas. These cells were named the islets of Langerhans after the physician who discovered them (**Figure 12-9**). B cells produce **insulin** (**IN**-suh-lin), which (1) promotes the utilization of glucose for energy in the cells, necessary for maintenance of normal levels of blood glucose (fasting blood sugar 70–100 mg/dL); (2) promotes fatty acid transport and fat deposition into cells; (3) promotes amino acid transport into cells; (4) facilitates protein synthesis; and (5) causes the liver to convert glucose into glycose. Lack of insulin secretion by the islet cells causes diabetes mellitus.

The alpha cells (A cells) contained in the islets of Langerhans secrete the hormone **glucagon** (**GLOO**-kah-gon). The action of glucagon may be antagonistic, or opposite, to that of insulin. Glucagon's function is to increase the level of glucose in the bloodstream. This is done by stimulating the conversion of liver glycogen to glucose. The control of glucagon secretion is achieved by negative feedback. Low glucose levels in the bloodstream stimulate the A cells to secrete glucagon, which quickly increases the glucose level in the bloodstream.

Media Link

View the **Dual Role of the Pancreas** animation on the Online Resources.

PINEAL GLAND

The **pineal gland** (**PIN**-ee-al) or body is a small, pinecone-shaped organ attached by a slim stalk to the roof of the third ventricle in the brain. The hormone produced by the pineal gland is called **melatonin** (mel-ah-**TOH**-nin). The pineal gland is stimulated by a group of nerve cells, called the suprachiasmatic (**soo**-prah-**KYE**-az-**mat**-ik) nucleus (SCN), located in the brain over the pathway of fibers of the optic nerve. The amount of light entering the eye stimulates the SCN, which then stimulates the pineal gland to release its hormone. The amount of light

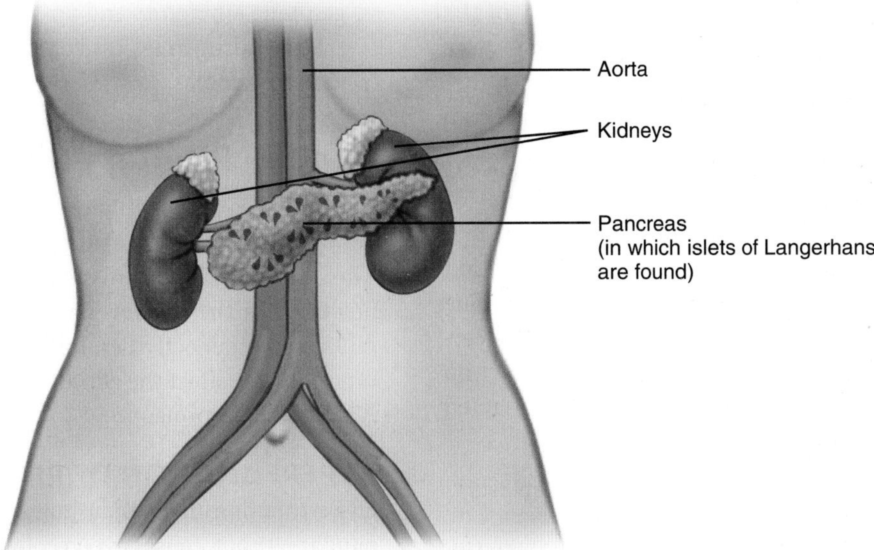

Figure 12-9 *Location of islets of Langerhans*

affects the amount of melatonin secreted. The darker it is, the more melatonin is produced; the lighter it is, the less melatonin is produced. Because winter has longer, darker days, more melatonin is produced, which can lead to a condition known as seasonal affective disorder (SAD). This condition produces lethargy, anxiety, and mood disorders. Scientists have discovered for those affected with SAD that symptoms disappear when light is introduced. The suggested treatment is bright light exposure for 0.5 to 3 hours per day.

There are no clear answers to the function of melatonin; however, melatonin causes body temperature to drop. For example, falling asleep is associated with lowered body temperature, whereas waking up is associated with rising body temperature.

Media Link

View the **Endocrine System** animation on the Online Resources.

DISORDERS OF THE ENDOCRINE SYSTEM

Endocrine gland disturbances may be caused by several factors: congenital disease of the gland itself, infections in other parts of the body, autoimmune disorders, and dietary deficiencies. Most disturbances result from (1) hyperactivity of the glands, causing

The Effects of Aging on the Endocrine System

The aging process affects nearly every gland in the endocrine system. The blood levels of some hormones increase, while others decrease. The hypothalamus is responsible for releasing hormones that stimulate the pituitary gland. During aging, either impaired secretion of some hypothalamic-releasing hormone or impaired pituitary response occurs. These changes, in turn, affect the homeostasis of the body.

With increasing age, the pituitary becomes smaller and production of the growth hormone is interfered with. The thyroid gland may become lumpy, and metabolism generally slows down with age. The parathormone blood levels change, which may contribute to osteoporosis. From the pancreas, there is a loss of insulin receptor cells, which may lead to type 2 diabetes. The aldosterone level from the adrenal cortex decreases, which can contribute to light-headedness and a drop in blood pressure upon sudden position change, called orthostatic hypotension.

The gonad glands are affected; aging men sometimes experience a slightly decreased level of testosterone, while women have decreased levels of estrogen and progesterone after menopause.

oversecretion of hormones; or (2) hypoactivity of the gland, resulting in undersecretion of hormones. See **Table 12-3** for more information on disorders of the endocrine glands. Health care providers most often see these patients in a physician's office.

Congenital Pituitary Disorders

During fetal development, there is a complex interaction between the fetal and maternal endocrine system. The failure to develop an endocrine system and the corresponding release of hormones has a cascading effect on other developing systems. Glands that may be affected with accompanying disease include the following:

- Pineal: hypoplasia associated with retinal dysplasia
- Pituitary: pituitary tumors
- Thyroid: congenital hypothyroidism associated with neurological abnormalities and incomplete ossification of bone and mineral matter; goiter cretinism
- Thymus: when the thymus doesn't completely descend, it can cause trachea and esophageal compression.
- Adrenal: congenital adrenal hypoplasia, a familial, inherited disorder of the adrenal steroidogenesis enzyme that impairs cortisol production by the adrenal cortex. Androgen excess can affect the structure and functions of the reproductive organs, leading to genitalia ambiguity and post-natal virilization in both sexes; androgen excess can often cause fertility problems in females.

General Pituitary Disorders

Disturbances of the pituitary gland may produce a number of body changes outside of embryonic development. This gland is chiefly involved in the growth function; however, as the master gland, the pituitary indirectly influences other activities. Both hyperfunction and hypofunction will affect the action of hormones such as TSH, ATH, LH, and prolactin.

HYPERFUNCTION OF PITUITARY

Hyperfunctioning of the pituitary gland, often due to a pituitary tumor, causes hypersecretion of the pituitary GH. This can lead to gigantism (giantism) and acromegaly.

Hyperfunctioning during preadolescence causes **gigantism** (jigh-**GAN**-tizm), an overgrowth of the long bones leading to excessive tallness.

If hypersecretion of the GH occurs during adulthood, **acromegaly** (ack-roh-**MEG**-ah-lee) is the result. This is an overdevelopment of the bones of the face, hands, and feet (**Figure 12-10**). In adults whose long bones have already matured, the GH attacks the cartilaginous regions and the bony joints. Thus, the chin protrudes, and the lips, nose, and extremities enlarge

Table 12-3 *Disorders of the Endocrine Glands*

GLAND	HORMONE	HYPERFUNCTION	HYPOFUNCTION
Anterior pituitary	Growth hormone	Gigantism Adults—acromegaly	Pituitary short stature (formerly known as *dwarfism*)
Posterior pituitary	ADH/vasopressin	Diabetes insipidus	
Thyroid	Thyroxine	Hyperthyroidism Graves' disease	Hypothyroidism Children—cretinism Adults—myxedema
Parathyroid	Parathormone	Possible kidney stones Possible bone fracture	Tetany
Thymus	Thymosin		T-lymphocyte deficiency
Adrenal	Cortisol cortisone	Cushing's syndrome	Addison's disease
Pancreas	Insulin		Diabetes mellitus
Gonads	See Chapter 22		

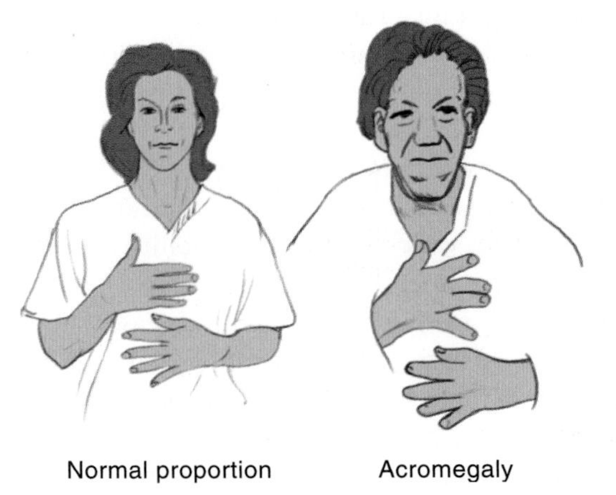

Figure 12-10 *Comparison of an individual of normal proportion with an individual with acromegaly*

disproportionately. Lethargy and severe headaches frequently set in as well.

Treatment of acromegaly and gigantism depends on the underlying cause. Treatment may include drug therapy, which inhibits GH; surgery; and radiation therapy.

HYPOFUNCTION OF PITUITARY

Hypofunction of the pituitary gland during childhood leads to **pituitary short stature** (formerly known as *dwarfism*). Growth of the long bones is abnormally decreased by an inadequate production of GH. Despite the small size, however, the body of an individual with pituitary short stature is normally proportioned and intelligence is normal. Unfortunately, the physique remains juvenile and sexually immature (**Figure 12-11**). Treatment involves early diagnosis and injections of human GH. The treatment period is five years or more.

> **Did You Know?**
>
> *There really was a dwarf known as "Tom Thumb," but this was a stage name created by P.T. Barnum. His real name was Charles Sherwood. He weighed 9 pounds 2 ounces at birth, but he stopped growing after his first birthday. At 18, Tom Thumb was 2 feet, 6½ inches.*

DIABETES INSIPIDUS

Another disorder caused by posterior lobe dysfunction is **diabetes insipidus** (dye-ah-**BEE**-teez in-**SIP**-ih-dus). In this condition, there is a drop in the amount of ADH, which causes an excessive loss of water and electrolytes. Symptoms include excessive thirst, or **polydipsia** (pol-ee-**DIP**-see-ah). Treatment is with medications.

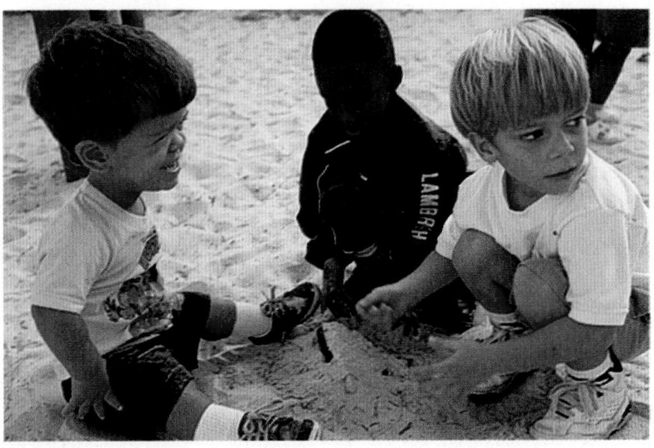

Figure 12-11 *The child on the left has pituitary short stature.*

Thyroid Disorders

Because the thyroid gland controls metabolic activity, any disorder will affect other structures besides the gland itself. Persons at risk include those with other immune system problems.

DIAGNOSTIC TESTS FOR THYROID

To diagnose thyroid function, a blood test is done; blood levels of TSH, T_3, and T_4 are checked to see if they are within normal limits.

Another diagnostic tool may be a thyroid nuclear test. There are two types: thyroid scan and radioactive iodine uptake (RAI). In both tests the patient takes a small amount of radioactive iodine.

- Thyroid scan measures the structure and function of the thyroid, checking for enlargement or nodules.
- RAI measures the amount of radioactivity in the thyroid gland. The amount of radioactive iodine detected in the thyroid corresponds to the amount of hormone the thyroid is producing.

HYPERTHYROIDISM

Hyperthyroidism (high-per-**THIGH**-roid-izm) is a result of overactivity by the thyroid gland. Hyperthyroidism can be caused by toxic nodules or medication. Most cases of hyperthyroidism are due to the autoimmune disorder Graves' disease. Too much thyroxine is secreted in a process called hypersecretion, leading to enlargement of the gland. People with hyperthyroidism consume large quantities of food yet lose body fat and weight. Symptoms include feeling too hot, fast-growing and rougher fingernails, and weakened muscles. They may suffer from increased blood pressure, rapid heartbeat, hand tremors, fatigue, perspiration, and irritability. In addition, the liver releases excess glucose into the bloodstream, increasing the blood sugar level. The most pronounced symptoms of hyperthyroidism include enlargement of the thyroid gland, or a **goiter** (**GOI**-ter); bulging of the eyeballs, or **exophthalmos** (eck-sof-**THAL**-mos); dilation of the pupils; and wide-opened eyelids (**Figure 12-12**). The immediate cause of exophthalmos is not completely known. It is not directly caused by hyperthyroidism because removal of the thyroid does not always cause the eyeballs to return to their normal state.

Treatment of hyperthyroidism includes total or partial removal of the thyroid and administration of drugs, such as propylthiouracil and methylthiouracil, to reduce thyroxine secretion. The use of radioactive iodine to suppress the activity of the thyroid gland is another treatment for hyperthyroidism.

If the level of thyroid hormones goes too high and symptoms become worse, the patient may be experiencing "thyroid storm." One indication is an extremely high body temperature between 40°C and 40.5°C (104°F and 105°F). Thyroid storm is a life-threatening condition, and emergency medical treatment is needed. Treatments include intravenous fluid and electrolytes as well as drugs to block the release and action of thyroid hormones.

HYPOTHYROIDISM

Hypothyroidism (high-poh-**THIGH**-roid-izm) is a condition in which the thyroid gland does not secrete sufficient thyroxine in a process called hyposecretion.

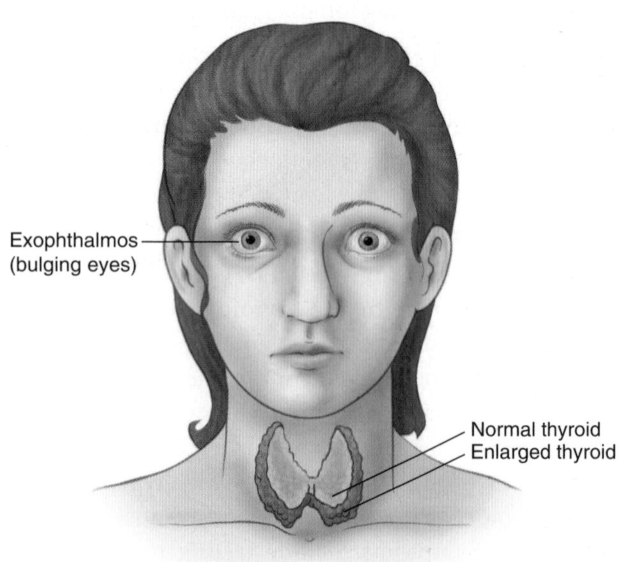

Figure 12-12 *Exophthalmos and an enlarged thyroid, two symptoms of Graves' disease*

General symptoms include fatigue, increased sensitivity to cold, dry skin, and weight gain. This is manifested by low T_3 or T_4 levels or increased TSH blood levels.

Adult hypothyroidism may occur because of iodine deficiency. The major cause is an inflammation of the thyroid, which destroys the ability of the gland to make thyroxine. This inflammation is caused by Hashimoto's disease, an autoimmune disease that attacks the body's own thyroid gland. Symptoms include dry and itchy skin, dry and brittle hair, constipation, and muscle cramps at night.

Depending on the time hypothyroidism strikes its victims, two different sets of disorders may occur: myxedema or cretinism.

MYXEDEMA. Adult onset is more common in people over the age of 60. The face becomes swollen, weight increases, and motivation and memory fail for a person experiencing **myxedema** (micks-eh-**DEE**-mah). Treatment is daily medication of thyroid hormone. It is important for the health care provider to ensure the patient understands the necessity of taking the medication. Follow-up tests to measure TSH blood levels are also important.

CRETINISM. Developing early in infancy or childhood, **cretinism** (**CREE**-tin-izm) is characterized by a lack of mental and physical growth, resulting in intellectual disability and malformation. The sexual development and physical growth of children affected by cretinism do not proceed beyond those of a 7- or 8-year-old child.

In treating cretinism, thyroid hormones or thyroid extract may restore a degree of normal development if administered in time. In most cases, however, normal development cannot be completely restored once the condition has set in.

THYROID CANCER

Thyroid cancer is the most common cancer of the endocrine system. It affects more females than males. Diagnosed in the early stages, survival rates are between 90% and 100%.

Parathyroid Disorders

The parathyroid glands regulate the use of calcium and phosphorus. Both of these minerals are involved in many of the body systems.

Hyperfunctioning of the parathyroid glands may cause an increase in the amount of blood calcium, increasing the tendency for calcium to crystallize in the kidneys as kidney stones. Excess amounts of calcium and phosphorus are withdrawn from the bones; this may lead to eventual deformity. Sometimes, so much calcium is removed from the bones that they become honeycombed with cavities. Afflicted bones become so fragile that even walking can cause fractures. Treatment may be the surgical removal of the parathyroid glands, as well as the use of supplemental medication—such as calcium or Vitamin D—or medication to suppress the parathyroid.

Hypofunctioning of the parathyroid glands may cause a decrease in calcium and an increase in serum phosphorus. Conditions that may occur are hypocalcemia or tetany. Hypocalcemia can cause a complex pattern of symptoms; these can be treated with calcium. In the case of **tetany**, severely diminished calcium levels affect the normal function of nerves. Convulsive twitching, or spasms, develop, and the afflicted person dies of spasms in the respiratory muscles. Treatment consists of administering vitamin D, calcium, and parathormone to restore a normal calcium balance.

Adrenal Disorders

The adrenal glands produce glucocorticoid hormones. Therefore, disorders of the adrenal glands result in either abundance or deficiency of these hormones.

Pheochromocytoma (fee-oh-**kroh**-moh-sye-**TOH**-mah) is a tumor of the adrenal gland that causes an excessive secretion of epinephrine, which may be fatal. This tumor is not cancerous and must be removed.

HYPERFUNCTION OF ADRENAL CORTEX

Cushing's syndrome (**KUSH**-ingz **SIN**-drohm) results from hypersecretion of glucocorticoid hormones from the adrenal cortex. This hypersecretion may be caused by an adrenal cortical tumor or the prolonged use of prednisone; more women than men tend to develop this endocrine disorder. Symptoms include high blood pressure, muscular weakness, obesity, poor healing of skin lesions, a tendency to bruise easily, hirsutism (excessive hair growth), menstrual disorders in women, and hyperglycemia. The most noticeable characteristics are a rounded "moon" face and a "buffalo hump" that develops from the redistribution of body fat (**Figure 12-13**). Therapy consists of surgical removal of the adrenal cortical tumor.

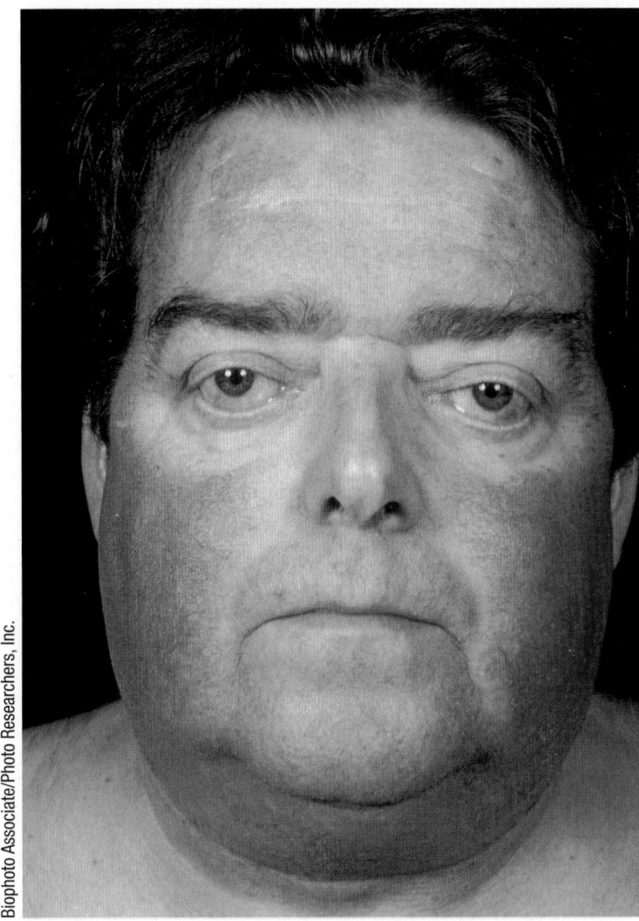

Biophoto Associate/Photo Researchers, Inc.

Figure 12-13 *Cushing's syndrome causes a characteristic rounded, red "moon" face.*

HYPOFUNCTION OF ADRENAL CORTEX

Hypofunctioning of the adrenal cortex leads to **Addison's disease** (**AD**-ih-sonz). People with this disease exhibit the following symptoms: excessive pigmentation, prompting the characteristic "bronzing" of the skin; decreased levels of blood glucose (hypoglycemia); low blood pressure, which falls further when standing; pronounced muscular weakness and fatigue; diarrhea; weight loss; vomiting; and a severe drop of sodium in the blood and tissue fluids, causing a serious imbalance of electrolytes.

The medical treatment of Addison's disease is focused on replacement of the deficient hormones and increasing salt intake, especially during heavy exercise.

Gonad Disorders

See Chapter 22 for disorders of the ovaries and testes.

Pancreatic Disorders

Diabetes mellitus (dye-ah-**BEE**-teez **MELL**-ih-tus) is a condition caused by decreased secretion of insulin from the islets of Langerhans cells or by the ineffective use of insulin. Insulin is necessary for the cells to use glucose to provide energy. Carbohydrate metabolism in diabetes mellitus is disturbed and thus has an adverse effect on protein and fat metabolism, leaving excess glucose (energy) in the blood.

According to the latest data from the American Diabetes Association (ADA), 34.2 million Americans have diabetes: 28 million are diagnosed, while 7.3 million are undiagnosed.

Diabetes is divided into two main types: type 1 and type 2. Type 1 is usually exhibited in children or young adults. The cause of type 1 is thought to be an autoimmune reaction, which involves genetic and viral factors that destroy the islets of Langerhans cells. Patients who have type 1 diabetes must take insulin and monitor daily blood glucose levels.

Symptoms of type 1 diabetes include the following:

- **Polyuria** (pol-ee-**YOU**-ree-ah)—excessive urination
- Polydipsia (pol-ee-**DIP**-see-ah)—excessive thirst
- **Polyphagia** (pol-ee-**FAY**-jee-ah)—excessive hunger
- Weight loss
- Blurred vision
- Possible diabetic coma
- Ketones in the urine
- Slow-to-heal sores

Insulin deficiency causes glucose to accumulate in the bloodstream, rather than be transported to the cells and converted into energy. Eventually the excess becomes too much for the kidneys to reabsorb, and the excess glucose is excreted in the urine. Excretion of excess glucose requires an accompanying excretion of large amounts of water.

Because sufficient glucose is not available for cellular oxidation in diabetes mellitus, the body starts to burn up protein and fats. A person with diabetes is constantly hungry and usually eats voraciously but loses weight nonetheless.

When fats are utilized as a fuel source, they are rapidly but incompletely oxidized. One product of this abnormal rate of fat oxidation is ketone bodies. Ketone bodies are highly toxic; the type most commonly formed is acetoacetic acid. These ketone acids accumulate in the blood, promoting the development of acidosis, giving the breath and urine an odor of "sweet" acetone. If acidosis is severe, diabetic coma and death may result. Prolonged diabetes leads to atherosclerosis, stroke, hypertension, blindness (diabetic retinopathy), cataracts, glaucoma, dental disease, kidney damage, amputations, and nerve damage. For type 1 diabetes, therapy consists of daily insulin injections, exercise, and a controlled diet. In some individuals, a pancreatic transplant may be possible. The use of insulin pumps and prefilled insulin pens may also be effective for some people.

Medical Highlights

HORMONE IMBALANCE: MENTAL HEALTH

An imbalance of any hormone in the body can cause physical and mental health problems. Symptoms can develop when hormone levels are too high or too low.

The hypothalamus and pituitary gland influence mood. Over- or underproduction affects personality.

Overproduction of thyroid hormones causes anxiety, tremors, and irritation. Underproduction from the thyroid can cause myxedema in adults, which leads to lethargy and memory failure. In infants and children, cretinism may develop.

In most cases of hormone imbalance, the treatment is diagnosis and medication to correct the condition.

Stress is one condition that results in increased production of epinephrine (adrenaline) secretions of the adrenal medulla. When the body is under stress, the sympathetic nervous system stimulates the adrenal medulla to produce epinephrine. If the parasympathetic nervous system is unable to balance the sympathetic nervous system, the result is chronic stress.

Chronic stress adversely affects brain function, especially memory; it increases blood pressure, increasing the risk of stroke, heart disease, sleep problems, depression, and digestive problems. Some literature states that as many as 90% of physicians' office visits are stress related. Treatment for chronic stress may include using relaxation response techniques, massage, biofeedback, and medication if necessary.

Table 12-4 *Signs of Hypoglycemia and Hyperglycemia*

	HYPOGLYCEMIA (↓ BLOOD SUGAR)	***HYPERGLYCEMIA (↑ BLOOD SUGAR)***
Onset	Sudden	Slow
Reason	Too much insulin Too much exercise Not enough food	Not enough insulin Not enough exercise Too much food
Skin	Pale, moist to wet Sweating	Flushed, dry, hot No sweating
Symptoms	Nervous, trembling, confused, irritable	Drowsy, lethargic, weak, lapsing into unconsciousness
Breath	Normal odor	Fruity odor
Respiration	Normal to rapid	Kussmaul breathing (air hunger)
Glycosuria	Little to none	High amount
Ketonuria	None	Present
Blood sugar	Low—below 80 mg/dL	High—above 150 mg/dL
Treatment	Rapid response; give sugar in form of soft drink or orange juice Glucagon (IM); glucose 50% IV	Slow response IV fluids Regular insulin

Patient education is critical in the treatment of diabetes. People with insulin-dependent diabetes require education in the signs of **hypoglycemia** (high-poh-glye-**SEE**-mee-ah)–low blood sugar and insulin shock–and **hyperglycemia** (high-per-gly-**SEE**-mee-ah)–high blood sugar and diabetic coma–as illustrated in **Table 12-4** and **Figure 12-14**.

Type 2 diabetes occurs when the body becomes resistant to insulin or when the pancreas does not produce enough insulin. Characteristics and symptoms of type 2 diabetes include the following:

- Gradual onset
- Most common in adults over age 45
- Feelings of tiredness or illness
- Frequent urination, especially at night
- Unusual thirst
- Frequent infections and slow healing of sores

Type 2 diabetes makes up 90% to 95% of diabetics. Diabetes is most common in adults over age 45, people who are overweight, individuals who have an immediate family member with diabetes, and people of certain ethnic groups. The incidence of type 2 diabetes in younger people is growing; scientists believe the major cause is obesity. In this condition, insulin is secreted but in lower amounts.

TREATMENT OF DIABETES MELLITUS

Diabetes is recognized as a leading cause of death and disability in the United States. Diabetes can damage the coronary arteries and blood vessels. People with diabetes also have higher cholesterol and triglyceride blood levels. This combination causes the damaged vessels to trap cholesterol from the blood; in time, the blood vessels fill with fatty buildup, leading to heart disease, high blood pressure, and poor circulation. Heart and blood vessel damage occurs three times more often and at an earlier age in people with diabetes. The treatment focus for type 2 diabetes is on diet modification, weight reduction, glucose monitoring, and medication.

Individuals who have diabetes may need instruction on how to use their glucose monitoring system, how to inject insulin, how to exercise, how to use their calculated diet, and how to take prescribed medications as instructed.

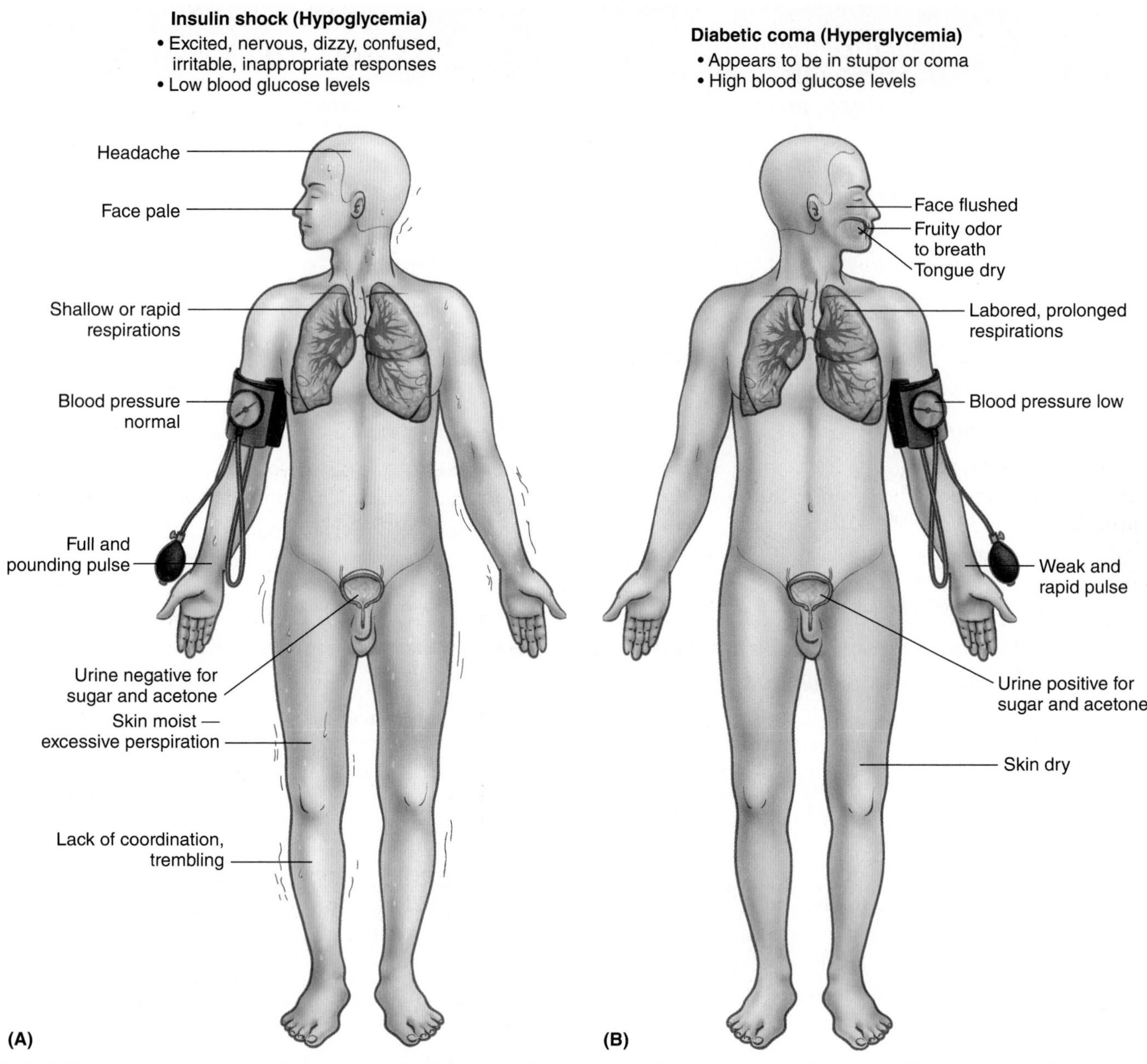

Figure 12-14 *(A) Insulin shock (hypoglycemia); (B) diabetic coma (hyperglycemia)*

Oral hypoglycemic and/or injectable agents used by those with type 2 diabetes can stimulate the pancreas to produce more insulin, block the action of stomach and intestinal enzymes that break down carbohydrates, increase the response of the cells to the insulin that is produced, and control liver production of glucose. Emerging treatment of diabetes is a closed-loop monitoring system known as an "artificial pancreas." It links a continuous glucose monitor to an insulin pump and automatically delivers the correct amount of insulin.

Patients with diabetes can lead normal, productive lives if they follow their treatment. They should wear a medic-alert bracelet stating that they have diabetes.

Research has shown that aggressive and intensive control of elevated blood sugar levels in patients with diabetes is very important when needed. It decreases the complications of kidney disease, nerve damage, and retinopathy, and may reduce the occurrence and severity of heart disease.

TESTS FOR DIABETES MELLITUS

The diagnostic tests to determine the presence of glucose are done on urine and blood samples. The most common test done is a finger prick to obtain a blood sample that is then measured in a glucometer (glucose monitor). This test may be done by the patient at home. For patients with diabetes, the ADA recommends a fasting blood sugar level of 70–130 mg/dL and an after-meal blood sugar level less than 180 mg/dL.

Another blood test for diabetes is the glycosylated hemoglobin (HbA1c) test. The glucose exposed to hemoglobin attaches itself to the protein in a way that reflects the average blood glucose concentration for the preceding two to three months. The test is done every three months. Normal HbA1c for those without diabetes is less than 5.7%; patients with diabetes may have 6.3% or higher. The goal in patients with diabetes is below 7%.

The glucose tolerance test (GTT) is an oral test for diabetes. The person fasts from midnight the night before the test. On the morning of the test, a fasting blood sugar will be drawn. The person then drinks a liquid containing a certain amount of glucose (usually 50 to 100 g of glucose). Blood will be drawn at one-hour intervals up to three hours after the glucose drink is taken. Above-normal blood sugar levels may indicate diabetes.

Urine can be tested for albumin or ketone bodies using a specifically coded dipstick. A urine sample is obtained, and then the tape is dipped into the urine and compared with the special coding bar on the outside of the dipstick container. Urine is also tested for microalbuminuria, which may indicate kidney damage from diabetes.

TRAUMA AND THE ENDOCRINE SYSTEM

Because of the location of the hypothalamus and the pituitary gland, a *traumatic brain injury* can cause disturbances to the endocrine system if any of the glands are damaged. Hormonal disturbances can occur immediately or months to years after the trauma. A common condition due to a traumatic brain injury is adrenal insufficiency. With this condition, the pituitary gland doesn't produce enough ADH, which can lead to diabetes insipidus.

Steroid Abuse in Sports

Some athletes of today have turned to the use of androgenic anabolic steroids to build bigger, stronger muscles and thus hope to achieve status in the world of sports. Anabolic steroids are variants of the male sex hormone testosterone.

The risks of taking steroids far outweigh any temporary improvement that an athlete may hope to gain. Effects on males who abuse steroids include liver changes, decrease in sperm production, atrophy of the testicles, breast enlargement, acne, and increased risk of cardiovascular disease. Effects on females include amenorrhea, or loss of menstrual cycle; abnormal placement of body hair; baldness; and voice changes.

In addition, both sexes complain of headaches; dizziness; hypertension; mood swings; and aggressiveness, commonly known as "roid rage." Sudden withdrawal of a high dosage of anabolic steroids can cause severe depression and possible suicide. Treatment includes meeting with psychologists and endocrine therapy.

One BODY

How the Endocrine System Interacts with Other Body Systems

INTEGUMENTARY SYSTEM

- Melanocyte-stimulating hormone influences production of melanin, which protects the skin from the harmful rays of the sun.
- Androgens cause activation of the sebaceous gland.
- Estrogen influences skin hydration.

SKELETAL SYSTEM

- Bone growth is influenced by growth hormone.
- Parathormone and calcitonin influence blood levels of calcium in the blood and storage in the bones.

MUSCULAR SYSTEM

- Growth hormone influences development of muscle.
- Thyroid hormones influence muscle metabolism and production of energy.

NERVOUS SYSTEM

- Many hormones influence the development of the nervous system.
- Thyroid hormones influence the metabolism of the nervous system.
- Epinephrine influences the sympathetic nervous system.

CIRCULATORY SYSTEM

- Epinephrine affects the heart rate and blood pressure.

LYMPHATIC SYSTEM

- Thymosin influences the production of T cells in the lymph system.
- Glucocorticoids depress the immune response and the inflammation process.

RESPIRATORY SYSTEM

- Epinephrine dilates the bronchioles to take in more oxygen.

DIGESTIVE SYSTEM

- Thyroid hormones influence the metabolism of the digestive process.
- Insulin and glucagon of the pancreas help control the blood glucose level.

URINARY SYSTEM

- Aldosterone and ADH influence water and electrolyte balance.

REPRODUCTIVE SYSTEM

- Anterior pituitary and gonad hormones influence the development of the reproductive system.
- The posterior pituitary hormone oxytocin influences childbirth.

Medical Terminology

acr/o	body extremity	**hyper-**	over, excessive
megaly	enlargement of	**glycemia**	blood sugar
acr/o/megaly	enlargement of body extremity	**hyper/glycemia**	excessive blood sugar
adren	toward the kidney	**hypo-**	deficient
-al	pertaining to	**hypo/glycemia**	deficient blood sugar
adren/al	pertaining to the kidney; above the kidney	**poly-**	many, much
		dipsia	thirst
dwarf	small by comparison	**poly/dipsia**	much or excessive thirst
-ism	abnormal condition of	**phagia**	eating
dwarf/ism	abnormal condition of smallness	**poly/phagia**	much or excessive eating
exo-	outside of	**uria**	urination
ophthalm	eye	**poly/uria**	much or excessive urination
-os	one who	**ster**	solid oil
exo/phthalm/os	one who has abnormal protrusion of the eyeball	**-oid**	resembles
		ster/oid	resembles a solid oil
gigant	largeness by comparison	**thyr**	shield
gigant/ism	abnormal condition of largeness by comparison	**thyr/oid**	resembles a shield

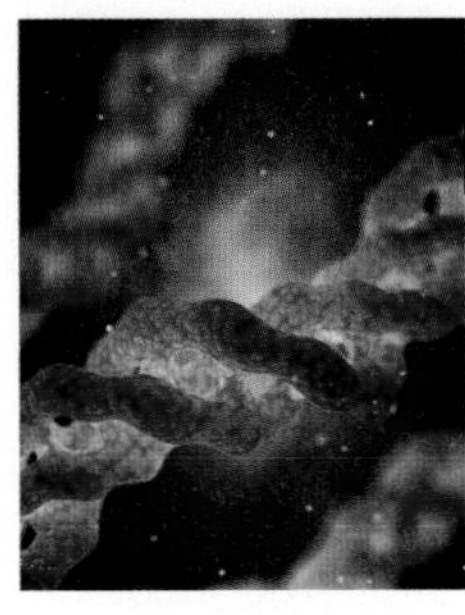

Study Tools

Workbook	Activities for Chapter 12
Online Resources	• PowerPoint presentations • Animations

REVIEW QUESTIONS

Select the letter of the choice that best completes the statement.

1. The master gland is known as the
a. pituitary.
b. thyroid.
c. adrenal.
d. ovary.

2. The pituitary hormone that is necessary to govern metabolism is
a. FSH.
b. MSH.
c. TSH.
d. ACTH.

3. The hormones that affect neuromuscular functioning, blood clotting, and holding the cells together are
a. thyroxine and calcitonin.
b. thyroxine and parathormone.
c. calcitonin and thymosin.
d. calcitonin and parathormone.

4. The gland that governs the production of lymphoid cells is the
a. thymus.
b. thyroid.
c. parathyroid.
d. pituitary.

5. The hormone that is responsible for stimulating ovulation is
a. TSH.
b. ICSH.
c. FSH.
d. LH.

6. The hormone that prepares us to fight or flee is
a. aldosterone.
b. epinephrine.
c. cortisol.
d. corticoid.

7. The secretions of the ovaries are
a. estrogen and LTH.
b. estrogen and LH.
c. progesterone and LTH.
d. progesterone and estrogen.

8. A decrease in the production of insulin causes
a. diabetes mellitus.
b. diabetes insipidus.
c. cretinism.
d. exophthalmos.

9. A hypofunction of the thyroid gland causes
a. exophthalmos.
b. glycosuria.
c. cretinism.
d. Graves' disease.

10. An oversecretion of the adrenal cortex is known as
a. myxedema.
b. Cushing's syndrome.
c. Addison's disease.
d. pituitary short stature.

COMPLETE THE CHART

Complete the following chart of the endocrine gland and its hormones.

GLAND	HORMONE	NORMAL FUNCTION	DISORDERS
Pituitary			
Pineal			
Thyroid			
Parathyroid			
Thymus			
Adrenals			
Gonads			
Pancreas			

MATCHING

Match each term in Column I with its correct description in Column II.

	COLUMN I	COLUMN II
______	**1.** ACTH	a. master gland of the endocrine system
______	**2.** adrenals	b. any gland of internal secretion
______	**3.** cortisone	c. a hormone secreted by adrenals
______	**4.** gonad	d. regulates use of calcium
______	**5.** endocrine	e. the secretion of any endocrine gland
______	**6.** hormone	f. helps body meet emergencies
______	**7.** insulin	g. sex gland
______	**8.** parathyroid	h. regulates body metabolism
______	**9.** pituitary	i. one of the hormones secreted by the pituitary gland
______	**10.** thyroid	j. necessary to maintain levels of blood glucose
		k. hypofunction of endocrine glands

APPLYING THEORY TO PRACTICE

1. You have a thermostat in your house that regulates the furnace or the air conditioner. When a certain temperature is reached, it automatically shuts off. Explain how this process relates to the thyroid gland.

2. A patient comes to the physician's office and tells the physician they are experiencing leg cramping, which they have heard is related to calcium. Explain to the patient how calcium is affected by the action of the thyroid and parathyroid glands.

3. When you arrive at your office in the health maintenance organization, a patient calls out to you. They are near hysteria; they tell you they are experiencing heart palpitations and feel like they are "jumping out of their skin." You check the records and note this patient has been on thyroid medication. Explain to the patient what you think may be the cause of the symptoms and what action should be taken.

4. Your brother wants to be a football player. He is 5'7"; he heard that steroids could help him. Explain the action of steroids and why they should not be used.

5. What effect might a traumatic brain injury have on the endocrine system? At what point after the injury would you expect to see problems? If a person develops diabetes insipidus, what gland may have been damaged during the traumatic brain injury?

CASE STUDY

Jack, age 52, has been feeling tired lately. He also gets up frequently during the night to urinate and is always thirsty. He makes an appointment with Shanice, the medical assistant, to see his physician. When Jack arrives at the office, Shanice takes his medical history and weighs and measures him. He is 5'9" and weighs 270 pounds. The physician suspects Jack is exhibiting signs of diabetes. The physician orders tests to be done and asks Jack to return in three days for a follow-up visit. Shanice makes the arrangements for the tests and follow-up visit.

1. Explain the tests that are done to diagnose diabetes mellitus.

When Jack returns to the office, the physician says his blood sugar is 240 mg/dL, and he makes the diagnosis of type 2 diabetes. The physician orders oral medication, prescribes an exercise program, gives Jack a calculated diet, and tells him he must monitor his blood glucose.

2. Describe diabetes mellitus and the cause of Jack's symptoms.

Shanice reviews the medication, glucose monitoring system, exercise program, and diet requirements with Jack.

3. What is the benefit of exercise for patients with diabetes?

4. How do the oral agents help lower the blood glucose level?

5. What are the major complications of diabetes?

6. Explain the relationship between diabetes and heart disease.

7. What do researchers say is the biggest cause of type 2 diabetes?

8. Researchers are making many advances in diabetic research. List the advances that may be most helpful to Jack.

LAB ACTIVITY 12-1 Microscopic Study of Endocrine Glands

- ***Objective:*** To observe the structure and function of the endocrine glands
- ***Materials needed:*** prepared slides of thyroid tissue, pancreas tissue, and adrenal tissue; microscope; paper; pencil; textbook

Step 1: Examine the thyroid tissue. Identify the follicle cells of the thyroid. Record their functions.

Step 2: Examine the slide of the pancreas tissue. Identify the islets of Langerhans. Record their functions.

Step 3: Before placing the adrenal tissue under the microscope, hold the slide up to a light to distinguish between the cortex and medulla areas.

Step 4: Examine the adrenal tissue. Is there a difference in the type of cells you are examining? Compare with the picture in your textbook. Record your answer.

Step 5: Compare the three types of endocrine tissue. Record any differences you see.

LAB ACTIVITY 12-2 Structure and Function of Endocrine Glands

- ***Objective:*** To observe and identify placement and function of the endocrine glands
- ***Materials needed:*** anatomical chart of the endocrine system, paper, pencil

Step 1: Locate the organs of the endocrine system.

Step 2: What is special about the location of the pituitary gland? What is the function of the pituitary gland? Record your answers.

Step 3: Make sketches of the thyroid, parathyroid, pancreas, and adrenal glands.

Step 4: Describe and record the function of each of these glands.

Step 5: What are the gonads? What is their function? Record your answer.

LAB ACTIVITY 12-3 Action of Adrenal Medulla

- ***Objective:*** To apply the responses of the adrenal medulla to real experiences
- ***Materials needed:*** chairs, desk, paper, pen, textbook

Step 1: Form groups of four. Decide amongst yourselves who would like to be student A, B, C, and D.

Step 2: Student A: describe and record your memory of a frightening experience.
Student B: describe and record your memory of a sudden fall.
Student C: describe and record your memory of a sudden quiz.
Student D: describe and record your memory of meeting your teacher for the first time.

Step 3: When every group member has finished their description, pass the descriptions around until every group member has read every description.

Step 4: Discuss as a group how the different responses connect to the fight or flight instinct. What feelings did people experience? What physical responses occurred?

Step 5: Find and record what causes the fight or flight response.

CHAPTER 13
Blood

Objectives

- List the important components of blood and their function.
- Describe the process of inflammation.
- Describe the process of blood clotting.
- Recognize the significance of the various blood types.
- Describe some disorders of the blood.
- Define the key words that relate to this chapter.

Key Words

abscess
agranulocytes
albumin
anemia
antibody
anticoagulants
antigen
antiprothrombin
antithromboplastin
aplastic anemia
basophils
B lymphocytes
carbon monoxide (CO) poisoning
clotting time
coagulation
Cooley's anemia
diapedesis
embolism
eosinophils
erythroblastosis fetalis
erythrocytes
erythropoiesis
erythropoietin
fibrin
fibrinogen
gamma globulin
globulin
granulocytes
hematocrit
hematopoiesis
hemoglobin
hemolysis
hemophilia
hemostasis
heparin
hereditary hemochromatosis
hereditary spherocytosis
inflammation
international normalized ratio (INR)
iron-deficiency anemia
leukemia
leukocytes
leukocytosis
leukopenia
lymphocytes
monocytes
multiple myeloma
myeloblasts
neutrophils
oxyhemoglobin
pathogenic
pernicious anemia
plasma
polycythemia
prothrombin
prothrombin time (PT)
pus
pyrexia
pyrogens
Rh factor
sedimentation rate
septicemia
stem cells
thalassemia
thrombin
thrombocytes
thrombocytopenia
thromboplastin
thrombosis
thrombus
T lymphocytes
universal donor
universal recipient

The average adult's body has 8 to 10 pints of blood. Loss of more than 2 pints at any one time leads to a serious condition.

FUNCTION OF BLOOD

Blood is the transporting fluid of the body. It carries nutrients from the digestive tract to the cells, oxygen from the lungs to the cells, waste products from the cells to the various organs of excretion, and hormones from secreting cells to other parts of the body. It aids in the distribution of heat formed in the more active tissues, such as the skeletal muscles, to all parts of the body. Blood also helps regulate the acid-base (pH) balance and protects against infection. Consequently, blood is a vital fluid to life and health (**Table 13-1**).

BLOOD COMPOSITION

Blood is made up of **plasma** (**PLAZ**-mah), the liquid portion of blood without its cellular elements. Serum is the name given to plasma after a blood clot is formed: serum = plasma − (fibrinogen + prothrombin). Blood also contains cellular elements, including **erythrocytes** (eh-**RITH**-roh-sights) or red blood cells (RBCs), **leukocytes** (**LOO**-koh-sights) or white blood cells (WBCs), and **thrombocytes** (**THROM**-boh-sights) or platelets (**Figure 13-1**).

Blood Plasma

Plasma is a straw-colored, complex liquid that comprises between 55% and 60% of the total blood volume.

Plasma contains the following six substances in solution:

1. *Water*—Water makes up about 92% of the total volume of plasma. This percentage is maintained by the kidneys and by water intake and output.
2. *Plasma proteins*—These three proteins are the most abundant of those found in plasma: fibrinogen, serum albumin, and serum globulin.
   a. **Fibrinogen** (figh-**BRIN**-oh-jen) is necessary for blood clotting. Without fibrinogen, the slightest cut or wound would bleed profusely. It is synthesized in the liver.
   b. **Albumin** (al-**BYOO**-men) is the most abundant of all the plasma proteins. A product of the liver, albumin helps maintain the blood's osmotic pressure and volume. It provides the "pulse pressure" needed to hold and pull water from the tissue fluid back into the blood vessels. Normally, plasma proteins do not pass through the capillary walls because their molecules are relatively large. They are colloidal substances; they can give up or take up water-soluble substances, thus regulating the osmotic pressure within the blood vessels.

Table 13-1 *Summary of the Various Functions of Blood*

FUNCTION	EFFECT ON THE BODY
Transport	1. Transports oxygen from the lungs to the tissues, and carbon dioxide from the tissues to lungs 2. Transports nutrient molecules (glucose, amino acids, fatty acids, and glycerol) from the small intestine or storage site to the cells of the body 3. Transports waste products, such as lactic acid, urea, and creatinine, from the cells to kidneys and sweat glands for excretion
Regulatory	1. Regulates hormones and other chemicals that control the functioning of organs and systems 2. Helps regulate the body pH through buffers and amino acids it carries; pH of blood is 7.4 3. Regulates body temperature by circulating excess heat to body surfaces and lungs 4. Regulates water content of cells through its dissolved sodium ion, thus playing a role in osmosis
Protection	1. Circulates antibodies and defensive cells to combat infection and disease 2. Produces clots to prevent excessive loss of blood

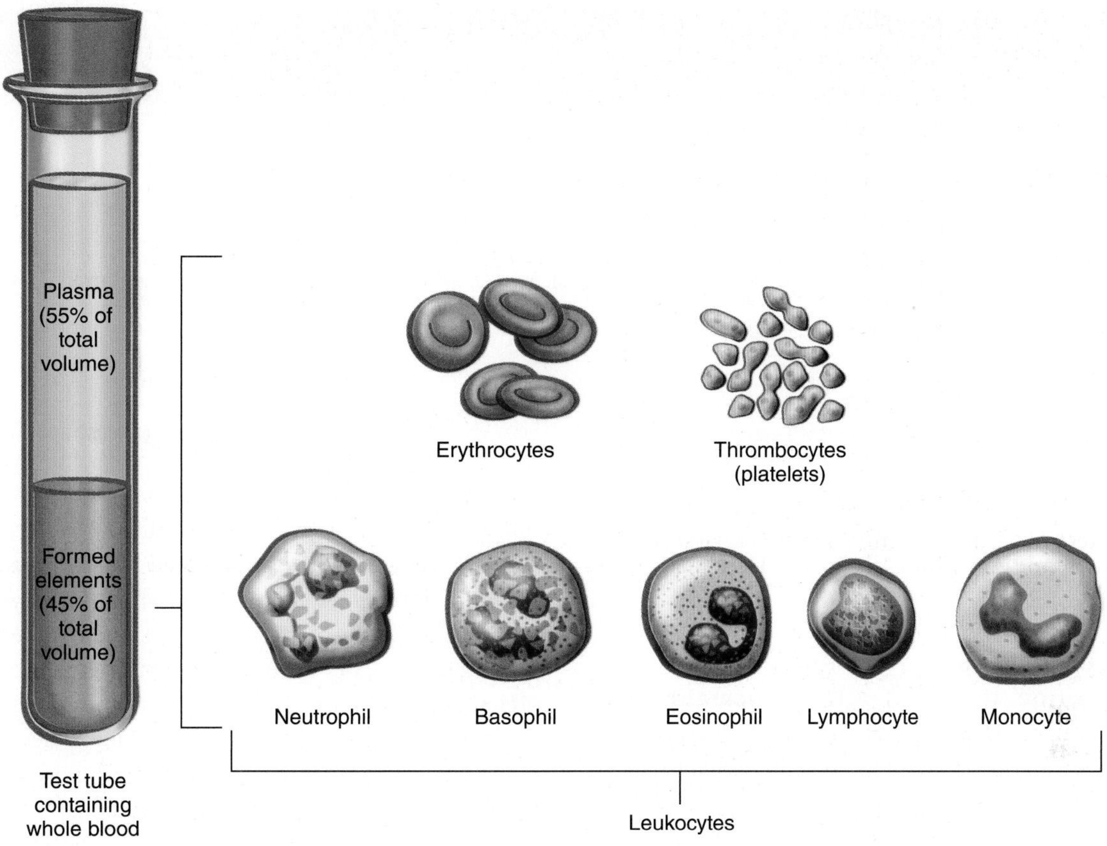

Figure 13-1 *The major fluid and formed components of blood. The formed elements include erythrocytes, thrombocytes, and leukocytes, which are 45% of the total volume.*

c. **Globulin** (**GLOB**-yoo-lin) is formed not only in the liver, but also in the lymphatic system (discussed in Chapter 16). **Gamma globulin** has been fractionated, or separated, from globulin. This portion helps in the synthesis of antibodies, which destroy or render harmless various disease-causing organisms. **Prothrombin** (proh-**THROM**-bin) is yet another globulin, formed continually in the liver, which helps blood coagulate. Vitamin K is necessary in aiding the process of prothrombin synthesis.

3. *Nutrients*—Nutrient molecules are absorbed from the digestive tract. Glucose, fatty acids, cholesterol, and amino acids are dissolved in the blood plasma.
4. *Electrolytes*—The most abundant electrolytes are sodium chloride and potassium chloride. These come from foods and chemical processes occurring in the body.
5. *Hormones, vitamins, and enzymes*—These three substances are found in very small amounts in the blood plasma. They generally help the body control its chemical reactions.
6. *Metabolic waste products*—All of the body's cells are actively engaged in chemical reactions to maintain homeostasis. As a result of this, waste products are formed and subsequently carried by the plasma to the various excretory organs, which is one of the major functions of plasma.

Changes in the Composition of Circulating Blood

The major substances added to and removed from the blood as it circulates through organs along the various sites of the circulatory system are outlined in **Table 13-2**. This table includes only the major changes in the blood as it passes through certain specialized organs or structures.

Formation of Blood Cells

Blood cells have different life spans. RBCs last about 120 days, whereas WBCs can last hours to days. The body needs to continuously replace these blood cells. On average, about one percent of blood cells are replaced daily. Hematopoiesis is the overall process of the production of

Table 13-2 *Changes in the Composition of Circulating Blood*

ORGANS	BLOOD LOSES	BLOOD GAINS
Digestive glands	Raw materials needed to make digestive juices and enzymes	Carbon dioxide
Kidneys	Water, urea, and mineral salts	Carbon dioxide
Liver	Excess glucose, amino acids, and worn-out red blood cells	Released glucose, urea, and plasma proteins
Lungs	Carbon dioxide and water	Oxygen
Muscles	Glucose and oxygen	Lactic acid and carbon dioxide
Small intestinal villi	Oxygen	End products of digestion (glucose and amino acids)

blood cells, while erythropoiesis is the differentiation and maturation of RBCs.

HEMATOPOIESIS

The formation of blood cells is **hematopoiesis** (**hem**-ah-toh-poy-**EE**-sis); this occurs in the red bone marrow, which is also known as myeloid tissue. All blood cells are produced by the red bone marrow. However, certain lymphatic tissues, such as the spleen, tonsils, and lymph nodes, produce some WBCs called agranular leukocytes. All blood cells develop from undifferentiated mesenchymal cells called **stem cells**, or hemocytoblasts.

ERYTHROPOIESIS

Erythropoiesis (eh-rith-roh-poy-**EE**-sis), or the manufacture of RBCs, occurs in the red bone marrow of essentially all bones until the individual reaches adolescence. As one grows older, the red marrow of the long bones is replaced by fat marrow; erythrocytes are thereafter formed only in the short and flat bones.

The rate of erythropoiesis is influenced by **erythropoietin**, a hormone produced largely in the kidneys. When the number of circulating RBCs decreases or when the oxygen transported by the blood diminishes, an unidentified sensor in the kidney detects the change, and the production of erythropoietin is increased. This substance is then transported through the plasma to the bone marrow, where it accelerates the production of RBCs.

Erythrocytes come from stem cells in the red bone marrow called hemocytoblasts. As the hemocytoblast matures into an erythrocyte, it loses its nucleus and cytoplasmic organelles. The hemocytoblast also becomes smaller, gains hemoglobin, develops a biconcave shape (refer to **Figure 13-1**), and enters the bloodstream. To aid in erythropoiesis, vitamin B_{12}, folic acid, copper, cobalt, iron, and proteins are needed.

Because erythrocytes are enucleated, meaning they contain no nucleus, they only live about 120 days. Destruction occurs as the cells age, rendering them more vulnerable to rupturing. They are broken down by the spleen and liver. Hemoglobin breaks down into globin and heme. Most of the iron content of heme is used to make new RBCs; the balance of the heme group is degraded to bilirubin and is stored in the liver. The normal count of RBCs ranges from 4.5 to 6.2 million/μL (million per microliter) venous blood for men and 4.2 to 5.4 million/μL venous blood for women.

Hemoglobin

Erythrocytes contain a red pigment (coloring agent) called **hemoglobin** (**hee**-moh-**GLOH**-bin), which provides their characteristic color. Hemoglobin is made of a protein molecule called globin and an iron compound called heme. A single blood cell contains several million molecules of hemoglobin. Hemoglobin is vital to the function of RBCs, allowing them to transport oxygen to the tissues and some carbon dioxide away from the tissues. Normal hemoglobin count is 14–19 gm/dL for males and 12–16 gm/dL for females.

FUNCTION

In the capillaries of the lung, erythrocytes pick up oxygen from the inspired air. The oxygen chemically combines with the hemoglobin, forming the compound **oxyhemoglobin**. The oxyhemoglobin-laden

Did You Know?

The red blood cell is a traveler. It makes about 250,000 round trips in the body before it heads to its destruction in the liver and spleen. The iron part of the hemoglobin cell gets to travel again because it gets recycled.

erythrocytes circulate to the capillaries of tissues. Here, oxygen is released to the tissues. The carbon dioxide that is formed in the cells is picked up by the plasma as bicarbonate. The RBCs circulate back to the lungs to give up the carbon dioxide and absorb more oxygen. Arteries carry blood away from the heart, and veins carry blood toward the heart. However, there are exceptions. Blood cells that travel in the arteries—except for pulmonary arteries—carry oxyhemoglobin, which gives blood its bright red color. Blood cells in the veins—except for pulmonary veins—contain carbaminohemoglobin (kahr-**bam**-ih-noh-**hee**-moh-**GLOH**-bin), which is responsible for the dark, reddish-blue color characteristic of venous blood.

Carbon monoxide (CO) poisoning is a serious and sometimes fatal condition. Carbon monoxide is an odorless gas present in the exhaust of gasoline engines. Carbon monoxide rapidly combines with hemoglobin and binds at the same site on the hemoglobin molecule as oxygen, thus crowding oxygen out. The cells are deprived of their oxygen supply. Symptoms may include headache, dizziness, drowsiness, and unconsciousness. Death may occur in severe cases of carbon monoxide poisoning. It is important to remember that carbon monoxide gas is odorless. Carbon monoxide is also present in the flue gases of furnaces, gas or oil-fired space heaters, and lanterns. A defective furnace or heater, as well as plugged chimneys and vents, can bring carbon monoxide into the home. Always be certain to allow for proper ventilation of home and work areas. Never allow a car to run in an unventilated garage. Commercial carbon monoxide detectors are available for home use.

HEMOLYSIS

A rupture or bursting of the RBC is called **hemolysis** (hee-**MOL**-ih-sis). This sometimes occurs as a result of a blood transfusion reaction or other disease processes.

White Blood Cells

WBCs are called leukocytes. They have nuclei and no pigment; they are called WBCs because they lack pigmentation. They are larger than erythrocytes and are either granular, meaning they have a grainy appearance, or agranular, meaning they do not have a grainy appearance. Leukocytes are manufactured in both red bone marrow and lymphatic tissue. Leukocytes are the body's natural defense against injury and disease.

TYPES OF LEUKOCYTES

Leukocytes are classified into two major groups of cells: **granulocytes** and **agranulocytes**. When stained with Wright's stain, the cytoplasm of a leukocyte will show the presence or absence of granules. In the laboratory, stains are applied to blood smears so that formed elements may be easily identified. Granulocytes are made in red bone marrow from cells called **myeloblasts**. Granulocytes are destroyed as they age and as a result of participating in bacterial destruction. The life span of WBCs is variable, but most granulocytes live only a few days.

There are three types of granulocytes: neutrophils, eosinophils, and basophils.

Neutrophils (**NEW**-troh-fills), also called polymorphonuclear leukocytes, phagocytize bacteria with lysosomal enzymes. They are usually the first cells of the immune system to respond to a bacterial or viral invasion. Phagocytosis is a process that surrounds, engulfs, and digests harmful bacteria. **Eosinophils** (ee-oh-**SIN**-oh-fills) phagocytize the remains of antibody-antigen reactions. They also increase in great numbers in allergic conditions, malaria, and in parasite or worm infestations.

Basophils (**BAY**-soh-fills) are activated during an allergic reaction or inflammation. Basophils produce histamine, a vasodilator, and heparin, an anticoagulant.

Agranulocytes are divided into lymphocytes and monocytes. **Lymphocytes** (**LIM**-foh-sights) are further subdivided into **B lymphocytes**, which are synthesized in the bone marrow, and **T lymphocytes** from the thymus gland. Still others are formed by the lymph nodes and spleen. Their life span ranges from a few days to several years. They basically help the body by synthesizing and releasing antibody molecules and by protecting against the formation of cancer cells.

Monocytes (**MON**-oh-sights) are formed in bone marrow and the spleen. They assist in phagocytosis and are able to leave the bloodstream to attach themselves to tissues; there they become tissue macrophages, or histiocytes. During inflammation, histiocytes help wall off and isolate the infected area.

The aforementioned types of leukocytes—basophils, neutrophils, eosinophils, and monocytes—that can perform phagocytosis are called phagocytes. Unlike erythrocytes, they can move through the intercellular spaces of the capillary wall into neighboring tissue. This process is known as **diapedesis** (dye-ah-ped-**EE**-sis).

A normal leukocyte count averages from 3,200 to 9,800/μL.

To summarize, leukocytes help protect the body against infection and injury. This is achieved through (1) phagocytosis and destruction of bacteria, (2) synthesis of antibody molecules, (3) "cleaning up" of cellular remnants at the site of inflammation, and (4) walling off of the infected area. See **Table 13-3** and **Table 13-4**.

Inflammation

If living tissue is damaged in any way, the body usually responds to the damage by either neutralizing or eliminating the cause of the damage. When this happens, the damaged body part goes through an inflammation process. **Inflammation** (in-flah-**MAY**-shun) occurs when tissues are subjected to chemical or physical trauma, such as a cut or heat, or invasion by **pathogenic** (path-oh-**JEN**-ic), or disease-causing microorganisms, such as bacteria, fungi, protozoa, and viruses.

The characteristic symptoms of inflammation are redness, local heat, swelling, and pain. Inflammation is due to irritation by bacterial toxins, increased blood flow, congestion of blood vessels, and the collection of blood plasma in the surrounding tissues, or edema (**Figure 13-2**). Histamine released from the basophil and other chemical substances increases blood flow to the injured area and also increases capillary permeability. Thus, large amounts of blood plasma and fibrinogen enter the damaged area. The damaged area is walled off as a result of the clotting action of fibrinogen on the damaged tissue and macrophage action.

Neutrophils move very quickly to the damaged area and are the first cells to arrive. The neutrophils move through the capillary walls by diapedesis and

Table 13-3 *Types of Leukocytes and Their Percentage*

MAJOR TYPES OF LEUKOCYTES	SPECIFIC KINDS OF LEUKOCYTES
Granulocytes 60%–70%	Neutrophils Eosinophils Basophils
Agranulocytes	Lymphocytes 20%–30% Monocytes 5%–8%

Table 13-4 *Characteristics and Functions of the Leukocytes*

LEUKOCYTE	WHERE FORMED	TYPE OF NUCLEUS	CYTOPLASM	FUNCTION
Agranular Leukocytes				
1. Lymphocyte	Lymph glands and nodes, bone marrow, spleen	One large, spherical nucleus; may be indented; sharply defined and stains dark blue	Cytoplasm stains pale blue and contains scattered violet granules	Helps form antibodies at a site of inflammation; protects against cancer
2. Monocyte (macrophage)	Lymph glands and nodes, bone marrow, spleen	One lobulated or horseshoe-shaped nucleus that stains blue	Abundant cytoplasm that stains gray-blue	Phagocytosis of cellular debris and foreign particles
Granular Leukocytes				
3. Neutrophil	Formed in bone marrow from neutrophilic myelocytes	Lobulated: contains 1 to 5 or more lobes; stains deep blue	Cytoplasm has a pink tinge with very fine granules	Displays marked phagocytosis toward bacteria during infections and inflammations; contributes to pus formation
4. Eosinophil	Formed in bone marrow from eosinophilic myelocytes	Irregularly shaped with two lobes; stains blue, but less deeply than neutrophil	Cytoplasm has a sky-blue tinge with many coarse, uniform, and round or oval bright-red granules	Marked increase in allergic conditions, malaria, and parasitic or worm infections; phagocytizes the remains of antibody-antigen reactions
5. Basophil (mast cell)	Formed in bone marrow from basophilic myelocytes	Centrally located, slightly lobulated nucleus; stains light purple and is hidden by granules	Cytoplasm has a mauve color with many large deep-purple granules	Phagocytosis; releases heparin and histamine and promotes the inflammatory response

begin phagocytosis of the pathogenic microorganisms. Macrophages also participate in phagocytosis.

In some cases of inflammation, a cream-colored liquid called **pus** forms. Pus is a combination of dead tissue, dead and living bacteria, dead leukocytes, and blood plasma. If the damaged area is below the epidermis, an **abscess** (**AB**-sess), or pus-filled cavity, forms. If it is on the skin or a mucosal surface, it is called an ulcer. In many inflammations, chemical substances called **pyrogens** are formed, which are circulated to the hypothalamus. In the hypothalamus, the pyrogens affect the temperature control center, which raises the body's temperature and causes fever, or **pyrexia** (pye-**REK**-see-ah). This higher temperature helps the body destroy pathogenic organisms.

During inflammation, the production of neutrophils by the bone marrow increases. If the WBC count exceeds 10,000 cells/μL, a condition called **leukocytosis** (loo-koh-sigh-**TOH**-sis) occurs. Following healing, the leukocyte count returns to normal. Sometimes a decrease in the number of WBCs occurs. This is called **leukopenia** (loo-koh-**PEE**-nee-ah). It can be caused by marrow-depressant drugs, pathologic conditions, or radiation.

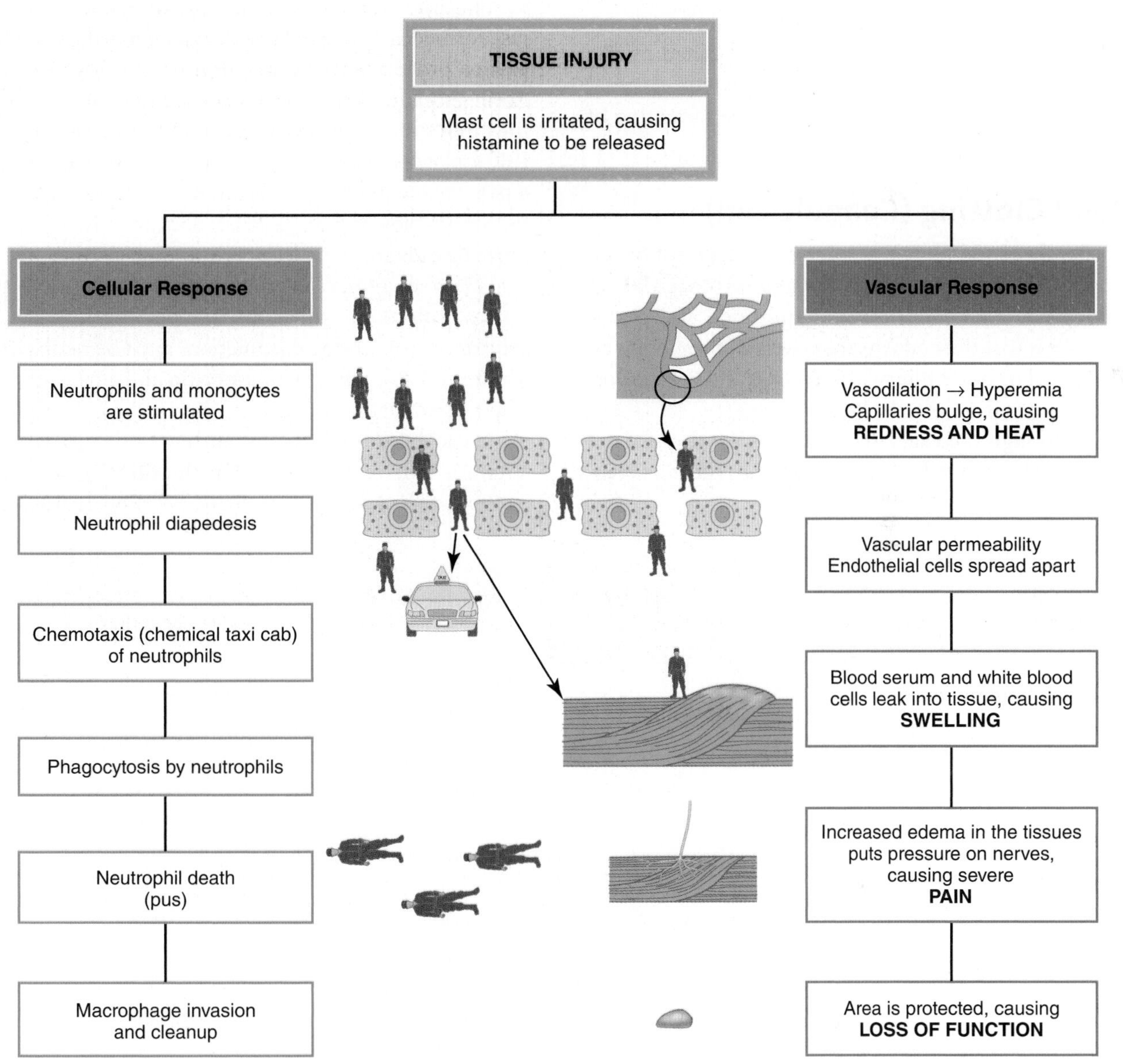

Figure 13-2 *Cellular and vascular response to inflammation*

Thrombocytes (Blood Platelets)

Thrombocytes are the smallest of the solid components of blood. They are ovoid-shaped structures that are synthesized from the larger megakaryocytes in red bone marrow. Thrombocytes are not cells but fragments of the megakaryocyte's cytoplasm (refer to **Figure 13-1**).

The normal blood platelet count ranges from 150,000 to 350,000/µL of blood. Platelets function in the initiation of the blood clotting process.

HEMOSTASIS

Hemostasis (hee-moh-**STAY**-sis) is a term that refers to stopping or controlling bleeding. This can be accomplished by vasoconstriction, which is initiated by the chemical serotonin released by the platelets; external pressure to block the flow of blood; or the formation of a blood clot.

Blood Clotting (Coagulation)

In order for blood to clot, a series of events must occur, like a cascade falling. Blood clotting, or **coagulation** (koh-ag-you-**LAY**-shun), is a complicated and essential process that depends in large part on thrombocytes. When a cut or other injury ruptures a blood vessel, clotting must occur to stop the bleeding.

Although the exact details of this process are not clear, there is a general agreement that the following reaction occurs.

When a blood vessel or tissue is injured, platelets and injured tissue release **thromboplastin** (throm-boh-**PLAS**-tin). An injury to a blood vessel makes the blood vessel lining rough; as blood platelets flow over the roughened area, they disintegrate, releasing thromboplastin. Thromboplastin is a complex substance that can only cause coagulation if calcium ions and prothrombin are present. Prothrombin is a plasma protein synthesized in the liver.

The thromboplastin and calcium ions act as enzymes in a reaction that converts prothrombin into **thrombin** (**THROM**-bin). This reaction occurs only in the presence of bleeding; normally, there is no thrombin in the blood plasma.

In the next stage of coagulation, the thrombin just formed acts as an enzyme, changing fibrinogen, a plasma protein, into **fibrin** (**FIGH**-brin). These gel-like fibrin threads layer themselves over the cut, creating a fine, mesh-like network. This fibrin network entraps the RBCs, platelets, and plasma, creating a blood clot. At first, a pale-yellow fluid called serum oozes out of the cut. As the serum slowly dries, a crust, or scab, forms over the fibrin threads and completes the common clotting process.

For coagulation to occur successfully, two **anticoagulants** (an-tih-koh-**AG**-you-lantz), which are substances preventing coagulation, must be neutralized. These are called **antithromboplastin** and **antiprothrombin (heparin)**; they are neutralized by thromboplastin.

Prothrombin is dependent on vitamin K. Vitamin K is manufactured in the body by a type of bacteria found in the intestines. See **Figure 13-3** and

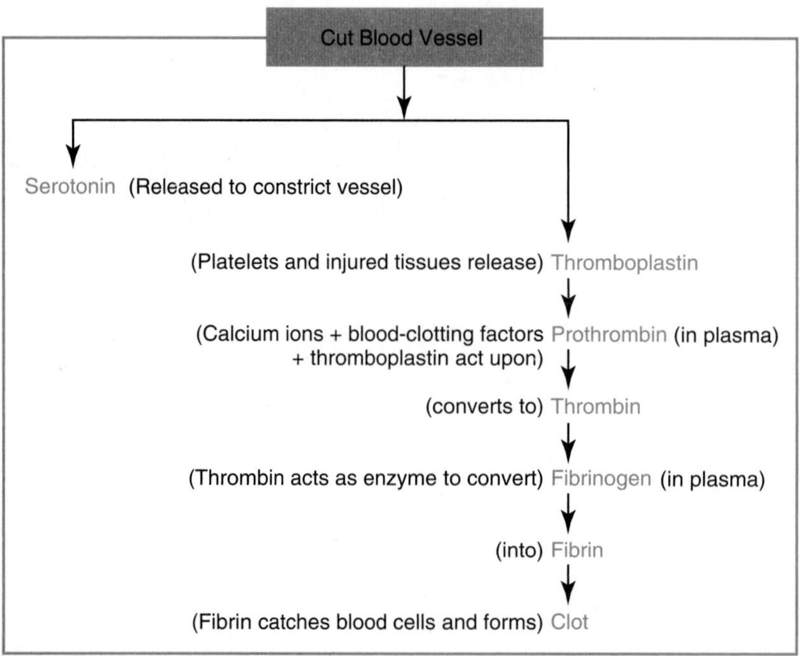

Figure 13-3 *Blood clotting process*

Figure 13-4 for a summary of the coagulation process. It is important to note that prothrombin and fibrinogen are plasma proteins manufactured in the liver; therefore, serious liver disease may interfere with the blood clotting process.

CLOTTING TIME

The time it takes for blood to clot is known as its **clotting time**. The clotting time for humans is from 8 to 15 minutes. This information is quite useful prior to surgery.

BLEEDING TIME

Bleeding time is a crude test of hemostasis; it indicates how well platelets interact with blood vessel walls to form clots. Normal bleeding time for humans is 3–10 minutes.

Media Link

View the **Blood** animation on the Online Resources.

BLOOD TYPES

There are four major groups or types of blood: A, B, AB, and O. Blood type is inherited from one's parents. It is determined by the presence—or absence—of the blood protein called agglutinogen, or **antigen** (**AN**-tih-jen), on the surface of the RBC. People with type A blood have the A antigen; type B blood have the B antigen; type AB have both A and B antigens; and type O have neither of the antigens.

There is a protein present in plasma known as agglutinin, or **antibody**. An individual with type A blood has B antibodies in the blood plasma. Type B blood possesses A antibodies; type O contains *both* A and B antibodies; and type AB contains *no* antibodies.

Knowledge of one's correct blood type is important when blood transfusions and surgery are required because antibodies react with antigens of the same type, causing the RBCs to clump together. The clumping of blood, a process known as agglutination, clogs the blood

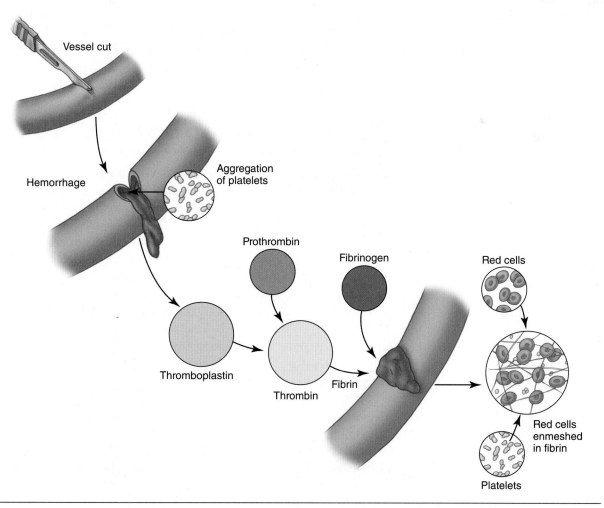

Figure 13-4 *Clotting*

Table 13-5 *Blood Type Crossmatches*

IF THE PATIENT'S BLOOD TYPE IS	THE DONOR'S BLOOD TYPE MUST BE
O+	O+, O−
O− (universal donor)	O−
A+	A+, A−, O+, O−
A−	A−, O−
B+	B+, B−, O+, O−
B−	B−, O−
AB+ (universal recipient)	AB+, AB−, A+, A−, B+, B−, O+, O−
AB−	AB−, A−, B−, O−

vessels, impeding circulation; this may cause death. By way of an example, if a person with type A blood needs a transfusion, they must only receive type A blood. Should the person receive type B blood, the B antigens of the type B blood would clump with the B antibodies of the person's type A blood.

To avoid a blood mismatch, a test known as a type and crossmatch is done before receiving a blood transfusion. This determines the blood type of both the recipient and donor.

A person with type O Rh-negative blood is considered a **universal donor** because that type of blood has no antigens for A or B blood and no antigens for the Rh factor. Therefore, that blood is compatible with all other types of blood. A person with type AB+ blood is considered a **universal recipient** because that type of blood has both A and B antigens and the Rh antigen. The Rh factors positive (+) and negative (−) are covered in more detail in the following section. See **Table 13-5** for blood type crossmatches.

Rh Factor

Human RBCs, in addition to containing antigens A and B, also contain the Rh antigen. It is known as the **Rh factor** because it was first found in the rhesus monkey. The Rh factor is found on the surface of RBCs. People possessing the Rh factor are said to be Rh positive (Rh+). Those without the Rh factor are Rh negative (Rh−).

About 85% of North Americans are Rh positive, and 15% are Rh negative. If an Rh-negative individual receives a transfusion of Rh-positive blood, they will develop antibodies to it. The antibodies take 2 weeks to develop. Generally, there is no problem with the first transfusion; but if a second transfusion of Rh-positive blood is given, the accumulated Rh antibodies will clump with the Rh antigen (agglutinogen) of the blood being received. So, both blood type and Rh factor must be considered for safe and successful transfusions.

The same problem arises when an Rh-negative mother is pregnant with an Rh-positive fetus. The mother's blood can develop anti-Rh antibodies to the fetus's Rh antigens. The first Rh-positive child will normally suffer no harmful effects; however, subsequent Rh-positive pregnancies will be affected because the mother's accumulated anti-Rh antibodies will clump the baby's RBCs. If the condition is left untreated, the baby will usually be born with a condition known as **erythroblastosis fetalis** (eh-**rith**-roh-blass-**TOH**-sis fee-**TAL**-is), or hemolytic disease of the newborn. This condition is rare today because of the use of the drug RhoGAM, which is a special preparation of immune globulin. RhoGAM is given to the Rh-negative mother within 72 hours after delivery of each baby if the baby is found to be Rh-positive. Some physicians also administer the drug to Rh-negative mothers during the last trimester of pregnancy. The antibodies in the RhoGAM will destroy any Rh-positive cells of the baby that may have entered the mother's bloodstream; therefore, the mother's immune system will not be stimulated to produce antibodies.

BLOOD NORMS

Tests have been devised to use physiological blood norms in diagnosing and following the course of certain diseases. Some of these norms are listed in **Table 13-6**.

A complete blood count (CBC) gives information about the kinds and numbers of cells, including RBC, WBC, WBC differential, hemoglobin (Hgb), hematocrit (Hct), and platelet counts.

Patients who are taking anticoagulant medications to prolong the clotting time of their blood must have an **international normalized ratio (INR)**, which measures the clotting time of blood; this is also known as **prothrombin time (PT)**. A normal INR level is 1.0 to 1.5. The dosage of the patient's medication is based on their clotting time.

Sedimentation rate is the time required for erythrocytes to settle to the bottom of an upright tube at room temperature. An elevated sedimentation rate indicates whether disease is present and is valuable in observing the progression of inflammatory conditions. A normal sedimentation rate is 0–22 mm/hour for males and 0–29 mm/hour for females.

Table 13-6 *Blood Tests*

TEST	*NORMAL RANGE*
Bleeding time	3–10 minutes
Coagulation time	8–15 minutes
Hemoglobin (Hgb) count	Men: 14–18 gm/dL Women: 12–16 gm/dL
Platelet count	150,000–350,000/μL
Prothrombin time (quick)	9.5–11 seconds
Sedimentation rate (Westergren) in first hour	Men: 0–22 mm/hour Women: 0–29 mm/hour
Red blood cell (RBC) count	Men: 4.5–6.2 million/μL Women: 4.2–5.4 million/μL
White blood cell (WBC) count	3200–9800/μL
Hematocrit (Hct)	Men: 47% (±5%) Women: 42% (±5%)
Cholesterol level	<200 mg/dL

Hematocrit (hee-**MAT**-oh-krit) is a blood test that measures the percentage of the volume of whole blood that is made up of RBCs. This measurement depends on the number of RBCs and their size. A normal hematocrit (Hct) in males is 47% (±5%), and a normal Hct in females is 42% (±5%). This test is done to diagnose abnormal states of fluid levels in the blood, as well as polycythemia and anemia.

DISORDERS OF THE BLOOD

Sickle cell anemia is a chronic blood disease inherited from both parents. The disease causes RBCs to form in an abnormal crescent shape (refer to **Figure 13-1**). These cells carry less oxygen and break easily, causing anemia. Sickle cells on average only live 10–20 days. The sickling trait, a less serious disease, occurs with inheritance from only one parent. The disease is the most common inherited disease in the United States, affecting 1 in 365 African Americans and 1 in 16,300 Hispanic Americans. (CDC Data and Statistics on Sickle Cell Disease, 2019). The rigid sickling of the RBCs causes a blockage called vaso-occlusion and results in painful crises. A painful crisis is a sudden attack of pain, often occurring in bones and joints in adults and children. Sickle cell complications include stroke, pulmonary hypertension, and structural damage to organs. Hydroxyurea taken daily prevents painful episodes about 50% of the time. Treatment for sickle cell anemia is bone marrow transplants to eligible candidates and blood transfusions when necessary.

Thalassemia (thal-ah-**SEE**-mee-ah) is an inherited blood disorder that causes mild to severe anemia. **Cooley's anemia**, also known as thalassemia major, is caused by a defect in hemoglobin formation. It affects people of Mediterranean descent. Treatment consists of antibiotics and frequent RBC transfusions. There is no natural way for the body to eliminate iron. The iron from the transfused blood may build up to produce a condition known as "iron overload," which becomes toxic to tissues and organs. Iron overload can result in early death from organ failure. Drugs known as iron chelators can help rid the body of excess iron, preventing or delaying problems related to iron overload.

Hereditary spherocytosis is an inherited blood disorder that is characterized by a mutation in a gene relating to a membrane protein that allows RBCs to change shape; instead of a biconcave disc shape, the cells are round like a sphere. The spherical RBCs are more fragile than the disc shape. Symptoms include anemia, jaundice, and an enlarged spleen. Treatment includes folic acid, the removal of the spleen, and blood transfusions.

Hereditary hemochromatosis (he-moe-kroe-muh-**TOE**-sis) is a genetic disorder that causes the body to absorb too much iron from food. Excess iron is stored in the liver, pancreas, and heart; this then leads to liver, pancreas, and heart disease. Signs and symptoms don't usually appear until midlife. Symptoms include darkening of the skin, abnormal heart rhythm, diabetes, fatigue, and joint pain. Treatment is phlebotomy, or removing approximately 1 pint of blood from the body, which then lowers iron levels.

Hemophilia (hee-moh-**FILL**-ee-ah) is a hereditary disease in which the blood clots slowly or abnormally. This causes prolonged bleeding with even minor cuts and bumps. Although sex-linked hemophilia occurs mostly in males, it is transmitted genetically by females to their sons. Treatment of the several types of hemophilia varies depending on the severity of the condition. Hemophilia A is caused by lack of blood factor VIII. Mild hemophilia A treatment involves an injection of the hormone desmopressin to stimulate a release of more of the clotting factor. Hemophilia B is caused by lack of blood factor IX. Moderate to severe hemophilia A or hemophilia B requires an injection of the clotting factor. A person with hemophilia is taught to avoid trauma, if possible, and report promptly any bleeding, no matter how slight.

Anemia (ah-**NEE**-mee-ah) is a deficiency in the number and/or percentage of RBCs and the amount of hemoglobin in the blood. Anemia may result from a large or chronic loss of blood, such as caused by a hemorrhage, or it may be one of many other causes; it is estimated that there are as many as 400 types of anemia. Anemia is divided into three groups: anemia caused by blood loss, anemia caused by a decrease in RBC production, or anemia caused by the destruction of RBCs. These conditions always cause some hemoglobin deficiency because there is not enough oxygen transported to the cells for cellular oxidation. Consequently, not enough energy is being released. Anemia is characterized by varying degrees of dyspnea, pallor, palpitation, and fatigue.

Iron-deficiency anemia is a condition that often exists in women, children, and adolescents. It is caused by a deficiency of adequate amounts of iron in the diet. This leads to insufficient hemoglobin synthesis in the RBCs. The condition is easily alleviated by ingestion of iron supplements or foods that contain iron, such as green, leafy vegetables; beans; cashews; and pistachios. Increasing the intake of vitamin C-rich foods, such as strawberries and tomatoes, can help improve iron absorption from iron-rich foods. It may take up to six months of treatment with iron supplements before iron levels are normal.

Pernicious anemia (per-**NISH**-us) is a form of anemia caused by a deficiency of vitamin B_{12} and/or lack of the intrinsic factor. Pernicious anemia is seen in association with some autoimmune endocrine diseases. The intrinsic factor produced by the stomach mucosa is necessary for the absorption and utilization of vitamin B_{12}, Vitamin B_{12}, and folic acid are necessary for the development of mature RBCs. Symptoms such as dyspnea, pallor, and fatigue are present, along with specific neurologic changes. Treatment for pernicious anemia involves medicating with vitamin B_{12} either by injection, pills, or nasal spray.

Aplastic anemia (ay-**PLAS**-tic) is a disease caused by the suppression of the bone marrow. Aplastic anemia has multiple causes. Some of these are idiopathic, meaning they occur sporadically for no known reason. Other causes are secondary, resulting from a previous illness. Acquired causes of the disease may be toxins, drugs, exposure to radiation or chemotherapy, inherited condition, or history of autoimmune disease. In this condition, bone marrow does not produce enough RBCs and WBCs. Treatment consists of the removal of the toxic substances or the discontinuation of the drugs, radiation, or chemotherapy. In severe conditions, a bone marrow transplant may be performed.

Polycythemia (pol-ee-sigh-**THEE**-mee-ah) is a condition in which too many RBCs are formed. This can be a temporary condition that occurs at high altitudes where less oxygen is present. Another type is polycythemia vera, which may be caused by a gene mutation. The increase in the number of RBCs causes a thickening of the blood, with possible blood clot formation. Treatment for this condition is phlebotomy, low-dose aspirin, and/or medication that suppresses the bone marrow's ability to produce blood cells.

An **embolism** (**EM**-boh-lizm) is a condition in which an embolus is carried by the bloodstream until it reaches an artery too small for passage. An embolus is a substance foreign to the bloodstream (**Figure 13-5**); it may be air, a blood clot, cancer cells, fat, bacterial clumps, a needle, or even a bullet that was lodged in tissue and breaks free.

Thrombosis (throm-**BOH**-sis) is the formation of a blood clot in a blood vessel. The blood clot formed is called a **thrombus** (**THROM**-bus) (**Figure 13-6**). It is caused by unusually slow blood circulation, changes in the blood or blood vessel walls, immobility, injury to a vessel, surgery, or a decrease in mobility. A thrombosis that travels through the circulatory system can cause blockages that seriously affect the structure and function of the body's systems.

Hematoma is a localized, clotted mass of blood found in an organ, tissue, or subdural space. It is caused by a traumatic injury, such as a blow, that causes a blood vessel to rupture.

Thrombocytopenia (throm-boh-sigh-toh-**PEE**-nee-ah) is a blood disease in which the number of platelets, or thrombocytes, is decreased. In this condition, blood will not clot properly. The cause may be a bone marrow disorder or an immune problem. Treatment is with blood platelets or medication to boost the platelet count.

Leukemia (loo-**KEE**-mee-ah) is a cancerous or malignant condition that affects the blood and bone marrow; in leukemia, the number of WBCs is greatly increased. The overabundant immature leukocytes replace the erythrocytes, thus interfering with the transport of oxygen to the tissues. They can also hinder the synthesis of new RBCs from bone marrow. Common symptoms of leukemia are anemia, fever, night sweats, headache, swollen lymph nodes, and bruising easily. Leukemia may be classified as acute or chronic. Acute forms commonly affect children, progress rapidly, and may be fatal. Chronic forms occur more often in older adults, who may be asymptomatic. It may not cause death. Leukemia is also classified as myelogenous, which means it affects the bone marrow, or lymphocytic, which means it affects the lymph nodes. The cause

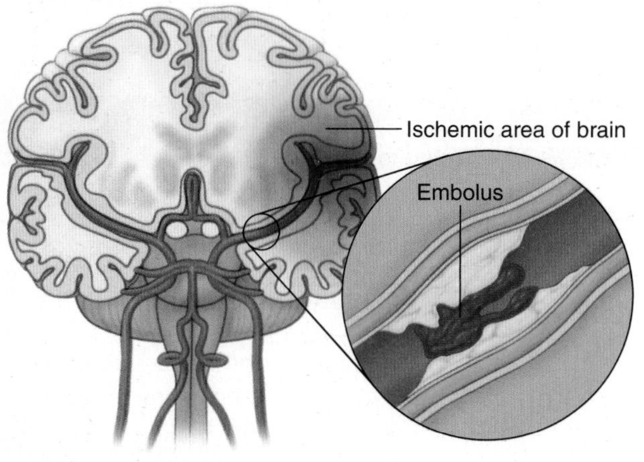

Figure 13-5 *An embolus is a foreign object circulating in the blood*

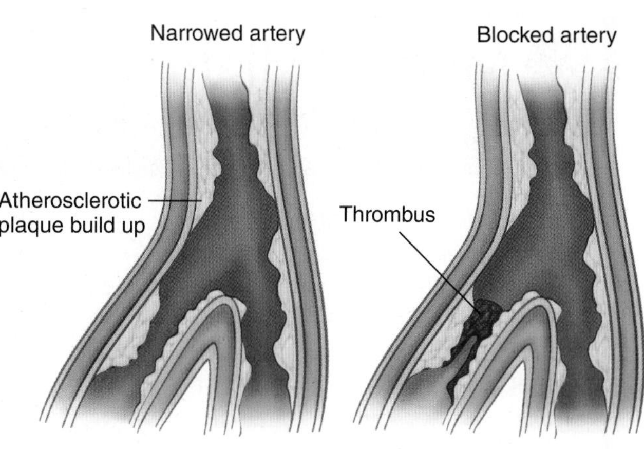

Figure 13-6 *A thrombus is a blood clot attached to the interior wall of an artery or vein*

is unknown. Treatment today consists of drug therapy; bone marrow, or stem cell, transplants; biologic therapy, or treatments that help the immune system recognize and attack leukemic cells; targeted therapy, or when drugs attack specific vulnerabilities within the cancer cell; and radiation therapy. Some of these treatments have given people with leukemia remission periods that last for several years.

Multiple myeloma (my-eh-**LOH**-mah) is a malignant neoplasm of plasma cells or B lymphocytes. The plasma cells multiply abnormally in the bone marrow, causing weakness in the bone that leads to pathologic fractures and bone pain. Overgrowth of plasma cells leads to a decrease in other blood components. The prognosis is poor; chemotherapy, stem cell transplant, and radiation may lead to a possible remission period.

Septicemia (sep-tih-**SEE**-mee-ah) describes the presence of pathogenic organisms or toxins in the blood. As a systemic response, the blood vessels dilate, which causes the blood pressure to drop and may cause the person to go into shock. Multiorgan failure is common due to the lack of blood flow to the major organs, such as the brain, lungs, kidneys, and liver. One of the early signs of sepsis is a change in mental status.

In almost all cases of septicemia, the individual must be hospitalized. Treatment is with intravenous antibiotics and therapy to support any other organ dysfunction.

Career Profile

CLINICAL LABORATORY TECHNICIAN AND CLINICAL LABORATORY TECHNOLOGIST

- Clinical laboratory testing plays a key role in the detection, diagnosis, and treatment of disease. Clinical laboratory personnel obtain and analyze body fluids, tissues, and cells.
- Clinical laboratory technicians perform routine tests in a medical laboratory and are able to discriminate and recognize factors that directly affect procedures and results. Clinical lab technicians have either an associate's degree or certification from a hospital or vocational-technical school. They work under the supervision of a medical technologist or physician.
- Clinical laboratory technologists physically and chemically analyze and culture all body fluids. Knowledge of specimen collection, anatomy and physiology, biochemistry, and laboratory equipment is essential. The education requirement is at least a bachelor's degree.

The American Society of Clinical Pathology is a professional organization that oversees credentialing and education in the medical laboratory profession.

Medical Highlights

BONE MARROW TRANSPLANT/STEM CELL TRANSPLANT

Bone marrow transplant, or *stem cell transplant*, is a procedure that transplants healthy bone marrow into a patient whose bone marrow is not functioning properly. It involves taking blood cells normally found in the bone marrow (stem cells), filtering these cells, and returning them either to the patient or to another person. The goal of bone marrow transplant is to lengthen the life of the patient, who would otherwise die.

Stem cell transplant has been used in the treatment of leukemia, aplastic anemia, and sickle cell anemia; it has also been used to replace the bone marrow and restore normal function after high doses of radiation have been given to treat malignancies.

Bone marrow is the medium for development and storage of about 95% of the body's blood cells. The blood cells that produce other blood cells are called stem cells. In the bone marrow, there is approximately one stem cell for every 100,000 blood cells. The different types of stem cell transplants are the following:

- *Autologous stem cell transplant*—the donor is the patient themself. Stem cells are taken from the patient by either bone marrow harvest or apheresis (**AFER**-ee-sis). In bone marrow harvest, stem cells are collected from the bone marrow by needle. In apheresis, the donor is connected to a special cell separation machine in which the blood is taken from one vein of one arm and circulated through the machine, which removes the stem cells and returns the remaining blood to the donor through another needle inserted into the opposite arm.
- *Allogenic stem cell transplant*—the donor shares the same genetic type as the patient. Stem cells are taken by bone marrow harvest or apheresis from a genetically matched donor, usually a brother or sister. Other possible donors for allogenic stem cell transplants include the following:
  1. Identical twin—this is a complete genetic match for a marrow transplant.
  2. Parent—the genetic match is at least half identical to the recipient.
  3. Unrelated stem cell transplant—the genetically matched marrow stem cells are from an unrelated donor.
- *Umbilical cord transplant*—a type of allogenic transplant in which stem cells are taken from umbilical cord blood immediately after the delivery of an infant. These stem cells reproduce mature functioning blood cells more quickly and efficiently than do stem cells taken from the bone marrow of a child or adult. The stem cells are tested, typed, counted, and frozen for storage until they are ready to be transplanted.

The decision to undergo a stem cell transplant will be based on the following factors: the patient's age; the extent of the disease; the availability of a donor; the tolerance of the patient for specific medications, procedures, and therapies; expectations for the course of the disease; and expectations for the course of the transplant.

In advance of the procedure, preparation for the patient includes

- an extensive evaluation, which includes all other treatment options. These are discussed and evaluated by the transplant team along with the patient;
- a complete medical examination; and
- a trip to the transplant center up to 10 days prior for hydration, evaluation, placement of the central venous catheter for administration of blood products and medications, and other preparations.

Stem cell transplant procedures vary depending on the type of transplant, the disease requiring the transplant, and tolerance to certain medications. Most often, high doses of chemotherapy or radiation are included in the preparation to effectively treat a malignancy and make room in the bone marrow for new cells to grow. The therapy is often called ablative because it stops the process of blood cell production and the marrow becomes empty. An empty marrow is necessary to make room for the new stem cells to grow and establish a new blood cell production system. After chemotherapy, the marrow transplant is administered through a central venous catheter. The stem cells find their way into the bone marrow and reproduce healthy new cells.

Engraftment is the period of time following the transplant. The marrow begins reproducing new blood cells, usually between the 15th and 30th day. Blood counts are performed to evaluate the progress of engraftment. Although the new bone marrow may begin making cells in the first 30 days following transplant, it may take months or years for the entire immune system to fully recover.

Complications that may occur with a stem cell transplant include infection, low platelet count, thrombocytopenia, anemia, pain, fluid overload, respiratory distress, organ damage, graft failure, graft-versus-host disease, and emotional and psychological stress.

Although complications may occur, an increased number of diseases call for this procedure. Medical advances have greatly improved the outcome for stem cell transplant patients.

Reference: Cleveland Clinic/Health Library

Medical Terminology

an-	without	**-oma**	tumor or swelling
-emia	blood	**hemat/oma**	swelling that contains blood
an/emia	without blood	**leuko**	white blood
coagul	clotting	**leuko/cyte**	white blood cell
-tion	process of	**lympho**	clear spring, water
coagul/a/tion	process of blood clotting	**lympho/cyte**	clear blood cell
edem	swelling	**mono-**	one
-a	presence of	**mono/cyte**	type of white blood cell with one large nucleus
edem/a	presence of swelling	**patho**	disease
embol	plug	**gen**	producing
-ism	condition of	**-ic**	refers to
embol/ism	condition of a plug or blockage	**patho/gen/ic**	refers to disease producing
erythro	red	**poly-**	many
-cyte	cell	**cyth**	cells
erythro/cyte	red blood cell	**poly/cyth/emia**	many blood cells
poiesis	formation of	**thrombo**	clot
erythro/poiesis	formation of red blood cell	**-sis**	condition of
hemat	blood	**thrombo/sis**	condition of a blood clot

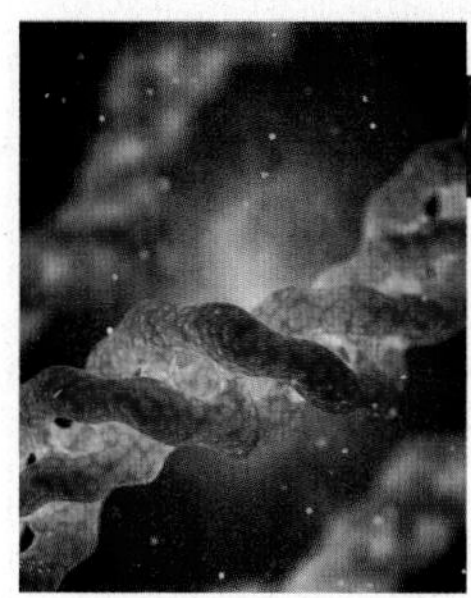

Study Tools

Workbook	Activities for Chapter 13
Online Resources	• PowerPoint presentations • Animations

REVIEW QUESTIONS

Select the letter of the choice that best completes the statement.

1. Blood from a universal donor is
   a. type B−.
   b. type A−.
   c. type AB−.
   d. type O−.

2. Blood from a universal recipient is
   a. type B+.
   b. type A+.
   c. type AB+.
   d. type O+.

3. Rh-negative blood is found in
   a. 5% of the population.
   b. 10% of the population.
   c. 15% of the population.
   d. 20% of the population.

4. The leukocytes that phagocytize bacteria with lysosomal enzymes are the
   a. eosinophils.
   b. basophils.
   c. neutrophils.
   d. monocytes.

5. The prothrombin in the blood clotting process is dependent on
   a. vitamin A.
   b. vitamin K.
   c. vitamin P.
   d. vitamin D.

6. Which of the following is not a blood cell?
   a. Erythrocyte
   b. Leukocyte
   c. Osteocyte
   d. Monocyte

7. Erythrocytes contain all but one of the following elements:
   a. Rh factor
   b. Leukocytes
   c. Hemoglobin
   d. Globin and heme

8. What characteristic is not true of normal thrombocytes?
   a. They average 4500 for each μL of blood.
   b. They are also called platelets.
   c. They are plate-shaped cells.
   d. They initiate the blood clotting process.

9. The normal leukocyte cell
   a. can only be produced in the lymphatic tissue.
   b. goes to the infection site to engulf and destroy microorganisms.
   c. is too large to move through the intercellular spaces of the capillary wall.
   d. exists in numbers that amount to an average of 12,000 cells per μL of blood.

10. The clotting process
    a. requires a normal platelet count, which is 5000 to 9000 for each μL of blood.
    b. is delayed by the rupture of platelets, which produces thromboplastin.
    c. occurs in less time with persons who have type O blood.
    d. requires vitamin K for the synthesis of prothrombin.

COMPARE AND CONTRAST

List the similarities and differences between the following terms:

1. Hematopoiesis and erythropoiesis
2. B lymphocyte and T lymphocyte
3. Pernicious anemia and aplastic anemia
4. Thrombosis and embolism
5. Leukemia and multiple myeloma

ANSWER THE FOLLOWING QUESTIONS

1. Name the three major types of blood cells.
2. What name is given to the liquid portion of blood?
3. What five proteins are contained in the blood and what is their function?
4. Describe the process of blood clot formation.
5. How can sickle cell anemia affect the structure and function of organs?

APPLYING THEORY TO PRACTICE

1. You hear that your friend has been in a car accident and needs a blood transfusion; you want to donate blood. You friend has type O+ blood, and you have type A+ blood. Can your blood be given to your friend? Explain the reason for your answer.
2. Why is blood considered the "gift of life"?
3. A female patient comes to the physician's office. She is pregnant and states that she is Rh negative and her husband is Rh positive. She has heard that there may be a problem with the baby. Explain to her about the Rh factor and how this situation is treated today.
4. In the hospital, you are caring for a 6-year-old child with leukemia. The mother says the physicians want to do a bone marrow (stem cell) transplant. She asks you to re-explain what a stem cell transplant is because she was so upset when the physician first told her. She wants to know if she can be a donor or if a friend can be a donor. Explain the types of stem cell transplants to the mother.
5. You are employed as a medical technologist. A patient comes to the lab and requires a complete blood count (CBC) and sedimentation rate. The patient asks you to explain these tests and their purpose.

CASE STUDY

John, age 24, is involved in an automobile accident. Ravi, a paramedic, arrives on the scene and does emergency first aid. John has multiple lacerations on his hands and arms; the laceration on his right arm is bleeding profusely. Ravi applies a pressure bandage and notes that John's blood pressure is 90/60. Ravi starts an intravenous line and transports John to the hospital. The emergency department physician examines John and notes he also has contusions near his liver. The physician has the medical technologist draw blood for CBC and to type and crossmatch the blood.

1. What condition may be caused by a severe loss of blood? What therapeutic technology should be used to treat it?
2. Name the blood components and their functions.
3. What is a normal blood count for John?
4. Why is the emergency department physician concerned about possible liver damage? How does liver damage relate to the blood?
5. Describe the role of a medical technologist.
6. Explain typing and crossmatching.
7. It is determined that John has type A+ blood. Can John receive blood from Ravi, who is O−?
8. In addition to blood loss, what other circulatory problems can be caused by the type of trauma John may have endured in the car accident? What body systems may be affected?

LAB ACTIVITY 13-1 Red Blood Cells (RBCs) and White Blood Cells (WBCs)

- ***Objective:*** To observe the structure of red and white blood cells
- ***Materials needed:*** prepared stained slides of blood cells, microscope, medicine dropper, disposable gloves, safety goggles, sharps or biohazard container, household bleach (1-part bleach to 10-parts water), textbook, paper, pencil
- ***Note:*** Remember to use all safety measures when in contact with blood or blood products and dispose of items according to standard precautions.

Step 1: Put on gloves and safety goggles.

Step 2: Examine the stained slide of blood cells under a microscope. Draw and describe the structure of an RBC. Which are more numerous: RBCs or WBCs? Record your answer.

Step 3: Identify the five types of WBCs. Compare their appearance with the diagram in the textbook. What are the differences among the types of WBCs? What is the function of each type of WBC?

Step 4: What are the differences between RBCs and WBCs? Record your answer.

Step 5: Dispose of the blood cell slides in an approved biohazard or sharps container.

Step 6: Clean all other equipment with household bleach (1-part bleach to 10-parts water).

Step 7: Remove goggles and gloves.

Step 8: Wash hands.

Simulated Blood Transfusion Compatibility

- ***Objective:*** To observe the transfusion reactions of blood
- ***Materials needed:*** at least five paper cups, food coloring, water, medicine dropper, marking pen, paper, pencil

POTENTIAL BLOOD TYPES FOR TRANSFUSION

BLOOD TYPE OF PATIENT	***GROUP A***	***GROUP B***	***GROUP AB***	***GROUP O***
Group A				
Group B				
Group AB				
Group O				

Note: Prepare chart for recording.

Step 1: Fill four cups about two-thirds full with water. Leave the fifth cup empty.

Step 2: Label the paper cups with water *Group A*, *Group B*, *Group AB*, and *Group O*. Label the empty cup *Patient*.

Step 3: Add red color to Cup A, blue to Cup B, and equal amounts of red and blue to cup AB; do not add food coloring to Cup O.

Step 4: Pour a small amount of liquid from Cup B into the Patient cup. The "patient" now has that type of "blood."

Step 5: Using a medicine dropper, transfer "blood" from Group A to the Patient cup. Did the color change in the Patient cup? Record your findings as either safe or unsafe.*

Step 6: Rinse the Patient cup. Add liquid from Group B to the Patient cup.

Step 7: Repeat step 5 using "blood" from the Group B cup.

Step 8: Repeat step 6.

Step 9: Repeat step 5 using "blood" from the Group AB cup.

Step 10: Repeat step 6.

Step 11: Repeat step 5 using liquid from the Group O cup.

Step 12: Which "blood" groups can safely give blood to the patient who had Group B blood?

*As long as the liquid in the Patient cup does not change color, the "transfusion" is safe.

13-3 Compare Normal and Genetically Mutated Red Blood Cells

- ***Objective***: To observe the difference between a normal red blood cell, a sickle cell, and a thalassemia cell
- ***Materials needed***: prepared slides of a red blood cell, a sickle cell, and a thalassemia cell; microscope; disposable gloves; goggles; paper; pen; biohazard or sharps container

Step 1: Wash hands and put on goggles and gloves.

Step 2: Examine the three prepared slides.

Step 3: Draw the shape of a normal red blood cell, a sickle cell, and a thalassemia cell.

Step 4: Is the shape of the sickle cell and the thalassemia cell different from the red blood cell? Describe and note the differences.

Step 5: What action is prevented by the shape of the sickle cell and thalassemia cell? Record your answer, and then compare your answer with the text.

Step 6: Dispose of or store the slides in the appropriate container. Remove gloves and goggles and wash hands.

CHAPTER 14

Heart

Objectives

- Describe the functions of the circulatory system.
- Describe the structure of the heart.
- Describe the functions of the various structures of the heart.
- Describe how blood is circulated through the heart to the lungs and body.
- Describe the conduction system of the heart.
- Discuss the diseases of the heart.
- Define the key words that relate to this chapter.

Key Words

angina pectoris
angioplasty
aorta
aortic semilunar valve
arrhythmia
ascites
atrial fibrillation
atrioventricular bundle
atrioventricular (AV) node
atrium
balloon surgery
bicuspid (mitral) valve
bradycardia
bundle of His
cardiac arrest
cardiac catheterization
cardiac output
cardiac stents
cardiopulmonary resuscitation (CPR)
cardiotonic
chordae tendineae
coarctation of aorta
conduction defect
congenital heart disease (CHD)
congestive heart failure
coronary artery disease (CAD)
coronary bypass
coronary sinus
defibrillator
diuretics
electrocardiogram (ECG or EKG)
endocarditis
endocardium
epicardium
heart block
heart failure
left ventricle
lubb dupp sounds
mitral valve prolapse
murmurs
myocardial infarction
myocarditis
myocardium
pacemaker
palpitations
pericarditis
pericardium
premature contractions
premature ventricular contractions (PVCs)
pulmonary artery
pulmonary semilunar valve
pulmonary veins
Purkinje fibers
rheumatic heart disease
right ventricle
septum
sinoatrial (SA) node
stethoscope
stroke volume
tachycardia
tetralogy of Fallot
transmyocardial laser revascularization (TMR)
tricuspid valve
vena cava
ventricular fibrillation

The circulatory system is the longest system of the body. If one were to lay all of the blood vessels in a single human body end to end, they would stretch one-fourth of the way from the earth to the moon: a distance of some 60,000 miles.[1]

FUNCTIONS OF THE CIRCULATORY SYSTEM

1. The heart is the pump that circulates oxygenated blood to all parts of the body.
2. Arteries, veins, and capillaries are the structures that take blood from the heart to the cells and return blood from the cells back to the heart. Blood will then be circulated to the lungs for oxygen.
3. Blood carries oxygen, hormones, and nutrients to the cells and carries the waste products away.
4. The lymphatic system (see Chapter 16) returns excess fluid from the tissues to the general circulation and is part of the circulatory system.

Organs of the Circulatory System

The organs of the circulatory system include the heart, arteries, veins, capillaries, and blood.

The heart is the muscular pump responsible for circulating the blood throughout the body.

[1] Sherman, Irwin W. and Vilia G. Sherman, *Biology: A Human Approach*. NY: Oxford University Press, 1979.

THE HEART

The blood's circulatory system is extremely efficient. The main organ responsible for this efficiency is the heart—a tough, simply constructed muscle about the size of a closed fist.

The heart is located in the thoracic cavity. This places the heart between the lungs, behind the sternum, and above the diaphragm. Although the heart is centrally located, its axis of symmetry is not along the midline (**Figure 14-1**). The heart's apex, or conical tip, lies on the diaphragm and points to the left of the body. It is at the apex that the heartbeat is most easily felt and heard through a **stethoscope.**

The adult human heart is about 5 inches long and 3.5 inches wide, and it weighs less than 1 pound, or 12 to 13 ounces (**Figure 14-2** and **Figure 14-3**). The importance of a healthy, well-functioning heart is obvious: to circulate life-sustaining blood throughout the body. When the heart stops beating, life stops as well. To explain further, if the blood flow to the brain ceases for 30–180 seconds, the individual loses consciousness. After 3 minutes, brain damage is more likely; after 4–5 minutes without blood flow, the brain cells are irreversibly damaged.

Try this simple demonstration: Place the disk of a stethoscope over the heart's apex. This is the area between the fifth and sixth ribs, along an imaginary line extending from the middle of the left clavicle. Because the heartbeat is felt and heard so easily at the apex, this gives rise to the popular but incorrect notion that the heart is located on the left side of the body.

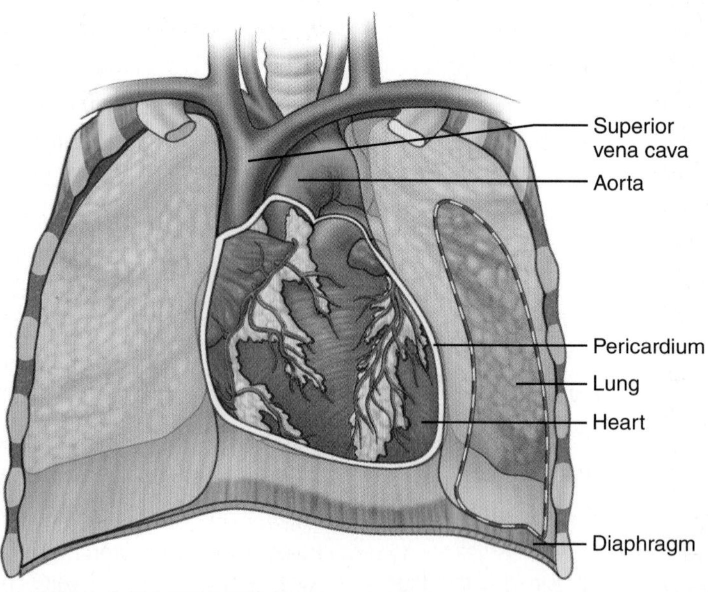

Figure 14-1 *The heart is located in the thoracic cavity between the lungs.*

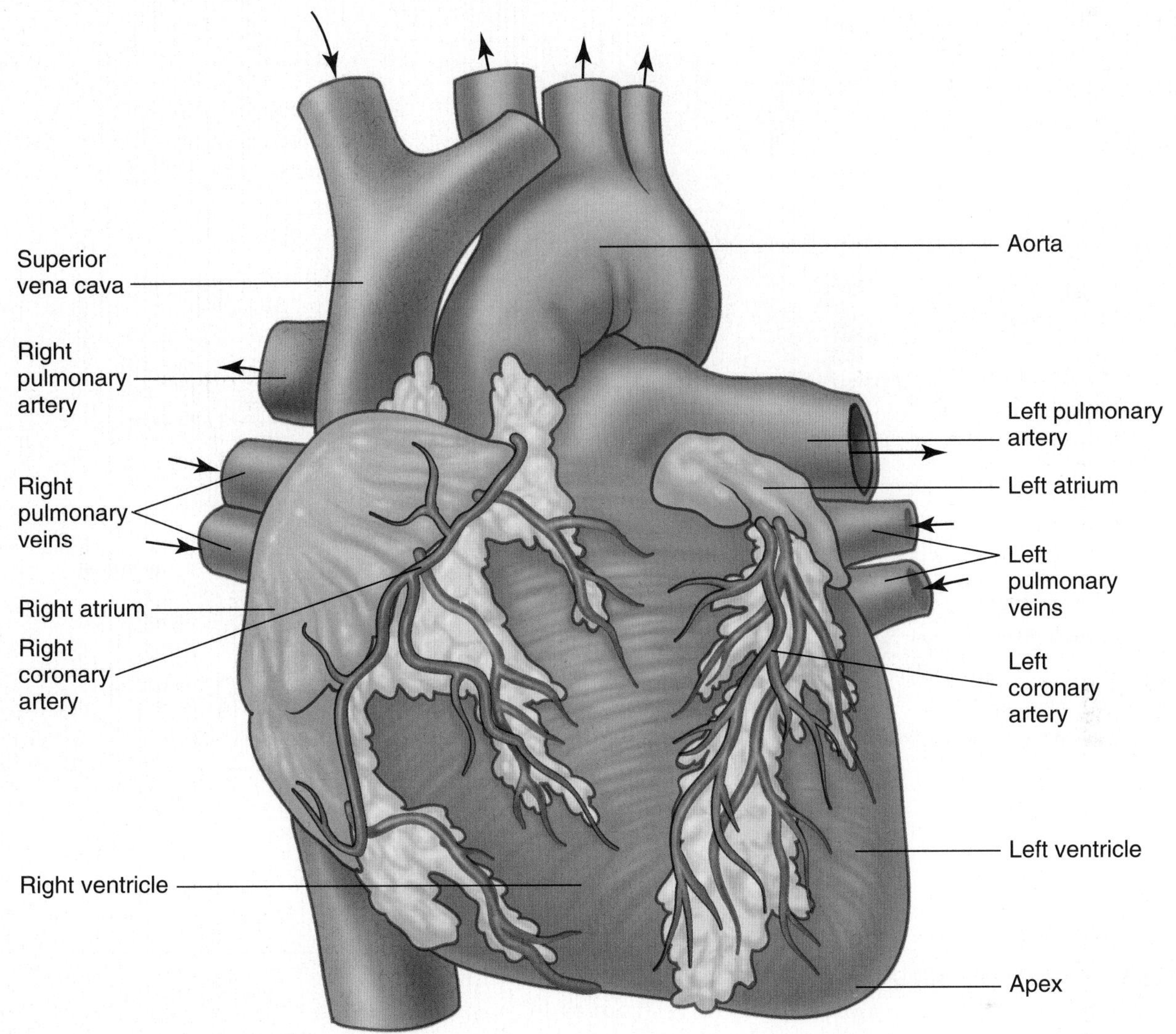

Figure 14-2 *Anterior external view of the heart*

Knowledge of the correct position of the heart can make all the difference in the treatment of **cardiac arrest** (when the heart stops beating) or a heart attack (when blood flow to heart is blocked). **Cardiopulmonary resuscitation (CPR)** is a lifesaving technique useful in many emergencies. This includes cardiac arrest or near drowning in which someone's heartbeat has stopped. CPR can keep oxygenated blood flowing to the brain and other vital organs until more definitive medical treatment can restore a normal heart rhythm. The American Heart Association recommends immediately starting heart compressions and calling 911.

All health care providers must have current and correct CPR certification. If performing CPR on an adult, the person should use two hands and their upper body weight to push straight down on, or compress, the chest at least 2 inches; they should push hard at the rate of 100 compressions per minute. Singing the song "Stayin' Alive" can help time the compressions. After every 30 compressions, the person should open the mouth and give two rescue breaths. If performing CPR on an infant, use two fingers instead of two hands and perform gentler compressions. If administering CPR to a stranger who is an adult, AHA does not recommend rescue breathing.

Structure of the Heart

The heart is a hollow, muscular, double pump that circulates the blood through the blood vessels to all parts of the body. Surrounding the heart is the **pericardium** (pehr-ih-**KAR**-dee-um), a double layer of serous and fibrous tissue (**Figure 14-1**). Between these two pericardial layers is a space filled with a lubricating fluid called *pericardial fluid*. This fluid prevents the two layers from rubbing against each other and creating friction. The thin innermost layer covering the heart is the visceral or serous pericardium, which is also referred to as the outer layer of the heart, or the **epicardium** (**ep**-ih-**KAR**-dee-um).

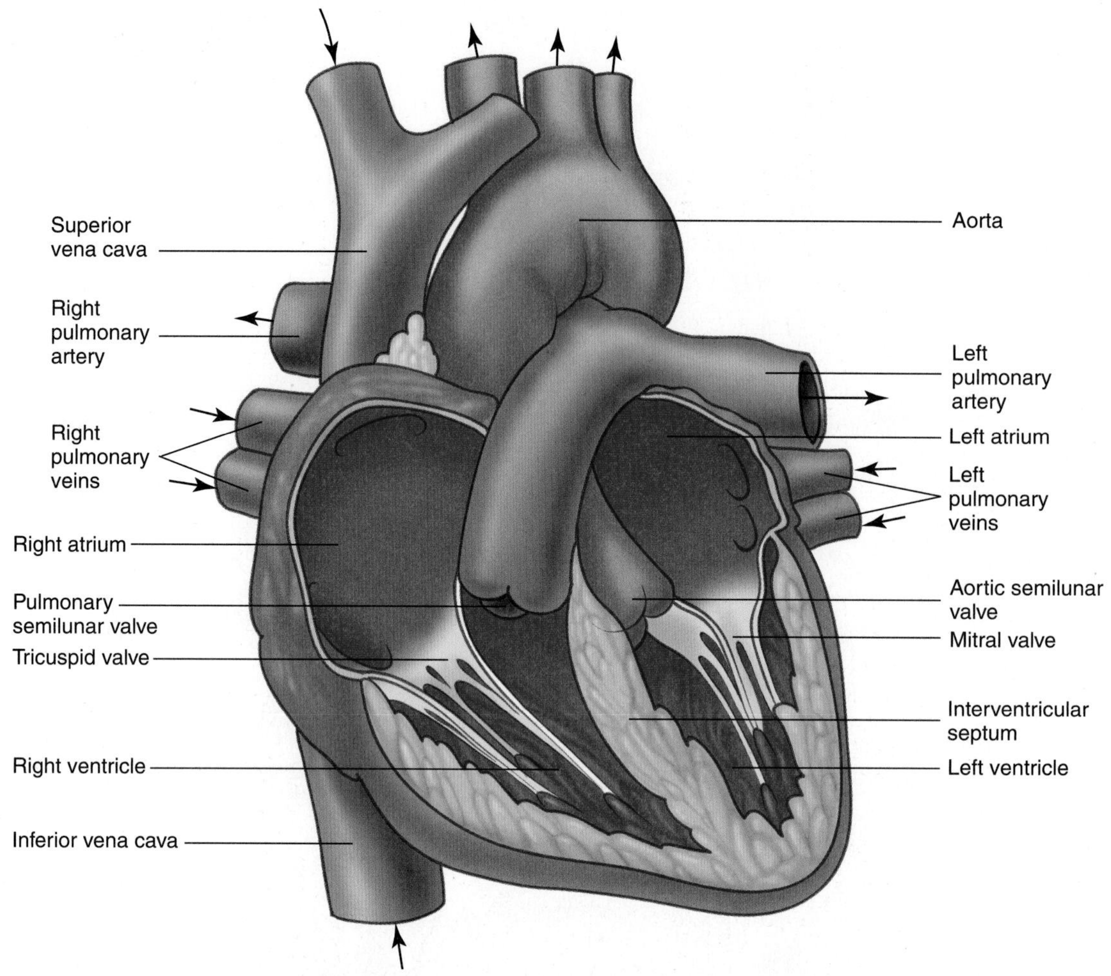

Figure 14-3 *Anterior cross-sectional view of the heart*

The tough outer membrane is the parietal or fibrous pericardium.

Cardiac muscle tissue, or **myocardium** (**my**-oh-**KAR**-dee-um), makes up the major portion of the heart (**Figure 14-4**). This specialized muscle tissue is capable of constant contraction and relaxation, which creates the pumping movement that is necessary to maintain the flow of blood through the body. The inner lining is smooth tissue called the **endocardium** (**en**-doh-**KAR**-dee-um). The endocardium covers the heart valves and lines the blood vessels, providing smooth transit for the flowing blood.

A frontal view of the human heart reveals a thick, muscular wall separating it into a right half and a left half. This partition, known as the interventricular **septum** (**SEP**-tum), completely separates the blood in the right half from that in the left half (**Figure 14-3**).

The structures leading to and from the heart are as follows (**Figure 14-3**):

- Superior and inferior **vena cava** (**VEE**-nah **KAY**-vah)—the large venous blood vessels that bring deoxygenated blood, which has lesser amounts of oxygen, to the right atrium from all parts of the body
- **Coronary sinus** (**KOR**-oh-nair-ee **SIGH**-nus)—a large venous channel located on the posterior wall of the heart that receives blood from the coronary veins and empties the blood into the right atrium
- **Pulmonary artery** (**PULL**-mah-nair-ee)—takes blood away from the right ventricle to the lungs for oxygen
- **Pulmonary veins**—bring oxygenated blood (oxygen-rich blood) from the lungs to the left atrium
- **Aorta** (ay-**OR**-tah)—takes blood away from the left ventricle to the rest of the body

CHAMBERS AND VALVES

The human heart is separated into right and left halves by the septum. Each half is divided further into two parts, creating four chambers. The two upper chambers are called the right atrium and the left atrium. The **atrium** (**AY**-tree-um) may be referred to as the auricle. The

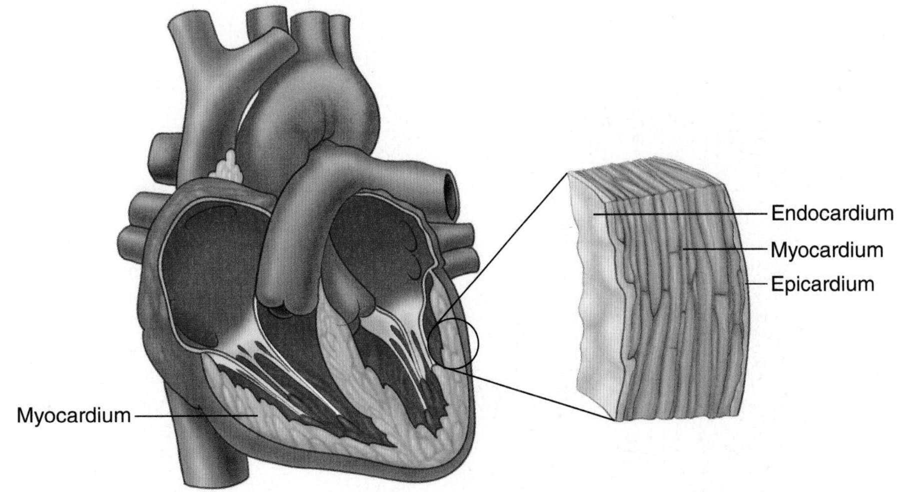

Figure 14-4 *A simplified view of the tissues of the heart walls*

lower chambers are the **right ventricle** (**VEN**-trih-kuhl) and the **left ventricle** (**Figure 14-3**).

The heart has four valves, which permit the blood to flow in one direction only. These valves open and close during the contraction of the heart, preventing the blood from flowing backward (**Figure 14-5**).

Atrioventricular (AV) valves are located between the atria and the ventricles. The following are the AV valves:

- The **tricuspid valve** (try-**KUSS**-pid) is positioned between the right atrium and the right ventricle. Its name comes from the fact that there are three points, or cusps, of attachment. The **chordae tendineae** (**KOR**-dee **TEN**-din-ee) are small fibrous strands that are projections of the myocardium and connect the edges of the tricuspid valve to the papillary muscle. When the right ventricle contracts, the papillary muscle contracts, pulling on the chordae tendineae to prevent inversion of the tricuspid valve. The tricuspid valve allows blood to flow from the right atrium into the right ventricle but not in the opposite direction.
- The **bicuspid (mitral) valve** (bye-**KUSS**-pid) (**MY**-tral) resembles a bishop's hat, called a miter. It is located between the left atrium and the left ventricle. Blood flows from the left atrium into the left ventricle; the mitral valve prevents backflow from the left ventricle to the left atrium (**Figure 14-5** and **Figure 14-6**).

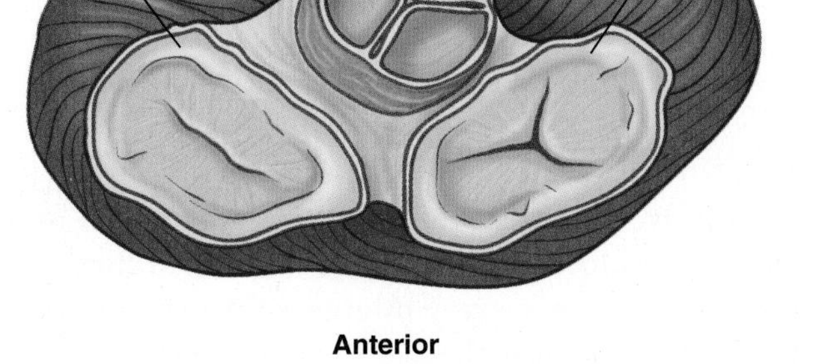

Figure 14-5 *The valves of the heart viewed from above with the atria removed*

The following semilunar valves are located where blood will leave the heart:

- The **pulmonary semilunar valve** (**PULL**-mah-nair-ee sem-ee-**LOO**-nar) is found at the orifice, or opening, of the pulmonary arteries. It allows blood to travel from the right ventricle into the right and left pulmonary arteries and then into the lungs (**Figure 14-5** and **Figure 14-6**).
- The **aortic semilunar valve** (ay-**OR**-tic) is at the orifice of the aorta. This valve permits the blood to pass from the left ventricle into the aorta but not backward into the left ventricle (**Figure 14-5** and **Figure 14-6**).

Media Link

View the **Anatomy of the Heart** video on the Online Resources.

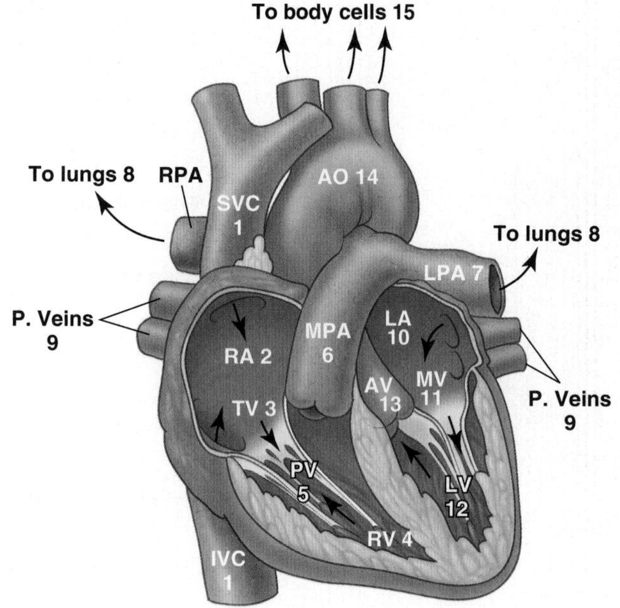

AO — Aorta
AV — Aortic semilunar valve
IVC — Inferior vena cava
LA — Left atrium
LPA — Left pulmonary artery
LV — Left ventricle
MPA — Main pulmonary artery
MV — Mitral valve
PV — Pulmonary semilunar valve
P. Veins — Pulmonary veins
RA — Right atrium
RPA — Right pulmonary artery
RV — Right ventricle
SVC — Superior vena cava
TV — Tricuspid valve

1. Blood reaches heart through superior vena cava (SVC) and inferior vena cava (IVC)
2. To right atrium
3. To tricuspid valve
4. To right ventricle
5. To pulmonary valve (semilunar)
6. To main pulmonary artery
7. To left pulmonary artery and right pulmonary artery
8. To lungs—blood receives O_2
9. From lungs to pulmonary veins
10. To left atrium
11. To mitral (bicuspid) valve
12. To left ventricle
13. To aortic valve (semilunar)
14. To aorta (largest artery in the body)
15. Blood with oxygen then goes to all cells of the body

Figure 14-6 *Physiology of the heart*

Circulation and Physiology of the Heart

The structure of the heart allows it to function as a double pump; think of the heart as having a right side and a left side. The following two major functions occur each time the heart beats (**Figure 14-6**):

- *Right heart*—Deoxygenated blood flows into the heart from the superior and inferior vena cava to the right atrium; travels on to the tricuspid valve; then to the right ventricle through the pulmonary semilunar valves; and finally to the right and left pulmonary arteries, which take blood to the lungs for oxygen.
- *Left heart*—Oxygenated blood flows from the lungs into the heart through the pulmonary veins to the left atrium; travels on through the bicuspid (mitral) valve; then to the left ventricle through the aortic semilunar valve; to the aorta; and finally to general circulation.

It is sometimes hard to imagine this idea of two pumping actions occurring at the same time. Each time the ventricles contract, blood leaves the right ventricle to go to the lungs, and blood leaves the left ventricle to go to the aorta.

HEART RATE AND CARDIAC OUTPUT

When an individual is at rest, the heart beats between 72 and 80 times per minute. With each beat, between 60 and 80 milliliters (mL) of blood are ejected from the ventricles. This is known as the **stroke volume**. The **cardiac output** is the total volume of blood ejected from the heart per minute:

$$\text{Stroke volume} \times \text{Heart rate} = \text{Cardiac output}$$

$$60 \text{ mL} \times 80 = 4800 \text{ mL/minute}$$

The average adult body contains about 5000 mL of blood. This means all the blood is pumped through the heart about once every minute. Exercise increases cardiac output because the heart rate is increased. During exercise, muscles receive about 60% of the cardiac output. At rest, the muscles receive only 27% of the cardiac output.

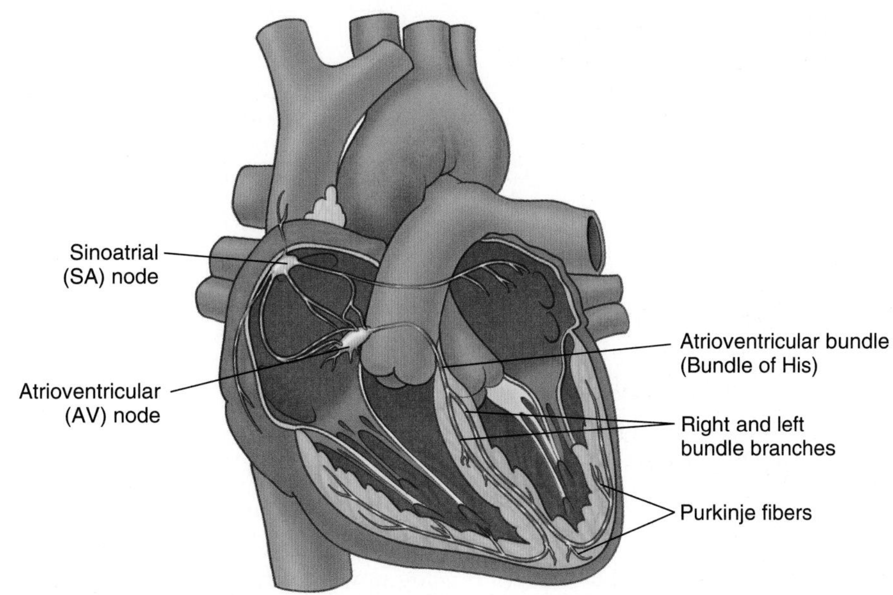

Figure 14-7 *An electrical impulse from the SA node travels to the AV node and causes the ventricle to contract.*

BLOOD SUPPLY TO THE HEART

The heart receives its blood supply from the coronary artery, which branches into right and left coronary arteries. (Further discussion on this subject can be found in Chapter 15.)

Heart Sounds

Physicians listen at specific locations on the chest wall to hear how the heart is functioning. During the cardiac cycle, the valves make a sound when they close. These are referred to as the **lubb dupp sounds**. The "lubb" sound is heard first and is made by the tripcuspid and bicuspid valves closing between the atria and ventricles. Physicians refer to it as the S_1 sound. It is heard loudest at the apex of the heart.

The "dupp" sound is the second sound heard and is shorter and higher pitched. It is caused by the semilunar valves in the aorta and the main pulmonary artery closing. Physicians refer to it as the S_2 sound.

Conduction System of Heart Contractions

The heart rate is affected by the endocrine and nervous systems. Impulses from the sympathetic nervous system can speed up the heart rate, and impulses from the parasympathetic system can slow it down. The thyroid hormone also affects the heart rate.

The cardiac conduction system is a group of specialized cardiac muscle cells in the walls of the heart that send signals to the heart muscle, causing it to contract. These cells are located at the opening of the superior vena cava into the right atrium and are known as the **sinoatrial (SA) node** (sigh-noh-**AY**-tree-ahl), or **pacemaker**. The SA node sends out an electrical impulse that begins and regulates the heart. The impulse spreads over the atria, making them contract (depolarize). This causes blood to flow downward from the upper atrial chamber to the atrioventricular openings. The electrical impulse eventually reaches the **atrioventricular (AV) node** (ay-tree-oh-ven-**TRICK**-you-lar), which is another conducting cell group located between the atria and ventricles.

From the AV node, the electrical impulse is carried to conducting fibers in the septum. These conducting fibers are known as the **atrioventricular bundle**, or the **bundle of His** (HISS). This bundle divides into a right and left branch; each branch then subdivides into a fine network of branches spreading throughout the ventricles called the Purkinje network. The electrical impulse shoots along the **Purkinje fibers** (per-**KIN**-jee) to the ventricles, causing them to contract. The heart then rests briefly and repolarizes. See **Figure 14-7**.

The combined action of the SA and AV nodes is instrumental in the cardiac cycle. The cardiac cycle

> **Did You Know?**
>
> *The human heart creates enough pressure to squirt blood 30 feet.*

Media Link

View the **Conduction System** animation on the Online Resources.

comprises one complete heartbeat, with both atrial and ventricular contractions:

1. The SA node stimulates the contraction of both atria. Blood flows from the atria into the ventricles through the open tricuspid and mitral valves. At the same time, the ventricles are relaxed, allowing them to fill with blood. At this point, because the semilunar valves are closed, blood cannot enter the main pulmonary artery or aorta.
2. The AV node stimulates the contraction of both ventricles so that the blood in the ventricles is pumped into the main pulmonary artery and the aorta through the semilunar valves, which are now open. At this point, the atria are relaxed, and the tricuspid and mitral valves are closed.
3. The ventricles relax; the semilunar valves are closed to prevent blood flowing back into the ventricles. The heart rests briefly for repolarization. The cycle begins again with the signal from the SA node.

Each cardiac cycle takes 0.8 seconds. The average person's heart rate is between 60–100 beats per minute.

ELECTROCARDIOGRAM (ECG OR EKG)

The **electrocardiogram (ECG or EKG)** (ee-**leck**-troh-**KAR**-dee-oh-gram) is a device used to record the electrical activity of the heart, which causes the contraction—systole—and the relaxation—diastole—of the atria and ventricles during the cardiac cycle.

The Effects of Aging on the Heart Muscle

The impact of aging on the heart influences the total circulatory system. The heart, as a muscular organ, changes as muscle fibers are replaced by fibrous tissue. This change leads to a diminished contractibility and filling capacity. Heart valves increase in thickness, modifying the normal closing of the valves and causing murmurs. Cardiac output decreases as a person ages. The diminished output becomes significant when an older adult is physically or mentally stressed by illness, strenuous physical activity, or disabilities.

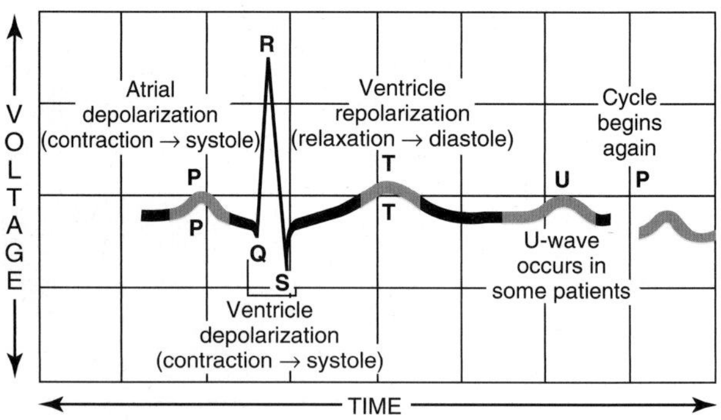

Figure 14-8 *ECG reading*

The baseline of the ECG is the flat line that separates the various waves. It is present when there is no current flowing in the heart. The waves are either deflecting upward, known as *positive deflection*, or deflecting downward, known as *negative deflection*. The P and T waves and QRS complexes recorded during the ECG represent the depolarization (contraction) and repolarization (relaxation) of the myocardial cells. The P wave represents atrial depolarization, the QRS complex represents ventricular depolarization, and the T wave represents ventricular repolarization. By observing the size, shape, and location of each wave, the physician can analyze and interpret the conduction of electricity through the cardiac cells, the heart's rate, the heart's rhythm, and the general health of the heart (**Figure 14-8**).

PREVENTION OF HEART DISEASE

Heart disease is the leading cause of death for both women and men in the United States. Risk factors for heart disease include family history, racial or ethnic predisposition, high blood pressure, high cholesterol, diabetes, current smoking, physical inactivity, and obesity.

Steps to lower risk or prevent heart disease include the following:

1. Prevention and control of high blood cholesterol levels and triglycerides. The test done to measure a person's cholesterol levels and triglycerides is called a lipid panel. A person's lipoprotein level is one of the determinants of transportation of fats throughout the body. Two important lipoproteins are very low-density lipoprotein (VLDL) and low-density lipoprotein (LDL). LDL carries newly formed cholesterol from the liver to the other tissues in the body; it is related to early narrowing of the arteries—atherosclerosis—and it is usually low in triglycerides. VLDL is responsible for moving fats and

cholesterol through the blood. VLDL contains high amounts of triglycerides and lower amounts of cholesterol. High-density lipoprotein (HDL) is called "good" cholesterol because it helps counter the fat buildup on artery walls. It does this by picking up the deposited fats and transporting them to the liver for elimination. Triglycerides are another form of fat, and high triglyceride levels increase the risk for coronary heart disease.

Desirable levels for persons with or without heart disease are as follows:

- Total cholesterol—less than 200 mg/dL (dL means deciliter, which is a unit of volume measurement used in the metric system; it is equal to one-tenth of a liter, or 3.3 ounces). Total cholesterol is the sum of the LDL, HDL, and one-fifth of the measured triglycerides.
- Low-density lipoprotein (LDL)—less than 100 mg/dL
- Very low-density lipoprotein (VLDL)—5 to 30 mg/dL
- High-density lipoprotein (HDL)—40 mg/dL or higher
- Triglycerides—less than 150 mg/dL

Medications called statins (Lipitor®, Crestor®, Zocor®) lower blood cholesterol levels. These drugs block the enzyme the liver needs to manufacture cholesterol, which depletes cholesterol in the liver and causes cells to remove cholesterol from the blood.

2. Prevention and control of high blood pressure using lifestyle changes, such as reducing stress and taking medication if necessary
3. Prevention and control of diabetes through weight loss and regular exercise
4. No tobacco usage
5. Moderate alcohol use: no more than two drinks a day for men and one a day for women
6. Maintaining a healthy weight: body mass index should be 18 to 24.9
7. Regular physical exercise: 30 minutes for most days of the week
8. Good sleeping habits, averaging 7–9 hours per night
9. Proper diet and nutrition: lots of fresh fruits and vegetables, decrease salt intake, and eat fewer foods that are high in saturated fats

DISEASES OF THE HEART

Heart diseases may begin during fetal development, or they may develop at another stage of life.

Congenital Diseases of the Heart

Congenital heart disease (CHD) is one of the most common birth defects, affecting 1% of births, or 40,000 births every year (Data and Statistics on Congenital Heart Defects, 2019). Down syndrome is one of the most widely known conditions that can cause CHD.

Coarctation of aorta (koh-ark-**TEY**-shun) is when a part of the aorta is narrower than usual. Symptoms include pale skin, heavy sweating, and a weaker pulse in the legs or groin than in the arms or neck. Treatment is a balloon angioplasty.

Tetralogy of Fallot means that there are four anatomical abnormalities:

- Pulmonary stenosis, or narrowing of the pulmonary valve and main pulmonary artery from the right ventricle
- Ventricular septal defect, or an opening between the right and left ventricle
- Right ventricular hypertrophy, or an enlargement of the right ventricle
- Aortic valve is enlarged and seems to open from both ventricles

Symptoms can vary from none to severe. When the baby cries or has a bowel movement, they turn very blue, have shortness of breath, become limp, and may lose consciousness.

Treatment may start with a temporary cardiac shunt, creating a passage so deoxygenated blood can flow to the lungs and return to the body with oxygen. When the infant is stable, further surgery will be done on all four defects. If the surgery is successful, most infants do well.

General Diseases of the Heart

One of the leading causes of death is cardiovascular disease. Common symptoms of heart disease are as follows:

- **Arrhythmia** (ah-**RITH**-mee-ah) is the term used to discuss any change or deviation from the normal rate or rhythm of the heart.
- **Bradycardia** (**brad**-ee-**KAR**-dee-ah) is the term used for a slow heart rate (fewer than 60 beats per minute).
- **Tachycardia** (**tack**-ee-**KAR**-dee-ah) is the term used for a rapid heart rate (more than 100 beats per minute).

- **Murmurs** indicate some defects in the valves of the heart. When valves fail to close properly, a gurgling or hissing sound will occur. Cardiac murmurs may be classified according to which valve is affected or according to the heart's cardiac cycle: if the murmur occurs when the heart is contracting, it is a systolic (sis-**TOL**-ick) murmur; if the heart is at rest, it is a diastolic (dye-ah-**STOL**-ick) murmur. Surgery can be done to replace the defective valve.
- **Mitral valve prolapse** is a condition in which the valve between the left atrium and the left ventricle closes imperfectly. Symptoms are thought to occur because of a response to stress. These symptoms include fatigue; **palpitations** (pal-pih-**TAY**-shunz), which is when the heart feels like it is racing; headache; chest pain; and anxiety. Exercise, restricted sugar and caffeine intake, adequate fluid intake, and relaxation techniques help alleviate symptoms.

Diseases of the Coronary Artery

Coronary artery disease (CAD) is a narrowing of the coronary arteries that supply oxygen- and nutrient-filled blood to the heart muscle. The narrowing, or atherosclerosis, usually results from the buildup of plaque on the artery walls. If the artery becomes completely blocked, a myocardial infarction may occur. To prevent CAD, a person should stop smoking, increase exercise, and reduce cholesterol levels. Angina is one of the most important symptoms of this disease.

Angina pectoris (an-**JYE**-nuh **PECK**-toh-ris), or angina, is a severe chest pain that arises when the heart does not receive enough oxygen. It is not a disease in itself, but a symptom of an underlying problem with coronary circulation, caused by narrowed coronary arteries. The chest pain radiates from the precordial area to the left shoulder, down the arm along the ulnar nerve. Women may experience "atypical" symptoms with angina. Many

Medical Highlights

DIAGNOSTIC TESTS FOR THE HEART

NONINVASIVE TESTS

The term *noninvasive* means that no surgery is done and no instrument is inserted into the body.

Angiography (**an**-jee-**OG**-rah-fee)—an X-ray that uses dye injected into the coronary arteries to study the circulation of the blood through the arteries. This test is often done along with a cardiac catheterization test.

Cardiac MRI—the use of radio waves and a strong magnetic field to provide remarkably clear and detailed pictures of the size and thickness of the chambers. The images can determine the extent of damage caused by a heart attack or progressive heart disease.

Coronary calcium scoring/heart scan—a test performed on a computed tomography machine that can help evaluate the risk of heart disease. It is able to detect calcified plaque in the arteries of the heart.

Echocardiography (**eck**-oh-**kar**-dee-**OG**-rah-fee)—an ultrasonic procedure that provides a direct view of any problems of the heart muscle, the pumping ability, and the state of the heart valves.

Electrocardiogram (EKG or ECG)—see prior information in this chapter.

Exercise stress test (treadmill test)—test given while the patient walks on a treadmill to see if exercise brings on changes in the ECG. The heart is stimulated to pump to maximum capacity.

Holter monitor—a portable, battery-operated ECG machine worn by the patient to record ECGs on tape over a period of 24 to 48 hours. At the end of the time period, the monitor is returned to the physician's office, and the tape is read and evaluated.

MUGA (multiple gated acquisition) scan—a test done to evaluate the right and left ventricles. This scan is performed by taking a sample of the patient's blood and attaching a radioactive marker to red blood cells. These red blood cells are then injected back into the patient's bloodstream. The patient is scanned by a special camera that detects the low-level radiation being given off by the red blood cells. As the red blood cells move through the heart, they produce a moving image of the beating heart and its chambers.

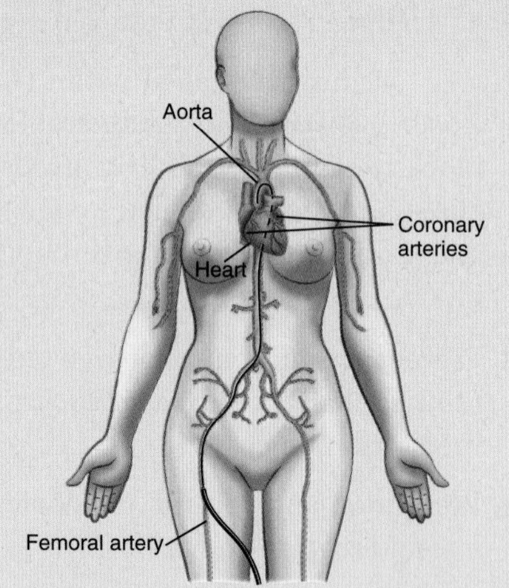

Figure 14-9 *Cardiac catheterization*

Nuclear perfusion testing—a specialized type of stress test. This test uses thallium or Cardiolite®, radioactive substances that are injected into the bloodstream when the maximum level of exercise is reached. The radioactive substance distributes itself throughout the cardiac muscle in proportion to the blood flow received by the muscle. An image of the heart is then made by a special camera that can "see" the thallium or Cardiolite. If one of the coronary arteries is blocked, not as much thallium or Cardiolite accumulates in the muscle supplied by the blocked artery.

INVASIVE TESTS

Cardiac catheterization—involves insertion of a catheter, usually into the femoral artery or vein. The catheter is fed into the chamber of the heart (**Figure 14-9**). Dye is inserted and pictures are taken as the fluid moves through the chambers. The patient may experience a warm or flushing sensation as the dye moves through the circulation, lasting only a few seconds. This test determines patency, or openness, of the coronary blood vessels as well as the efficiency of the structures of the heart.

Intravascular coronary ultrasound (IVUS)—a combination of echocardiography and cardiac catheterizations. An IVUS uses sound waves to produce an image of the coronary arteries and to see their condition. The sound waves are sent through a catheter, which is threaded through an artery and into the heart. This allows the physician to look inside the blood vessels. It is usually done at the same time that an angioplasty is done.

Transesophageal echocardiography (TEE)—an invasive technique that is performed by placing an instrument in the esophagus to measure sound waves. The TEE procedure provides a closer look at the heart's valves and chambers without interference from the lungs or ribs.

Health care providers must be certain to ask the patient scheduled to receive any type of dye for a test if they are allergic to any substance, especially fish.

BLOOD TESTS

Blood tests help diagnose heart disease, identify risk factors, help establish that a heart attack has occurred, and measure the extent of damage to the heart.

Arterial blood gases—measures the amount of oxygen in the blood, which should be high, and the amount of carbon dioxide, which should be low.

BNP—measures the level of the protein B-type natriuretic peptide that is produced by the heart and blood vessels to help eliminate body fluids. The BNP has been shown to rise if one has angina or a heart attack.

Cardiac troponin T—a protein normally only found in heart cells that is released into the blood after a heart attack, usually within 8 to 12 hours. Patients are monitored at intervals during this period to see if the level of this protein is rising. Patients with elevated levels may have suffered heart damage.

C-reactive protein—produced in the liver in response to infection and inflammation. Low risk level for developing heart disease is below 1 mg/L; average risk level is 1–3 mg/L; and high risk level is over 3 mg/L. It may indicate heart disease in conjunction with other findings.

Lipid panel—measures cholesterol, LDL, VLDL, HDL, and triglyceride levels.

women report a hot or burning sensation or even tenderness to touch in the back, shoulders, arms, or jaw. Often, they have no chest discomfort at all. Victims often experience a feeling of impending death. Angina pectoris occurs quite suddenly; it may be brought on by stress or physical exhaustion. Immediate treatment is to call 911, lie down, and chew an aspirin. For others with a history of angina, it may be treated with sublingual nitroglycerine, which helps dilate the coronary arteries.

Myocardial infarction (my-oh-**KAR**-dee-al in-**FARK**-shun) is commonly known as an "MI," or "heart attack," and is caused by a lack of blood supply to the myocardium. This may be due to the blocking of the coronary artery by a blood clot; a narrowing of the coronary artery as a result of arteriosclerosis, which is a loss of elasticity and thickening of the wall; or atherosclerosis, caused by plaque buildup in the arterial walls (**Figure 14-10**). The heart muscle becomes damaged due to lack of blood supply. The amount of tissue affected depends on how much of the heart area is deprived of blood. A symptom is crushing, severe chest pain that radiates to the left shoulder, arm, neck, and jaw. Patients may also complain of nausea; increased perspiration; fatigue; *dyspnea* (**DISP**-nee-ah), or difficulty breathing; sweating; and dizziness. Myocardial infarcts may act differently in women. Women and some people with diabetes frequently have no chest pain and are said to have "silent" MIs.

Mortality is highest when treatment is delayed; therefore, immediate medical care is critical. Arrhythmias are the leading cause of death in the first hours of a heart attack. Arrhythmias may be treated with cardioversion, which is done with a defibrillator, or medication. Blood thinners will be given intravenously

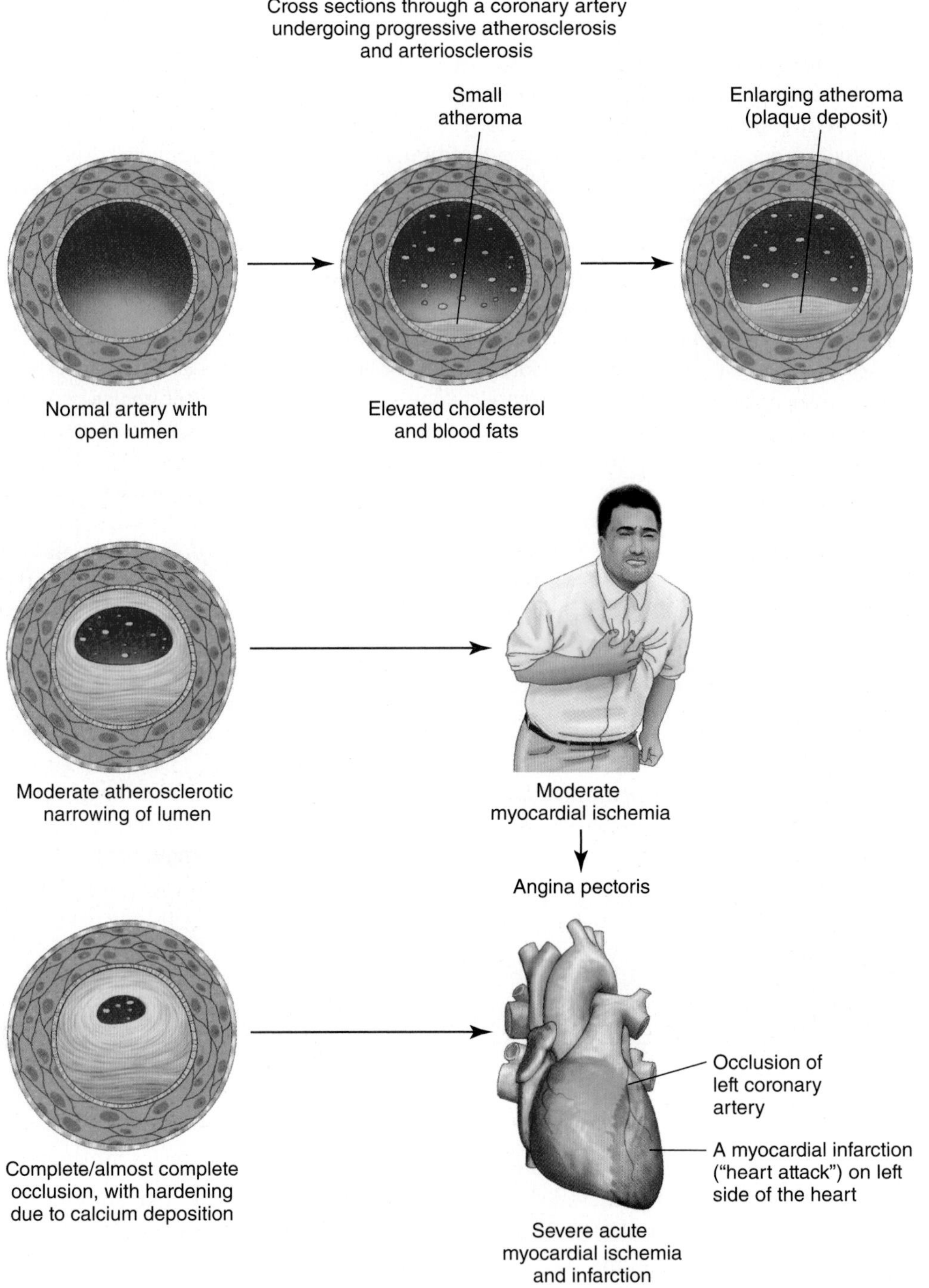

Figure 14-10 *Progressive atherosclerosis*

during the first 12 hours. Treatment consists of bed rest, oxygen, and medications. Morphine or Demerol® is given to alleviate the pain, drugs such as tissue plasminogen activator (tPA) are used to dissolve the blood clot, and **cardiotonic** (**kard**-ee-oh-**TON**-ik) drugs such as digitalis are used to slow and strengthen the heartbeat. Other medications include vasodilators; beta-blockers; and **diuretics** (**dye**-you-**RET**-icks), which are drugs that reduce the amount of fluid in the body. Anticoagulant therapy is used to prevent further clots from forming. Angioplasty and bypass surgery may also be necessary.

Infectious Diseases of the Heart

A bacteria or virus is usually the cause of infectious diseases of the heart. Some of these conditions may be treated with antibiotic therapy.

- **Pericarditis** (pehr-ih-kar-**DYE**-tis) is an inflammation of the outer membrane covering the heart. This type of inflammation causes a pleural effusion, or an accumulation of fluid within the pericardial sac that restricts the contraction of the heart and reduces the ability of the heart to pump blood throughout the body. The most common symptom is a sharp, stabbing chest pain felt behind the sternum or in the left side of the chest. The pain may travel into the left shoulder and neck. The other symptoms are cough, dyspnea, rapid pulse, and fever.
- **Myocarditis** (my-oh-kar-**DYE**-tis) is an inflammation of the heart muscle. The symptoms may be the same as for pericarditis.
- **Endocarditis** (en-doh-kar-**DYE**-tis) is an inflammation of the membrane that lines the heart and covers the valves. This causes the formation of rough spots in the endocardium, which may lead to the development of a fatal thrombus, or blood clot.
- **Rheumatic heart disease** (roo-**MAT**-ick) may be a result of a person having frequent strep throat infections during childhood; these infections may lead to rheumatic fever. The antibodies that form to protect the child from the strep throat or rheumatic fever may also attack the lining of the heart, especially the bicuspid or mitral valve. The valve becomes inflamed and the structure may be scarred, which leads to narrowing of the valve. This affects the function of the mitral valve; it is unable to close properly, which interferes with the blood flow from the left atrium to the left ventricle. It is very important for children who have streptococcal infections to be treated with antibiotic therapy.

Heart Failure

Heart failure is a chronic disease that occurs when the ventricles of the heart are unable to contract effectively and blood pools in the heart. Different symptoms can arise depending on which ventricle fails to beat properly. If the left ventricle fails, dyspnea occurs. If the right ventricle fails, engorgement of organs with venous blood occurs; as well as *edema* (eh-**DEE**-mah), which is excessive fluid in tissues; and **ascites** (ah-**SIGH**-teez), which is the abnormal accumulation of serous fluid in the abdominal cavity. Other symptoms may include lung congestion and coughing. Treatment is with drugs such as angiotensin-converting enzyme (ACE) inhibitors, which are vasodilators that lower blood pressure and improve blood flow; beta-blockers that slow the heart and reduce blood pressure; diuretics; and digitoxin, which slows and strengthens the heart muscle.

CONGESTIVE HEART FAILURE

Congestive heart failure is similar to heart failure. In left-sided failure, fluid accumulates in the lungs, backing blood up into the lung vessels; this is known as pulmonary edema. In right-sided failure, fluid builds up throughout the body. Because of the pressure of gravity, the edema is first noticed in the legs and feet. Treatment consists of cardiotonics and diuretics. Other drugs which may be useful include ACE inhibitors and beta-blockers.

Media Link

View the **Congestive Heart Failure** animation on the Online Resources.

Rhythm/Conduction Defects

A **conduction defect**, or rhythm defect, is said to occur when the conduction system of the heart is affected.

- **Heart block** is the interruption of the AV node message from the SA node. The abnormal patterns are seen on an electrocardiograph. *First-degree block* is characterized by a momentary delay at the AV node before the impulse is transmitted to the ventricles. *Second-degree block* can be of two forms. One occurs in cycles of delayed impulses until the SA node fails to conduct to the AV node, and then returns to near normal. A second form is characterized by a pattern of only every second, third, or fourth impulse being conducted to the ventricles. *Third-degree block* is known as "complete heart block." No impulse is carried over from the pacemaker. Because the heart is essential to life, there is a built-in safety factor. The atria continue to beat 72 beats per minute while the ventricles contract independently at about half the atrial rate. This action is adequate to sustain life but results in a severe decrease in cardiac output. Conduction defects may be treated by medications and/or the use of an artificial pacemaker.
- **Premature contractions** refer to an arrhythmia disorder that occurs when an area of the heart known as an ectopic (abnormal place) pacemaker, not the SA node, sparks and stimulates a contraction of the myocardium. The three types are identified by their

Career Profile

EMERGENCY MEDICAL TECHNICIAN AND PARAMEDIC

Emergency medical technicians (EMTs) and paramedics respond to medical emergencies. Once they arrive at a scene, they determine the nature and extent of the patient's condition. They also try to determine if the patient has a preexisting medical problem. Following strict procedures, they give appropriate emergency care and then transport the patient to a medical facility. Guidance for handling complicated problems is given by a physician via radio or phone to EMTs in the field.

EMTs and paramedics may use special equipment, such as defibrillators. The specific responsibilities of EMTs and paramedics depend on their level of qualification and training. The National Registry of Emergency Medical Technicians registers emergency medical services at four levels: First Responder, EMT-Basic, EMT-Intermediate, and EMT-Paramedic.

EMTs and paramedics work both indoors and outdoors in all types of weather. They are required to do considerable kneeling, bending, and heavy lifting. Formal training and certification are needed in all states to become an EMT or paramedic. Job outlook is good—demand is expected to grow faster than average.

location: atrial, ventricular, or AV junctional. Premature contractions are usually of no clinical significance and are caused by stress, caffeine, or fatigue.

- **Atrial fibrillation** occurs when abnormal impulses from the atria bombard the AV node. The AV node blocks many of the extra signals from reaching the ventricles, but some do get through. This action causes the ventricles to beat faster, and a person can experience tachycardia. Treatment is medication to prevent blood clots and cardioversion to reset the heart rate. If no improvement is noted, a procedure called radio-frequency ablation is done. A catheter is inserted through the groin to the heart and used to burn areas of the tissue within the atria. This causes scarring of the tissue and stops the erratic heartbeats. Radio-frequency ablation can stop atrial fibrillation for about one year.
- **Premature ventricular contractions (PVCs)** originate in the ventricles and cause contractions ahead of the next anticipated beat. They can be benign, or they can be deadly when they result in ventricular fibrillation. If frequent (five or six per minute) or in pairs, they may require immediate intervention to decrease the irritability of the cardiac muscle and maintain cardiac output.
- In **ventricular fibrillation** (fih-brih-**LAY**-shun) the rhythm breaks down and muscle fibers contract at random, without coordination. This results in ineffective heart action and is a life-threatening condition. An electrical device called a **defibrillator** (dee-**fib**-rih-**LAY**-tor) is used to discharge a strong electrical current through the patient's heart via electrode paddles held against the bare chest wall. The shock interferes with the uncoordinated action and attempts to shock the SA node into resuming its control.

HEART TRAUMA OR INJURY

Heart trauma or injury occurs when a blunt chest injury causes contusion to the heart muscle. The cause is usually a motor vehicle accident, fall, or direct blow. The heart muscle will usually heal within 2–4 weeks if no further damage occurs. At that time, any structure and functions of the organ and the circulatory system affected by the trauma should be back to normal.

TYPES OF HEART SURGERY

Angioplasty (**AN**-jee-oh-**plas**-tee), or **balloon surgery**, is a procedure to help open clogged vessels. A small, deflated balloon is threaded into the coronary artery; when it reaches the blocked area, the balloon is inflated. The balloon is then opened and closed a few times until the blockage is pushed against the arterial wall and the area is unblocked. The balloon is then deflated and removed (**Figure 14-11**).

Cardiac stents are tiny, webbed, stainless steel devices that hold arteries open after an angioplasty (**Figure 14-12**). About 25% of the patients who receive stents develop restenosis, in which scar tissue or another

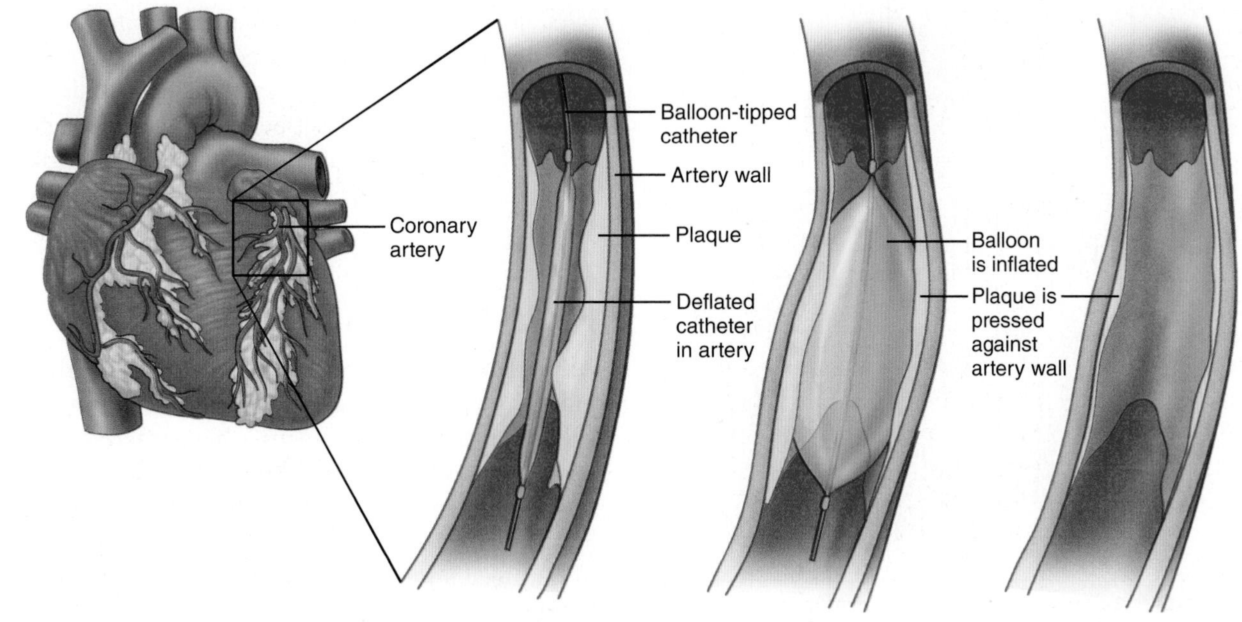

Figure 14-11 *Balloon angioplasty is a procedure performed to reopen a blocked coronary artery.*

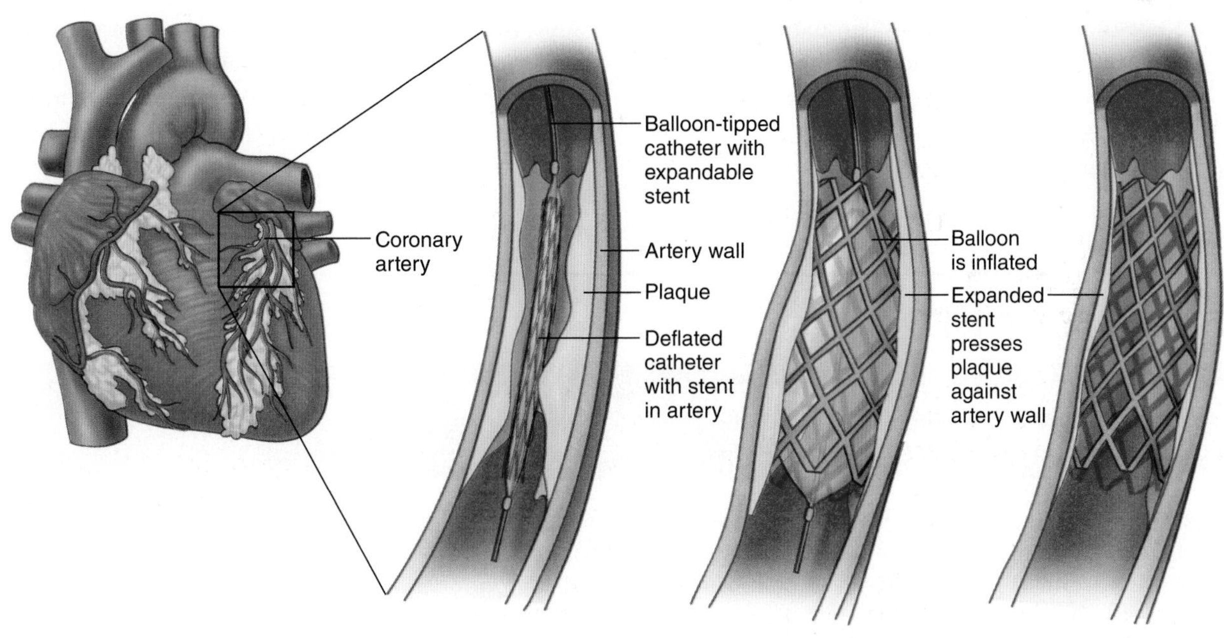

Figure 14-12 *A cardiac stent is put in place to hold arteries open after an angioplasty and to prevent restenosis.*

blockage forms inside the stent and reclogs the arteries. To prevent restenosis, drug-coated stents are used. The drugs used successfully discourage tissue regrowth in the artery. These drug-coated stents may have a tendency to form blood clots.

Coronary bypass involves surgically providing a detour, or bypass, to allow the blood supply to go around the blocked area of a coronary artery (**Figure 14-13**). A healthy blood vessel, usually from the mammary artery or saphenous (sah-**FEE**-nus) vein, is used for this purpose. The blood vessel is inserted before the blocked area and provides another route for the blood supply to the myocardium.

Transmyocardial laser revascularization (TMR) is the use of lasers to puncture holes in the heart muscle to improve blood flow. This procedure benefits patients

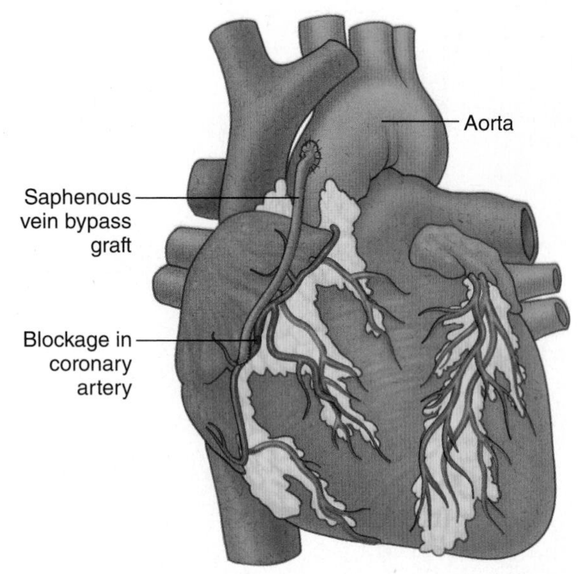

Figure 14-13 *Coronary bypass provides a detour, or bypass, to allow the blood supply to go around a blocked coronary artery.*

who are not candidates for bypass or angioplasty surgery. The laser instrument is placed on the heart muscle around a blocked artery, and the heart muscle is zapped. The laser's energy creates a tiny hole about 1 mm in size through the heart wall to the blood-filled chamber. The outside of the hole heals in a matter of minutes, but the channel created remains. The new channel allows blood from the heart chamber to reach the heart muscle. The trauma caused by the laser beam stimulates the growth of new blood vessels. The full effect of the TMR does not take place until about two weeks to six months after surgery.

Heart Transplants

A heart transplant is needed in cases when the individual's own heart can no longer function properly. This happens when someone has suffered repeated heart attacks and irreparable damage has been done to the heart muscle, valves, or blood vessels leading to and from the heart. Occasionally, a baby or young child might need a heart transplant because of a congenital heart defect.

Problems always arise even after the most "successful" of heart transplants, however. The problems involve histocompatibility, or matching of tissue type, and organ rejection.

Medical science has counteracted organ rejection by developing chemicals called immunosuppressants. But suppressing the recipient's immune system indefinitely is not medically wise because they will be more susceptible to disease and infection.

 Career Profile

CARDIOVASCULAR TECHNOLOGIST AND TECHNICIAN/ECG TECHNICIAN

Cardiovascular technologists and technicians assist physicians in diagnosing and treating cardiac and peripheral vascular disease. Cardiovascular technicians may also be known as ECG technicians because they take electrocardiograms. More skilled technicians may also do Holter monitor and stress testing. Cardiovascular technologists who specialize in cardiac catheterization procedures are called cardiology technologists.

Education to prepare a technician for ECG, Holter, and stress testing usually requires a one-year certificate program. Training for cardiology technologists involves a two-year program, which is dedicated to core courses and clinical practice.

The job prospects for cardiology technologists are excellent. However, cardiovascular technologists' job prospects are not as good because nurses and others may also be trained to do procedures such as ECGs and stress testing.

Medical Highlights

PACEMAKERS, DEFIBRILLATORS, AND HEART PUMPS

PACEMAKER

A *pacemaker* is a surgically implanted, battery-operated electronic device that sends electrical impulses to regulate the rhythm of the heart (**Figure 14-14**). Pacemakers have two main parts: the generator and the leads. The battery life is about 5 to 10 years. The lead is a flexible, insulated wire. Most pacemakers use two leads: one end of each lead is attached to the generator, and the other ends are attached to the right atrium and the right ventricle. These leads transmit the electrical impulses to keep the heart rate from dropping too low. They also maintain coordination between the atrium and ventricle. Two types of pacemakers are demand pacing and fixed-rate pacing. The demand pacemaker sends electrical prompts to the heart when the heart rate falls outside of a predetermined rate or skips a beat. The fixed-rate pacemaker sends electrical impulses at a steady rate that does not respond to the activity of the heart.

Cardiac resynchronization therapy (CRT) uses a specialized type of pacemaker to recoordinate the action of both the right and left ventricles of the heart, pacing both ventricles simultaneously.

DEFIBRILLATOR

An *implantable cardioverter-defibrillator (ICD)* is a device implanted under the skin and attached to the heart with small wires. The ICD monitors the heart rhythm; if the heart starts beating at a dangerous rhythm, such as ventricular tachycardia or ventricular fibrillation, the ICD shocks it back into normal rhythm. Many ICDs record the heart's electrical patterns whenever an abnormal heartbeat occurs. Physicians can review the record and plan future treatment options.

An *automated external defibrillator* is an easy-to-use, extremely portable defibrillator kept with emergency equipment and often located where people work and play.

Proper operation of pacemakers and defibrillators may be at risk from external devices. These include MRI machines, other medical devices, electronic surveillance systems, and possibly cell phones. It is recommended that patients with pacemakers not place cell phones over the area of the pacemaker. If a patient has to clear security in an airport, for instance, the patient should make sure to tell security personnel they have an electronic medical device and ask that the handheld metal detector not be held near the device any longer than necessary. They should also carry identification that shows others they have a pacemaker or defibrillator.

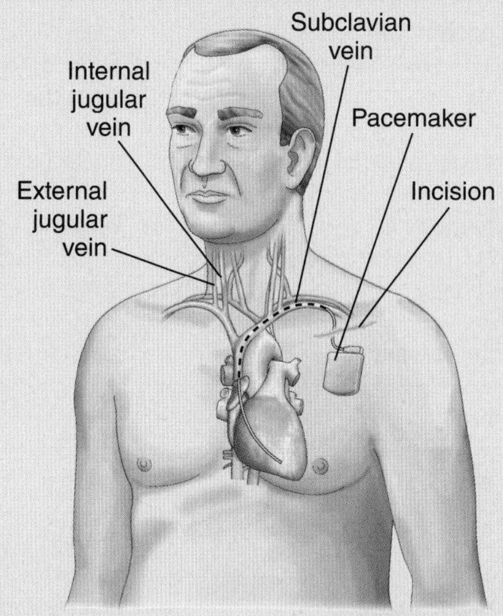

Figure 14-14 *Pacemaker*

HEART PUMPS

Heart pumps are called *left ventricular assist devices (LVADs)*. They are implanted in the abdomen and attached to a weakened heart to help it pump. LVADs are now being considered as an alternative to a heart transplant. Implanted heart pumps significantly extend and improve the lives of some people with end-stage heart failure and those who are not eligible to undergo a heart transplant.

Medical Terminology

angin-	tightness with pain	**endo-**	within, inner
-a	presence of	**-itis**	inflammation of
pector	chest	**endo/card/itis**	inflammation within the heart
-is	presence	**myo**	muscle
angin/a pector/is	presence of pain in the chest	**-al**	presence of
angio	vessel	**myo/cardi/al**	presence of heart muscle
-plasty	surgical repair	**infarct**	area of tissue death
angio/plasty	surgical repair of vessels	**myo/cardi/al infarct**	area of tissue death in the heart muscle
brady-	slow	**peri-**	around
card	heart	**peri/card/itis**	inflammation around the heart
-ia	condition of	**sept**	wall, partition
brady/card/ia	condition of slow heart	**-um**	presence of
ton	strength	**sept/um**	presence of partition
-ic	pertaining to	**steth/o**	chest
cardio/ton/ics	pertaining to heart strengthener	**scope**	instrument used to examine
electro	electric current or activity	**steth/o/scope**	instrument used to examine the chest
-gram	recording of	**tachy-**	rapid, fast
electro/cardio/gram	recording of electric activity of the heart	**tachy/cardia**	rapid or fast heart rate

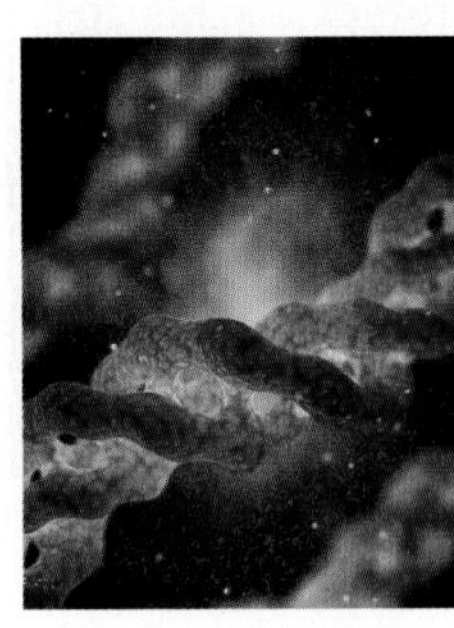

Study Tools

Workbook	Activities for Chapter 14
Online Resources	• PowerPoint presentations • Animations

REVIEW QUESTIONS

Select the letter of the choice that best completes the statement.

1. The outer layer of the heart is called the
   a. myocardium.
   b. endocardium.
   c. pericardium.
   d. pleural lining.

2. The muscle layer of the heart is called the
   a. myocardium.
   b. endocardium.
   c. pericardium.
   d. pleural lining.

3. When administering CPR, the first action to take after calling 911 is to
   a. begin chest compressions.
   b. check patient's airway.
   c. turn patient on their side.
   d. give two quick breaths.

4. The valve between the right atrium and the right ventricle is called the
   a. tricuspid valve.
   b. aortic semilunar valve.
   c. bicuspid valve.
   d. pulmonary semilunar valve.

5. The blood vessel that brings blood to the right atrium is called the
   a. pulmonary vein.
   b. aorta.
   c. pulmonary artery.
   d. vena cava.

6. The pacemaker of the heart is known as the
   a. SA node.
   b. AV node.
   c. bundle branches.
   d. Purkinje fibers.

7. The heart's conducting system normally follows this sequence:
   a. AV bundle branches, AV node, and SA node.
   b. AV node, AV bundle branches, and SA node.
   c. SA node, AV node, and AV bundle branches.
   d. AV bundle branches, SA node, and AV node.

8. The device used to measure the electrical activity of the heart is called an
   a. EEG.
   b. MRI.
   c. ECG.
   d. EMG.

9. A heart rate below 60 is called
   a. bradycardia.
   b. tachycardia.
   c. arrhythmia.
   d. murmur.

10. An inflammation of the inner layer of the heart is called
    a. pericarditis.
    b. myocarditis.
    c. endocarditis.
    d. phlebitis.

11. The term "heart attack" is another name for
    a. rheumatic heart disease.
    b. myocardial infarction.
    c. heart block.
    d. congestive heart failure.

12. The leading cause of death in the first hours after a heart attack is
    a. dyspnea.
    b. arrhythmia.
    c. edema.
    d. electrolyte imbalance.

13. The treatment for a heart attack may include all but
    a. angioplasty.
    b. antibiotics.
    c. coronary bypass.
    d. anticoagulants.

14. The desirable blood level for high density lipoprotein (HDL) is
    a. 40 mg/dL.
    b. 20 mg/dL.
    c. 30 mg/dL.
    d. 10 mg/dL.

15. The treatment for conduction defect may include
    a. coronary bypass.
    b. cardiotonics.
    c. insertion of a pacemaker.
    d. angioplasty.

LABELING

Study the following diagram of the heart and label the various structures, including valves and vessels. Trace blood from the right atrium to the aorta.

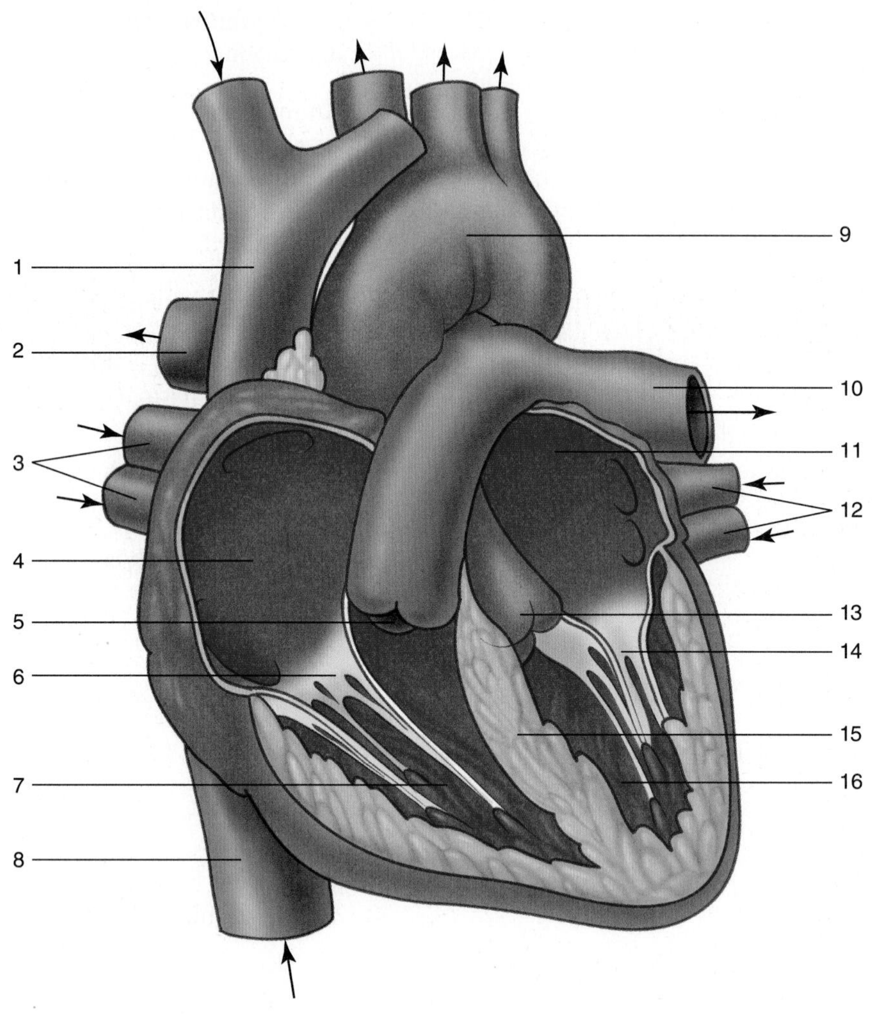

1. ____________________
2. ____________________
3. ____________________
4. ____________________
5. ____________________
6. ____________________
7. ____________________
8. ____________________
9. ____________________
10. ____________________
11. ____________________
12. ____________________
13. ____________________
14. ____________________
15. ____________________
16. ____________________

MATCHING

Match each term in Column I with its correct description or function in Column II.

	COLUMN I	COLUMN II
______	**1.** pulmonary artery	a. vein that carries freshly oxygenated blood from the lung to the heart
______	**2.** lymphatic system	b. circulation route that carries blood to and from the heart and lungs
______	**3.** pulmonary vein	c. divides the heart into right and left sides
______	**4.** septum	d. artery that carries deoxygenated blood from the heart to the lung
______	**5.** pulmonary circulation	e. system that consists of lymph and tissue fluid derived from the blood
______	**6.** left ventricle	f. blood from the pulmonary vein that reenters the heart through the left atrium
______	**7.** general circulation	g. artery that carries blood with nourishment, oxygen, and other materials from the heart to all parts of the body
______	**8.** right ventricle	h. ventricle from which the aorta receives blood
______	**9.** aorta	i. circulation that carries blood throughout the body
		j. ventricle from which the pulmonary artery leaves the heart

APPLYING THEORY TO PRACTICE

1. Pretend you are a blood cell that has just arrived in the right atrium. Trace the journey you will take to get to the aorta.

2. A child has chronic strep throat. If this condition is not treated, what heart disease can occur? Describe what happens to the heart's structure and function. How can this be prevented?

3. Your 70-year-old neighbor, Mr. Michael, tells you he has been diagnosed with a second-degree heart block. The physician told him he would need a pacemaker implanted. He knows you are a registered nurse, and he wants you to explain pacemakers to him. He is worried he will not be able to use a cell phone. Describe for Mr. Michael what a pacemaker is and what precautions, if any, he has to take after the pacemaker is implanted.

4. A 50-year-old female patient comes into the physician's office and states, "people in my family all start to die at 50 from heart disease." What guidelines can you give her to help prevent the disease? Explain to the patient the symptoms a woman may experience if she is having a heart attack. Also explain the diagnostic technologies that may be used to check her heart.

5. A 20-year-old patient comes into the emergency room with trauma to the chest. She was in a boxing match, received a direct blow to the chest, and fainted. She is sitting up and smiling by the time she gets to the hospital. How long will it take her heart to heal? How will the structure and functions of her organs and circulatory system be affected? What should she avoid doing for the time being?

CASE STUDY

Mr. Vincent is a 45-year-old salesman who is overweight. At work he suddenly develops severe chest pain and is nauseated. A coworker takes him to a nearby hospital. The emergency department physician orders an immediate ECG. A diagnosis of acute myocardial infarct is made. Mr. Vincent is scheduled for a balloon angioplasty with insertion of a cardiac stent. The surgery is completed, and Mr. Vincent recovers with no complications. Margaret, the nurse clinician, is assigned to educate Mr. Vincent regarding his procedure and follow-up care.

1. What is the cause of a myocardial infarct?
2. What is the function of the heart?
3. Describe the cardiac cycle and how it would be affected by a myocardial infarct.
4. Describe the therapeutic technology used—the balloon angioplasty—and the rationale for the insertion of a cardiac stent.
5. How can Mr. Vincent prevent a future heart attack from occurring?
6. Explain the effect of cholesterol and triglycerides on the arteries. What blood level values for cholesterol and triglycerides should Mr. Vincent maintain?
7. What are the effects of cardiotonics and anticoagulant medication?

LAB ACTIVITY 14-1 Heart Structure

- ***Objective:*** To observe the structure of the heart
- ***Materials needed:*** anatomical model of human heart, preserved sheep heart, dissecting kit, disposable gloves, paper, pencil

Step 1: Put on disposable gloves.

Step 2: Rinse the sheep heart in cold water to remove preservatives.

Step 3: Locate the apex of the heart. Compare the size and shape of the sheep heart with the anatomical model of the human heart. Is there a difference? List any differences you see.

Step 4: Describe and record the appearance of the pericardium of the sheep heart. Using a scalpel, carefully pull the pericardium away from the myocardium.

Step 5: Using a scalpel, carefully scrape away any accumulation of fat that may surround the heart. This will help you see the heart chambers and coronary blood vessels.

Step 6: Locate and describe coronary arteries. Record your observations.

Step 7: Identify the right and left atria. Describe and record their appearance.

Step 8: Locate the ventricles of the heart. Feel both ventricle chambers. Is there a difference between the right and left chambers? Record your answer.

Step 9: Locate and describe the pulmonary artery and aorta. Draw and label a simple sketch to show these features.

Step 10: Using a scalpel or scissors, carefully cut through the aorta and locate the aortic semilunar valve (see Figure 14-3). Describe the appearance of the semilunar valve.

Step 11: Examine the heart on its posterior side. Locate and identify the superior and inferior vena cava.

Step 12: Using a scalpel or scissors, carefully cut through the wall of the superior vena cava to view the right atrium. Observe the right tricuspid valve. Sketch the tricuspid valve.

Step 13: Continue to cut carefully through the right atrium into the right ventricle. Observe the walls of the right ventricle and locate the pulmonary semilunar valve. Record your observations.

Step 14: Using a scalpel or scissors, carefully cut through the aorta into the left atrium. Observe the bicuspid valve. Record your observations.

Step 15: Continue to cut carefully into the left ventricle. Observe the walls of the left ventricle. Record your observations.

Step 16: Is there a difference between the structures of the right and left ventricles? Record your answer.

Step 17: Dispose of the sheep heart in the appropriate laboratory container.

Step 18: Clean all equipment.

Step 19: Remove gloves and wash hands.

Step 20: Use the anatomical model of a human heart and record any heart features you found in steps 6 through 14.

LAB ACTIVITY 14-2 Read an EKG

- ***Objective:*** Compare the differences between a normal EKG, an EKG that shows a myocardial infarct, and an EKG that shows a second-degree heart block
- ***Materials needed:*** copies of normal EKG, myocardial infarct EKG and second-degree heart block EKG; textbook; paper; pen; Internet access

Step 1: Identify the P and T waves and QRS complex on the normal EKG.

Step 2: Identify the P and T waves and QRS complex on the myocardial infarct EKG and the second-degree heart block.

Step 3: Describe what the P wave, QRS complex, and T wave mean. Check with your textbook to confirm your response.

Step 4: Describe the major differences between the normal EKG, myocardial infarct EKG, and second-degree heart block EKG. What is the reason for the differences in the complexes? Write down your answer.

Step 5: Use the Internet to research various heart diseases and what their EKG tracings would look like. Investigate the diagnostic and therapeutic technologies that would be used to diagnose and treat each of these conditions.

LAB ACTIVITY 14-3 Heart Sounds

- ***Objective:*** Compare and contrast the heart sounds of a normal heart and the heart sounds of persons affected with disorders of the heart
- ***Materials needed:*** Internet access, paper, pencil, a quiet room, headphones if available

Step 1: With a partner, access the Internet in a quiet room. If available, use headphones to help you listen more closely.

Step 2: Find a website with normal heart sounds. Have each partner listen carefully to the heart sounds for one minute and record their observations. When both partners have had a turn, describe the rate and rhythm you observed to your partner.

Step 3: Find a website that has the sounds of a heart murmur. Have each partner listen carefully to the heart sounds for one minute and record their observations. When both partners have had a turn, describe the rate and rhythm you observed to your partner.

Step 4: Find a website with the heart sound of someone with a second-degree heart block. Have each partner listen carefully to the heart sounds for one minute and record their observations. When both partners have had a turn, describe the rate and rhythm you observed to your partner.

Step 5: Compare the sounds of the normal heart with the heart murmur and the heart block sounds. Is there a difference? Describe the difference.

Step 6: Report any differences you and your partner observed to the class. Discuss to see if all are in agreement.

CHAPTER 15

Circulation and Blood Vessels

Objectives

- Trace the path of cardiopulmonary circulation.
- Name and describe the specialized circulatory systems.
- Trace the blood in fetal circulation.
- List the types of blood vessels.
- Identify the principal arteries and veins of the body.
- Describe some disorders of the circulation and blood vessels.
- Define the key words that relate to this chapter.

Key Words

aneurysm
aphasia
arteries
arterioles
arteriosclerosis
atherosclerosis
atrial septal defect
atrioventricular septal defect
brachial artery
capillaries
cardiopulmonary circulation
carotid artery
cerebral hemorrhage
cerebral vascular accident (CVA)
common carotid artery
coronary artery
coronary circulation
cyanosis
diastolic blood pressure
dorsalis pedis artery
ductus arteriosus
ductus venosus
dysphasia
femoral artery
fetal circulation
foramen ovale
gangrene
hemiplegia
hepatic vein
hypertension
hypoperfusion
hypotension
peripheral vascular disease (PVD)
phlebitis
popliteal artery
portal circulation
portal vein
pulse
pulse pressure
radial artery
shock
stroke
systemic circulation
systolic blood pressure
temporal artery
transient ischemic attacks (TIAs)
tricuspid atresia

(continues)

Key Words *continued*

tunica adventitia (externa)
tunica intima
tunica media
valves
varicose veins
vascular malformations
veins
venipuncture
ventricular septal defect
venules
white-coat hypertension

Blood vessels circulate blood through two major circulatory systems (**Figure 15-1**):

1. *Cardiopulmonary circulation*—blood from the heart to the lungs for oxygen and back to the heart
2. *Systemic circulation*—blood from the heart to the tissues and cells and back to the heart

   Specialized systemic routes are as follows:

a. *Coronary circulation*—brings blood from the heart to the myocardium

b. *Portal circulation*—takes blood from the organs of digestion to the liver through the portal vein

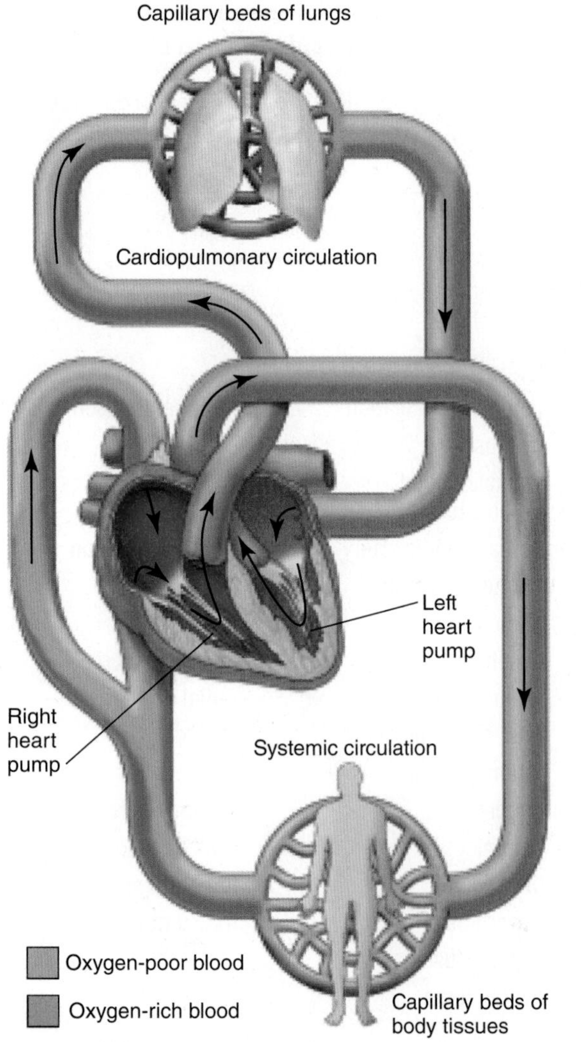

Figure 15-1 *Cardiopulmonary circulation*

c. *Fetal circulation*—occurs in the pregnant female; fetus obtains oxygen and nutrients from the mother's blood

CARDIOPULMONARY CIRCULATION

Cardiopulmonary circulation takes deoxygenated blood from the heart to the lungs, where carbon dioxide is exchanged for oxygen. The oxygenated blood returns to the heart. As stated in Chapter 14, blood enters the right atrium, which contracts, forcing the blood through the tricuspid valve into the right ventricle.

The right ventricle contracts to push the blood through the pulmonary valve into the main pulmonary artery. The main pulmonary artery bifurcates, or divides in two. It branches into the right pulmonary artery, bringing blood to the right lung, and into the left pulmonary artery, bringing blood to the left lung (**Figure 15-2**).

Inside the lungs, the pulmonary arteries branch into countless small arteries called **arterioles** (ar-**TEE**-ree-ohlz). The arterioles connect to dense beds of capillaries lying in the alveoli lung tissue. Here, gaseous exchange takes place by the process of diffusion. Carbon dioxide leaves the red blood cells and is discharged into the air in the alveoli, to be excreted from the lungs. Oxygen from air in the alveoli combines with hemoglobin in the red blood cells. From these capillaries, the blood travels into small veins or **venules** (**VEN**-youls) (**Figure 15-3**). Venules from the right and left lungs form large pulmonary veins. These veins carry oxygenated blood from the lungs back to the heart and into the left atrium.

The left atrium contracts, sending blood through the bicuspid, or mitral valve, into the left ventricle. This chamber then acts as a pump for newly oxygenated blood. When the left ventricle contracts, it sends oxygenated blood through the aortic semilunar valve into the aorta.

SYSTEMIC CIRCULATION

The function of the general circulation, or **systemic circulation**, is fourfold: it circulates chemicals, such as nutrients, oxygen, water, and secretions, to the tissues

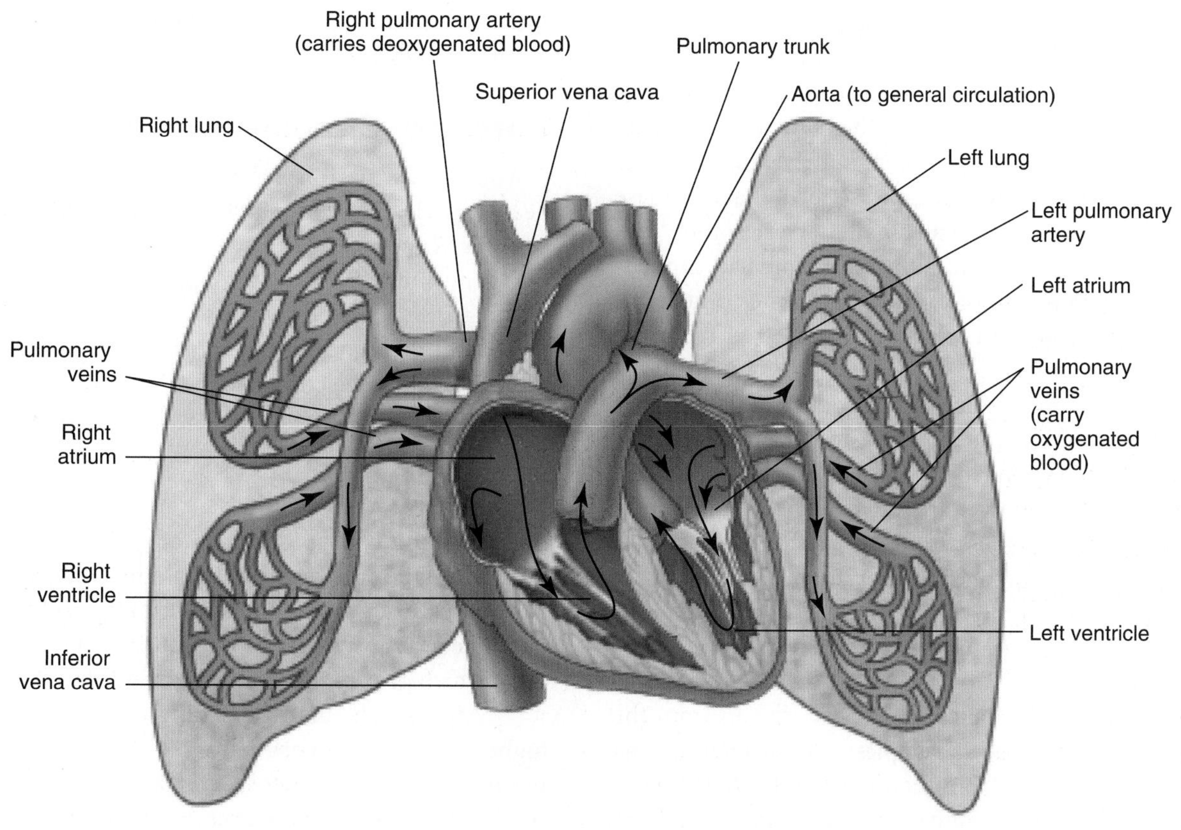

Figure 15-2 *Cardiopulmonary circulation*

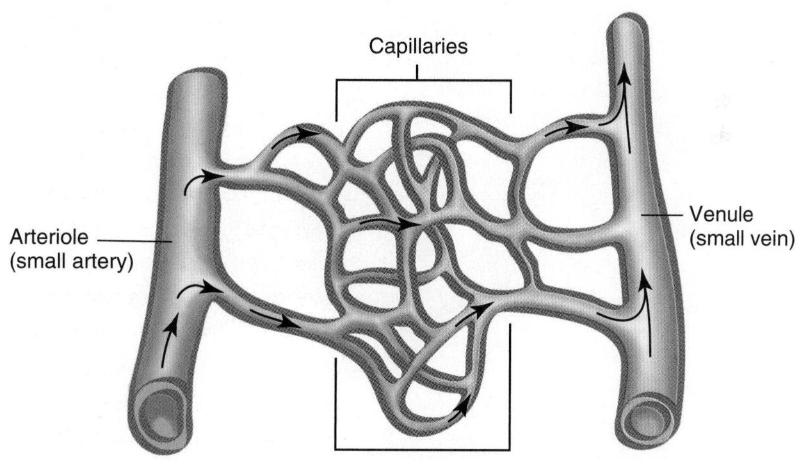

Figure 15-3 *Arterioles deliver oxygenated blood to the capillaries. After the oxygen has been extracted, the oxygen-poor blood is returned to circulation as venous blood.*

and back to the heart; it carries products, such as carbon dioxide and other dissolved wastes, away from the tissues; it helps equalize body temperature; and it aids in protecting the body from harmful bacteria.

The aorta is the largest artery in the body. The first branch of the aorta is the **coronary artery**, which takes blood to the myocardium, or cardiac muscle. As the aorta emerges from the anterior portion of the heart, known as an ascending aorta, it forms an arch. This arch is known as the aortic arch. Three branches come from this arch: the *brachiocephalic artery* (**bray**-kee-oh-seh-**FAL**-ick), the **common carotid artery** (kah-**ROT**-id), and the left *subclavian arteries*. These arteries and their branches carry blood to the arms, neck, and head.

From the aortic arch, the aorta descends along the middorsal wall of the thorax and abdomen. Many arteries branch off from the descending aorta, carrying oxygenated blood throughout the body.

As the descending aorta proceeds posteriorly, it sends off additional branches to the body wall, stomach, intestines, liver, pancreas, spleen, kidneys, reproductive organs, urinary bladder, legs, and so forth. Each of these arteries subdivides into still smaller arteries, then into arterioles, and finally into numerous capillaries embedded in the tissues. This is where hormones, nutrients, oxygen, and other materials are transferred from the blood into the tissue.

In turn, metabolic waste products, such as carbon dioxide and nitrogenous wastes, are picked up by the blood capillaries. Hormones and nutrients from the small intestines and liver are also absorbed by the blood capillaries. Blood goes from the capillaries first into venules, through increasingly larger veins, and finally into one or more of the veins that exit the organ. Eventually the blood empties into one of the two largest veins in the body.

Deoxygenated, or venous, blood returns from the lower parts of the body and empties into the inferior vena cava. Venous blood from the upper body parts (arms, neck, and head) passes into the superior vena cava. Both the inferior and superior vena cava empty their deoxygenated blood into the right atrium.

Coronary Circulation

Coronary circulation brings oxygenated blood to the heart muscle. The coronary artery has a right and left branch. These branches encircle the heart muscle with many tiny branches going to all parts of it. The blood circulates to the capillaries, where the exchange of gases takes place, and then goes to the veins. Deoxygenated blood returns through the coronary veins to the *coronary sinus*. This is a trough in the posterior wall of the right atrium.

Portal Circulation

Portal circulation is a branch of general circulation. Veins from the pancreas, stomach, small intestine (superior mesenteric vein), colon (inferior mesenteric vein), and spleen empty their blood into the hepatic **portal vein**, which goes to the liver (**Figure 15-4**).

After meals, blood reaching the liver contains a higher than normal concentration of glucose. The liver removes the excess glucose, converting it to glycogen. In the event of vigorous exercise, work, or prolonged

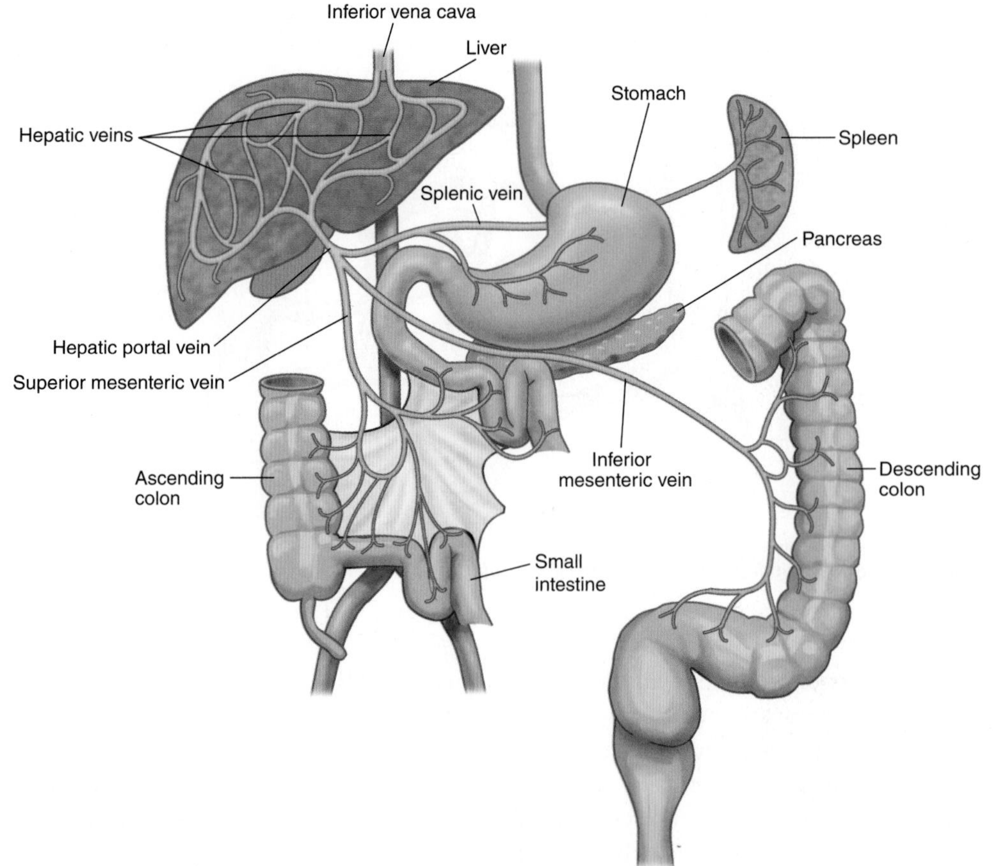

Figure 15-4 *Portal circulation*

periods without nourishment, glycogen reserves will be changed back into glucose for energy. The liver ensures that the blood's glucose concentration is kept within a relatively narrow range.

Deoxygenated venous blood leaves the liver through the **hepatic vein** (heh-**PAT**-ick), which carries it to the inferior vena cava. From the inferior vena cava, blood enters the right atrium (**Figure 15-4**).

Fetal Circulation

Fetal circulation occurs in the fetus. Instead of using its own lungs and digestive system, the fetus obtains oxygen and nutrients from the mother's blood. Fetal and maternal bloods do not mix. The exchange of gases, food, and waste takes place in the structure known as the placenta, located in the pregnant uterus (**Figure 15-5**).

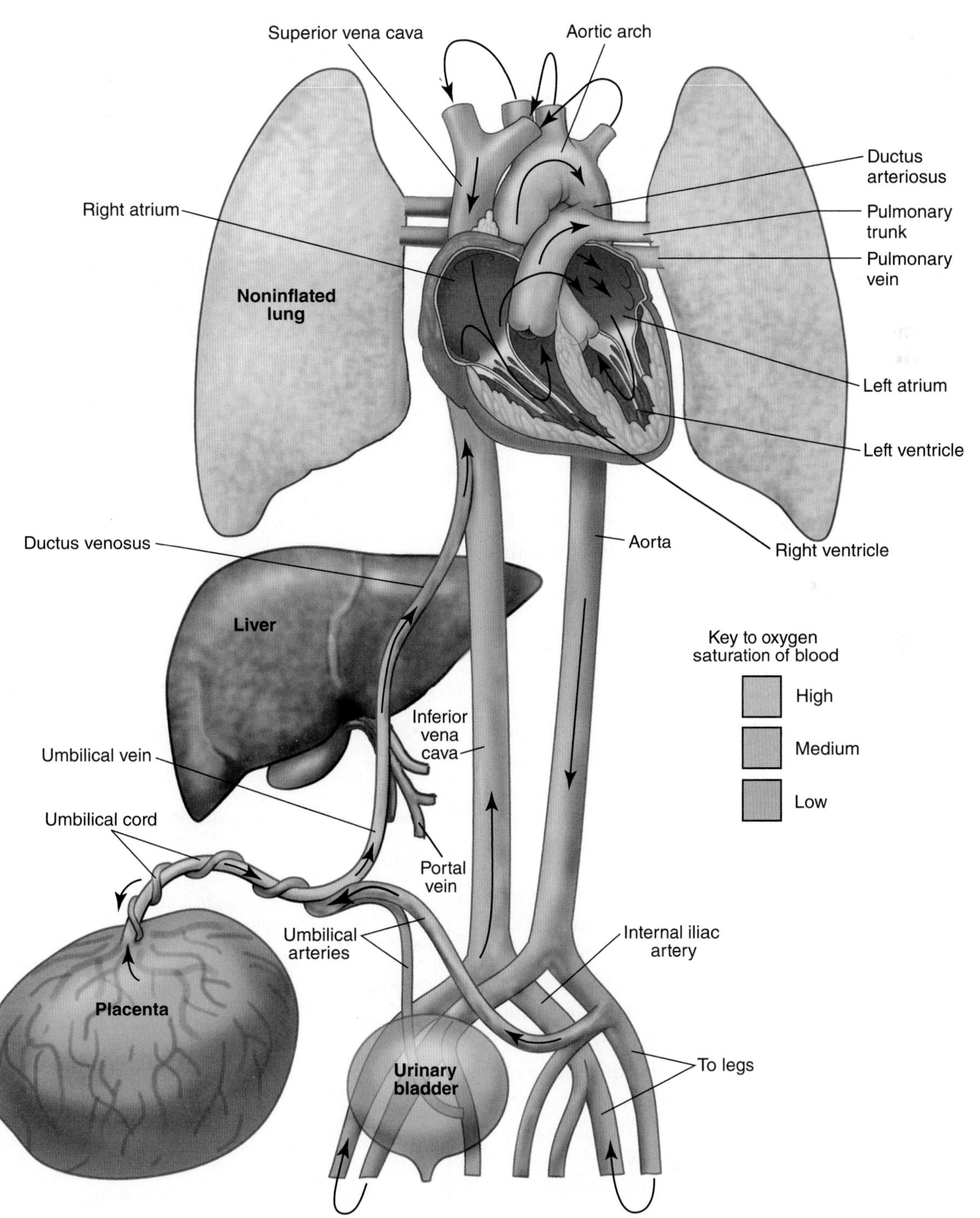

Figure 15-5 *Fetal circulation*

In fetal circulation, oxygenated blood comes through the placenta of the mother to the fetus via the umbilical vein. Most of the blood joins the inferior vena cava by way of a small vessel called the **ductus venosus** (**DUK**-tus vee-**NO**-sus) and goes to the right atrium. The remaining blood goes to the liver. The blood in the right atrium goes through an opening in the atrial septum called the **foramen ovale** (for-**AY**-men oh-**VAL**-ee) and then goes into the left atrium. The blood bypasses the right ventricle and the pulmonary circuit. Some blood goes into the right ventricle and is pumped into the pulmonary artery. The purpose of the blood circulating through the heart is to give the heart and blood vessels oxygen and nutrients to grow. However, most of this blood shunts directly into the systemic circulation through the **ductus arteriosus** (**DUK**-tus ar-tier-ee-**OH**-sis), which connects the main pulmonary artery to the aorta. Blood returns to the placenta through the umbilical arteries. At birth, these adaptations, which include the ductus venosus and the ductus arteriosus, close within 30 minutes, and the foramen ovale completely closes within one year. Normal cardiopulmonary circulation begins at birth. Congenital heart defects may occur if these structures do not properly close. The most common symptom of a congenital heart defect is **cyanosis** (sigh-ah-**NOH**-sis), a discoloration to the skin and mucous membrane caused by a lack of oxygen in the blood. This discoloration varies with skin tone: on light skin, it appears bluish; on dark skin, it may appear more gray or white; and on skin with yellow undertones, it may appear gray-green.

BLOOD VESSELS

The heart pumps the blood to all parts of the body through a remarkable system of three types of blood vessels: arteries, capillaries, and veins.

Arteries

Arteries carry oxygenated blood away from the heart to the capillaries; there is one exception: the pulmonary arteries carry deoxygenated blood from the heart to the lungs.

As seen in **Figure 15-6**, the arterial walls consist of three layers. The outer layer is called the **tunica adventitia (externa)** (**TYOO**-nih-kah ad-ven-**TISH**-ee-ah). This layer consists of fibrous connective tissue with bundles of smooth muscle cells that lend great elasticity to the arteries. This elasticity allows the arteries to withstand sudden large increases in internal pressure created by the large volume of blood forced into them at each heart contraction.

The **tunica media** is the middle arterial layer. It consists of muscle cells arranged in a circular pattern. This layer controls the artery's diameter by dilation and constriction, which regulates the flow of blood through the artery. This keeps the blood flow steady and even and reduces the heart's work.

An inner layer, the **tunica intima** (**IN**-tih-mah), consists of three smaller layers of endothelium that give the artery a smooth lining to allow for the free flow of blood.

The arteries transport blood under very high pressure; they are elastic, muscular, and thick walled. The thickness of the arteries makes them the strongest of the three types of blood vessels.

The aorta leads away from the heart and branches into smaller arteries. These smaller arteries, in turn, branch into arterioles, which still have some smooth muscle in the walls. Arterioles give rise to the capillaries.

Table 15-1 lists and **Figure 15-7** illustrates the principal arteries and the areas they serve.

Table 15-1 ***Principal Arteries***

PRINCIPAL ARTERY	*AREA(S) SERVED*
Common carotid	Head and face
Internal carotid	Brain
External carotid	Face (*pulse point*)
Vertebral	Spinal column and brain
Brachiocephalic	Right arm, head, and shoulder
Subclavian	Shoulder
Axillary	Axilla
Brachial	Upper arm and elbow area (*pulse point*)
Radial	Arm, wrist (*pulse point*)
Thoracic aorta	Chest cavity
Splenic	Spleen
Hepatic	Liver
Superior mesenteric	Small intestines and colon
Renal	Kidney
Common iliac	Lower abdomen
Internal iliac	Pelvis and bladder
External iliac	Groin and lower leg
Femoral	Groin (*pulse point*)
Popliteal	Knee area (*pulse point*)
Anterior tibial	Anterior lower leg
Posterior tibial	Posterior lower leg
Dorsalis pedis	Ankle (*pulse point*)

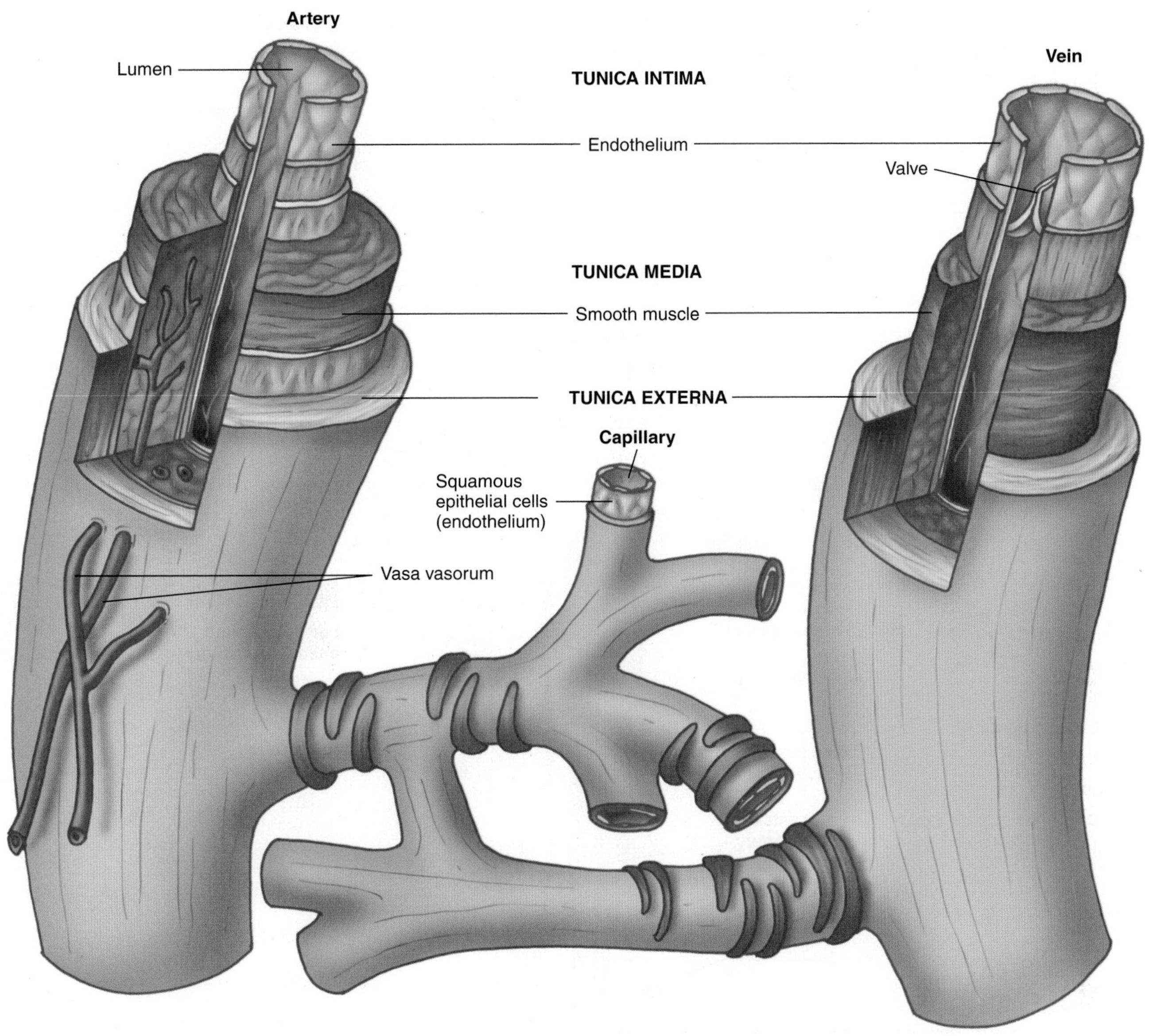

Figure 15-6 *The three layers of the walls of the arteries and veins: tunica intima, tunica media, and tunica adventitia*

Capillaries

Capillaries (**KAP**-ih-lay-reez) are the smallest blood vessels and can only be seen through a compound microscope. Capillaries connect the arterioles with venules. Capillaries are branches of the finest arteriole divisions, known as metarterioles. The metarterioles lose most of their connective tissue and muscle layers until they disappear. There remains only a simple endothelial cell layer; this endothelial cell layer constitutes the capillaries. Tissue such as skeletal muscle, liver tissue, and kidney tissue have extensive capillary networks because they are metabolically active. The epidermis of the skin and the lens and cornea of the eye have no capillary networks.

The capillary walls are extremely thin to allow for the selective permeability of various cells and substances. Nutrient molecules and oxygen pass out of the capillaries and into the surrounding cells and tissues by diffusion. Metabolic waste products, such as carbon dioxide and nitrogenous wastes, pass back from the cells and tissues into the bloodstream for excretion at their proper sites (lungs and kidneys).

Tiny openings in the capillary walls allow white blood cells to leave the bloodstream and enter the tissue spaces to help destroy invading bacteria. In the capillaries, some of the plasma diffuses out of the bloodstream and into the tissue spaces. This fluid is called interstitial fluid and is returned to the bloodstream in the form of lymph via the lymphatic vessels.

Blood flow through the capillaries is influenced by hydrostatic pressure. Hydrostatic pressure is the force exerted by a fluid pressing against a wall. In

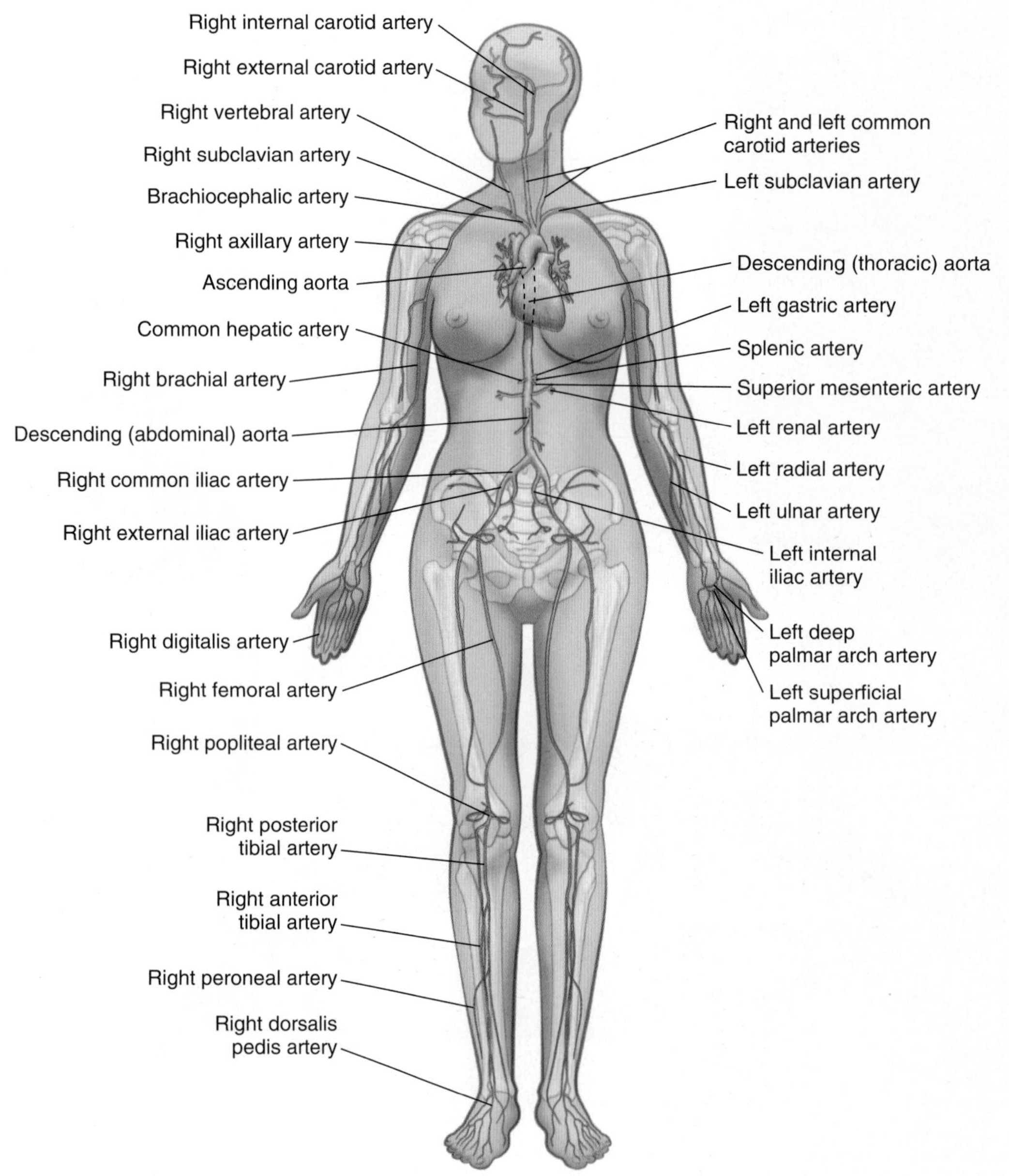

Figure 15-7 *The principal arteries of the body*

capillaries, hydrostatic pressure is the same as capillary blood pressure—the pressure exerted by blood on the capillary wall. This pressure tends to force fluid through the capillary walls, leaving behind cells and most proteins.

Capillaries are ultimately responsible for transporting blood to all tissues. Not all capillaries are open simultaneously. This system allows for regulation of blood flow to active tissues. For example, if a person is running, the skeletal muscles need more oxygen, while the digestive organs need less oxygen. This may explain why, a person runs after a heavy meal, they get indigestion or have abdominal cramps.

Veins

The **veins** carry deoxygenated blood away from the capillaries to the heart. The smallest vein, the venule, is hardly larger than a capillary, but it contains a muscular layer that is not present within capillaries. **Table 15-2** lists and **Figure 15-8** illustrates the principal veins and the areas they serve.

The veins are composed of three layers: the tunica adventitia, also known as the tunica externa; tunica media; and tunica intima. Veins are considerably less elastic and muscular than arteries. The walls of the veins are much thinner than those of the arteries because they

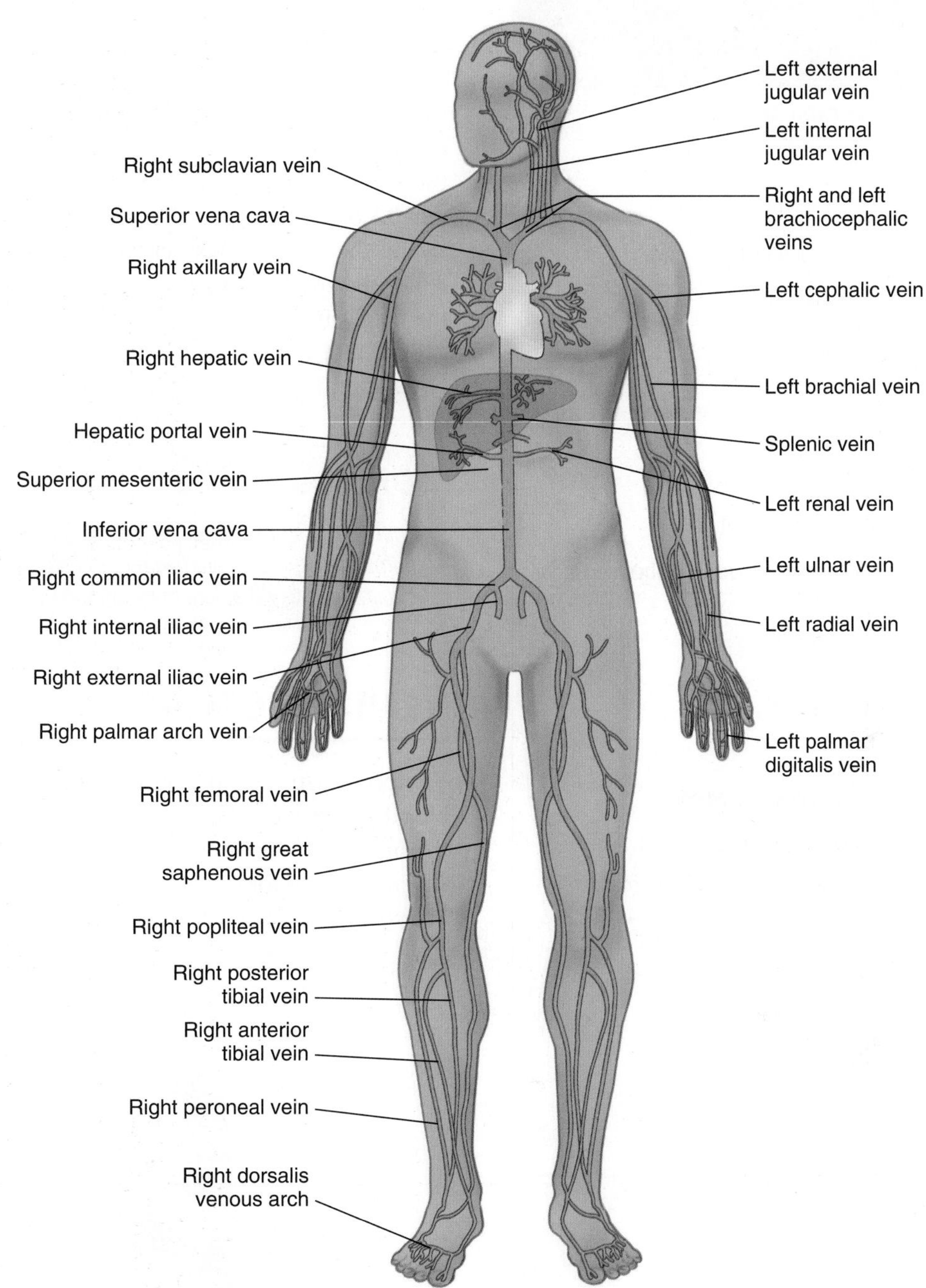

Figure 15-8 *The principal veins of the body*

do not have to withstand such high internal pressures. The pressure from the heart's contraction is greatly diminished by the time the blood reaches the veins for its return journey. Thus, the thinner-walled veins can collapse easily when not filled with blood. Finally, veins have **valves** along their length. These valves allow blood to flow in only one direction: toward the heart. This prevents reflux, or backflow, of blood toward the capillaries (**Figure 15-9**). Valves are found in abundance in veins where there is a greater chance of reflux. There are many valves in the lower extremities where blood has to oppose the force of gravity.

Eventually, all of the venules converge to make up larger veins, which ultimately form the body's largest veins, the vena cavae. Venous blood from the upper part of the body returns to the right atrium via the superior vena cava; blood from the lower body parts is conducted to the heart via the inferior vena cava.

Table 15-2 *Principal Veins*

PRINCIPAL VEIN	AREA(S) SERVED
External jugular	Face
Internal jugular	Head and neck
Subclavian	Shoulder and upper limbs
Brachiocephalic	Head and shoulder
Cephalic	Shoulder and axilla
Axillary	Axilla
Brachial	Upper arm
Radial	Lower arm and wrist
Superior vena cava	Upper part of body
Inferior vena cava	Lower part of body and abdomen area
Hepatic	Liver
Renal	Kidney
Hepatic portal	Organs of digestion
Splenic	Spleen
Superior mesenteric	Small intestine and colon
Common iliac	Lower abdomen and pelvis
Internal iliac	Bladder and reproductive organs
External iliac	Lower limbs
Great saphenous	Upper leg
Femoral	Upper leg and groin
Popliteal	Knee
Posterior tibial	Posterior leg
Dorsal venous arch	Foot

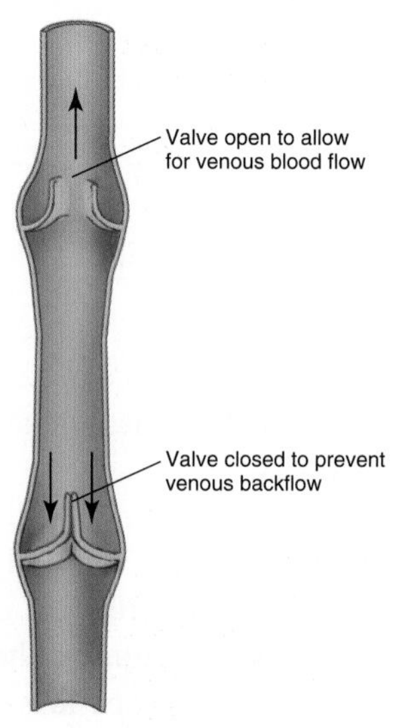

Figure 15-9 *Valves in the veins*

VENOUS RETURN

In addition to valves, the skeletal muscles contract to help push the blood along its path. In the abdominal and thoracic cavity, pressure changes occur when a person breathes; this also helps bring the venous blood back to the heart. Think about an individual sitting for a long period of time, especially on a car ride. Think how sleepy they start to get. The reason may be that blood is not getting back to the heart for oxygen. To reduce the drowsiness, they should pull over, stop the car, and get out and walk around for a while. This will improve circulation and the drowsiness should pass.

Did You Know?

There are about 62,000 miles of blood vessels in the human body. If they are laid end to end, they would encircle the world at least two and one-half times.

VENIPUNCTURE

Venipuncture, or phlebotomy, is a method of drawing blood using a needle to access a vein for intravenous therapy or to get a sampling of venous blood for testing. The major sites for venipuncture are the cephalic, basilic, and the median cubital veins (**Figure 15-10**).

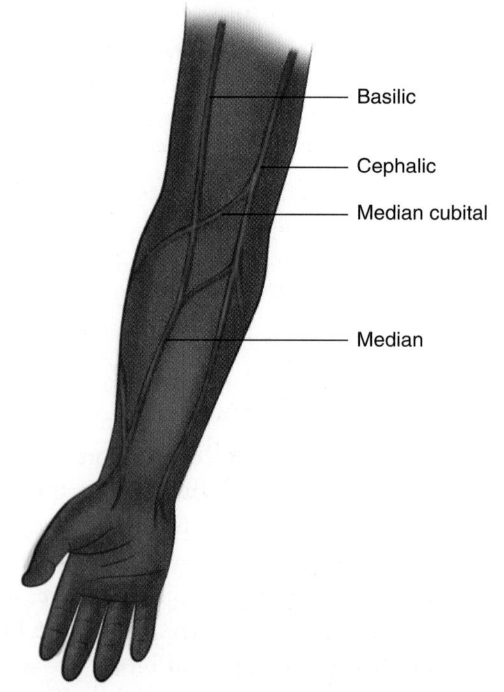

Figure 15-10 *The major sites for venipuncture*

BLOOD PRESSURE

When the heart pumps blood into the arteries, the surge of blood filling the vessels creates pressure against their walls. The pressure measured at the moment of contraction is the **systolic blood pressure** (sis-**TOL**-ick). The lessened force of the blood measured when the ventricles relax is called the **diastolic blood pressure** (dye-ah-**STOL**-ick). This force is measured using a sphygmomanometer (**sfig**-moh-mah-**NOM**-eh-ter), a manual blood pressure device or a digital device. When using a manual sphygmomanometer, a stethoscope is used to listen to the sounds. The systolic pressure is the first sound heard, and the diastolic is the last sound heard (**Figure 15-11**).

The average systolic pressure measured in the upper arm is 120 mm Hg. The average diastolic pressure in an adult is 80 mm Hg. The blood pressure is recorded as a ratio with the systolic over the diastolic, such as 120/80. Factors that influence blood pressure include the following:

- Volume of the blood—changes in the volume, such as a loss of blood, means there will be less blood for the heart to pump.
- Cardiac output—any factor that increases the heart rate, stroke volume, or both, will increase the blood pressure. Factors include adrenaline, thyroid hormones, and an increase in calcium ions.
- Blood viscosity—the thicker the blood, the harder the heart has to pump.
- Total peripheral resistance—if the area through which the blood has to pump is reduced, such as in atherosclerosis, the heart has to pump harder.
- Elasticity of vessel walls—the capacity of the blood vessel to return to its normal shape after stretching and compressing. If artery walls are rigid and unable to expand and recoil, their resistance to blood flow will greatly increase, and blood pressure will rise to even higher levels.
- Stressors—these cause the muscles around the blood vessels to constrict, making the heart pump harder, thereby increasing blood pressure.

Pulse pressure is the difference between the systolic and diastolic. For example, if the blood pressure is 120/80, the pulse pressure is 40. A pulse pressure of more than 60 mm Hg, especially in older adults, indicates a higher risk of cardiac problems.

The Effects of Aging on Circulation and Blood Vessels

The arteries that are pliable and elastic when young become less elastic, dilated, and elongated with age. These physiological changes mean the heart has to work harder to push blood through the arteries. Arterial changes appear to be widespread and result in diminished circulation to organs and tissues.

A frequent cardiovascular measure is blood pressure (BP). It is debatable how aging affects this measure of cardiovascular status. Some researchers believe normal BP for older people is typically 140 mm Hg systolic and 90 mm Hg diastolic (140/90).

Some researchers think systolic increases are due to reduced aortic elasticity; others believe that peripheral resistance in the vessels causes an increase in systolic and diastolic pressures.

In the carotid arteries, the baroreceptors—the neural receptors sensitive to blood pressure—become rigid and less sensitive to pressure changes with aging. This results in a slow response to postural changes, which may cause dizziness and fainting. This hypotensive response is called orthostatic hypotension. Under normal circumstances, the heart continues to adequately supply blood to all parts of the body. However, an aging heart may be less able to increase workloads from effects such as illness, stress, infections, or extreme physical exertion.

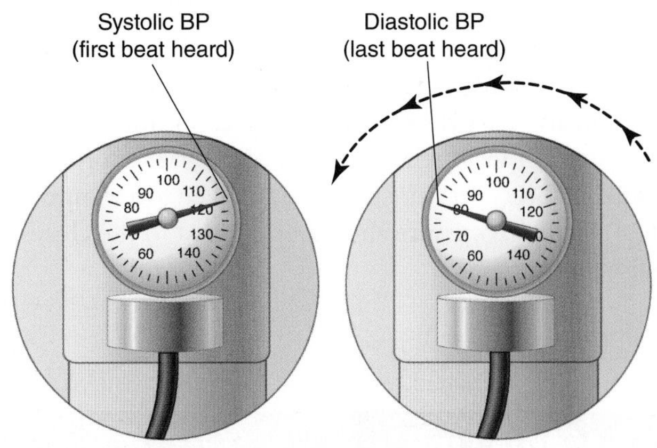

Figure 15-11 *Using a stethoscope to measure blood pressure (BP), the systolic blood pressure is the first sound heard, and the diastolic blood pressure is the last sound heard.*

Pulse

If a person touches certain areas that are pulse points of the body, such as the radial artery at the wrist, they will feel alternating, beating throbs. These throbs represent the body's pulse. A **pulse** is the alternating expansion and contraction of an artery as blood flows through it. The pulse rate is usually the same as the heart rate. The pulse provides information about heart rate as well as strength and rhythm of the heart.

Try this simple demonstration: Place the fingertips—excluding the thumb, which has its own pulse point—over an artery that is near the surface of the skin and over a bone. The seven paired locations where the pulse can be conveniently felt are as follows (**Figure 15-12**):

1. **Temporal artery**—slightly above the outer edge of the eye
2. **Carotid artery** (kah-**ROT**-id)—in the neck, along the front margin of the sternocleidomastoid muscle, near the lower edge of the thyroid cartilage
3. **Brachial artery** (**BRAY**-kee-al)—at the crook of the elbow, along the inner border of the biceps muscle
4. **Radial artery** (**RAY**-dee-al)—at the wrist, on the same side as the thumb; most common site for taking pulse
5. **Femoral artery** (**FEM**-or-al)—in the inguinal or groin area
6. **Popliteal artery** (pop-lit-**EE**-al)—behind the knee; may be hard to palpate
7. **Dorsalis pedis artery** (dor-**SAY**-lis **PED**-is)—on the anterior surface of the foot, below the ankle joint

A pressure point is where the main artery lies near the skin surface over a bone. The seven locations where a person can feel their pulse may also serve as pressure points. If direct pressure cannot be applied to a wound to stop bleeding, pressure should be applied to the closest pulse point.

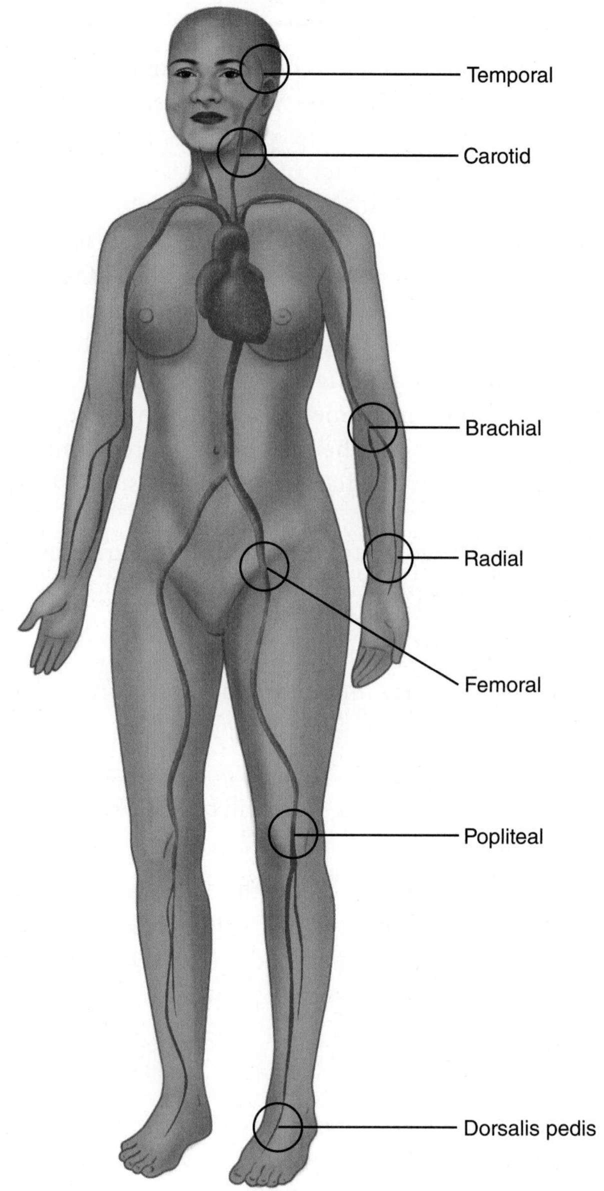

Figure 15-12 *Pulse points of the body*

Media Link

View the **Blood Pressure and Pulse** video on the Online Resources.

DISORDERS OF CIRCULATION AND BLOOD VESSELS

Disorders of circulation or blood vessels can be present at birth, or they can develop at a different stage of life.

Congenital Circulation and Blood Vessel Disorders

There are several congenital disorders that may occur. A child may be born with an **atrial septal defect** opening between both atria. This can be corrected with open heart surgery using a patch, a device to plug the opening, or sewing it closed. An **atrioventricular septal defect,** an opening between the atria and ventricles and the valves between the atria and ventricles, may also occur. It is treated by surgical closing using patches; the valves are also replaced. A **ventricular septal defect** is the most common type of congenital heart disease and consists of an opening between both ventricles. Surgical closing using patches is done to treat it.

Tricuspid atresia (try-cusp-id uh-**TREE**-zuh) is the third most common form of congenital heart defect. The tricuspid valve between the right atria and right ventricle does not form. Temporary surgery is done to increase the blood flow from the heart to the lungs. The surgery includes atrial septostomy, which enlarges the heart's atria to allow more blood flow between the right and left atrium. A cardiac shunting is done to create a bypass from a main blood artery vessel leading out of the heart to a blood vessel leading to the lungs. When the child is between 2–3 years of age, a more permanent treatment is done. The surgeon creates a path for the blood returning to the heart through the inferior vena cava to flow directly to the pulmonary artery, which takes blood to the lungs. Follow-up care is necessary throughout life.

Vascular malformations are a type of birthmark or growth often present at birth composed of blood vessels that can cause functional or cosmetic problems. Infantile hemangioma is a strawberry birthmark usually on the face. Arteriovenous malformation (AVM) is a congenital malformation of blood vessels that includes arteries, veins, capillaries, and lymph vessels. Treatment of vascular malformation depends on the type and location. It may include surgery or focused radiation therapy.

General Circulation and Blood Vessel Disorders

An **aneurysm** (AN-you-rizm) is the ballooning out of an artery accompanied by a thinning arterial wall, caused by a weakening of the blood vessel—almost like having a bubble on a tire. Aneurysms often occur in the aorta, brain, back of the knee, intestines, or spleen. The aneurysm pulsates with each systolic beat. The symptoms are pain and pressure, but sometimes there are no symptoms. Treatment varies from watchful waiting to emergency surgery. For treatment of a brain aneurysm, physicians may use interventional radiology (IR). MRI and CT scans take three-dimensional color pictures, which reveal the anatomy of the brain in minute detail. Physicians then use IR to reach the aneurysm. They insert a wire catheter into the groin, guide it to the brain aneurysm, and then release tiny coils that provide scaffolding to reinforce the artery and prevent the aneurysm from bursting.

Arteriosclerosis (ar-**tee**-ree-oh-skleh-**ROH**-sis) occurs when the arterial walls thicken because of a loss of elasticity as aging occurs. Sometimes it is referred to as hardening of the arteries. **Atherosclerosis** (**ath**-er-oh-skleh-**ROH**-sis) occurs when deposits of fat form along the walls of the arteries (see Chapter 14).

In both arteriosclerosis and atherosclerosis, there is a narrowing of the blood vessel opening. This interferes with the blood supply to the body parts and causes hypertension. Symptoms develop where circulation is impaired and may include numbness and tingling of the lower extremities or a loss of memory (**Figure 15-13**). Treatment involves medications to lower the blood pressure and lifestyle changes, such as no smoking, limited alcohol intake, and regular exercise. In some cases, an endarterectomy may be done to remove plaque. Thrombolytic therapy using intravenous clot-dissolving medications may also be used.

Gangrene (**GANG**-green) is death of body tissue due to an insufficient blood supply caused by disease or physical trauma. Symptoms depend on the location and cause of gangrene. Treatment requires that the dead tissue be removed—in some cases, this may be an amputation—to allow healing and to prevent further infection.

Phlebitis (fleh-**BYE**-tis), or thrombophlebitis, is an inflammation of the lining of a vein accompanied by clotting of blood in the vein. Symptoms include edema of the affected area, pain, and redness along the length of the vein. Treatment may be with warm compresses to the affected area, compression stockings, and anti-inflammatory medication (**Figure 15-14**).

Hemorrhoids (**HEM**-oh-royds) are varicose veins in the walls of the lower rectum and the tissues around the anus. Conservative treatment for hemorrhoids includes sitz baths, which are warm baths for the buttocks, and over-the-counter topical ointments. In more severe cases, rubber band ligation or hemorrhoidectomy may be done.

Cerebral hemorrhage (**SER**-eh-bral **HEM**-eh-rij) refers to bleeding from blood vessels within the brain. It can be caused by arteriosclerosis; disease; or physical trauma, such as a blow to the head. A cerebral hemorrhage may lead to a stroke.

Peripheral vascular disease (PVD) (per-**IF**-er-al) is caused by blockage of the arteries, usually in the legs. Symptoms are pain or cramping in the legs or buttocks while walking. Such cramping subsides when the person stands still. This is called *intermittent*

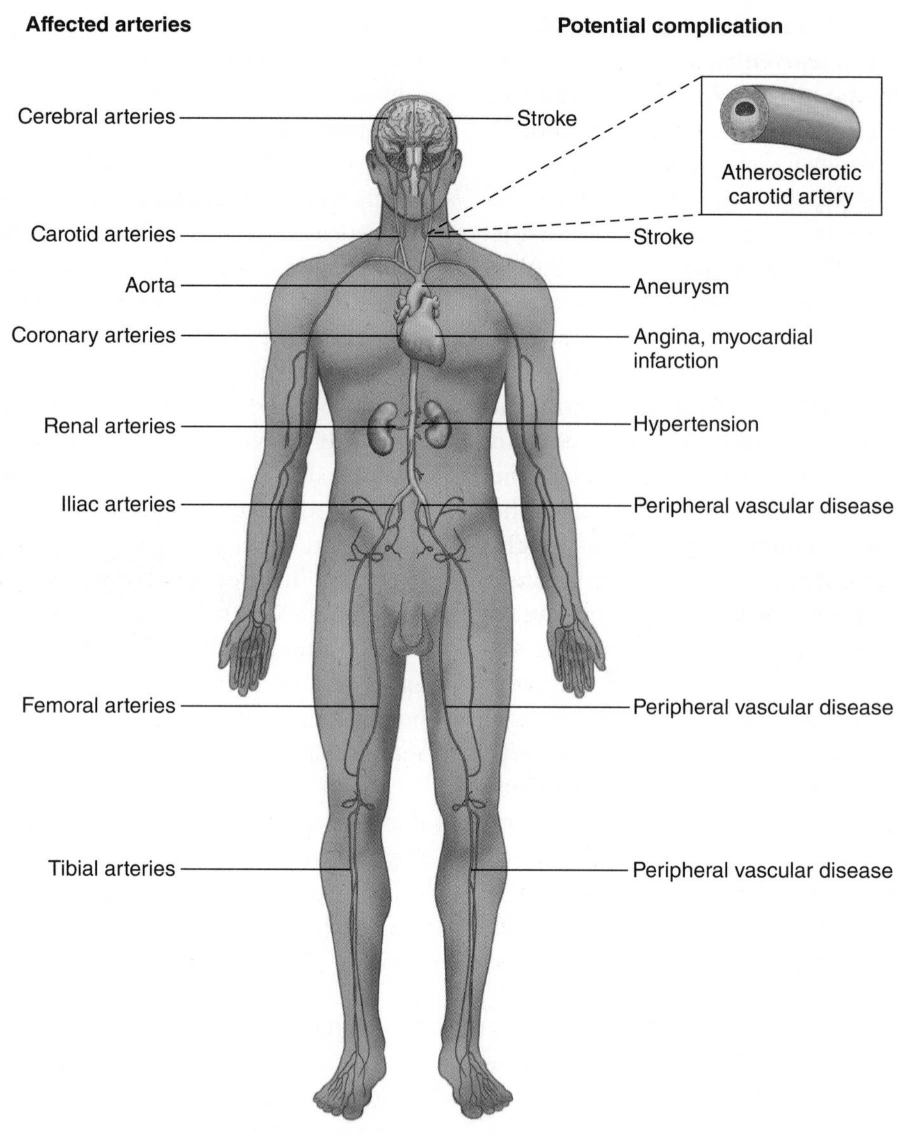

Figure 15-13 *Arteries affected by atherosclerosis (left column) and the potential complications of this condition (right column)*

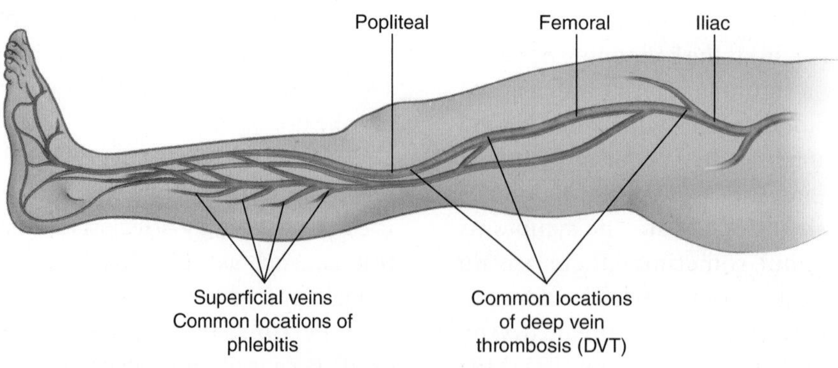

Figure 15-14 *Common locations for the development of phlebitis and deep vein thrombosis*

claudication (klaw-dih-**KAY**-shun). As the condition worsens, symptoms may include pain in the toes or feet while at rest, numbness, paleness, and cyanosis in the foot or leg. The condition must be treated or amputation may be necessary. Treatments include medication to reduce cholesterol, improved or modified diet, and other treatments to improve circulation. A percutaneous balloon angioplasty, which is a balloon angioplasty performed through the groin using a needle and guide wire, may be done to enlarge the blocked or damaged artery.

Varicose veins (**VAR**-ih-kohs **VAYNS**) are swollen veins that result from a slowing of blood flow back to the heart (**Figure 15-15**). The weight of the stagnant blood distends the valves; the continued pooling of blood then causes distention and inelasticity of the vein walls. The veins most commonly affected are in the legs and feet. This condition may develop due to hereditary weakness or as a result of prolonged periods of standing. Age and pregnancy are other factors responsible for varicose veins. Treatment includes avoiding excess standing, making sure to exercise, elevating the legs when sleeping, and wearing support hose. People need to avoid high heels and tight clothing, especially around the waist. A procedure known as *sclerotherapy* (**skler**-oh-**THAIR**-ah-pee) may be done, in which a sclerosing solution is injected into the vein. The solution causes the vein to scar and close. Other options include laser therapy or vein stripping.

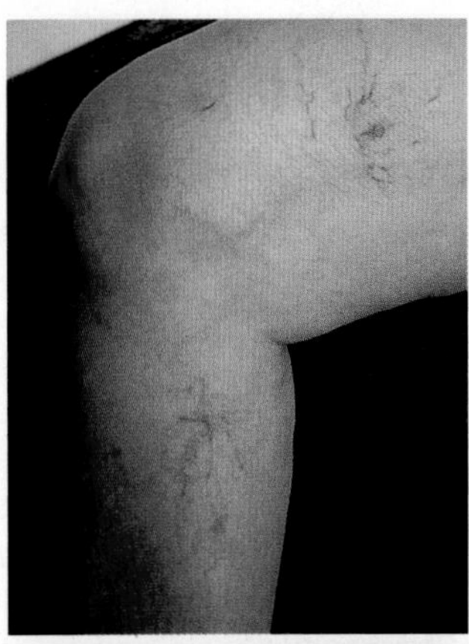

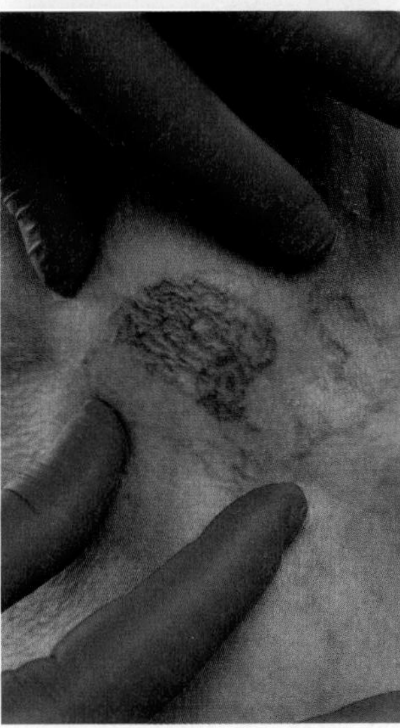

Figure 15-15 *Varicose veins*

Career Profile

REGISTERED NURSE AND NURSE PRACTITIONER

Registered nurses (RNs) provide for the physical, mental, and emotional needs of their patients. They observe, assess, and record symptoms, reactions, and progress; they also assist physicians during treatments and examinations, administer medications, and assist in convalescence and rehabilitation. Registered nurses develop nursing care plans, instruct patients and their families in proper care, and help individuals and groups improve and maintain their health.

Registered nurses work in hospitals, the home, offices, nursing homes, public health services, and industries.

In all states, students must graduate from an accredited school of nursing and pass a national licensing examination to become an RN. There are three major educational paths to nursing: associate's degree in nursing (ADN) programs take two years, Bachelor of Science in nursing (BSN) degree programs take four years, and diploma programs given in hospitals last 2–3 years.

The employment outlook is expected to be above average in the coming years. The job outlook is best for the nurse with a BSN.

A nurse practitioner or nurse clinician is an RN with a master's degree and clinical experience in a particular branch of nursing. The nurse practitioner has acquired expert knowledge in a specific medical specialty. Nurse practitioners are employed by physicians in private practice or clinics, or they sometimes practice independently, especially in rural areas.

Career Profile

LICENSED PRACTICAL NURSE

Licensed practical nurses (LPNs), or licensed vocational nurses (LVNs) as they are called in Texas and California, care for people who are sick, injured, convalescing, or handicapped under the direction of a physician or registered nurse.

Most LPNs provide basic bedside care. They take vital signs, treat bedsores, prepare and give injections, and administer some treatments. They collect laboratory specimens, observe patients, and report any adverse reactions. They help patients with activities of daily living, keep them comfortable, and care for their emotional needs. In states where the law allows, they may administer prescribed medicines.

Licensed practical nurses in nursing homes also evaluate residents' needs, develop care plans, and supervise nursing aides.

All states require LPNs to graduate from an accredited practical nursing program and pass a national licensing examination.

The job outlook for the practical nurse is good and is expected to increase faster than the average during the next few years.

Hypertension, or high blood pressure, is frequently called the "silent killer" because there are usually no symptoms of the disease. Hypertension is classified as either essential hypertension—also known as primary hypertension—or secondary hypertension. About 90% to 95% of cases are essential, which means high blood pressure with no obvious cause. The remaining 5% are secondary and are caused by conditions that affect the kidneys, arteries, heart, or endocrine system. Hypertension leads to strokes, heart attacks, aneurysm, memory problems, and kidney failure. Most people discover that they have the condition during a routine physical. There are several categories of hypertension (American Heart Association, 2019), including:

- Normal: systolic less than 120 mm Hg, diastolic less than 80 mm Hg
- Stage 1 hypertension: systolic greater than 130–139 mm Hg, diastolic greater than 89 mm Hg
- Stage 2 hypertension: systolic greater than 139 mm Hg, diastolic greater than 89 mm Hg
- Hypertensive crisis: systolic over 180 mm Hg, diastolic over 120 mm Hg

Nearly 45% of Americans have hypertension, and only about one out of four have their condition under control (Blood Pressure Facts, 2020). Risk factors for hypertension are stress, smoking, obesity, diets high in fat and/or sodium, and a family history of the disease. Having prehypertension or diabetes are also risk factors. Treatment consists of relaxation techniques, reducing fat and sodium in the diet, exercise, weight loss, and medication to control blood pressure. In the treatment of hypertension, patients often do not understand the disease and its risks. They frequently stop taking their medication because of costs and side effects. Health care providers must realize that better education and communication will lead to more effective treatment and a higher level of compliance by patients.

White-coat hypertension is so called because it is an increase in a patient's blood pressure that occurs only when a medical professional—the "white coat"—takes the blood pressure. The thought is that the stress of a medical examination causes the blood pressure to rise, resulting in an inaccurate diagnosis of hypertension. Blood pressure medication does not help the problem. The best way to differentiate between white-coat hypertension and true hypertension is to ask the patient to wear a device that measures the blood pressure over a 24-hour period.

Hypotension is low blood pressure. The reading is usually less than 90/60. Chronic low blood pressure is almost never serious. Health problems may occur if blood pressure drops suddenly and the brain is deprived of an adequate blood supply, leading to dizziness. It most commonly occurs when rising from a prone or sitting position to a standing position. This is known as postural hypotension or orthostatic hypotension.

Increasing water intake will increase blood volume and help raise blood pressure.

Transient ischemic attacks (TIAs) (iss-**KEE**-mick) are temporary interruptions of the blood flow, known as ischemia, to the brain. The cause is usually a narrowing of the carotid artery due to an accumulation of fat. Patients may experience stroke-like symptoms, such as dizziness, weakness, or temporary paralysis that lasts only a few minutes. Most symptoms of a TIA disappear within an hour, although they may persist for up to 24 hours. About one-third of people who have had a TIA will have an acute stroke sometime in the future. Many strokes can be prevented by heeding the warning signs given by a TIA and treating the underlying risk factors.

Cerebral vascular accident (CVA) (**ser**-eh-bro-**VAS**-kyou-lar) or **stroke** is the sudden interruption of the blood supply to the brain. This results in a loss of oxygen to brain cells, causing impairment of the brain tissue or death (**Figure 15-16**). Stroke is the fifth leading cause of death in the United States but the first leading cause of disability. Fewer Americans are dying from stroke today because of education, better use of medical therapies, and lifestyle changes.

Risk factors include smoking, hypertension, diabetes, heart disease, and family history. About 90% of strokes are caused by fat deposits accumulating in the carotid arteries, or blood clots becoming lodged in the carotid arteries, choking the blood supply to the brain. The remaining 10% of strokes, called hemorrhagic

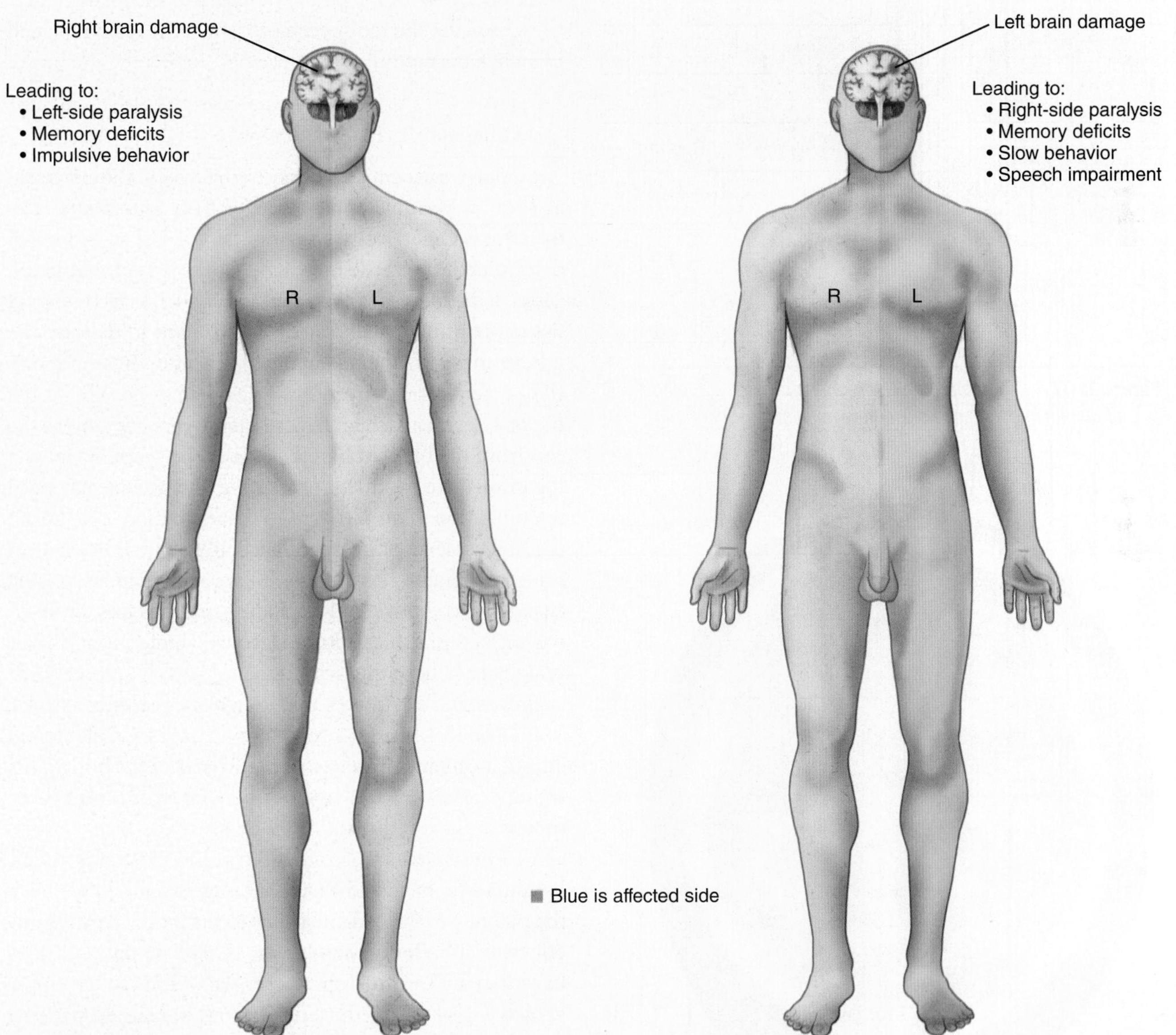

Figure 15-16 *The location of the damage caused by a cerebral vascular accident depends on which side of the brain is affected.*

strokes, are caused when blood vessels within the brain rupture. See **Figure 15-17** and **Figure 15-18**.

Symptoms depend on the side of the brain that has had its blood supply interrupted. Loss of blood supply to the right cerebrum can affect spatial and perceptual abilities and cause weakness or **hemiplegia** (**hem**-ee-**PLEE**-jee-ah), which is paralysis on the left side of the body. Loss of blood supply to the left cerebrum will result in **aphasia** (ah-**FAY**-zee-ah), a loss of speech and memory, as well as right-sided hemiplegia. No two stroke patients will experience the same injuries or disabilities.

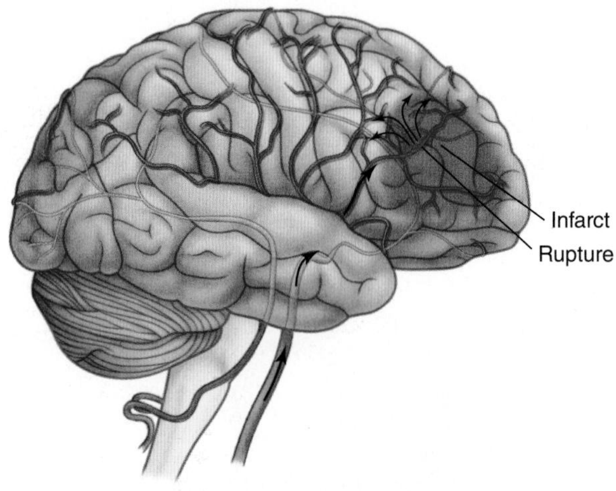

Figure 15-17 *In a hemorrhagic stroke, the rupture of a blood vessel results in decreased blood flow to an area of the brain tissue.*

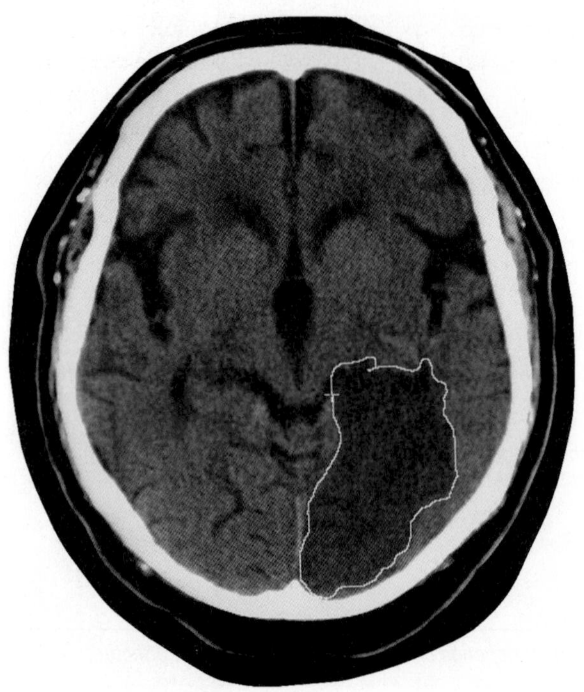

Figure 15-18 *On this magnetic resonance image of a brain, the area of a bleed is visible in the lower right.*

Symptoms common to many stroke patients include vision problems; sudden severe headache; trouble walking or staying balanced; communication difficulties; **dysphasia** (dis-**FAY**-zee-ah), which is the inability to say what one wishes to say; emotional lability, or uncontrolled, unexplained displays of crying, anger, or laughter; depression; coma; and possible death.

An acronym to help assess whether someone is having a stroke is F-A-S-T:

1. Face—Ask the person to smile and see if one side of the face droops down.
2. Arms—Ask the person to raise both arms; watch to see if one arm drifts down.
3. Speech—Ask the person to repeat a simple sentence; check for slurred speech or if the sentence is repeated back correctly.
4. Time—If any symptoms are present, call for emergency help immediately.

For treatment to be most effective, it should begin as soon as possible and no longer than four hours after the stroke. On arrival at the hospital, a CT scan is done to determine if the cause is a blood clot or a ruptured blood vessel. If the cause is a blood clot, a drug such as recombinant tissue plasminogen activator (tPA) is used to dissolve the clot, restoring the blood supply to the brain. In addition to tPA, a stent retriever procedure may be done. This is the use of a device attached to a catheter to directly remove the clot from the blocked blood vessel in the brain. Following the immediate treatment of a stroke, the patient may need rehabilitation. The purpose of rehabilitation is to reach the highest level of independence. Rehabilitation involves physical therapy; relearning self-care skills; and addressing changes in cognitive skills including memory loss, problem solving, communication, and social interaction. Today, physicians believe that with ongoing rehabilitation efforts, improvements can be seen up through one year after a stroke.

Rehabilitation programs today use electrical stimulation machines to stimulate parts of the body. This therapy offers a wide array of benefits from improving motor skills to reducing numbness.

Physicians are exploring ways to prevent strokes. Patients who have had TIAs are being examined to check the patency of the carotid artery to see if they would benefit from a balloon angioplasty. In 39% of patients who have had a TIA, one aspirin per day seems to have prevented a stroke. To reduce risk factors, encourage patients to stop smoking, get exercise, and control hypertension. A stroke occurs suddenly, and a patient who wakes up paralyzed and unable to speak will be very frightened. A health care provider must be very supportive of patients who have had strokes.

Hypoperfusion/Shock

Hypoperfusion means inadequate flow of blood carrying oxygen to the organs and body systems. This can be caused by excessive blood or fluid loss. Hypoperfused tissue is no longer being given enough oxygen and will stop working optimally. The most sensitive organ to a decrease in blood supply and oxygenation is the brain. After just four minutes of decreased blood flow to the brain, brain cells will be irreversibly damaged. Another cause of hypoperfusion is due to a change in the size of the arteries and veins. Blood vessels may become too dilated, and there is not enough pressure to move blood through the blood vessels. The main cause of hypoperfusion is inadequate pumping of the heart. Hypoperfusion leads to **shock**. The body will attempt to compensate for hypoperfusion by increasing the respiratory rate, increasing the heart rate, or sacrificing blood supply to organs to protect blood flow to the brain. Treatment includes blood transfusion, medications, and in cases of septic shock, antibiotics.

One BODY — How the Cardiovascular System Interacts with Other Body Systems

The cardiovascular system plays a role in the maintenance of all body systems by carrying oxygen, nutrients, and hormones to all cells and carrying away cellular waste products and carbon dioxide for excretion by the body.

INTEGUMENTARY SYSTEM

- The capillary network in the skin helps maintain body temperature.

SKELETAL SYSTEM

- Red bone marrow produces blood cells.
- The bones of the thoracic cavity protect the heart and major blood vessels.

MUSCULAR SYSTEM

- The action of the muscles helps return venous blood to the heart.

NERVOUS SYSTEM

- The autonomic nervous system influences the heart rate and blood pressure.

ENDOCRINE SYSTEM

- The blood serves as the transport medium for hormones produced by the endocrine system.
- The hormones adrenaline and thyroxine affect the heart rate.

LYMPHATIC SYSTEM

- Lymphocytes are carried by the blood to sites of infection and inflammation.

RESPIRATORY SYSTEM

- The exchange of gases between carbon dioxide and oxygen takes place in the capillary network of the lungs.

DIGESTIVE SYSTEM

- Blood picks up the end products of digestion for distribution to other organs of the body.

URINARY SYSTEM

- Blood pressure affects the filtration rate in the kidneys.
- As the blood is filtered through the kidneys, waste products, excess electrolytes, and excess fluid are removed; this action preserves blood volume.

REPRODUCTIVE SYSTEM

- Estrogen maintains vascular health in women.
- Engorgement of the blood vessels in the male maintains erection of the penis.

Medical Terminology

a-	without	**dys-**	difficult
phas	speech	**dys/phas/ia**	pertaining to difficulty in speech
-ia	abnormal condition of	**embol**	plug or clot
a/phas/ia	abnormal condition of being without speech	**-ism**	condition of
arterio	arteries	**embol/ism**	condition of having a blood clot
-sclerosis	hardening	**hemi-**	half
arterio/sclerosis	hardening of the arteries	**-plegia**	paralysis
athero	fatty	**hemi/plegia**	condition of paralysis on one side or half
athero/sclerosis	hardening of the arteries by fat	**hyper-**	over or excessive
cerebr	main brain	**tens**	condition of tension or pressure
-al	pertaining to	**-ion**	process of
vascular	blood vessels	**hyper/tens/ion**	condition of excessive blood pressure
cerebr/al vascular accident	accident pertaining to the blood vessels in the main brain	**hypo-**	under or low
cyan	blue	**hypo/tens/ion**	condition of low blood pressure
-osis	process of becoming	**phleb**	vein
cyan/osis	process of becoming blue	**-itis**	inflammation of
diastol	relaxation	**phleb/itis**	inflammation of a vein
-ic	pertaining to	**systole**	contraction
diastol/ic pressure	pertaining to the relaxation phase of the heart cycle	**systol/ic pressure**	pertaining to the contraction phase of the heart cycle

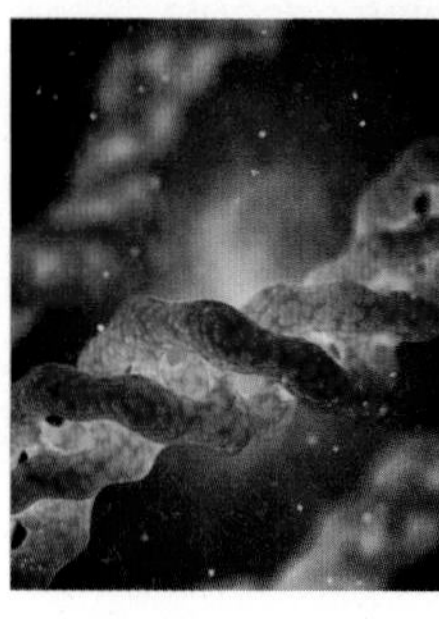

Study Tools

Workbook	Activities for Chapter 15
Online Resources	• PowerPoint presentations • Animations

REVIEW QUESTIONS

Select the letter of the choice that best completes the statement.

1. The name of the blood vessel that supplies the myocardium is the
   a. coronary artery.
   b. brachial artery.
   c. aorta.
   d. subclavian artery.

2. Special circulation that collects blood from the organs of digestion and takes it to the liver is the
   a. coronary.
   b. fetal.
   c. cardiopulmonary.
   d. portal.

3. The most common site for taking a pulse is the
   a. popliteal artery.
   b. dorsalis pedis artery.
   c. radial artery.
   d. temporal artery.

4. The blood vessel that carries blood away from the heart to the lungs is called the
   a. pulmonary artery.
   b. pulmonary vein.
   c. coronary sinus.
   d. coronary artery.

5. The inner layer of the artery is called the
   a. tunica adventitia.
   b. tunica intima.
   c. tunica media.
   d. externa.

6. The blood supply to the brain is carried by the
   a. external carotid artery.
   b. popliteal artery.
   c. internal carotid artery.
   d. coronary artery.

7. The blood supply returns from the legs through the
   a. saphenous vein.
   b. external jugular vein.
   c. superior vena cava.
   d. hepatic vein.

8. A buildup of fat in the arterial walls can cause the disease of
   a. gangrene.
   b. atherosclerosis.
   c. arteriosclerosis.
   d. aneurysm.

9. An inflammation of the lining of the vein is called
   a. hemorrhoid.
   b. thrombus.
   c. embolism.
   d. phlebitis.

10. The thinning and ballooning of an artery is called
   a. aneurysm.
   b. arteriosclerosis.
   c. phlebitis.
   d. atherosclerosis.

MATCHING

Match each term in Column I with its correct description in Column II.

	COLUMN I	COLUMN II
______	1. capillaries	a. small arteries that lead to capillaries
______	2. valves	b. deposit of fatty substances in the arteries
______	3. arterioles	c. blood pressure over 140/90
______	4. aorta	d. permit blood flow in only one direction
______	5. coronary	e. goes to the liver from the small intestine
______	6. hypertension	f. blood vessels that carry blood back to the heart
______	7. atherosclerosis	g. largest artery in the body
______	8. portal vein	h. loss of elasticity in the arteries
______	9. superior and inferior vena cavae	i. connect arterioles with venules
______	10. arteriosclerosis	j. arteries that nourish the heart

APPLYING THEORY TO PRACTICE

1. You are a red blood cell, and you are leaving the arch of the aorta. Trace your journey to the right great toe. Name all the blood vessels through which you will travel.
2. You are a red blood cell in the left finger. You need oxygen, and you must get to the lungs. Trace your journey from the finger to the lungs. Name the blood vessels and structures through which you will travel. What other chemicals are transported by the circulatory system?
3. Your grandmother has symptoms of peripheral vascular disease that the physicians say is a result of arteriosclerosis. Explain PVD and arteriosclerosis to your grandmother. If the following arteries are affected—carotid artery, coronary artery, renal artery, and femoral artery—what complication may occur? Describe how arteriosclerosis can be prevented.
4. The fetal heart is unique. Why is it different? Describe the structures of the fetal heart that change at birth.
5. Take the pulse and blood pressure of a 20-year-old, a 40-year-old, and a 70-year-old. Compare and contrast the results; if they are different, why are they different?
6. Why is hypertension called the "silent killer"? What is considered normal blood pressure? What are the complications of hypertension?

CASE STUDY

Mrs. William arrives in the emergency department with her son George. She cannot speak, and there is weakness and numbness on her right side. She is seen by Victoria, the nurse practitioner, who also notices a drooping on the right side of Mrs. William's face. George states that his mother was fine and eating her breakfast when this occurred. Victoria checks the woman's blood pressure, and it is 200/100. The emergency department physician and Victoria examine the patient, and the physician makes the diagnosis of a cerebral vascular accident (CVA).

1. Describe what a CVA is. What is the other name given to a CVA?
2. What is the correlation between Mrs. William's blood pressure and her CVA?
3. What other body systems will be affected because of the CVA?
4. What is the major cause of strokes?
5. Investigate and report the simple diagnostic tests Victoria will do to determine Mrs. William's state of paralysis.
6. Mrs. William cannot speak. Which side of her brain was affected?
7. Investigate and report some of the therapeutic technologies that will help Mrs. William in her recovery.
8. Explain some of the actions people can take to avoid a CVA.

LAB ACTIVITY 15-1 Structure of Blood Vessels

- ***Objective:*** To observe the structure of the various blood vessels in the human body
- ***Materials needed:*** microscopic slides of cross sections of a normal artery, vein, and an atherosclerotic artery; microscope; textbook; disposable gloves; biodegradable container for slides; household bleach; paper; pencil

Step 1: Put on gloves.

Step 2: Observe the slide of the structure of the normal artery. Record a brief description of the features you see.

Step 3: Observe the slide of the structure of a vein. Record a brief description of the features you see.

Step 4: What is the difference between the artery and the vein? Record your observations.

Step 5: Observe the slide of the atherosclerotic artery. Compare with the diagram in the textbook. Record your observations. Contrast the appearance of the normal artery with the appearance of the atherosclerotic artery.

Step 6: Place slides in the biodegradable container for disposal.

Step 7: Clean all equipment with household bleach.

Step 8: Remove gloves and wash your hands.

LAB ACTIVITY 15-2 Principal Arteries and Veins

- ***Objective:*** To locate and identify the principal arteries and veins within the body
- ***Materials needed:*** unlabeled anatomical charts of the principal arteries and veins; magnetic labels with the names of the arteries and veins; textbook; paper; pencil

Step 1: Locate and name the arteries on the anatomical chart that supply the following organs or body regions with blood: brain, face, pectoral girdle, upper arm, radius, ulna, heart, lungs, liver, stomach, spleen, kidney, intestines, femur, tibia, fibula, and pelvic girdle. Place the names of the arteries in their appropriate places on the chart.

Step 2: Compare your answers to the diagrams in this chapter of the textbook.

Step 3: Locate and name the veins that return the blood to the heart from the following organs or body regions: brain, face, pectoral girdle, upper arm, radius, ulna, heart, lungs, liver, stomach, spleen, kidney, intestines, femur, tibia, fibula, and pelvic girdle. Place the names of the veins in their appropriate places on the chart.

Step 4: Compare your answers to the diagrams in this chapter of the textbook.

Step 5: Do the arteries and veins that supply these locations have the same or similar names? Record your answer.

LAB ACTIVITY 15-3 Vital Signs

- ***Objective:*** To determine the pulse points in the body and to take a pulse
- ***Materials needed:*** wristwatch with second hand, textbook, paper, pencil
- ***Note:*** This activity must be done with a lab partner.

Step 1: Have your lab partner sit with the wrist resting on a table.

Step 2: Locate your partner's radial pulse with the pads of your first three fingers. (Remember: Do not use the thumb because it has its own pulse.)

Step 3: Gently compress the radial artery to feel the pulse.

Step 4: Count the pulse for 1 full minute. Take notice of the rhythm and volume. Record the pulse and describe any irregularities you notice.

Step 5: On your lab partner, locate and take the pulse at the following pulse points: temporal, carotid, brachial, popliteal, and dorsalis pedis. Compare locations with the diagram in this chapter of the textbook. Record the count at each pulse point. Record and explain any differences in your answer.

Step 6: Switch places with your lab partner and repeat steps 1–5.

CHAPTER 16

The Lymphatic and Immune Systems

Objectives

- Describe the lymphatic system and its function.
- Describe the function of interstitial fluid and lymph.
- Describe the organs of the lymphatic system and their function.
- Describe the disorders of the lymphatic system.
- Describe immunity and the defense mechanisms of the body.
- Describe autoimmune diseases.
- Describe the cause, symptoms, and treatment of AIDS.
- Define the key words that relate to this chapter.

Key Words

acquired immunity
acquired immunodeficiency syndrome (AIDS)
active acquired immunity
adenoids
allergen
anaphylaxis (anaphylactic shock)
antigen
artificial acquired immunity
autoimmune disorder
autoimmunity
cellular immunity
chemokines
complement cascade
congenital myasthenia gravis
cytokines
edema
germinal center
Guillain-Barré syndrome
herd immunity
hereditary lymphedema
Hodgkin's disease
human immunodeficiency virus (HIV)
humoral immunity
hypersensitivity
immunity
immunization
immunoglobulins
infectious mononucleosis
interferon
interleukins
interstitial fluid
lacteals
lingual
lupus
lymph
lymph nodes
lymph vessels
lymphadenitis
lymphatic malformations
lymphatic system
lymphatics
lymphedema
lymphokines
lymphoma
macrophages
natural acquired immunity
natural immunity
palatine

(*continues*)

Key Words *continued*

passive acquired immunity
Peyer's patches
right lymphatic duct
scleroderma
Sjögren's syndrome
spleen
thoracic duct (left lymphatic duct)
tonsillitis
tonsils
trabeculae
tumor necrosis factor (TNF)

The **lymphatic system** can be considered a supplement to the circulatory system. It is composed of lymph, lymph vessels, lymph nodes, tonsils, spleen, thymus gland, Peyer's patches, lacteals, and lymphoid tissue (see *Medical Highlights*). Unlike the circulatory system, it has no muscular pump or heart (**Figure 16-1**).

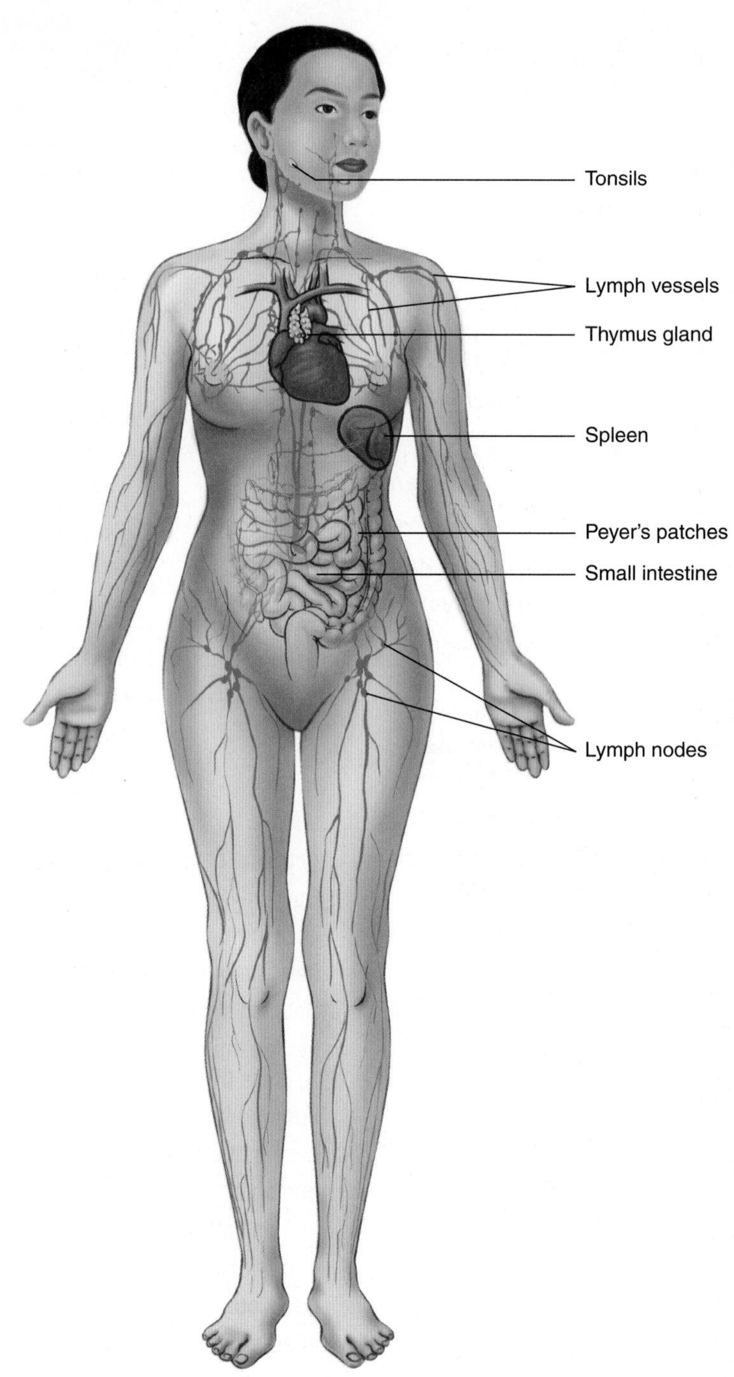

Figure 16-1 *The vessels and organs of the lymphatic system*

FUNCTIONS OF THE LYMPHATIC SYSTEM

1. Interstitial fluid and lymph act as an intermediary between the blood in the capillaries and the tissue.
2. Lymph vessels transport the excess filtered tissue fluid back into the circulatory system.
3. Lymph tissue in bone marrow produces lymphocytes.
4. Lymph nodes and organs produce lymphocytes to destroy invading bacteria.
5. Lacteals, specialized lymph capillaries in villi of the small intestine, absorb digested fat and fat-soluble vitamins.
6. Spleen produces lymphocytes and monocytes.
7. Thymus gland produces T lymphocytes.

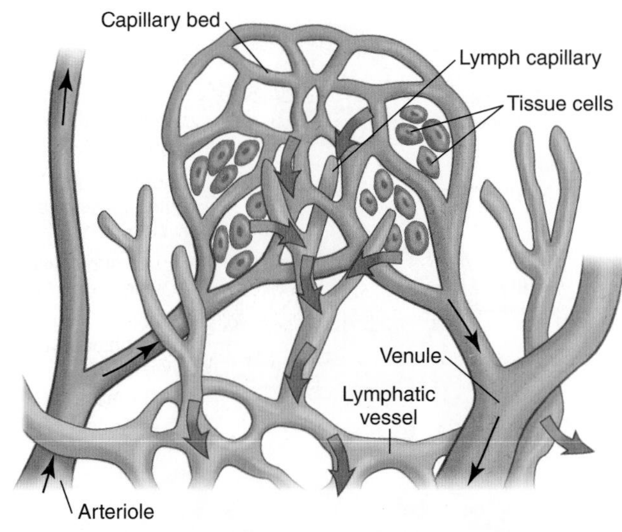

Figure 16-2 *Lymph capillaries begin as blind-ended tubes.*

Interstitial Fluid and Lymph

Interstitial fluid (in-ter-**STISH**-al) acts as the intermediary between the blood in the capillaries and the tissue. It is similar in composition to blood plasma; it diffuses from the capillaries into the tissue spaces. Because the fluid fills the surrounding spaces between tissue cells, it is also referred to as intercellular fluid. Interstitial fluid is composed of water, lymphocytes, some granulocytes, digested nutrients, hormones, salts, carbon dioxide, and urea. It does not contain red blood cells or protein molecules, which are too large to diffuse through the capillaries.

Interstitial fluid also carries metabolic waste products, such as carbon dioxide and urea, away from the cells and back into the capillaries for excretion. Most of the fluid is reabsorbed into the capillaries by differences in osmotic pressure. Some of the fluid enters lymph capillaries and is now called **lymph** (**LIMF**). However, some of the fluid may not get reabsorbed, which results in swelling of the tissue and the resulting condition known as **edema** (eh-**DEE**-mah).

Because the lymphatic system has no pump, other factors operate to push lymph through the lymph vessels. The contractions of the skeletal muscles against the lymph vessels cause the lymph to surge forward into larger vessels. The breathing movements of the body also cause lymph to flow. Valves located within lymph capillaries prevent backward lymph flow.

Lymph Vessels

The lymph capillaries begin as blind-ended tubes (**Figure 16-2**). Lymph enters the cells of the capillary walls and flows into progressively larger vessels. **Lymph vessels** accompany and closely parallel the veins. They form an extensive, branchlike system throughout the body. This system may be considered an auxiliary to the circulatory system.

Lymph vessels are located in almost all of the tissues and organs that have blood vessels. Lymph capillaries are not in the cartilage, central nervous system, red bone marrow, epidermis, eyeball, inner ear, hair, nails, or spleen.

The interstitial fluid surrounding tissue cells enters small lymph vessels. These, in turn, join to form larger lymph vessels called **lymphatics** (lim-**FAT**-iks). They continue to unite, forming larger and larger lymphatics, until the lymph flows into one of two large, main lymphatics. They are the **thoracic duct (left lymphatic duct)** and the **right lymphatic duct**.

The thoracic duct receives lymph from the left side of the chest, head, neck, abdominal area, and lower limbs. Lymph in the thoracic duct is carried to the left subclavian vein. From there, it is carried to the superior vena cava and the right atrium. In this manner, lymph carrying digested nutrients and other materials can return to the systemic circulation. Lymph from the right arm, right side of the head, and upper trunk enters the right lymphatic duct. From there, it enters the right subclavian vein at the right shoulder and then flows into the superior vena cava (**Figure 16-3**).

Unlike the circulatory system, which travels in closed circuits through the blood vessels, lymph travels in only one direction: from the body organs to the heart. It does not flow continually through vessels forming a closed circular route.

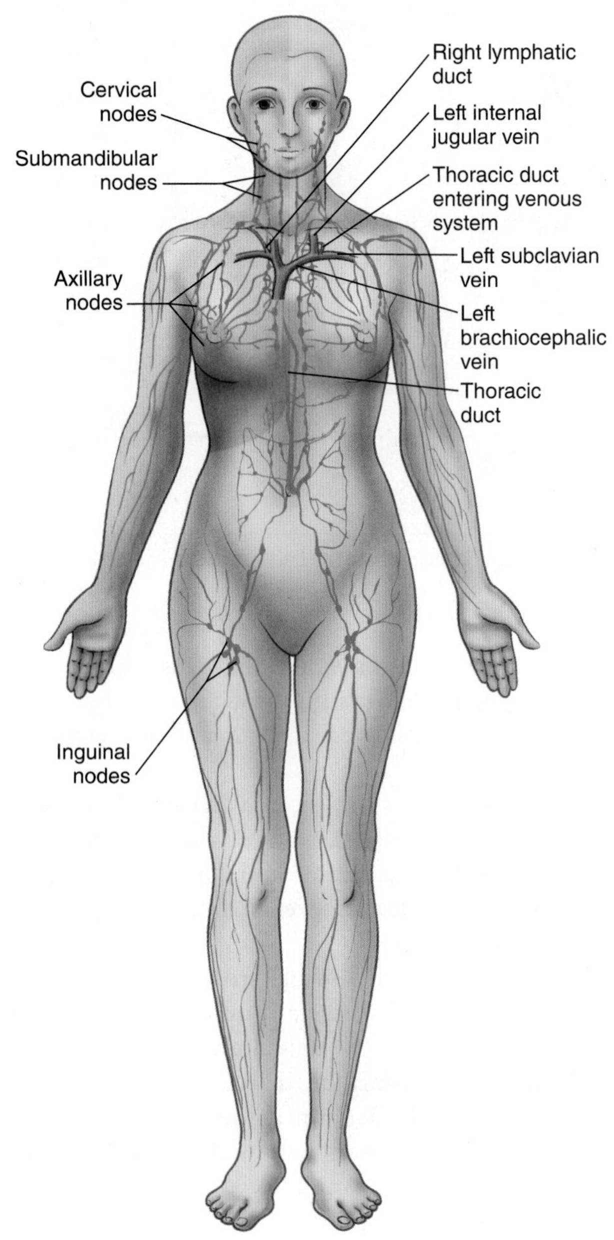

Figure 16-3 *Lymphatics pass their lymph into two main collecting ducts: the thoracic duct and the right lymphatic duct.*

ORGANS OF THE LYMPHATIC SYSTEM

The organs that make up the lymphatic system consist of the lymph nodes, tonsils, spleen, thymus gland, Peyer's patches, and lacteals.

Lymph Nodes

Lymph nodes are tiny, oval-shaped structures ranging from the size of a pinhead to that of an almond. Red bone marrow produces all red blood cells and platelets and around 60%–70% of lymphocytes in adults. A lymph node contains a slight depression on one side called the hilum (**HIGH**-lum), where the efferent (**EFF**-er-ent) lymphatic vessels and a nodal vein leave the node and a nodal artery enters. Each lymph node is covered by a capsule of fibrous connective tissue that extends into the node. These capsular extensions are called **trabeculae** (trah-**BEK**-you-lee). They divide the node into a series of compartments that contain lymphatic sinuses and lymphatic tissue. Lymphatic vessels enter the node at various sites, called afferent (**AFF**-er-ent) lymphatic vessels (**Figure 16-4A**). The lymphatic tissue of the node consists of different kinds of lymphocytes and other cells that make up dense masses of tissue called lymph nodules. The lymph nodule surrounds a **germinal center** that produces the lymphocytes. Lymph sinuses are spaces between the lymph tissues that contain a network of fibers and **macrophages**. The capsule, trabeculae, and hilum make up the framework, or stroma, of the lymph node.

As lymph enters the node through the afferent vessel, the immune response is activated. Any microorganisms or foreign substances in the lymph stimulate the germinal centers to produce lymphocytes, which are then released into the lymph (**Figure 16-4B**). Eventually, the lymphocytes reach the blood and produce antibodies against the microorganisms. The macrophages will remove the dead microorganisms and foreign substances by phagocytosis.

Lymph nodes are located alone or grouped in various places along the lymph vessels throughout the body. If the harmful substances occur in such large quantities that they cannot be destroyed by the lymphocytes, the node becomes inflamed. This causes a swelling in the lymph glands, a condition known as **lymphadenitis** (limf-ad-eh-**NIGH**-tis).

Spleen

The **spleen** (**SPLEEN**) is a saclike mass of lymphatic tissue. It is located near the upper left area of the abdominal cavity, just beneath the diaphragm. The spleen forms lymphocytes and monocytes. Blood passing through the spleen is filtered, as in any lymph node.

The spleen stores large amounts of red blood cells. During excessive bleeding or vigorous exercise, the spleen contracts, forcing the stored red blood cells into circulation. It also destroys and removes old or fragile red blood cells and forms erythrocytes in the embryo.

Thymus Gland

The *thymus gland* is located in the upper anterior part of the thorax, above the heart. Its function is to produce and mature lymphocytes. These lymphocytes are called T

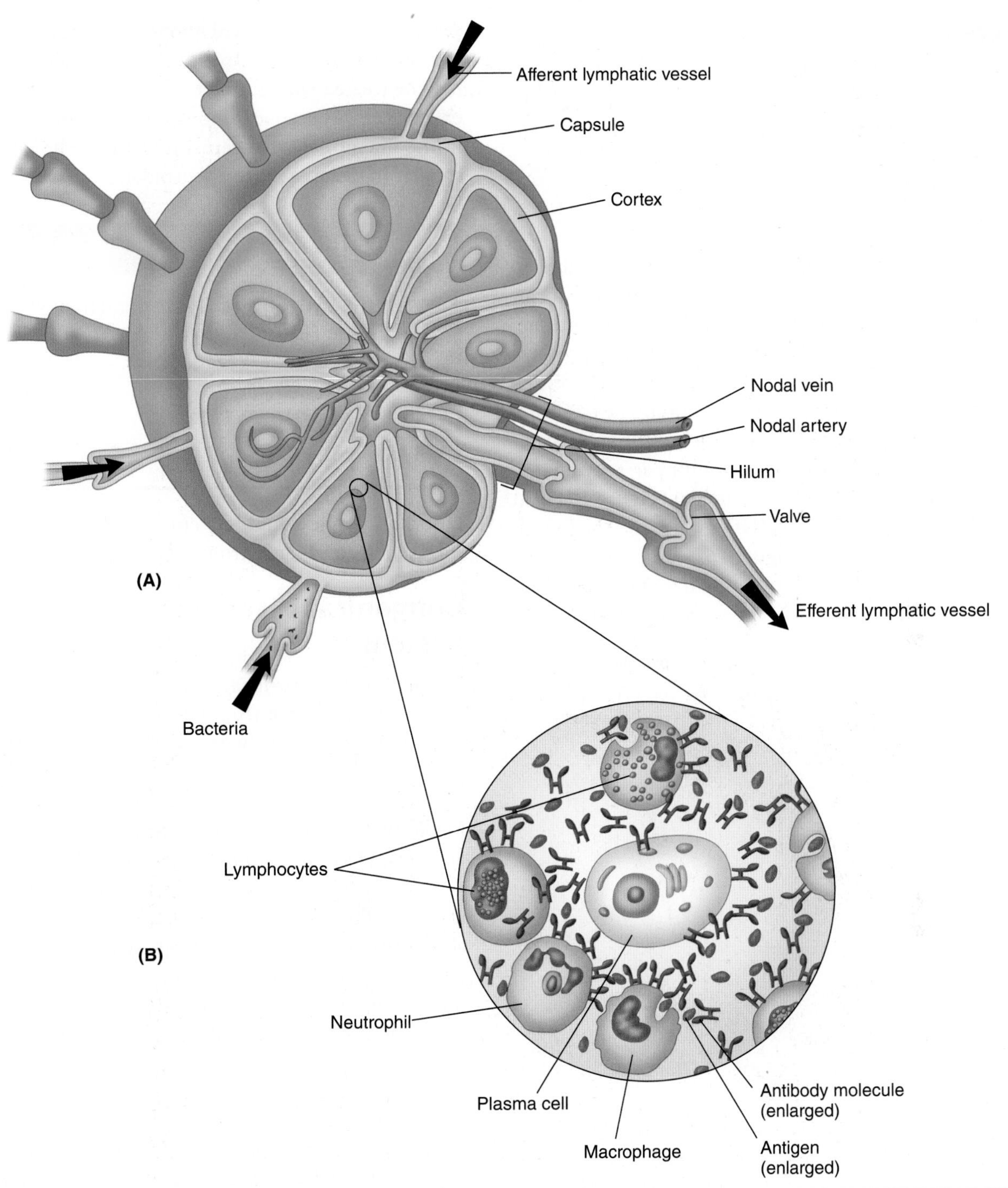

Figure 16-4 *Lymph node: (A) section through a lymph node, showing the flow of lymph; (B) microscopic detail of bacteria being destroyed within the lymph node*

lymphocytes. The thymus is classified with the lymphatic organs because it consists largely of lymphatic tissue. It is also considered an endocrine gland because it secretes a hormone called thymosin, which stimulates the maturation of lymphocytes into T cells. The manufacturing of T cells ends after puberty, so there are a finite number of T cells in the adult body.

Peyer's Patches

Peyer's patches (**PYE**-erz), also known as aggregated lymphatic follicles, are found in the walls of the small intestines. They resemble tonsils and produce macrophages. The macrophages destroy bacteria and prevent bacteria from penetrating the walls of the small intestines **(Figure 16-4B)**.

Lacteals

Lacteals (**LACK**-tee-ahlz) are specialized lymph capillaries in the villi of the small intestine. They absorb digested fat and fat-soluble vitamins and carry them to the general circulation.

Tonsils

Tonsils are masses of lymphatic tissues that are capable of producing lymphocytes and filtering bacteria. They form a protective ring of reticuloendothelial cells around the back of the nose and upper throat, which help prevent pathogens from entering the respiratory tract. Tonsils are subdivided according to their location. The **palatine** (**PAL**-ah-tine) are located on the sides of the soft palate (**Figure 16-5A**). The tonsils in the upper part of the throat are the **adenoids** (**AD**-eh-noids) (**Figure 16-5B**). The third pair, the **lingual** (**LING**-gwall), may be found at the back of the tongue (**Figure 16-5A**).

During childhood, the tonsils frequently become infected and enlarged, causing difficulty swallowing, severe sore throat, elevated temperature, and chills. This condition is known as **tonsillitis** (ton-sih-**LYE**-tis), which may be treated with antibiotics. Surgery is done only in extreme cases because the tonsils have an important role in the line of defense against infection. The tonsils get smaller in size as a person gets older.

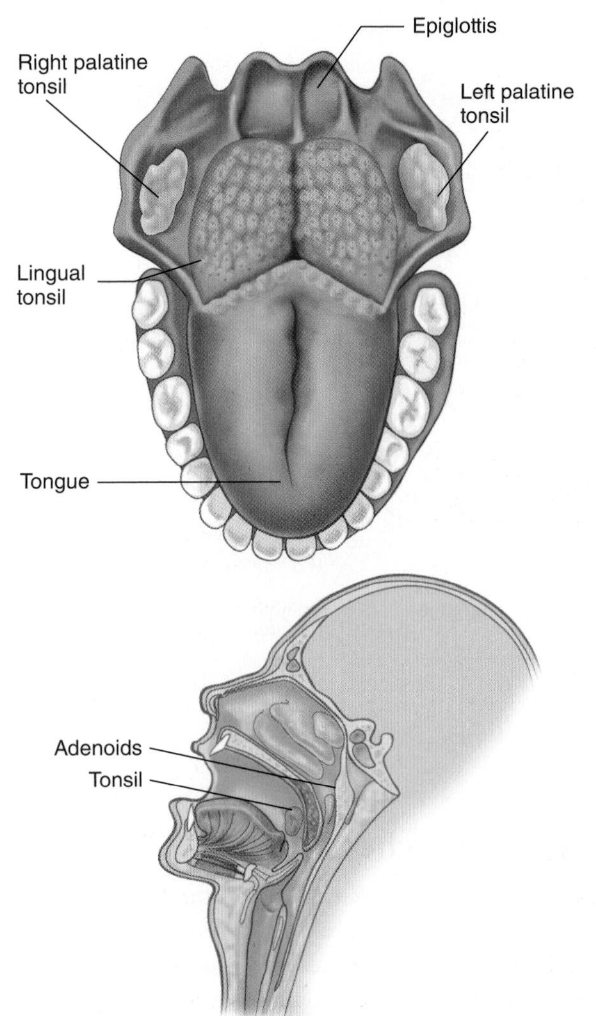

Figure 16-5 *Tonsils*

> **Media Link**
>
> View the **Lymphatic System** animation on the Online Resources.

DISORDERS OF THE LYMPH SYSTEM

Disorders of the lymph system can happen at different stages in life and range from very mild to serious.

Congenital Disorders of the Lymph System

Congenital myasthenia gravis is caused by a genetic disorder and can occur anytime between birth and adulthood. Symptoms include an abnormally large thymus gland that produces antibodies that block or destroy muscle cell receptor sites. This causes the muscles to become weak and tired. Medications may help the communication between muscles and nerves (see also *myasthenia gravis* in Chapter 8).

Lymphatic malformations are rare. Nonmalignant masses of fluid-filled channels or spaces occur because of abnormal development of the lymph system. The symptoms and severity vary depending on the size and location of the mass. These masses occur at birth or appear by the age of 2. Treatment may be percutaneous drainage surgery, sclerotherapy, laser, radio-frequency ablation, or medical therapy. Percutaneous drainage surgery is done by making an incision through the skin and draining the fluid from the mass. Sclerotherapy is injecting a sclerosing agent directly into the mass. The solution causes scarring in the malformation and eventually causes it to shred. *Cystic hygroma* is a type of lymphatic malformation meaning "water tumor." This is a fluid-filled abnormal mass that occurs on the baby's head or neck. During pregnancy, an ultrasound may diagnose this condition. After birth, caution must be taken to protect the baby's head with extra bedding. Treatment is sclerotherapy and other types of surgery to drain the cystic mass.

Hereditary lymphedema is swelling that occurs in certain parts of the body. The following are the three different types of hereditary lymphedema:

- Milroy disease: characterized by edema at birth, usually of the legs and feet

- Merge disease: usually occurs around puberty
- Lymphedema Tarde: occurs after the age of 35, usually of the legs

Treatment is aimed at reducing swelling through the use of compression bandages or compression garments.

General Disorders of the Lymph System

Lymphedema (lim-feh-**DEE**-mah) is swelling of the tissues due to an abnormal collection of lymph. It is caused by damage to the lymph system that prevents lymph from draining properly. Lymphedema is most commonly caused by removal or damage to lymph nodes as part of cancer treatment. Treatment includes light exercise, compression garments, and massage to help manually promote lymph drainage. Surgery may be done, including moving healthy lymph nodes from one part of the body to the affected area. A lymphaticovenous anastomosis may also be done, which consists of connecting existing lymph vessels to tiny veins in the area.

Lymphoma (lim-**FOH**-mah) is a tumor of the lymphatic tissue, usually malignant. It begins as a large mass with no associated pain. The enlarged nodes will compress surrounding structures and cause complications. The immune system becomes suppressed, and the individual is susceptible to opportunistic infections. Lymphomas are classified as non-Hodgkin's lymphomas or **Hodgkin's disease**. Hodgkin's disease is distinguished from other lymphomas by the presence of large cancerous lymphocytes known as Reed-Sternberg cells. Treatment of non-Hodgkin's lymphomas and Hodgkin's disease with chemotherapy, stem cell therapy, and radiation produces good results.

Infectious mononucleosis (in-**FECK**-shus mon-oh-new-klee-**OH**-sis) is a disease caused by the Epstein-Barr virus. It frequently occurs in young adults and children. This disease is spread by oral contact and is frequently called the "kissing disease" or "mono." The symptoms are enlarged lymph nodes, fever, and physical and mental fatigue. There is a marked increase in the number of leukocytes. This illness is treated symptomatically, meaning the symptoms are treated as they appear. Bed rest is essential in the treatment of mono. In some cases, it may cause an enlargement of the spleen; the liver may also be affected, and hepatitis can result.

TRAUMA TO THE LYMPH SYSTEM

A ruptured spleen may be caused by a blow to the left upper abdomen or left lower chest, such as might occur in contact sports, bike accidents, or car crashes. Symptoms include severe pain in the upper abdomen on the left side, possible pain in the left shoulder, dizziness, and lower blood pressure. Treatment depends on the severity; hospitalization is necessary if the blood pressure is falling, and the patient may need a blood transfusion. In severe cases if the blood pressure keeps falling and vital signs are unstable, the spleen is removed from the lymphatic system altogether. Removal of the spleen can cause a compromised immune system that increases the chance of infection.

The Effects of Aging on the Lymphatic and Immune Systems

With advancing age, the lymphatic system becomes less effective at combating disease and fighting infections. The ultimate consequence of any age-related decline in the immune function is an increase in the incidence and severity of infectious diseases, such as pneumonia, gastrointestinal diseases, urinary tract infections, skin infections, and cancers. The major problem with aging in the immune system appears to be the loss of the ability of the specific immune system cells—T cells and B cells—to undergo rapid cell division. As a result, the immune system has trouble keeping up with the rate of cell division performed by harmful bacteria and viruses. Older adults are encouraged to get tetanus immunizations every 10 years. In addition, a physician may recommend the hepatitis vaccine, flu vaccine, and Pneumovax® to prevent pneumonia.

FUNCTION OF THE IMMUNE SYSTEM

The immune system's function is to protect the body from harmful substances, such as allergens, toxins, malignant cells, and pathogens, which are disease-producing microorganisms. It initiates an immune response to help defend against the invading harmful substances.

Immunity

Immunity (im-**YOO**-nih-tee) is the body's ability to resist infections from pathogens, foreign substances, and toxic chemicals. Humoral immunity, so-called because it involves substances found in body fluids or humors, and cellular immunity derive from the body's lymphoid tissue. The B lymphocytes are cells that produce antibodies and provide **humoral immunity**. The T lymphocytes are responsible for providing **cellular immunity**. Individuals differ in their ability to resist infection. In addition, an individual's resistance varies at different times.

Normal Defense Mechanisms

The individual's immune system serves as the normal defense mechanism against the attack of infectious agents. A unique feature of the immune system is its ability to recognize which agents are not consistent with the genetic makeup of the host. These agents are called antigens. Antibodies will form to protect the body against antigens. The immune defenses are categorized as specific and nonspecific.

ANTIGEN-ANTIBODY REACTION

The antigen-antibody reaction is the immune reaction that involves binding antigens to antibodies. An **antigen** (**AN**-tih-jen) is identified as any substance that the body regards as being foreign; this includes viruses, bacteria, toxins, and transplanted tissue. The immune system responds to the presence of an antigen by producing a disease-fighting protein called an antibody (**AN**-tih-bod-ee) (**Figure 16-6**).

NONSPECIFIC IMMUNE DEFENSE

The nonspecific immune defense mounts a response to protect the individual from all microorganisms; it is not dependent on prior exposure to the antigens. Nonspecific immune defenses include the following:

- Skin and normal flora serve as a physical barrier against infectious agents.
- Mucous membranes entrap infectious agents and contain antibodies, lactoferrin, and lysozyme, which inhibit bacterial growth.
- Sneezing, coughing, and tearing reflexes physically expel mucus and microorganisms with force. Tears continually flush away microorganisms. They also contain bactericides, which are bacteria-killing chemicals.
- Elimination and an acidic environment prevent microbial growth of pathogenic organisms. These include resident flora of the large intestine, acidity of the urine, and normal vaginal flora. The mechanical process of defecation evacuates the bowel of feces and microorganisms. The flushing action of urination prevents microorganisms from ascending the urinary tract.
- Inflammation is the nonspecific response to cellular injury. Tissue injury releases multiple substances that produce dramatic changes in the injured tissue. The intensity of the inflammatory process is usually in proportion to the degree of tissue injury. For a detailed description on inflammation, see Chapter 13.
- **Immunoglobulins** (im-you-noh-**GLOB**-you-lins) are a class of specialized proteins that function as antibodies. The five classes of immunoglobulins are the following:

1. *Immunoglobulin G (IgG):* The most abundant class of antibodies, IgG is found in the blood serum and lymph. These antibodies are active against bacteria, fungi, viruses, and foreign particles.
2. *Immunoglobulin A (IgA):* The IgA class of antibodies is produced against ingested antigens. These antibodies are found in the respiratory tract, digestive system, saliva, sweat, or tears, and they function to prevent the attachment of viruses and bacteria to the epithelial cells that line most organs.
3. *Immunoglobulin M (IgM):* This class of antibodies is found in circulating body fluids. The IgM antibodies are the first antibodies to appear in response to an initial exposure to an antigen.
4. *Immunoglobulin D (IgD):* This class of antibodies is found only on B cells and are important in B-cell activation.
5. *Immunoglobulin E (IgE):* The IgE class of antibodies is produced in the lungs, skin, and mucous membranes. These antibodies are responsible for allergic reactions.

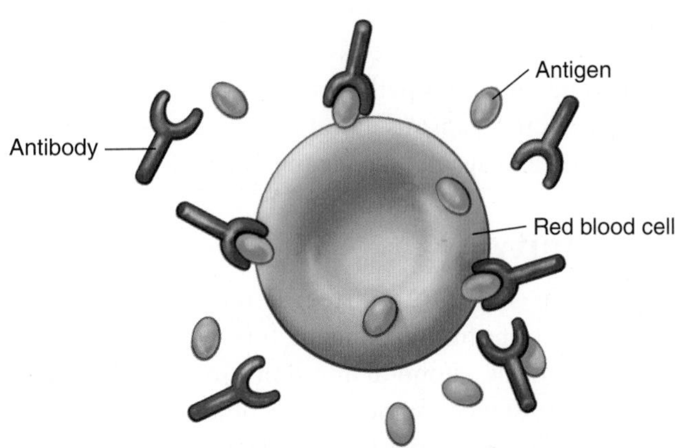

Figure 16-6 *Antigen-antibody reaction: An antigen is a substance foreign to the body. Antibodies will form to protect the body from the foreign substance.*

SPECIFIC IMMUNE DEFENSE

A specific immune defense is triggered by antigens. The lymphocytes of the body are the precursors of a whole range of cells that are involved in the immune response. A list of those cells and their functions follows (**Figure 16-7**):

- *B cells*—lymphocytes found in the lymph node, spleen, and other lymphoid tissue where they replicate. Their clones form plasma cells and memory cells.

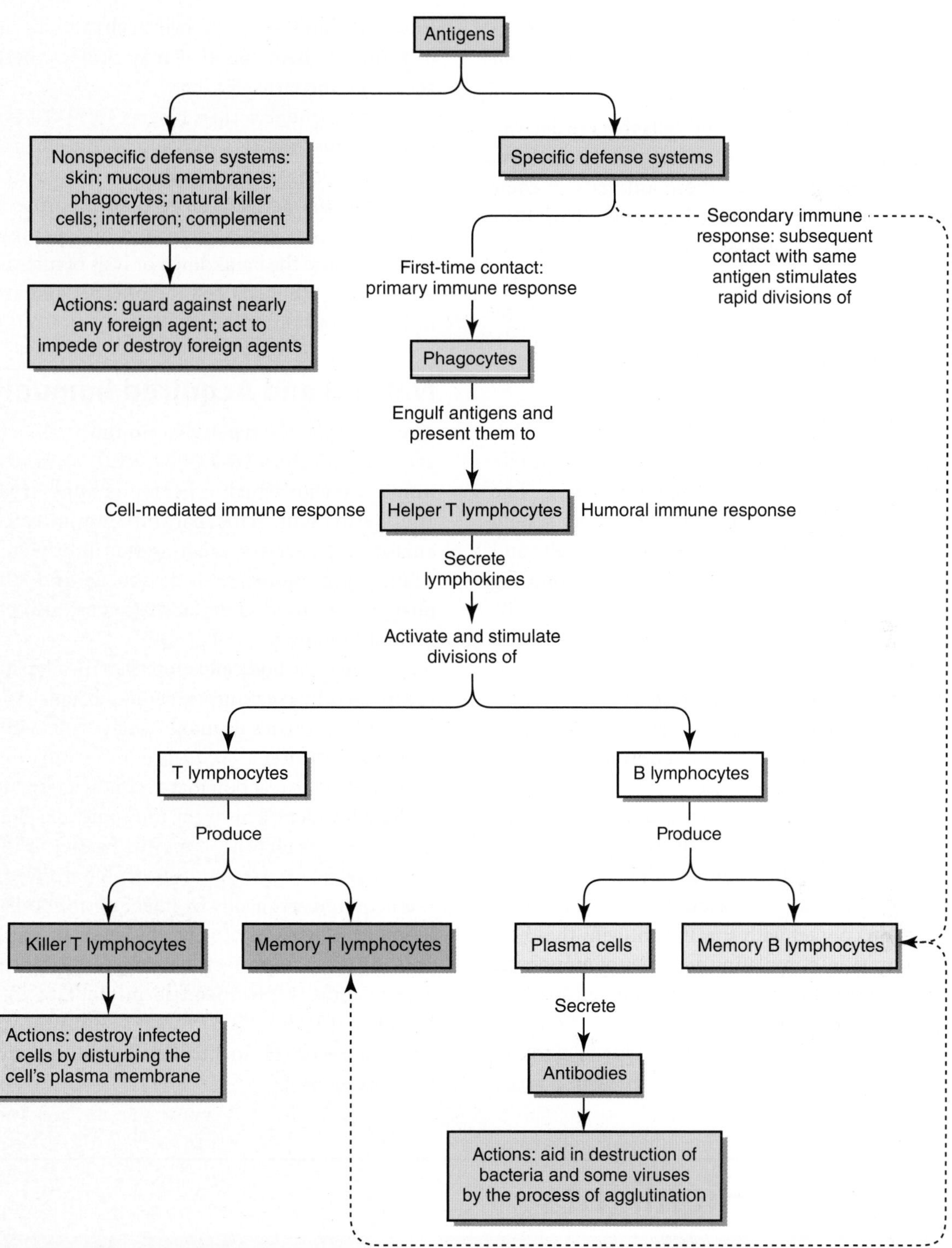

Figure 16-7 *Overview of the body's defense mechanisms*

- *Plasma cells*—cells formed by B cells that produce huge quantities of the same antibody or immunoglobulin.
- *Helper T cells*—T cells from the thymus gland that bind with specific antigens presented by macrophages. They stimulate the production of killer T cells and more B cells to fight the invading pathogens. They release lymphokines.
- *Killer T cells,* or cytotoxic T cells—cells that kill virus-invaded body cells and cancerous body cells. They are also involved in transplant rejection.
- *Suppressor T cells*—cells that slow down the activities of B and T cells once an infection has been controlled.
- *Memory cells*—cells that are descendants of activated T and B cells produced during an initial immune

response. They will exist in the body for years, enabling it to respond quickly to any future infection from the same pathogen.

- *Macrophages*—large white blood cells that are derived from monocytes that engulf and digest the antigen. They then present parts of these antigens in their cell membrane for recognition by T cells. This antigen presentation function is critical for normal T-cell response.

Chemicals and the Immune Response

Cytokines are a broad category of loose proteins, which are signaling molecules that mediate and regulate immunity, inflammation, and hematopoiesis. The cytokines are directly released into the circulatory system or into the tissue. They locate target immune cells and interact with receptors on the immune cells by binding to them. This interaction triggers or stimulates a specific response by the target cells. The classes of cytokines include chemokines, lymphokines, interleukins, interferon, and the tumor necrosis factor (TNF).

Chemokines play an important role while introducing chemokine receptors in antitumor immune response and autoimmune disease.

Lymphokines (LIM-foh-kynz) and interleukins (in-ter-**LOO**-kinz) are released by lymphocytes and macrophages as part of the immune response.

Lymphokines are released by the sensitized lymphocytes on contact with specific antigens that help cellular immunity by stimulating activities of monocytes and macrophages. **Interleukins** regulate cell growth, differentiation, and motility.

Interferon (in-ter-**FEAR**-on) is a type of cytokine produced by cells that have been infected with a virus. Interferon binds to neighboring cells and stimulates them to produce chemicals that may protect these adjacent cells from the virus.

The **tumor necrosis factor (TNF)** is a cytokine that stimulates macrophages and causes cell death in cancer cells.

Complement cascade is a complex series of reactions that activate more than 20 proteins that are usually inactive unless activated by a pathogen. These proteins cause the breakdown or lysis of microorganisms, attack the pathogen's cell membrane, and enhance the inflammatory process.

Natural and Acquired Immunities

The two general types of immunity are natural and acquired (**Table 16-1**). **Natural immunity** is the immunity with which a person is born. It is inherited and permanent. This natural immunity consists of anatomical barriers, such as the unbroken skin, and cellular secretions, such as mucus and tears. Blood phagocytes and local inflammation are also part of one's natural immunity.

When the body encounters an invader, it tries to kill the invader by creating a specific substance to combat it. The body also tries to make itself permanently resistant to these intruders. **Acquired immunity** or adaptive immunity is the reaction that occurs as a result of exposure to these invaders. This is the immunity developed during an individual's lifetime. It may be passive or active.

Passive acquired immunity is borrowed immunity. It is acquired artificially by injecting antibodies from the blood or plasma of other individuals or animals into a person's body to protect them against a specific disease. The immunity produced is immediate in its effect; however, it lasts only 3–5 weeks. After this period, the antibodies will be inactivated by the individual's own macrophages.

Medical Highlights

MUCOSA-ASSOCIATED LYMPHOID TISSUE (MALT)

A significant quantity of lymphoid tissue is associated with human mucosa. Mucosa-associated lymphoid tissue (MALT) is scattered along mucosa linings. These surfaces protect the body from an enormous quantity and variety of antigens. Examples include the tonsils, Peyer's patches in the small intestines, and the vermiform appendix. The nomenclature incorporates their location:

- GALT: gut-associated lymphoid tissue
- BALT: bronchotracheal-associated lymphoid tissue
- NALT: nose-associated lymphoid tissue
- VALT: vulvovaginal-associated lymphoid tissue

MALT protects passages that are open to the exterior of the body and the never-ending onslaught of foreign matter entering them.

Table 16-1 *Types of Immunity*

NATURAL IMMUNITY	*ACQUIRED IMMUNITY/ADAPTIVE IMMUNITY*	
Lasts a lifetime; born with it; inherited	Reaction as a result of exposure	
	ACTIVE: Lasts a long time	**PASSIVE:** Borrowed; lasts a short time
	Natural—Get the disease and recover; get a mild form of disease with no symptoms and recovery	*Natural*—Baby gets from the placenta or human breast milk
	Artificial—Vaccination; immunization	*Artificial*—Serum from another; immunoglobulin; antitoxin

Because it is immediate, passive immunity is used when one has been exposed to a virulent disease, such as measles, COVID-19, tetanus, or infectious hepatitis, and has not acquired active immunity to that disease. The borrowed antibodies will confer temporary protection.

A baby has temporary passive immunity from their mother's antibodies. These antibodies pass through the placenta to enter the baby's blood. In addition, human breast milk also offers the baby some passive immunity. Thus, a newborn infant may be protected against poliomyelitis, measles, and mumps. Measles and mumps immunity may last for nearly a year. Then the child must develop their own active immunity.

Active acquired immunity is preferable to passive immunity because it lasts longer. There are two types of active acquired immunity: natural acquired immunity and artificial acquired immunity.

- **Natural acquired immunity** is the result of having had and recovered from a disease. For example, a child who has had measles and has recovered will not ordinarily get measles again because the child's body has manufactured antibodies. This form of immunity is also acquired by having a series of unnoticed or mild infections. For example, a person who has had a mild form of a disease one or more times and has fought it off, sometimes unnoticed, is later immune to the disease.
- **Artificial acquired immunity** comes from being inoculated with a suitable vaccine, antigen, or toxoid. For example, a child vaccinated for measles has been given a very mild form of the disease; the child's body will thus be stimulated to manufacture its own antibodies.

Immunization is the process of increasing an individual's resistance to a particular infection by artificial means. An antigen may be a substance that is injected to stimulate production of antibodies. For example, toxins produced by bacteria, dead or weakened bacteria, viruses, and foreign proteins are examples of antigens. These weakened toxins stimulate the body to produce antibodies. Refer to **Table 16-2A** and **Table 16-2B** for the U.S. government's current immunization schedules.

Herd immunity, or community immunity, is a form of indirect protection from infectious disease. This occurs when a sufficient percentage of a population has become immune to an infection through vaccination or previous infection, making the spread of the disease from person to person unlikely. Individuals who are not vaccinated are offered some protection because the disease has little opportunity to spread within the community.

Autoimmunity

Autoimmunity is when a person's own immune system mistakenly targets the normal cells, tissues, and organs of their own body. This is known as an **autoimmune disorder** (**aw**-toh-ih-**MYOUN**). The cause of autoimmune disease is unknown. Because the incidence of autoimmune disease is increasing, researchers suspect environmental factors like infection or exposure to chemicals might be involved.

There are many different autoimmune diseases, and they can affect the body in different ways. For example, the autoimmune reaction is directed against the nervous system in multiple sclerosis and the skin in psoriasis. In other autoimmune diseases, such as systemic lupus erythematosus, affected tissues and organs may vary among individuals with the same disease. One person with lupus may have affected skin and joints, whereas another may be affected with blood-clotting problems.

Causes of autoimmune disease can include a genetic familial predisposition, viruses, or even sunlight, which can act as a trigger for lupus.

The following are examples of autoimmune diseases and the chapters in which they are discussed in more detail:

- Addison's disease, Chapter 12
- Type 1 diabetes mellitus, Chapter 12

- Guillain-Barré syndrome, Chapter 16
- Hypothyroidism/hyperthyroidism, Chapter 12
- Lupus/systemic lupus erythematosus, Chapter 16
- Multiple sclerosis, Chapter 9
- Myasthenia gravis, Chapter 8
- Pernicious anemia, Chapter 13
- Psoriasis, Chapter 6
- Rheumatoid arthritis, Chapter 7
- Scleroderma, Chapter 16
- Sjögren's syndrome, Chapter 16
- Irritable bowel syndrome/Crohn's disease, Chapter 19

Guillain-Barré syndrome is a rare disease that attacks the nerves that control the muscles in the legs or sometimes the arms and upper body. Filtering the blood with a procedure called plasmapheresis is the main treatment. Symptoms are weakness and tingling, which quickly spread throughout the body and can cause paralysis. The person must be hospitalized to receive treatment.

Lupus, also known as *systemic lupus erythematosus (SLE)* (sis-**TEM**-ik **LOO**-pus er-ih-thee-mah-**TOH**-sus), is a chronic inflammatory autoimmune disease. Patients with lupus most commonly experience profound fatigue, a butterfly-shaped rash on the face, and joint pain. In severe cases, the immune system may attack and damage several organs, such as the kidney, brain, blood, or lung, severely impairing their function. For many individuals, symptoms and damage from the disease can be controlled with anti-inflammatory medication and symptomatic prescribed medication. Most people with lupus have a mild disease characterized by episodes called "flares," or when signs and symptoms get worse and then improve or disappear. Treatment is with non-steroid anti-inflammatory drugs, corticoid steroids, biologics, or immune suppressant drugs.

Scleroderma (sklehr-oh-**DER**-mah), or systemic sclerosis, is a disease that results in the thickening of the skin and blood vessels (**Figure 16-8**). Overproduction of collagen is a characteristic of this disease. Almost every patient with scleroderma has Raynaud's disease, which is a spasm of the blood vessels of the fingers and toes. Symptoms of Raynaud's include increased sensitivity of the fingers and toes to the cold, changes in skin color, pain, and occasionally ulcers of the fingertips or toes. For people with scleroderma, the thickening of skin and blood vessels can result in loss of movement and dyspnea. Treat or slow skin changes by using creams or oils to reduce swelling and joint pain. Current treatment for scleroderma includes a wide variety of medications that help control the sclerosis and prevent complications, such as anti-inflammatory drugs, immune suppressant drugs, and vasodilators.

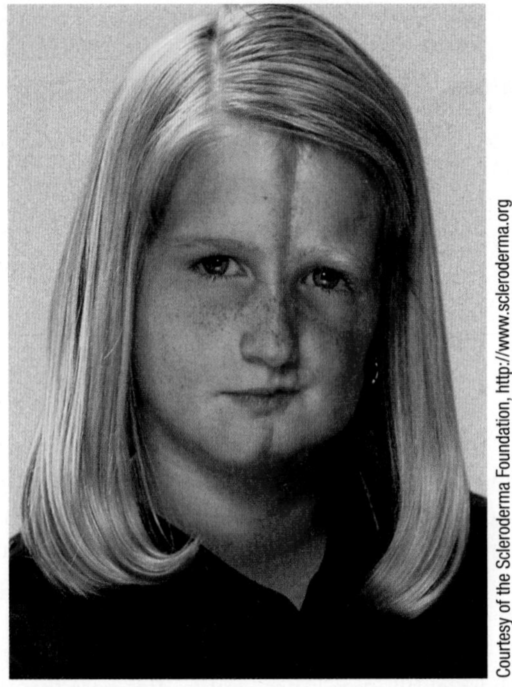

Figure 16-8 *Scleroderma*

Sjögren's syndrome (**SHOW**-grun) is a disease that affects the moisture-producing glands. Classic symptoms are dry eyes and dry mouth; it may also affect the joints, kidneys, blood vessels, and central nervous system. Because it affects almost 4 million people, it is one of the most prevalent autoimmune diseases. There is no cure; treatment with over-the-counter and prescription drugs may improve symptoms and prevent complications.

Hypersensitivity

Hypersensitivity occurs when the body's immune system fails to protect itself against foreign material. Instead, the antibodies that are formed irritate certain body cells. A hypersensitive or allergic individual is generally more sensitive to certain allergens than most people are.

An **allergen** (**AL**-er-jen) is an antigen that causes allergic responses. Examples of allergens include grass, ragweed pollen, ingested food, peanuts, penicillin and other antibiotics, and bee and wasp stings. Such allergens

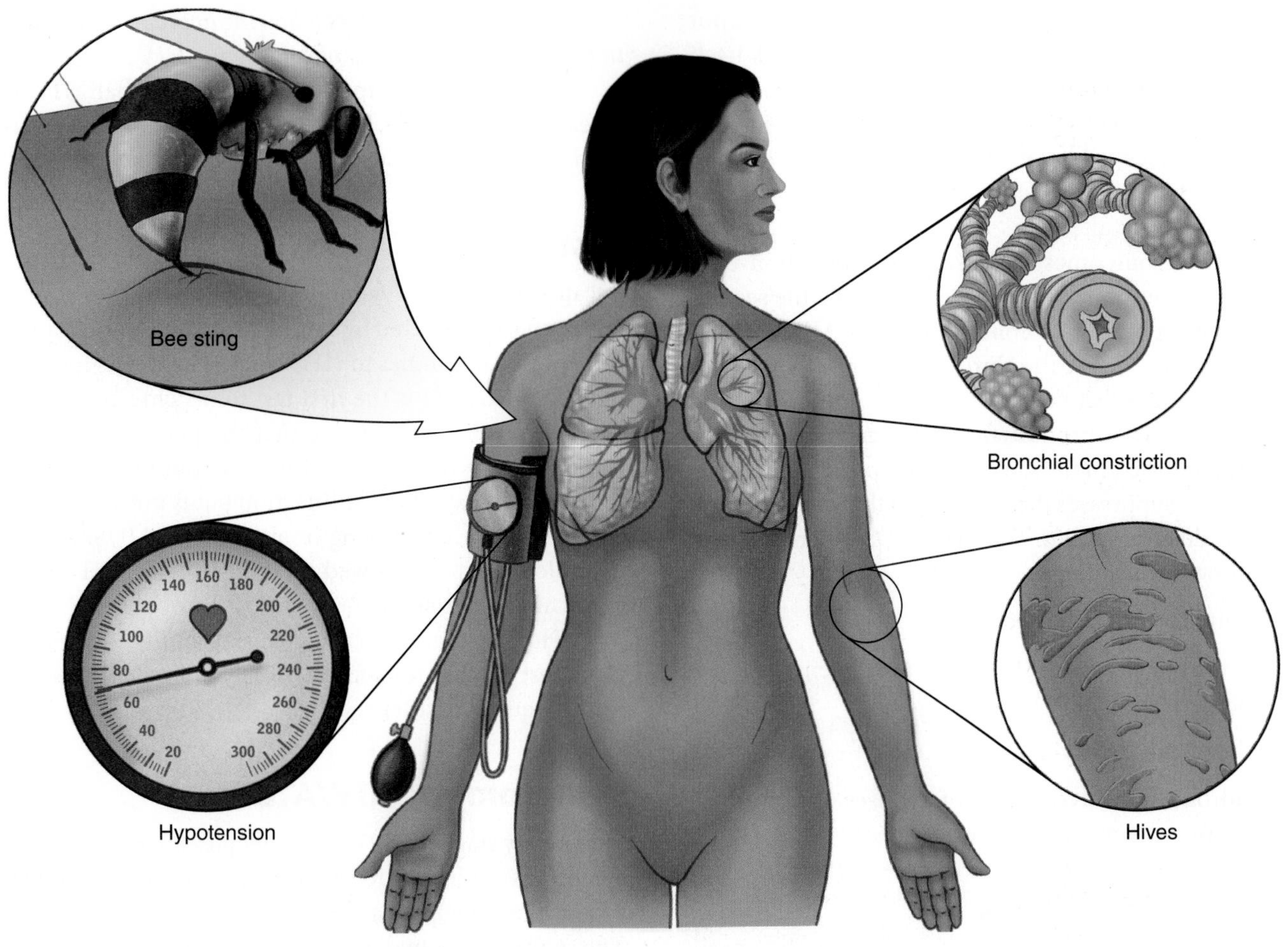

Figure 16-9 *An anaphylactic allergic reaction is the result of a severe antigen-antibody reaction and requires prompt treatment. As an example, the many systems that may respond to someone allergic to a bee sting are shown.*

stimulate antibody formation, some of which are known as the IgE antibodies. Antibodies are found in individuals who are allergic, drug sensitive, or hypersensitive. The antibodies bind to certain cells in the body, causing a characteristic allergic reaction.

In asthma, the IgE antibodies bind to the bronchi and bronchioles; in hay fever, they bind to the mucous membranes of the respiratory tract and eyes, causing a runny nose and itchy eyes. In hives and rashes, they bind to the skin cells and cause reactions across the structure of the skin organ.

An even more severe and sometimes fatal allergic reaction is called **anaphylaxis (anaphylactic shock)** (an-ah-fih-**LACK**-sis). This is the result of an antigen-antibody reaction that stimulates a massive secretion of histamine. Anaphylaxis can be caused by insect stings or by drugs such as penicillin. A person suffering from anaphylaxis experiences breathing problems, headache, facial swelling, falling blood pressure, stomach cramps, and vomiting. The antidote is an injection of either adrenaline or antihistamine. If proper care is not given immediately, death may occur in minutes (**Figure 16-9**).

Health care providers should always ask patients if they are sensitive to any allergens or drugs. This precaution is necessary to prevent negative and sometimes fatal allergic responses to injected drugs. People with such hypersensitivities should wear a medical alert tag to alert health professionals in the event of an emergency. Such tags have saved the lives of patients rendered unconscious or otherwise unable to communicate.

HIV/AIDS

The **human immunodeficiency virus (HIV)** (im-you-noh-dee-**FISH**-en-see) causes AIDS, which is the last stage of HIV infection. HIV progressively destroys the body's T4-helper lymphocyte cells. HIV cannot reproduce on its own; instead, the virus attaches itself to a T-helper cell and fuses with it. The virus action does not cause symptoms. Individuals diagnosed with AIDS are susceptible to *opportunistic infections*, an infection that occurs because of a weakened immune system. It may take years before a person develops AIDS. Three outcomes can result from infection with HIV: one is the actual development of AIDS, the second is the development of a condition called AIDS-related complex (ARC), and the third condition is known as asymptomatic infection.

Acquired immunodeficiency syndrome (AIDS) was first reported in the United States in 1981. AIDS is a disease that suppresses the function of the body's natural immune defense system. The term AIDS is derived from the following meanings:

- *A*cquired—not inherited
- *I*mmune—refers to the body's natural defenses against cancers, disease, and infections
- *D*eficiency—lacks cellular immunity
- *S*yndrome—involves the set of diseases or conditions that are present to signal the diagnosis

HIV/AIDS Statistics

According to the Centers for Disease and Control and Prevention (CDC), about 38,000 people in the United States become infected with HIV annually. More than 1.1 million people in the United States are living with HIV. The overall growth of the epidemic has stabilized in recent years because of the significant number of people receiving antiretroviral therapy (ART).

Transmission of AIDS

The transmission of AIDS occurs in the following ways:

1. Sexual contact with an infected partner; can enter the body through the lining of the vagina, vulva, penis, rectum, or mouth during sex
2. Sharing hypodermic needles among infected intravenous drug users
3. From a mother who is infected to her baby; can occur in utero, at birth, or through breast-feeding
4. Transfusion of contaminated blood or donations of semen, skin grafts, or organ transplants from an infected individual

Health care providers following appropriate precautions are at very little risk of contracting HIV.

Scientists have found no evidence that HIV is spread through sweat, tears, urine, or feces. The virus is fragile and does not survive outside the body.

Tests for HIV/AIDS

Because early HIV infection often causes no symptoms, it is primarily detected by testing a person's blood for the presence of antibodies to HIV. Nucleic acid tests look for the actual virus; it is the first test to become positive. A positive result may indicate that the person has either fought off the infection and is now immune to AIDS, the person is carrying the infection but is not sick, or the person may be developing or already has AIDS. A positive blood test is followed up by another blood test to confirm the diagnosis.

Home testing can be done using an FDA-approved rapid test kit that tests for the HIV antibody. A positive result requires further testing.

Symptoms of HIV/AIDS

In the early stages of HIV, some people experience flulike symptoms, including fever, headache, malaise, a blotchy red rash on the upper torso, and enlarged lymph glands. The symptoms then disappear within a week to a month. More persistent symptoms may not occur for another 10 years; however, even though no symptoms are present, the HIV-infected person can still transmit the disease. If no treatment is given, other symptoms will appear, such as enlarged lymph glands, lack of energy, weight loss, frequent fevers, night sweats, persistent yeast infections (thrush), persistent skin rashes, flaky skin, chronic diarrhea, cough, shortness of breath, and some mental impairment.

As the disease progresses, the untreated person with HIV will develop AIDS and become susceptible to opportunistic infections. Opportunistic conditions include the following (**Figure 16-10**):

- Bacterial infections such as tuberculosis, which is the most common infection
- Cancers, especially Kaposi's sarcoma, cervical cancer, or lymphomas
- Parasitic infections such as *Pneumocystis carinii* pneumonia and toxoplasmosis
- Fungal infections

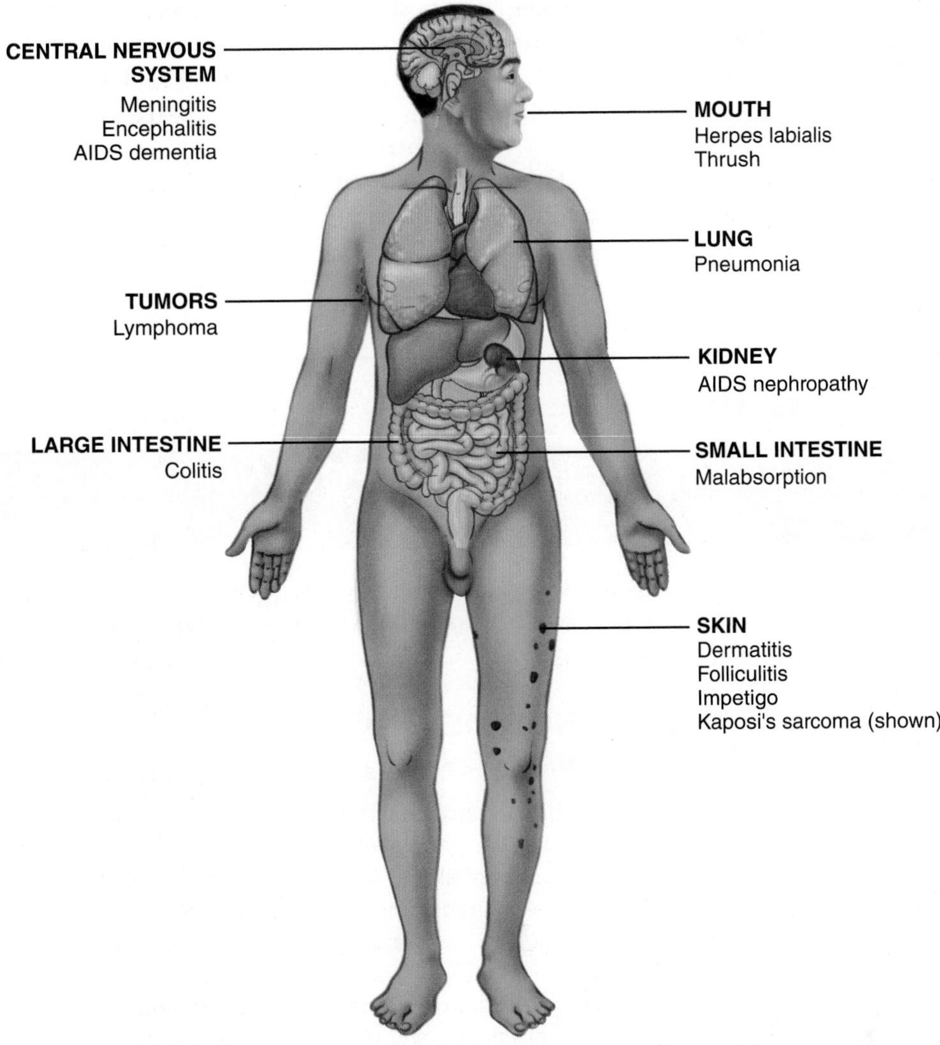

Figure 16-10 *Conditions associated with AIDS*

- Viral infections such as cytomegalovirus (CMV) disease, herpes simplex, hepatitis B, and non-A and non-B hepatitis

Treatment of AIDS

There is no cure for AIDS; however, a variety of drugs can be used in combination to control the virus. ART drugs are used to control the virus by blocking viral replication. Drugs such as protease inhibitors disable the protein protease that HIV needs to replicate itself. Other drugs also inhibit the virus from replicating itself or prevent the virus from inserting itself into the T4 cell. Combinations of these drugs, called the AIDS cocktail, have extended the lives of many infected individuals by stemming the growth and activity of the virus. Pre-exposure prophylaxis (PrEP) treatment may be used by people who are at high risk of being exposed to HIV. It contains the same antiviral drugs that can stop the virus from replicating itself.

Measures to Prevent Transmission of AIDS

The following measures will help prevent transmission of the disease:

- Limit the number of sexual contacts.
- Use treatment as a preventative; a person should take their prescribed drugs.
- Use condoms. The CDC recommends that people use male latex condoms when having oral, anal, or vaginal sex. A person should use a new condom every time they have sex.
- Do not share hypodermic needles or syringes.

- Ensure that soiled articles, materials, and surfaces are cleaned with soap and hot water after incidents involving bleeding.
- Tell their partner if they have HIV.

If a person thinks they have been exposed to HIV, they should notify a physician immediately and start PEP (post-exposure prophylaxis) ART drugs within the first 72 hours; this will greatly reduce the risk of HIV. The ART drugs are taken for 28 days.

To prevent health care providers from contracting AIDS and other diseases, the CDC has published guidelines called standard precautions (see Chapter 17).

Career Profile

CERTIFIED PATIENT CARE TECHNICIAN (CPCT), NURSING AIDE, AND PSYCHIATRIC AIDE

CERTIFIED PATIENT CARE TECHNICIAN (CPCT)

There is an increasing need today for health care providers trained in many diverse areas. Certified Patient Care Technicians (CPCTs) can meet that need. CPCTs take vital signs, perform ECGs and venipunctures (blood drawings), and assist in hemodialysis. CPCTs also take care of the activities of daily living for their patient.

Under the supervision of a registered nurse (RN), CPCTs care for patients in all health care settings including hospitals, skilled nursing facilities, doctor's offices, and health care clinics.

Training can be completed in less than one year. National certification for patient care technicians is required. A competency examination is administered by the National Center for Competency Testing of National Health Career Association.

Growth outlook in this field has a better than average employment rate.

NURSING AIDE AND PSYCHIATRIC AIDE

Nursing aides and psychiatric aides help care for people who are physically or mentally ill; injured; disabled; or confined to hospitals, nursing homes, or residential care facilities.

Nursing aides work under the supervision of nursing and medical staff. They answer call bells; deliver messages; serve meals; make beds; and help patients eat, dress, and bathe. Nursing aides may also provide skin care, take vital signs, and assist patients in and out of bed. They observe patients' physical, mental, and emotional states and report any changes to the nursing or medical staff. Nursing aides employed in nursing homes are often the principal caregivers, having far more contact with the residents than other staff members do.

Psychiatric aides care for people with mental impairments and work under a health care team. In addition to helping patients with the activities of daily living, they socialize with patients and lead them in educational and recreational activities. Because they have the closest contact with patients, psychiatric aides have a great deal of influence on patients' outlook and treatment.

Most states require a nursing aide to have training. Nursing aides employed in nursing homes must complete a minimum of 75 hours of mandatory training and pass a competency examination within four months of employment. Aides who complete the course are placed on the state registry of nursing aides.

In response to the aging population, the job outlook for these types of aides is good and is expected to grow faster than the average.

Table 16-2A *Recommended Immunization Schedule for Persons Aged 0 through 18 Years—United States 2020 (for those who fall behind or start late, see the catch-up schedule [Table 16-2B])*

These recommendations must be read with the notes that follow. For those who fall behind or start late, provide catch-up vaccination at the earliest opportunity as indicated by the green bars. To determine minimum intervals between doses, see the catch-up schedule (Table 16-2B). School entry and adolescent vaccine age groups are shaded in gray.

Vaccine	Birth	1 mo	2 mos	4 mos	6 mos	9 mos	12 mos	15 mos	18 mos	19–23 mos	2–3 yrs	4–6 yrs	7–10 yrs	11–12 yrs	13–15 yrs	16 yrs	17–18 yrs
Hepatitis B (HepB)	1st dose	2nd dose			←—— 3rd dose ——→												
Rotavirus (RV): RV1 (2-dose series), RV5 (3-dose series)			1st dose	2nd dose	See Notes												
Diphtheria, tetanus, acellular pertussis (DTaP <7 yrs)			1st dose	2nd dose	3rd dose			←— 4th dose —→				5th dose					
***Haemophilus influenzae* type b (Hib)**			1st dose	2nd dose	See Notes		←3rd or 4th dose, See Notes→										
Pneumococcal conjugate (PCV13)			1st dose	2nd dose	3rd dose		←— 4th dose —→										
Inactivated poliovirus (IPV <18 yrs)			1st dose	2nd dose	←—— 3rd dose ——→							4th dose					
Influenza (IIV)					Annual vaccination 1 or 2 doses								Annual vaccination 1 dose only				
or													or				
Influenza (LAIV)											Annual vaccination 1 or 2 doses		Annual vaccination 1 dose only				
Measles, mumps, rubella (MMR)					See Notes		←— 1st dose —→					2nd dose					
Varicella (VAR)							←— 1st dose —→					2nd dose					
Hepatitis A (HepA)					See Notes		2-dose series, See Notes										
Tetanus, diphtheria, acellular pertussis (Tdap ≥7 yrs)														Tdap			
Human papillomavirus (HPV)													*	See Notes			
Meningococcal (MenACWY-D ≥9 mos, MenACWY-CRM ≥2 mos)			See Notes											1st dose		2nd dose	
Meningococcal B													See Notes				
Pneumococcal polysaccharide (PPSV23)											See Notes						

Legend:
- Range of recommended ages for all children
- Range of recommended ages for catch-up immunization
- Range of recommended ages for certain high-risk groups
- Recommended based on shared clinical decision-making or *can be used in this age group
- No recommendation/not applicable

Source: The Centers for Disease Control and Prevention

Table 16-2B Catch-Up Immunization Schedule for Persons Aged 4 Months through 18 Years Who Start Late or Who Are More Than 1 Month Behind—United States 2020

The table below provides catch-up schedules and minimum intervals between doses for children whose vaccinations have been delayed. A vaccine series does not need to be restarted, regardless of the time that has elapsed between doses. Use the section appropriate for the child's age. **Always use this table in conjunction with Table 16-2A and the notes that follow.**

Vaccine	Minimum Age for Dose 1	Minimum Interval Between Doses			
		Dose 1 to Dose 2	Dose 2 to Dose 3	Dose 3 to Dose 4	Dose 4 to Dose 5
Children age 4 months through 6 years					
Hepatitis B	Birth	**4 weeks**	**8 weeks *and* at least 16 weeks after first dose.** Minimum age for the final dose is 24 weeks.		
Rotavirus	6 weeks Maximum age for first dose is 14 weeks, 6 days	**4 weeks**	**4 weeks** Maximum age for final dose is 8 months, 0 days.		
Diphtheria, tetanus, and acellular pertussis	6 weeks	**4 weeks**	**4 weeks**	**6 months**	**6 months**
Haemophilus influenzae type b	6 weeks	**No further doses needed** if first dose was administered at age 15 months or older. **4 weeks** if first dose was administered before the 1st birthday. **8 weeks (as final dose)** if first dose was administered at age 12 through 14 months.	**No further doses needed** if previous dose was administered at age 15 months or older. **4 weeks** if current age is younger than 12 months ***and*** first dose was administered at younger than age 7 months ***and*** at least 1 previous dose was PRP-T (ActHib, Pentacel, Hiberix) or unknown. **8 weeks *and* age 12 through 59 months (as final dose)** if current age is younger than 12 months ***and*** first dose was administered at age 7 through 11 months; OR if current age is 12 through 59 months ***and*** first dose was administered before the 1st birthday ***and*** second dose administered at younger than 15 months; OR if both doses were PRP-OMP (PedvaxHIB, Comvax) ***and*** were administered before the 1st birthday.	**8 weeks (as final dose)** This dose only necessary for children age 12 through 59 months who received 3 doses before the 1st birthday.	
Pneumococcal conjugate	6 weeks	**No further doses needed** for healthy children if first dose was administered at age 24 months or older. **4 weeks** if first dose was administered before the 1st birthday. **8 weeks (as final dose for healthy children)** if first dose was administered at the 1st birthday or after.	**No further doses needed** for healthy children if previous dose administered at age 24 months or older. **4 weeks** if current age is younger than 12 months and previous dose was administered at <7 months old. **8 weeks (as final dose for healthy children)** if previous dose was administered between 7–11 months (wait until at least 12 months old); OR if current age is 12 months or older and at least 1 dose was given before age 12 months.	**8 weeks (as final dose)** This dose only necessary for children age 12 through 59 months who received 3 doses before age 12 months or for children at high risk who received 3 doses at any age.	
Inactivated poliovirus	6 weeks	**4 weeks**	**4 weeks** if current age is < 4 years. **6 months (as final dose)** if current age is 4 years or older.	**6 months** (minimum age 4 years for final dose).	
Measles, mumps, rubella	12 months	**4 weeks**			
Varicella	12 months	**3 months**			
Hepatitis A	12 months	**6 months**			
Meningococcal ACWY	2 months MenACWY-CRM 9 months MenACWY-D	**8 weeks**	See Notes	See Notes	
Children and adolescents age 7 through 18 years					
Meningococcal ACWY	Not applicable (N/A)	**8 weeks**			
Tetanus, diphtheria; tetanus, diphtheria, and acellular pertussis	7 years	**4 weeks**	**4 weeks** if first dose of DTaP/DT was administered before the 1st birthday. **6 months (as final dose)** if first dose of DTaP/DT or Tdap/Td was administered at or after the 1st birthday.	**6 months** if first dose of DTaP/DT was administered before the 1st birthday.	
Human papillomavirus	9 years	**Routine dosing intervals are recommended.**			
Hepatitis A	N/A	**6 months**			
Hepatitis B	N/A	**4 weeks**	**8 weeks *and* at least 16 weeks after first dose.**		
Inactivated poliovirus	N/A	**4 weeks**	**6 months** A fourth dose is not necessary if the third dose was administered at age 4 years or older and at least 6 months after the previous dose.	A fourth dose of IPV is indicated if all previous doses were administered at <4 years or if the third dose was administered <6 months after the second dose.	
Measles, mumps, rubella	N/A	**4 weeks**			
Varicella	N/A	**3 months** if younger than age 13 years. **4 weeks** if age 13 years or older.			

Source: The Centers for Disease Control and Prevention

Table 16-2B (*Continued*)

Always use this table in conjunction with Table 16-2A and the notes that follow.

VACCINE	INDICATION									
	Pregnancy	Immunocompromised status (excluding HIV infection)	HIV infection CD4+ count[1]		Kidney failure, end-stage renal disease, or on hemodialysis	Heart disease or chronic lung disease	CSF leaks or cochlear implants	Asplenia or persistent complement component deficiencies	Chronic liver disease	Diabetes
			<15% and total CD4 cell count of <200/mm3	≥15% and total CD4 cell count of ≥200/mm3						
Hepatitis B										
Rotavirus		SCID[2]								
Diphtheria, tetanus, & acellular pertussis (DTaP)										
Haemophilus influenzae type b										
Pneumococcal conjugate										
Inactivated poliovirus										
Influenza (IIV) or Influenza (LAIV)						Asthma, wheezing: 2–4yrs[3]				
Measles, mumps, rubella										
Varicella										
Hepatitis A										
Tetanus, diphtheria, & acellular pertussis (Tdap)										
Human papillomavirus										
Meningococcal ACWY										
Meningococcal B										
Pneumococcal polysaccharide										

- Vaccination according to the routine schedule recommended
- Recommended for persons with an additional risk factor for which the vaccine would be indicated
- Vaccination is recommended, and additional doses may be necessary based on medical condition. See Notes.
- Not recommended/contraindicated—vaccine should not be administered
- Precaution—vaccine might be indicated if benefit of protection outweighs risk of adverse reaction
- Delay vaccination until after pregnancy if vaccine indicated
- No recommendation/not applicable

1 For additional information regarding HIV laboratory parameters and use of live vaccines, see the General Best Practice Guidelines for Immunization, "Altered Immunocompetence," at www.cdc.gov/vaccines/hcp/acip-recs/general-recs/immunocompetence.html and Table 4-1 (footnote D) at www.cdc.gov/vaccines/hcp/acip-recs/general-recs/contraindications.html.

2 Severe Combined Immunodeficiency

3 LAIV contraindicated for children 2–4 years of age with asthma or wheezing during the preceding 12 months.

(Continues)

Table 16-2B (*Continued*)

Notes — Recommended Child and Adolescent Immunization Schedule for ages 18 years or younger, United States, 2020

For vaccine recommendations for persons 19 years of age or older, see the Recommended Adult Immunization Schedule.

Additional information

- Consult relevant ACIP statements for detailed recommendations at www.cdc.gov/vaccines/hcp/acip-recs/index.html.
- For information on contraindications and precautions for the use of a vaccine, consult the General Best Practice Guidelines for Immunization at www.cdc.gov/vaccines/hcp/acip-recs/general-recs/contraindications.html and relevant ACIP statements at www.cdc.gov/vaccines/hcp/acip-recs/index.html.
- For calculating intervals between doses, 4 weeks = 28 days. Intervals of ≥4 months are determined by calendar months.
- Within a number range (e.g., 12–18), a dash (-) should be read as "through."
- Vaccine doses administered ≤4 days before the minimum age or interval are considered valid. Doses of any vaccine administered ≥5 days earlier than the minimum age or minimum interval should not be counted as valid and should be repeated as age-appropriate. The repeat dose should be spaced after the invalid dose by the recommended minimum interval. For further details, see Table 3-1, Recommended and minimum ages and intervals between vaccine doses, in General Best Practice Guidelines for Immunization at www.cdc.gov/vaccines/hcp/acip-recs/general-recs/timing.html.
- Information on travel vaccine requirements and recommendations is available at www.cdc.gov/travel/.
- For vaccination of persons with immunodeficiencies, see Table 8-1, Vaccination of persons with primary and secondary immunodeficiencies, in General Best Practice Guidelines for Immunization at www.cdc.gov/vaccines/hcp/acip-recs/general-recs/immunocompetence.html, and Immunization in Special Clinical Circumstances (In: Kimberlin DW, Brady MT, Jackson MA, Long SS, eds. *Red Book: 2018 Report of the Committee on Infectious Diseases*. 31st ed. Itasca, IL: American Academy of Pediatrics; 2018:67–111).
- For information regarding vaccination in the setting of a vaccine-preventable disease outbreak, contact your state or local health department.
- The National Vaccine Injury Compensation Program (VICP) is a no-fault alternative to the traditional legal system for resolving vaccine injury claims. All routine child and adolescent vaccines are covered by VICP except for pneumococcal polysaccharide vaccine (PPSV23). For more information, see www.hrsa.gov/vaccinecompensation/index.html.

Diphtheria, tetanus, and pertussis (DTaP) vaccination (minimum age: 6 weeks [4 years for Kinrix or Quadracel])

Routine vaccination

- 5-dose series at 2, 4, 6, 15–18 months, 4–6 years
  - **Prospectively:** Dose 4 may be administered as early as age 12 months if at least 6 months have elapsed since dose 3.
  - **Retrospectively:** A 4th dose that was inadvertently administered as early as 12 months may be counted if at least 4 months have elapsed since dose 3.

Catch-up vaccination

- Dose 5 is not necessary if dose 4 was administered at age 4 years or older and at least 6 months after dose 3.
- For other catch-up guidance, see Table 2.

***Haemophilus influenzae* type b vaccination** (minimum age: 6 weeks)

Routine vaccination

- **ActHIB, Hiberix, or Pentacel:** 4-dose series at 2, 4, 6, 12–15 months
- **PedvaxHIB:** 3-dose series at 2, 4, 12–15 months

Catch-up vaccination

- **Dose 1 at 7–11 months:** Administer dose 2 at least 4 weeks later and dose 3 (final dose) at 12–15 months or 8 weeks after dose 2 (whichever is later).
- **Dose 1 at 12–14 months:** Administer dose 2 (final dose) at least 8 weeks after dose 1.
- **Dose 1 before 12 months and dose 2 before 15 months:** Administer dose 3 (final dose) 8 weeks after dose 2.
- **2 doses of PedvaxHIB before 12 months:** Administer dose 3 (final dose) at 12–59 months and at least 8 weeks after dose 2.
- **Unvaccinated at 15–59 months:** 1 dose
- **Previously unvaccinated children age 60 months or older** who are not considered high risk do not require catch-up vaccination.
- For other catch-up guidance, see Table 2.

Special situations

- **Chemotherapy or radiation treatment:**
  12–59 months
  - Unvaccinated or only 1 dose before age 12 months: 2 doses, 8 weeks apart
  - 2 or more doses before age 12 months: 1 dose at least 8 weeks after previous dose

  Doses administered within 14 days of starting therapy or during therapy should be repeated at least 3 months after therapy completion.
- **Hematopoietic stem cell transplant (HSCT):**
  - 3-dose series 4 weeks apart starting 6 to 12 months after successful transplant, regardless of Hib vaccination history
- **Anatomic or functional asplenia (including sickle cell disease):**
  12–59 months
  - Unvaccinated or only 1 dose before age 12 months: 2 doses, 8 weeks apart
  - 2 or more doses before age 12 months: 1 dose at least 8 weeks after previous dose

  Unvaccinated persons age 5 years or older*
  - 1 dose
- **Elective splenectomy:**
  Unvaccinated persons age 15 months or older*
  - 1 dose (preferably at least 14 days before procedure)
- **HIV infection:**
  12–59 months
  - Unvaccinated or only 1 dose before age 12 months: 2 doses, 8 weeks apart
  - 2 or more doses before age 12 months: 1 dose at least 8 weeks after previous dose

  Unvaccinated persons age 5–18 years*
  - 1 dose
- **Immunoglobulin deficiency, early component complement deficiency:**
  12–59 months
  - Unvaccinated or only 1 dose before age 12 months: 2 doses, 8 weeks apart
  - 2 or more doses before age 12 months: 1 dose at least 8 weeks after previous dose

**Unvaccinated* = Less than routine series (through 14 months) OR no doses (15 months or older)

Table 16-2B (*Continued*)

Notes Recommended Child and Adolescent Immunization Schedule for ages 18 years or younger, United States, 2020

Hepatitis A vaccination
(minimum age: 12 months for routine vaccination)

Routine vaccination
- 2-dose series (minimum interval: 6 months) beginning at age 12 months

Catch-up vaccination
- Unvaccinated persons through 18 years should complete a 2-dose series (minimum interval: 6 months).
- Persons who previously received 1 dose at age 12 months or older should receive dose 2 at least 6 months after dose 1.
- Adolescents 18 years and older may receive the combined HepA and HepB vaccine, **Twinrix®**, as a 3-dose series (0, 1, and 6 months) or 4-dose series (0, 7, and 21–30 days, followed by a dose at 12 months).

International travel
- Persons traveling to or working in countries with high or intermediate endemic hepatitis A (www.cdc.gov/travel/):
  - **Infants age 6–11 months**: 1 dose before departure; revaccinate with 2 doses, separated by at least 6 months, between 12 and 23 months of age
  - **Unvaccinated age 12 months and older**: Administer dose 1 as soon as travel is considered.

Hepatitis B vaccination
(minimum age: birth)

Birth dose (monovalent HepB vaccine only)
- **Mother is HBsAg-negative:** 1 dose within 24 hours of birth for **all** medically stable infants ≥2,000 grams. Infants <2,000 grams: Administer 1 dose at chronological age 1 month or hospital discharge.
- **Mother is HBsAg-positive:**
  - Administer **HepB vaccine** and **hepatitis B immune globulin (HBIG)** (in separate limbs) within 12 hours of birth, regardless of birth weight. For infants <2,000 grams, administer 3 additional doses of vaccine (total of 4 doses) beginning at age 1 month.
  - Test for HBsAg and anti-HBs at age 9–12 months. If HepB series is delayed, test 1–2 months after final dose.
- **Mother's HBsAg status is unknown:**
  - Administer **HepB vaccine** within 12 hours of birth, regardless of birth weight.
  - For infants <2,000 grams, administer **HBIG** in addition to HepB vaccine (in separate limbs) within 12 hours of birth. Administer 3 additional doses of vaccine (total of 4 doses) beginning at age 1 month.
  - Determine mother's HBsAg status as soon as possible. If mother is HBsAg-positive, administer **HBIG** to infants ≥2,000 grams as soon as possible, but no later than 7 days of age.

Routine series
- 3-dose series at 0, 1–2, 6–18 months (use monovalent HepB vaccine for doses administered before age 6 weeks)
- Infants who did not receive a birth dose should begin the series as soon as feasible (see Table 2).
- Administration of **4 doses** is permitted when a combination vaccine containing HepB is used after the birth dose.
- **Minimum age** for the final (3rd or 4th) dose: 24 weeks
- **Minimum intervals:** dose 1 to dose 2: 4 weeks / dose 2 to dose 3: 8 weeks / dose 1 to dose 3: 16 weeks (when 4 doses are administered, substitute "dose 4" for "dose 3" in these calculations)

Catch-up vaccination
- Unvaccinated persons should complete a 3-dose series at 0, 1–2, 6 months.
- Adolescents age 11–15 years may use an alternative 2-dose schedule with at least 4 months between doses (adult formulation **Recombivax HB** only).
- Adolescents 18 years and older may receive a 2-dose series of HepB (**Heplisav-B®**) at least 4 weeks apart.
- Adolescents 18 years and older may receive the combined HepA and HepB vaccine, **Twinrix**, as a 3-dose series (0, 1, and 6 months) or 4-dose series (0, 7, and 21–30 days, followed by a dose at 12 months).
- For other catch-up guidance, see Table 2.

Special situations
- Revaccination is not generally recommended for persons with a normal immune status who were vaccinated as infants, children, adolescents, or adults.
- **Revaccination** may be recommended for certain populations, including:
  - **Infants born to HBsAg-positive mothers**
  - **Hemodialysis patients**
  - **Other immunocompromised persons**
- For detailed revaccination recommendations, see www.cdc.gov/vaccines/hcp/acip-recs/vacc-specific/hepb.html.

Human papillomavirus vaccination
(minimum age: 9 years)

Routine and catch-up vaccination
- HPV vaccination routinely recommended at **age 11–12 years (can start at age 9 years)** and catch-up HPV vaccination recommended for all persons through age 18 years if not adequately vaccinated
- 2- or 3-dose series depending on age at initial vaccination:
  - **Age 9 through 14 years at initial vaccination**: 2-dose series at 0, 6–12 months (minimum interval: 5 months; repeat dose if administered too soon)
  - **Age 15 years or older at initial vaccination**: 3-dose series at 0, 1–2 months, 6 months (minimum intervals: dose 1 to dose 2: 4 weeks / dose 2 to dose 3: 12 weeks / dose 1 to dose 3: 5 months; repeat dose if administered too soon)
- If completed valid vaccination series with any HPV vaccine, no additional doses needed

Special situations
- **Immunocompromising conditions, including HIV infection:** 3-dose series as above
- **History of sexual abuse or assault:** Start at age 9 years.
- **Pregnancy:** HPV vaccination not recommended until after pregnancy; no intervention needed if vaccinated while pregnant; pregnancy testing not needed before vaccination

Influenza vaccination
(minimum age: 6 months [IIV], 2 years [LAIV], 18 years [recombinant influenza vaccine, RIV])

Routine vaccination
- Use any influenza vaccine appropriate for age and health status annually:
  - 2 doses, separated by at least 4 weeks, for **children age 6 months–8 years** who have received fewer than 2 influenza vaccine doses before July 1, 2019, or whose influenza vaccination history is unknown (administer dose 2 even if the child turns 9 between receipt of dose 1 and dose 2)
  - 1 dose for **children age 6 months–8 years** who have received at least 2 influenza vaccine doses before July 1, 2019
  - 1 dose for **all persons age 9 years and older**
- For the 2020–21 season, see the 2020–21 ACIP influenza vaccine recommendations.

Special situations
- **Egg allergy, hives only**: Any influenza vaccine appropriate for age and health status annually
- **Egg allergy with symptoms other than hives** (e.g., angioedema, respiratory distress, need for emergency medical services or epinephrine): Any influenza vaccine appropriate for age and health status annually in medical setting under supervision of health care provider who can recognize and manage severe allergic reactions
- **LAIV should not be used** in persons with the following conditions or situations:
  - History of severe allergic reaction to a previous dose of any influenza vaccine or to any vaccine component (excluding egg, see details above)
  - Receiving aspirin or salicylate-containing medications
  - Age 2–4 years with history of asthma or wheezing
  - Immunocompromised due to any cause (including medications and HIV infection)
  - Anatomic or functional asplenia
  - Cochlear implant
  - Cerebrospinal fluid-oropharyngeal communication
  - Close contacts or caregivers of severely immunosuppressed persons who require a protected environment
  - Pregnancy
  - Received influenza antiviral medications within the previous 48 hours

(*Continues*)

Table 16-2B *(Continued)*

Notes — Recommended Child and Adolescent Immunization Schedule for ages 18 years or younger, United States, 2020

Measles, mumps, and rubella vaccination (minimum age: 12 months for routine vaccination)

Routine vaccination
- 2-dose series at 12–15 months, 4–6 years
- Dose 2 may be administered as early as 4 weeks after dose 1.

Catch-up vaccination
- Unvaccinated children and adolescents: 2-dose series at least 4 weeks apart
- The maximum age for use of MMRV is 12 years.

Special situations

International travel
- **Infants age 6–11 months:** 1 dose before departure; revaccinate with 2-dose series with dose 1 at 12–15 months (12 months for children in high-risk areas) and dose 2 as early as 4 weeks later.
- **Unvaccinated children age 12 months and older:** 2-dose series at least 4 weeks apart before departure

Meningococcal serogroup A,C,W,Y vaccination (minimum age: 2 months [MenACWY-CRM, Menveo], 9 months [MenACWY-D, Menactra])

Routine vaccination
- 2-dose series at 11–12 years, 16 years

Catch-up vaccination
- Age 13–15 years: 1 dose now and booster at age 16–18 years (minimum interval: 8 weeks)
- Age 16–18 years: 1 dose

Special situations

Anatomic or functional asplenia (including sickle cell disease), HIV infection, persistent complement component deficiency, complement inhibitor (e.g., eculizumab, ravulizumab) use:
- **Menveo**
  - Dose 1 at age 8 weeks: 4-dose series at 2, 4, 6, 12 months
  - Dose 1 at age 7–23 months: 2-dose series (dose 2 at least 12 weeks after dose 1 and after age 12 months)
  - Dose 1 at age 24 months or older: 2-dose series at least 8 weeks apart
- **Menactra**
  - **Persistent complement component deficiency or complement inhibitor use:**
    - Age 9–23 months: 2-dose series at least 12 weeks apart
    - Age 24 months or older: 2-dose series at least 8 weeks apart
  - **Anatomic or functional asplenia, sickle cell disease, or HIV infection:**
    - Age 9–23 months: Not recommended
    - Age 24 months or older: 2-dose series at least 8 weeks apart
    - **Menactra** must be administered at least 4 weeks after completion of PCV13 series.

Travel in countries with hyperendemic or epidemic meningococcal disease, including countries in the African meningitis belt or during the Hajj (www.cdc.gov/travel/):
- Children less than age 24 months:
  - **Menveo (age 2–23 months):**
    - Dose 1 at 8 weeks: 4-dose series at 2, 4, 6, 12 months
    - Dose 1 at 7–23 months: 2-dose series (dose 2 at least 12 weeks after dose 1 and after age 12 months)
  - **Menactra (age 9–23 months):**
    - 2-dose series (dose 2 at least 12 weeks after dose 1; dose 2 may be administered as early as 8 weeks after dose 1 in travelers)
- Children age 2 years or older: 1 dose **Menveo** or **Menactra**

First-year college students who live in residential housing (if not previously vaccinated at age 16 years or older) or military recruits:
- 1 dose **Menveo** or **Menactra**

Adolescent vaccination of children who received MenACWY prior to age 10 years:
- **Children for whom boosters are recommended** because of an ongoing increased risk of meningococcal disease (e.g., those with complement deficiency, HIV, or asplenia): Follow the booster schedule for persons at increased risk (see below).
- **Children for whom boosters are not recommended** (e.g., those who received a single dose for travel to a country where meningococcal disease is endemic): Administer MenACWY according to the recommended adolescent schedule with dose 1 at age 11–12 years and dose 2 at age 16 years.

Note: Menactra should be administered either before or at the same time as DTaP. For MenACWY **booster dose recommendations** for groups listed under "Special situations" and in an outbreak setting and for additional meningococcal vaccination information, see www.cdc.gov/vaccines/hcp/acip-recs/vacc-specific/mening.html.

Meningococcal serogroup B vaccination (minimum age: 10 years [MenB-4C, Bexsero; MenB-FHbp, Trumenba])

Shared clinical decision-making
- **Adolescents not at increased risk** age 16–23 years (preferred age 16–18 years) based on shared clinical decision-making:
  - **Bexsero:** 2-dose series at least 1 month apart
  - **Trumenba:** 2-dose series at least 6 months apart; if dose 2 is administered earlier than 6 months, administer a 3rd dose at least 4 months after dose 2.

Special situations

Anatomic or functional asplenia (including sickle cell disease), persistent complement component deficiency, complement inhibitor (e.g., eculizumab, ravulizumab) use:
- **Bexsero:** 2-dose series at least 1 month apart
- **Trumenba:** 3-dose series at 0, 1–2, 6 months

Bexsero and **Trumenba** are not interchangeable; the same product should be used for all doses in a series.

For MenB **booster dose recommendations** for groups listed under "Special situations" and in an outbreak setting and for additional meningococcal vaccination information, see www.cdc.gov/vaccines/acip/recommendations.html and www.cdc.gov/vaccines/hcp/acip-recs/vacc-specific/mening.html.

Pneumococcal vaccination (minimum age: 6 weeks [PCV13], 2 years [PPSV23])

Routine vaccination with PCV13
- 4-dose series at 2, 4, 6, 12–15 months

Catch-up vaccination with PCV13
- 1 dose for healthy children age 24–59 months with any incomplete* PCV13 series
- For other catch-up guidance, see Table 2.

Special situations

High-risk conditions below: When both PCV13 and PPSV23 are indicated, administer PCV13 first. PCV13 and PPSV23 should not be administered during the same visit.

Chronic heart disease (particularly cyanotic congenital heart disease and cardiac failure), chronic lung disease (including asthma treated with high-dose, oral corticosteroids), diabetes mellitus:

Age 2–5 years
- Any incomplete* series with:
  - 3 PCV13 doses: 1 dose PCV13 (at least 8 weeks after any prior PCV13 dose)
  - Less than 3 PCV13 doses: 2 doses PCV13 (8 weeks after the most recent dose and administered 8 weeks apart)
- No history of PPSV23: 1 dose PPSV23 (at least 8 weeks after any prior PCV13 dose)

Age 6–18 years
- No history of PPSV23: 1 dose PPSV23 (at least 8 weeks after any prior PCV13 dose)

Cerebrospinal fluid leak, cochlear implant:

Age 2–5 years
- Any incomplete* series with:
  - 3 PCV13 doses: 1 dose PCV13 (at least 8 weeks after any prior PCV13 dose)
  - Less than 3 PCV13 doses: 2 doses PCV13 (8 weeks after the most recent dose and administered 8 weeks apart)
- No history of PPSV23: 1 dose PPSV23 (at least 8 weeks after any prior PCV13 dose)

Age 6–18 years
- No history of either PCV13 or PPSV23: 1 dose PCV13, 1 dose PPSV23 at least 8 weeks later
- Any PCV13 but no PPSV23: 1 dose PPSV23 at least 8 weeks after the most recent dose of PCV13
- PPSV23 but no PCV13: 1 dose PCV13 at least 8 weeks after the most recent dose of PPSV23

Table 16-2B (*Continued*)

Notes Recommended Child and Adolescent Immunization Schedule for ages 18 years or younger, United States, 2020

Sickle cell disease and other hemoglobinopathies; anatomic or functional asplenia; congenital or acquired immunodeficiency; HIV infection; chronic renal failure; nephrotic syndrome; malignant neoplasms, leukemias, lymphomas, Hodgkin disease, and other diseases associated with treatment with immunosuppressive drugs or radiation therapy; solid organ transplantation; multiple myeloma:

Age 2–5 years

- Any incomplete* series with:
  - 3 PCV13 doses: 1 dose PCV13 (at least 8 weeks after any prior PCV13 dose)
  - Less than 3 PCV13 doses: 2 doses PCV13 (8 weeks after the most recent dose and administered 8 weeks apart)
- No history of PPSV23: 1 dose PPSV23 (at least 8 weeks after any prior PCV13 dose) and a 2nd dose of PPSV23 5 years later

Age 6–18 years

- No history of either PCV13 or PPSV23: 1 dose PCV13, 2 doses PPSV23 (dose 1 of PPSV23 administered 8 weeks after PCV13 and dose 2 of PPSV23 administered at least 5 years after dose 1 of PPSV23)
- Any PCV13 but no PPSV23: 2 doses PPSV23 (dose 1 of PPSV23 administered 8 weeks after the most recent dose of PCV13 and dose 2 of PPSV23 administered at least 5 years after dose 1 of PPSV23)
- PPSV23 but no PCV13: 1 dose PCV13 at least 8 weeks after the most recent PPSV23 dose and a 2nd dose of PPSV23 administered 5 years after dose 1 of PPSV23 and at least 8 weeks after a dose of PCV13

Chronic liver disease, alcoholism:

Age 6–18 years

- No history of PPSV23: 1 dose PPSV23 (at least 8 weeks after any prior PCV13 dose)

**Incomplete series* = Not having received all doses in either the recommended series or an age-appropriate catch-up series See Tables 8, 9, and 11 in the ACIP pneumococcal vaccine recommendations at www.cdc.gov/mmwr/pdf/rr/rr5911.pdf for complete schedule details.

Poliovirus vaccination
(minimum age: 6 weeks)

Routine vaccination

- 4-dose series at ages 2, 4, 6–18 months, 4–6 years; administer the final dose at or after age 4 years and at least 6 months after the previous dose.
- 4 or more doses of IPV can be administered before age 4 years when a combination vaccine containing IPV is used. However, a dose is still recommended at or after age 4 years and at least 6 months after the previous dose.

Catch-up vaccination

- In the first 6 months of life, use minimum ages and intervals only for travel to a polio-endemic region or during an outbreak.
- IPV is not routinely recommended for U.S. residents 18 years and older.

Series containing oral polio vaccine (OPV), either mixed OPV-IPV or OPV-only series:

- Total number of doses needed to complete the series is the same as that recommended for the U.S. IPV schedule. See www.cdc.gov/mmwr/volumes/66/wr/mm6601a6.htm?s_cid=mm6601a6_w.
- Only trivalent OPV (tOPV) counts toward the U.S. vaccination requirements.
  - Doses of OPV administered before April 1, 2016, should be counted (unless specifically noted as administered during a campaign).
  - Doses of OPV administered on or after April 1, 2016, should not be counted.
  - For guidance to assess doses documented as "OPV," see www.cdc.gov/mmwr/volumes/66/wr/mm6606a7.htm?s_cid=mm6606a7_w.
- For other catch-up guidance, see Table 2.

Rotavirus vaccination
(minimum age: 6 weeks)

Routine vaccination

- **Rotarix:** 2-dose series at 2 and 4 months
- **RotaTeq:** 3-dose series at 2, 4, and 6 months
- If any dose in the series is either **RotaTeq** or unknown, default to 3-dose series.

Catch-up vaccination

- Do not start the series on or after age 15 weeks, 0 days.
- The maximum age for the final dose is 8 months, 0 days.
- For other catch-up guidance, see Table 2.

Tetanus, diphtheria, and pertussis (Tdap) vaccination
(minimum age: 11 years for routine vaccination, 7 years for catch-up vaccination)

Routine vaccination

- **Adolescents age 11–12 years:** 1 dose Tdap
- **Pregnancy:** 1 dose Tdap during each pregnancy, preferably in early part of gestational weeks 27–36
- Tdap may be administered regardless of the interval since the last tetanus- and diphtheria-toxoid-containing vaccine.

Catch-up vaccination

- **Adolescents age 13–18 years who have not received Tdap:** 1 dose Tdap, then Td or Tdap booster every 10 years
- **Persons age 7–18 years not fully vaccinated* with DTaP:** 1 dose Tdap as part of the catch-up series (preferably the first dose); if additional doses are needed, use Td or Tdap.
- **Tdap administered at 7–10 years:**
  - **Children age 7–9 years** who receive Tdap should receive the routine Tdap dose at age 11–12 years.
  - **Children age 10 years** who receive Tdap do not need to receive the routine Tdap dose at age 11–12 years.
- **DTaP inadvertently administered at or after age 7 years:**
  - **Children age 7–9 years:** DTaP may count as part of catch-up series. Routine Tdap dose at age 11–12 years should be administered.
  - **Children age 10–18 years:** Count dose of DTaP as the adolescent Tdap booster.
- For other catch-up guidance, see Table 2.
- For information on use of Tdap or Td as tetanus prophylaxis in wound management, see www.cdc.gov/mmwr/volumes/67/rr/rr6702a1.htm.

**Fully vaccinated* = 5 valid doses of DTaP OR 4 valid doses of DTaP if dose 4 was administered at age 4 years or older

Varicella vaccination
(minimum age: 12 months)

Routine vaccination

- 2-dose series at 12–15 months, 4–6 years
- Dose 2 may be administered as early as 3 months after dose 1 (a dose administered after a 4-week interval may be counted).

Catch-up vaccination

- Ensure persons age 7–18 years without evidence of immunity (see www.cdc.gov/mmwr/pdf/rr/rr5604.pdf) have 2-dose series:
  - **Age 7–12 years:** routine interval: 3 months (a dose administered after a 4-week interval may be counted)
  - **Age 13 years and older:** routine interval: 4–8 weeks (minimum interval: 4 weeks)
  - The maximum age for use of MMRV is 12 years.

1/29/2020 Centers for Disease Control and Prevention | Recommended Child and Adolescent Immunization Schedule, United States, 2020

(Continues)

Career Profile

HOME HEALTH AIDE

Home health aides help older adults or people who are disabled or ill and live in their homes instead of in a health care facility.

Home health aides provide housekeeping services, personal care, and emotional support for their clients. Aides may plan meals, shop for food, and cook. Home health aides take vital signs, help get clients in and out of bed, and assist with medication routines. Occasionally, they change nonsterile dressings, use special equipment such as hydraulic lifts, and give massages.

Home health aides also provide psychological support. They assist with toilet training for children with severe mental handicaps or just listen to clients talk about their problems. In home care agencies, aides are supervised by a registered nurse, a physical therapist, or a social worker who assigns them specific duties.

The federal government has enacted guidelines for home health aides who receive reimbursement from Medicare. Federal law requires home health aides to pass a competency test covering 12 areas. Federal law suggests at least 75 hours of classroom and practical training supervised by a registered nurse. Home health aides who do not receive reimbursement under Medicare provisions may receive on-the-job training.

The job outlook for this profession is good—it is expected to be one of the fastest growing careers in the years ahead.

One BODY How the Lymphatic and Immune Systems Interact with Other Body Systems

INTEGUMENTARY SYSTEM

- The intact skin and normal flora of the skin is the body's first line of defense against invading pathogens.

SKELETAL SYSTEM

- Red bone marrow is where the lymphocytes are produced.

MUSCULAR SYSTEM

- Action of the muscles is necessary for the return of lymph through the capillaries, vessels, and ducts to the veins of the circulatory system. Inflammation is a nonspecific response to an injury of the muscular system.

NERVOUS SYSTEM

- Excess stress diminishes the action of the immune system.

ENDOCRINE SYSTEM

- The hormone thymosin stimulates the production of lymphoid cells.

CARDIOVASCULAR SYSTEM

- Blood capillaries and lymph capillaries maintain the blood and tissue volume of fluid in the body. Lymphocytes and other antibodies travel through the circulatory system to the site of infection.

RESPIRATORY SYSTEM

- Tonsils present in the throat defend against invading bacteria and other pathogens from entering the lungs.

DIGESTIVE SYSTEM

- The lacteals of the lymph absorb fat and fat-soluble vitamins. Normal flora of intestines destroys invading pathogens.

URINARY SYSTEM

- Interstitial fluid carries away the waste products from the urinary system. The flushing action of urination prevents microorganisms from ascending through the urinary tract.

REPRODUCTIVE SYSTEM

- Vaginal flora prevents microbial growth of pathogens. A baby receives temporary passive immunity from the mother's antibodies during pregnancy that usually lasts about six months after birth.

Medical Terminology

hyper-	over
sensitive	sensitive
-ity	condition of
hyper/sensitiv/ity	condition of being oversensitive
immun	not serving disease
immun/ity	condition of not serving disease; represents protection against disease
-tion	process of
immun/iza/tion	process of protection against disease
inter-	between
-stitial	tissues
inter/stitial	between the body tissues
lymph	clear white fluid
-oma	tumor
lymph/oma	tumor of the lymph
aden	glands
-itis	inflammation of
lymph aden/itis	inflammation of the lymph glands
sclero	hard
-derma	skin
sclero/derma	condition in which the skin hardens

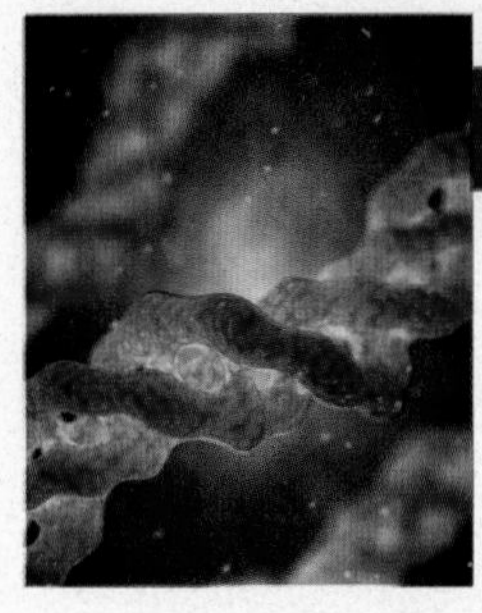

Study Tools

Workbook	Activities for Chapter 16
Online Resources	• PowerPoint presentations • Animations

REVIEW QUESTIONS

Select the letter of the choice that best completes the statement.

1. Lymph fluid contains all but
a. red blood cells.
b. hormones.
c. salts.
d. digested nutrients.

2. The function of the lymph nodes is to produce
a. platelets.
b. lymphocytes.
c. basophils.
d. erythrocytes.

3. The name of the vessel through which lymph finally rejoins general circulation is called the
a. thoracic duct.
b. left lymphatic duct.
c. superior vena cava.
d. right lymphatic duct.

4. The organ composed of lymphatic tissue that filters blood and produces white blood cells is called the
a. spleen.
b. liver.
c. kidney.
d. stomach.

5. The ability of the body to resist disease is known as
a. sensitivity.
b. resistance.
c. immunity.
d. noninfection.

6. Peyer's patches are lymphoid tissue found in the
a. lymph vessels.
b. lymph nodes.
c. intestinal walls.
d. thymus gland.

7. B lymphocyte cells form
a. helper T cells.
b. plasma cells.
c. killer T cells.
d. macrophages.

8. The type of immunoglobulin found in saliva is
a. IgA.
b. IgG.
c. IgD.
d. IgM.

9. The protective ring of lymphoid tissue around the back of the nose and throat is formed by the
a. lacteals.
b. villi.
c. tonsils.
d. lymph nodes.

10. An opportunistic infection that is frequently associated with HIV is
a. Hodgkin's disease.
b. Kaposi's sarcoma.
c. Myasthenia gravis.
d. Graves' disease.

MATCHING

Match each term in Column I with its correct description in Column II.

	COLUMN I	*COLUMN II*
__________	**1.** natural immunity	a. immunization
__________	**2.** acquired active immunity	b. immunoglobulin
__________	**3.** acquired passive immunity	c. inherited
__________	**4.** acquired active artificial immunity	d. obtained through human breast milk
__________	**5.** acquired passive artificial immunity	e. have the disease and recover

FILL IN THE BLANKS

1. A person who is highly sensitive to an allergen is said to be ______________.
2. An antigen that causes an allergic response is a(n) ______________.
3. A fatal allergic response is ______________.
4. A cancer of the lymph nodes is called ______________.
5. A lymph organ that is also part of the endocrine gland is the ______________.

APPLYING THEORY TO PRACTICE

1. Riley, age 5, is entering school and must have his immunizations completed. Explain to his mother, Mrs. Ayers, what this means. List the immunizations necessary for a 5-year-old child.
2. Your friend says, "I think I have the kissing disease. What is it?" Explain the disease to your friend and how the disease is transmitted and treated.
3. Your grandmother, age 79, says her physician is recommending she get a flu shot. She states she has always been healthy and asks you why she needs this shot.
4. Keisha is a medical assistant working in a pediatrician's office. Mrs. Romano brings her daughter Molly, age 7, to the office for her usual checkup. Molly is going to camp. She is concerned that Molly may be exposed to all kinds of infections or disease. What will Keisha explain to Mrs. Romano about the body's normal defense mechanisms and how they will protect Molly?
5. Many diseases are being classified as autoimmune. Explain autoimmunity and lupus to a new patient with no knowledge of autoimmune diseases and share other examples of autoimmune diseases.
6. Multiple systems work together to keep the body in homeostasis. Compare and contrast the circulatory system and the lymphatic system, highlighting how they are similar and how they are different.

CASE STUDY

Mrs. Gonzalez is a volunteer at a hospice. She is confused about the disease HIV/AIDS and is afraid if she comes in contact with a patient's urine or feces, she may get the disease. Lauren, the licensed practical nurse, tries to reassure Mrs. Gonzalez she cannot get HIV/AIDS in this manner. The following actions may help Lauren explain the disease to Mrs. Gonzalez.

1. Explain what HIV and AIDS are.
2. Describe the modes of transmission.
3. Discuss the early and later symptoms.
4. Describe opportunistic infections.
5. Investigate and report the diagnostic technologies used to detect it.
6. Investigate and report the therapeutic technologies used to treat HIV and the treatment's limitations.
7. Explain how to prevent HIV/AIDS.

LAB ACTIVITY 16-1 Lymph Nodes

- ***Objective:*** To observe the structure of the lymph nodes
- ***Materials needed:*** microscopic slide of lymph node, microscope, paper, pencil

Step 1: Examine the slide of the lymph node.

Step 2: Describe and identify the parts: trabeculae, germinal center, lymphocytes, and antigen. Record your findings.

LAB ACTIVITY 16-2 Lymph Vessels

- ***Objective:*** To observe the location and function of the lymph vessels
- ***Materials needed:*** unlabeled chart of lymph vessels, textbook, paper, pencil

Step 1: On the unlabeled chart of the lymph vessels, locate the right lymphatic duct and the left lymphatic duct. Compare with the diagram in the textbook.

Step 2: Into what structures do the lymph fluids from the right and left lymphatic ducts empty? Record your answer.

LAB ACTIVITY 16-3 Lymphatic System Flow

- ***Objective:*** To demonstrate how lymph fluid is drained from the body
- ***Materials needed:*** Two plastic rubber gloves, needle, water, pen, paper, textbook

Step 1: Take a plastic glove and poke holes with the needle in the fingertips of the glove.

Step 2: Fill the glove with water. Observe how the fluid drains through the openings.

Step 3: Fill the other glove with water and see how the fluid remains in the glove.

Step 4: What is the difference between step 2 and 3? Write down your observation.

Step 5: Report to a partner how lymph collects in the body, and how it reaches the superior vena cava. Check any differences you report against the textbook.

CHAPTER 17

Infection Control and Standard Precautions

Objectives

- Describe six types of pathogenic microorganisms.
- Explain the infectious process and the chain of infection.
- Describe methods to break the chain of infection.
- Describe the stages of infection.
- Explain standard precautions.
- Define the key words that relate to this chapter.

Key Words

airborne transmission
asepsis
bacteria
biological agents
C. diff
carriers
chemical agents
cleansing
compromised host
contact transmission
convalescent stage
disinfection
droplet transmission
flora
fomites
fungi
helminths
hospital-acquired infection (HAI)
host
illness stage
incubation stage
infectious agent
means of transmission
MRSA
nosocomial infection
parasites
pathogenicity
physical agents
portal of entry
portal of exit
prodromal stage
protozoa
reservoir
resident flora
reverse isolation
rickettsia
spores
sterilization
susceptible host
transient flora
vector-borne transmission
vehicle transmission
virulence
viruses
VISA
VRSA

Health care providers are responsible for providing care that utilizes infection control principles to provide a safe environment. The practice of medical **asepsis** (ay-**SEP**-sis) is to reduce and prevent the spread of pathogens. This chapter discusses infection control principles as they relate to microorganisms; pathogens; infection and colonization; chain of infection; stages of the infectious process; and nosocomial infections (nos-oh-**KOH**-mee-al in-**FECK**-shuns), also known as hospital-acquired infections (HAIs).

FLORA

Flora are microorganisms that occur in, or have adapted to live in, a specific environment, such as the intestine, skin, vagina, or oral cavity. There are two types of flora: resident and transient. **Resident flora**, or normal flora, are always present. They prevent the overgrowth of harmful microorganisms. Only when the homeostasis is upset does disease result. An example is *Propionibacterium* found on the skin, which can cause acne. **Transient flora** occur in periods of limited duration. An example is *Staphylococcus aureus,* which causes skin infections. Transient flora attach to the skin for a brief time but do not continually live on the skin. Vigorous handwashing with soap and water or the use of an alcohol-based hand sanitizer is an effective means of removing most flora.

Pathogenicity and Virulence

Most microorganisms found in the environment do not cause disease or infection, but some do. Disease may be defined in several ways. It may be defined as a change in structure or function within the body that is considered abnormal, or it may be defined as any change from normal. It usually refers to a condition in which abnormal symptoms occur and a pathological state is present. Both of these definitions have one thing in common: an alteration of homeostasis. Disease-producing microorganisms are called pathogens; **pathogenicity** (**path**-oh-jeh-**NIS**-ih-tee) refers to the ability of a microorganism to produce disease. **Virulence** (**VEER**-you-lens) refers to the frequency with which a pathogen causes disease. Factors affecting virulence are the strength of the pathogen to adhere to healthy cells, the ability of the pathogen to damage cells or interfere with the body's normal systems, and the ability of the pathogen to evade the action of the white blood cells. There are six types of pathogenic microorganisms: bacteria, viruses, fungi, protozoa, rickettsia, and helminths (**Table 17-1**).

BACTERIA

Bacteria (back-**TEER**-ee-uh) are small, one-celled microorganisms that lack a true nucleus or mechanism to provide metabolism. Bacteria need an environment that will provide food for survival. Although most bacteria multiply by simple cell division, some forms of bacteria

Table 17-1 *Some Common Infections Caused by Microorganisms in Humans*

BACTERIA	***VIRUSES***	***FUNGI***
Staphylococcus	Common cold	Ringworm *(Tinea)*
Streptococcus	Herpes simplex	Athlete's foot
E. coli	Mononucleosis	*Candidiasis*
Klebsiella	HIV	Thrush *(Candida albicans)*
Pseudomonas	Measles	Vaginitis
Shigella	Mumps	Histoplasmosis
Salmonella	Rubella	*Coccidioidomycosis*
	Influenza (flu)	
RICKETTSIA	***PROTOZOA***	***HELMINTHS***
Rocky Mountain spotted fever	Malaria	Roundworms
		Flatworms
Lyme disease	Giardiasis	Pinworms
Typhus		Tapeworms

produce **spores**, a resistant stage that withstands an unfavorable environment. When proper environmental conditions return, spores germinate and form new cells. Spores are resistant to heat, drying, and disinfectants.

Pathogenic bacteria cause a wide range of illnesses including diarrhea, pneumonia, sinusitis, urinary tract infections, and gonorrhea (**Figure 17-1**).

Rickettsia (rih-**KET**-see-uh) infections are caused by multiple types of bacteria that can only live inside the cells of another organism. Infection from rickettsia is spread through the bites of fleas, ticks, mites, and lice. Common infections are Lyme disease, Rocky Mountain spotted fever, and typhus.

Did You Know?

What causes "stinky feet"? Blame the foot odor, called bromohidrosis, on normal bacterial flora. The foot sweat is broken down into small noxious fatty acids by the bacteria on the feet, namely, Corynebacteria *and* Micrococci. *These bacteria thrive on moisture, warmth, and darkness.*

VIRUSES

Viruses (**VYE**-ruh-sez) are organisms that can live only inside cells. They cannot get nourishment or reproduce outside the cell. Viruses contain a core of DNA or RNA surrounded by a protein coating. Some viruses have the ability to create an additional coating called an envelope. This envelope protects the virus from attack by the immune system. Viruses damage the cell they inhabit by blocking the normal protein synthesis and by using the cell's mechanism for metabolism to reproduce themselves.

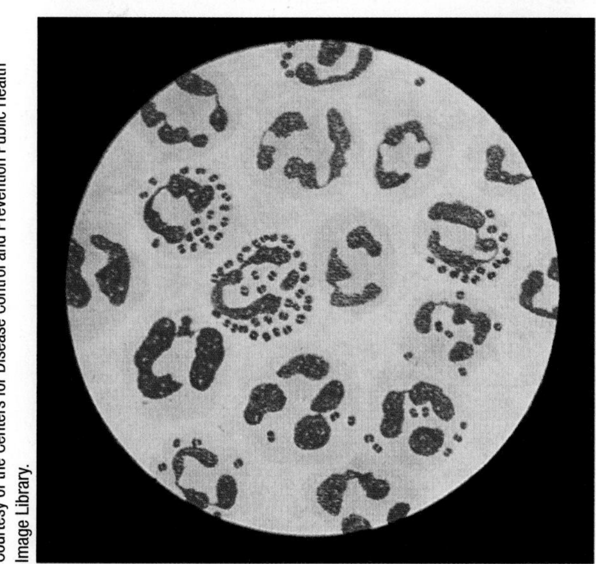

Courtesy of the Centers for Disease Control and Prevention Public Health Image Library.

Figure 17-1 *Neisseria Gonorrhoeae*

The same viral infection may cause different symptoms in different individuals. Some viruses will immediately trigger a disease response, while others may remain latent for many years. Viral infections include the common cold, influenza, measles, hepatitis, genital herpes, HIV, COVID-19, Ebola virus, and the West Nile virus (**Figure 17-2**).

FUNGI

Fungi (**FUN**-jye) are microscopic, plantlike organisms that may cause disease. The diseases are referred to as mycoses. Yeast is a single-cell form of fungi. Some fungi and yeast are beneficial to humans; fungi obtain food from living organisms or organic matter. Disease from fungi is found mainly in individuals who are immunologically impaired. Fungi can cause infections of the hair, skin, nails, and mucous membranes; for example, thrush and athlete's foot are both fungal infections (**Figure 17-3**).

PARASITES

Parasitic infections are caused by **parasites**, which are organisms that live off of another organism to survive. The deadliest parasitic infection in the world is malaria.

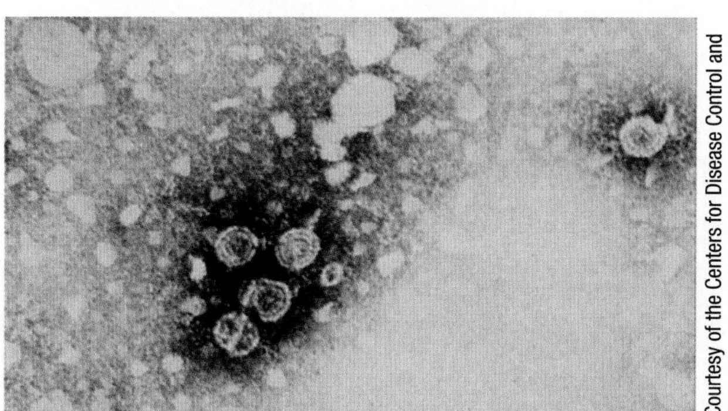

Courtesy of the Centers for Disease Control and Prevention Public Health Image Library.

Figure 17-2 *Electron micrograph of hepatitis B virus*

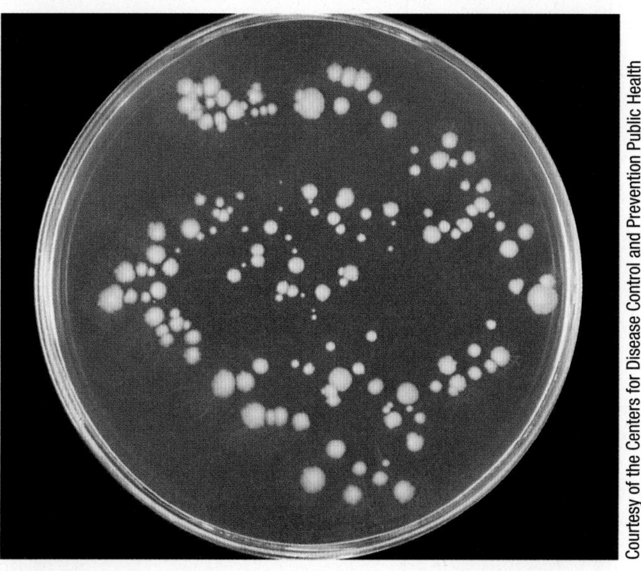

Courtesy of the Centers for Disease Control and Prevention Public Health Image Library.

Figure 17-3 *Candida albicans, commonly known as thrush*

Although malaria is not a disease found in the United States, there are other parasites in the United States that can cause illness such as giardiasis, toxoplasmosis, trichomoniasis, and cryptosporidiosis.

PROTOZOA

Protozoa (pro-toh-**ZOH**-uh) are single-celled, parasitic organisms with the ability to move (**Figure 17-4**). Most protozoa obtain their food from dead or decaying organic matter. Infection is spread through ingestion of contaminated food or water or through insect bites. Common infections are malaria, gastroenteritis, and vaginal infections.

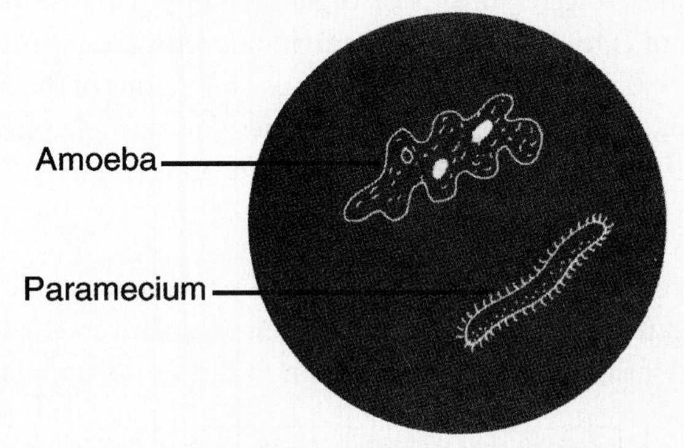

Figure 17-4 *Protozoa*

HELMINTHS

Helminths (**HELL**-minths), or parasitic worms, are any one of the roundworms, flatworms, pinworms, or tapeworms. They are the most common helminths. Pinworms cause anal itching but do not cause serious illness. Tapeworms may cause intestinal disease in humans. Tapeworms can be acquired by eating contaminated food or water or by eating uncooked or inadequately cooked meat.

CHAIN OF INFECTION

The chain of infection describes the elements of an infectious process. It is an interactive process that involves an agent, a host, and the environment. This process must include several essential elements, or "links in the chain," for the transmission of microorganisms to occur. **Figure 17-5** identifies the six essential links in the chain of infection. Without the transmission of microorganisms, the infectious process cannot occur. Knowledge about the chain of infection facilitates control or prevention of disease by breaking the links in the chain. This is achieved by altering one or more of the interactive processes of agent, host, or environment.

Medical Highlights

COVID-19

The world came to a halt in 2020 because of the first global pandemic in more than 100 years. This pandemic was caused by the expeditious spread of coronavirus-2 (SARS-CoV-2), the virus that causes COVID-19. COVID-19 is an acute respiratory syndrome that first emerged in Wuhan, China, in December of 2019. Public health experts were not prepared for the rapid spread. It wasn't until March 11, 2020, that the World Health Organization declared a global pandemic. There have been other coronaviruses identified, but COVID-19 is very different and much less predictable. The long list of symptoms is unusual, with the main point of attack being the lungs, but other symptoms have been identified.

COVID-19 affects individuals differently. The effects of the virus include being asymptomatic, severely ill, or death. The incubation period of the virus is 2–14 days, with symptoms appearing within this time period. Most individuals display symptoms 4–5 days after exposure. Mild to moderate symptoms include fever, chills, cough, shortness of breath or difficulty breathing, fatigue, muscle or body aches, headache, loss of taste and/or smell, sore throat, congestion or runny nose, nausea and/or vomiting, and diarrhea. Severe symptoms include trouble breathing, persistent pain or pressure in the chest, confusion, inability to wake or stay awake, and cyanotic lips and face.

Mild cases of COVID-19 are treated at home by self-isolating, keeping hydrated, using over-the-counter pain medication, and resting. Severe cases may require hospitalization for supportive care. Because there are no FDA-approved treatments at this time for COVID-19, doctors are using various therapies including antiviral drugs, steroids, convalescent plasma, and disease-modifying antirheumatic drugs (DMARDS). A vaccine has been approved by the FDA that will help slow the spread of the virus and prevent individuals from becoming sick.

Prevention of COVID-19 spread includes wearing masks in any public setting, frequent handwashing, use of alcohol-based hand sanitizer with 60%–95% alcohol, and social distancing at a minimum of six feet.

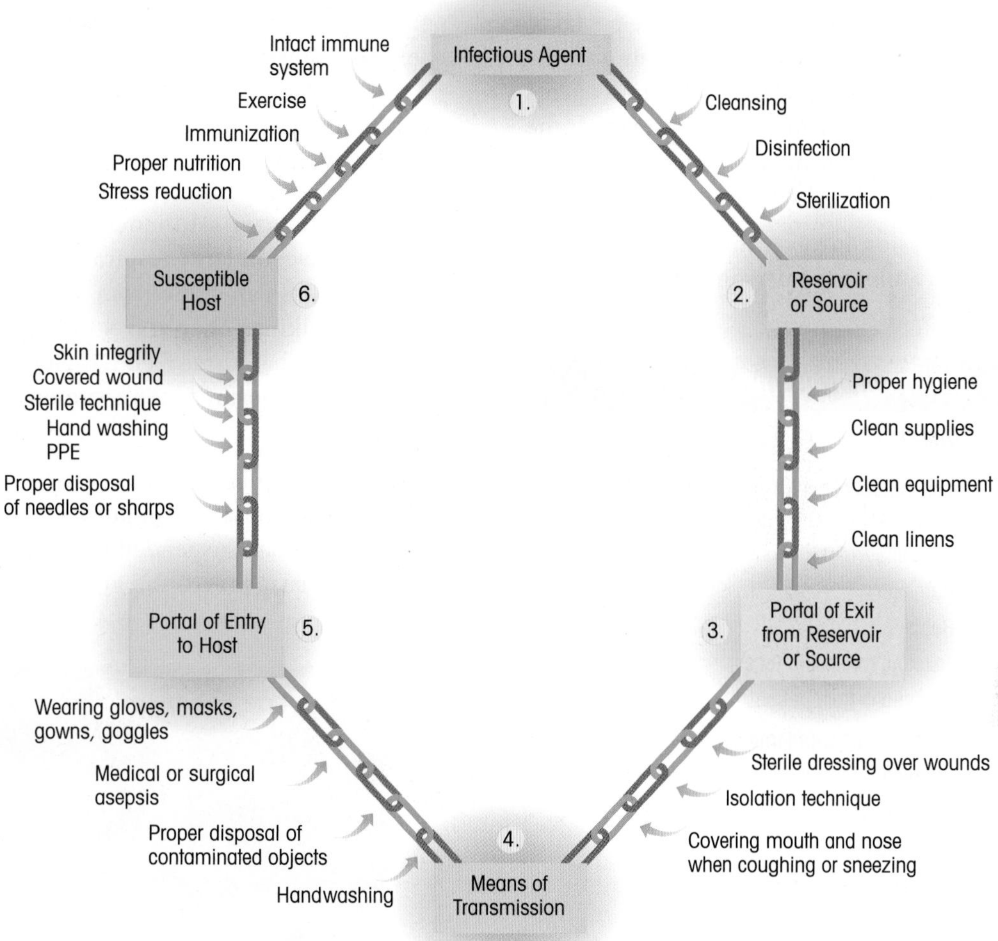

Figure 17-5 *The chain of infection: preventive measures follow each link of the chain*

Infectious Agent

An **infectious agent** is an entity that is capable of causing disease. Agents that cause disease include the following:

- **Biological agents**—living organisms that invade the host, such as bacteria, viruses, fungi, protozoa, and rickettsia
- **Chemical agents**—substances that can interact with the body, such as pesticides, food additives, medications, and industrial chemicals
- **Physical agents**—factors in the environment, such as heat, light, noise, and radiation

In the chain of infection, the main concern is biological (infectious) agents and their effect on the host.

RESERVOIR

The **reservoir** (**REH**-zer-vwor) is a place where the biological agent can survive. Colonization and reproduction take place while the agent is in the reservoir. A reservoir that promotes growth of pathogens must contain the proper nutrients, such as oxygen and organic matter; maintain proper temperature; contain moisture; maintain a compatible pH level; and maintain the proper amount of light exposure. The most common reservoirs are humans, animals, the environment, and **fomites** (**FOE**-mih-teez). Fomites are objects, such as instruments or dressings, that have been contaminated with an infectious agent.

Humans and animals can have symptoms of the infectious agents, or they can be strictly carriers of the agent. **Carriers** have the infectious agent but are symptom free. The agent can be spread to others in both instances.

PORTAL OF EXIT. The **portal of exit** is the route by which an infectious agent leaves the reservoir to be transferred to a host. The agent leaves the reservoir through body secretions, such as sputum, semen, vaginal secretions, urine, saliva, feces, blood, draining wounds, coughing, and sneezing.

MEANS OF TRANSMISSION. The **means of transmission** is the process that bridges the gap between the portal of exit of the infectious agent from the reservoir and the portal of entry of the "new" host. Most infectious agents have a usual means of transmission; however, some microorganisms may be transmitted by more than one means (**Table 17-2**).

- **Contact transmission** involves the physical transfer of an agent from an infected person to an uninfected person through direct contact with the infected person. Direct contact can occur as a result of exposure to sexually transmitted diseases, colds, or the flu.

  Contact with the infected person through contaminated secretions is called indirect contact (**Figure 17-6**).

- **Droplet transmission** occurs when bacteria or viruses travel on large respiratory droplets. Respiratory droplets can come from a sneeze, cough, drip, or from talking. They travel only short distances—usually less than three feet. Droplet transmission can occur from exposure to the common cold, for example (**Figure 17-7**).

- **Airborne transmission** occurs when a susceptible person contacts contaminated droplets or dust particles that are suspended in the air. The longer the particle is suspended, the greater the chance it will

Table 17-2 *Means of Transmission*

MEANS	***EXAMPLES***
Contact	Direct contact of health care provider with patient: • Touching • Bathing • Rubbing • Toileting (urine and feces) • Secretions from patient Indirect contact with fomites: • Clothing • Bed linens • Dressings • Health care equipment • Instruments used in treatments • Specimen containers used for laboratory analysis • Personal belongings • Personal care equipment • Diagnostic equipment
Droplet	Direct contact with an infected person within 3 feet: • Coughing • Sneezing • Dripping
Airborne	Inhaling microorganisms carried by moisture or dust particles in air: • Coughing • Talking • Sneezing
Vehicle	Contact with contaminated inanimate objects: • Water • Blood • Drugs • Food • Urine
Vector-borne	Contact with contaminated animate hosts: • Animals • Insects

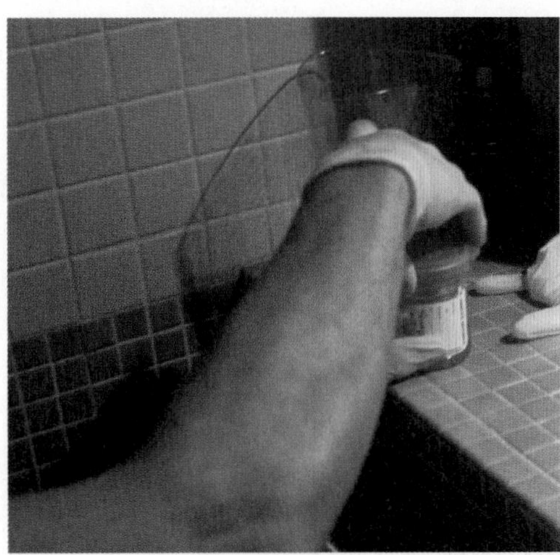

Figure 17-6 *Care must be taken in the handling of body fluids to prevent the transfer of infectious agents through contact with secretions.*

Science Source/Photo Researchers, Inc.

Figure 17-7 *The width of the area that droplet nuclei from a sneeze can encompass.*

find an available port of entry in the human host. An example of an organism that relies on airborne transmission is measles. Spores of anthrax are also transmitted in an airborne powder form.

- **Vehicle transmission** occurs when the agent is transferred to a susceptible host by contaminated inanimate objects, such as water, food, meat, drugs, and blood (**Figure 17-8**). An example is salmonellosis transmitted through contaminated food.
- **Vector-borne transmission** occurs when an agent is transferred to a susceptible human host by animate means, such as mosquitoes, fleas, ticks, lice, and other animals (**Figure 17-9**). Lyme disease is an example.

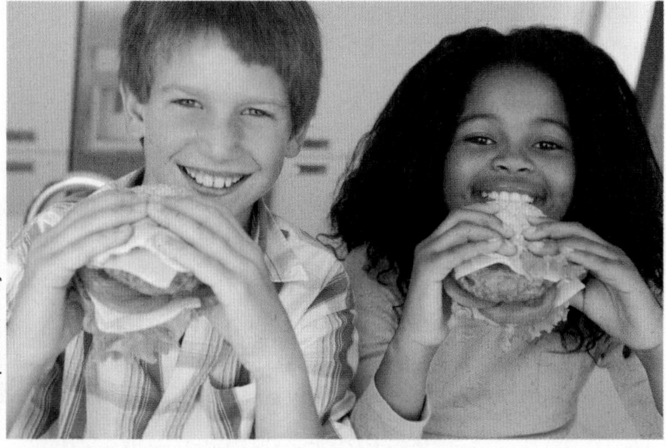

Figure 17-8 *Vehicle transmission can occur through contaminated food such as meat.*

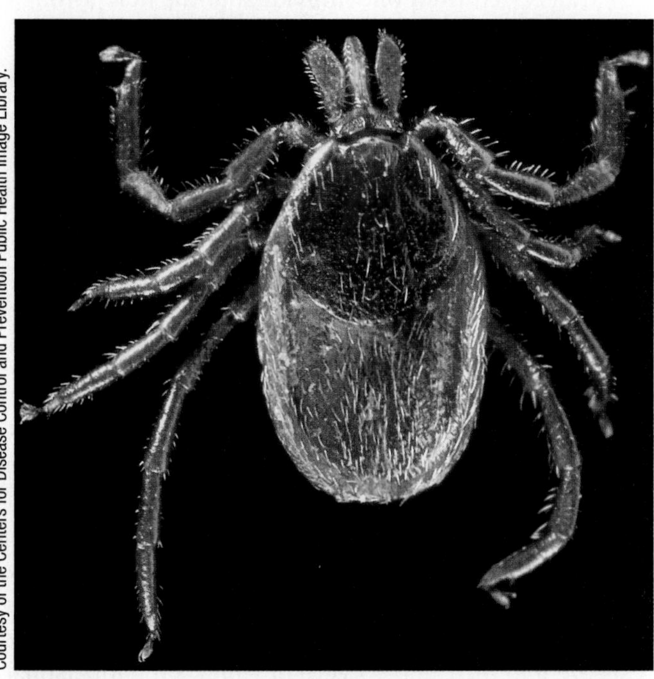

Figure 17-9 *Lyme disease is transmitted by the bite of a deer tick.*

PORTAL OF ENTRY. A **portal of entry** is the route by which an infectious agent enters the host. Portals of entry include the following:

- Integumentary system—through a break in the skin or mucous membrane
- Respiratory tract—by inhaling contaminated droplets
- Genitourinary tract—through contamination with infected vaginal secretions, semen, or catheter access sites
- Gastrointestinal tract—by ingesting contaminated food or water
- Circulatory system—through the bite of insects or rodents
- Transplacental—through transfer of a microorganism from mother to fetus via the placenta and umbilical cord

Host

A **host** is a simple or complex organism that can be affected by an agent. As the term is used here, a host is an individual who is at risk of contracting an infectious disease. A **susceptible host** is a person who lacks resistance to an agent and is vulnerable to a disease. A **compromised host** is a person whose normal defense mechanisms are impaired and who is therefore more susceptible to infection.

The following characteristics of the host influence the susceptibility to and severity of infections:

- Age—as a person ages, immunity declines
- Concurrent disease—the existence of other diseases indicates susceptibility
- Weakened immune system—because of certain types of medications, such as chemotherapy drugs
- Stress—a person experiencing a compromised emotional state has lower defense mechanisms
- Immunization/vaccination status—people who are not fully immunized
- Lifestyle—practices such as having multiple sex partners, sharing needles, or tobacco or drug use can alter defenses
- Career—certain forms of employment involve an increased exposure to pathogenic sources, such as needles or chemical agents
- Nutritional status—people who maintain targeted weight for height and body frame are less prone to illness
- Heredity—some people are genetically more susceptible to infections than others are

BREAKING THE CHAIN OF INFECTION

Health care providers focus on breaking the chain of infection by applying proper infection control practices to interfere with the spread of microorganisms. Specific strategies can be directed at breaking or blocking the transmission of infection from one link in the chain to the next.

Between Agent and Reservoir

The keys to eliminating infection between the infectious agent and reservoir in the chain are cleansing, disinfection, and sterilization. These tactics serve to prevent the formation of a reservoir and environment within which infectious agents can live and multiply.

- **Cleansing** is the removal of soil or organic matter from equipment used in providing care to someone. To reduce the amount of contamination and loosen material on reusable objects, the objects are cleansed prior to sterilization and disinfection. The steps for proper cleansing are as follows:
  1. Rinse the object under cold water, because warm water causes proteins in organic material to coagulate and stick.
  2. Apply detergent and scrub the object under running water with a soft brush.
  3. Rinse the object under warm water.
  4. Dry the object before sterilization.
- **Disinfection** is the elimination of pathogens, except spores, from inanimate objects. Disinfectants are chemical solutions used to clean inanimate objects. Common disinfectants are alcohol and sodium hypochlorite. In the home, Lysol® and bleach are common disinfectants. In some cases, the standard disinfectants may not be effective; if in doubt, check with the Centers for Disease Control and Prevention.
- **Sterilization** is the total elimination of all microorganisms, including spores. Methods of achieving sterilization are moist heat or steam, radiation, chemicals, and ethylene oxide gas. The method of sterilization depends on the type and amount of contamination and the object to be sterilized. Boiling water is not a totally effective sterilization method because some spores and viruses can survive boiling water; however, boiling water is still the best and most common method of sterilization in the home.

If the reservoir is an already infected individual, that individual may need to be isolated. The individual's infectious condition needs to be vigorously treated to reduce the reservoir of infectious material or eliminate the agent.

Media Link

View the **Controlling Disease** video on the Online Resources.

Between Reservoir and Portal of Exit

Promoting proper hygiene, maintaining clean dressings and linens, and ensuring the use of clean equipment in a patient's care can break the chain between the reservoir and the portal of exit. The aim is to eliminate the reservoir for the microorganism before the pathogen can escape to a susceptible host.

- *Proper hygiene*—Health care providers must teach the importance of maintaining the cleanliness and integrity of the skin and mucous membranes. Bathing and handwashing are the best means of eliminating the potential for infection.
- *Clean supplies*—An open injury represents a potential reservoir for infectious agents or portal of exit for a pathogen to be transferred to another individual. Dressings on open or oozing wounds must be changed and cleansed regularly.
- *Clean linens*—Dressing gowns, linens, or towels are catchalls for body secretions. Infectious agents can be transferred from one individual to the next through contact with linens.
- *Clean equipment*—All equipment used in the care of a patient must be cleansed and disinfected after each use. To protect themselves, health care providers may wear gloves and masks when cleaning equipment to avoid being splashed with contaminated waste products or secretions.

Between Portal of Exit and Means of Transmission

The goal in breaking the chain between the portal of exit and the means of transmission is to block the exit of the infectious agent. The health care provider must maintain clean dressings on all injuries or wounds. People should be encouraged to cover the mouth and nose when sneezing or coughing, and the health care provider must do so as well. If a tissue is not available, a person should cough or sneeze into their upper sleeve or elbow. Proper personal protective equipment (PPE) must be worn when caring for a person who may have infectious secretions and care must be taken to properly dispose of contaminated articles. Isolation must be carried out in certain situations.

Between Means of Transmission and Portal of Entry

The goal is to break the chain of infection between the means of transmission and portal of entry. Health care providers must always wash their hands between care cases, which may involve contact with contaminated items and use of gloves, masks, gowns, and goggles as appropriate. Proper disposal must be used for all contaminated material and for needles and sharps.

Between Portal of Entry and Host

The goal is to break the chain between portal of entry and susceptible host, preventing the transmission of infection to an uninfected person, including the health care provider. This is accomplished by maintaining skin integrity, such as covering wounds and the sterilization technique; hand washing; PPE; and proper disposal of needles or sharps.

Between Host and Agent

To break the chain between susceptible host and agent means eliminating infection before it begins. Proper nutrition, stress reduction, exercise, and immunization allow an individual to maintain an intact immune system, thus preventing infection.

STAGES OF THE INFECTIOUS PROCESS

Activation of the immune response indicates the occurrence of infection. Infection results from tissue invasion and damage by an infectious agent. There are two types of infectious responses:

- Localized infection, which is limited to a defined area or single organ with symptoms that resemble inflammation, such as an ear infection
- Systemic infection, which affects the entire body and involves multiple organs, such as influenza

Localized and systemic infections progress through four stages: incubation, prodromal, illness, and convalescence.

The **incubation stage** (**in**-kyoo-**BAY**-shun) is the time interval between entry of an infectious agent into the host and the onset of symptoms. During this time, the infectious agent invades the tissue and begins to multiply to produce an infection; the patient is typically infectious to others during this period. The incubation period for chickenpox is 2–3 weeks; the infected person is contagious from five days before the skin eruptions to no more than six days after the eruptions appear.

The **prodromal stage** (pro-**DRO**-muhl) is the time interval from the onset of nonspecific symptoms until specific symptoms begin to appear. During this period, the infectious agent continues to invade and multiply in the host. A patient may also be infectious to others during this period.

The **illness stage** is the time period when the patient is manifesting specific signs and symptoms of an infectious process.

The **convalescent stage** (**kon**-vuh-**LEH**-sint) is the period of time from the beginning of disappearance of acute symptoms until the patient returns to the previous state of health.

NOSOCOMIAL OR HOSPITAL-ACQUIRED INFECTIONS (HAIS)

A **nosocomial infection** or **hospital-acquired infection (HAI)** is acquired in a hospital or other health care facility and was not present or incubating at the time of the patient's admission. These types of infections typically fall into one of four categories: urinary tract, surgical wounds, pneumonia, or septicemia.

Nosocomial infections include those that become symptomatic after the patient is discharged, as well as infections passed among health care providers. Personnel who fail to follow proper handwashing principles transmit most infections.

Hospitalized patients are at risk for infections because the environment provides exposure to a variety of virulent organisms to which the patient has not typically been exposed. Therefore, the patient has not developed resistance to these organisms.

Individuals in long-term care facilities often have multiple illnesses, which decreases their resistance to infection.

Examples of these types of infections are MRSA, VRSA, VISA, and *C. diff*:

- Methicillin-resistant *Staphylococcus aureus* (**MRSA**) is a type of staph bacteria that is resistant to the methicillin antibiotic drug class because of a genetic change. *Staphylococcus aureus* is a common bacterium found in the nose and on the skin of about one-third of the population. The bacteria are usually harmless unless they enter a skin wound. In healthy individuals, *Staphylococcus aureus* may cause a minor infection. In people who are immunocompromised, the bacteria can cause a severe infection. This infection is found mainly in a health care setting; however, it can also be found in the general population. MRSA symptoms appear as red bumps on the site of entry; in this case, the site of entry is a wound. The tissue surrounding the wound can be swollen, painful, warm to the touch, and contain pus. A fever may also be prevalent. More severe life-threatening symptoms of MRSA occur in the health care setting because of the weakened condition of the patient. Severe symptoms of MRSA include infection of the blood or organs, chills, cough, chest pain, rashes, muscle aches, and fever. MRSA is spread through direct contact. Treatment is symptomatic and with specific antibiotics.

- Like MRSA, **VRSA**, or vancomycin-resistant *Staphylococcus aureus*, is a type of staph bacteria that has genetically changed and is resistant to the vancomycin family of antibiotics. If there is only some resistance to the antibiotic, it is called vancomycin-intermediate *S. aureus*, or **VISA**. Signs and symptoms of VRSA are the same as MRSA.
- *C. diff*, or *Clostridium difficile*, is a highly contagious bacterial disease and the number one cause of diarrhea in the health care setting. *C. diff* occurs because of the prolonged use of antibiotics. The intestines contain millions of bacteria, normal flora that help protect the body from infection. Antibiotics used over time may destroy the normal flora—and without healthy bacteria, *C. diff* quickly grows out of control. *C. diff* bacteria attack the intestines. The bacteria produce spores that may persist for weeks or months. Symptoms include diarrhea, fever, loss of appetite, and dehydration. *C. diff* is transmitted through direct contact with infected feces or via direct contact with fomites, such as from bedrails, remote controls, linens, and so on. Most at risk are older adults, especially those who are hospitalized or live in a health care setting. Complications are severe dehydration, kidney failure, perforation of the large intestine, and death. Treatment is symptomatic and requires specific antibiotics such as Flagyl®, vancomycin, or probiotics. Probiotics are organisms that help restore the normal flora.

To prevent the spread of MRSA, VRSA, VISA, or C. diff, *the health care provider must follow strict anticontamination guidelines for infections. Handwashing is critical; alcohol-based sanitizers should not be used for* C. diff *because they may not effectively destroy spores.*

Medical Highlights

CHANGES OCCURRING IN INFECTIOUS DISEASES

In the world today, infectious diseases have been drastically reduced. Factors that have led to this decline include the following:

- Improved sanitation and hygiene
- Vaccine development and distribution
- Development of drugs, especially antibiotics
- Better communication and collaboration by scientists and physicians through the Internet, informing people of health risks and preventive measures
- Development of public health departments worldwide

During the past 25 years, much progress has been made in expanding scientific knowledge. A major area of advance has been in the field of genetics. *Emerging infectious diseases* are any infectious diseases that have recently appeared or that have existed in the past. These diseases are now rapidly increasing in frequency, geographic range, or both. Some of the factors involved are infectious agents that are capable of rapid genetic mutation and evolution. This leads to new strains of old microbes against which current drugs are not effective. There has been a huge increase in the ability of people to travel quickly over long distances. This factor and others have given microbes an enhanced opportunity for genetic change and worldwide dispersal.

Emerging infectious disease threats include the following:

1. Influenza—the virus causing the most common form of this disease is not new, but every winter new strains of the influenza virus emerge through mutations. While effective flu vaccines exist, they need to be reformulated every year to match the changes.
2. Ebola disease—caused by the Ebola virus, Ebola disease is transmitted through direct contact with blood or other bodily fluids from an infected person, alive or dead. Symptoms appear 2–21 days following exposure. The disease starts with fever, fatigue, loss of appetite, vomiting, diarrhea, and headache. In a later stage, patients may have profuse external or internal bleeding, which may lead to death. No vaccine or validated treatment is available at present. The Ebola virus is easily killed by soap, bleach, high temperature, and sunlight. Follow recommended standard precautions and use special strong protective gear when taking care of patients, living or deceased.
3. HIV/AIDS—this is among the worst pandemic diseases (see Chapter 16).

(continues)

4. "Superbugs"—a growing number of bacteria are now resistant to even the newest, most powerful antibiotics, and hence pose a large threat to the general public. Antibiotic-resistant bugs are created when bacteria develop genetic traits that make them resistant to treatment with current antibiotic drugs. An example of a superbug is methicillin-resistant *Staphylococcus aureus* (MRSA).
5. Whooping cough and measles—these are occurring in some children under age three who have not received all the vaccinations recommended by the Department of Health and Human Services.

Part of the changing infectious disease landscape includes recent scientific and medical progress. Areas of advancement include the following:

- Laboratory technology and techniques—new genetic laboratory tests provide, in only a few days, microbe identification tests that once took a month or more. There are laboratory methods to measure the amount of a virus in a blood or tissue sample. This allows the physician to gauge the extent of a viral disease or the effectiveness of a drug treatment.
- Infectomics—this is the study of microbe genes that has led to the mapping of numerous microbe genomes. A genome, made of all the genes of an organism, is an instruction book for the hereditary traits of the specific organism. By studying microbe genes, scientists can identify all of the fundamental functions that allow the microbe to survive and cause disease. The understanding of which nucleic acids and proteins are involved in each step of a microbe infection makes it possible to develop treatments or vaccines to target those components of the infection.
- Cell biology—among many immune system insights has been the identification of a group of immune chemicals called cytokines. The discovery of how to produce and harvest these agents has led to the production of new drugs such as interferon, the interleukins, and growth factors. These drugs have been used in the treatment of hepatitis C, cancer, and diseases that are present when a person has a weakened immune system caused by AIDS.
- Epidemiology—instead of months, it now takes only weeks or days to discover how a disease is spreading.

Health care providers must stay alert to emerging threats, assess the risks, and take appropriate preventive steps to contain the infectious disease.

References: "The Changing Face of Infectious Disease," Medical Essay, Mayo Foundation for Medical Education and Research, Rochester, MN (October 2003); and "Super Bugs Pose Bigger Threat than SARS," WebMD.

STANDARD PRECAUTIONS

Standard precautions are guidelines to be used during routine patient care and cleaning duties. Standard precautions combine the major features of universal precautions and body substance isolation. Universal precautions help control contamination from blood-borne viruses such as HIV and hepatitis virus. They must be used when a person expects to have contact with blood, body fluids (except sweat), mucous membranes, and nonintact skin. See **Table 17-3** and **17-4**.

Methods to prevent the spread of infections are standardized in recommendations from the Centers for Disease Control and Prevention (CDC).

Handwashing

Handwashing is the single most effective way to prevent infection.

1. Wash hands after touching blood, body fluids, secretions, excretions, and contaminated items, *regardless of if gloves are worn.*
2. Wash hands immediately after removing gloves, between patient contacts, and when otherwise indicated to avoid transfer of microorganisms to other patients or the surrounding environment.
3. Use a plain, nonantimicrobial soap for handwashing.
4. Wash hands for a minimum of 20 seconds.
5. Be certain to dry hands; if hands remain damp after washing, bacteria are more readily transferred to other surfaces. Use paper towels to dry.

Alcohol-based hand sanitizers can be used instead of soap and water except in cases where microbes produce spores. Apply enough sanitizer to cover all of both hands; rub hands together until dry. Do not rinse or wipe dry.

Table 17-3 Examples of Personal Protective Equipment in Common Health Care Provider Tasks

TASK	GLOVES	GOWN	GOGGLES/ FACE SHIELD	SURGICAL MASK
Controlling bleeding with squirting blood	Yes	Yes	Yes	Yes
Emptying a catheter bag	Yes	Yes	Yes	Yes
Serving a meal tray	No	No	No	No
Giving oral care	Yes	No	Yes	Yes
Helping the dentist with a procedure	Yes	Yes	Yes	Yes
Cleaning a resident and changing the bed after an episode of diarrhea	Yes	Yes	No	No
Taking an oral temperature	No	No	No	No
Taking a rectal temperature	No	No	No	No
Taking a blood pressure	No	No	No	No
Cleaning soiled medical equipment, such as bedpans	Yes	Yes	Yes	Yes
Shaving a patient with a disposable razor	Yes*	No	No	No
Giving eye care	Yes	No	No	No
Giving special mouth care to an unconscious patient	Yes	No, unless coughing is likely	No, unless coughing is likely	No, unless coughing is likely
Washing a patient's genital area	Yes	Yes	No	No
Washing a patient's arms and legs when the skin is intact	No	No	No	No

*Because of the high risk of this procedure for contact with blood

Media Link

View the **Proper Handwashing** video and the **Infection Control** animation on the Online Resources.

Gloves

Wear gloves—clean, nonsterile gloves are adequate—when coming in contact with blood, body fluids, secretions, excretions, and contaminated items. Put on clean gloves just before touching mucous membranes and nonintact skin. Remove gloves after use and wash hands.

Mask, Eye Protection, and Face Shield

Wear a mask and eye protection or a face shield to protect mucous membranes of the eyes, nose, and mouth during procedures and patient care activities that are likely to generate splashes or sprays of blood, body fluids, secretions, or excretions.

Gown

Wear a clean, nonsterile gown to protect skin and prevent soiling of clothing during procedures and patient care activities that are likely to generate splashes or sprays of blood, body fluids, secretions, or excretions, or cause soiling of clothing. Remove a soiled gown promptly and wash hands to avoid transfer of microorganisms to other patients or to the surrounding environment.

Patient Care Equipment

Handle used patient care equipment soiled with blood, body fluids, secretions, or excretions in a manner that prevents skin and mucous membrane exposures, contamination of clothing, and transfer of microorganisms to other patients and environments. Be certain that

Table 17-4 *Recommendations for Application of Standard Precautions for the Care of all Patients in all Health Care Settings*

COMPONENT	RECOMMENDATIONS
Hand hygiene (washing)	After touching blood, body fluids, secretions, excretions, contaminated items; immediately after removing gloves; between patient contacts
Personal protective equipment (PPE) Gloves Gown Mask, eye protection (goggles), face shield*	For touching blood, body fluids, secretions, excretions, contaminated items; for touching mucous membranes and nonintact skin During procedures and patient care activities when contact with clothing or exposed skin with blood, body fluids, secretions, and excretions is anticipated During procedures and patient care activities likely to generate splashes or sprays of blood, body fluids, and secretions, especially suctioning (endotracheal intubation)
Soiled patient care equipment	Handle in a manner that prevents transfer of microorganisms to others and to the environment; wear gloves if visibly contaminated; perform hand hygiene
Environmental control	Develop procedures for routine care, cleaning, and disinfection of environmental surfaces, especially frequently touched surfaces in patient care areas
Textiles and laundry	Handle in a manner that prevents transfer of microorganisms to others and to the environment
Needles and other sharps	Do not recap, bend, break, or hand-manipulate used needles; if recapping is required, use a one-handed scoop technique only; use safety features when available; place used sharps in puncture-resistant container
Patient resuscitation	Use mouthpiece, resuscitation bag, or other ventilation devices to prevent contact with mouth and oral secretions
Patient placement	Prioritize for single-patient room if patient is at increased risk of transmission, is likely to contaminate the environment, does not maintain appropriate hygiene, or is at increased risk of acquiring infection or developing adverse outcome following infection
Respiratory hygiene/cough etiquette (source containment of infectious respiratory secretions in symptomatic patients, beginning at initial point of encounter (e.g., triage and reception areas in emergency departments and physician offices)	Instruct symptomatic persons to cover mouth/nose when sneezing/coughing; use tissues and dispose in no-touch receptacle. If tissues are not available, cough or sneeze into upper sleeve or elbow. Observe hand hygiene after soiling of hands with respiratory secretions; wear surgical mask if tolerated or maintain spatial separation >3 feet if possible

*During aerosol-generating procedures on patients with suspected or proven infections transmitted by respiratory aerosols (e.g., Ebola virus), wear a fit-tested N95 or higher respirator in addition to gloves, gown, and face/eye protection.

Courtesy of the Centers for Disease Control and Prevention; See Sections II.D.-II.J. and III.A.1

reusable equipment is properly cleaned and reprocessed before it is used on another patient. Single-use items must be discarded properly.

Linens

Handle, transport, and process used linens soiled with blood, body fluids, secretions, or excretions in a manner that prevents skin and mucous membrane exposures and contamination of clothing and avoids transfer of microorganisms to other patients and environments.

Occupational Health and Blood-borne Pathogens

1. Take care to prevent injuries from needles, scalpels, and other sharp instruments or devices when handling these instruments after procedures, when cleaning used instruments, and when disposing of used needles. *Never recap used needles or use any technique that involves directing the point of the needle toward any part of the body.* Place used disposable syringes,

needles, scalpels, and other sharp items in appropriate puncture-resistant containers located as close as practical to the area in which the items were used.

2. Use mouthpieces, resuscitation bags, or other ventilation devices as an alternative to mouth-to-mouth resuscitation methods in areas where there is a need for resuscitation.

Patient Placement

Place a patient who contaminates the environment or who does not assist in maintaining appropriate hygiene or environmental precautions in a private room or other relatively isolated area if possible.

ISOLATION

The CDC guidelines condensed the former disease-specific precautions into three sets of precautions based on the route of transmission: airborne (**Figure 17-10**), contact (**Figure 17-11**), or droplet (**Figure 17-12**). These transmission-based precautions are to be used *in addition* to the standard precautions. Transmission-based precautions are practices designed for patients documented as or suspected of being infected with highly transmissible or epidemiologically important pathogens for which additional precautions beyond the standard precautions are required to interrupt transmission in hospitals (refer to **Table 17-5**).

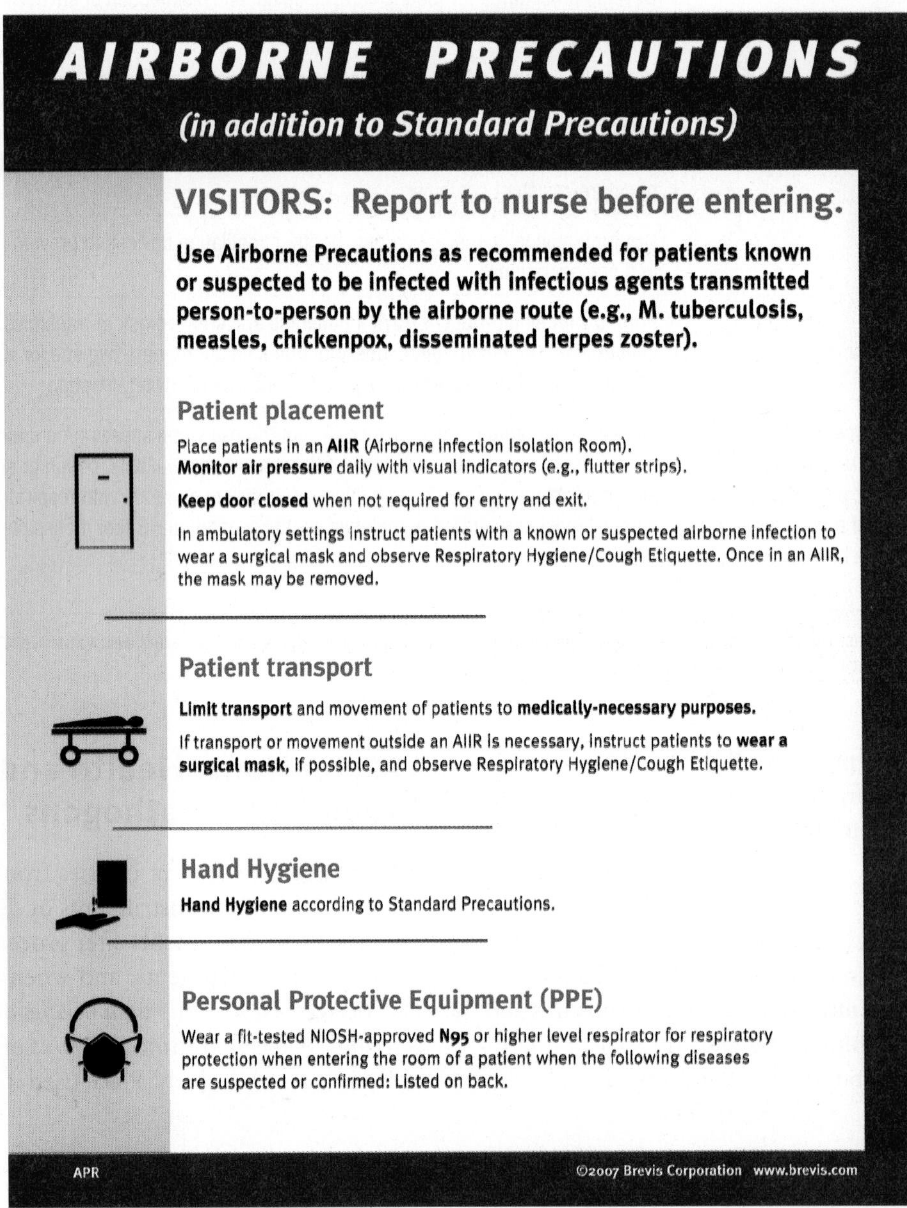

Figure 17-10 *Airborne precautions*

Transmission-based precautions are also used in the event of suspicious infections and with patients who are immunosuppressed from either disease or chemotherapy. More than one of the transmission-based precautions are used at the same time for patients with certain infections or conditions.

Patients requiring isolation should be placed in a private room with adequate ventilation and should have their own supplies. Personal belongings should be kept to a minimum, and health care providers should use disposable supplies and equipment when possible. All articles leaving the room, such as soiled linens and collected specimens, should be labeled and either placed in impermeable bags or double bagged. Health care providers must understand that physical protection gained from the barrier isolation may have a negative psychological impact on the patient. Be supportive of patients in isolation and allow them to verbalize their feelings. Be sure to provide information regarding the necessity for isolation.

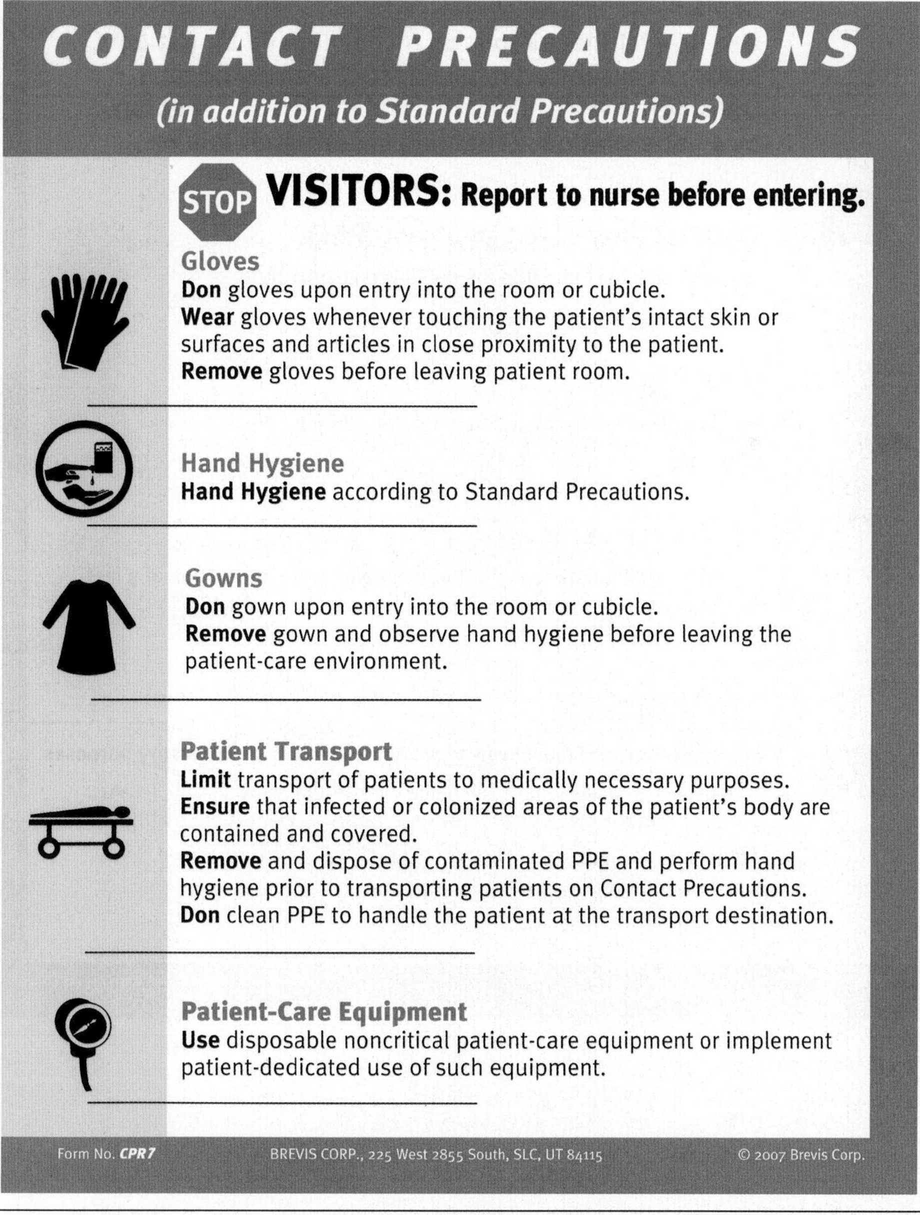

Figure 17-11 *Contact precautions*

Reverse isolation, also known as protective isolation, is a barrier protection designed to prevent infection in patients who are severely compromised and highly susceptible to infection. This includes patients who are taking immunosuppressive medications; are receiving chemotherapy or radiation therapy; have diseases such as leukemia, which depress resistance to infectious organisms; or have extensive burns, dermatitis, or other skin impairments that prevent adequate coverage with dressings.

These patients are at increased risk for infection from their own microorganisms; contact with health care

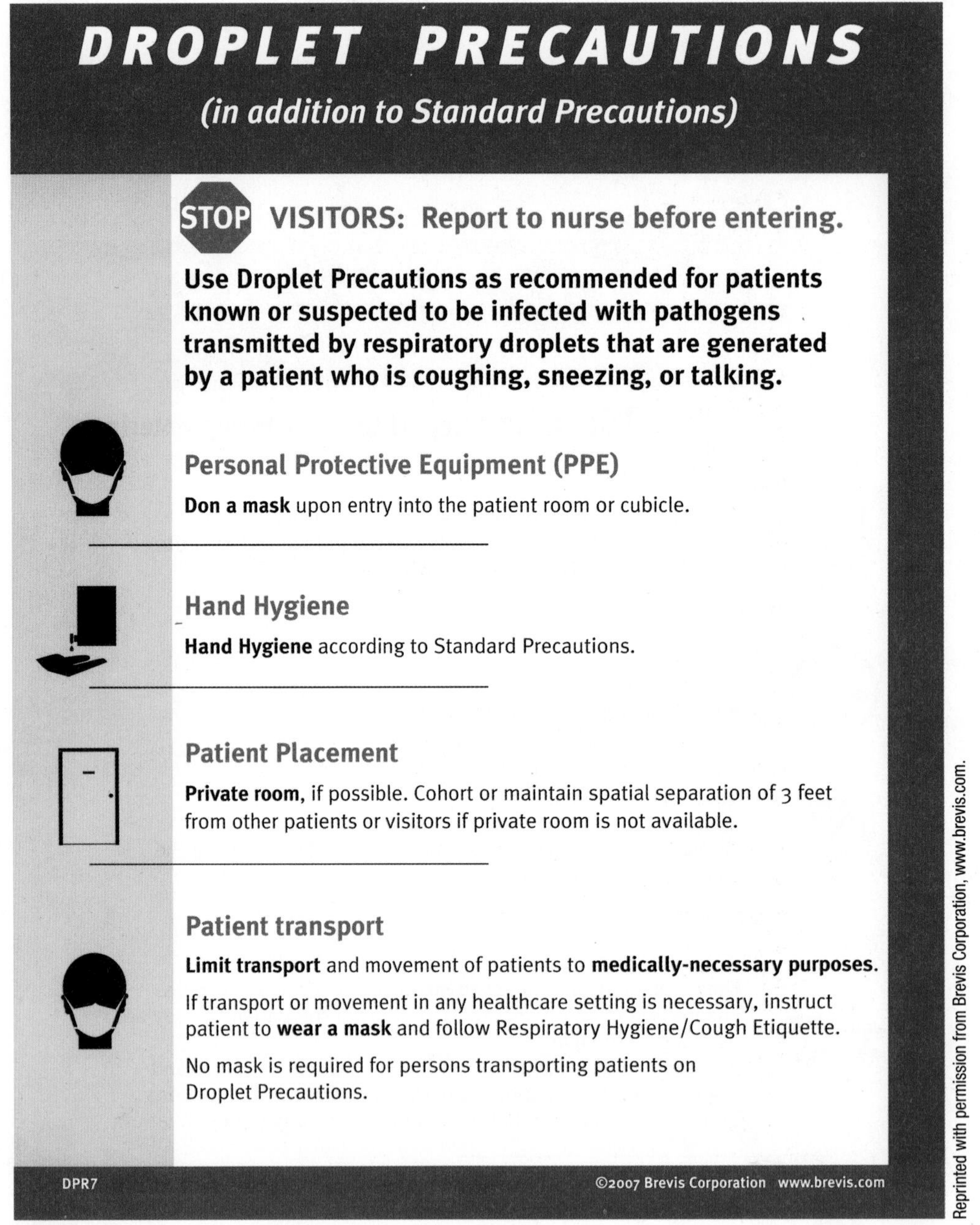

DROPLET PRECAUTIONS
(in addition to Standard Precautions)
STOP
VISITORS: Report to nurse before entering.
Use Droplet Precautions as recommended for patients known or suspected to be infected with pathogens transmitted by respiratory droplets that are generated by a patient who is coughing, sneezing, or talking.
Personal Protective Equipment (PPE)
Don a mask upon entry into the patient room or cubicle.
Hand Hygiene
Hand Hygiene according to Standard Precautions.
Patient Placement
Private room, if possible. Cohort or maintain spatial separation of 3 feet from other patients or visitors if private room is not available.
Patient transport
Limit transport and movement of patients to medically-necessary purposes.
If transport or movement in any healthcare setting is necessary, instruct patient to wear a mask and follow Respiratory Hygiene/Cough Etiquette.
No mask is required for persons transporting patients on Droplet Precautions.
DPR7
©2007 Brevis Corporation www.brevis.com

Figure 17-12 *Droplet precautions*

providers whose hands have not been properly washed; and exposure to improperly disinfected and nonsterile items, such as air, food, water, and equipment. Responsibilities toward these patients include ensuring that everyone entering the patients' rooms has completed a meticulous handwashing and is properly attired in gown, gloves, and mask; ensuring that the patients' environment is as clear of pathogens as possible; and knowing the institutional policy regarding caring for patients requiring reverse isolation.

Table 17-5 ***Precautions Related to Type of Disease***

PRECAUTION	***TYPE OF DISEASE***	
Standard Precautions	All patients, regardless of disease or condition	
Airborne Precautions	In addition to standard precautions, used for patients known to have or suspected of having serious illnesses spread by airborne droplet nuclei, including • Measles • Varicella • Tuberculosis	
Contact Precautions	In addition to standard precautions, used for patients known to have or suspected of having serious illnesses easily spread by direct patient contact or contact with fomites, including • Wound infections • Gastrointestinal infections • Respiratory infections • Skin infections, including Herpes simplex Impetigo Major abscesses, cellulitis, or pressure ulcers Pediculosis Scabies Varicella (zoster) • Viral hemorrhagic infections (*Ebola*)	
Droplet Precautions	In addition to standard precautions, used for patients known to have or suspected of having illnesses spread by large particle droplets, including • Meningitis • Pneumonia • Diphtheria • Pertussis • Scarlet fever	 • Adenovirus • Influenza • Mumps • Rubella • Parvovirus 19

Current Infection control guidance from Centers for Disease Control and Prevention/Hospital Infection Control Practices Advisory Committee, 2008 ©. Available at cdc.gov/hipac/pubs.html
From Table 1, Synopsis of Types of Precautions and Patients Requiring Precautions, Centers for Disease Control and Prevention/Hospital Infection Control Practices Advisory Committee, 1997 ©. Available at http://www.cdc.gov/ncidod/hip/isolat/isotab_1.htm.

Medical Terminology

coloniz	group of microorganisms living together
-tion	process of
coloniza/tion	process of microorganisms living together
dis-	removal of
infect	contamination with pathogenic microorganisms
dis/infect/ion	process of removal of pathogenic microorganisms
infect/ion	process of contamination with pathogenic microorganisms
nosocomi	hospital or infirmary
-al	pertaining to
nosocomi/al	pertaining to hospital or infirmary
pathogen	producing disease
-ic	relating to or characterized by
pathogen/ic	relating to producing disease
prodrom	early symptom
prodrom/al	pertaining to early symptoms of disease
steriliz	free from microorganisms
steriliza/tion	process of being free of microorganisms
vir-	poison; venom; pathogen
-ulent	quantity; frequency; full of
-ulence	forms a noun out of a word ending in -ent
vir/ulence	something full of pathogens that cause disease

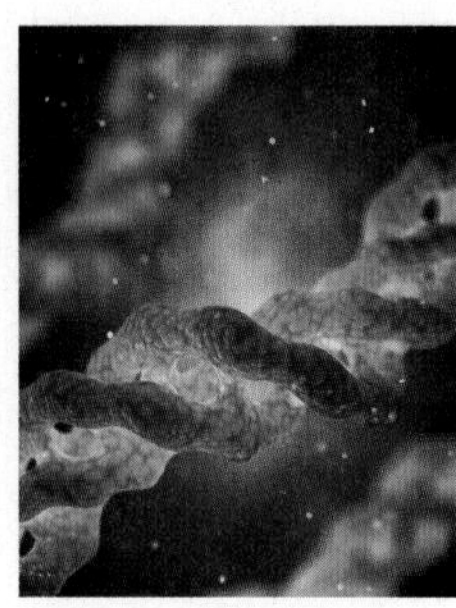

Study Tools

Workbook	Activities for Chapter 17
Online Resources	• PowerPoint presentations • Animation

REVIEW QUESTIONS

Select the letter of the choice that best completes the statement.

1. The type of microorganisms that are always present, especially on the skin, are called
   a. transient flora.
   b. rickettsia.
   c. resident flora.
   d. viruses.

2. Organisms that can only live inside the cells are called
   a. flora.
   b. bacteria.
   c. viruses.
   d. fungi.

3. Infections that are spread by fleas and ticks are caused by
   a. viruses.
   b. fungi.
   c. protozoa.
   d. rickettsia.

4. In the chain of infection, colonization and reproduction take place while the agent is in the
   a. portal of entry link.
   b. reservoir link.
   c. portal of exit link.
   d. transmission link.

5. Salmonellosis, a disease caused by contaminated food, is transmitted by
   a. vehicle transmission.
   b. vector-borne transmission.
   c. contact transmission.
   d. airborne transmission.

6. The processes used for the total elimination of all microorganisms, including spores, include all but
   a. steam.
   b. radiation.
   c. ethylene gas.
   d. alcohol.

7. Using clean linens and equipment will help break the chain of infection
   a. between reservoir and portal of entry.
   b. between portal of exit and means of transmission.
   c. between agent and reservoir.
   d. between portal of entry and host.

8. In an infectious process, the time interval from the onset of nonspecific symptoms until specific symptoms appear is called the
   a. incubation stage.
   b. prodromal stage.
   c. illness stage.
   d. convalescent stage.

9. A person is most infectious to other people during the
   a. incubation stage.
   b. prodromal stage.
   c. illness stage.
   d. convalescent stage.

10. The most common endemic nosocomial infection involves all but the
    a. respiratory system.
    b. circulatory system.
    c. integumentary system.
    d. digestive system.

FILL IN THE BLANKS

1. Some forms of bacteria produce resistant forms called _______________.
2. Malaria is caused by a group of parasitic organisms known as _______________.
3. _______________ of disease may have the infectious agent but are symptom free.
4. A _______________ host is a person whose normal defense mechanisms are impaired and who is therefore more susceptible to infection.
5. When medical equipment is cleaned, it first must be rinsed with _______________ water.
6. Forms of barrier protection include gloves, masks, gowns, and _______________.
7. Nosocomial infections are referred to as _______________ acquired infections.
8. Nosocomial infections are passed among _______________.
9. Reverse isolation may be used for a person receiving _______________.
10. Alcohol-based hand sanitizers may be used instead of washing with soap and water in all cases except if the bacteria may produce _______________.

APPLYING THEORY TO PRACTICE

1. Dominick, age 12, lives in a wooded area. He gets a tick bite. Dominick is taken to the emergency department by his mother, Lara. What diseases do ticks carry? How are the diseases carried by ticks transmitted? Define the term "means of transmission" and the types involved.
2. Mrs. Ozuna, 90 years old, is admitted to the hospital for the repair of a hip fracture. To monitor her urinary output, a Foley catheter is inserted prior to surgery. A few days after surgery, Mrs. Ozuna develops a urinary tract infection. What is the probable cause of her infection, and how is it related to her hospitalization?
3. Health care providers may be exposed to diseases because of their occupations. List at least five diseases to which health care providers may be exposed and how they would contract these diseases. Describe methods by which health care providers can prevent each of these diseases.
4. Lawson is a doctor at a hospital. Lawson is going to examine a patient who has an open sore and has been diagnosed with AIDS. List the standard precautions they need to take and explain why each is necessary.
5. Your father, age 74, has been recovering in a rehabilitation center from a hip fracture. He has developed *C. diff* and is in isolation. Explain the process of how this disease occurs. What preventive measures need to be taken to prevent the spread of this condition? What are probiotics?

CASE STUDY

Sage is a mother of a toddler. She began feeling ill a few days after visiting a friend. She is experiencing congestion, sore throat, body aches, cough, and fever.

1. Identify, in the chain of infection, the possible means of transmission in which she could have been infected.
2. What are some possible illnesses that she could have?
3. What types of tests can be run to identify the exact virus or bacteria causing the infection?
4. Before Sage receives a diagnosis, what precautions should she adhere to so she doesn't infect others?
5. If she has a bacterial infection, what are some treatment options?
6. If she has the COVID-19 virus, what are some treatment options?

17-1 Pathogenic Microorganisms

- ***Objective:*** Describe the different types of pathogenic microorganisms
- ***Materials needed:*** slides of bacteria, fungi, protozoa, rickettsia, and helminths; textbook; biohazardous container; paper; pencil

Step 1: Examine a slide of typical bacteria. Record your observations.

Step 2: Examine a slide of typical fungi. Record your observations.

Step 3: Examine a slide of typical protozoa. Record your observations.

Step 4: Examine a slide of typical rickettsia. Record your observations.

Step 5: Examine a slide of typical helminths. Record your observations.

Step 6: Discuss your observations with your lab partner.

Step 7: Record the differences between the pathogenic organisms.

Step 8: Make a list of at least one illness caused by each different type of organism.

Step 9: Compare your list with the textbook regarding types of illnesses caused by each of the pathogenic organisms.

Step 10: Be certain to dispose of the slides using the correct method and place them in the approved container after the lab activity.

Step 11: Wash hands.

LAB ACTIVITY 17-2 Tracking the Virus

- ***Objective:*** Demonstrate disease transmission and practice contact tracing methods
- ***Materials needed:*** test tubes, droppers, distilled water, NaOH, phenolphthalein solution, white board, white board markers, paper, pencil

Step 1: Your teacher will prepare test tubes before class by filling all test tubes (one for each student) with distilled water. One test tube will be filled halfway with NaOH and the rest of the way with water. Choose your test tube.

Step 2: Walk around the room and transfer a dropper of water from your test tube into another student's test tube and receive a dropper of water from that same student. Record the student's name with whom the liquid was exchanged.

Step 3: After you have exchanged with a total of four students and recorded each of their names, the teacher will add a few drops of phenolphthalein solution to your test tube.

Step 4: If you were "infected," meaning your water turned pink, write your name on the board.

Step 5: Trace the source of the infection by asking your classmates questions. Sample questions could be:

- If the test tube remained clear, who did they exchange water with?
- If they were "infected," who did they exchange water with?

Step 6: Wash test tubes and place on a test tube drying rack.

Step 7: Wash hands and dry thoroughly.

Step 8: Answer the following questions:

- Were you able to narrow down the list of people who caused the infection to 4–5 people? If yes, how did you do this?
- Why would it be important to be able to identify where an illness began?
- What are some diseases that spread through contact and droplet transmission?
- What are some measures you can take to avoid disease transmission for contact and droplet transmission?

CHAPTER 18

Respiratory System

Objectives

- Describe the functions of the respiratory system.
- Describe the structures and functions of the organs of respiration.
- Explain the breathing and respiratory process.
- Discuss how breathing is controlled by neural and chemical factors.
- Discuss respiratory disorders.
- Define the key words that relate to this chapter.

Key Words

alveolar sac
alveoli
anthrax
apnea
asbestosis
asthma
atelectasis
bronchioles
bronchitis
bronchogenic cysts
bronchoscopy
bronchus
cancer of the larynx
cancer of the lung
cellular respiration (oxidation)
chronic obstructive pulmonary disease (COPD)
common cold
congenital cystic adenomatoid malformation (CCAM)
congenital lung disorder
congenital malformation of the diaphragm
coughing
diaphragm
diphtheria
dyspnea
emphysema
epiglottis
eupnea
expiration
expiratory reserve volume (ERV)
external nares
external respiration
functional residual capacity
glottis
Hering-Breuer reflex
hiccoughs
hyperpnea
hyperventilation
influenza
inspiration
inspiratory reserve volume (IRV)
internal respiration
Kussmaul respiration
laryngitis
larynx
mediastinum
nasal septum
olfactory nerves
orthopnea
pertussis (whooping cough)
pharyngitis

(*continues*)

Key Words *continued*

pharynx
pleura
pleural fluid
pleurisy
pneumonia
pneumothorax
primary cilia dyskinesia
pulmonary embolism
rales
residual volume
respiratory distress syndrome
respiratory syncytial virus (RSV)
silicosis
sinuses
sinusitis
sneezing
spirometer
sudden infant death syndrome (SIDS)
surfactant
tachypnea
thoracentesis
tidal volume
total lung capacity
trachea
tuberculosis
turbinates
ventilation
vital lung capacity
wheezing
yawning

The respiratory system obtains oxygen for use by the millions of body cells and eliminates carbon dioxide and water that is produced in cellular respiration. Oxygen and nutrients stored in the cells combine to produce heat and energy. Oxygen must be in constant supply for the body to survive.

FUNCTIONS OF THE RESPIRATORY SYSTEM

1. Provides the structures for the exchange of oxygen and carbon dioxide in the body through respiration, which is subdivided into external respiration, internal respiration, and cellular respiration (**Figure 18-1**).
2. Protects from dust and microbes entering the body through mucus production, cilia, and coughing.
3. Responsible for the production of sound; the larynx contains the vocal cords. When air is expelled from the lungs, it passes over the vocal cords and produces sound.

Respiration

Respiration is the physical and chemical processes by which the body supplies its cells and tissues with the oxygen needed for metabolism and relieves them of the carbon dioxide formed in the energy-producing reactions. Respiration is subdivided into external respiration, which takes place in the lungs; internal respiration, which is between the cells of the body and the blood by way of the fluid bathing the cells; and cellular respiration, which occurs within the cells of the body.

External respiration is also known as *breathing*, or *ventilation*. The air that people breathe contains a mixture of gases—nitrogen, oxygen, and carbon dioxide. Nitrogen is an inert gas that does not interact with anything in the body. It is important because it is a support gas that keeps the lungs open with constant volume. Oxygen is in a higher concentration than carbon dioxide in the air people take in.

External respiration is the exchange of oxygen and carbon dioxide between the lungs, the body, and the outside environment. The breathing process consists of inspiration, or inhalation, and expiration, or exhalation. On inspiration, air enters the body and is warmed, moistened, and filtered as it passes to the air sacs of the lungs called alveoli (al-**VEE**-oh-lye). The concentration of oxygen in the alveoli is greater than in the bloodstream. Oxygen diffuses from the area of greater concentration (the alveoli) to an area of lesser concentration (the bloodstream), then into the red blood cells. At the same time, the concentration of carbon dioxide in the blood is greater than in the alveoli, so it diffuses from the blood to the alveoli. Expiration expels the carbon dioxide from the alveoli of the lungs. Some water vapor is also given off in the process.

Internal respiration includes the exchange of carbon dioxide and oxygen between the cells and the lymph surrounding them, plus the oxidative process of energy in the cells. After inspiration, the alveoli are rich with oxygen and transfer the oxygen into the blood. The resulting greater concentration of oxygen in the blood diffuses the oxygen into the tissue cells. At the same time, the cells build up a higher carbon dioxide concentration. This concentration increases to a point that exceeds the level in the blood. This causes the carbon dioxide to diffuse out of the cells and into the blood, where it is then carried away to be eliminated.

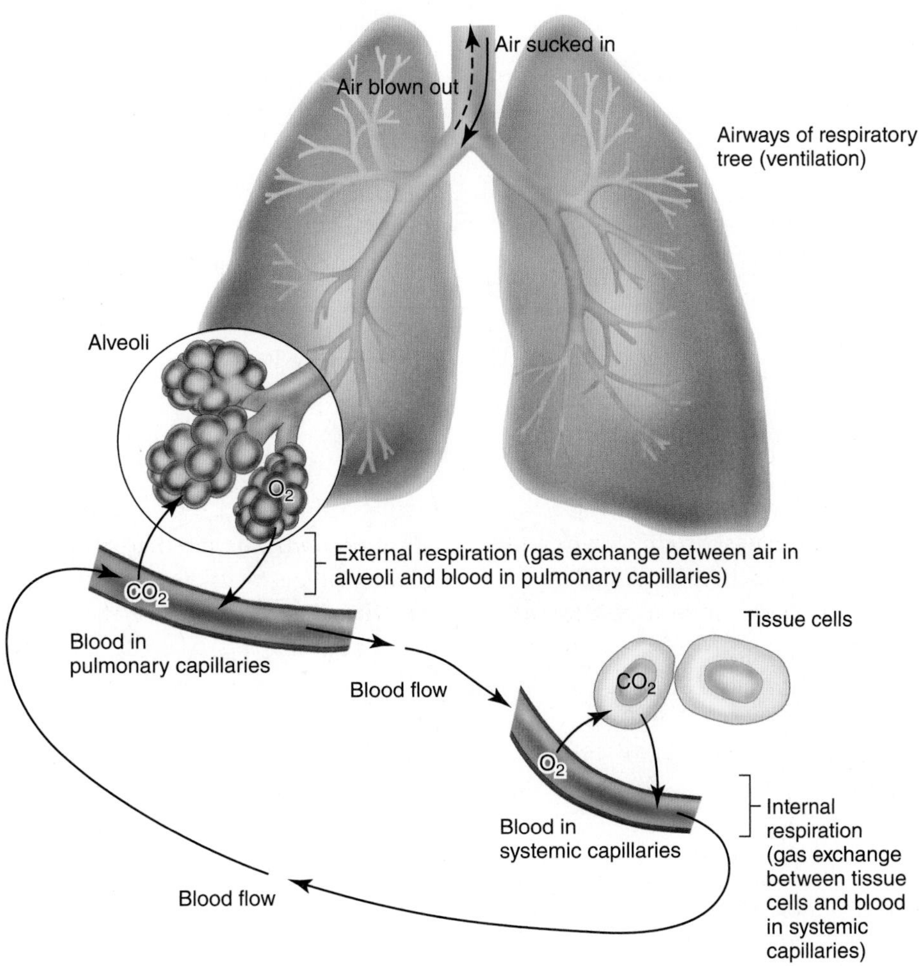

Figure 18-1 *Respiration*

Deoxygenated blood, produced during internal respiration, carries carbon dioxide in the form of bicarbonate ions. These ions are transported by both blood plasma and red blood cells. Exhalation expels carbon dioxide from the red blood cells and the plasma, and bicarbonate ions decompose to carbon dioxide and water.

Media Link

View the **Respiration** animation on the Online Resources.

Cellular respiration (oxidation) involves the use of oxygen to release energy stored in nutrient molecules, such as glucose into adenosine triphosphate (ADP), and release waste products. Just as wood gives off energy in the form of heat and light when burned, or oxidized, so too does food give off energy when it is burned, or oxidized, in the cells. Much of this energy is released in the form of heat to maintain body temperature. Some of it, however, is used directly by the cells for such work as contraction of muscle cells. It is also used to carry on other vital processes.

Food, when oxidized, gives off waste products, including carbon dioxide and water vapor. These waste products are carried away through the process of internal respiration.

RESPIRATORY ORGANS AND STRUCTURES

Air moves into the lungs through several passageways. The following structures are included: nasal cavity, pharynx, larynx, trachea, bronchi and bronchioles, alveoli, lungs, pleura, diaphragm, and mediastinum (**Figure 18-2**).

Nasal Cavity

In humans, air enters the respiratory system through two oval openings in the nose. They are called the nostrils, or **external nares** (**NAIR**-eez). From here, air enters

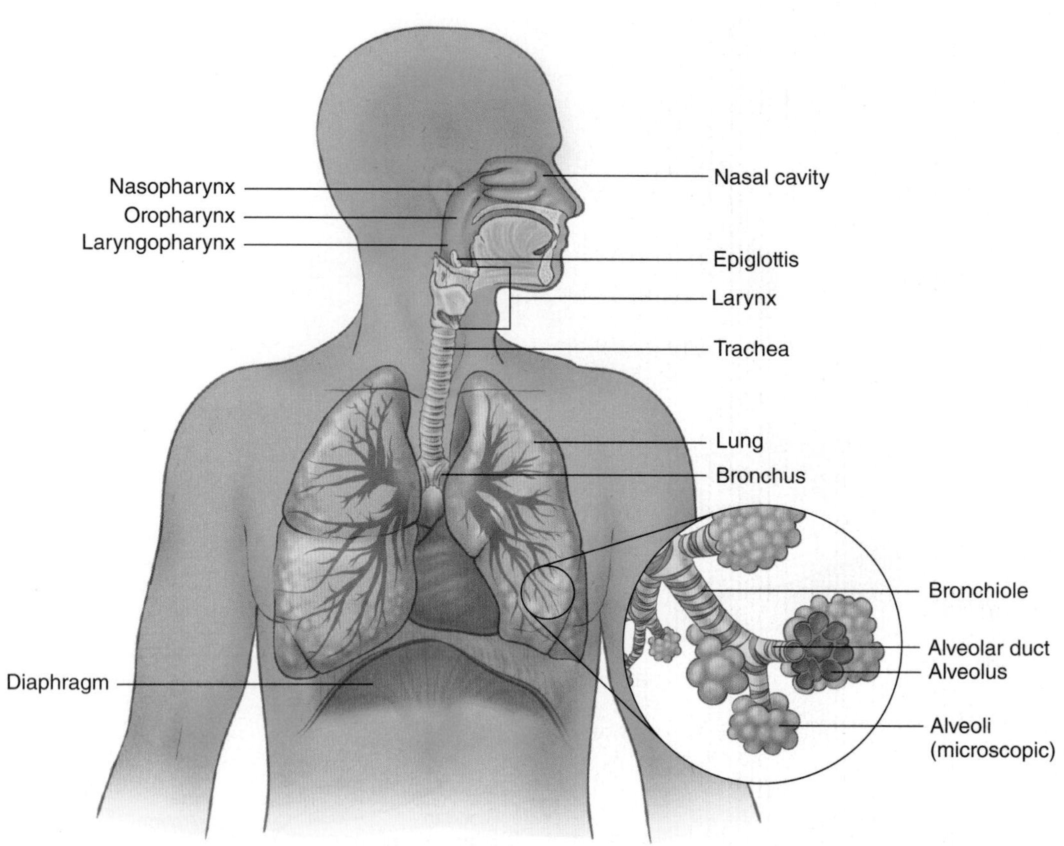

Figure 18-2 *The organs of the respiratory system and the pathway of external respiration: Air enters through the nasal cavity → pharynx → larynx → trachea → bronchus → bronchiole → alveoli*

the nasal cavity, which is divided into a right and left chamber by a partition known as the **nasal septum**. Both cavities are lined with mucous membranes.

Protruding into the nasal cavity are three **turbinates**, or nasal conchae bones; refer to **Figure 7-4** in Chapter 7 for more detail on the bones of the nose. These three scroll-like bones—the superior, middle, and inferior concha—divide the nasal cavity into three narrow passageways. The turbinates increase the surface area of the nasal cavity, causing turbulence in the flowing air, which moves in various directions before exiting the nasal cavity. As it moves through the nasal cavity, air is filtered of dust and dirt particles by the mucous membranes lining the conchal and nasal cavities. The air is also moistened by the mucus and warmed by the blood vessels that supply the nasal cavity. At the front of the nares are small hairs, or *cilia* (**SIH**-lee-ah), which entrap and prevent the entry of larger dirt particles. By the time the air reaches the lungs, it has been warmed, moistened, and filtered. **Olfactory nerves** providing the sense of smell are located in the mucous membrane, in the upper part of the nasal cavity.

The **sinuses**, named frontal, ethmoidal, sphenoidal, and maxillary, are cavities of the skull filled with air in and around the nasal region (**Figure 18-3**). Short ducts connect the sinuses with the nasal cavity. Sinuses help lighten the bones of the skull. The mucous membrane lines the sinuses and helps warm and moisten air passing through them. The sinuses also give resonance to the voice. The unpleasant voice sound of a nasal cold results from the blockage of sinuses.

Pharynx

After air leaves the nasal cavity, it enters the **pharynx** (**FAR**-inks), commonly known as the throat. The pharynx serves as a common passageway for air and food. It is about 5 inches long and is subdivided into the nasopharynx, the oropharynx, and laryngopharynx. The nasopharynx lies above and behind the soft palate. The left and right eustachian tubes open directly into the nasopharynx, connecting it with each middle ear. Because of this connection, nasopharyngeal inflammation can lead to middle ear infections. The oropharynx is also called the oral part

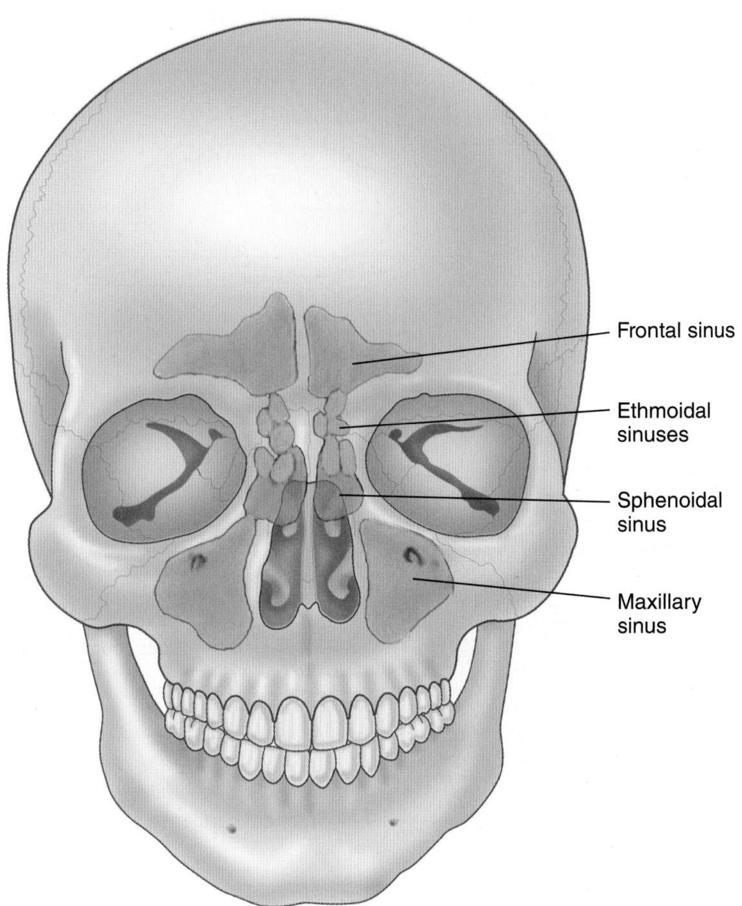

Figure 18-3 *Nasal sinuses*

of the mouth; it extends from the soft palate, behind the mouth, to just above the hyoid bone. The laryngopharynx is located below the oropharynx and superior to the larynx. Air travels down the pharynx on its way to the lungs; food travels this route on its way to the stomach.

The **epiglottis** (ep-ih-**GLOT**-is) is the flap of cartilage lying behind the tongue and in front of the entrance to the larynx. At rest the epiglottis is upright and allows air to pass through the larynx and to the lungs. During swallowing, it folds back to cover the entrance to the larynx, preventing food and drink from entering the trachea. The larynx draws upward and forward to close the trachea. At the end of each swallow, the epiglottis moves up again, the larynx returns to rest, and the flow of air into the trachea continues (**Figure 18-4**).

Larynx

The **larynx** (**LAR**-inks), or voice box, is a triangular chamber found below the pharynx. The laryngeal walls are composed of nine fibrocartilaginous plates. The largest of these is commonly called the Adam's apple. During puberty, the vocal cords become larger in the male, making the Adam's apple more prominent.

The larynx is lined with a mucous membrane, continuous from the pharyngeal lining above to the tracheal lining below. Within the larynx are the characteristic vocal cords. There is a space between the vocal cords known as the **glottis** (**GLOT**-is). When air is expelled from the lungs, it passes the vocal cords (**Figure 18-5**). This sets off a vibration, creating sound. The action of the lips and tongue on this sound produces speech.

Trachea

The **trachea** (**TRAY**-kee-ah), or windpipe, is a tubelike passageway some 11.2 centimeters (about 4.5 inches) in length. It extends from the larynx, passes in front of the esophagus, and continues to form the two bronchi, one for each lung. The walls of the trachea are composed of alternate bands of membranes and 15 to 20 C-shaped rings of hyaline cartilage (**Figure 18-6**). These C-shaped rings are virtually noncollapsible, keeping the trachea open for the passage of oxygen into the lungs. However, the trachea can be obstructed by large pieces of food, tumorous growths, or the swelling of inflamed lymph nodes in the neck.

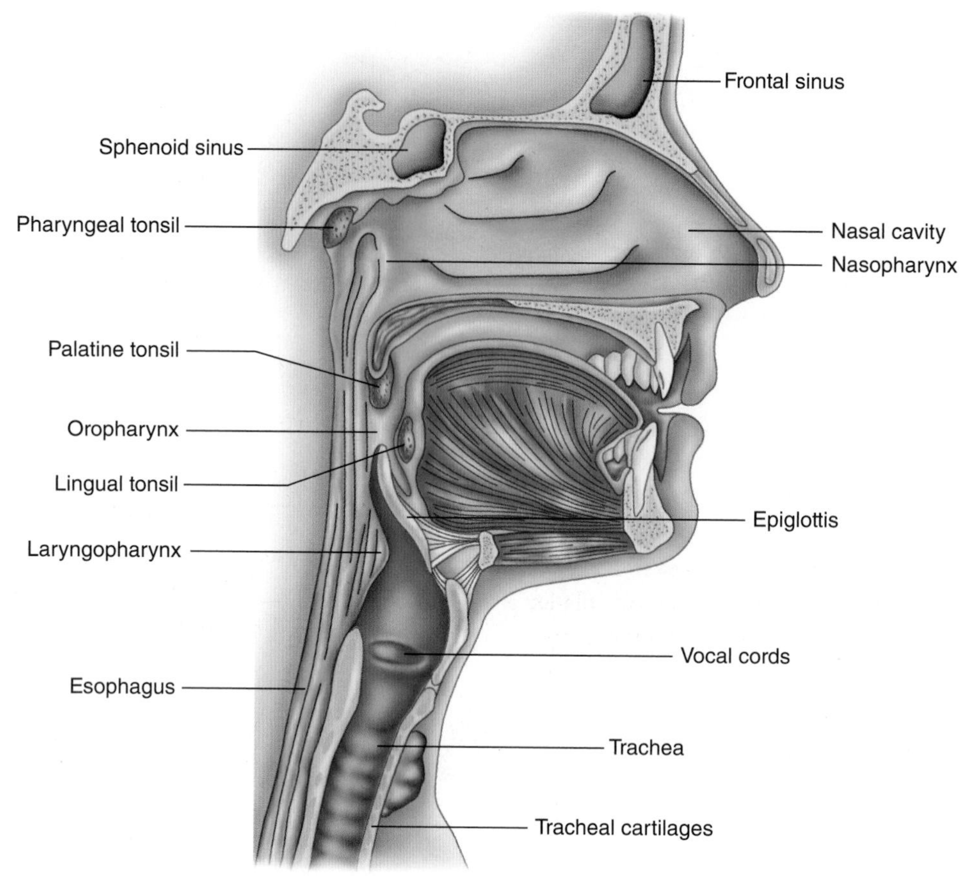

Figure 18-4 *Sagittal section of the face and neck*

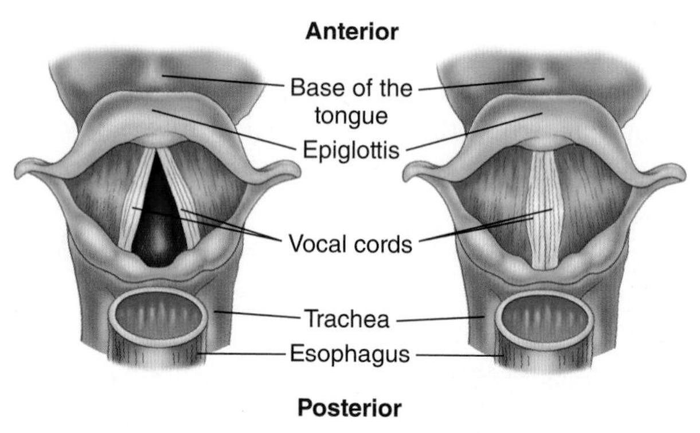

Figure 18-5 *View of the larynx and vocal folds: Shown on the left, the vocal folds are open during breathing. On the right, the vocal folds vibrate together during speech.*

The walls of the trachea are lined with both mucous membrane and ciliated epithelium. The function of the mucus is to entrap inhaled dust particles; the cilia then sweep such dust-laden mucus upward to the pharynx. Coughing and expectoration dislodge and eliminate the dust-laden mucus from the pharynx.

Smoking creates a constant irritation to the trachea. Over time, this irritation causes the epithelium of the trachea to change from a pseudostratified ciliated columnar to a stratified squamous epithelium lacking the cilia.

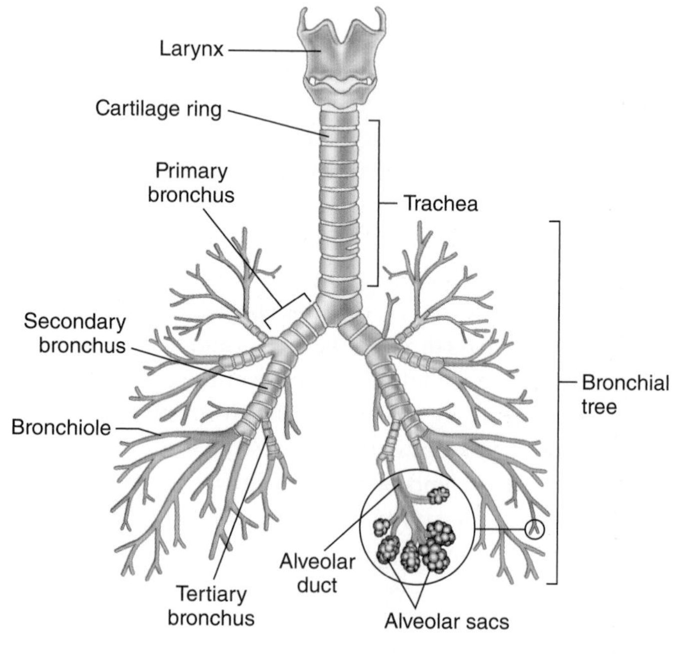

Figure 18-6 *Larynx, trachea, and bronchial tree*

Without cilia, the epithelium cannot clear the passageway of mucus and debris. The mucus and debris provide an ideal environment for the growth of microorganisms, leading to respiratory infections. This constant irritation and respiratory inflammation triggers the cough reflex, resulting in what is called a "smoker's cough."

Bronchi and Bronchioles

The lower end of the trachea separates into the right **bronchus** (**BRONG**-kus) and the left bronchus. There is a slight difference between the two bronchi; the right bronchus is somewhat shorter, wider, and more vertical in position (**Figure 18-6**).

As the bronchi enter the lung, they subdivide into bronchial tubes and smaller **bronchioles** (**BRONG**-kee-ohlz). The divisions are Y-shaped in form. The two bronchi are similar in structure to the trachea; their walls are also lined with ciliated epithelium and ringed with hyaline cartilage. However, the bronchial tubes and smaller bronchi are ringed with cartilaginous plates instead of incomplete C-shaped rings. As they become smaller, the bronchioles lose their cartilaginous plates and fibrous tissue. Their thinner walls are made from smooth muscle and elastic tissue lined with ciliated epithelium. At the end of each bronchiole is an alveolar duct, which ends in a saclike cluster called an **alveolar sac** (**Figure 18-6**).

Alveoli

The alveolar sacs consist of many **alveoli,** which have a single layer of epithelial tissue. An adult lung has about 500 million alveoli, about three times the amount necessary to sustain life. Each alveolus forming a part of the alveolar sac possesses a globular shape. Their inner surfaces are covered with a lipid material known as **surfactant** (sir-**FAK**-tant). The surfactant helps stabilize the alveoli, preventing their collapse. Each alveolus is encased by a network of blood capillaries.

It is through the moist walls of both the alveoli and the capillaries that rapid exchange of carbon dioxide and oxygen occurs. In the blood capillaries, carbon dioxide diffuses from the erythrocytes, through the capillary walls, into the alveoli, and is exhaled through the mouth and nose.

The opposite process occurs with oxygen, which diffuses from the alveoli into the capillaries, and from there into the erythrocytes.

Lungs

The lungs are fairly large, cone-shaped organs filling up the two lateral chambers of the thoracic cavity (**Figure 18-7**). They are separated from each other by

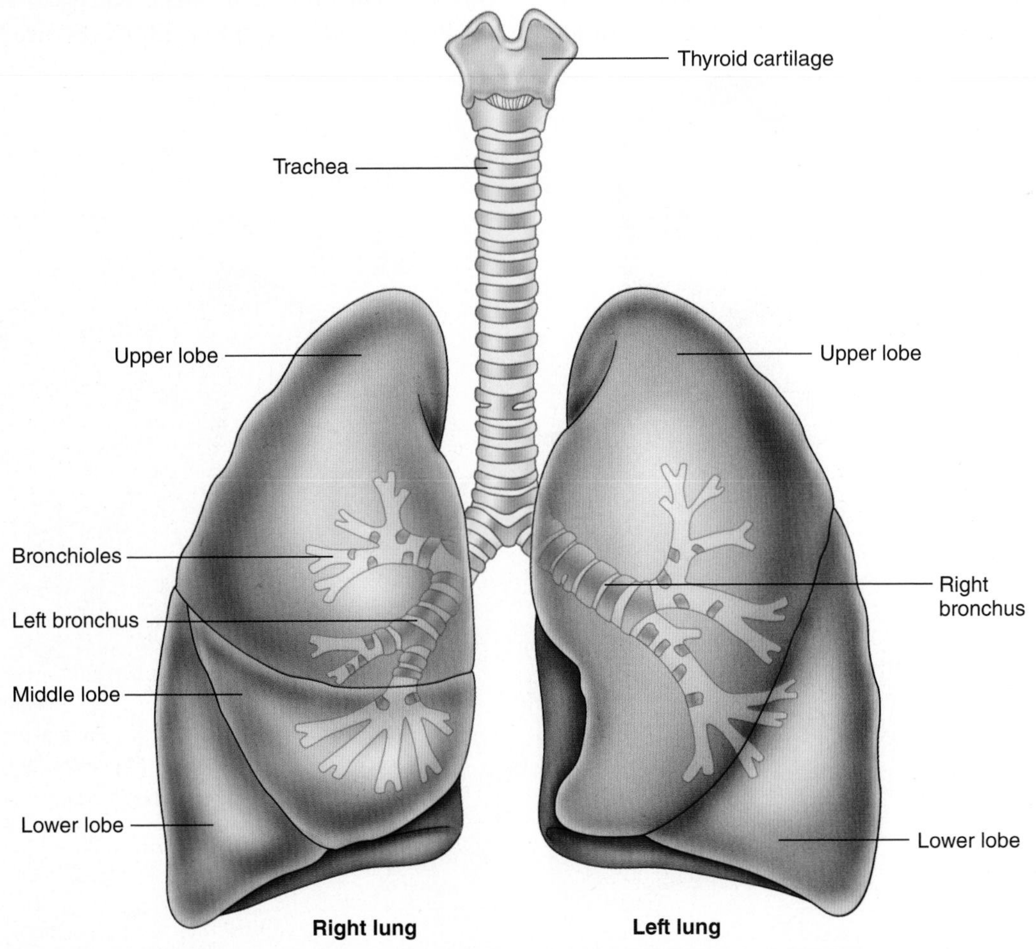

Figure 18-7 *External view of the lungs: The right lung has three lobes, and the left lung has two lobes.*

the mediastinum and the heart. The upper part of the lung, underneath the collarbone, is the apex; the broad lower part is the base. Each base is concave, allowing it to fit snugly over the convex part of the diaphragm.

Lung tissue is porous and spongy, due to the alveoli and the tremendous amount of air it contains. If a specimen of a cow lung were placed into a tankful of water, for example, it would float quite easily.

The right lung is larger and broader than the left because the heart inclines to the left side. The right lung is also shorter due to the diaphragm's upward displacement on the right to accommodate the liver. The right lung is divided by fissures, or clefts, into three lobes: upper, middle, and lower.

The left lung is smaller, narrower, and longer than its counterpart. It is subdivided into two lobes: upper and lower.

Pleura

The lungs are covered with a thin, moist, slippery membrane of tough endothelial cells, or **pleura** (**PLOOR**-ah). There are two pleural membranes. The one covering the lungs and dipping between the lobes is the pulmonary, or visceral, pleura. Lining the thoracic cavity and the upper surface of the diaphragm is the parietal pleura (**Figure 18-8**). Consequently, each lung is enclosed in a double-walled sac. **Pleurisy** (**PLOOR**-ih-see) is an inflammation of this lining.

The space between the two pleural membranes is the pleural cavity, filled with serous fluid called **pleural fluid.** This fluid is necessary to prevent friction between the two pleural membranes as they rub against each other during each breath.

The pleural cavity may, on occasion, fill up with an enormous quantity of serous fluid. This occurs when there is an inflammation of the pleura. The increased pleural fluid compresses and sometimes even causes parts of the lung to collapse. This makes breathing extremely difficult. To alleviate the pressure, a **thoracentesis** (**thoh**-rah-sen-**TEE**-sis) may be performed. This procedure entails the insertion of a hollow, tubelike instrument through the thoracic cavity and into the pleural cavity to drain the excess fluid.

Diaphragm

The **diaphragm** (**DYE**-ah-fram) is a dome-shaped sheet of muscle that separates the thoracic cavity from the abdomen. It is the contraction and relaxation of this muscle with the intercostal muscles that makes breathing possible. The phrenic nerves (**FREN**-ick) stimulate the diaphragm and cause it to contract (**Figure 18-9**).

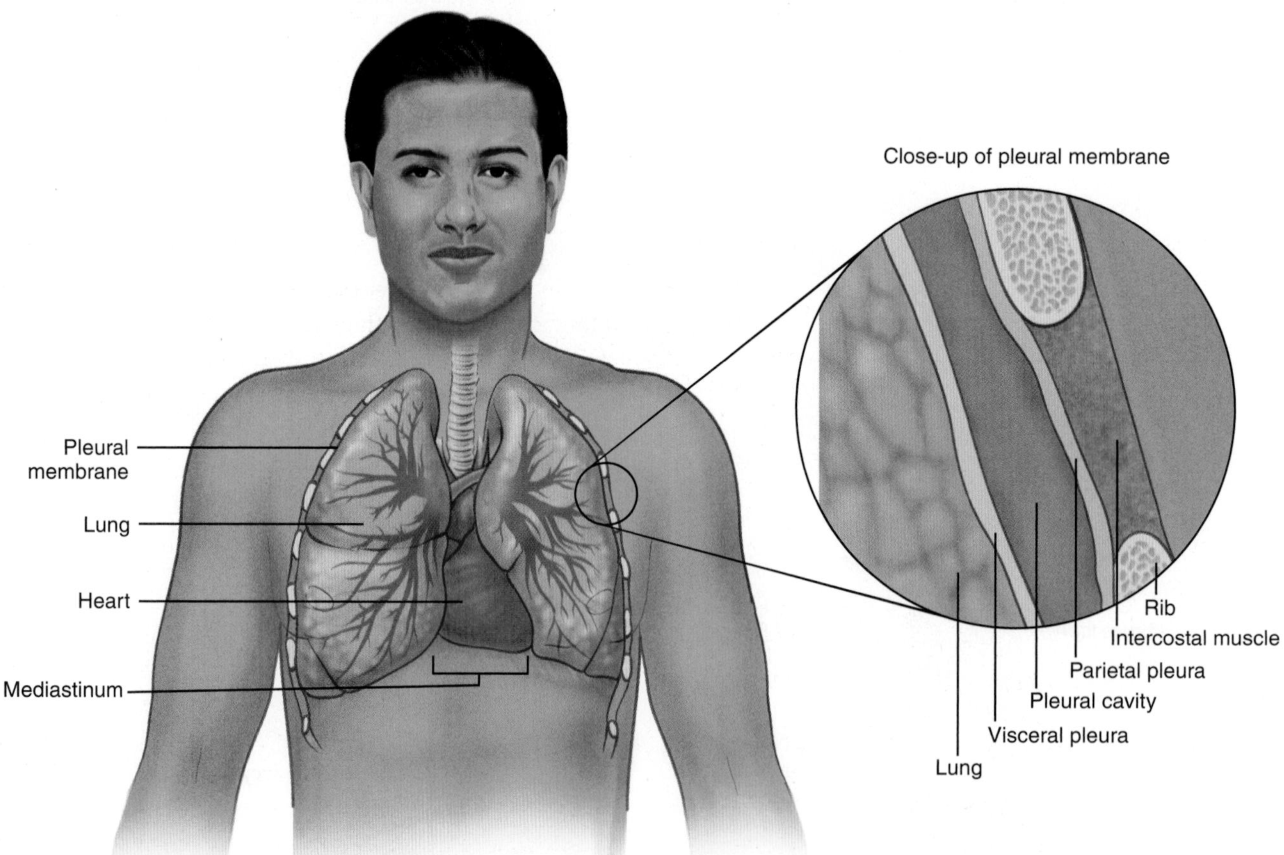

Figure 18-8 *The pleural fluid between the parietal and visceral pleura allows the lungs to move smoothly within the chest.*

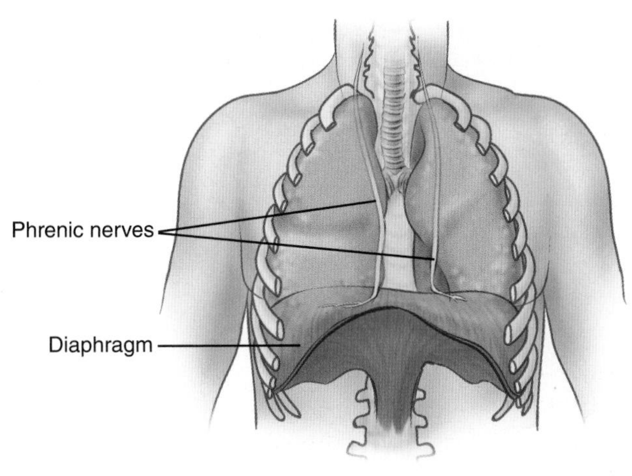

Figure 18-9 *The phrenic nerves control the diaphragm and the intercostal muscles.*

Mediastinum

The **mediastinum** (mee-dee-as-**TYE**-num), also called the interpleural space, is situated between the lungs along the median plane of the thorax. It extends from the sternum to the vertebrae. The mediastinum contains the thoracic viscera: the thymus gland, heart, aorta and its branches, pulmonary arteries and veins, superior and inferior vena cava, esophagus, trachea, thoracic duct, lymph nodes, and vessels.

MECHANICS OF BREATHING

Respiration, or breathing, is the process in the lungs where oxygen is exchanged for carbon dioxide. A single breath consists of one inhalation and one expiration. **Ventilation** is another term for moving air in and out of the lungs (breathing). Pulmonary ventilation of the lungs is due to changes in pressure that occur within the chest cavity. The normal pressure within the pleural space is always negative, meaning it is less than atmospheric pressure. The negative pressure helps keep the lungs expanded. The variation in pressure is brought about by cellular respiration and mechanical breathing movements.

The Breathing Process

Pulmonary ventilation allows the exchange of oxygen between the alveoli and erythrocyte, and eventually between the erythrocyte and cells.

INHALATION/INSPIRATION

There are two groups of intercostal muscles: external intercostals and internal intercostals. Their muscle fibers cross each other at an angle of 90 degrees. During inhalation, or **inspiration,** the external intercostals lift the ribs upward and outward (**Figure 18-10**). This increases the volume of the thoracic cavity. Simultaneously, the sternum rises along with the ribs and the dome-shaped diaphragm contracts and becomes flattened, moving downward. As the diaphragm moves downward, pressure is exerted on the abdominal viscera. This causes the anterior muscles to protrude slightly, increasing the space within the chest cavity in a vertical direction. As a result, there is a decrease in pressure. Because atmospheric pressure is now greater, air rushes in all the way down to the alveoli, resulting in inhalation.

EXHALATION/EXPIRATION

In exhalation, or **expiration,** just the opposite takes place (**Figure 18-10**). Expiration is a passive process; the contracted diaphragm and intercostal muscles relax. The ribs move down; the diaphragm moves up. In addition, the surface tension of the fluid lining the alveoli reduces the elasticity of the lung tissue and causes the alveoli to collapse. This action, coupled with the relaxation of contracted respiratory muscles, relaxes the lungs; the space within the thoracic cavity decreases, thus increasing the internal pressure. Increased pressure forces air from the lungs, resulting in exhalation.

The lungs are extremely elastic. They are able to change capacity as the size of the thoracic cavity is altered. This ability is known as compliance. When lung tissue becomes diseased and fibrotic, the lung's compliance decreases and ventilation decreases.

Respiratory Movements and Frequency of Respiration

The rhythmic movement of the rib cage where air is drawn in and expelled from the lungs makes up the respiratory movements. Inspiration and expiration combined is counted as one respiratory movement. The normal rate in quiet breathing for an adult is about 14 to 20 breaths per minute. This rate is changeable. The respiratory rate can be increased by muscular activity, increased body temperatures, and in certain pathological disorders, such as hyperthyroidism. It changes with sex, females having the higher rate at 16 to 20 breaths per minute. Age will also change the respiratory rate. For example, at birth the rate is 40 to 60 breaths per minute; at 5 years, the rate is 24 to 26 breaths. The body's position also affects the respiration rate. When the body is asleep or prone, the rate is 12 to 14 breaths per minute. In a sitting position, it is 18, and in a standing position, it is 20 to 22 breaths per minute. Emotions play a role in decreasing or increasing the respiratory rate, probably through the action of the hypothalamus and pons (see Chapter 9).

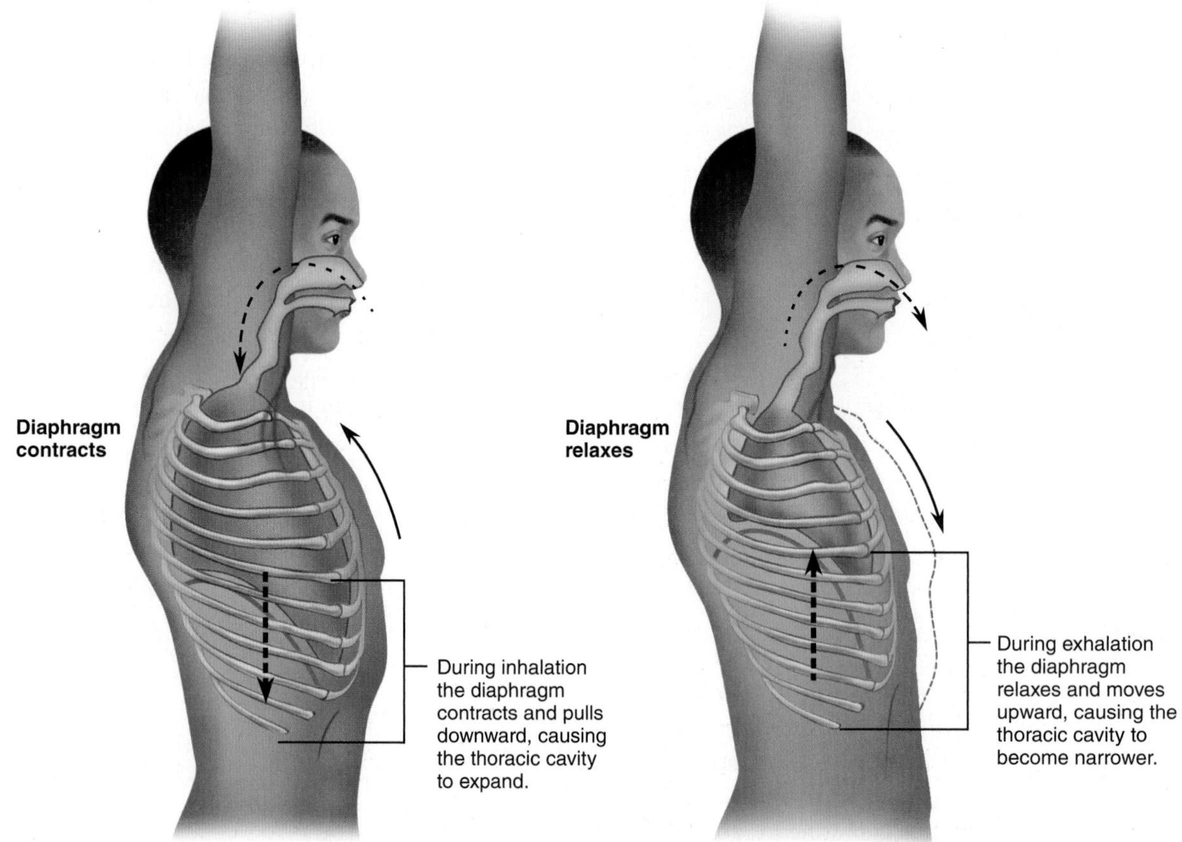
Diaphragm
contracts
During inhalation
the diaphragm
contracts and pulls
downward, causing
the thoracic cavity
to expand.
Diaphragm
relaxes
During exhalation
the diaphragm
relaxes and moves
upward, causing the
thoracic cavity to
become narrower.

Figure 18-10 *Mechanics of breathing: Movements of the diaphragm and thoracic cavity produce inhalation (left) and exhalation (right).*

Other situations that can affect the respiratory rate are the following:

- **Coughing**—a deep breath is taken, followed by a forceful exhalation from the mouth to clear the lower respiratory tract.
- **Hiccoughs,** or hiccups—these are caused by spasms of the diaphragm and a spasmodic closure of the glottis. This is believed to be the result of an irritation to the diaphragm or the phrenic nerve.
- **Sneezing**—this occurs like a cough, except air is forced through the nose to clear the upper respiratory tract.
- **Yawning**—a deep, prolonged breath that fills the lungs is taken, believed to be caused by the need to increase oxygen within the blood or flush out excess carbon dioxide.

Did You Know?

Yawning may occur if a person is not breathing deeply enough throughout the day, allowing carbon dioxide to build up in their system. Yawning expands the lungs to flush out carbon dioxide. A person may also yawn when they see someone else yawn. Researchers suspect this has to do with the mirror neuron system, the part of the brain that causes people to mimic expressions and behavior.

CONTROL OF BREATHING

The rate of breathing is controlled by neural (nervous) and chemical factors. Although both have the same goal—that of respiratory control—they function independently of one another.

Neural Factors

The respiratory center is located in the medulla oblongata in the brain. It is subdivided into two centers: one to regulate inspiration, the other for expiratory control.

The upper part of the medulla contains a grouping of cells that is the seat of the respiratory center. An increase of carbon dioxide or lack of oxygen in the blood will trigger the respiratory center.

Two neuronal pathways are involved in breathing. One group of phrenic nerves leads to and stimulates the diaphragm and intercostal muscles. The other nerve pathway carries sensory impulses from the nose, larynx, lungs, skin, and abdominal organs via the vagus nerve in the medulla.

The rhythm of breathing can be changed by stimuli originating within the body's surface membranes. For example, a sudden drenching with cold water can make

The Effects of Aging on the Respiratory System

With aging, lung tissue loses elasticity, the rib cage becomes less flexible, and muscle strength decreases. The number of functioning alveoli decreases. These factors compromise the oxygen-carbon dioxide exchange, reducing the oxygen content in the blood. The combination of a less efficient heart pump and reduced oxygen causes characteristic signs of activity intolerance.

The sensory receptors in the airway that produce the cough reflex lose their sensitivity with age. The diminished cough reflex enables debris and irritants to reach the deep lung tissues, which can cause respiratory tract infections.

Breathlessness is the most common physiological response to exercise of a person who is sedentary. It is estimated that the maximum oxygen consumption rate during stress and moderate exercise can increase nine times for an older adult. Lung capacity exhibits changes also. There is an increase in residual volume and functional residual capacity and a decrease in vital capacity and expiratory airflow.

The presence of respiratory disease tends to be higher in older adults. Experts recommend that older adults take the flu vaccine annually and get a one-time vaccination for pneumonia.

a person gasp, while irritation to the nose or larynx can make a person sneeze or cough.

Although the medulla's respiratory center is primarily responsible for respiratory control, it is not the only part of the brain that controls breathing. A lung reflex called the **Hering-Breuer reflex**—named after K. Ewald Hering (1834–1918), a German physiologist, and Josef Breuer (1842–1925), an Austrian physician—is involved in preventing the overstretching of the lungs. When the lungs are inflated, the nerve endings in the walls are stimulated. A nerve message is sent from the lungs to the medulla by way of the vagus nerve, inhibiting inspiration and stimulating expiration. This mechanism prevents overinflation of the lungs, keeping them from being ripped apart like an overinflated balloon.

Chemical Factors

Chemical control of respiration is dependent on the level of carbon dioxide in the blood. When blood circulates through active tissue, it receives carbon dioxide and other metabolic waste products of cellular respiration. As blood circulates through the respiratory center, the respiratory center senses the increased carbon dioxide in the blood and increases the respiratory rate. For example, a person performing vigorous exercise or physical labor breathes more deeply and quickly to cope with the need for more oxygen and to rid the body of excess carbon dioxide produced.

Other chemical regulators of respiration are the chemoreceptors, which are found in carotid arteries and the aorta. As the arterial blood flows around these carotid and aortic bodies, the chemoreceptors are particularly sensitive to the amount of oxygen present. If oxygen declines to very low levels, impulses are sent from the carotid and aortic bodies to the respiratory center, which will stimulate the rate and depth of respiration. The respiratory center can be affected by drugs such as depressants, barbiturates, and morphine.

LUNG CAPACITY AND VOLUME

A person can measure how much air their lungs can hold (their lung capacity) by using a device called a **spirometer** (spih-**ROM**-eh-ter). A spirometer measures the volume and flow of air during inspiration and expiration. By comparing the reading with the norm for a person's age, height, weight, and sex, it can be determined if any deficiencies exist. Disease processes affect total lung capacity (**Figure 18-11**).

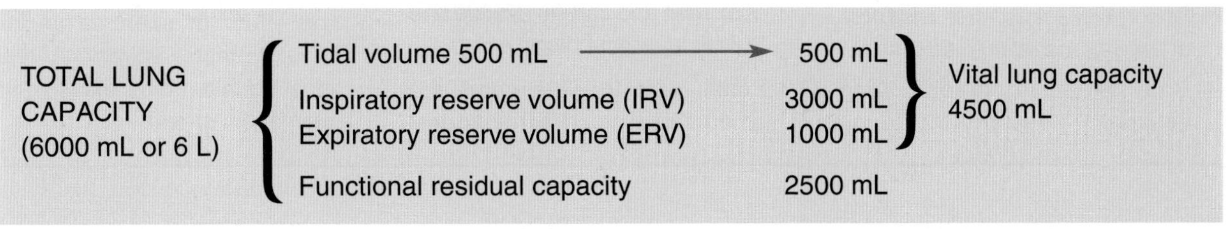

Figure 18-11 *Lung capacity and volume*

The following are the different lung capacity measurements:

- **Tidal volume** is the amount of air that moves in and out of the lungs with each breath. The normal amount is about 500 mL.
- **Inspiratory reserve volume (IRV)** is the amount of air a person can take in over and above the tidal volume. The normal amount is 2100 to 3000 mL.
- **Expiratory reserve volume (ERV)** is the amount of air a person can exhale over and above the tidal volume. The normal amount is 1000 mL.
- **Vital lung capacity** is the total amount of air involved with tidal volume, inspiratory reserve volume, and expiratory reserve volume. The normal vital capacity is 4500 mL.
- **Residual volume** is the amount of air that cannot be voluntarily expelled from the lungs. This allows for the continuous exchange of gases between breaths. The normal residual volume is 1500 mL.
- **Functional residual capacity** is the sum of the expiratory reserve volume plus the residual volume. The normal amount is 2500 mL.
- **Total lung capacity** includes tidal volume, inspiratory reserve, expiratory reserve, and residual air. The normal amount is 6000 mL.

TYPES OF RESPIRATION

The health care provider should be aware of the various changes to the respiratory rate and sounds of human respiration. These changes can be alerts to an abnormal respiratory condition in a patient. The following conditions describe various kinds and conditions of respiration:

- **Apnea** is the temporary stoppage of breathing movements.
- **Dyspnea** (**DISP**-nee-ah) is difficult, labored, or painful breathing, usually accompanied by discomfort and breathlessness.
- **Eupnea** (youp-**NEE**-ah) is normal or easy breathing with the usual quiet inhalations and exhalations.
- **Hyperpnea** (high-perp-**NEE**-ah) is an increase in the depth and rate of breathing accompanied by abnormal exaggeration of respiratory movements.
- **Orthopnea** (or-**THOP**-nee-ah) is difficult or labored breathing when the body is in a horizontal position. It is usually corrected by sitting or standing.
- **Tachypnea** (**tack**-ihp-**NEE**-ah) is an abnormally rapid and shallow rate of breathing.

Medical Highlights

PULMONARY FUNCTION TESTS

Pulmonary function tests are a group of tests that measure the volume and flow of air and/or the concentration of oxygen in the blood. These tests are measured against a norm for the individual's age, height, and sex. These types of tests include the following:

- Spirometry—measures the force of expirations using a spirometer
- Body plethysmographic lung volumes—measures total lung capacity. The patient sits in a small, airtight room called a plethysmograph and breathes through a mouthpiece where pressure and airflow data are measured and collected.
- Arterial blood gases—a blood gas analyzer measures the amounts of oxygen and carbon dioxide and the pH in arterial blood samples.
- Pulse oximetry—estimates the amount of oxygen in the arterial blood by shining a light through the fingertip. Because blood that holds a lot of oxygen is a different color than less oxygenated blood, the device is able to estimate the blood's oxygen content.
- Exercise stress tests—evaluates the effect of exercise on lung function tests. Spirometry readings are taken after exercise and then again at rest.

- **Hyperventilation** (**high**-per-**ven**-tih-**LAY**-shun) is a condition that can be caused by disease or stress. Rapid breathing occurs, which causes the body to lose carbon dioxide too quickly. The lowered blood level of carbon dioxide leads to alkalosis. Symptoms are dizziness and possible fainting. To correct this condition, the person should breathe into a paper bag. The exhaled air contains more carbon dioxide; the air breathed in will have higher levels of carbon dioxide from the paper bag, so this activity will restore the normal blood levels of carbon dioxide.
- **Kussmaul respiration** is caused by ketoacidosis (**KEE**-toh-as-ah-DOH-sis), which occurs when there is too much glucose (energy) accompanied by high levels of ketones in the blood, most commonly caused by diabetes. Kussmaul respirations are rapid, shallow breaths that become deeper and more labored as ketoacidosis gets more severe.

Medical Highlights

SLEEP APNEA

Sleep is essential to health and well-being. Sleep apnea (**AP**-nee-ah) is a serious sleep disorder in which breathing repeatedly stops and starts. Breathing pauses can last from a few seconds to minutes; this may occur as many as 30 times a night. According to a Mayo Clinic report, an estimated 18 million Americans have this potentially serious disorder. Signs of sleep apnea include loud snoring, having more than a few breathing pauses during a night's sleep witnessed by another person, and daytime drowsiness. There are three kinds of sleep apnea: obstructive sleep apnea; central sleep apnea; and complex sleep apnea, which occurs when a person has both obstructive sleep apnea and central sleep apnea.

Obstructive sleep apnea is caused by a blockage in the back of the throat that prevents air from reaching the lungs. In obstructive sleep apnea, the muscles that normally keep the airway open relax and sag during sleep, causing the tongue and palate to repeatedly block the breathing for about 20 seconds. This lowers the level of oxygen in the blood. The brain senses this decrease and briefly rouses the person from sleep so that the airway can be reopened.

With central sleep apnea, the brain fails to send proper signals to the muscles that keep a person breathing adequately. As a result, they may awaken within 20 seconds because of a lack of air.

In addition to the decline in well-being and the danger associated with major sleep deprivation, sudden drops in blood oxygen levels may increase blood pressure and strain the cardiovascular system. Obstructive sleep apnea is more common in individuals who are male; overweight; are smokers; or who consume alcohol, sedatives, or tranquilizers. Evaluation of the condition is done at a sleep center, which involves overnight monitoring of lung, brain, and heart activity; breathing patterns; and blood oxygen levels using a test called a polysomnography.

Treatment includes losing weight, quitting smoking, sleeping on the side or stomach, or using a nasal strip that helps keep the air passages open. For a moderate to severe sleep apnea problem, the physician may recommend a nasal continuous positive airway pressure (CPAP) machine. Through a mask placed over the nose, a CPAP machine delivers air at a pressure greater than the surrounding air. The pressure increase is just enough to keep the airway passages open. Other devices include an expiratory positive airway pressure (EPAP) device, an adaptive servo ventilation (ASV) device, a hypoglossal nerve stimulator, and oral appliances. An EPAP is placed over each nostril before going to sleep. The valve allows air to move in freely, but when a person exhales, the air must go through small holes in the valve. This increases pressure in the airway and keeps it open. ASV is a device that learns a person's normal breathing pattern and stores the information in a computer. After a person falls asleep, the machine uses pressure to normalize their breathing pattern and prevent pauses in their breathing. Hypoglossal nerve stimulation (HNS) is done by a small device surgically implanted in the chest that can be turned on and off by the patient. The device monitors breathing and stimulates the nerves to keep the airway open. An oral appliance is any appliance designed to keep the mouth open.

DISORDERS OF THE RESPIRATORY SYSTEM

Upper respiratory infections account for just under 50% of all acute illnesses (**Figure 18-12**). Most disorders of the upper respiratory tract are not life threatening. Bacterial infections, which are less common, may be treated with antibiotics.

The majority of the disorders of the respiratory system involve some defect or problem that prevents outside air from reaching the alveoli. In many cases, the cause of the problem is an infection within the upper or lower airway. The problem may also stem from a mechanical obstruction of the airway.

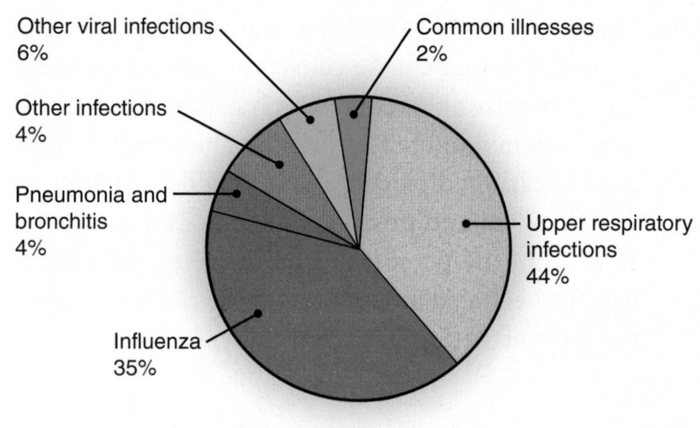

Figure 18-12 *Frequency of infectious diseases*

Infectious Causes

The respiratory system is subject to various infections and inflammations caused by bacteria, viruses, and irritants. Many of these viruses are forms of the coronavirus. Severe acute respiratory distress syndrome (SARS) was an example of a highly contagious coronavirus that quickly spread in 2002; it has rarely been seen since 2004. COVID-19 is a coronavirus that caused a global pandemic in 2020 (see Chapter 17).

The **common cold** is usually caused by a virus that is highly contagious. There are several hundred strains of viruses that can cause the common cold. Symptoms include runny nose, watery eyes, sneezing, head congestion and sore throat. This infection is responsible for the greatest loss of production hours each year. It is often the basis for more serious respiratory diseases. It lowers the body's resistance to infection. A person who has a cold should get lots of rest, drink fluids, and eat wholesome, nourishing foods. Handwashing or the use of hand sanitizers is the best preventive measure against the common cold. A person with a cold should cough or sneeze into a tissue and discard it, or cough or sneeze into their upper shirt sleeve, completely covering the nose and mouth. Over-the-counter cold remedies may ease the symptoms. Colds generally run their course in seven days.

Pharyngitis (**fah**-rin-**JIGH**-tis) is a red, inflamed throat that may be caused by one of several bacteria or viruses. Bacterial causes include the *Streptococcus*

 Career Profile

RESPIRATORY THERAPIST

Respiratory therapists evaluate, treat, and care for patients with breathing or other cardiopulmonary disorders. Practicing under the direction of a physician, advanced-level respiratory therapists assume responsibility for all respiratory care procedures. Entry-level respiratory therapists follow specific, well-defined respiratory care procedures under the direction of a physician or an advanced-level respiratory therapist.

Both levels of therapists may have the same education; however, the advanced respiratory therapist has demonstrated advanced competency. Advanced respiratory therapists consult with physicians and staff to help plan and modify patient care plans. They also provide complex therapies that require considerable judgment, such as caring for patients on life support in the intensive care unit. Respiratory therapists perform physical examinations and conduct diagnostic tests measuring lung capacity and oxygen and carbon dioxide levels.

Formal training is necessary for this career. Programs vary in length and in the credential or degree awarded. Associate's degree or bachelor's degree programs prepare graduates for jobs as Registered Respiratory Therapists (RRTs). Shorter programs award certificates and lead to entry-level jobs as Certified Respiratory Therapists (CRTs).

Respiratory therapists usually work 40 hours a week. Therapists may work around the clock in a hospital setting. Many therapists also work outside the hospital in respiratory therapy clinics, nursing homes, or in a physician's practice. The job outlook is excellent.

organism, which causes "strep throat"; this is treated with antibiotics. Pharyngitis also occurs as a result of irritants such as too much smoking or speaking. It is characterized by painful swallowing and extreme dryness of the throat. Treatment is rest, drinking warm fluids, and gargling with warm saltwater.

Laryngitis (lar-in-**JIGH**-tis) is an inflammation of the larynx, or voice box. It is often secondary to other respiratory infections. Laryngitis can also be caused by overusing the voice, such as with excessive talking, singing, or shouting. It can be recognized by the incidence of hoarseness or loss of voice. Treatment includes resting the voice, using a humidifier, and drinking plenty of fluids.

Sinusitis (sigh-nus-**EYE**-tis) is an infection of the mucous membrane that lines the sinus cavities. It is usually the result of a viral or bacterial infection. One or several of the cavities may be infected. Pain and nasal discharge are symptoms of this infection. To reduce congestion in the sinuses, use a humidifier and nasal saline rinses, drink plenty of fluids to thin the mucus, and apply a warm moist cloth to the face several times a day.

Bronchitis (brong-**KYE**-tis) is an inflammation of the mucous membrane of the trachea and the bronchial tubes that produces excessive mucus. Bronchitis can be caused by bacteria or a virus. It may be acute or chronic and often follows infections of the upper respiratory tract. Acute bronchitis can be caused by the spreading of an inflammation from the nasopharynx, or by inhalation of irritating vapors. This condition is characterized by a cough; fever; substernal pain; and **rales** (**RAHLZ**), or a rattling sound in the lungs. Treatment is symptomatic. Acute bronchitis may become chronic after many episodes.

Influenza (in-flew-**EN**-zah) or "flu" is a viral infection characterized by inflammation of the mucous membrane of the respiratory system. The infection is accompanied by fever, a mucopurulent discharge, muscular pain, and extreme exhaustion. Complications such as bronchopneumonia, neuritis, otitis media (middle ear infection), and pleurisy often follow influenza. Treatment is aimed at the symptoms. A yearly vaccine is recommended for infants over 6 months of age, older adults, pregnant women, and those with certain health conditions. The vaccine is also recommended for people who work or live with others at high risk.

H1N1, also called the swine flu, is a subtype of the influenza A virus; it is called the swine flu because it originally appeared to be associated with pigs, which later proved to be incorrect. Symptoms are similar to influenza symptoms. A yearly flu vaccine is recommended because it includes protection against the swine flu virus.

Respiratory syncytial virus (RSV) is a very common and contagious virus that leads to mild cold-like symptoms. In infants and young children, symptoms can be more severe. Many children have had it by the age of two. Treatment includes over-the-counter drugs, nasal saline rinses, and suctioning if necessary. Individuals diagnosed with RSV usually recover within 1–2 weeks.

Pneumonia (new-**MOH**-nee-ah) is an infection of the lung. It may be caused by bacteria or a virus. In this condition, the alveoli become filled with a thick fluid called exudate, which contains both pus and red blood cells. The symptoms of pneumonia are cough with phlegm, chest pain, fever, chills, and dyspnea. Treatment may require the administration of oxygen and antibiotics. The pneumonia vaccine is recommended for older adults.

Tuberculosis (too-ber-kew-**LOH**-sis) or TB is an infectious, contagious disease of the lungs caused by the tubercle bacillus, *Mycobacterium tuberculosis*. The organs usually most affected in TB are the lungs; however, the organism may also affect the kidney, bones, and lymph nodes. In pulmonary TB, lesions called tubercles form within the lung tissue. Symptoms of TB are cough, low-grade fever in the afternoon, weight loss, and night sweats. The diagnostic test for TB is the Mantoux test, a skin test that is read within 48 to 72 hours by a health care provider. A positive skin test must be followed by a chest X-ray and sputum sample. There is an increased risk for TB in patients with HIV/AIDS due to a compromised immune system.

The incidence of TB had been declining because of early detection, treatment with drugs, and patient education. There is a new strain of the TB bacteria that is resistant to treatment. People with TB must stay on the drug isoniazid for a long time. Many people stop taking the drug when they start to feel better, which then can lead to drug-resistant organisms.

Diphtheria (dif-**THEE**-ree-ah) is a very infectious disease caused by the bacillus *Corynebacterium diphtheriae*. As part of the normal immunization process, children receive a vaccine, DTaP, which is effective against diphtheria. These vaccines protect against diphtheria, tetanus, and pertussis.

Pertussis (whooping cough) (per-**TUS**-sis) is a highly contagious infection caused by the bacterium *Bordetella pertussis*. Pertussis is characterized by severe coughing attacks that end in a "whooping" sound and dyspnea. Treatment is with antibiotics. Officials at the CDC state that pertussis is at the highest level in 50 years. This is due to the lack of vaccination in children and adolescents. It is recommended that all children receive the DTaP vaccine, and that preteens, teens, and adults receive the Tdap vaccine.

Anthrax is a disease-causing organism that can create a potential health hazard. The bacterium *Bacillus anthracis* and its spores cause anthrax. The inactive microscopic spores normally reside in the soil. In a favorable environment the

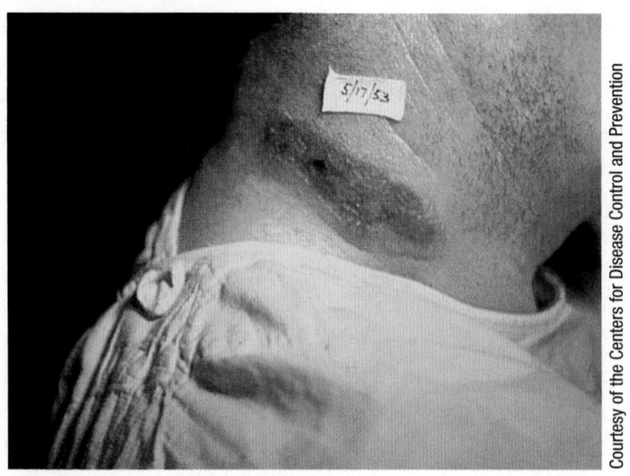

Figure 18-13 *Cutaneous anthrax*

spores can change into the anthrax bacteria, which produce a toxin that can be fatal to humans and animals.

Anthrax spores are invisible, odorless, and tasteless. The amount needed to make a person ill is smaller than a speck of dust. The disease has three forms:

- *Cutaneous anthrax*, in which spores enter a cut and cause a local infection (**Figure 18-13**).
- *Intestinal anthrax*, in which the bacteria is ingested from contaminated meat. Symptoms include diarrhea and vomiting of blood.
- *Inhalation anthrax* is when the anthrax spores can be manipulated through the air.

Noninfectious Causes

Respiratory ailments unrelated to infectious causes sometimes develop in the respiratory system.

CONGENITAL RESPIRATORY DISORDERS

Congenital lung disorder, also known as cystic lung disorder or congenital lung malformations, are most often discovered on ultrasound. The following are examples of congenital lung disorder:

- **Cystic fibrosis** affects multiple body systems. The cells in the lungs absorb too much sodium and water, causing the secretions to become thick, build up, and lead to frequent infections. For more information on cystic fibrosis, see Chapter 23.
- **Bronchogenic cysts** can be found in the trachea or the lower lobes of the lung; the cysts compromise the airway. These can be removed surgically.
- **Congenital cystic adenomatoid malformation** (CCAM) is when a baby's lung tissue grows more than normal. CCAM can cause cysts that fill with fluid or solid masses in the lung. The cysts prevent the alveoli from developing normally. Treatment is surgical aspiration of the fluid.
- **Primary cilia dyskinesia** is characterized by structural defects of the cilia that result in ineffective clearing of mucous particles, including bacteria. A common complication is bronchitis. There is no cure, just attempts to slow the progress of the disease.
- **Congenital malformation of the diaphragm** is when the diaphragm fails to close properly during prenatal development. This opening allows contents of the abdomen to move into the chest, impacting the growth and development of the lungs. Treatment is surgical repair after delivery.
- **Respiratory distress syndrome** occurs in premature infants in whom there is a deficiency of surfactant, or the substance that keeps the alveoli open. Lungs do not start producing surfactant until the 24th–28th week of pregnancy. By the 35th week, most babies have developed adequate levels of surfactant. Symptoms include tachycardia, nasal flares, tachypnea cyanosis, and grunting. Treatment is with surfactant medication, oxygen, and—if necessary—a ventilator.

GENERAL RESPIRATORY DISORDERS

Rhinitis (rye-**NIGH**-tis) is inflammation of the nasal mucous membrane, causing swelling and increased secretions. Various forms include acute rhinitis and allergic rhinitis caused by any allergen, more commonly known as hay fever.

Asthma (**AZ**-mah) is a disease in which the airway becomes constricted or encounters resistance due to the bronchioles' inflammatory response to a stimulus (**Figure 18-14**). The stimulus may be an allergen or psychological stress. About 5% of the American population has asthma, which is an increase over the past 10 years. The symptoms include difficulty exhaling; dyspnea; **wheezing**, or the sound produced by a rush of air through a narrowed passageway; tightness in the chest; and coughing. Treatment is with anti-inflammatory drugs; bronchodilators, which are rescue inhalers that treat the symptoms; and controller inhalers that prevent the symptoms.

Media Link

View the **Asthma** animation on the Online Resources.

Atelectasis (at-ee-**LEK**-tah-sis) is a condition in which a part of the lung fails to expand normally due to a blockage of the air passages. It is the most common respiratory complication after surgery due to the effects of anesthesia. Treatment consists of breathing exercises and medication.

Pneumothorax (new-moh-**THOH**-racks) is a total collapse of the lung. This condition occurs if there is a

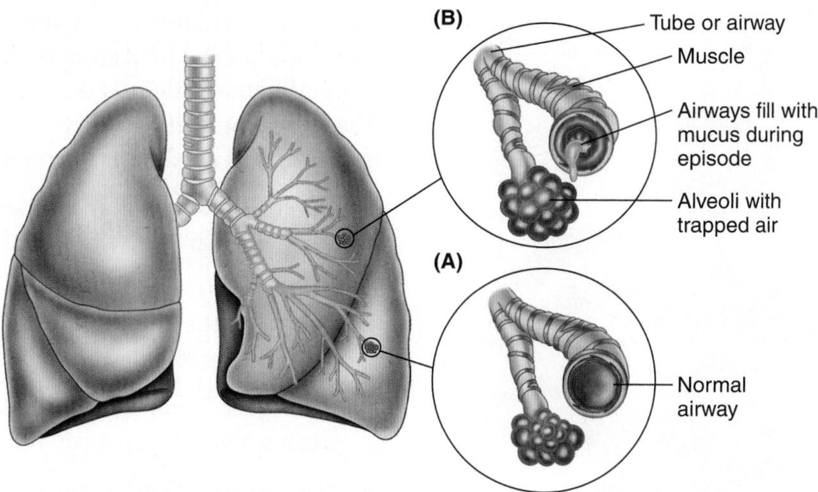

Figure 18-14 *During an asthma episode, the airways become blocked. (A) Before the episode, the muscles are relaxed, and the airways are open. (B) During the episode, the muscles tighten, and the airways fill with mucus.*

buildup of air within the pleural cavity on one side of the chest. The cause of a pneumothorax may be blunt trauma to the chest, which may occur in a car accident or physical assault. The excess air increases pressure on the lung, causing the organ to collapse. Breathing is not possible with a collapsed lung, but the unaffected lung can still continue the breathing process; however, the exchange of lung gases and the regular breathing pattern will be affected. Treatment is having the air removed either with a syringe or a chest tube connected to a one-way valve system.

A **pulmonary embolism** occurs when a blood clot (embolism) travels to the lung. This condition may occur after surgery or if a person has been on bed rest. Symptoms include a sudden, severe pain in the chest and dyspnea. Diagnosis is confirmed by a lung scan. Treatment includes anticoagulant therapy; thrombolytic therapy, which are drugs administered to dissolve clots; and compression stockings. To prevent an embolism, early ambulation, or walking, after surgery is important.

Sudden infant death syndrome (SIDS), also known as "crib death," usually occurs in infants between two weeks and one year of age. The infant stops breathing during sleep. The exact cause of SIDS is unknown. Researchers believe SIDS is caused by several different factors, including problems with sleep arousal, which is the brain's ability to wake up, or a disturbance of the respiratory control center in the brain. The most important way to reduce the risk of SIDS is for all infants under one year of age to be placed on their backs to sleep. It is also very important to keep the baby from being overheated.

Environmental Causes

Asbestosis (ass-beh-**STOH**-sis) is a respiratory disease caused by inhaling asbestos fibers, which can result in the formation of scar tissue, or fibrosis, inside the lung. Scarred lung tissue does not expand and contract normally. Symptoms appear usually 10–20 years after initial exposure and include shortness of breath on exertion, cough, tightness in the chest, and chest pain. No cure is available. Supportive treatment of symptoms includes treatment to remove secretions from the lungs, aerosol medication to thin secretions, and oxygen therapy. Professionals today are less likely to get asbestosis-related diseases because of government regulations.

Silicosis (sill-ih-**KOH**-sis) is caused by breathing dust containing silicon dioxide over a long period of time. The lungs become fibrosed, which results in a reduced capacity for expansion. Silicosis is also called miner's asthma or miner's disease. There is no cure, but the use of bronchodilators helps relax the bronchi and decrease inflammation.

Chronic obstructive pulmonary disease (COPD) refers to a group of lung diseases that limit airflow as a person exhales and make it increasingly difficult to breathe out. COPD is the third leading cause of death in the United States. Approximately 16.4 million U.S. adults have been diagnosed with COPD (American Lung Association, 2020).

Smoking is the primary risk factor for COPD. Smoking causes irritation of the larynx and trachea, and swelling and narrowing of the lung airways, reducing lung function and causing permanent damage to the lungs and alveoli. Other lung irritants, such as pollution, dust, or chemicals, may cause or contribute to COPD.

The following are the two main forms of COPD:

- Chronic bronchitis involves a long-term cough with mucus.
- **Emphysema** (em-fih-**SEE**-mah) occurs when the alveoli of the lung become over dilated, lose their elasticity, and cannot rebound (**Figure 18-15**). The alveoli may eventually rupture. In this process, air becomes trapped in the alveoli, it is difficult to exhale,

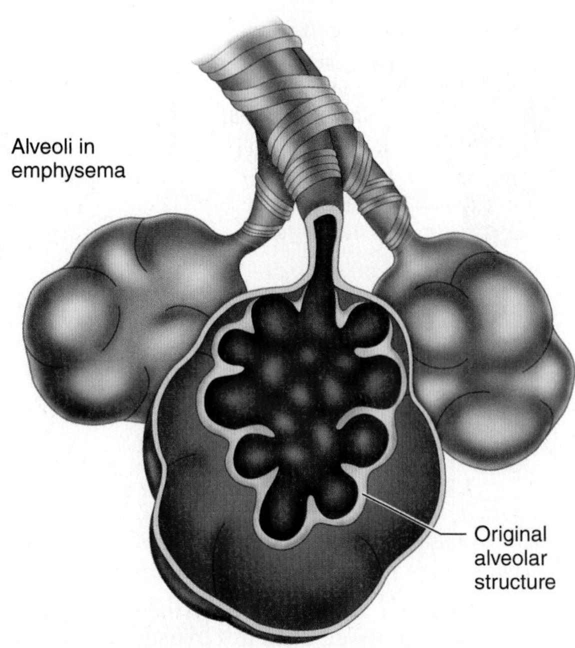

Figure 18-15 *Emphysema: The alveoli become overdistended as the disease progresses.*

forced exhalation is required, and there is a reduced exchange of carbon dioxide and oxygen. The patient with emphysema has a chronic cough that produces copious amounts of sputum, wheezing, and cyanosis of the fingers. The patient experiences dyspnea that becomes more severe as the disease progresses.

The goal of treatment in COPD is to alleviate the symptoms as much as possible. Medications used to treat COPD include bronchodilators, inhaled steroids, and anti-inflammatory agents, as well as the use of supplemental oxygen. Persons with COPD need to reduce their exposure to respiratory irritants, stop smoking, prevent infections, and restructure their activity to minimize their need for oxygen. The practice of pursed-lip breathing may help empty the lungs before each breath. Severe cases may require surgery such as a lung transplant or removal of parts of the diseased lung to help other lung areas work better.

When a patient is receiving oxygen therapy, take the following safety precautions (remember that oxygen supports and accelerates burning):

- Never use oxygen near an open flame, including candles and cigarettes.
- Do not use electrical equipment, such as hair dryers or razors; the equipment might create a spark. Use battery-operated devices.
- Do not use flammable products, such as oil-based face creams, aerosol sprays, rubbing alcohol, and lubricants.
- Place oxygen devices, such as a compressor, in well-ventilated areas and away from a heat source.

Cancer of the lung is the most common cause of cancer death in the United States.

Ninety percent of all lung cancers are caused by smoking. One type is known as small cell, which spreads rapidly to other organs. This type is found mainly in people who are smokers (**Figure 18-16**). The other types of lung cancer, also known as non-small cell cancer, are squamous cell or adenocarcinoma, which do not spread as rapidly. Symptoms include cough, shortness of breath, coughing up a small amount of blood, and weight loss. Diagnosis is made by X-ray and white-light **bronchoscopy** (brong-**KOS**-koh-pee). A small, flexible tube is passed through the mouth or nose, and a piece of tissue is obtained for study. It is important for the health care provider to know that the throat may be anesthetized for this procedure and that the cough reflex must have returned before the person can have fluid or food. Treatment of lung cancer can include surgery, chemotherapy, immunotherapy, laser therapy, cryosurgery, and/or radiation.

Cancer of the larynx is curable if early detection is made of the disorder. It is found most frequently in men over age 50.

(A)

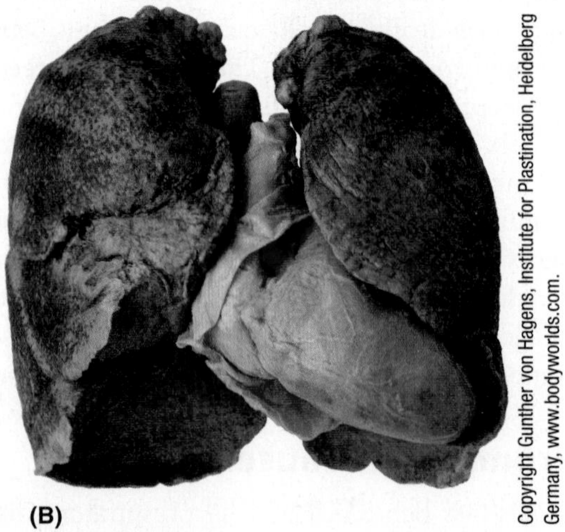

(B)

Figure 18-16 *Changes in the lung: (A) healthy lungs of a nonsmoker; (B) damaged lungs of a smoker*

One BODY How the Respiratory System Interacts with Other Body Systems

INTEGUMENTARY SYSTEM

- The skin forms a barrier against invading microorganisms.
- Sensory receptors send stimuli to the brain to react to severe temperature changes, such as cold, which may alter the respiratory rhythm.

SKELETAL SYSTEM

- The ribs and sternum protect the organs of respiration in the thoracic cavity.

MUSCULAR SYSTEM

- The diaphragm and intercostal muscles are necessary for the mechanics of breathing.

NERVOUS SYSTEM

- The medulla oblongata is responsible for the rate and depth of respiration.
- Phrenic nerves stimulate the diaphragm and intercostal muscles.
- Stretch receptors in the lung initiate the Hering-Breuer reflex.

ENDOCRINE SYSTEM

- Epinephrine dilates the bronchioles to prepare for fight or flight.
- Excess thyroid hormone increases respirations.

CIRCULATORY SYSTEM

- Exchange of gases takes place in the lungs by means of blood capillaries.
- Blood vessels are necessary for transporting oxygen to the cells and taking away the waste products of metabolism (carbon dioxide).
- Chemoreceptors in the carotid arteries and aorta are sensitive to the blood oxygen level; a low level of oxygen increases the respiratory rate.

LYMPHATIC SYSTEM

- Normal defense mechanisms respond to any respiratory infections.

DIGESTIVE SYSTEM

- The pharynx is the common passageway for food and air.
- Oxygen supplied by the respiratory system and nutrients from the digestive system provide the cells with energy.

URINARY SYSTEM

- The kidney assists the respiratory system in maintaining the pH level in the blood.

REPRODUCTIVE SYSTEM

- Respiratory rate increases with sexual activity.
- The fetus obtains oxygen through the placenta of the mother.

Medical Terminology

alveol	tiny cavity	**spira**	breathe
-i	pertaining to	**-tion**	process
alveol/i	pertaining to tiny cavity	**ex/pira/tion**	process of breathing out
a-	without	**in-**	in
-pnea	breathing	**in/spira/tion**	process of breathing in
a/pnea	without breathing	**laryng**	voice box
bronch	major branch of windpipe	**laryng/itis**	inflammation of voice box
-itis	inflammation of	**pharyng**	throat
bronch/itis	inflammation of major branch of windpipe	**pharyng/itis**	inflammation of the throat
scope	instrument used to examine	**pneumon**	lungs, air
-y	pertaining to	**-ia**	abnormal condition of
bronch/o/scop/y	pertaining to examination of windpipe	**pneumon/ia**	abnormal condition of the lungs
dys-	difficult	**-thorax**	chest
dys/pnea	difficult breathing	**pneumo/thorax**	air in the chest cavity
emphysem	blowing up with air	**spiro**	breath
-a	pertaining to	**-meter**	instrument used to measure
emphysem/a	pertaining to blowing up with air	**spiro/meter**	instrument used to measure breaths
ex-	out	**tachy-**	fast
		tachy/pnea	fast breathing

Study Tools

Workbook	Activities for Chapter 18
Online Resources	• PowerPoint presentations • Animations

REVIEW QUESTIONS

Select the letter of the choice that best completes the statement.

1. The exchange of oxygen and carbon dioxide between the body and the air breathed in is called
   a. cellular respiration.
   b. external respiration.
   c. internal respiration.
   d. breathing.

2. Oxygen moves from an area of higher concentration through a process called
   a. active transport.
   b. osmosis.
   c. diffusion.
   d. filtration.

3. When air travels through the nose it is filtered and
   a. warmed and moistened.
   b. warmed and exchanged for carbon dioxide.
   c. cooled and exchanged for carbon dioxide.
   d. cooled and moistened.

4. The structure responsible for giving tone to the voice is the
   a. nares.
   b. nasal septum.
   c. glottis.
   d. concha.

5. This structure contains 15 to 20 cartilage rings and serves as a passageway for air; it is known as the
   a. nasopharynx.
   b. trachea.
   c. pharynx.
   d. larynx.

6. The structure at the end of the bronchial tree where the exchange between oxygen and carbon dioxide occurs is the
   a. alveolar duct.
   b. alveolus.
   c. bronchiole.
   d. bronchial tree.

7. Collapse of the lung is called
   a. pleurisy.
   b. pneumonia.
   c. pneumothorax.
   d. thoracentesis.

8. The rate of breathing is affected by the
   a. cerebrum.
   b. medulla.
   c. cerebellum.
   d. frontal lobe.

9. Difficult or labored breathing is known as
   a. eupnea.
   b. dyspnea.
   c. orthopnea.
   d. hyperpnea.

10. Pharyngitis is inflammation of the
    a. throat.
    b. voice box.
    c. windpipe.
    d. upper nose.

11. An inflammation of the lining of the lung is called
    a. pneumonia.
    b. pleurisy.
    c. sinusitis.
    d. tuberculosis.

12. The vaccine used to protect children against whooping cough is
    a. MMR.
    b. Mantoux.
    c. DTaP.
    d. Salk.

13. Chronic obstructive pulmonary disease means the person has
    a. asthma.
    b. pneumonia.
    c. atelectasis.
    d. emphysema.

14. A respiratory disorder with wheezing and dyspnea is known as
    a. acute bronchitis.
    b. atelectasis.
    c. asthma.
    d. SIDS.

15. A respiratory disease that has shown a marked increase in the past 50 years is
    a. asthma.
    b. asbestosis.
    c. COPD.
    d. pertussis.

COMPARE AND CONTRAST

Compare and contrast the following terms:

1. Tachypnea and hyperventilation
2. Vital lung capacity and total lung capacity
3. Common cold and influenza
4. Acute bronchitis and chronic bronchitis
5. Pneumonia and asthma

MATCHING

Match each term in Column I with its function or description in Column II.

	COLUMN I	*COLUMN II*
________	**1.** respiratory control	a. inspiratory reserve volume
________	**2.** oxidation	b. amount of air that moves in and out with each breath
________	**3.** vagus nerve	c. cellular respiration
________	**4.** breathing	d. located in the medulla
________	**5.** tachypnea	e. ventilation
________	**6.** diaphragm	f. difficult or labored breathing
________	**7.** intercostal muscles	g. becomes flattened and moves downward during inhalation
________	**8.** tidal volume	h. air that cannot be forcibly removed
________	**9.** residual volume	i. less than atmospheric pressure
________	**10.** pressure in the pleural space	j. abnormal rapid shallow breathing
		k. muscles between the ribs that contract during inhalation
		l. inhibits inspiration and stimulates expiration

APPLYING THEORY TO PRACTICE

1. a. You are a molecule of oxygen, floating in the air. Suddenly you feel a whoosh, and you are in a dark tube with little hairs tickling you: the nose! Trace your journey from there to the alveoli of the lung; you will recognize it, as it looks like a bunch of grapes. Name the structures along the way.
   b. After you arrive at the alveoli, squeeze into the capillary around the alveolus and get to the pulmonary vein. You can now begin a new journey to the left knee; trace that journey. Name the structures and vessels along the way.
2. You have a cold, sinusitis, and your voice sounds different. What is happening? How do you explain it?
3. Take a breath; now breathe steadily deeper. Let the breath out; force more and more air out until you gasp. Name the processes you have experienced.
4. Many people about to have their first baby worry about sudden infant death syndrome, or SIDS. How do you explain SIDS to parents? What is the best advice you can give?
5. Jog in place. Note how the body's activity affects the rate of breathing. How does exercise change breathing? Why?
6. Your brother has been diagnosed with diabetes. You are concerned because he begins hyperventilating. Describe how the excess glucose (energy) in his blood may be the culprit.

7. The digestive and respiratory system work together to maintain homeostasis. Compare and contrast the systems. What would happen if either of these systems was out of homeostasis?

8. Construct a model to illustrate the chemical processes of cellular respiration. Identify where food and oxygen molecules are broken down and new bonds are formed. Show the net transfer of energy.

9. A friend was in a car accident and suffered a collapsed lung. What is the medical term for this, what causes this condition, and how is this trauma to the lung treated?

CASE STUDY

Alan has been smoking for 20 years and has shortness of breath and a cough. His physician, Dr. Shah, suspects COPD and orders diagnostic tests.

1. Explain what diagnostic technologies Dr. Shah would use to investigate Alan's illness.

After the results of the test, Dr. Shah tells Alan he has COPD in the form of emphysema.

2. Explain COPD and emphysema.
3. Name the organs of external respiration.
4. What structural changes occur with emphysema?
5. What other body systems are affected by emphysema?
6. Describe the therapeutic technologies used to treat emphysema. Is there any surgical procedure for emphysema?
7. How does this disease affect a person's lifestyle?

18-1 Breathing Process

- ***Objective:*** To observe the mechanism involved in breathing
- ***Materials needed:*** bell jar, Y-tube, two balloons of the same size, rubber stopper to fit top of bell jar, another small rubber stopper, scissors, string, rubber bands, piece of thin rubber sheeting, textbook, paper, pencil

Step 1: Carefully insert a glass Y-tube into the opening of a rubber stopper.

Step 2: Fasten the two balloons to the two arms of the Y-tube. Use string or rubber bands to keep balloons in place.

Step 3: Place the rubber stopper in the bell jar so that the balloons are inside the jar.

Step 4: Using the scissors, carefully cut a piece of rubber sheeting large enough to fit over the bottom of the bell jar.

Step 5: From the center of the rubber sheeting, pinch a fingerful of the sheeting, being careful not to pierce it. Place a small rubber stopper into this area.

Step 6: Take the string and tie it around the rubber stopper so you may now use this place as a handle to grasp the rubber sheeting.

Step 7: Take another piece of string and secure the rubber sheeting to the bottom of the bell jar.

Step 8: Grasp the "handle" and pull down on the rubber sheeting. Observe what is happening to the balloons. Record your observations and describe how this relates to the breathing process.

Step 9: Grasp the "handle" and push up on the rubber sheeting. Observe what is happening to the balloons. Record your observations and describe how this relates to the breathing process.

Step 10: Compare the bell jar apparatus with Figure 18-2 in the textbook. What structures do the bell jar, balloons, top of the Y-tube, branches of the Y-tube, and rubber sheeting represent?

LAB ACTIVITY 18-2 Breath Sounds

- ***Objective:*** To observe the sounds made during breathing
- ***Materials needed:*** stethoscope, alcohol wipes, paper, pencil
- ***Note:*** This activity is done with a lab partner.

Step 1: Clean the earpieces of the stethoscope with alcohol wipes.

Step 2: Place the diaphragm of the stethoscope on the throat of your lab partner, just above the sternum. Listen to the sounds made during inspiration and expiration. Are the sounds similar? Describe and record the sounds you hear.

Step 3: Move the stethoscope down until you no longer hear breath sounds.

Step 4: Place the stethoscope at two different intercostal spaces. Describe and record the sounds you hear.

Step 5: Place the stethoscope on the upper area of the back and listen to the sounds on inspiration and expiration. Describe and record the sounds you hear.

Step 6: Remove the stethoscope and clean the earpieces.

Step 7: Switch places with your lab partner and repeat steps 2 through 5.

Step 8: Compare your results.

LAB ACTIVITY 18-3 Lung Tissue

- ***Objective:*** To observe the normal and abnormal structure of the lung tissue
- ***Materials needed:*** slide of normal lung tissue, slides of pathological lung tissue showing (1) emphysema and (2) lung cancer, microscope, textbook, paper, pencil

Step 1: Examine the slide of normal lung tissue. Identify, if possible, a bronchiole and alveolus. Draw a sketch of what you see. What is the function of the alveolus?

Step 2: Examine the slide of tissue showing emphysema. Compare the slide to the normal lung tissue. Describe and record your observations.

Step 3: Examine the slide of cancer tissue. Compare the slide to the normal lung tissue. Describe and record your observations.

CHAPTER 19

Digestive System

Objectives

- Describe the general function of the digestive system.
- List the structures and the functions of the digestive system.
- Describe the action of the enzymes on carbohydrates, fats, and proteins.
- Trace food from the beginning of the digestive process to the end.
- Describe common disorders of the digestive system.
- Define the key words that relate to this chapter.

Key Words

abdominal pain
absorption
acini cells
alimentary canal
anus
appendicitis
atresia
Barrett's syndrome
bicuspids
bile
bolus
buccal cavity
cancer of stomach
canines
canker sores
cardiac sphincter
cecum
celiac disease
cholecystitis
chyme
cirrhosis
cleft lip
cleft palate
colon
colon cancer
colorectal cancer
colostomy
common bile duct
constipation
crown
cystic duct
deciduous
defecation
dentin
diarrhea
digestion
diverticulitis
diverticulosis
duodenum
enamel
enteritis
esophageal sphincter
esophagus
feces
flatulence
gallbladder
gallstones
gastritis
gastroenteritis
gastroesophageal reflux disease (GERD)
gingivae
gingivitis
greater omentum

(continues)

Key Words *continued*

haustra
heartburn
hemoccult
hemorrhoids
hepatic duct
hepatitis
hiatal hernia
Hirschsprung's disease
histamine
ileocecal valve
ileum
incisors
inflammatory bowel disease (IBD)
irritable bowel syndrome (IBS)
jejunum
lingual frenulum
lipase
liver
masticated
Meckel's diverticulum
mesentery
molars
nausea
neck
oral thrush
pancreatitis
parotid glands
peptic ulcers
periodontal membrane
periodontitis
peristalsis
peritonitis
protease
ptyalin
pulp cavity
pyloric sphincter
pyloric stenosis
pylorospasm
rectum
root
rugae
salivary amylase
salivary glands infection
segmented movement
sigmoid colon
stomach
stomatitis
sublingual glands
submandibular gland
taste buds
ulcer
uvula
vermiform appendix
villi
vomiting
wisdom teeth

All food that is eaten must be changed into a soluble, absorbable form within the body before it can be used by the cells. This means that certain physical and chemical changes must take place to change the insoluble complex food molecules into simpler soluble ones. These can then be transported by the blood to the cells and absorbed through the cell membranes. The process of changing complex solid foods into simpler soluble forms that can be absorbed by the body cells is called **digestion**. It is accomplished by the action of various digestive juices containing enzymes. Enzymes (**EN**-zimez) promote chemical reactions in living things, although they themselves are unaffected by the chemical reactions.

Digestion is performed by the digestive system, which includes the alimentary canal and accessory digestive organs. The **alimentary canal** is also known as the digestive tract, or gastrointestinal (GI) tract. The alimentary canal consists of the oral cavity (mouth), pharynx (throat), esophagus (gullet), stomach, small intestine, large intestine (colon), and anus (**Figure 19-1**). It is a continuous tube some 30 feet (9 meters) in length from the mouth to the anus.

The accessory organs of digestion are the tongue, teeth, salivary glands, pancreas, liver, and gallbladder.

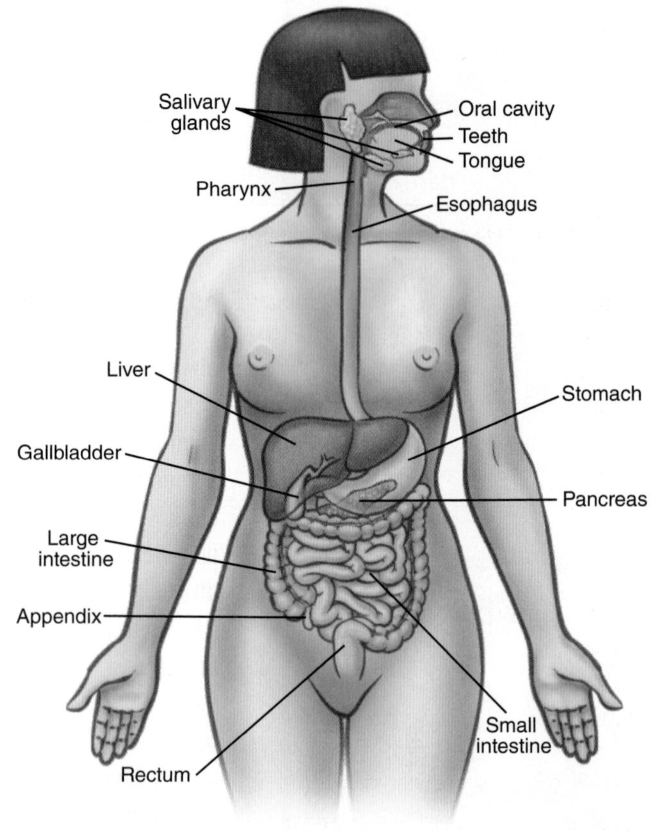

Figure 19-1 *Structures of the digestive system*

FUNCTIONS OF THE DIGESTIVE SYSTEM

The purposes of the digestive system are to change food into forms that the body can use for energy and to eliminate waste products. The specific functions are as follows:

1. Take food into the mouth, where it is mechanically and chemically broken down.
2. Use **peristalsis** (pehr-ih-**STALL**-sis), which are rhythmic muscular contractions, and **segmented movement**, which is when single segments of the intestine alternate between contraction and relaxation, to push food through the digestive tract.
3. Use both mechanical (teeth) and chemical mechanisms (digestive enzymes) to break down food into the end products of fat, carbohydrates, and proteins.
4. Absorb nutrients into the blood capillaries and lacteals of the small intestines for use in the body.
5. Eliminate waste products through excretion.

LAYERS OF THE DIGESTIVE SYSTEM

The walls of the alimentary canal are composed of four layers: (1) The innermost lining, called the mucosa, is made of epithelial cells. (2) The submucosa, where absorption takes place, consists of connective tissue with fibers, blood vessels, and nerve endings. (3) The third layer, the muscularis of the mouth, pharynx, and first part of the esophagus, consists of skeletal muscle that allows the voluntary act of swallowing. The rest of the tract consists of smooth muscle, which helps break down food and propels it through the tract by peristalsis. (4) The fourth layer is a serous layer called the serosa, which is also known as the visceral peritoneum.

The mucosa secretes slimy mucus. In some areas, it also produces digestive juices. This slimy mucus lubricates the alimentary canal, aiding in the passage of food. It also insulates the digestive tract from the effects of powerful enzymes while protecting the delicate epithelial cells from abrasive substances within the food.

LINING OF THE DIGESTIVE CAVITY

The abdominal cavity is lined with a serous membrane called the peritoneum. This is a two-layered membrane with the outer, or parietal, side lining the abdominal cavity and the inner, or visceral, side covering the outside of each organ in the abdominal cavity. An inflammation or infection of the lining of this cavity caused by disease-producing organisms is called **peritonitis** (pehr-ih-toh-**NIGH**-tis). Peritonitis is caused by a leaking hole in the intestines or a burst appendix.

There are two specialized layers of peritoneum. The peritoneum that attaches to the posterior wall of the abdominal cavity is called the **mesentery** (**MEZ**-in-tehr-ee). The small intestines are attached to this layer. In the anterior portion of the abdominal cavity, a double fold of peritoneum extends down from the greater curvature of the stomach. This hangs over the abdominal organs like a protective apron. This layer contains large amounts of fat and is called the **greater omentum** (oh-**MEN**-tum). The peritoneal structure between the liver and stomach is called the lesser omentum.

ORGANS OF DIGESTION

Many organs contribute to digestion. Each serves a specific function in the process. Digestion starts to take place before food enters the mouth; the smell of food activates the salivary glands.

Mouth

Food enters the digestive tract through the mouth, also called the oral cavity or **buccal cavity** (**BUCK**-ull). The lips protect the opening to the mouth. The inside of the mouth is covered with a mucous membrane. Its roof consists of a hard and a soft palate. The hard palate is hard because it is formed from the maxillary and palatine bones, which are covered by mucous membrane. Behind the hard palate is the soft palate, made from a movable mucous membrane fold. The soft palate is an arch-shaped structure, separating the mouth from the nasopharynx. Hanging from the middle of the soft palate is a cone-shaped flap of tissue called the **uvula** (**YOO**-vyuh-luh). The uvula prevents food from entering the nasal cavity when swallowing (**Figure 19-2**).

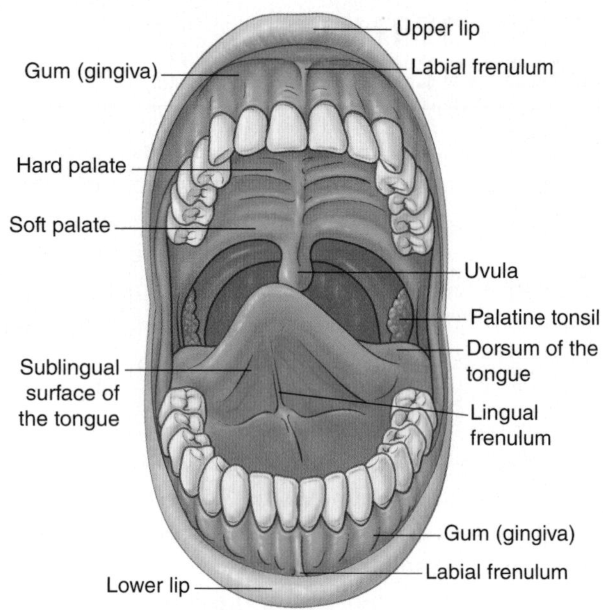

Figure 19-2 *Structures of the tongue and oral cavity*

Tongue: Accessory Organ of Digestion

The tongue and its muscles are attached to the floor of the mouth, helping in both chewing and swallowing. The tongue is made from skeletal muscles that lie in many different planes. Because of this, the tongue can be moved in various directions. It is attached to four bones: the hyoid, the mandible, and two temporal bones. On the tongue's epithelial surface are projections called papillae. There are nerve endings located in many of these papillae, forming the sense organs of taste, or **taste buds**. These taste buds respond to bitterness, saltiness, sweetness, sourness, and umami in foods. They are also sensitive to cold, heat, and pressure.

For food to be tasted, it must be in solution. The solution passes through the taste bud openings, stimulating the nerve endings in the taste cells.

The sensation of taste is coupled with the sense of smell. When a person experiences an odor, it stimulates the olfactory nerve endings in the upper part of the nasal cavity. They may confuse the odor of a food with its flavor. A bad cold with nasal congestion frequently impedes the ability to taste the flavor of foods because the increased mucous secretions cover the olfactory nerve endings.

The **lingual frenulum** (**LING**-gwall **FRIN**-yoo-lum) is a band of tissue that attaches the tongue to the floor of the mouth. The frenulum limits the motion of the tongue. A condition known as tongue-tie or ankyloglossia (an-ky-lo-**GLOS**-sia) may occur if there is congenital shortness of the frenulum, which interferes with speech.

Salivary Glands

Saliva is 99.5% water, which provides a medium for dissolving foods. The remainder is chlorides, which activate **salivary amylase** (**SAL**-ih-vehr-ee **AM**-eh-layz); mucin; and the enzyme lysozyme, which destroys bacteria. Salivary amylase begins the breakdown of complex carbohydrates into simple sugars.

Saliva is secreted into the oral cavity by three pairs of salivary glands: the parotid, the submandibular, and the sublingual (**Figure 19-3**). The **parotid glands** (pah-**RAH**-tid) are found on both sides of the face, in front of and below the ears. They are the largest salivary glands, the ones that become inflamed during an attack of mumps. If a person has mumps, chewing is painful because the motion squeezes these tender, inflamed glands. To treat mumps, it is important to drink plenty of fluids and apply ice packs to swollen glands.

A parotid duct carries its secretions into the mouth; these secretions consist almost entirely of **ptyalin** (**TYE**-ah-lin), or salivary amylase. It opens on the inner surface of the cheeks, opposite the second molar of the upper jaw.

Below the parotid salivary gland and near the angle of the lower jaw is a **submandibular gland** (sub-man-**DIB**-yoo-lar). This gland is about the size of a walnut and its secretions contain both mucin, which forms mucus, and salivary amylase. The secretions enter the buccal cavity via the submandibular duct at the anterior base of the tongue.

The final pair of salivary glands are the **sublingual glands** (sub-**LING**-gwuhl), the smallest of the three. They are found under the sides of the tongue. Their secretion consists mainly of mucus and contains no salivary amylase.

In addition to the major salivary glands, there are hundreds of minor salivary glands found in the lining of the lips, the tongue, and the roof of the mouth.

Teeth: Accessory Organ of Digestion

The **gingivae** (**JIN**-jih-vee), or gums, support and protect the teeth. They are made up of fleshy tissue covered with mucous membrane. This membrane surrounds the narrow portions of the teeth, also called the cervix or neck, and covers the structures in the upper and lower jaws.

Food must be thoroughly chewed, or **masticated** (**MASS**-tih-kay-ted), by the teeth. Teeth help break food down into very small morsels, increasing the food's surface area. This activity enables the digestive enzymes to digest the food more efficiently and quickly. During normal growth and development, the human mouth develops two sets of teeth: (1) the deciduous or milk teeth, which are later replaced by (2) the permanent teeth.

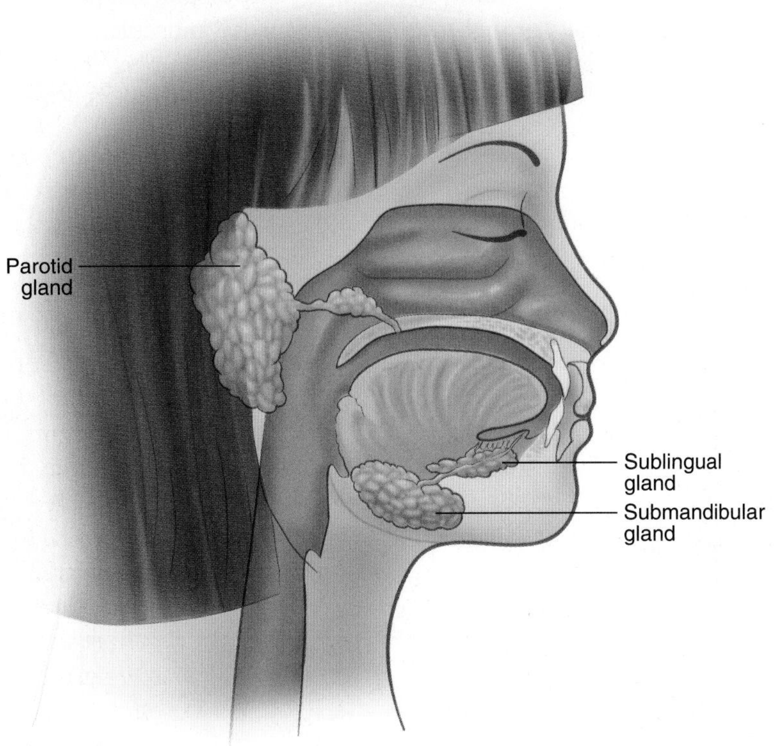

Figure 19-3 *Salivary glands*

Deciduous (dih-**SIH**-joo-us) teeth start to erupt at about 6 months and continue until around two years of age. In total, 20 deciduous teeth are cut during the first two years. They include four incisors, two canines, and four molars in each jaw. This relationship is expressed in the dentition formula, as shown in **Figure 19-4A**. The **incisors** (in-**SIGH**-zerz) have sharp edges for biting, the **canines** (**KAY**-ninez) are pointed for tearing, and the **molars** (**MOH**-lerz) have ridges designed for crushing and grinding. There are no premolars among the deciduous teeth. Deciduous teeth may last up to the age of 12.

Permanent teeth begin developing at about the age of six, pushing out their deciduous predecessors. The last permanent teeth to emerge are the third molars, or **wisdom teeth**, which may appear anywhere from 17–25 years of age. In some cases, the wisdom teeth cannot erupt because of a lack of space in the mouth. This causes the wisdom teeth to become impacted and need to be removed. In total, the adult mouth develops 32 teeth, 16 in each jaw (**Figure 19-4B**).

The adult mouth has eight premolars, or **bicuspids** (bye-**KUS**-pidz): four in the upper jaw and four in the lower. Bicuspids are broad, with two ridges on each crown, and have only two roots. Their design is ideal for grinding food. **Figure 19-4** shows the arrangement of the deciduous and permanent teeth.

STRUCTURE OF A TOOTH

Each tooth may be divided into three major parts: the crown, the neck, and the root (**Figure 19-4C**). The **crown** is the part of the tooth that is visible; the **neck** is where the tooth enters the gumline; the **root** is embedded in the alveolar processes of the jaw.

Inside the tooth is the **pulp cavity**, which contains the nerves and blood supply. The pulp cavity is surrounded by calcified tissue called **dentin**. In the crown portion, the dentin is covered by **enamel**. Enamel is the hardest substance in the body. If the enamel wears down on the surface of the tooth, bacteria may enter and caries, or cavities, will develop. Narrow extensions of the pulp cavity called root canals project into the root. At the base of each root canal is the opening for blood vessels and nerves. The dentin of the root is covered with a substance called cementum, which attaches the root to the **periodontal membrane** (**pehr**-ee-oh-**DON**-tal **MEM**-brain), anchoring the tooth in place.

Esophagus

When food is swallowed, it enters the upper portion of the esophagus. The **esophagus** (ih-**SOF**-ah-gus) is a muscular tube about 25 centimeters (10 inches) long. It begins at the lower end of the pharynx, behind the trachea.

It continues downward through the mediastinum, in front of the vertebral column, and passes through the diaphragm. From there, the esophagus enters the upper part, or cardiac portion, of the stomach.

The esophageal walls have four layers: the mucosa, submucosa, muscular, and external serous layers. The muscles in the upper third are voluntary and the lower portion is smooth muscle, or involuntary.

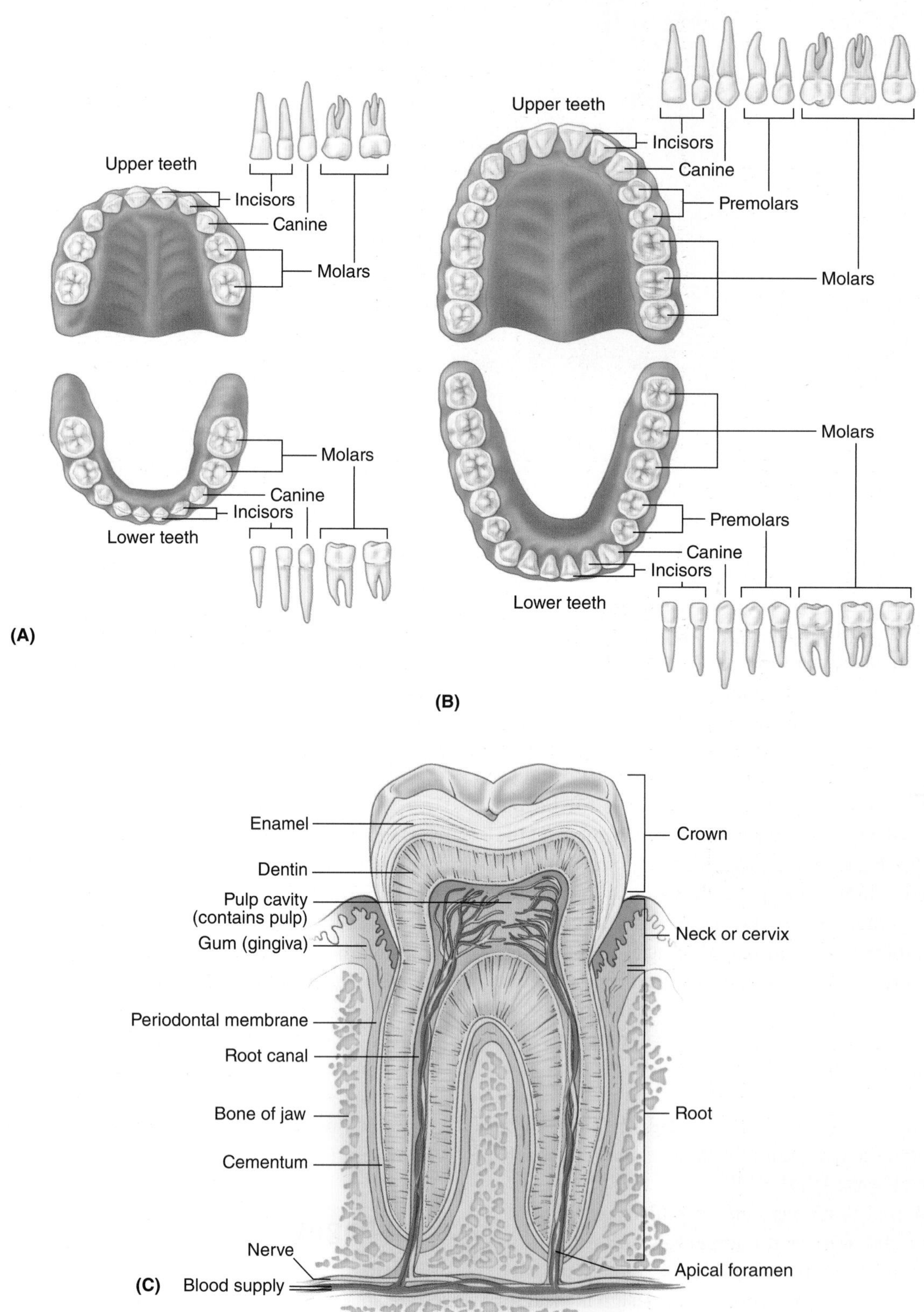

Figure 19-4 *(A) Deciduous teeth; (B) permanent teeth; (C) anatomy of a typical tooth*

The muscular ring between the esophagus and stomach is called the esophageal or cardiac sphincter. During swallowing, it relaxes to allow food to enter the stomach and closes to prevent the stomach contents from regurgitating into the esophagus. Regurgitation means "to flow backward."

Stomach

The **stomach** is found in the upper part of the abdominal cavity, just to the left of and below the diaphragm. The shape and position are determined by several factors. These include the amount of food contained within the stomach, the stage of digestion, the position of a person's body, and the pressure exerted on the stomach from the intestines below. The stomach and small intestines have more nerve cells than the spinal cord does, causing some experts to call it a "mini-brain."

The stomach is divided into three portions: the upper part or fundus, the middle section called the body or greater curvature, and the lower portion called the pylorus (**Figure 19-5A**). The opening from the esophagus into the stomach is through a circle of muscle called the **cardiac sphincter**, or **esophageal sphincter**. It is called the cardiac sphincter because of its proximity to the cardiac portion of the stomach. Toward the other end of the stomach lies the **pyloric sphincter** (pye-**LOR**-ick) valve, which regulates the entrance of food into the **duodenum** (dew-uh-**DEE**-num), the first part of the small intestine. Sometimes the pyloric sphincter valve fails to relax in infants. In such cases, food remaining in the stomach does not get completely digested and eventually is vomited. This condition is called **pylorospasm** (pye-**LOR**-uh-spaz-em).

The stomach wall consists of four layers: mucosa, submucosa, muscularis, and serosa (**Figure 19-5B**).

1. The mucosa layer is the innermost layer. It is a thick layer made up of small gastric glands embedded in connective tissue. When the stomach is not distended with food, the gastric mucosa is thrown into folds called **rugae** (**ROO**-gee) (**Figure 19-5A**).
2. The submucosa layer is made of loose areolar connective tissue.
3. The muscularis layer consists of three layers of smooth muscle: the outer longitudinal layer, a middle circular layer, and an inner oblique layer (**Figure 19-5A**). These muscles help the stomach perform peristalsis, which pushes food into the small intestine.
4. The serosa is the thick outer layer covering the stomach. It is continuous with the peritoneum. The serosa and peritoneum meet at certain points, surrounding the organs around the stomach and holding them in a kind of sling.

GASTRIC JUICES

The gastric mucosa contains millions of gastric glands that secrete the gastric juices necessary for digestion (**Figure 19-5B**).

- Enteroendocrine glands secrete gastrin, which in turn stimulates cells to produce hydrochloric acid (HCl) and pepsinogen.
- Parietal cells produce HCl acid, which converts pepsinogen into pepsin and destroys bacteria and microorganisms that enter the stomach. It is the body's natural sterilizer.
- Parietal cells also produce the intrinsic factor, an element necessary for the absorption of vitamin B_{12}; without it, a condition known as pernicious anemia develops.
- Chief-type cells produce pepsinogen, which converts to pepsin. The enzyme pepsin breaks down protein into smaller pieces called proteoses and peptone.
- Mucous cells secrete alkaline mucus, which helps neutralize the effects of HCl acid and the other digestive juices. Mucous cells in the stomach replace themselves every 3–5 days.
- Rennin, a digestive enzyme, is found in infants and children, but not adults. It prepares milk proteins for digestion by other enzymes.

When food leaves the stomach, it is known as **chyme (KIME)**, which is a pulpy mass of gastric juice and partially digested food.

Small Intestine

The small intestine has the same four layers as the stomach: the mucosa, submucosa, muscularis, and serosa. The mucosa of the small intestine is in numerous folds called plicae circularis (**PLYE**-kee).

The final preparation of food to be absorbed occurs in the small intestine. This coiled portion of the alimentary canal can be as long as 20 feet. The small intestine is divided into three sections: the duodenum, the **jejunum** (jih-**JOO**-num), and the **ileum** (**ILL**-ee-um). The small intestine is held in place by the mesentery. The small intestine lining secretes digestive juices and is covered with villi, which absorb the end products of digestion.

The first segment of the small intestine is the duodenum. This 12-inch structure curves around the head of the pancreas. A few inches into the duodenum is the hepatopancreatic ampulla, which is the site where the pancreatic duct and the common bile duct of the liver enter. The pancreatic duct empties the digestive juices of the pancreas and the common bile duct empties bile from the liver.

The next sections of the small intestine are the jejunum, which is about 8 feet long, and the ileum, which is 10–12 feet long (**Figure 19-6**).

Did You Know?

The surface area of a human small intestine is about equal to that of a tennis court.

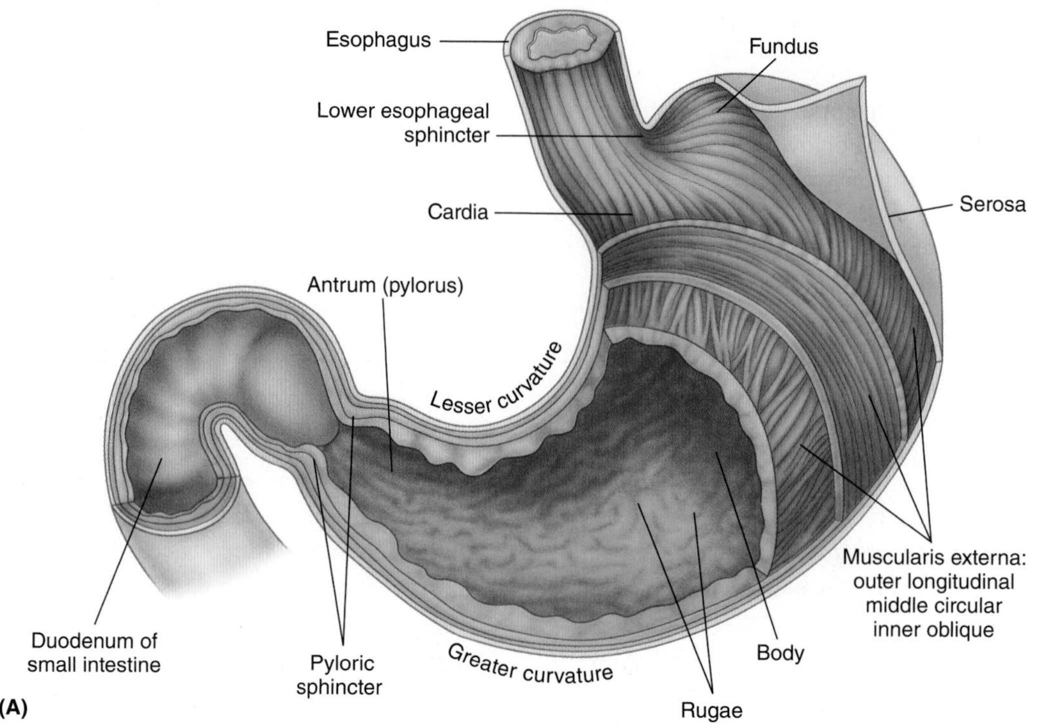

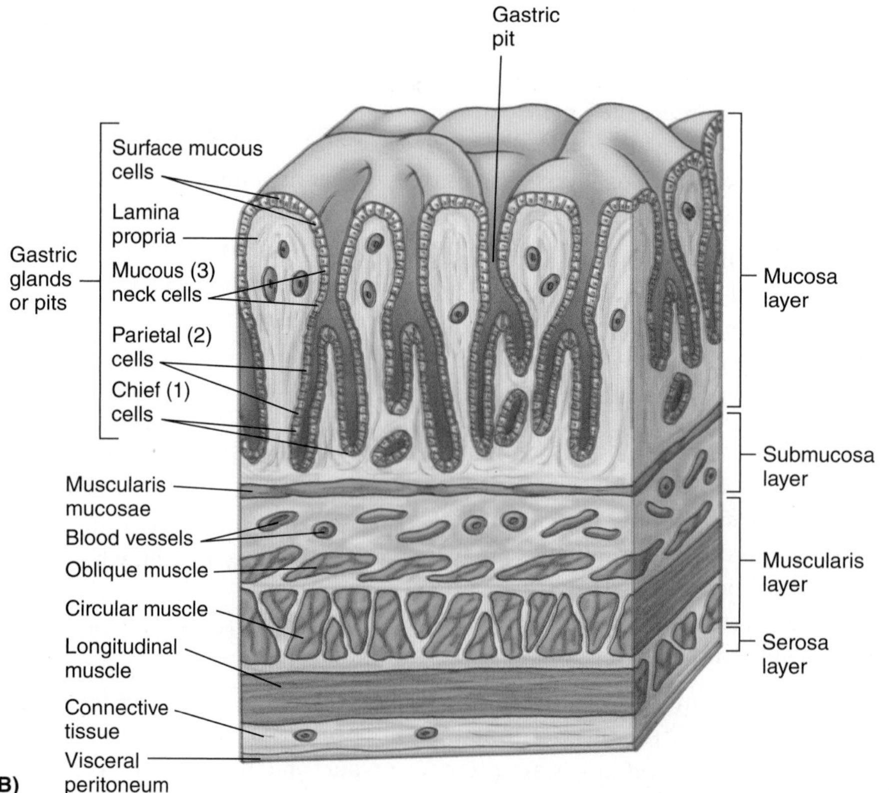

Figure 19-5 *(A) The parts of the stomach; (B) the four layers of the stomach and the three secreting glands of the gastric mucosa*

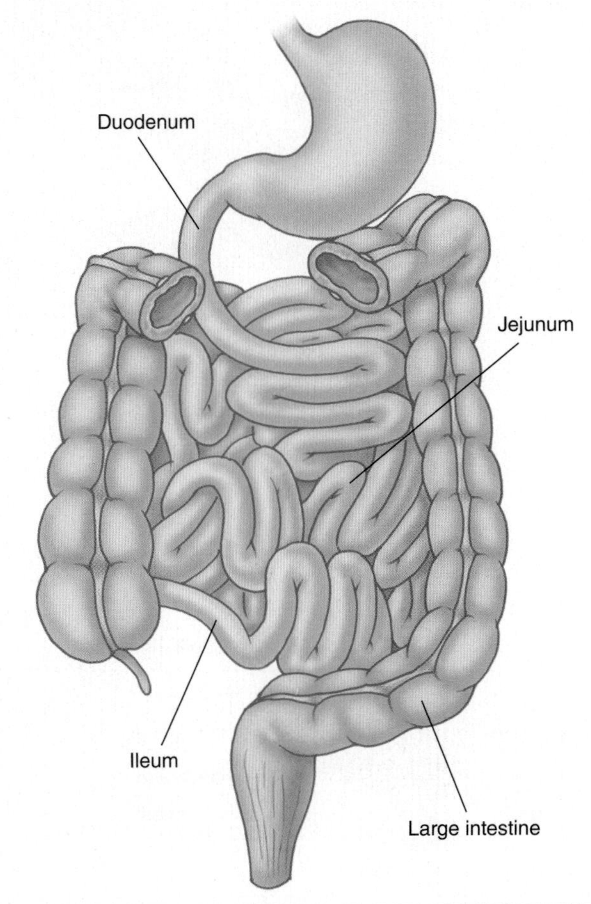

Figure 19-6 *Small intestine*

DIGESTIVE JUICES IN THE SMALL INTESTINES

- Enzymes, secretin (sih-**KREE**-tin), and cholecystokinin (**koh**-lee-sis-toh-**KYE**-nin) stimulate the digestive juices of the pancreas, liver, and gallbladder.
- Pancreatic juices include **protease** (**PRO**-tee-ace), which breaks down protein to amino acids; the amylase carbohydrase, which breaks down starches to glucose; and **lipase** (**LIP**-ace), which breaks down fats to fatty acids and glycerol. The pancreatic juices also contain sodium bicarbonate, which neutralizes the food content of the stomach, which is high in acid (**Figure 19-7**).
- Liver produces **bile**, which is necessary to break down or emulsify fat into smaller fat globules to be digested by lipase.
- Intestinal juices secreted by the cells of the small intestine—including maltase, lactase, and sucrase—change starch into glucose, peptidase changes proteoses and peptone into amino acids, and lipase changes fat into fatty acids and glycerol.
- Brunner's glands produce a rich, alkaline mucus containing bicarbonate. These secretions, in combination with bicarbonates from the pancreas, neutralize stomach acid.

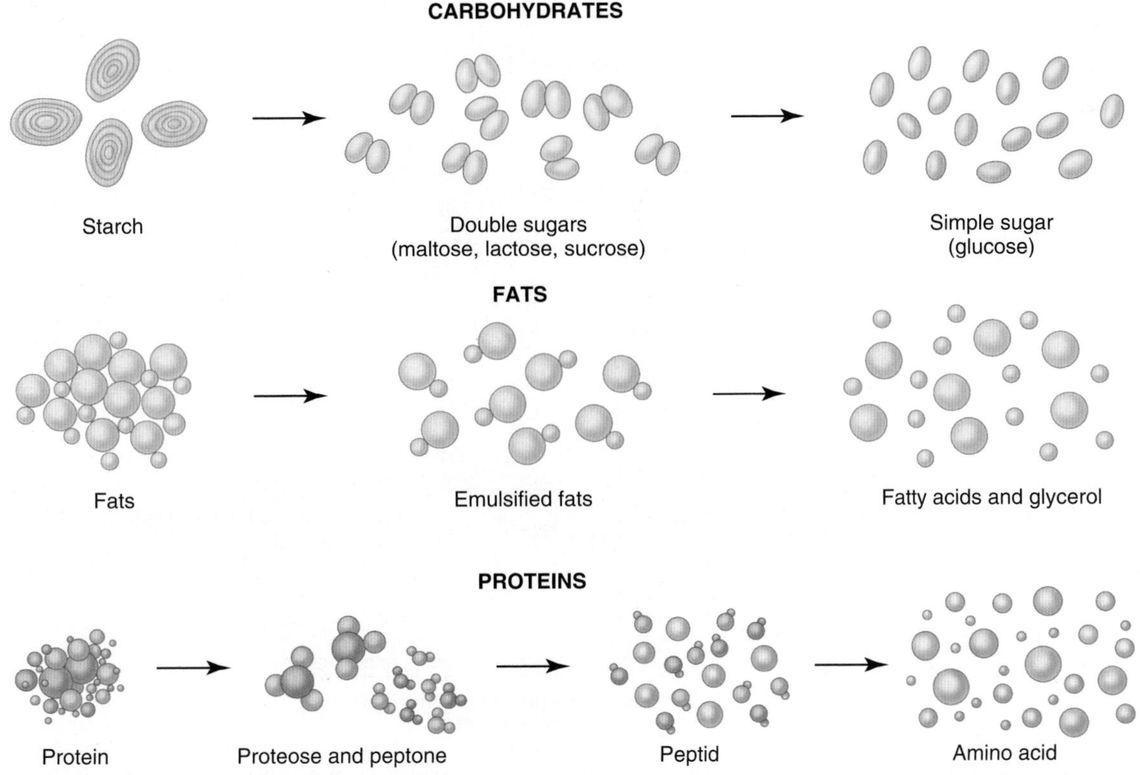

Figure 19-7 *Phases in the digestion of starch, fat, and protein*

The combined action of pancreatic juice, bile, and intestinal juice completes the process of changing carbohydrates first into starch then into glucose, protein into amino acids, and fats into fatty acids and glycerol (**Figure 19-7**). The end products of digestion are now ready for absorption. See **Table 19-1**.

ABSORPTION IN THE SMALL INTESTINE

Absorption (**UB**-sorp-shun) is possible because the lining of the small intestine is not smooth. It is covered with millions of tiny projections called **villi** (**VILL**-eye). Each microscopic villus contains a network of blood and lymph capillaries called lacteals (**Figure 19-8**). The digested portion of the food passes through the villi, into the bloodstream and lacteals, and on to the body cells. The undigestible portion passes on to the large intestine.

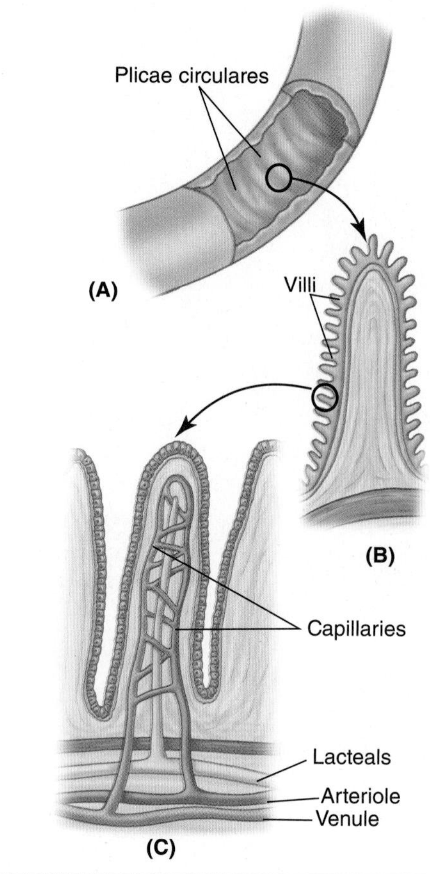

Figure 19-8 *Structures of absorption in the small intestines: (A) plicae circulares; (B) villi; (C) capillaries*

Pancreas: Accessory Organ of Digestion

The pancreas is a feather-shaped organ located behind the stomach (**Figure 19-9**). It functions as both an exocrine gland and an endocrine gland. Internally, the pancreas consists of groups of cell clusters. The islets of Langerhans, one group of cells, are part of the endocrine system and produce insulin and glucagon (refer to Chapter 12). The **acini cells** (**ASS**-uh-nigh) produce

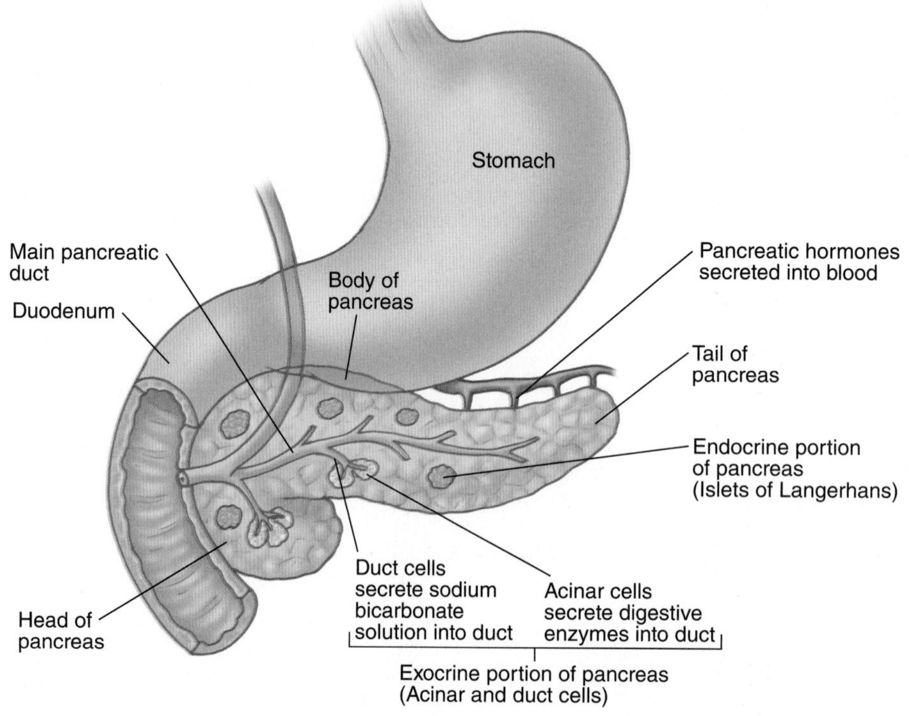

Figure 19-9 *The structure of the pancreas, showing both exocrine and endocrine portions; note that the glandular portions of the pancreas are grossly exaggerated.*

the digestive juices, and the duct cells secrete sodium bicarbonate, which helps neutralize the acidic content of food as it leaves the stomach. These secretions leave the pancreas through a large duct called the pancreatic duct. The secretions unite with the common bile duct of the liver and enter the duodenum in a common duct called the hepatopancreatic ampulla, or the ampulla of Vater.

Liver: Accessory Organ of Digestion

The **liver** is the largest organ in the body. It is located below the diaphragm, in the upper right quadrant of the abdomen (**Figure 19-10**). The liver is divided into two principal lobes: the right lobe and the left lobe. The lobes of the liver are made up of many functional units called lobules. The portal vein carries the products of digestion from the small intestine to the liver.

Some of the liver's many functions are as follows:

- Manufactures bile, a yellow to green fluid that is necessary for the digestion of fat. About 800–1000 mL of bile is produced daily. Bile contains bile salts; bile pigments, mainly bilirubin (bill-ih-**ROO**-bin), which comes from the breakdown of the hemoglobin molecule; cholesterol; phospholipids; and some electrolytes. The **hepatic duct** (hih-**PAA**-tik) from the liver joins with the **cystic duct** (**SIS**-tik) of the gallbladder to form the **common bile duct**, which carries the bile to the duodenum. If this duct is blocked, bile may then enter

Table 19-1 *Summary of Digestive Enzymes Involved in Human Digestion*

ORGAN	*JUICE*	*GLAND*	*ENZYME(S)*	*ACTION*	*ADDITIONAL FACTS*
Mouth	Saliva	Salivary	Amylase found in ptyalin	Starch → Maltose	Physical and chemical hydrolysis Mucous flow starts here and continues throughout digestive tract
Esophagus	Mucus	Mucus	None	Lubrication of food	Peristalsis begins here
Stomach	Gastric juice along with HCl acid	Gastric	Protease, pepsin	Proteins → peptones and proteoses	Gastrin activates the gastric glands HCl supplies an acidic medium and kills bacteria Temporary food storage
Small intestine	Intestinal	Intestinal	Peptidases	Peptones and proteoses into amino acids	Absorption of end products occurs in small intestine
			Maltase	Maltose → glucose	Villi facilitate absorption
			Lactase	Lactose → glucose and galactose	
			Sucrase	Sucrose → glucose and fructose	
			Lipase	Fats → fatty acids and glycerol	
	Bile	Liver	None	Emulsifies fat	Neutralizes stomach acid
	Pancreatic	Pancreas	Protease	Proteins → peptones and amino acids	Secretin stimulates the flow of pancreatic juice
			Amylase	Starch → maltose	
			Lipase	Fats → fatty acids and glycerol	
			Nucleases	Nucleic acids	

the bloodstream, causing jaundice (**JAWN**-dis), which gives the skin and sclera of the eyes a yellow color.

- Produces and stores glucose in the form of glycogen or converts excess glucose to fat. When needed, the liver can then transform glycogen and fat into glucose.
- Detoxifies alcohol, drugs, and other harmful substances.
- Filters blood coming from the digestive tract.
- Manufactures blood proteins such as fibrinogen and prothrombin, which are necessary for blood clotting; heparin, an anticoagulant; albumin, which is needed for fluid balance in the cells; and globulin, which is necessary for immunity.
- Turns ammonia that is produced when protein is broken down during the digestive process into urea; an increase in ammonia levels is toxic to the body. The urea is secreted into the blood stream, then goes to the kidneys and is excreted in the urine.
- Stores vitamins A, D, E, and K.
- Breaks down hormones no longer useful to the body.
- Removes worn-out red blood cells from circulation and recycles the iron content.

Gallbladder: Accessory Organ of Digestion

The **gallbladder** is a small, green organ in the inferior surface of the liver (**Figure 19-10**). It stores and concentrates bile when the bile is not needed by the body. When food high in fat enters the duodenum, bile is released by the gallbladder through the cystic duct.

Large Intestine (Colon)

The ileum empties its intestinal chyme into the sidewall of the large intestine through an opening called the **ileocecal valve** (**il**-ee-oh-**SEE**-kul). This valve permits passage of the chyme to the large intestine and prevents the backflow of chyme into the ileum.

The large intestine is about 5 feet long and 2 inches in diameter. Large amounts of mucus are secreted by the colon mucosa. This lubrication eases the passage of fecal material. The longitudinal smooth muscle layer is in three bands called the teniae coli (**TEE**-nee-ee **KO**-lye). The colon is gathered to fit these bands, giving it a puckered appearance. These little puckers, or pockets, are called **haustra** (**HAW**-struh) and provide more surface area in the colon. The **colon**, as the large intestine is also called, frames the abdomen (**Figure 19-11**).

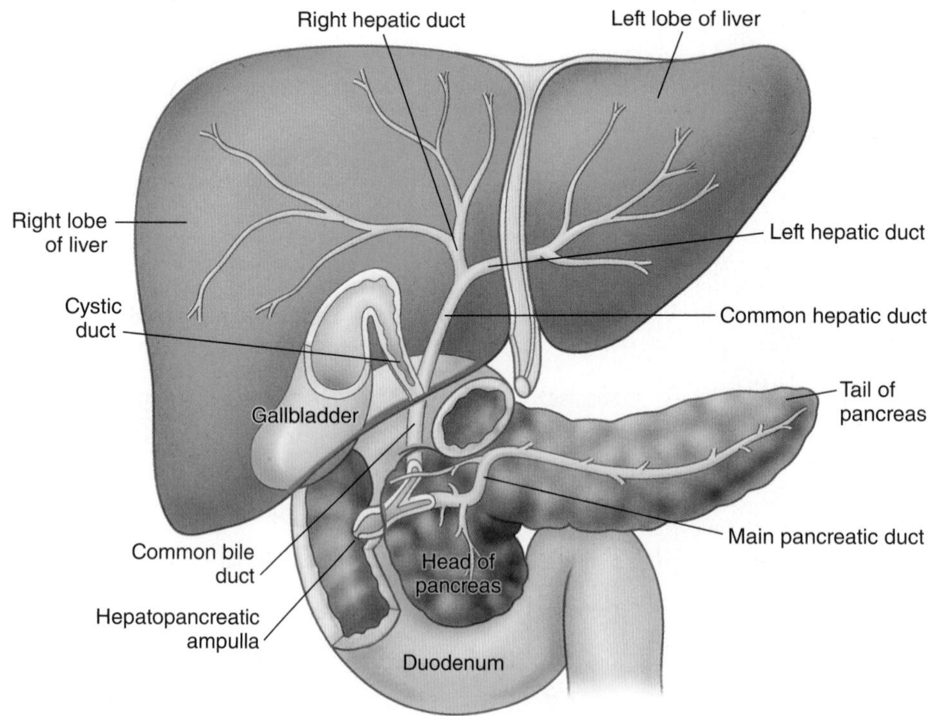

Figure 19-10 *Liver, gallbladder, and pancreas*

Cecum and Appendix

Located slightly below the ileocecal valve, in the lower right portion of the abdomen, is a blind pouch called the **cecum** (**SEE**-kum).

To the lower left of the cecum is the **vermiform appendix** (**VER**-mih-form). The appendix is a fingerlike projection protruding into the abdominal cavity (**Figure 19-11**). It has massive amounts of lymphoid tissue and plays a role in immunity.

Ascending, Transverse, and Descending Colon

The colon continues upward along the right side of the abdominal cavity to the hepatic flexure at the underside of the liver, forming the *ascending colon*. Then it veers to the left of the abdominal cavity, across the abdominal cavity, and to a point below the spleen called the splenic flexure, forming the *transverse colon*. The *descending colon* travels down from the splenic flexure on the left side of the abdominal cavity. As the descending colon reaches the left iliac region, it enters the pelvis in an S-shaped bend. This section is known as the **sigmoid colon** (**SIG**-moyd). The last 7–8 inches of the digestive tract is known as the **rectum** (**RECK**-tum) (**Figure 19-11**).

Anal Canal

The terminal one inch of the rectum is the anal canal. Its external opening is the **anus.** The anus is guarded by two anal sphincter muscles: one is an internal sphincter of smooth, involuntary muscle; the other is an external sphincter of striated, voluntary muscle. Both of these remain contracted to close the anal opening until **defecation** (def-ih-**KAY**-shun), the evacuation of the large intestine, takes place. The mucous membrane lining the anal canal is folded into vertical folds called rectal columns, which contain a network of arteries and veins.

GENERAL OVERVIEW OF DIGESTION

Food enters the GI tract via the mouth. In the oral cavity, food is mechanically broken down by the cutting, ripping, and grinding action of the teeth. Chemical digestion of carbohydrates is initiated by the secretion of saliva containing a digestive enzyme. Then, the action of the saliva and rolling motion of the tongue turn the food into a soft, pliable ball called a **bolus** (**BOH**-lus). The bolus slides down to the throat, or pharynx, to be swallowed. Next it travels through the esophagus into the stomach.

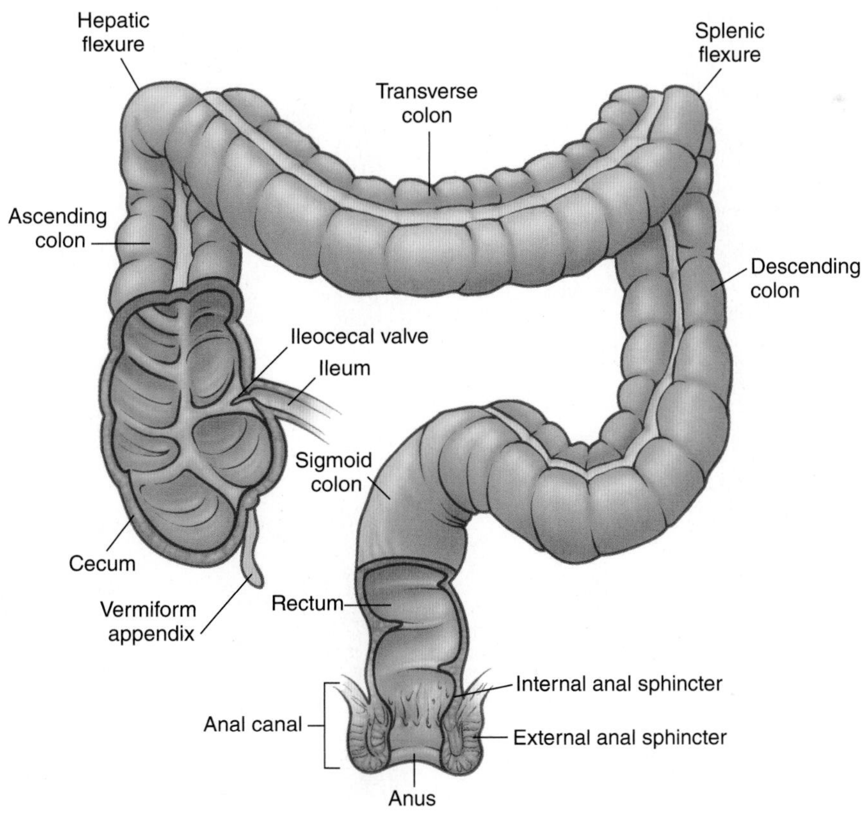

Figure 19-11 *The structure of the colon*

Food is pushed along the esophagus by peristalsis. From the stomach, peristaltic contractions continue to push the food into the small intestine. The nervous system stimulates gland activity and peristalsis.

Each part of the alimentary canal contributes to the overall digestive process. Protein digestion, for instance, is initiated by the stomach. Then the small intestine starts and finishes fat digestion and completes the digestion of carbohydrates and proteins. Numerous digestive glands are located in the stomach and small intestine, which secrete digestive juices containing powerful enzymes to chemically digest the food. Due to digestion, insoluble food becomes a soluble fluid substance. This substance is then transported across the small intestinal wall into the bloodstream and lacteals.

Circulated and absorbed through the blood capillaries into the interstitial fluid and finally into the body cells, the soluble food molecules are utilized for energy, repair, and production of new cells. The remaining undigested substances will pass into the large intestine as feces (**FEE**-seez) and leave the alimentary canal via the anus.

Action in the Mouth

Food is broken down by the teeth and mixed with saliva. Saliva contains salivary amylase, which converts the starches in carbohydrates into simple sugars. For example, if a person places a cracker in their mouth for a few minutes, it will have no taste because it is being broken down into glucose. Saliva is affected by the nervous system; just thinking of food will cause the mouth to water, or the opposite effect can occur—a dry mouth may result when a person is nervous or frightened.

Action in the Pharynx

Food leaves the mouth and travels to the pharynx. This structure serves as the common passageway for food and air.

SWALLOWING

Swallowing is a complex process involving the constrictor muscles of the pharynx. It begins as a voluntary process, changing to an involuntary process as the food enters the esophagus. When a person swallows, the tip of the tongue arches slightly and moves backward and upward. This action forces the food against the hard palate; simultaneously, the soft palate and the uvula shut off the opening to the nasopharynx. Food is thus prevented from entering the nasopharynx (**Figure 19-12A**).

In swallowing, the constrictor muscles of the pharynx contract, pushing food into the upper part of the esophagus. At the same time, other pharyngeal muscles raise the larynx, causing the epiglottis to cover the trachea (windpipe) (**Figure 19-12B**) and prevent

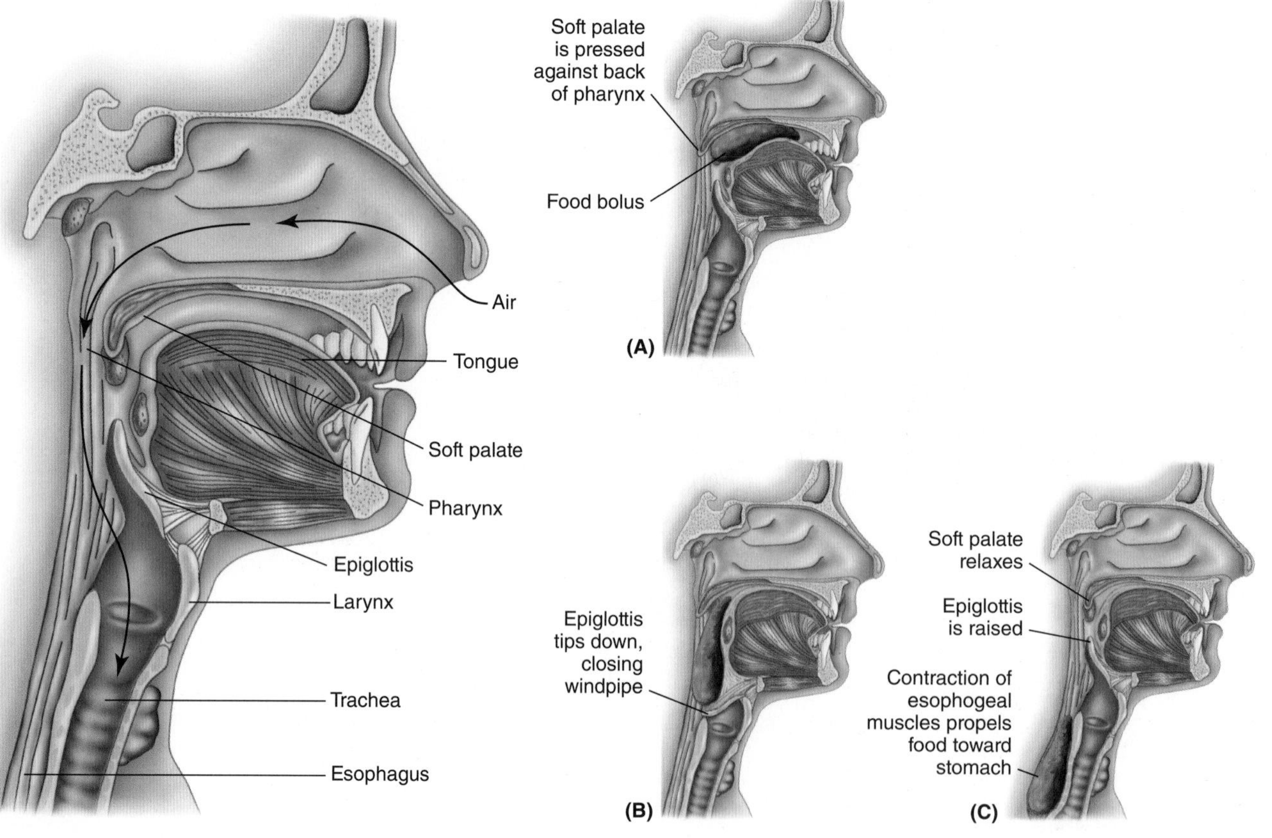

Figure 19-12 *The swallowing sequence into the esophagus*

food from entering it. If a person talks while eating, the epiglottis may not close and food may enter the trachea.

The act of swallowing is voluntary, but as a bolus of food passes over the posterior part of the tongue and stimulates receptors in the walls of the pharynx, swallowing becomes an involuntary reflex action. With the contraction of the pharyngeal muscles, followed by the contraction of the muscles lining the esophagus, food passes down into the stomach (**Figure 19-12C**). When a person swallows, they can place their fingers near the trachea to feel the structure move upward.

Action in the Esophagus

Food is pushed through the esophagus by peristalsis. This action explains why a person can swallow even while standing on their head; once food enters the esophagus, it goes to the stomach and is not affected by gravity.

Action in the Stomach

When food reaches the stomach, the cardiac sphincter relaxes and allows the food to enter. About 2–3 quarts of digestive juices are produced daily, which may explain the gurgling noises that can be heard from the stomach at times. When food enters, the gastric juices are released and begin to work on proteins. Salivary amylase continues its work in the stomach.

The action of the gastric juices is helped by the churning of the stomach walls. The semiliquid food is called chyme. The chyme leaves the stomach through the pyloric sphincter, which acts as a gatekeeper. This action allows a small squirt of chyme into the duodenum from time to time. Food takes about 2–4 hours to leave the stomach. Food moves through the stomach by peristalsis; vomiting is an action that occurs because of reverse peristalsis. The only known substances to be absorbed in the stomach are alcohol and some medications.

Action in the Small Intestine

In the small intestine, the process of digestion is completed and absorption occurs. Bile emulsifies fat to prepare it for digestion by pancreatic and intestinal juices. Pancreatic juices neutralize the acidic chyme and complete the digestion of carbohydrates, fats, and proteins (**Figure 19-13**). The end products of digestion are as follows:

- Carbohydrates are converted to simple sugars, such as glucose.
- Proteins are broken down into amino acids.
- Fats are changed into fatty acids and glycerol.

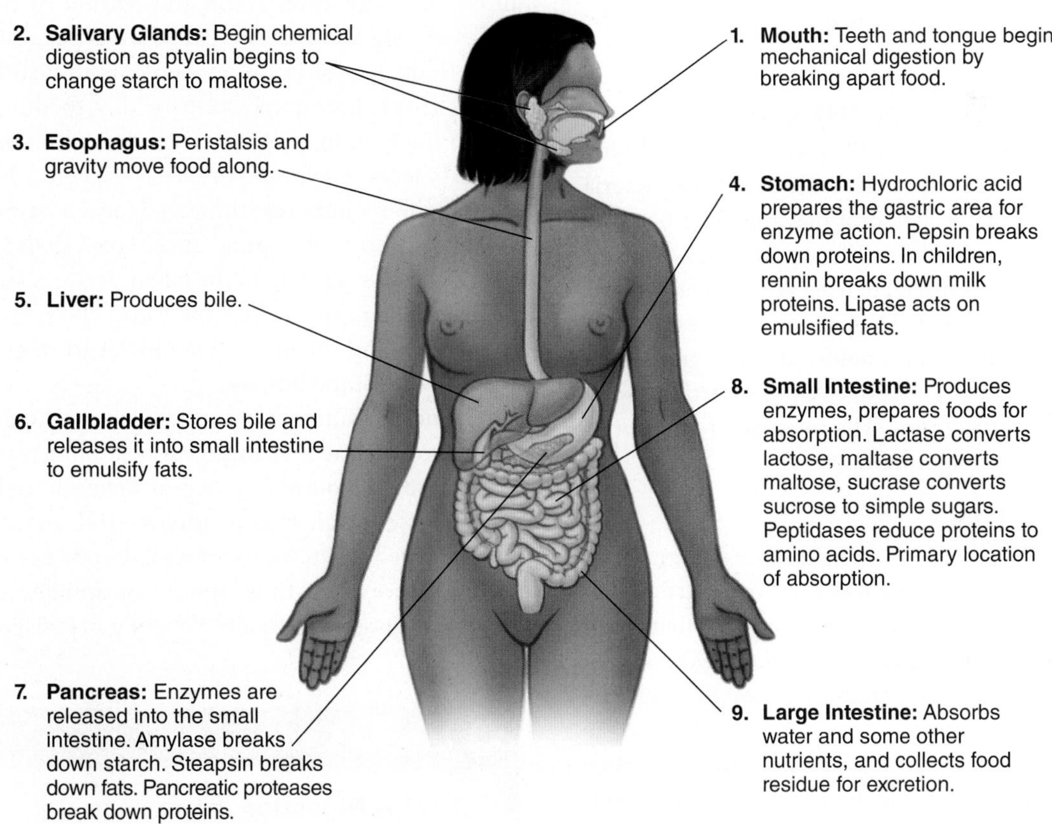
1. Mouth: Teeth and tongue begin mechanical digestion by breaking apart food.
2. Salivary Glands: Begin chemical digestion as ptyalin begins to change starch to maltose.
3. Esophagus: Peristalsis and gravity move food along.
4. Stomach: Hydrochloric acid prepares the gastric area for enzyme action. Pepsin breaks down proteins. In children, rennin breaks down milk proteins. Lipase acts on emulsified fats.
5. Liver: Produces bile.
6. Gallbladder: Stores bile and releases it into small intestine to emulsify fats.
7. Pancreas: Enzymes are released into the small intestine. Amylase breaks down starch. Steapsin breaks down fats. Pancreatic proteases break down proteins.
8. Small Intestine: Produces enzymes, prepares foods for absorption. Lactase converts lactose, maltase converts maltose, sucrase converts sucrose to simple sugars. Peptidases reduce proteins to amino acids. Primary location of absorption.
9. Large Intestine: Absorbs water and some other nutrients, and collects food residue for excretion.

Figure 19-13 *Overview of digestion*

The glucose, amino acids, fatty acids, and glycerol are then absorbed through the villi of the small intestine into the blood and lymph capillaries, or lacteals. The portal vein transports the blood from the small intestine and takes it to the liver, where it is distributed to the organs of the body.

The passage of food through the small intestine occurs because of peristalsis and segmented movement. Because inactive segments exist between active ones, the food is moved forward and backward—it is mixed as well as propelled. It takes about 6–8 hours for food to go through the small intestine; undigested foods then reach the ileocecal valve and enter the large intestine.

Action in the Large Intestine

The large intestine is responsible for water absorption, bacterial action, fecal formation, gas formation, and defecation. The purpose of these functions is to regulate the body's water balance, keeping the body at homeostasis, while storing and excreting the waste products of digestion (**Figure 19-13**).

ABSORPTION

The large intestine aids in the regulation of the body's water balance by absorbing large quantities of water back into the bloodstream. The water is drawn from the undigested food and indigestible material (cellulose) that pass through the colon. The large intestine absorbs vitamins B complex and K, which have been formed by the action of the intestinal bacteria.

BACTERIAL ACTION IN THE COLON

A few hours following the birth of an infant, the lining of the colon starts to accumulate bacteria. These bacteria persist throughout the person's lifetime. The bacteria multiply rapidly to form the bacterial population, or flora, of the colon. The intestinal bacteria are nonpathogenic, or harmless, to their host. They act on undigested food remains, turning them into acids, amines, gases, and other waste products. These decomposed products are excreted through the colon. The overuse of antibiotics can destroy this normal flora.

GAS FORMATION

Most people produce 1–3 pints of gas per day and pass it through the rectum in the form of **flatulence** (**FLA-chuh-lens**). Gas is produced by swallowed air that is not belched or burped up and by the normal breakdown of food. The unpleasant odor of flatulence comes from bacteria in the large intestine.

Research has not been able to show why some foods produce gas in one person and not in another. Some foods that produce gas are beans, vegetables such as broccoli and cabbage, fruit, whole grains, and milk and milk products.

Diet modification may reduce the amount of gas produced; however, it is important to remember that some of the foods that produce gas provide essential nutrients.

FECAL FORMATION

Initially, the undigested or indigestible material in the colon contains a lot of water and is in a liquid state. Due to water absorption and bacterial action, it is subsequently converted into a semisolid form called feces.

Feces consist of bacteria, waste products from the blood, acids, amines, inorganic salts, gases, mucus, and cellulose. Amines are waste products of amino acids. The gases are ammonia, carbon dioxide, hydrogen, hydrogen sulfide, and methane. The characteristically foul odor of feces derives from these substances.

Cellulose is the fibrous part of plants that humans are unable to digest. It contributes to the bulk of the feces. This bulk stimulates the muscular activity of the colon, resulting in defecation. Regular defecation (regularity) can be promoted by exercising daily and eating foods containing cellulose, such as whole-grain cereals, fruits, and vegetables. These foods supply the necessary bulk to initiate bowel movements.

DEFECATION

Once approximately every 12 hours, the fecal material moves into the lower colon and rectum by means of a series of long contractions called mass peristalsis. However, frequency of bowel movements in healthy people varies from three movements a day to three a week. When the rectum becomes distended with the accumulation of feces, a defecation reflex is triggered. Nerve endings in the rectum are stimulated, and a nerve impulse is transmitted to the spinal cord. From the spinal cord, nerve impulses are sent to the colon, rectum, and internal anal sphincter. This causes the colon and rectal muscles to contract and the internal sphincter to relax, resulting in emptying of the bowels.

For defecation to occur, the external anal sphincter must also be relaxed. The external anal sphincter surrounds and guards the outer opening of the anus and is under conscious control. Due to this control, defecation can be prevented when inconvenient, despite the defecation reflex. However, if this urge is continually ignored, it lessens or disappears totally, resulting in constipation.

Media Link

View the **Digestion** animation on the Online Resources.

METABOLISM

After digestion and absorption, nutrients are carried by the blood to the cells of the body. Within the cells, nutrients are changed into energy through a complex process called metabolism. During aerobic metabolism, nutrients are combined with oxygen within each cell. This process is known as oxidation. Oxidation ultimately reduces carbohydrates to carbon dioxide and water; proteins are reduced to carbon dioxide, water, and nitrogen. Anaerobic metabolism reduces fats without the use of oxygen. The complete oxidation of carbohydrates, proteins, and fats is commonly called the Krebs cycle.

As nutrients are oxidized, energy is released. When this released energy is used to build new substances from simpler ones, the process is called anabolism. When released energy is used to reduce substances to simpler ones, the process is called catabolism. This building up and breaking down of substances is called metabolism and is a continuous process within the body and requires a continuous supply of nutrients.

Metabolism is governed primarily by the hormones secreted by the thyroid gland.

The Effects of Aging on the Digestive System

It is normal for the digestive processes to slow down with age. Changes most frequently associated with aging are a decrease in the sensory ability of the taste buds and a reduction in the production of saliva. Dry mouth is also a side effect of more than 400 commonly used medicines, including drugs for high blood pressure, antidepressants, and antihistamines. In addition, there is sometimes a loss of teeth due to gum disease or decay. These changes lead to a loss of appetite and result in poor nutrition.

There is a slowdown in peristalsis from the esophagus to the colon, which may lead to the development of diverticulosis and chronic constipation. With age, the intestines become more susceptible to overgrowth of certain bacteria, which can impair absorption of nutrients and cause bloating and gas.

The liver becomes smaller and may lose some of its ability to detoxify drugs, alcohol, and other harmful substances. As a result, it takes longer for the body to rid itself of medication. Although this may seem like a lot of changes, the impact on digestion is generally mild.

COMMON SYMPTOMS OF DIGESTIVE DISORDERS

Common symptoms of digestive disorders include the following:

- **Abdominal pain** (visceral pain)—may come from organs within the abdominal cavity.
- **Nausea**—feeling of sickness in the stomach accompanied by a loathing for food and an involuntary impulse to vomit.
- **Vomiting** (emesis)—the expelling of undigested food or fluid through the mouth. Vomiting is a violent act in which the stomach turns itself inside out, forcing itself into the lower part of the esophagus.
- **Diarrhea** (dye-ah-**REE**-ah)—characterized by loose, watery, and frequent bowel movements. It may result from irritation of the colon's lining by bacterial, viral, or parasitic infections. It can also be caused by poor diet, nervousness, reaction to medications, toxic substances, or irritants in food. Excessive diarrhea may lead to dehydration and electrolyte imbalance. This can be a life-threatening situation in the very young and the very old.
- **Constipation**—condition in which defecation is delayed. The colon absorbs excessive water from the feces, rendering them dry and hard. When this occurs, defecation (evacuation) becomes difficult. Constipation can also be caused by emotions such as anxiety or fear. Headaches and other symptoms that frequently accompany constipation result from the distention of the rectum, as opposed to toxins from the feces. Treatment usually consists of eating proper foods containing fiber, fruits, and vegetables; drinking plenty of fluids; getting enough exercise; setting regular bowel habits; and avoiding tension as much as possible.

Lifestyle and the Environment

There are external factors that may influence GI problems. What individuals ingest can affect how the digestive system functions. Alcohol, tobacco, low fiber diets, and inactive lifestyles can have a direct result on digestive health. Many external influences can be controlled and corrected before diseases such as cancer, pancreatitis, gallstones, and cirrhosis occur.

Stress and Trauma to the Digestive System

Psychological trauma can have devastating effects on the GI system. Stress is a common result of psychological trauma that many people experience. Stress can cause nausea, vomiting, bloating, constipation, diarrhea, and pain. In more severe cases, stress hormones can alter the lining of the gut, which in turn can cause inflammation and allow bacteria to enter the bloodstream, causing septicemia.

CONGENITAL DISORDERS OF THE DIGESTIVE SYSTEM

Most congenital digestive disorders result in some type of intestinal obstructions. These problems are frequently manifested by feeding problems, abdominal distention, and emesis at birth or within 1–2 days.

Cleft lip occurs when the tissue that forms a baby's lip or mouth does not close properly during the early stages of fetal development.

Cleft palate occurs when the palate of the mouth does not close properly during the early stages of fetal development. Both cleft lip and cleft palate interfere with feeding and speech and must be surgically repaired.

Atresia is the congenital absence or the pathological closure of an opening, passage, or cavity. Most cases of atresia require surgical repair. Some types of atresia include the following:

- Esophageal atresia and fistula esophagus occur when the esophagus ends in a blind pouch rather than attaching to the stomach or when there is an abnormal opening between the esophagus and trachea.
- Gastric atresia is a complete blockage of the pyloric outlet of the stomach.
- Duodenal atresia is the absence or complete closure of portions of the duodenum.
- Biliary atresia is when one or more of the bile ducts are abnormally large or absent.

Hirschsprung's disease, or congenital megacolon, occurs when nerve cells stop growing and are missing at the end of the colon. The body can't sense when waste material reaches the anus. It is typically diagnosed in newborns, and symptoms include no bowel movement in the first 48 hours. Two surgical procedures may be considered. The first of these is pull-through surgery, in which the diseased part of the colon is removed, and the normal section is pulled through the colon from the inside to attach to the anus. The second surgery that may be considered is an ostomy; an ostomy requires the surgeon to remove the diseased tissue and move the healthy tissue to an opening created on the abdomen. Both surgeries may alter the structure of the organ by removing the diseased tissue, but connecting the normal tissue allows the digestive system to work as intended.

Meckel's diverticulum is the most common congenital defect of the digestive system. It is an outpouching, or bulge, in the lower part of the small intestine, and it is caused by leftover umbilical cord. Symptoms occur in only a few cases and include GI bleeding and abdominal pain. Surgery is done to remove the disorder.

Pyloric stenosis (pye-**LOR**-ick sten-**OH**-sis) is a narrowing of the pyloric sphincter at the lower end of the stomach. It is often found in infants. Projectile vomiting may result; surgery is often necessary.

GENERAL DISORDERS OF THE DIGESTIVE SYSTEM

In times of stress, it is not unusual to have "butterflies" in the stomach, nausea, or another type of distress associated with the digestive system. However, sometimes symptoms associated with the digestive system are a reflection of disease.

Stomatitis (stoh-mah-**TYE**-tis) is an inflammation of the soft tissues of the mouth cavity. Pain and salivation may occur also.

Canker sores are small, painful ulcers that appear periodically on the tongue or mouth. The cause is unknown and is not contagious. Treatment is with mouthwashes and over-the-counter medications.

Oral thrush is caused by *Candida albicans;* it causes creamy-white lesions on the tongue and/or the inner lining of the mouth. It may occur at any age. It is seen more often in people with suppressed immune systems. Treatment is with antifungal medications.

Salivary glands infection or swollen glands are most commonly caused by salivary stones, which are buildups of crystallized saliva deposits that result in pain and swelling. Treatment is taking antibiotics, drinking 8–10 glasses of water daily with lemon to stimulate salivary glands, and keeping the mouth clean.

Did You Know?

Butterflies in the stomach, a fluttery feeling, is caused by a reduction in blood flow to the stomach. When the body is under stress, adrenaline is released, causing increased blood flow to the vital organs and less blood flow to the digestive system.

Gingivitis (jin-jih-**VYE**-tis) or gum disease begins with plaque, which is a colorless film that hardens into tartar. Plaque must be removed every day by brushing and flossing the teeth, otherwise the gums become swollen and bleed easily. Tartar can only be removed by professional cleaning. If left untreated, gum disease advances into periodontal disease.

Periodontitis, or periodontal disease, is a chronic bacterial infection of the gums and surrounding tissue that causes oral bacteria by-products to enter the bloodstream. This condition leads to the inner layer of gum and bone receding from the teeth and forming pockets that collect bacteria and become infected. As the disease progresses, the teeth lose their anchors and may become loose and fall out. Scientists are studying the effects that periodontal disease has on the cardiovascular system. It is important to take care of the teeth and gums to prevent diseases in other parts of the body. Treatment is professional cleaning and good brushing and flossing.

Gastroesophageal reflux disease (GERD) (**GAS**-troh-eh-sof-ah-**JEE**-al **REE**-flucks) is a chronic disorder that affects the cardiac sphincter muscle connecting the esophagus with the stomach. In GERD, the sphincter muscle is weak or relaxes inappropriately, allowing the stomach's acid contents to flow up into the esophagus, causing heartburn and possible chest pain. Treatment includes lifestyle changes and antacids.

Barrett's syndrome is when the mucosa of the esophagus becomes damaged by acid reflux inflaming the lining. It may increase the risk of cancer.

Heartburn, or acid indigestion, results from a backflow of the highly acidic gastric juice into the lower end of the esophagus, which has many causes other than GERD. This irritates the lining of the esophagus, causing a burning sensation in the lower chest. Heartburn may be experienced on a daily basis by some people. Twenty-five percent of all pregnant women experience heartburn.

Temporary relief from heartburn and GERD can be obtained by doing the following:

- Avoid smoking and drinking alcohol.
- Take prescription medications to block the production of acid.
- Avoid lying down for 2–3 hours after eating.
- Sleep with two pillows to elevate the head.
- Avoid foods that trigger heartburn, such as coffee, chocolate, citrus fruits and juices, fried and fatty foods, spicy foods, and tomato products.
- Avoid tight-fitting clothing.

Hiatal hernia (high-**AY**-tal **HER**-nee-ah), or rupture, occurs when the stomach protrudes above the diaphragm through the esophagus opening (**Figure 19-14**). Hiatal hernia is not uncommon in people over the age of 50. Changes in diet may relieve the symptoms, as well as the use of antacids. Surgery is not usually required.

Gastritis (gas-**TRY**-tis) is an acute or chronic inflammation of the stomach lining caused by bacteria, a virus, or certain foods. Gastritis produces discomfort, nausea, and vomiting. Treatment depends on the cause; antibiotics may help.

Gastroenteritis (**gas**-troh-en-ter-**EYE**-tis) is the inflammation of the mucous membrane lining of the stomach and intestinal tract. A common cause is a virus, which causes diarrhea and vomiting for 24 to 36 hours. Avoid contaminated food and water and wash hands frequently. If this condition persists, dehydration may occur. Treatment is symptomatic.

Enteritis is inflammation of the intestine that may be caused by a bacterial, viral, or protozoan infection. Enteritis can also be caused by an allergic reaction to certain foods or food poisoning. Treatment is with antibiotics and anti-diarrheal drugs.

Celiac disease or coeliac disease is a genetic autoimmune disease in which the ingestion of gluten leads to damage to the lining of the small intestine. This damage prevents the absorption of nutrients. Treatment is to follow a strict gluten-free diet.

An **ulcer** (**UL**-ser) is a sore or lesion that forms in the mucosal lining of the stomach or duodenum, where acid and pepsin are present. Ulcers found in the stomach are called

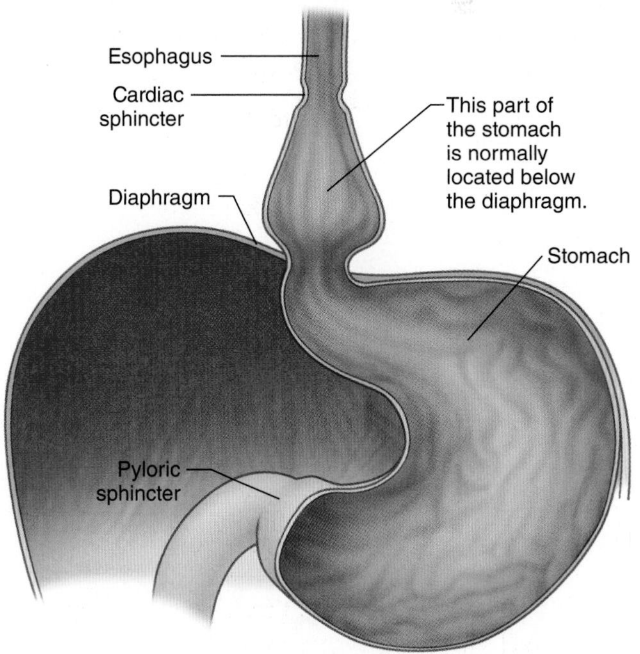

Figure 19-14 *Hiatal hernia: In a hiatal hernia, part of the stomach protrudes above the diaphragm.*

gastric ulcers; those in the duodenum are called duodenal ulcers. In general, both types are referred to as **peptic ulcers** (**PEP**-tick) (**Figure 19-15**). Research shows that most ulcers develop as a result of an infection with bacteria called *Helicobacter pylori* (*H. pylori*). Some ulcers may occur because of the use of nonsteroidal anti-inflammatory drugs. Stress, spicy foods, and the use of alcohol do not cause ulcers. They can, however, make ulcers worse.

The most common symptom of an ulcer is a burning pain in the abdomen between the sternum and navel. The pain occurs between meals and at night. It may be relieved by eating or taking an antacid. Ulcers are diagnosed by X-ray and by testing for *H. pylori* bacteria. If the cause is *H. pylori*, antibiotics are the treatment. Elimination of the bacteria allows the ulcer to heal and not reoccur.

Treatment for ulcers from other causes include the use of histamine (H_2) blockers. These drugs reduce the amount of acid the stomach produces by blocking **histamine** (**HISS**-tah-mean), a powerful stimulant of acid secretion.

Additional treatment includes the use of proton pump inhibitors (PPI), which are drugs that reduce the production of stomach acid; the use of mucosal protective medications; and lifestyle changes.

Cancer of stomach, also called gastric cancer, can develop in any part of the stomach and may spread throughout the stomach and to other organs. Stomach cancer is hard to diagnose. Often there are no symptoms in the early stages and, in many cases, the cancer has spread before it is found. The initial symptoms of stomach cancer are much like those of other digestive disorders: heartburn, loss of appetite, unexplained persistent nausea, a feeling of bloated discomfort after eating, and occasional mild stomach pain. Later symptoms include traces of blood in the feces, pain, weight loss, and vomiting.

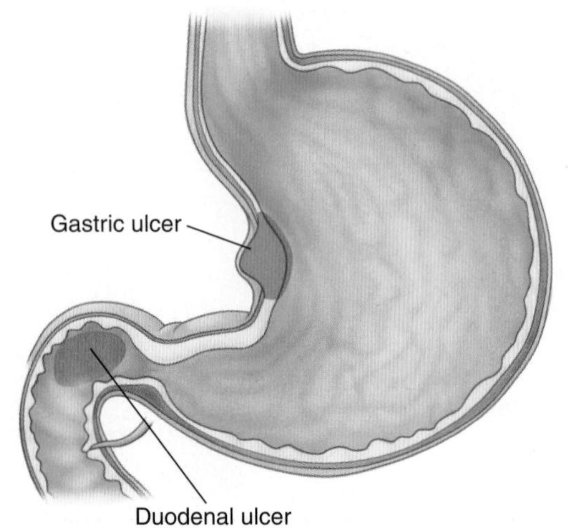

Figure 19-15 *The location of peptic ulcers: gastric and duodenal*

Treatment involves surgical removal of the stomach tumor as soon as possible. Depending on the size and extent of growth of the tumor, part or all of the stomach may have to be removed. If the cancer has spread, chemotherapy is prescribed. Targeted therapy is done, which uses drugs that attack specific abnormalities within the cancer cell.

Irritable bowel syndrome (IBS) is a group of symptoms that affect the large intestine. Unlike irritable bowel disease, it does not cause inflammation, ulcers, or other damage to the large intestine. Signs and symptoms include cramping, abdominal pain, bloating, gas, and diarrhea or constipation (sometimes both). It is a chronic condition that affects quality of life, and it needs to be managed long term. Symptoms can be triggered by food, stress, or hormones. Treatment includes dietary changes, medication, relaxation techniques, and counseling.

Inflammatory bowel disease (IBD) is an umbrella term used to describe disorders that involve chronic inflammation of the digestive tract. This disorder affects about three million Americans. The most prominent symptom is chronic diarrhea. Crohn's disease and ulcerative colitis are two separate inflammatory diseases that are considered IBDs. Both diseases have one strong feature in common: they are considered autoimmune diseases (see Chapter 16). More details on these two diseases follow:

- Crohn's disease can occur anywhere in the digestive tract. It may occur simultaneously in different locations. Crohn's disease generally penetrates every layer of tissue in the affected area. Patients with Crohn's may have remissions and exacerbations, or flare-ups.
- Ulcerative colitis is typically found in the colon and rectum. It usually affects only the innermost lining of the colon and rectum.

The cause of IBD is unknown. The symptoms of these two diseases under the IBD umbrella are similar: chronic diarrhea, vomiting, abdominal cramping, blood in the stool, weight loss, and fatigue. Diagnosis may involve blood tests and examinations of the digestive tract, including a possible colonoscopy. Treatment is mainly drug therapy, including anti-inflammatory drugs, immune modulators, and antibiotics. Life treatments include diet modification, high fluid intake, and the reduction and management of stress. Inflammatory bowel disease takes an emotional toll; there is constant anxiety about the need to run to the toilet. Support groups such as the Crohn's and Colitis Foundation of America are available. Surgical treatment may be necessary for about 70% of patients with Crohn's disease and for about 20% of patients with ulcerative colitis.

Appendicitis (ah-**pen**-dih-**SIGH**-tis) occurs when the vermiform appendix becomes inflamed. A classical symptom is pain that begins near the belly button and moves to the lower right side. If it ruptures, the bacteria from the appendix can spread to the peritoneal cavity, causing peritonitis. Treatment is with antibiotics and surgery.

Hepatitis (hep-ah-**TYE**-tis) is an inflammation of the liver. Clinical symptoms are fever, nausea, anorexia, clay-colored stool, dark urine, and jaundice. The different strains of viral hepatitis include hepatitis A, B, C, D, and E. Standard precautions are followed in the care of all hepatitis patients. More details about the different strains of hepatitis follow:

- Hepatitis A is caused by the hepatitis A virus (HAV). This viral infection of the liver is often spread through water or food contaminated with the feces of an infected person. The hepatitis A vaccine is recommended for young children living in areas with a high incidence of the disease or anyone traveling to countries where hepatitis A is endemic, which means restricted or peculiar to a certain area.
- Hepatitis B is a viral infection caused by the hepatitis B virus (HBV), found in highest concentrations in the blood and lower concentrations in other body fluids. It is transmitted through contact with the blood and body fluids of an infected person. Hepatitis B is spread through having sex with an infected partner without using a condom, by sharing needles, by getting a tattoo or piercing with tools that were not cleaned properly, from mother to child through placenta, or through accidental needle sticks or sharps exposure in the health care field. Treatment is with antiviral drugs. A vaccine is available for hepatitis B and is recommended for all ages; babies are now receiving the vaccine as part of their regular immunization program.
- Hepatitis C is a viral infection caused by the hepatitis C virus (HCV). Intravenous drug use is the single biggest risk factor for hepatitis C. Hepatitis C currently affects more than three million Americans. Most patients with hepatitis C are unaware of their infection because they can be symptom free for as long as a decade. The United States Preventive Task Force recommends that all adults between the ages of 18 and 75 be screened for Hepatitis C. The disease can be life threatening; the consequences are a severely damaged liver. Treatment is with direct acting antiviral drugs, and the treatment may be as short as 6–8 weeks.
- Hepatitis D virus (HDV) requires coinfection with the B type. There is currently no cure or vaccine.
- Hepatitis E virus (HEV) is transmitted through the fecal-oral route. The most common source of infection is fecal contamination of water. There is no cure or vaccine available. It usually resolves on its own within 4–6 weeks.

Cirrhosis (sih-**ROH**-sis) is a chronic, progressive, inflammatory disease of the liver characterized by replacement of normal tissue with fibrous connective tissue. Three-fourths of cirrhosis cases are caused by excessive alcohol consumption. Viral hepatitis may also cause cirrhosis.

Many people with early stages of cirrhosis have no symptoms. As scar tissue replaces healthy liver cells, the organ's functions start to fail and a person may experience the following symptoms: fatigue; nausea; weight loss; itchy skin; abdominal pain; and spider angiomas, which are spiderlike blood vessels that develop on the skin. The complications of cirrhosis include edema and ascites, or fluid in the abdomen; bruising and bleeding; jaundice; gallstones; splenomegaly, or an enlarged spleen; toxins in the blood and brain; sensitivity to medication; portal hypertension; and varices, which are enlarged blood vessels in the esophagus and stomach (**Figure 19-16**).

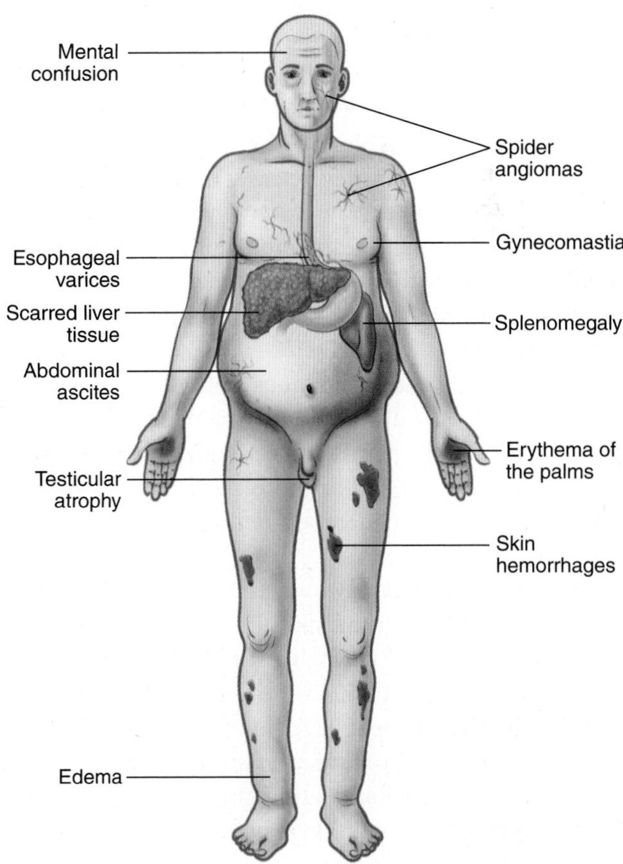

Figure 19-16 *Various clinical conditions associated with cirrhosis of the liver*

Liver damage caused by cirrhosis cannot be reversed, but treatment can delay or reduce complications. Treatment depends on the underlying causes. In all cases, regardless of the cause, following a healthy diet and avoiding alcohol are essential. If the liver becomes too damaged from scarring, a liver transplant may be necessary.

Cholecystitis (**koh**-lee-sis-**TYE**-tis) is inflammation of the gallbladder. This condition may cause blockage of the cystic duct, which would inhibit the release of stored bile. Treatment is with antibiotics, analgesics, and drugs to dissolve gallstones.

Gallstones, or cholelithiasis, are collections of crystallized cholesterol in the gallbladder. These are combined with bile salts and bile pigments. Gallstones can block the bile duct, causing pain and digestive disorders; pain may also occur in the back between the shoulder blades. In such cases, bile cannot flow into the small intestine. Most gallstones are small and may pass with undigested food. However, the larger and obstructive ones must be surgically removed. The gallstones or gallbladder may be removed through laparoscopic surgery (see *Medical Highlights—Minimally Invasive Surgery: Laparoscopy*).

Pancreatitis (pan-kree-ah-**TYE**-tis) is inflammation of the pancreas. The pancreas can become edematous, hemorrhagic, or necrotic. One-third of pancreatitis cases are due to unknown causes; however, acute pancreatitis is usually caused by gallstones or heavy use of alcohol. In acute pancreatitis, the patient complains of severe pain in the upper abdomen that may reach to the back. Severe cases may cause dehydration and low blood pressure. Acute pancreatitis usually improves on its own. Treatment includes fasting for several days to allow the pancreas to recover. Medication is given for the pain, and intravenous fluids may be given for dehydration.

Chronic pancreatitis does not resolve itself and results in the slow destruction of the pancreas. Medication is given for the pain. Pancreatic enzymes and insulin are given if they are not being secreted by the pancreas. In severe cases, the pancreas is removed by laparoscopic surgery.

Diverticulosis (dye-ver-tick-you-**LOH**-sis) is a condition in which little sacs, called diverticula, develop in the wall of the colon (**Figure 19-17**). The majority of people over the age of 60 in the United States have this condition. Most people have no symptoms and would not know they had diverticulosis unless an X-ray was taken or intestinal examination conducted. About 20% of people with this condition may develop **diverticulitis**, which is an inflammation in the wall of the colon. People with this condition must follow a restricted diet and take antibiotics.

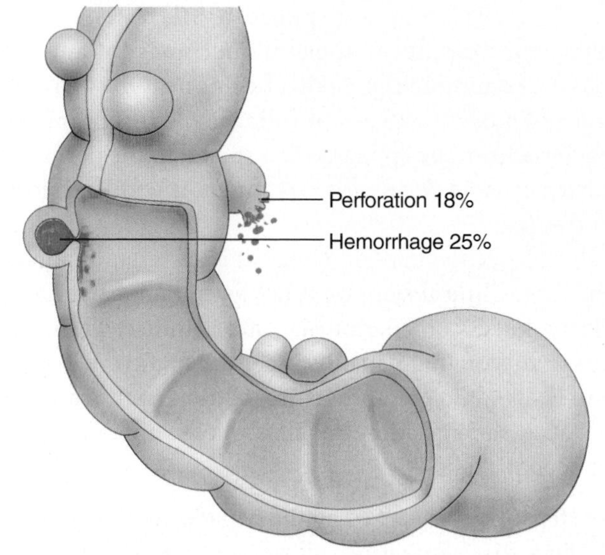

Figure 19-17 *Diverticulosis*

Hemorrhoids (**HEH**-muh-roidz) are a condition in which the veins around the anus or lower rectum become swollen and inflamed. This condition results from straining during defecation. Other contributing factors include aging, chronic constipation, pregnancy, and diarrhea. Hemorrhoids are either located internally within the anus or externally under the skin around the anus. Symptoms include painful swelling and possible bleeding covering the stool in the case of an internal hemorrhoid.

Measures to reduce symptoms include application of a hemorrhoid cream or suppository. To reduce the size of the hemorrhoid, rubber band ligation may be done. In severe cases, a hemorrhoidectomy may be done. To prevent hemorrhoids, it is important to exercise, drink ample fluids, and increase fiber intake. These measures help reduce constipation and straining when defecating.

Colorectal cancer involves the large intestine, while rectal cancer involves only the rectum; collectively they are known as colorectal cancers. The number of cases has declined over the years because of the growing awareness of screening methods that allow precancerous tissue changes in the colon to be removed before they become cancerous growths. Most cancers begin as small, benign growths of tissue known as polyps. Over time, these polyps can become cancerous. Early detection is critical. It is recommended that people of average risk begin the following screenings at age 45 and continue screening every ten years until the age of 75:

- A stool slide specimen looking for hidden blood, or **hemoccult**, every year.

- In a colonoscopy—the screening tool of choice—a flexible, slender tube equipped with a camera is used to view the entire length of the colon. This instrument allows the physician to remove polyps and take tissue samples if needed. If any polyps are found during a colonoscopy, then the test should be conducted every three years instead of 10.
- In a virtual colonoscopy, a computerized tomography (CT) scanner is used to create hundreds of images of the colon without the use of a colonoscope. This procedure is noninvasive but does not allow for the removal of polyps. It is recommended every five years.

A colon resection may be performed in a patient with **colon cancer**. Sometimes it may be necessary to perform a **colostomy**. In this procedure, an opening is made through the abdomen into the colon, the cancerous tissue is removed, and the healthy tissue is brought out through the opening onto the skin. A pouch is worn to collect the body's wastes. This procedure causes stress and anxiety. The health care provider must be supportive of a patient with this condition. For rectal cancer, surgery is the first method of treatment. Colorectal cancers may be followed up with chemotherapy and radiation.

Medical Highlights

MINIMALLY INVASIVE SURGERY: LAPAROSCOPY

Advances in computer imaging have paved the way for less invasive procedures than those that used to call for large incisions and long recovery times. *Minimally invasive surgery* is an approach to surgery that minimizes trauma, maximizes outcomes, and enables patients to return quickly to their homes.

Laparoscopy is such a type of surgery. A laparoscopy is the direct visualization of the abdominal cavity through a tube, called a laparoscope, which is placed through a small incision. The instrument is like a miniature telescope with a fiberoptic system that brings light into the abdomen.

Carbon dioxide gas is put into the abdomen through a special needle inserted below the navel. This gas helps separate the organs inside the abdominal cavity, making it easier to see the internal organs. The gas is removed at the end of the procedure. The laparoscope may be fitted with miniature surgical devices to enable the physician to correct abnormal conditions.

Laparoscopic surgery is used when possible to remove gallstones, repair a hiatal hernia, perform appendectomies, remove the pancreas, and perform other medical procedures.

Career Profile

DENTIST

Dentists diagnose, prevent, and treat problems of the teeth and tissues of the mouth. They also perform corrective surgery of the gums and supporting bones in gum disease. Dentists extract teeth and make molds and measurements for dentures to replace missing teeth. They may administer anesthetics and write prescriptions for medications. Most dentists are general practitioners. Other dentists may practice in several specialty areas.

Dentists should have good visual memory, excellent judgment of space and shape, and a high degree of manual dexterity. High school students who wish to become dentists should take courses in biology, chemistry, physics, health, and mathematics.

All dentists must be licensed. To qualify for a license, a candidate must graduate from a dental school accredited by the American Dental Association Commission on Dental Accreditation and pass written and practical examinations.

Employment is expected to grow at an average rate, but may increase as the population ages.

Career Profile

DENTAL HYGIENIST

Dental hygienists clean teeth, provide other preventive dental care, and teach patients how to practice good oral hygiene. Hygienists examine teeth, remove plaque, take and develop X-rays, remove sutures, and smooth and polish restorations.

Dental hygienists should have good manual dexterity because they use dental instruments with little room for error within the patient's mouth.

Dental hygienists must be licensed by the state in which they practice. To qualify for licensure, a candidate must graduate from an accredited dental hygiene school and pass both a written test and a clinical examination. Some programs lead to a bachelor's degree, but most grant an associate's, degree.

Employment is expected to grow faster than the average for dental hygienists in response to increasing demand for dental care and the greater substitution of hygienists for services previously performed by a dentist.

Career Profile

DENTAL ASSISTANT

Dental assistants perform a variety of patient care and laboratory duties. They work at the chairside to assist while the dentist examines and treats patients. Assistants keep patients' mouths dry and clear by using suction or other devices.

Assistants must be a dentist's "third hand"; therefore, dentists look for people who are reliable, can work with others, and have good manual dexterity.

Those with laboratory duties make casts of the teeth and mouth from impressions taken by the dentist. Dental assistants with office duties schedule and confirm appointments, receive patients, keep treatment records, send bills, receive payments, and order supplies and materials.

Programs in dental assisting take one year or less and are offered at community colleges, vocational schools, and technical institutes.

The employment outlook is good. Population growth and greater retention of natural teeth by middle-aged and older people will fuel the demand for dental services.

Career Profile

DENTAL LABORATORY TECHNICIAN

Dental laboratory technicians fill prescriptions from dentists for crowns, bridges, dentures, and other dental prosthetics.

A high degree of manual dexterity, good vision, and the ability to recognize very fine color shadings and variations in shape are necessary.

Training in dental laboratory technology is available through community colleges, vocational schools, and technical institutes. Programs vary in length.

The employment outlook for dental laboratory technicians is excellent.

One BODY How the Digestive System Interacts with Other Body Systems

INTEGUMENTARY SYSTEM

- Provides nutrients for energy, growth, and repair of the system.
- Helps form vitamin D, which is necessary for the absorption of calcium.
- Supplies fat for insulation in the dermal and epidermal layers.

SKELETAL SYSTEM

- Provides nutrients for energy, growth, and repair of the system.
- Protects some of the organs of the digestive system.

MUSCULAR SYSTEM

- Provides nutrients for energy, growth, and repair of the system.
- Skeletal muscles of the face help chew and grind food.
- Peristalsis moves food through the digestive tract.
- Smooth muscle of the stomach churns food.

NERVOUS SYSTEM

- Provides nutrients for energy, growth, and repair of the system.
- Sensory receptors in the nose initiate the salivary response.
- Sensory receptors send messages to the brain for swallowing, peristalsis, and defecation.
- Stress and anxiety often affect the digestive system, causing nausea, vomiting, feeling of "butterflies" in the stomach, diarrhea, or constipation.

ENDOCRINE SYSTEM

- Provides nutrients for energy, growth, and repair of the system.
- The hormones secretin and cholecystokinin stimulate the digestive juices of the pancreas, liver, and gallbladder.
- Insulin and glucagon maintain glucose metabolism.

CIRCULATORY SYSTEM

- Provides nutrients for energy, growth, and repair of the system.
- Capillaries in the small intestines pick up the products of digestion and distribute them to all parts of the body.
- Liver manufactures plasma proteins necessary for blood clotting.

LYMPHATIC SYSTEM

- Provides nutrients for energy, growth, and repair of the system.
- Hydrochloric acid in the stomach, normal flora of the intestines, and lymphoid tissue in the appendix help control bacteria.

RESPIRATORY SYSTEM

- Provides nutrients for energy, growth, and repair of the system.
- Provides the oxygen necessary for metabolism to occur in the cells.

URINARY SYSTEM

- Provides nutrients for energy, growth, and repair of the system.
- Liver converts the end products of amino acids into urea to be excreted by the kidneys.

REPRODUCTIVE SYSTEM

- Provides nutrients for energy, growth, and repair of the system.

Medical Terminology

Term	Meaning	Term	Meaning
absorp	being absorbed	**gingiv**	gums
-tion	process of	**-ae**	pertaining to
absorp/tion	process of being absorbed	**gingiv/ae**	pertaining to the gums
aliment	food	**hemo**	blood
-ary	pertaining to	**-occult**	hidden
aliment/ary canal	pertaining to the food canal	**hem/occult**	hidden blood
appendic	appendix, attachment	**hepat**	liver
-itis	inflammation of	**hepat/itis**	inflammation of the liver
appendic/itis	inflammation of the appendix	**intrins**	within a body or organ
cec	blind	**-ic**	relating to
-um	presence of	**intrins/ic factor**	a factor within a body or organ
cec/um	presence of a blind pouch	**mastic**	chew or gnash
cholecyst	gallbladder	**-ate**	process of
cholecyst/itis	inflammation of the gallbladder	**mastic/ate**	process of chewing
cirrh	reddish, yellowness	**peri-**	around
-osis	condition of	**odont**	teeth
cirrh/osis	condition of yellowness indicating disease of the liver	**-al**	pertaining to
colo	colon	**peri/odont/al**	pertaining to membrane around teeth
-ostomy	opening into	**ton/e**	strength, stretching
col/ostomy	opening into the colon	**-um**	presence of
dia-	through	**peri/ton/e/um**	a type of serous membrane that is stretched around structures
-rrhea	flow or discharge		
dia/rrhea	excessive flow-through of liquid fecal material	**pylor**	gatekeeper to the small intestines
diverticul	offshoot, bypass	**pylor/ic sphincter**	muscle that is the gatekeeper to the small intestines
-osis	condition of		
diverticul/osis	condition of bypass, abnormal outpouching in the intestines	**sigm**	means an S
		-oid	resembling
gastro	stomach	**sigm/oid colon**	resembling an S shape
-enteri	small intestines	**stoma**	mouth
gastro/enter/itis	inflammation of the stomach and small intestines	**stoma/titis**	inflammation of the mouth

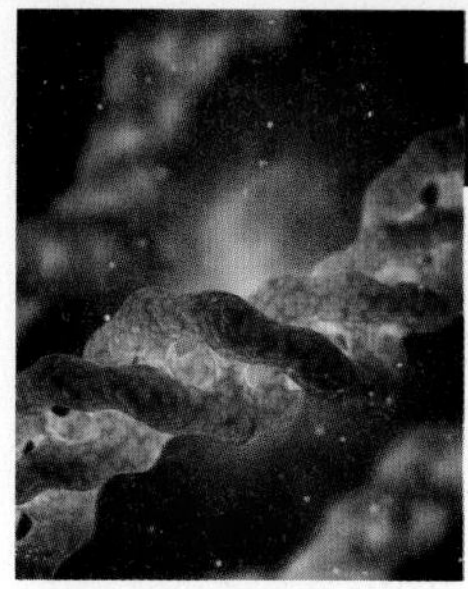

Study Tools

Workbook	Activities for Chapter 19
Online Resources	• PowerPoint presentations • Animations

REVIEW QUESTIONS

Select the letter of the choice that best completes the statement.

1. The process of changing complex foods into simpler substances to be absorbed is called
   a. metabolism.
   b. cellular respiration.
   c. peristalsis.
   d. digestion.

2. The walls of the digestive tube that contain mucus are called
   a. submucosa.
   b. mucosa.
   c. circular muscle.
   d. visceral peritoneum.

3. The accessory organs of the alimentary canal are the tongue, teeth, salivary glands, pancreas, liver, and
   a. stomach.
   b. esophagus.
   c. gallbladder.
   d. colon.

4. The taste buds are found on projections called
   a. papillae.
   b. parotid.
   c. palatine.
   d. pharynx.

5. The involuntary muscle action of the alimentary canal is called
   a. pushing.
   b. peristalsis.
   c. stenosis.
   d. contraction.

6. The opening from the esophagus to the stomach is called the
   a. ileocecal valve.
   b. cardiac sphincter.
   c. pyloric sphincter.
   d. cystic duct.

7. The lining of the abdominal cavity is called
   a. pleural.
   b. peritoneal.
   c. submucosal.
   d. epithelial.

8. The pancreatic enzyme that breaks down starches to glucose is called
   a. protease.
   b. lipase.
   c. secretin.
   d. amylase.

9. The enzyme that stimulates the liver to produce bile is called
   a. lipase.
   b. amylase.
   c. secretin.
   d. protease.

10. Food is absorbed in the small intestine in the
    a. villi.
    b. submucosa.
    c. peritoneal lining.
    d. colon.

TRUE OR FALSE

Read each statement carefully and determine if it is true or false. Correct any false statements.

T F **1.** The large intestine is called the colon.

T F **2.** The large intestine is 20 feet long and 2 inches wide.

T F **3.** The cecum is located where the small intestine joins the large intestine.

T F **4.** The function of the appendix is unknown.

T F **5.** The large intestine stores and eliminates the waste products of digestion.

T F **6.** Regulation of water balance occurs in the large intestine because its lining absorbs water.

T F **7.** Constipation may be overcome by intensive and long periods of work and exercise.

T F **8.** Bulk foods such as whole-grain cereals, fruits, and vegetables may help avoid constipation.

T F **9.** The rectum is an extension of the descending colon.

T F **10.** The hepatic flexure of the colon is at a point below the spleen.

MATCHING

Match each term in Column I with its correct description in Column II.

	COLUMN I	COLUMN II
______	**1.** enamel	a. substances that promote chemical reactions in living things
______	**2.** gingivae	b. a small, soft structure suspended from the soft palate
______	**3.** enzyme	c. gums that protect and support the teeth
______	**4.** salivary amylase	d. hardest substance in the body
______	**5.** uvula	e. the enzyme manufactured by the salivary glands
______	**6.** duodenum	f. frequent liquid bowel movements
______	**7.** gastroenteritis	g. narrowing of the sphincter in the stomach
______	**8.** diarrhea	h. the first section of the intestines
______	**9.** cholecystitis	i. inflammation of the stomach and intestinal lining
______	**10.** pyloric stenosis	j. inflammation of the gallbladder

COMPARE AND CONTRAST

Compare and contrast the following terms:

1. Gingivitis and periodontal disease

2. Gastroesophageal disease and heartburn

3. Hepatitis and cirrhosis

4. Crohn's disease and ulcerative colitis

5. Peptic ulcer and hiatal hernia

LABELING

Study the following diagram of a tooth and label the structures.

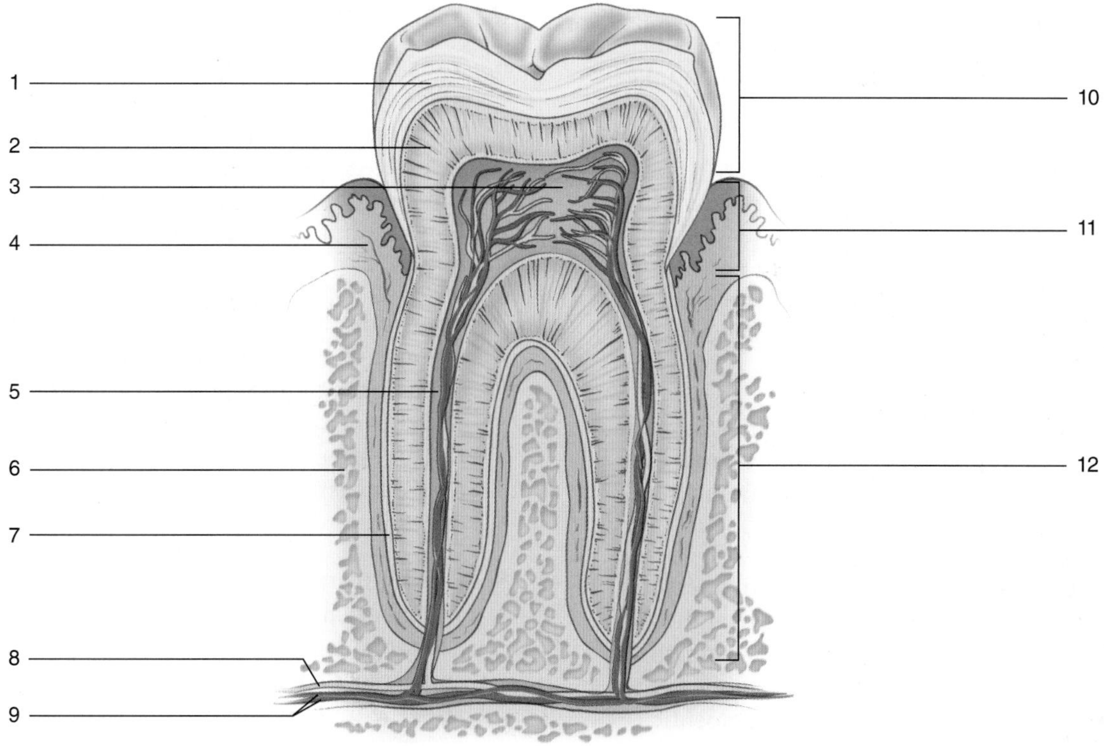

1. ______________________________

2. ______________________________

3. ______________________________

4. ______________________________

5. ______________________________

6. ______________________________

7. ______________________________

8. ______________________________

9. ______________________________

10. ______________________________

11. ______________________________

12. ______________________________

APPLYING THEORY TO PRACTICE

1. You have just eaten a slice of pizza for lunch. In about 12 hours, that slice of pizza may be ready for absorption in the villi of the small intestine. Trace the journey of the pizza, naming all the enzymes involved; where the action takes place; and the end products of carbohydrate, protein, and fat metabolism. Would you consider pizza a nutritious food? Explain your answer.
2. Enzymes secreted by the stomach are high in acid content. Explain the reason the lining of the digestive system does not become ulcerated.
3. Dental checkups make you nervous. Why is it a good health practice to see your dentist every six months? Describe the health careers in the dental profession.
4. In the emergency room, a 40-year-old person is complaining of a sharp pain between their shoulder blades on the right side. What is this symptomatic of? Report your suspicions to the patient and explain the different therapeutic technologies that may be necessary to treat this condition.
5. Your dad, who is 55, goes to the physician, who recommends he have a colonoscopy. He wants to know what a colonoscopy is, why it was recommended, and what this test can prevent. What would you tell him?
6. Your friend eats a high fat, low fiber diet and plays video games most of the day. Explain to them how this type of lifestyle could lead to diseases of their gastrointestinal system. Give them examples of the types of diseases that may result, and explore ways to minimize the effects.
7. Your mom has been under an immense amount of stress at work. She starts complaining about bloating, cramping, and diarrhea. Explain to her how stress can affect the digestive processes and how this trauma to the digestive system can cause long-term effects.
8. Create an amusement park using the digestive system organs as the rides. All organs and accessory organs must be included in your amusement park. The "rides" should be named after the function of the organ. For example, the mouth could be named the "Tunnel of Grub." The amusement park could be designed as a poster, booklet, or brochure. The names of the rides can be funny, but they should not be derogatory.

CASE STUDY

Kieran, age 40, goes to his health maintenance organization (HMO) to see his physician, Dr. Watson. He is complaining of general fatigue and loss of appetite, and he notices the whites of his eyes look yellowish. Dr. Watson orders blood tests for Kieran. The blood tests reveal he has hepatitis C.

1. Explain what hepatitis C is.
2. What is unusual about the appearance of symptoms in hepatitis C?
3. What functions of the liver may be impaired?
4. Are any other body systems affected by hepatitis?
5. What is the prognosis for Kieran?

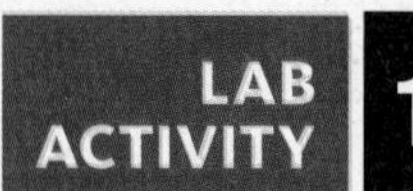

LAB ACTIVITY 19-1 Organs of Digestion

- ***Objective:*** To observe the location of the abdominal organs within the abdominal cavity
- ***Materials needed:*** anatomical torso with organs of digestion removed, organs of digestion, textbook, paper, pencil

Step 1: Place the organs of digestion in their correct anatomical place in the torso. Compare your results with **Figure 19-1** in the textbook.

Step 2: Describe the location of each of the organs of digestion using anatomical terminology.

LAB ACTIVITY 19-2 Stomach, Small Intestines, and Large Intestines

- ***Objective:*** To observe and compare the structure of the stomach, small intestines, and large intestines
- ***Materials needed:*** slides of cross sections of stomach, jejunum (including villi), and colon; microscope; textbook; paper; pencil

Step 1: Identify and describe the mucosa layers and gastric gland cells of the stomach. Compare your observations with **Figure 19-5B**.

Step 2: Sketch and identify the mucosa and villi of the jejunum. Compare your observations with **Figure 19-8**. Compare with the slides of the mucosa of the stomach. Describe and record the differences.

Step 3: Examine the slide of the mucosa of the colon. Compare with the slide of the jejunum. Describe and record the differences between the colon and jejunum.

LAB ACTIVITY 19-3 Action of Bile

- ***Objective:*** To observe the function of bile in digestion
- ***Materials needed:*** two small jars with lids labeled A and B, 2 tablespoons of oil, 2 tablespoons of water, container with liquid soap (simulated bile), tablespoon, teaspoon, paper, pencil

Step 1: Place about 1 teaspoon of vegetable oil in each jar.

Step 2: Place about 1 teaspoon of water in each jar.

Step 3: Add a few drops of liquid soap to jar A.

Step 4: Put the covers on both jars and shake gently.

Step 5: Observe what has happened in both jars. Record your results.

Step 6: What role does bile play in digestion? Is bile involved in mechanical or chemical digestion? Record your answer.

LAB ACTIVITY 19-4 Crohn's Disease

- ***Objective:*** To compare and contrast a healthy colon with a diseased colon
- ***Materials needed:*** slide of cross section of normal colon, pathologic slides of colon cross section showing Crohn's disease, microscope, disposal unit for the slides, paper, pencil, textbook

Step 1: Examine the slide of cross section of normal colon. Describe color and appearance of mucosa.

Step 2: Examine the slide of cross section of colon showing Crohn's disease. Describe color and appearance.

Step 3: Compare slides. Record the differences you observed.

Step 4: Return the slides to the approved container. Wash hands.

Step 5: Using the textbook or a medical website, identify various technologies used to diagnose Crohn's disease.

Step 6: With a lab partner, discuss how Crohn's disease is diagnosed and the importance of these technologies. Make a list of the technologies you discussed.

Step 7: With a lab partner, describe all therapeutic options for Crohn's disease. Make a list of the therapeutic options you discussed.

Step 8: Together with your partner, report your lists to another partner team. Add any additional information. Continue to report and exchange information with partner teams until you have met with all the other teams and your lists are complete.

CHAPTER 20

Nutrition

Objectives

- Define the term *nutrients*.
- Describe the function(s) of the different types of nutrients.
- Differentiate between fat-soluble and water-soluble vitamins.
- List the recommendations of the Dietary Guidelines for Americans.
- Explain BMR and BMI.
- Define the key words that relate to this chapter.

Key Words

anorexia nervosa
basal metabolic rate (BMR)
binge eating disorder
body mass index (BMI)
bulimia nervosa
calorie
complete proteins
essential amino acids
fiber
HDL
incomplete proteins
kilocalorie
LDL
mineral
nutrients
nutrition
obesity
organic food
roughage
trace elements
trans fat
triglycerides
vitamin
VLDL

The pace of an active daily life can be hectic and stressful. This can cause people to eat "on the run," "grab a bite" at a fast food restaurant, or forget to eat nutritiously.

The food people eat and the beverages they drink may or may not be nutritious. **Nutrition** (new-**TRIH**-shun) is the process by which the body assimilates food and uses it for energy, growth, and repair of tissue. Good nutrition prevents disease and promotes health. For food to be nutritious, it must contain the materials needed by the individual cells for proper cell functioning. These materials, or **nutrients** (**NEW**-tree-ents), are the following:

- Water
- Carbohydrates, including fiber
- Lipids
- Proteins
- Minerals and trace elements
- Vitamins

WATER

Water is an essential component of all body tissues. It has several important functions:

- Acts as a solvent for all biochemical reactions, helping to dissolve minerals and nutrients to make them more accessible to the body
- Serves as a transport medium for substances by carrying nutrients and oxygen to the cells
- Functions as a lubricant for joint movement and the digestive tract
- Helps control body temperature by evaporation from the pores of the skin
- Serves as a cushion for body organs, such as the lungs and brain
- Lessens burdens on the kidneys and liver by flushing out waste material

Water makes up between 55% and 65% of a person's total body weight. The body is continually losing water through evaporation, excretion, and respiration. This water loss must be replaced. Most foods people eat also contain some water. Exercise, sweating, urination, and vomiting increase the body's need for water. A person should drink water before they feel thirsty; thirst is a signal the body is on its way to dehydration. The Food and Nutrition Board recommends 13 to 15 8-ounce cups of fluid per day for a male and 9 to 11 8-ounce cups of fluid per day for a female.

CALORIE

A **calorie** (**KAL**-or-ee) is a unit that measures the amount of energy contained within the chemical bonds of different foods. The small calorie "c" is defined as the amount of heat required to raise the temperature of one gram of water by 1° Celsius. A **kilocalorie**, or large calorie "C", is equal to 1000 small calories and is the common measurement for dietary and nutritional calories. The calorie content of food is determined by measuring the amount of heat released when food is burned. The energy content of fat, 9 kilocalories per gram, is slightly more than twice that of carbohydrate or protein, which are both 4 kilocalories per gram.

A normal adult usually requires between 1600 and 3000 calories a day depending on age, sex, body weight, and degree of physical activity. Degree of activity falls into three levels: sedentary, a lifestyle that includes only light physical activity associated with activities of daily living; moderately active, a lifestyle that includes physical activity equal to walking 1.5–3 miles per day; and active, a lifestyle that includes physical activity equal to walking 3–5 miles per day. An excess number of calories, which is equivalent to energy required, results in an individual being overweight.

Newborn infants and young children have higher energy requirements per unit of body weight than adults do because of the high energy expenditure of growth. See **Table 20-1** for daily recommended energy (calorie) intakes.

CARBOHYDRATES

Carbohydrates include simple sugars, such as monosaccharides like glucose ($C_6H_{12}O_6$). Depending on the number of simple sugars found in the carbohydrate, they are classified as monosaccharides, disaccharides, or polysaccharides (starch). Only the monosaccharides are small enough to be absorbed and eventually taken into the cells. Disaccharides and polysaccharides are broken down by digestion into the smallest possible molecular subunits prior to absorption.

Carbohydrates are the main source of energy for the body. Excess carbohydrates are converted into fat and stored in fat tissue. Nutritionists recommend that carbohydrates comprise between 50% and 60% of the daily intake of calories.

Some carbohydrates should be avoided or minimized in the daily diet, such as candies, cakes, and so on. They supply calories but little else. Energy obtained from such foods is commonly referred to as "empty calories." Intake of these foods can also contribute to tooth decay.

Table 20-1 *Estimated Calorie Needs per Day by Age, Sex, and Physical Activity Level*[a]

Estimated amounts of calories needed to maintain calorie balance for various sex and age groups at three different levels of physical activity. The estimates are rounded to the nearest 200 calories. An individual's calorie needs may be higher or lower than these average estimates.

		PHYSICAL ACTIVITY LEVEL[b]		
SEX	**AGE (YEARS)**	**SEDENTARY**	**MODERATELY ACTIVE**	**ACTIVE**
Child (female and male)	2–3	1000–1200[c]	1000–1400[c]	1000–1400[c]
Female[d]	4–8	1200–1400	1400–1600	1400–1800
	9–13	1400–1600	1600–2000	1800–2200
	14–18	1800	2000	2400
	19–30	1800–2000	2000–2200	2400
	31–50	1800	2000	2200
	51+	1600	1800	2000–2200
Male	4–8	1200–1400	1400–1600	1600–2000
	9–13	1600–2000	1800–2200	2000–2600
	14–18	2000–2400	2400–2800	2800–3200
	19–30	2400–2600	2600–2800	3000
	31–50	2200–2400	2400–2600	2800–3000
	51+	2000–2200	2200–2400	2400–2800

[a]Based on Estimated Energy Requirements (EER) equations, using reference heights (average) and reference weights (healthy) for each age/sex group. For children and adolescents, reference height and weight vary. For adults, the reference man is 5 feet 10 inches tall and weighs 154 pounds. The reference woman is 5 feet 4 inches tall and weighs 126 pounds. EER equations are from the Institute of Medicine. Dietary Reference Intakes for Energy, Carbohydrate, Fiber, Fat, Fatty Acids, Cholesterol, Protein, and Amino Acids. Washington (DC): The National Academies Press; 2002.

[b]Sedentary means a lifestyle that includes only the light physical activity associated with typical day-to-day life. Moderately active means a lifestyle that includes physical activity equivalent to walking about 1.5 to 3 miles per day at 3 to 4 miles per hour, in addition to the light physical activity associated with typical day-to-day life. Active means a lifestyle that includes physical activity equivalent to walking more than 3 miles per day at 3 to 4 miles per hour, in addition to the light physical activity associated with typical day-to-day life.

[c]The calorie ranges shown are to accommodate needs of different ages within the group. For children and adolescents, more calories are needed at older ages. For adults, fewer calories are needed at older ages.

[d]Estimates for females do not include women who are pregnant or breastfeeding.

Source: From U.S. Department of Agriculture and U.S. Department of Health and Human Services. *Dietary Guidelines for Americans,* 2010, 7th Edition, Washington, DC: U.S. Government Printing Office, December 2010.

Foods containing starches and cellulose are a healthier source of carbohydrates. These foods, besides providing energy, can provide needed minerals, roughage, and vitamins.

FIBER

Dietary **fiber**, also known as **roughage** (**RUFF**-aj) or bulk, includes all parts of plant food that is not digested or absorbed. Fiber is found in whole-grain breads, cereals, corn (popcorn is a good source of fiber), beans, peas, and other vegetables and fruits. Eating a variety of fiber-containing plant foods is important for proper bowel function; reduces the symptoms of chronic constipation, diverticula disease, and hemorrhoids; and may lower the risk of heart diseases and some cancers. However, some of the health benefits associated with a high-fiber diet may come from other components present in these foods, not just from fiber itself. For this reason, fiber is best obtained from foods rather than a supplement.

LIPIDS

Lipids, or fats, are a group of compounds containing fatty acids combined with an alcohol. They can be subdivided into two groups: simple lipids (fats, oils, and waxes) and compound lipids (phospholipids, glycolipids, and sterols). Like carbohydrates, fats are a source of energy. The same amount of fat can release more than twice as many calories as the same amount of carbohydrate or protein. The human body stores reserves of energy as fat cells in adipose tissue. Likewise, any excess carbohydrate and protein in the diet is transformed into fat and stored along with any excess fat.

Fats are an essential nutrient to the maintenance of the human body. Stored fats provide a supply of energy during emergencies such as sickness or during deficient caloric intakes. Fats in the form of adipose tissue also cushion the internal organs and serve as insulation against the cold. Fats are components of the cell membrane and contribute to the formation of bile and steroid hormones, such as the sex hormones. Fats also contain certain kinds of vitamins called fat-soluble vitamins, which are an important part of a person's daily diet. It is therefore essential to have a diet that contains fats but does not exceed the body's calorie needs. Total daily dietary fat intake should not exceed 25% to 30% of the daily caloric intake.

Cholesterol (koh-**LES**-ter-ol) is a fat found in animal products such as meat, eggs, and cheese. Cholesterol is a white, waxlike substance used to build cells and make hormones. It is also manufactured by the liver. The cholesterol that one eats is not digested. There are no calories in cholesterol, but it is difficult for the body to get rid of it once it is in the body. Fats and oils in food are called **triglycerides** (try-**GLIS**-er-ides). The body turns excess calories into triglycerides, which are stored throughout the body in adipose tissue.

Triglycerides and cholesterol must be carried through the blood plasma by special proteins called lipoproteins: **HDL**, **LDL**, and **VLDL**. (See the Prevention of Heart Disease section in Chapter 14.)

The two most important steps a person can take to lower their blood cholesterol are (1) to reduce their intake of foods high in saturated fat, and (2) to lose weight if they are overweight. Fats are defined as follows:

- Saturated fat—oil from animal products that are solid at room temperature, such as butter, cheese, and meat fat
- Polyunsaturated fat—oil from vegetable products that are liquid at room temperature lowers blood cholesterol if used in moderation; includes safflower oil and sunflower oil
- Monounsaturated fat—oil from other vegetable products that are liquid at room temperature lowers blood cholesterol; includes olive oil and peanut oil
- **Trans fat**—the result of hydrogenation of vegetable oils. Manufacturers add hydrogen to vegetable oils. Hydrogenation increases the shelf life and flavor stability of foods containing these fats. Trans fat raises the LDL levels.

All food labels carry the amount of saturated fat and cholesterol in the product. If a label states the food is "cholesterol free," that does not necessarily mean it is healthy. Look carefully—many products with no cholesterol have saturated fats in them. Substitute foods high in saturated fat with skim milk, low-fat cheese, poultry, margarine, and low-fat ice cream. Some of the foods that help lower cholesterol include garlic, fresh fruits and vegetables, oat bran, wheat bran, and prunes.

PROTEINS

Proteins are structurally more complex than carbohydrates and lipids and contain an amino (NH_2) group. They are synthesized in the cell cytoplasm from constituent molecules called amino acids.

Proteins serve many different functions in the body. Some are enzymes and regulate the rate of chemical reactions; others are important in growth and repair of tissues. When necessary, proteins can also be used as a source of energy. In addition, contractile systems, such as muscles, hormonal systems, plasma transport systems, clotting systems, and defense systems (antibodies), are all dependent on proteins.

The body can synthesize some amino acids but not all. The amino acids that cannot be produced by the body are **essential amino acids**, which must be obtained from dietary sources. Proteins that contain all of the essential amino acids are known as **complete proteins**. Sources of such complete proteins are eggs, meat, milk, and milk products. Proteins that do not contain all the essential amino acids are called **incomplete proteins**. Vegetables contain incomplete proteins; however, a varied diet including vegetables will supply all the necessary complete proteins. For example, beans and wheat eaten alone will not provide all of the necessary complete proteins. When eaten together, however, they will complement each other and supply the necessary complete proteins.

The human body is unable to store excess amino acids. Any unused amino acids are broken down by the liver, and the amino group is excreted as a nitrogenous waste product called urea. The remainder of the amino acid may be burned for immediate energy or stored as fat or glycogen, a polysaccharide.

Protein synthesis cannot occur without all essential amino acids present at the same time. Therefore, it is important to include some complete protein with the foods eaten during the day. The daily intake of calories from proteins should be no more than 15% to 20%.

MINERALS AND TRACE ELEMENTS

A **mineral** is a chemical element that is obtained from inorganic compounds in food. The knowledge of the role of the essential minerals and trace elements is incomplete.

Many are notably necessary for normal human growth and maintenance.

Among the most important of essential minerals are sodium, potassium, calcium, iron, phosphorus, magnesium, and chloride.

Trace elements are minerals that are present in the body in very small amounts. These include zinc, copper, iodine, cobalt, manganese, selenium, chromium, molybdenum, and fluorine.

Table 20-2 summarizes the most important minerals and trace elements in the human diet. The toxic limits of some trace elements are extremely close to the recommended levels. This means that there is a critical difference between toxicity, health, and deficiency.

Table 20-2 *Essential Minerals and Trace Elements Needed in the Human Diet*

MINERAL	FOOD SOURCES	FUNCTION(S)	DEFICIENCIES
Calcium	Milk, cheese, dark green vegetables, dried legumes, sardines, shellfish	Bone and tooth formation Blood clotting Nerve transmission	Stunted growth Rickets Osteoporosis Convulsions
Chlorine	Common table salt, seafood, milk, meat, eggs	Formation of gastric juices Acid-base balance	Muscle cramps Mental apathy Poor appetite
Chromium	Fats, vegetable oils, meats, clams, whole-grain cereals	Involved in energy and glucose metabolism	Impaired ability to metabolize glucose
Copper	Drinking water, liver, shellfish, whole grains, cherries, legumes, kidney, poultry, oysters, nuts, chocolate	Constituent of enzymes Involved with iron transport	Anemia Incidence of disease rare
Fluorine	Drinking water, tea, coffee, seafood, rice, spinach, onions, lettuce	Maintenance of bone and tooth structure	Higher frequency of tooth decay
Iodine	Marine fish and shellfish, dairy products, many vegetables, iodized salt	Constituent of thyroid hormones	Goiter (enlarged thyroid)
Iron	Liver, lean meats, legumes, whole grains, dark green vegetables, eggs, dark molasses, shrimp, oysters	Constituent of hemoglobin Involved in energy metabolism	Iron-deficiency anemia
Magnesium	Whole grains, green leafy vegetables, nuts, meats, milk, legumes	Involved in muscle and nerve function Helps with heart rhythm Maintenance of bone strength	Behavioral disturbances Weakness Spasms Growth failure Cardiac arrhythmias
Phosphorus	Milk, cheese, meat, fish, poultry, whole grains, legumes, nuts	Bone and tooth formation Acid-base balance Involved in energy metabolism	Weakness Demineralization of bone
Potassium	Meats, milk, fruits, legumes, vegetables, sweet potatoes	Acid-base and water balance Nerve transmission Helps control blood pressure	Muscular weakness Paralysis
Selenium	Fish, poultry, meats, grains, milk, vegetables (depending on amount in soil)	Necessary for vitamin E function Regulates thyroid hormone	Anemia Deficiency is rare
Sodium	Common table salt, seafood, most other foods except fruit	Acid-base balance Body water balance Nerve transmission	Muscle cramps Mental apathy
Sulfur	Meat, fish, poultry, eggs, milk, cheese, legumes, nuts	Constituent of certain tissue proteins	Related to deficiencies of sulfur-containing amino acids
Zinc	Milk, liver, shellfish, herring, wheat bran	Supports nerve and immunity functions Necessary for vitamin A metabolism	Growth failure Lack of sexual maturity Impaired wound healing

Most of the essential minerals and trace elements are already present in the average normal American diet in sufficient concentrations, and supplementation is only indicated for special conditions of disease, during pregnancy, and advanced age. However, government surveys indicate that females in the United States might be consuming less than optimal daily intakes of calcium and iron.

Age-related osteoporosis is one of the most severely debilitating diseases in the United States. Although the question of whether osteoporosis is a nutritional disorder remains unanswered, there is much convincing evidence that calcium deficiency accelerates the age-related loss of bone. Menopause results in diminished calcium absorption in the intestines.

Females of childbearing age have a tendency to have low iron levels because of blood loss during the menstrual flow. Fatigue and iron-deficiency anemia in these women can usually be corrected by taking iron supplements.

VITAMINS

A **vitamin** is defined as a biologically active organic compound, often functioning as a coenzyme, that is necessary for normal health and growth. Most enzymatic activity relies on the presence of coenzymes. Although most vitamins must be obtained from the diet, some vitamins, such as vitamins K, D, and niacin, are also synthesized in the body. A dietary deficiency of a vitamin results in specific disorders. Vitamins are transported by the circulatory system to all the tissues of the body.

Recent evidence indicates that certain vitamins actually behave like hormones physiologically. As an example, vitamin D is synthesized in the body in inadequate amounts, but when the skin is exposed to sunlight, a chemical reaction occurs. The kidneys convert the chemical to an active form of a hormone. The fat-soluble vitamins A, D, E, and K are readily stored in the body; within the cell, they demonstrate many similarities to the steroid hormones estrogen, testosterone, and cortisol. The water-soluble vitamins are B_1, B_2, B_3, B_6, B_{12}, pantothenic acid, folic acid, biotin, and vitamin C. An excessive intake of water-soluble vitamins results in increased excretion rather than additional storage.

Certain conditions, such as pregnancy, disease, emotional stress, and advanced age must be considered when determining daily individual vitamin requirements. **Table 20-3** summarizes the major vitamins needed in the human diet.

EFFECTS OF ENERGY DEFICIENCIES AND EXCESSES

Energy availability is the amount of energy needed to fuel the body after use and the calories needed for activities of daily living. This energy is derived from the nutrients absorbed from food. *Energy deficiency* results in fatigue, problems with concentration, decreases in activity level, and other health issues. The cause of this deficiency may be malabsorption, or when nutrients are not absorbed in the digestive system. Many diseases can lead to this condition, such as anemia, Crohn's disease, celiac disease, and AIDS.

Excess energy intake correlates with the development of metabolic disorders, such as obesity and diabetes. Excess calories above and beyond what is required to maintain homeostasis are converted into glycogen molecules and can be stored in the body as adipose tissue. Many people today consume large amounts of energy drinks, soda, coffee, and tea, which can contain large amounts of caffeine, an organic compound. Excess amounts of caffeine can cause heart palpitations, high blood pressure, and type 2 diabetes. The FDA has suggested that a healthy adult consume no more than 400 milligrams of caffeine a day.

RECOMMENDED DAILY DIETARY ALLOWANCES

Developing universal minimum daily requirements that apply to everyone is an extremely difficult task. Nutritional requirements among individuals might vary for several reasons. Malabsorption disorders sometimes require that an individual receive greater than the average daily intake of certain nutrients. Differences in the microbial environment of the intestine and genetic factors influencing biochemical reactions must also be considered. People experiencing psychological or physical stress often require a greater amount of certain nutrients to help the body maintain homeostasis.

In recognition of individual variations in nutritional requirements, a table of Recommended Dietary Allowances (RDA), Adequate Intakes (AI), and Upper Limits (UL) has been approved by the Food and Nutrition Board of the National Academy of Sciences. It contains the daily recommendations for protein, fat-soluble vitamins, water-soluble vitamins, and minerals.

The allowances are intended to provide for individual variations among most individuals as they live in the United States under usual environmental stresses. For further information on dietary guidelines, visit http://www.usda.gov.

Did You Know?

Blue food is a rare occurrence in nature, except for blueberries and blue-purple potatoes. Consequently, there is no automatic appetite response to blue food.

Table 20-3 *Major Vitamins Needed in the Human Diet*

VITAMIN	FOOD SOURCES	FUNCTION(S)	DEFICIENCIES
A (fat soluble)	Butter, fortified margarine, green and yellow vegetables, milk, eggs, liver	Night vision Healthy skin Proper growth and repair of body tissues	Night blindness Dry skin Slow growth Poor gums and teeth
B_1 (thiamine) (water soluble)	Chicken, fish, meat, eggs, enriched bread, whole-grain cereals	Promotes normal appetite and digestion Needed by nervous system	Loss of appetite Nervous disorders Fatigue Severe deficiency causes beriberi
B_2 (riboflavin) (water soluble)	Cheese, eggs, fish, meat, liver, milk, cereals, enriched bread	Needed in cellular respiration	Eye problems Sores on skin and lips General fatigue
B_3 (niacin) (water soluble)	Eggs, fish, liver, meat, milk, potatoes, enriched bread	Needed for normal metabolism Growth Proper skin health	Indigestion Diarrhea Headaches Mental disturbances Skin disorders
B_{12} (cyanocobalamin) (water soluble)	Milk, liver, brain, beef, egg yolk, clams, oysters, sardines, salmon	Red blood cell synthesis Nucleic acid synthesis Nerve cell maintenance	Pernicious anemia Nerve cell malfunction
Folic acid (water soluble)	Liver, yeast, green vegetables, peanuts, mushrooms, beef, veal, egg yolk	Nucleic acid synthesis Needed for normal metabolism and growth	Anemia Growth retardation
C (ascorbic acid) (water soluble)	Citrus fruits, cabbage, green vegetables, tomatoes, potatoes	Needed for maintenance of normal bones, gums, teeth, and blood vessels	Weak bones Sore and bleeding gums Poor teeth Bleeding in skin Painful joints Severe deficiency results in scurvy
D (fat soluble)[a]	Beef, butter, eggs, milk	Needed for normal bone and teeth development Controls calcium and phosphorus metabolism	Poor bone and teeth structure Soft bones Rickets
E (tocopherol) (fat soluble)	Margarine, nuts, leafy vegetables, vegetable oils, whole wheat	Used in cellular respiration Protects red blood cells from destruction Acts as an antioxidant	Anemia in premature infants No known deficiency in adults
K (fat soluble)	Synthesized by colon bacteria Green, leafy vegetables; cereal	Essential for normal blood clotting	Slow blood clotting

[a]The role of vitamin D in insulin resistance, hypertension, and immune function is under investigation by researchers.

The Effects of Aging on Nutrition

Many factors affect the diet of older adults. As stated previously, gum disease, loss of teeth, and decrease in sensitivity of taste buds can lead to poor nutrition. Other factors, such as chronic disease; medication use; and societal, economic, physical, and emotional factors also affect nutrition. An example of a medication is digoxin, which treats heart failure but suppresses the appetite. Seniors may also lose their taste for meat, fish, or beans, which provide the protein that is necessary to build and repair tissue. Arthritis, heart disease, or other ailments make cooking a physical challenge. The loss of a spouse or partner may make an individual not want to bother cooking for oneself. Economic factors and availability also influence the choices people make regarding the buying of food.

Health care providers should be aware of the nutritional needs of the elderly, especially the need for vitamin D, calcium, and vitamin B12.

BASAL METABOLIC RATE

The **basal metabolic rate (BMR)** is the measure of the total energy utilized by the body to maintain the body processes necessary for life, the minimum level of heat produced by the body at rest. BMR is the number of calories needed to keep the heart pumping, to keep breathing, carry out all activities of daily living, and maintain homeostasis.

The purpose in determining the BMR is to calculate basic caloric needs for a person. A way to estimate the BMR is as follows:

For females: $661 + (4.38 \times \text{weight in pounds}) + (4.33 \times \text{height in inches}) - (4.7 \times \text{age}) = \text{BMR}$

For males: $67 + (6.24 \times \text{weight in pounds}) + (12.7 \times \text{height in inches}) - (6.9 \times \text{age}) = \text{BMR}$

Next, estimate the total number of calories the body needs per day by multiplying the BMR by the appropriate factor, as shown:

- 1.2 for an inactive person
- 1.3 for a moderately active person, or someone who exercises three times per week
- 1.7 for a very active person
- 1.9 for an extremely active person, such as a runner or swimmer

This method is one way to measure the number of calories burned each day. If the metabolic rate is lower than the calories supplied by food, the excess calories are converted to fats and weight increases. If all the food calories are burned, weight is maintained. Burning more calories than are supplied by the diet will result in weight loss.

BODY MASS INDEX

Obesity is a major health concern for the American people. Sixty-nine percent of Americans are overweight or obese. Obesity has direct links to serious illnesses such as type 2 diabetes, high blood pressure, coronary artery disease, and stroke. With obesity, the liver makes more triglycerides and less HDL. There is an increased risk of gallstones and sleep apnea. Researchers have found that body fat is a better predictor of health than body weight. **Body mass index (BMI)** relates body weight with health risks of being overweight. A method to determine BMI is as follows:

1. Multiply height $\times$ height in inches (e.g., 5 feet $\times$ 5 feet $=$ 60 inches $\times$ 60 inches $=$ 3600).
2. Divide weight by the previous answer (e.g., 150 lbs divided by 3600 $=$ 0.041).
3. Multiply the answer by 703 (e.g., $0.041 \times 703 = 28.28$). The result is the BMI.

Generally, a healthy BMI ranges from 19 to 25.

DIETARY GUIDELINES FOR AMERICANS

The dietary guidelines for most Americans are based on MyPlate, the current nutrition guide from the United States Department of Agriculture (USDA) Center for Nutrition Policy and Promotion (**Figure 20-1**). The plate idea allows a person to visualize how the food and the amount should appear on their plate. It is divided into five areas: fruits, vegetables, grains, proteins, and dairy. In general, the USDA food patterns provide recommended intake from each food group at various calorie levels. For example, at the 2000-calorie level, the recommended daily intake includes 2 cups of fruit, 2.5 cups of vegetables, 6 ounces of grains, 5.5 ounces of protein, and 3 cups of dairy.

The guidelines do not include a specific recommendation for water intake. The guidelines do state, however, that healthy individuals in general should have an adequate intake of water to meet their individual

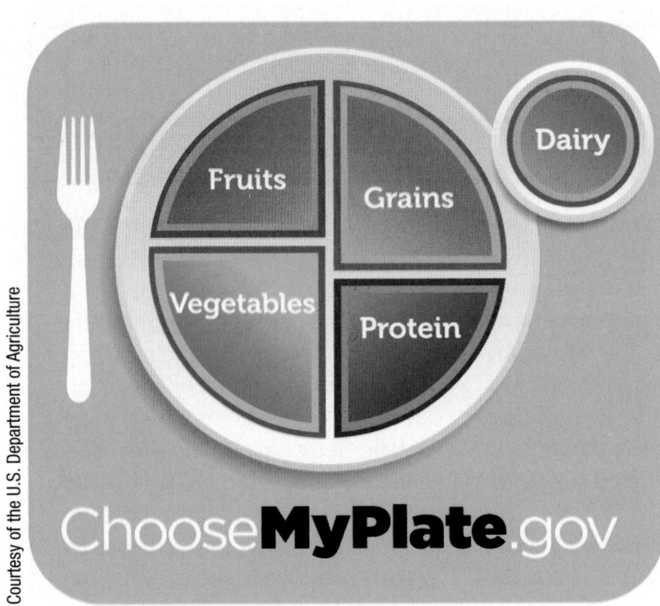

Figure 20-1 *The MyPlate icon is a visual tool to help individuals make better food choices.*

needs. The combination of thirst and typical behaviors such as drinking liquids with meals should provide sufficient water intake.

To build a healthy plate, the following is suggested:

- Make half the plate fruits and vegetables—eat red, orange, and dark green vegetables. Eat fruits, vegetables, or unsalted nuts as snacks.
- Switch to skim or 1% milk.
- Make at least half the grains whole grains, such as whole-grain cereals, breads, rice, and pasta.
- Vary the protein food choices. Twice a week, make seafood the protein; beans are also a natural source of fiber and protein. Keep meat and poultry portions small and lean.

Central to the MyPlate guidelines is the idea of individuality. To view individual dietary needs, go to the MyPlate website (http://www.choosemyplate.gov).

The importance of consuming a diet of a variety of foods to provide the essential nutrients at a caloric level to maintain desirable body weight is emphasized.

Some suggestions for an overall healthy lifestyle include the following:

- Let the MyPlate icon guide food choices.
- Be physically active. It is recommended that the average adult get the following amount of exercise per week: 150 minutes of moderate exercise, 75 minutes of vigorous exercise, or a combination.
- Keep food safe to eat.
- Choose a diet that is low in saturated fat, trans fat, and cholesterol, and that is moderate in total fat.
- Choose and prepare foods with less salt.
- If an adult chooses to drink alcoholic beverages, they should be consumed in moderation.

ORGANIC FOODS

Organic food is produced by farmers who emphasize the use of renewable resources and the conservation of soil and water. Organic meat, poultry, eggs, and dairy products come from animals that are given no antibiotics or growth hormone. Organic food is produced without using most conventional pesticides, fertilizers made with synthetic ingredients or sewage sludge, bioengineering, or ionizing radiation. Before a product is labeled "organic," a government-approved certifier inspects the farm where the food is grown to make sure the farmer is following all the rules necessary to meet USDA organic standards. Companies that handle or process organic food must also be certified. The use of the organic seal is voluntary; however, anyone who sells a product as "organic" that does not meet the USDA standards can be fined up to $11,000 for each violation.

The terms "natural" and "organic" are not interchangeable. Natural food has undergone zero or minimal processing and contains no additives, such as preservatives or artificial coloring.

PLANT-BASED FOODS

Plant-based food plans are diets consisting mostly of foods derived from plants, including vegetables, grains, and nuts, with few or no animal products. A plant-based diet is not necessarily a vegetarian diet, since it allows for some meat products. A plant-based diet focuses on healthy whole food rather than processed foods.

Recent offerings of plant-based burgers, such as Beyond Meat® and Impossible Burgers®, are designed to look and taste like meat and serve as an alternative source of protein. These products have ample protein, no cholesterol, and essential dietary minerals; however, they are processed, which means they may have more calories and sodium than red meat.

A plant-based diet is associated with a lower risk of heart disease, but it is important to remember to focus on the quality of plant food.

Career Profile

DIETITIAN AND NUTRITIONIST

Dietitians and nutritionists are both food and nutrition experts, but the two titles should not be used interchangeably. Dietitians may plan nutrition programs and supervise the preparation and serving of meals. They help prevent and treat illnesses by promoting healthy eating habits, scientifically evaluating patients' diets, and suggesting modifications such as reduced fat and sugar for those who are overweight.

Dietitians run food service systems for institutions such as hospitals and schools, and they also promote sound eating habits through education and research. Dietitians are also employed at large retail food stores to provide nutrition counseling when necessary.

A nutritionist is someone who has studied nutrition and may have a graduate degree. Some health care providers may have completed extra study in the area of nutrition and may practice *clinical nutrition*, which is often considered a type of alternative medicine.

Dietitians are considered nutritionists, but not all nutritionists are dietitians. Both dieticians and nutritionists can be meal planners.

According to the Academy of Nutrition and Dietetics, a registered dietitian must have a bachelor's degree with a major in dietetics, food, nutrition, or food service systems management. The dietitian must have completed an accredited practice program in a health care setting, community agency, or food service corporation and passed a national examination administered by the Commission on Dietetic Registration. The American Dietetic Association (ADA) awards the Registered Dietitian credential to those who pass the certification exam. According to the U.S. Bureau of Labor Statistics, 46 states have laws governing the dietetics profession. Only 33 states require dietitians to have a license.

Employment is expected to grow about as fast as the average for dietitians and nutritionists.

NUTRITION LABELING

The U.S. Food and Drug Administration (FDA) mandates nutrition labeling for most foods offered for sale and regulated by the FDA (**Figure 20-2**). The nutrition label is required to include the following information in order: calories; the amount of calories from total fat, such as saturated fat and trans fat; cholesterol; sodium; total carbohydrate; dietary fiber; total sugars, including added sugars; protein; vitamin D; calcium; iron; and potassium. The information on the package is to represent the packaged product prior to consumer preparation.

This final rule establishes a standard format for nutrition information on food labels, consisting of the following:

1. Quantitative amount per serving of each nutrient
2. Amount of each nutrient as a percent of the daily value for a 2000-calorie diet
3. Footnote with reference values for selected nutrients based on a 2000-calorie diet

On the food product label, the ingredients are listed in order of prominence, with the ingredients present in the greatest amount first, followed in descending order by those present in smaller amounts. Sometimes, the first ingredient may be water.

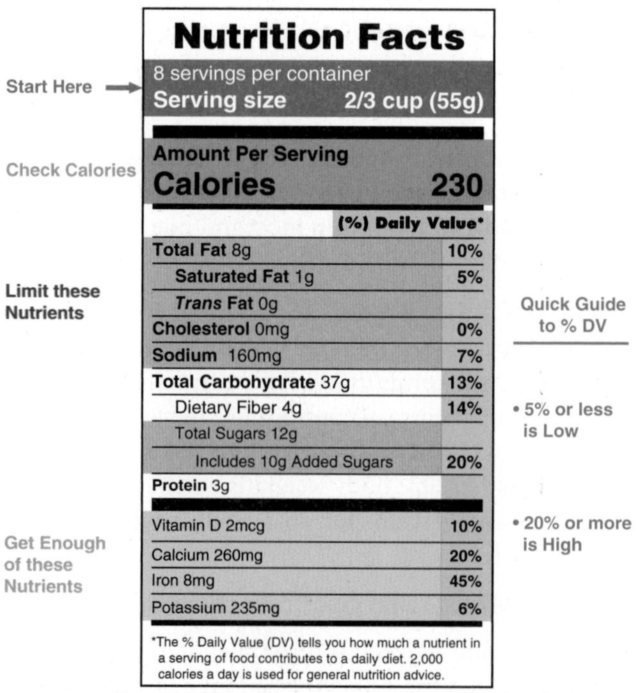

Figure 20-2 *Sample nutrition label*

FOOD SAFETY AND POISONING

According to the Centers for Disease Control, foodborne microorganisms cause 48 million illnesses every year in the United States. In most cases, symptoms of food poisoning resemble intestinal flu and last a few hours to several days. Microscopic organisms can grow undetected in food because they do not produce an odor or a difference in color or texture. These microbes can be prevented from contaminating food by proper storing and handling. The single most important thing is thorough handwashing before handling food. Scrub the hands with soap and water for at least 20 seconds after handling raw meat, fish, or poultry. Alcohol-based, rinse-free hand sanitizers should be used when handwashing with soap is not possible. Clean all cooking surfaces and utensils properly. For food safety, thoroughly cook all meat, poultry, eggs, and shellfish. Don't leave food at room temperature for more than two hours. Refrigerate food below 4°C (40°F). This will help stop the growth of most organisms that cause illness. Perishable frozen food should never be thawed on the kitchen counter or in hot water; frozen food should thaw in the refrigerator where it will remain in a safe, consistent temperature. Observe "sell by" and "use by" dates on labels.

OBESITY

Obesity (oh-**BEE**-sih-tee) is one of the most common "nutritional diseases." The most recent data suggest that 78 million adults and 12 million children are obese, figures many regard as an epidemic. The National Institutes of

Medical Highlights

ANTIOXIDANTS

Antioxidants in the form of certain vitamins, minerals, or plant chemicals (phytonutrients) that act like antioxidants may prevent, delay, or repair some types of cell damage.

Natural cell metabolism happens 24 hours per day. In the process of oxidation, where cells take in oxygen to release energy, there may be the formation of unstable by-products, called free radicals. While the body metabolizes oxygen efficiently, 1% to 2% may turn into free radicals. Free radicals can get into the DNA of a cell and create the seeds for a disease. Antioxidants work to stop this damage, causing a chain reaction that the free radicals have started. Antioxidants either prevent the chain reaction or stop it once it has started.

Daily food choices present opportunities to score some healthy, flavorful, and antioxidant-positive nutrition points. Among these choices are the following:

- *Coffee*—chemicals in a cup of coffee, including caffeine, behave as antioxidants. Antioxidant activity associated with coffee has been linked to protective effects on multiple diseases, including cancer and cardiovascular diseases.
- *Tea*—produced from the leaves of the camellia sinensis bush, tea is loaded with flavonoids and other polyphenols, which work as antioxidants. Drinking black tea may lower the risk of heart attack and atherosclerosis. Drinking green or black tea may lower the risk of several cancers.
- *Berries*—blueberries and strawberries are rich in anthocyanins, which appear to have heart-healthy effects. All berries offer high levels of antioxidants.
- *Dark Chocolate*—dark chocolate has more antioxidants than blueberries and raspberries and reduces the risk of heart disease.
- *Curcumin*—this substance is found in the spice turmeric, which is the main spice in curry. Curcumin is thought to have antioxidant properties that may decrease swelling and inflammation.
- *Cruciferous vegetables*—these include broccoli, cauliflower, Swiss chard, kale, and turnips. Much of the research on these foods demonstrates their protective effects in preventing certain types of cancer. Kale is a great source of calcium, which helps to keep bones healthy.
- *Pecans*—these nuts raise the antioxidants levels in the blood.
- *Beans (legumes)*—these are great protein substitutes, plus they have phytonutrients. They have antioxidant and anti-inflammatory properties, which support cardiovascular health.
- *Spinach*—spinach has antioxidant properties that help combat damage to the eyes.

There is overwhelming evidence that a diet rich in plant-based foods—fruits, vegetables, legumes, nuts, seeds, and whole grains—offers health benefits in addition to antioxidant content.

Mayo Clinic. "Antioxidants: Reap the Benefits from Food Sources." Mayo Clinic Health Letter. November 01, 2013.

Health defines overweight as a BMI of between 25 and 29.9 and obesity as a BMI of 30 or above.

Obesity can affect physical and mental health. Heart disease, high blood pressure, and noninsulin-dependent diabetes mellitus are more common in significantly overweight people than in those closer to their ideal body weight.

Many experts believe that obesity is due to environmental factors. Some of these contributing factors are the following:

- *Overconsumption*—During the past 20 years, society has begun to supersize everything. Muffins went from being 3 ounces to 6 ounces. An average portion of meat should be 3 ounces; instead, meat portions are 7–8 ounces.
- *Inactivity*—Nearly 60% of adults lead a sedentary lifestyle. The prevalence of obesity is four times greater among individuals who watch 21 hours or more of TV per week.

Steps to correct this problem include the following:

- Replace junk food with fresh fruits and vegetables.
- Keep a record of foods eaten and know the size of a typical portion.
- Increase the amount of exercise each day.

By reducing daily caloric intake by 500 calories per day, in one week a person will lose one pound (3500 calories equals 1 pound of weight). Individuals should consult their doctors before beginning any type of diet.

See **Table 20-4** for a list of popular diets.

Table 20-4 *Popular Diets*

NAME OF DIET	BASIC PRINCIPLES
Atkins® Diet	Based on a very low carbohydrate intake that avoids foods with refined flour or sugar and encourages the use of lean protein and low-starch vegetables; divided into four phases, with each phase becoming less restrictive
DASH Diet (Dietary Approaches to Stop Hypertension)	Focuses on eating low-fat dairy, fruits, vegetables, lean meats, and whole grains while limiting salty foods and processed foods; promoted as a heart-healthy diet to lower hypertension, or high blood pressure
Gluten-free Diet	Used as treatment for individuals with celiac disease or a sensitivity to gluten; others promote it as a diet to improve health, lose weight, and increase energy; eliminates gluten-grains such as wheat, barley, rye, and triticale; allows fruits, vegetables, beans, nuts, eggs, lean meats, fish, poultry, and most low-fat dairy products
Ketogenic Diet	Focuses on eating a high-fat, adequate protein, but very low-carbohydrate diet to lose weight; forces body into ketosis where fats and proteins are burned for energy instead of carbohydrates; encourages avoiding foods such as low-fat dairy, most fruits, grains, pasta, rice, starchy vegetables, beans, and foods high in sugar or sweeteners
Mediterranean Diet	Promoted as a heart-healthy diet that emphasizes vegetables, fruits, whole grains, beans, nuts, fish, and olive oil; encourages moderate comsumption of poultry, cheese, and eggs; suggests eating red meat and sweets only on special occasions
Nutrisystem® Diet	Commercial diet that allows clients to choose prepared meals from a list; the number and types of meals are selected based on the client's body build and weight loss goal; meals stress healthy carbohydrates and lean proteins; more expensive than other diets
Paleo Diet®	Also called the Caveman Diet because it concentrates on basic foods eaten by early humans; encourages eating fresh fruits and vegetables, seafood, lean meat, and healthy fats
Probiotic Diet	Encourages foods rich in probiotics or healthy bacteria; main sources are cultured foods such as yogurt and buttermilk, fermented vegetables such as pickled beets and sauerkraut, and microalgae or ocean-based plants such as chlorella and blue-green algae; promoted as a diet that can improve digestion, boost the immune system, and possibly reduce the risk of cancer
SlimFast® Diet	Commercial diet that provides shakes and meal bars for two meals and two or three snacks each day; only one additional 500-calorie meal is eaten per day
South Beach Diet®	Focuses on replacing bad carbohydrates and fats with good carbohydrates and fats; encourages eating lots of vegetables, fish, eggs, lean meats such as chicken, whole grains, and low-fat dairy products; divided into three phases, with each phase becoming less restrictive
Volumetrics Diet	Based on classifying foods into four density levels that range from low-density foods in level 1 such as fruits, vegetables, and nonfat milk to high-density foods in level 4 such as cookies, butter, and candy; encourages concentrating on foods in the lower density levels
WW (formerly Weight Watchers®)	Uses a point system to track foods eaten; encourages a balanced diet but stresses fruits, vegetables, lean meat, and low-fat dairy; provides group support to help an individual achieve their weight goal
Zone Diet®	Stresses five to six meals with smaller quantities; suggests that each meal contains 40% carbohydrates, 30% protein, and 30% healthy fat

Weight Loss Surgery/Bariatric Surgery

Doctors only recommend this type of surgery if an individual has a BMI of 40 or more—this would be equal to being about 100 pounds overweight. Some people with a lower BMI may have the surgery if they have a risk for heart disease or diabetes.

Gastric bypass surgery or bariatric surgery helps by the following:

1. *Restriction*—surgery is used to physically limit the amount of food the stomach can hold. A normal stomach holds 4 cups of food; after surgery, the stomach can hold only 1 cup of food.
2. *Malabsorption*—surgery shortens or bypasses part of the small intestines, which lowers the amount of calories and nutrients the body absorbs.

Types of surgery are gastric bypass, gastric banding, sleeve gastrectomy, and duodenal switch with biliopancreatic diversion.

One of the most common side effects is "dumping syndrome," which occurs in about 50% of people who have the surgery. The dumping syndrome occurs because food moves too quickly through the digestive system. Symptoms include nausea, sweating, fainting, and diarrhea after eating. Avoiding foods that are high in carbohydrates and replacing them with foods that are high in fiber may help prevent this syndrome.

EATING DISORDERS

Most people with eating disorders have distortions of their body image that can cause them to change exercise or eating habits. They may be underweight, overweight, or normal weight.

The three major types of eating disorders are the following:

- **Anorexia nervosa** (an-oh-**REK**-see-ah ner-**VOH**-sah) encompasses behaviors like dieting; fasting; overexercising; taking diet pills, diuretics, or laxatives; and vomiting. Characteristic symptoms include constant disdain for their bodies, maintaining a body weight 15% below the average for their age and height, missing three menstrual cycles in women, and an intense fear of gaining weight. It is important to note that anorexia nervosa can affect men and boys as well as woman and girls. Approximately one out of three individuals suffering from this disorder is male.
- **Bulimia nervosa** (byou-**LIM**-ee-ah) is characterized by episodic binge eating followed by purging behavior, such as self-induced vomiting, excessive exercise, and laxative abuse. Unlike patients with anorexia, patients with bulimia may be overweight, underweight, or normal weight.
- **Binge eating disorder** occurs when a person repeatedly eats unusually large amounts of food in a short period of time. This behavior occurs two times a week for six months or more. People with binge eating disorder do not purge. Characteristics are eating in secret and intense satisfaction when eating, followed by remorse and mood swings.

Treatment for eating disorders is difficult and lengthy. Treatment includes a combination of psychotherapy, nutrition education, and medical monitoring; in some cases, medications may be used. The goals are restitution of normal nutrition and resolution of the underlying psychological and physical problems.

Medical Terminology

BMI	body mass index	**VLDL**	very low density lipoprotein
BMR	basal metabolic rate	**milli-**	thousand
HDL	high-density lipoprotein	**milligram**	one-thousandth of a gram
LDL	low-density lipoprotein		

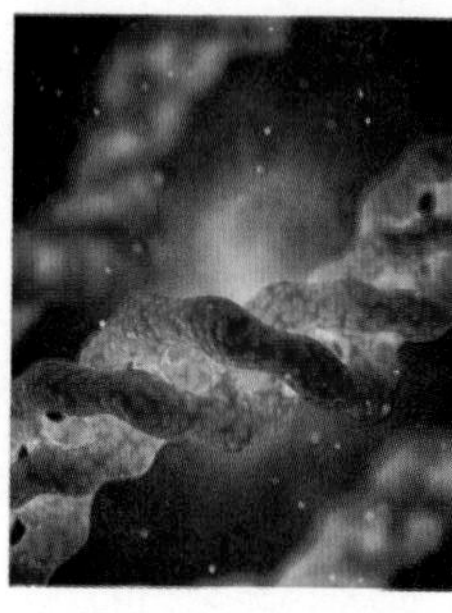

Study Tools

Workbook	Activities for Chapter 20
Online Resources	PowerPoint presentations

REVIEW QUESTIONS

Select the letter of the choice that best completes the statement.

1. Materials needed by the individual cells for proper cell function are
a. proteases.
b. enzymes.
c. amylases.
d. nutrients.

2. A gram of fat contains
a. 9 Calories.
b. 4 Calories.
c. 5 Calories.
d. 7 Calories.

3. The main source of energy for the body is provided by
a. fats.
b. carbohydrates.
c. proteins.
d. water.

4. To build and repair body tissue, you need
a. fats.
b. carbohydrates.
c. proteins.
d. water.

5. The most common bone disease is
a. osteomyelitis.
b. fracture.
c. osteoporosis.
d. bone cancer.

6. The minerals necessary to build bone and teeth are
a. iodine and calcium.
b. calcium and potassium.
c. calcium and phosphorus.
d. fluorine and calcium.

7. Iodine is required for the formation of the
a. adrenal hormone.
b. thyroid hormone.
c. parathyroid hormone.
d. pituitary hormone.

8. A vitamin needed to prevent night blindness is
a. vitamin A.
b. vitamin K.
c. vitamin C.
d. vitamin D.

9. The vitamin essential for blood clotting is
a. vitamin A.
b. vitamin K.
c. vitamin C.
d. vitamin D.

10. A loss of 1 pound per week can be accomplished by reducing daily calorie intake by
a. 100 calories.
b. 200 calories.
c. 400 calories.
d. 500 calories.

APPLYING THEORY TO PRACTICE

1. Go to MyPlate.gov and make a 3-day meal plan, including between-meal snacks, that will meet both recommended calorie intake and dietary allowances for a 12-year-old male with a height of 62 inches. Adjust this diet to meet the needs of a 70-year-old female with a height of 60 inches.
2. A patient has anemia. List the minerals and vitamins that will assist in the formation of red blood cells.
3. Research local restaurants to see if they offer healthy alternatives. Try to obtain the nutrition information or ingredient list and determine if they really are healthy meals.
4. A physician orders a diet of 20 grams of protein, 300 grams of carbohydrate, and 80 grams of fat. What are the total calories, and how much caloric value is there in protein, carbohydrate, and fat? Calculate the percentage of protein, carbohydrate, and fat.
5. Determine your individual BMR and BMI. Consult and use the method in this chapter. After calculating your BMR and BMI, go to MyPlate.gov and make a list of the information available to help someone maintain food nutrition and healthy weight. What does MyPlate recommend for you?

CASE STUDY

Victoria is a 60-year-old female who weighs 200 pounds and is 5 feet 5 inches tall. She has recently been diagnosed with type 2 diabetes. Her physician states that it is necessary for her to lose weight and sends her to see Rami, the health maintenance organization (HMO) nutritionist. Rami first discusses with Victoria what makes up a balanced diet. In addition, Rami will help Victoria determine her BMR and her caloric needs to reach and maintain a healthy weight.

1. Discuss the role of carbohydrates, fats, and proteins in the diet.
2. What is the importance of healthy cholesterol and triglyceride blood levels?
3. Victoria eats many Filipino dishes. Research foods often found in Filipino cuisine. What familiar foods can she incorporate into her diet? What foods should she avoid?
4. What is the recommended weight for Victoria according to weight and height scales?
5. Explain the term "Recommended Dietary Allowances."
6. What is Victoria's BMR? What factors influence the BMR?

LAB ACTIVITY 20-1 Test for Starch

- ***Objective:*** To observe the presence of starch in common foods
- ***Materials needed:*** paper plates, tincture of iodine solution, medicine dropper, gloves, slices of potato and white bread, cracker, pat of butter, sugar, paper, pencil
- ***Note:*** Tincture of iodine is a poison; it may cause burns to the skin, and it permanently stains clothing. Starch is a type of carbohydrate that will turn blue-black in the presence of an iodine solution.

Step 1 Arrange food samples on the paper plate.

Step 2 Put on gloves and, using the medicine dropper, carefully place a drop of iodine solution on each of the food samples.

Step 3 Observe the change in color of the food samples.

Step 4 Wash the medicine dropper, remove the gloves, and wash your hands.

Step 5 Record your observations.

Step 6 Which food samples contain starch? Record your answers.

LAB ACTIVITY 20-2 Test for Lipids or Fats

- ***Objective:*** To observe the presence of fat in common foods
- ***Materials needed:*** brown paper, slice of raw potato, teaspoon of cooking oil, butter, corn kernels, bean seeds, a light source, paper, pencil

Step 1 Place a few drops of cooking oil on brown paper. Hold the paper up to a light and look through the spot. Record your observations.

Step 2 Repeat the process with the butter, bean seeds, raw potato, and corn kernels by rubbing these foods onto the brown paper. Hold the paper up to the light.

Step 3 Record your observations.

Step 4 Which food samples contain lipids or fats? Record your answers.

LAB ACTIVITY 20-3 Test for Proteins

- ***Objective:*** To observe the presence of proteins in common foods
- ***Materials needed:*** biuret solution, test tubes, medicine dropper, white paper, small pieces of raw potato, teaspoon of cooking oil, butter, corn kernels, bean seeds, marking pen, labels, protective gloves, paper, pencil
- ***Note:*** Biuret solution is dangerous. It will burn clothing and skin, and it turns violet in the presence of protein.

Step 1 Place food samples in test tubes and label.

Step 2 Put on protective gloves.

Step 3 Using a medicine dropper, carefully add 10 drops of biuret solution to each test tube.

Step 4 Hold each test tube against the sheet of white paper.

Step 5 Record the color changes, if any, that occur.

Step 6 Properly dispose of contents of test tubes.

Step 7 Remove the protective gloves.

Step 8 What food samples contain protein? Record your answers.

CHAPTER 21

Urinary System

Objectives

- Explain the function of the urinary system.
- Describe the structure and function of the organs in the urinary system.
- Explain how the kidneys regulate water balance.
- List and describe some common disorders of the urinary system.
- Define the key words that relate to this chapter.

Key Words

acute glomerulonephritis
acute kidney failure
afferent arteriole
aldosterone
anuria
Bowman's capsule
calyces
chronic glomerulonephritis
chronic renal failure
collecting tubule
congenital anomalies of kidney and urinary tract (CAKUT)
cortex
cystitis or urinary tract infection
dialysis
dialyzer
distal convoluted tubule
dysuria
efferent arteriole
extracorporeal shock wave lithotripsy (ESWL)
filtrate
glomerulonephritis
glomerulus
hematuria
hemodialysis
hilum
hydronephrosis
hypospadias
incontinence
kidney dysplasia
kidney stones (renal calculi)
kidneys
loop of Henle
medulla
micturition
nephron
neurogenic bladder
nocturia
oliguria
osmoreceptors
overactive bladder
peritoneal dialysis
proximal convoluted tubule
pyelitis
pyelonephritis
pyuria
renal columns
renal fascia
renal papilla
renal pelvis
renal pyramids
renin
retroperitoneal
threshold
uremia
ureters
urethra
urinalysis
urinary bladder
urinary meatus
urine
vesicoureteral reflux (VUR)

After the cells of tissue have used the food and oxygen needed for growth and repair, the waste products must be removed and excreted from the body. The excretory organs through which elimination takes place include the lungs, kidneys, skin, and intestines. The urinary system functions largely as an excretory agent of nitrogenous wastes, salts, and water. The lungs give off carbon dioxide and water vapor during exhalation, and the skin excretes dissolved wastes present in perspiration, mostly dissolved salts. The indigestible residue, water, and bacteria are excreted by the intestines. The excretion of waste products is described and summarized in **Table 21-1**.

Table 21-1 *Elimination of Waste Products*

ORGAN	*PRODUCTS OF EXCRETION*	*PROCESS OF ELIMINATION*
Lungs	Carbon dioxide and water vapor	Exhalation
Kidneys	Nitrogenous wastes and salts dissolved in water to form urine	Urination
Skin	Dissolved salts	Perspiration
Intestines	Solid wastes and water	Defecation

URINARY SYSTEM

The urinary system performs the main part of the excretory functions in the body (**Figure 21-1**). The most important excretory organs are the **kidneys**. If the kidneys fail to function properly, toxic wastes start to accumulate in the body. Toxic wastes accumulating in the cells cause them to "suffocate" and literally poison themselves.

The urinary system consists of two kidneys, which form the urine; two ureters; a bladder; and a urethra. Each kidney has a long, tubular ureter that carries urine to the urinary bladder. This is a temporary storage sac for urine, from which urine is excreted through the urethra.

Functions of the Urinary System

The following are the functions of the urinary system:

- Performs excretion, which is the process of removing nitrogenous waste material, certain salts, and excess water from the blood
- Aids in maintaining acid-base, or electrolyte, balance by evaluating elements in the blood and selectively reabsorbing water and other substances to maintain the pH balance

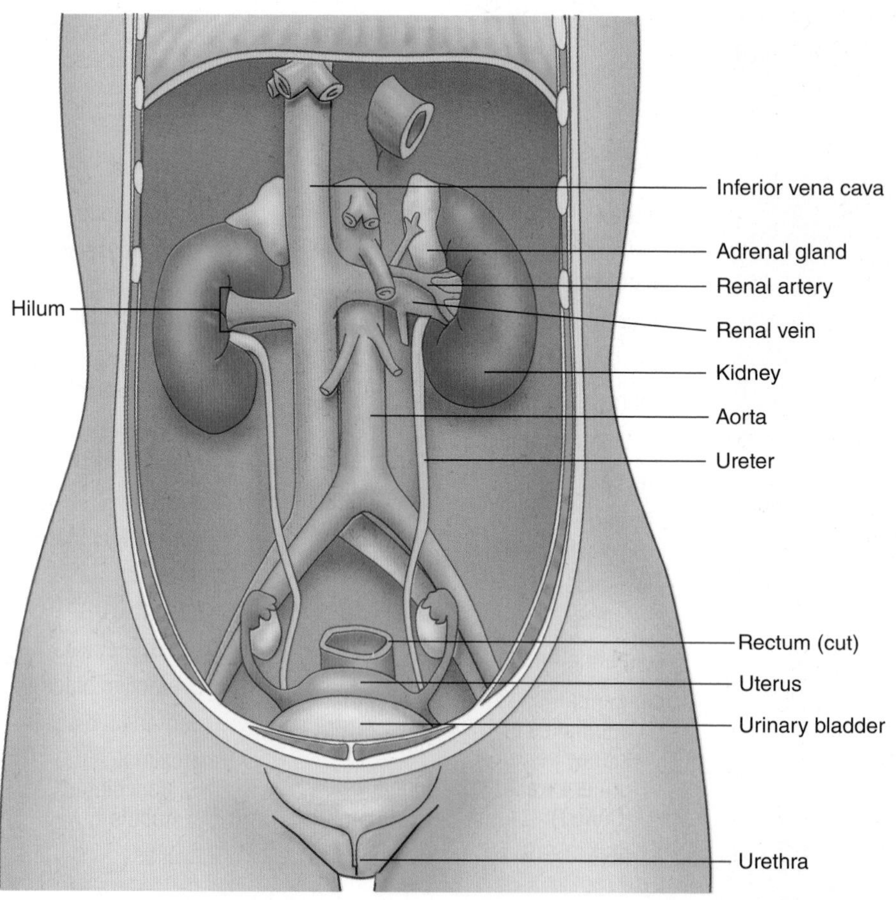

Figure 21-1 *The organs of the urinary system of a female*

- Produces the enzyme renin (**REN**-in), which helps maintain blood pressure through the filtration process
- Produces erythropoietin (eh-**rith**-roh-**POY**-eh-tin), a hormone that stimulates red blood cell production in the red bone marrow
- Helps with bone health by controlling calcium and phosphorous
- Secretes waste products in the form of urine
- Eliminates urine from the bladder where it is stored

Kidneys

The kidneys are bean-shaped organs resting high against the dorsal wall of the abdominal cavity. They lie on either side of the vertebral column, between the peritoneum and the back muscles. Because the kidneys are located behind the peritoneum, they are said to be **retroperitoneal** (**ret**-roh-**per**-ih-toh-**NEE**-al). They are positioned between the twelfth thoracic and the third lumbar vertebrae. The right kidney is situated slightly lower than the left due to the large area occupied by the liver.

Each kidney and its blood vessels are enclosed within a mass of fat tissue called the adipose capsule. In turn, each kidney and adipose capsule is covered by a tough, fibrous tissue called the **renal fascia** (**REE**-nal **FASH**-ee-ah). The term *renal* means "pertaining to the kidney."

There is an indentation along the concave medial border of the kidney called the **hilum** (**HIGH**-lum) (**Figure 21-1**). The hilum is a passageway for the lymph vessels, nerves, renal artery and vein, and the ureter. At the hilum, the fibrous capsule continues downward, forming the outer layer of the ureter. Cutting the kidney in half lengthwise will reveal its internal structure. The upper end of each ureter flares into a funnel-shaped structure known as the **renal pelvis** (**Figure 21-2**).

The kidneys have the potential to work harder than they actually do. Under ordinary circumstances, only a portion of the nephron, the functional unit of the kidney, is used. Should one kidney become diseased or have to be removed, such as in the case of a kidney organ affected by diabetes, more nephrons and tubules open up in the second kidney to assume the work of the nonfunctioning or missing kidney.

MEDULLA AND CORTEX

The kidney is divided into two layers: an outer, granular layer called the **cortex** (**KOR**-tex) and an inner, striated layer called the **medulla** (meh-**DULL**-ah). The bulk of the red medulla consists of radially striated cones called the **renal pyramids**. The base of each renal pyramid faces the cortex, while its apex, the **renal papilla** (pah-**PILL**-ah), empties into surrounding cuplike cavities called minor **calyces** (**KAY**-lik-sees). Each minor calyx collects urine from the ducts of the pyramids. Minor calyces join to

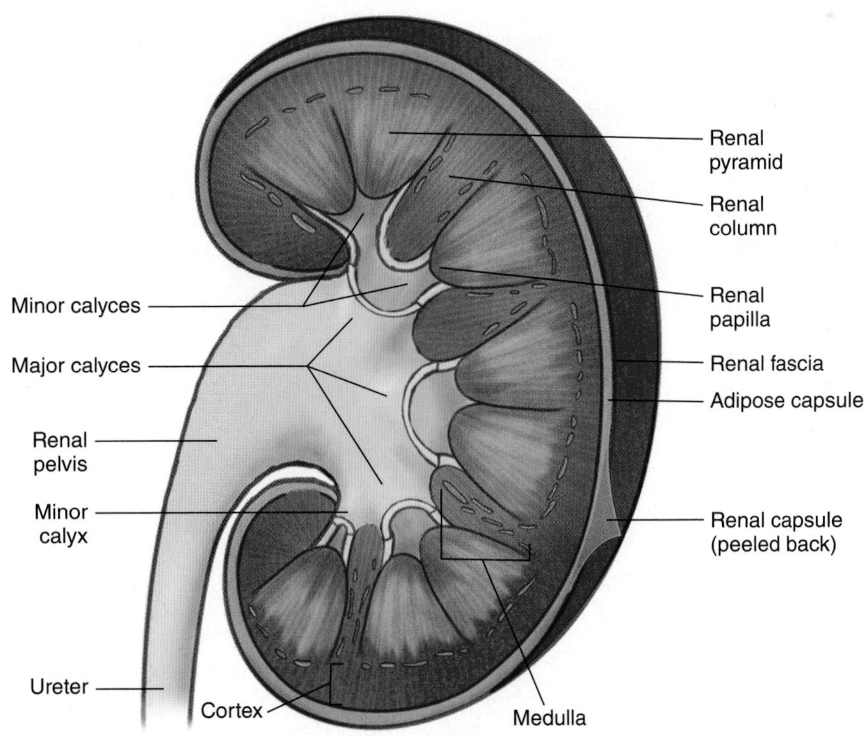

Figure 21-2 *The internal structure of a kidney*

form major calyces. The major calyces join to form the large collecting funnel: the renal pelvis.

The cortex is reddish brown and consists of millions of microscopic, functional units of the kidney called *nephrons*. Cortical tissue is interspersed between renal pyramids, separating and supporting them. These are called **renal columns**. Renal columns and renal pyramids alternate with one another (**Figure 21-2**).

NEPHRON

The **nephron** (**NEFF**-ron) is the basic structural and functional unit of the kidney. Each kidney has more than one million nephrons, which altogether comprise 140 miles of filters and tubes.

A nephron begins with the **afferent arteriole**, which carries blood from the renal artery. The afferent arteriole enters a double-walled hollow capsule called the **Bowman's capsule**, named for the English anatomist Sir William Bowman (1816–1892). Within the capsule, the afferent arteriole finely divides, forming a knotty ball called the **glomerulus** (gloh-**MARE**-you-lus) that contains some 50 separate capillaries. The combination of the Bowman's capsule and the glomerulus is known as the renal corpuscle. The Bowman's capsule sends off a highly convoluted (twisted) tubular branch referred to as the **proximal convoluted tubule** (**PROK**-sih-mal con-voh-**LOO**-ted **TOO**-byool).

The proximal convoluted tubule descends into the medulla to form the **loop of Henle**. In **Figure 21-3**, observe that the loop of Henle has a straight descending limb, a loop, and a straight ascending limb. When the ascending limb of Henle's loop returns to the cortex, it turns into the **distal convoluted tubule**. Eventually the proximal convoluted tubule and several distal convoluted tubules open into a larger, straight vessel known as the **collecting tubule**. The collecting tubule empties into the papillae, then the calyces, the renal pelvis, and finally the ureter.

The walls of the renal tubules are surrounded by capillaries. After the afferent arteriole branches out to form the glomerulus, it leaves the Bowman's capsule as the **efferent arteriole**. The efferent arteriole branches to form the peritubular capillaries surrounding the renal tubules. All of these capillaries eventually join to form a small branch of the renal vein, which carries blood from the kidney.

Did You Know?

The kidneys use more energy than the heart. The kidneys use 12% of a person's oxygen, yet the heart uses only 7%.

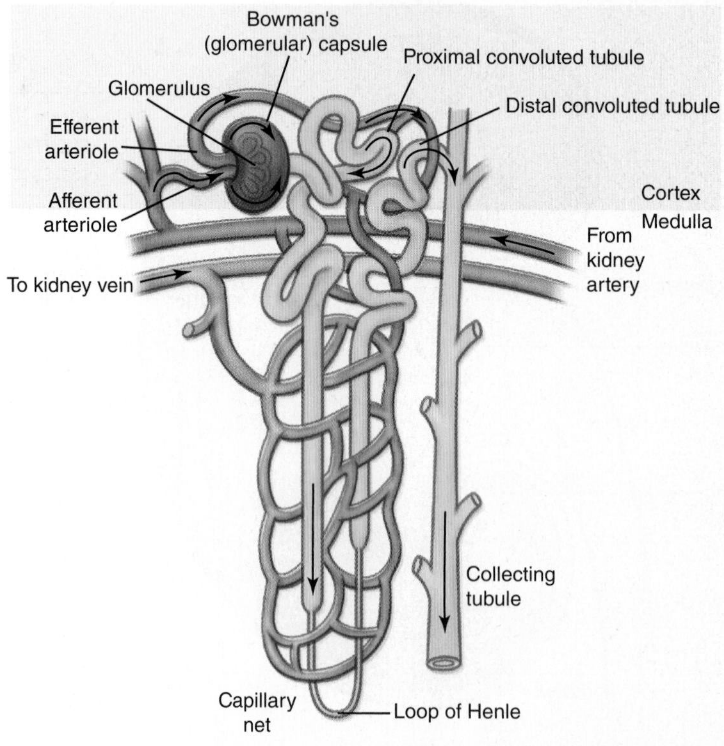

Figure 21-3 *Structure of a nephron*

THE PATH OF THE FORMATION OF URINE

Blood enters the afferent arteriole → passes through the glomerulus → to Bowman's capsule → now it becomes filtrate (blood minus the red blood cells, white blood cells, and plasma proteins) → continues through the proximal convoluted tubule → to the loop of Henle → to the distal convoluted tubule → to the collecting tubule (at this time, about 99% of the filtrate has been reabsorbed) → approximately one mL of urine is formed per minute → the urine goes to the renal papillae → minor calyces → major calyces → renal pelvis → ureter → urinary bladder → urethra → urinary meatus (**Figure 21-4**).

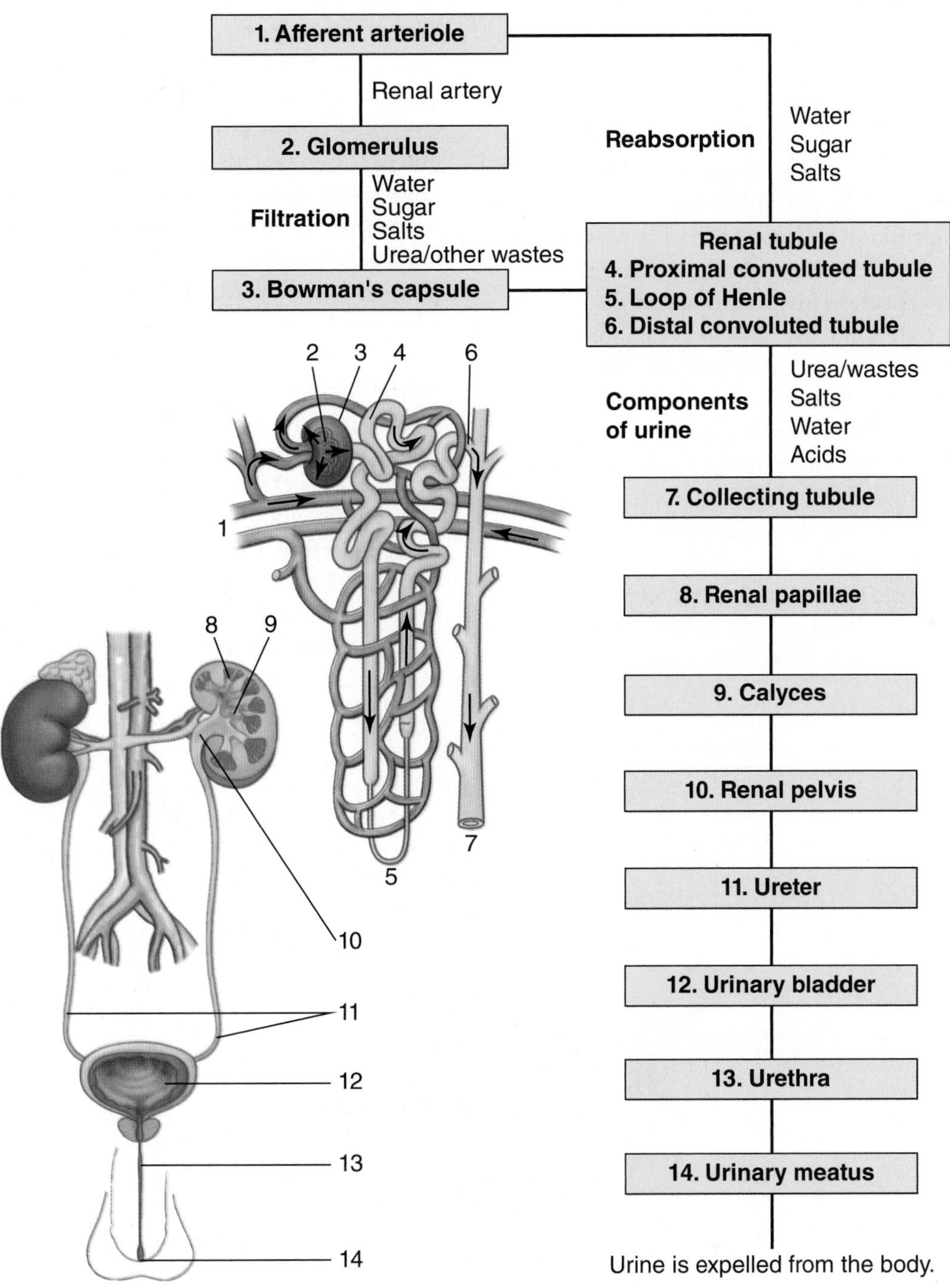

Figure 21-4 *Pathway of urine formation*

Urine Formation in the Nephron

The kidney nephrons form urine by three processes: (1) filtration of the blood by the glomerulus, (2) reabsorption by the renal tubules, and (3) secretion by the peritubular capillaries returning select substances to the glomerular filtrate (**Figure 21-5**).

FILTRATION

The first step in urine formation is filtration. In this process, blood from the renal artery enters the smaller afferent arteriole, which in turn enters the even smaller capillaries of the glomerulus. As the blood from the renal artery travels this course, the blood vessels grow narrower and narrower. This results in an increase in blood pressure. In most of the capillaries throughout the body, blood pressure is about 25 millimeters of mercury (mm Hg); in the glomerulus, it is between 60 and 90 mm Hg.

This high blood pressure forces a plasma-like fluid to filter from the blood in the glomerulus into the Bowman's capsule. This fluid is called the **filtrate**. It consists of water, glucose, amino acids, some salts, and urea. The filtrate does not contain plasma proteins or red blood cells because they are too large to pass through the pores of the capillary membrane. The Bowman's capsule filters out 125 mL of fluid from the blood in a single minute. In one hour, 7500 mL of filtrate leave the blood; this amounts to around 180 liters in a 24-hour period.

As the nephric filtrate continues along the tubules, 99% of this fluid is reabsorbed back into the bloodstream; therefore, only 1–2 liters (1000–2000 mL) of urine are excreted per day.

REABSORPTION

This process includes the reabsorption of useful substances from the filtrate within the renal tubules into the peritubular capillaries, which are the capillaries around the tubules. These include water; glucose; amino acids; vitamins; bicarbonate ions (HCO_3^-); and the chloride salts of calcium, magnesium, sodium, and potassium. Reabsorption starts in the proximal convoluted tubules, and it continues through the loop of Henle, the distal convoluted tubules, and the collecting tubules.

The proximal tubules reabsorb approximately 80% of the water filtered out of the blood in the glomeruli. Water absorbed through the proximal tubules constitutes obligatory water absorption, which is the amount necessary for cell function. Simultaneously, glucose, amino acids, vitamins, and some sodium ions are actively transported back into the blood. However, when levels exceed normal limits, the selective cells lining the tubules no longer reabsorb substances such as glucose but allow them to remain in the tubule to be eliminated in the urine. The term used to describe the limit of reabsorption is the **threshold**. Passing this level is referred to as "spilling over the threshold." For example, people who have diabetes spill sugar frequently, so sugar can be found in their urine (glycosuria). Another example occurs when a person is taking medications. The tubules will only reabsorb a certain amount of the drug; therefore, the medication may have to be taken every 4–6 hours to maintain a therapeutic dosage of the drug in the blood.

In the distal convoluted tubules, about 10% to 15% of water is reabsorbed into the bloodstream, depending

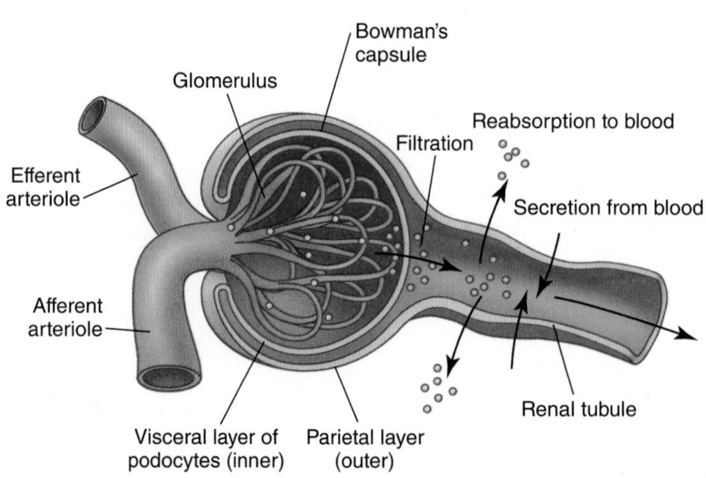

Figure 21-5 *Filtration, reabsorption, and secretion are the main functions of nephrons.*

on the needs of the body. This type of water absorption is called optional reabsorption. It is controlled by the antidiuretic hormone (ADH) and aldosterone. ADH and aldosterone help maintain the balance of body fluids (**Figure 21-6**).

SECRETION

The process of secretion is the opposite of reabsorption. Some substances are actively secreted into the tubules. Secretion transports substances from the blood in the peritubular capillaries into the glomerular filtrate in the distal and collecting tubules. Substances secreted into the glomerular filtrate include ammonia, creatinine, hydrogen ions (H^+), potassium ions (K^+), and some drugs. The electrolytes are selectively secreted to maintain the body's acid-base balance.

At the completion of these processes—filtration, reabsorption, and secretion—the glomerular filtrate is now called **urine**.

Media Link

View the **Urine Formation** animation on the Online Resources.

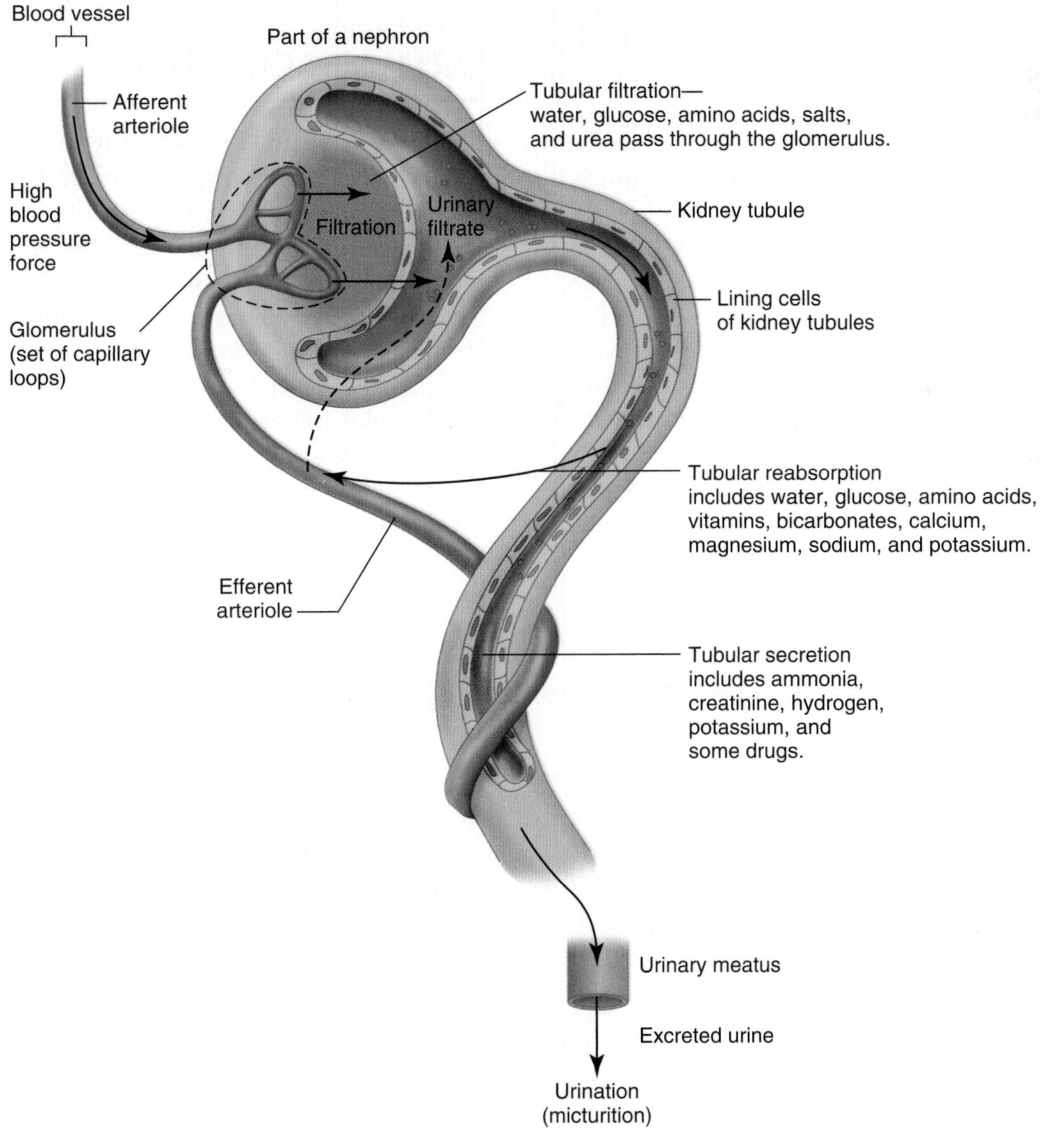

Figure 21-6 *Processes and structures of a nephron*

CONTROL OF URINARY SECRETION

The secretion of urine is under both chemical and nervous control.

Chemical Control

The reabsorption of water in the distal convoluted tubules and the collecting ducts is influenced by ADH, which helps increase the size of the cell membrane pores in the epithelial cells of the distal tubule and collecting ducts by increasing their permeability to water. The secretion and regulation of ADH is under the control of the hypothalamus. In the hypothalamus, highly sensitive receptor cells called **osmoreceptors** (oz-moh-ree-**SEP**-tors) are sensitive to the osmotic pressure of blood plasma. An increase in the osmotic blood pressure due to salt retention causes an increase in ADH secretion. This will inhibit normal urine formation, and water may also be held in the tissues. **Figure 21-7** shows the effect of salt retention on human tissues.

Other hormones are also involved in the reabsorption process. **Aldosterone** (al-**DOS**-tah-rown), which is secreted by the adrenal cortex, promotes the excretion of potassium and hydrogen ions and the reabsorption of sodium ions; chloride ions and water are also absorbed. As the blood passes through the glomerulus to Bowman's capsule, specialized cells are able to detect a drop in blood pressure. An enzyme called **renin** is released by the kidneys into the bloodstream. Renin stimulates the release of aldosterone by the adrenal cortex and constricts the blood vessels. In the absence of aldosterone, sodium and water are excreted in large amounts, and potassium is retained. Any dysfunction of the adrenal cortex produces pronounced changes in the salt and water content of body fluids.

Diuretics increase urinary output by inhibiting the reabsorption of water. Alcohol and caffeine are examples of common diuretics. Alcohol inhibits the secretion of ADH from the pituitary gland; this increases urinary output and may cause dehydration. Caffeine increases the loss of sodium ions, thus increasing the loss of water.

Nervous Control

The nervous control of urine secretion is accomplished directly through the action of nerve impulses on the blood vessels leading to the kidney and on those within the kidney leading to the glomeruli. Indirect nerve control is achieved through the stimulation of certain endocrine glands, whose hormonal secretions will control urinary secretion.

URINARY OUTPUT

The amount of urinary output is between 1000 and 2000 mL every 24 hours, with an average of 1500 mL per day. Volume will vary with diet, fluid intake, temperature, and physical activity. Another factor regulating secretion is the amount of solutes in the filtrate. Normal solutes found in urine include urea, creatinine, uric acid, ketone bodies, potassium, sodium, and chloride. Urine is usually a transparent clear fluid, but the color will vary from pale yellow to dark amber depending on its concentration. Cloudy urine is usually considered abnormal and may be the result of the presence of blood, pus, or bacteria.

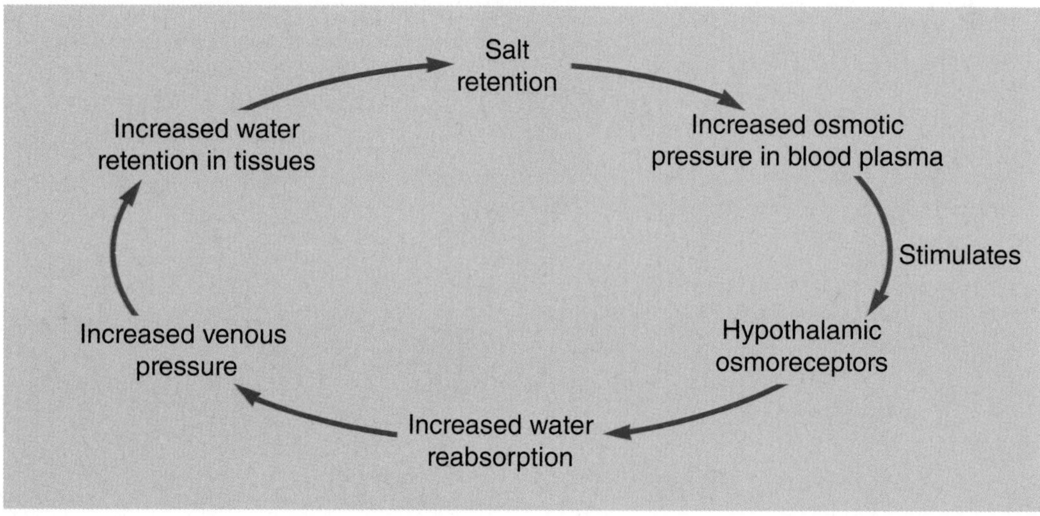

Figure 21-7 *Effects of salt retention on water retention in tissues*

Table 21-2 *Urinalysis Values*

URINALYSIS	*NORMAL VALUES*	*ABNORMAL RESULTS*
Color	Clear amber	Very light or very dark; cloudy
Odor	Pleasantly aromatic	Offensive, unpleasant
Albumin (protein)	Negative	Albuminuria
Acetone	Negative	Ketonuria
Red blood cells	2–3/HPF (high power field)	Hematuria
White blood cells	4–5/HPF (high power field)	White, cloudy urine
Bilirubin	Negative	Bilirubinuria
Glucose	Negative	Glycosuria
Specific gravity	1.005–1.030	Higher or lower than normal
Bacteria	Negative	Present
Casts	Rare	Present; several to many
pH	4.6–8.0	Higher or lower than normal

Urinalysis (you-rih-**NAL**-ih-sis), an examination of the urine, can determine the presence of blood cells; bacteria; acidity level; specific gravity or weight; and physical characteristics such as color, clarity, and odor. A urinalysis is the most common noninvasive diagnostic test done. **Table 21-2** shows normal values and abnormal results for a routine urinalysis.

Ureters

Urine passes from each kidney out of the collecting tubules into the renal pelvis, down the ureter, and into the urinary bladder. There are two **ureters** (**YOO**-reh-terz), one from each kidney, that carry urine from the kidneys to the urinary bladder. They are long, narrow tubes that are less than ¼-inch wide and 10–12 inches long. Mucous membrane lines both the renal pelvis and the ureters. Beneath the mucous membrane lining of the ureters are smooth muscle fibers. When the muscles contract, peristalsis is initiated, and the urine is pushed down the ureter into the urinary bladder.

Urinary Bladder

The **urinary bladder**, a hollow, muscular organ made of elastic fibers and involuntary muscle, acts like a reservoir (**Figure 21-8**). It stores the urine until about 1 pint (500 mL) is accumulated. The average capacity of the bladder is 500 mL. When the amount of urine reaches 200 to 400 mL, impulses are sent to the lower portion of the spinal cord to expel urine.

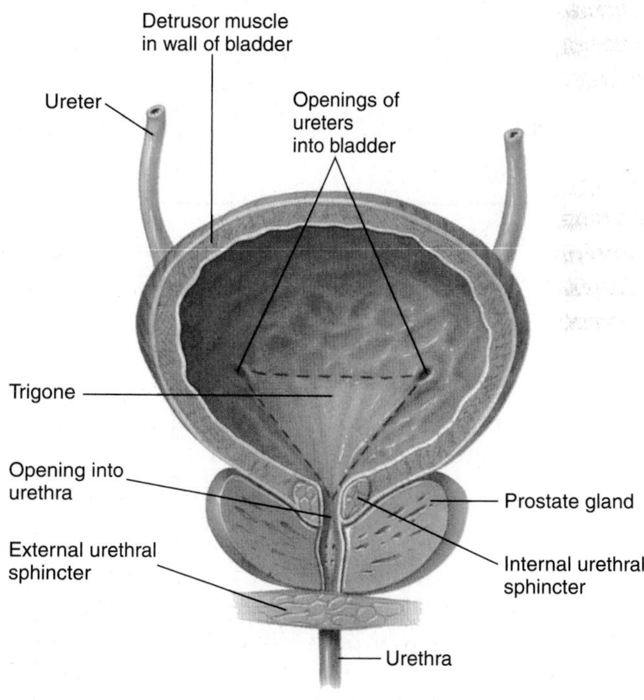

Figure 21-8 *The anatomy of the urinary bladder in a male*

Urethra

The **urethra** (you-**REE**-thrah) is the thin-walled tube leading from the bladder to the outside of the body. It transports urine by peristalsis. There are two urinary sphincters, one located at each end. These muscular rings control the flow of urine from the bladder into the urethra. The **urinary meatus** (mee-**AY**-tus) or urethral meatus is the external opening of the urethra. The male and female urethras are different. The female urethra is approximately 1.5 inches long, and the meatus is located between the clitoris and the opening of the vagina. The female urethra conveys only urine. The male urethra is approximately 8 inches long and the opening is located at the tip of the penis. The male urethra transports both urine and semen, but it cannot transport both at the same time.

Urination

Urination is also known as **micturition** (mik-tyoo-**RIH**-shun) or voiding and is the normal process of emptying the bladder. When urine in the bladder reaches between 200 and 400 mL, stretch receptors in the bladder wall transmit nerve impulses to the spinal cord, relaying the conscious desire to urinate. Urination requires the coordinated contraction of the bladder muscles and the relaxation of the sphincters. The external urinary sphincter, formed of skeletal muscles, surrounds the urethra as it leaves the bladder. The sphincter must open to release urine from the bladder to go through the urethra and to the urinary meatus outside.

DISORDERS OF THE URINARY SYSTEM

Some **congenital anomalies of the kidney and urinary tract (CAKUT)** may be observed while the pregnant mother is having an ultrasound. With CAKUT, the baby is born with kidney or urinary tract structures that develop differently; some common differences are underdeveloped kidneys, the absence of one kidney, and an extra ureter leading to the kidneys. This is the most common cause of end-stage renal disease in children. Details about specific conditions are as follows:

- **Kidney dysplasia** is when cysts, or fluid-filled sacs, grow inside one or both kidneys. If this condition occurs in only one kidney, there are few health problems. If this occurs in both kidneys, the child will need dialysis and a kidney transplant.
- **Vesicoureteral reflux** (**VUR**) is the abnormal upward flow of urine (reflux) from the bladder to the ureters. Treatment is surgery to fix the defective valve.
- **Hypospadias** is characterized by the location of the urethral opening on the underside of the penis. The child will experience difficulty when urinating, and they may have trouble directing the stream of urine.

Acute kidney failure may be sudden in onset. Causes may be nephritis (the inflammation of the nephron), shock, injury or trauma, bleeding, sudden heart failure, or poisoning. The symptoms of acute kidney failure include **oliguria** (**ol**-ih-**GOO**-ree-ah), which is scanty or diminished production of the urine; **anuria** (ah-**NEW**-ree-ah), which is absence of urine formation; fluid retention in the legs, ankles, and feet; shortness of breath; fatigue; confusion; and irregular heartbeat. Suppression of urine formation is dangerous. Unless anuria is relieved, **uremia** (you-**REE**-mee-ah), or excessive amounts of urea in the blood, will develop. Symptoms of uremia are headaches, dyspnea, nausea, vomiting, and, in extreme cases, coma and death. It is important to treat the underlying cause that originally damaged the kidneys. Treatment includes restoring fluid balance in the body, using diuretics and medications to control blood potassium levels, restoring calcium levels, and dialysis.

Chronic renal failure is the condition in which there is a gradual loss of function of the nephrons.

Glomerulonephritis (gloh-**mer**-you-loh-neh-**FRY**-tis) is an inflammation of the glomerulus of the nephron. The filtration process is affected. Plasma proteins are filtered through, and protein is found in the urine as albumin (albuminuria [**al**-byou-mih-**NEW**-ree-ah]); the protein in the urine may cause the urine to be foamy. Red blood cells may also pass through the filter. The urine will be slightly red in color, a condition known as **hematuria** (**hee**-mah-**TOO**-ree-ah), or blood in the urine. Other symptoms are hypertension and edema of the face and eye. The goal of treatment is to prevent kidney failure.

Acute glomerulonephritis occurs in some children about 1–3 weeks after a bacterial infection, usually strep throat. The illness is treated with antibiotics and recovery takes place.

Chronic glomerulonephritis occurs when the filtration membrane has been permanently affected. There is diminished function of the kidney, which may result in kidney failure.

Hydronephrosis (**high**-droh-neh-**FROH**-sis) occurs when the renal pelvis and calyces structures of the kidneys become distended due to an accumulation of fluid (**Figure 21-9**). This may also be a congenital condition. The urine backs up because of a blockage in the ureter or pressure on the outside of the ureter, which may narrow the passageway. The blockage may be caused by a kidney stone. Other conditions that can cause hydronephrosis are pregnancy or an enlarged prostate gland, which causes pressure on the ureters or bladder. The treatment for this condition is removal of the obstruction.

Pyelitis (pye-eh-**LYE**-tis) is an inflammation of the renal pelvis (**Figure 21-10A**). **Pyelonephritis**

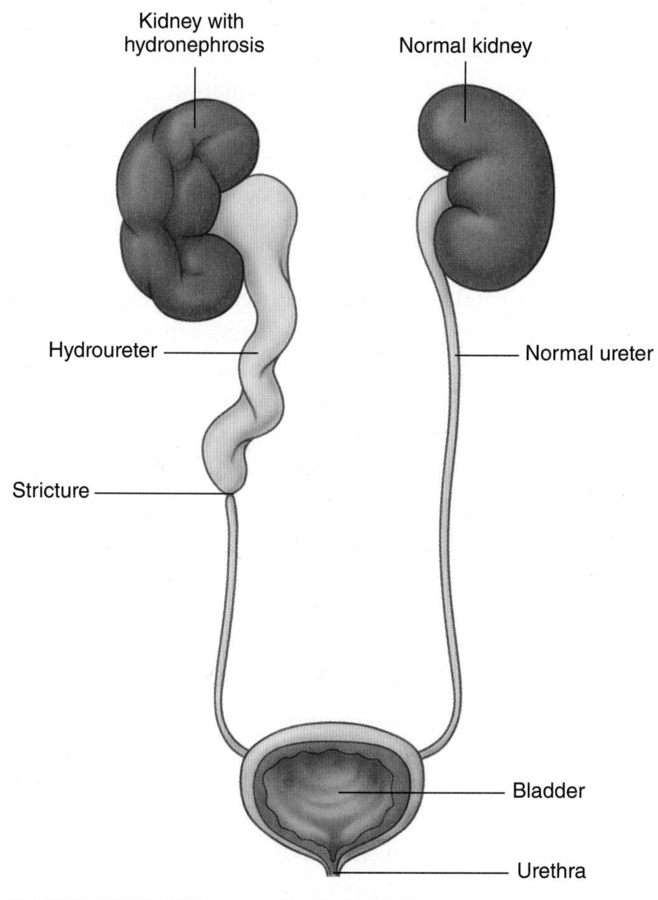

Figure 21-9 *Hydronephrosis*

(**pye**-eh-loh-neh-**FRY**-tis) is the inflammation of the kidney tissue and the renal pelvis (**Figure 21-10B**). This condition generally results from an infection that has spread from the ureters. One of the symptoms is **pyuria** (pye-**YOO**-ree-ah), the presence of pus in the urine. The course of treatment includes the administration of antibiotics.

Kidney stones (renal calculi) (**CAL**-cyou-lye) are stones formed in the kidney. Some materials contained in urine are only slightly soluble in water. Therefore, when stagnation occurs, microscopic crystals of calcium phosphate, along with uric acid and other substances, may clump together to form kidney stones. These kidney stones slowly grow in diameter. They eventually fill the renal pelvis and obstruct urine flow in the ureter. Usually, the first symptom of a kidney stone is extreme pain, which occurs suddenly in the kidney area or lower abdomen and moves to the groin. Other symptoms include nausea and vomiting, burning, frequent urge to void, chills, fever, and weakness. There may also be hematuria. Diagnosis is made by symptoms, ultrasound, and X-rays such as intravenous pyelogram (IVP) and kidney, ureter, and bladder (KUB). Treatment includes an increase in fluids, which will increase urinary output. This may help flush out the stone. Medications are given to help to dissolve the stone. If this is not successful, a urethroscope, or lithotripsy, may be done (see *Medical Highlights—Kidney Stone Removal*).

Cystitis or urinary tract infection (sis-**TIGH**-tis) is the inflammation of the mucous membrane lining of the urinary bladder (**Figure 21-10C**). The most common

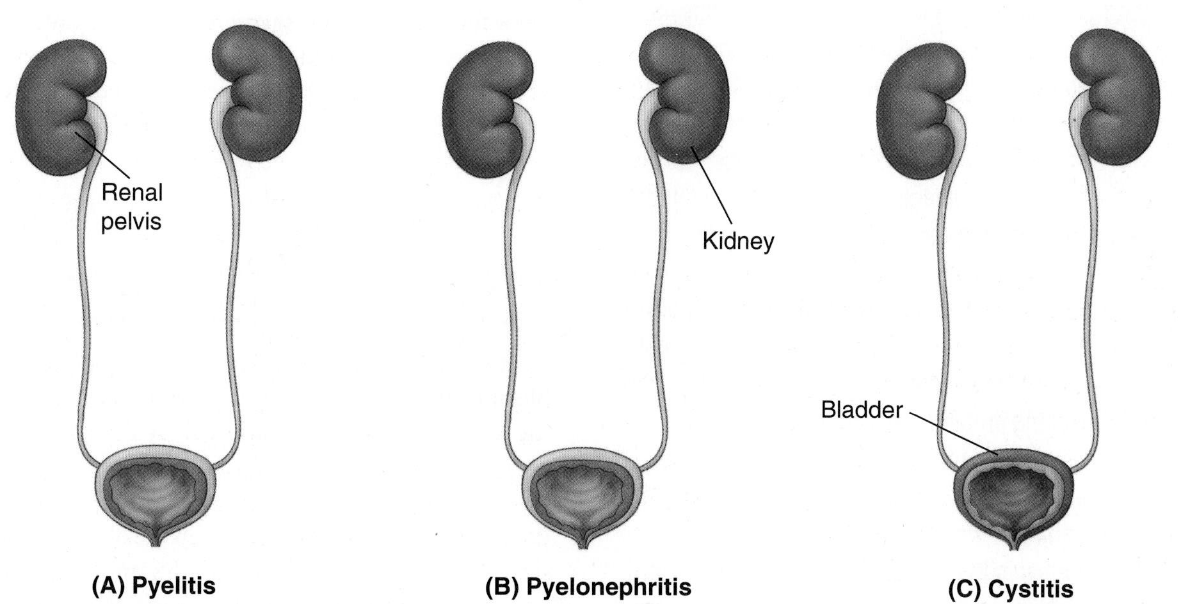

Figure 21-10 *Infection area of the urinary tract, indicated in green: (A) pyelitis (renal pelvis); (B) pyelonephritis (renal pelvis and tissue); and (C) cystitis (bladder)*

cause of cystitis is *E. coli*, an organism normally found in the rectum. Cystitis leads to painful urination (**dysuria** [dis-**YOO**-ree-ah]) or frequent urination. Older patients may present with confusion or personality changes. Cystitis is more common in females. The length of the female urethra is about 1.5–2 inches. Organisms can easily enter the urethra and then move into the bladder from the outside of the body. The treatment of cystitis involves antibiotics and urinary antiseptics with increased fluids. Drug-resistant urinary tract infection (UTI) UTIs are occurring more frequently. The organism survives exposure to a particular antibiotic and passes this survival advantage along. Urine for culture and sensitivity will help determine which antibiotic is most effective. The patient with cystitis must be reminded to complete the prescribed amount of medication to prevent reinfection. The female patient should be taught proper hygienic practices, such as wiping from front to back, avoiding douching, and avoiding harsh cleaning powders because the skin may become damaged. Some physicians recommend urination after sexual intercourse.

The Effects of Aging on the Urinary System

With advancing age, the kidneys shrink due to a loss of nephrons within the corticular region of the kidney. There is also evidence of collapsing glomeruli and sclerotic changes in the larger renal blood vessels. The end result is a decrease in renal blood flow. This change in flow compromises the ability of the kidney to eliminate unwanted substances from the bloodstream. There is also a decline in the glomerular filtration rate. Decrease in this rate means that, for drugs excreted by the kidney, the dose of drugs needs to be adjusted to compensate for the age-related decrease in kidney function. If adjustments are not made, there is an age-related risk for drug overdose. Other functions such as glucose reabsorption also decrease as people age, with the resulting problem of hyperglycemia.

There is a loss of muscle tone in the urinary bladder, which frequently causes **nocturia**, or frequent nighttime urination. Weakening of the bladder and the sphincters reduces the ability to maintain continence. Urinary incontinence, the involuntary loss of urine, is frequently observed in older adults.

Incontinence (in-**KON**-tih-nents) is also known as involuntary micturition. Incontinence is when an individual loses voluntary control over urination. Incontinence occurs in babies prior to toilet training because they lack control over the external sphincter muscle of the urethra. Thus, urination occurs when the bladder fills. Similarly, a person who has suffered a stroke or whose spinal cord has been severed may have no bladder control. In these conditions, a patient may require an indwelling catheter. This is a tube inserted into the neck of the bladder through the urethra. It directs the urine into a sterile urinary drainage bag.

Neurogenic bladder (new-roh-**JEN**-ick) is a condition caused by damaged nerves that control the urinary bladder; it may be the result of Parkinson's disease, multiple sclerosis, or diabetes. This results in dysuria, urinary retention, which is the inability to empty the bladder completely, and incontinence. Treatment may be medication and intermittent cauterization.

Overactive bladder is a condition that affects 35 million Americans and often interferes with the activities of daily living. There is a problem with bladder storage that causes a sudden urge to urinate, which may lead to incontinence. This condition occurs when the nerve signals between the bladder and brain are not coordinated. It may also occur when the muscles in the bladder are overactive. Conditions that contribute to overactive bladder are stroke, multiple sclerosis, and diabetes. Symptoms include urinating eight or more times daily and awakening two or more times at night to urinate. Treatment includes behavioral strategies such as pelvic exercises; fluid intake scheduling; timed voiding; and double voiding, which means after urinating, waiting a few minutes and trying to urinate again. Medications may also be used that relax the bladder, which can be helpful in relieving symptoms and reducing episodes of urge incontinence. Botox® injections may also help to relax the muscle.

TRAUMA TO THE KIDNEYS

Trauma to the kidneys can occur from outside sources. Blunt trauma is caused by an impact that does not break the skin. Examples include car accidents, falls, or a direct hit to the kidneys. There may or may not be outside bruising on the skin where the kidneys are located. Penetrating trauma occurs when an object pierces the skin and punctures the kidney. Examples include any object that perforates the skin, such as a knife or bullet. The most common sign of trauma to the kidneys is hematuria.

DIALYSIS

Dialysis (dye-**AL**-ih-sis) is a treatment used for kidney failure. Dialysis involves the passage of blood through a semipermeable membrane to rid the blood of harmful wastes, extra salt, and water. Dialysis devices serve as a substitute kidney. The two forms of dialysis are hemodialysis and peritoneal dialysis. Patients on dialysis must follow a strict treatment schedule; take medications regularly; and carefully monitor intake of fluids, protein, sodium, potassium, and phosphorous.

Hemodialysis (**hee**-moh-dye-**AL**-ih-sis) is a process for purifying blood by passing it through thin membranes and exposing it to a solution that continually circulates around the membrane. The solution is called a dialysate. Substances in the blood pass through the membranes into the lesser concentrated dialysate in response to the laws of diffusion. The part of the unit that actually substitutes for the kidney is a glass tube called a **dialyzer**, which is filled with thousands of minute hollow fibers attached firmly at both ends (**Figure 21-11**). Blood from the patient flows through the fibers, which are surrounded by circulating dialysate. The dialysate is individualized for each patient to provide the appropriate levels of sodium, bicarbonate, and other substances. These cross the membrane and enter the blood. At the same time, extra water and waste products leave the blood to enter the dialysate.

The patient is connected to the dialysis unit by means of needles and tubing that take blood from the patient to the machine and return it to the patient. A fistula (opening between an artery and a vein) or a graft (vein inserted between the artery and a vein) is surgically constructed to provide a site for inserting the needles. Artificial veins may last from 3–5 years. Most patients are assigned to a dialysis center for periodic treatment; however, treatment can also be done in the home if the patient and family are willing to assume responsibility. It is usually done 2–3 times a week, and each treatment lasts from 2–4 hours.

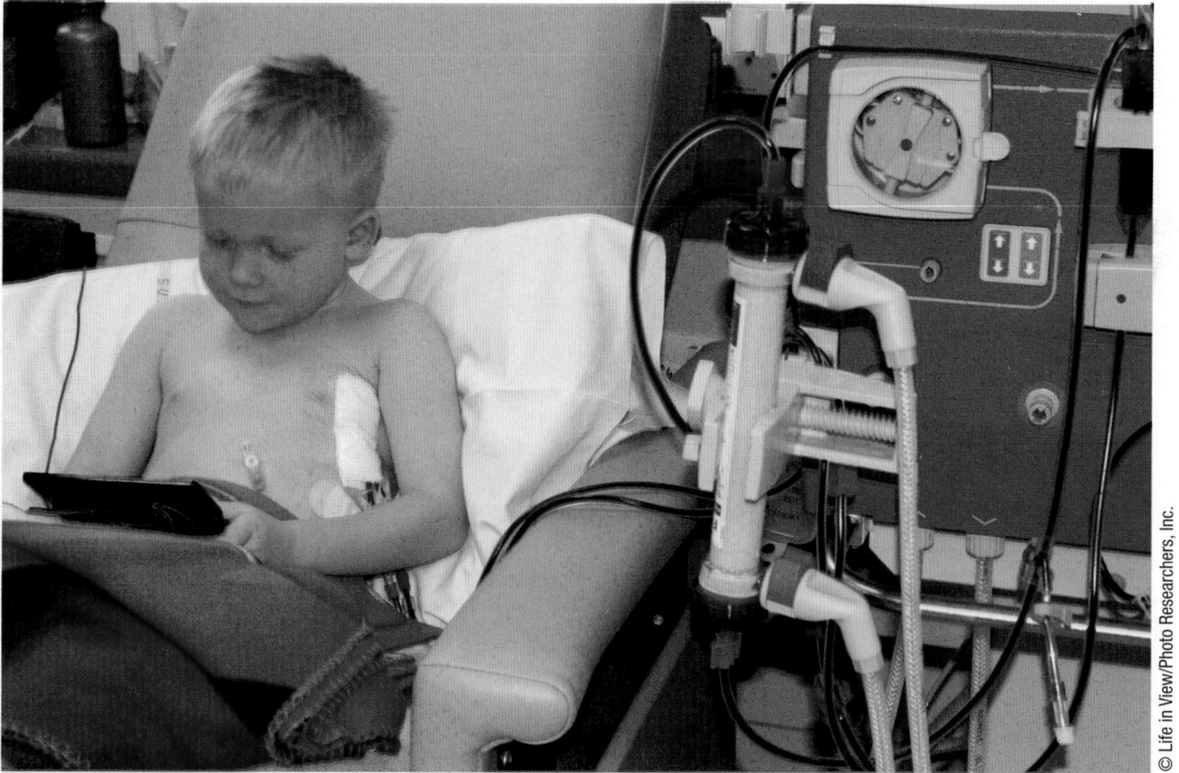

Figure 21-11 *In hemodialysis, waste is filtered from the patient's blood through a dialyzer, and the filtered blood is then returned to the patient's body.*

Peritoneal dialysis (pehr-ih-toh-**NEE**-al dye-**AL**-ih-sis) uses the patient's own peritoneal lining instead of a dialyzer to filter the blood. A cleansing solution called dialysate travels through a catheter implanted into the abdomen. Fluid, wastes, electrolytes, and chemicals pass from tiny blood vessels in the peritoneal membrane into the dialysate. After several hours, the dialysate is drained from the abdomen, taking the wastes from the blood with it. The abdomen is filled with fresh dialysate, and the cleaning procedure begins again. The most common type of peritoneal dialysis is continuous ambulatory peritoneal dialysis (CAPD). The dialysate stays in the abdomen for 4–6 hours. The process of draining the dialysate and replacing it with fresh solution takes about 30 minutes. Most people change the solution four times a day (**Figure 21-12**).

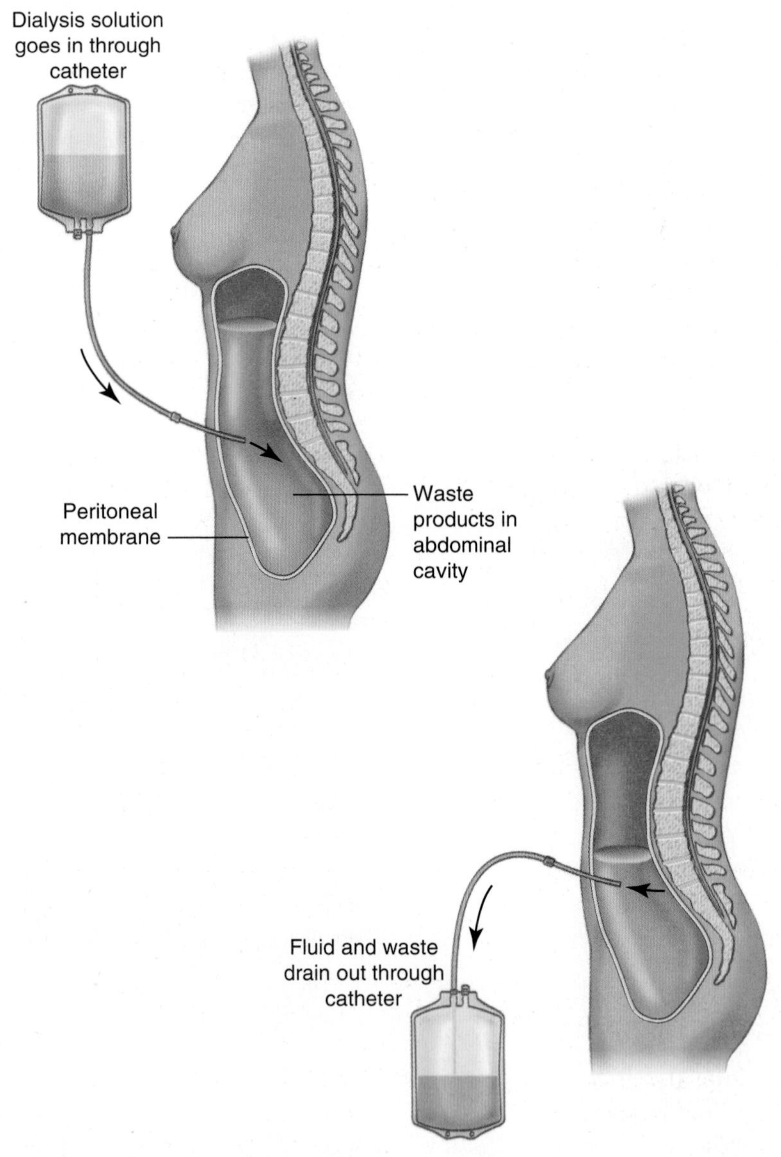

Figure 21-12 *Peritoneal dialysis uses the patient's own peritoneal lining as the dialyzer.*

 Medical Highlights

KIDNEY STONE REMOVAL

EXTRACORPOREAL SHOCK WAVE LITHOTRIPSY

A surgical procedure called **extracorporeal shock wave lithotripsy (ESWL)** (**ecks**-trah-kor-**POUR**-ee-al **LITH**-oh-**trip**-see) may be done to remove kidney stones located high in the ureters or the renal pelvis. ESWL uses shock waves created outside the body to travel through the skin and body tissues until the waves hit the dense stones. The stones become fractured by the sound waves, become sandlike, and are passed through the urinary tract. Several ESWL devices are in use. One device positions the patient in a water bath while the shock waves are transmitted. Most devices use either X-ray or ultrasound to help the surgeon locate the stone during the treatment.

This procedure can be done on an outpatient basis. Recovery time is short, and most people resume normal activities in a few days. Some complications may occur, such as hematuria, bruising, and minor discomfort on the back or abdomen. In addition, the shattered stone fragments may cause discomfort as they pass through the urinary tract.

PERCUTANEOUS NEPHROLITHOTOMY

In the percutaneous nephrolithotomy (**per**-kyou-**TAY**-nee-us **nef**-roh-lih-**THOT**-oh-mee) procedure, a surgeon makes a tiny hole in the patient's back and creates a tunnel directly into the kidney. Using an instrument called a nephroscope, the surgeon locates and removes the stones. For larger stones, an ultrasonic energy probe may be needed to break the stone into smaller pieces. One advantage of this procedure over ESWL is that the surgeon removes the stone fragments instead of relying on their natural passage from the kidney.

URETEROSCOPIC STONE REMOVAL

Ureteroscopic stone removal is done for mid and lower stones. A surgeon passes a small, fiberoptic instrument called a ureteroscope through the urethra and bladder into the ureter. The surgeon then locates the stone and either removes it with a cagelike device or shatters it with a special instrument that produces a form of shock wave.

Automated peritoneal dialysis, a type of peritoneal dialysis that can be done at night while the patient is asleep, takes 6–8 hours.

The main complication of peritoneal dialysis is peritonitis, an inflammation of the peritoneal lining.

KIDNEY TRANSPLANTS

Kidney transplants are done in cases of prolonged chronic debilitating diseases and renal failure involving both kidneys. Usually the patient has been on dialysis for a long period of time, waiting for a compatible organ.

There are two kinds of kidney transplants: those that use a kidney from a living donor—a family member is usually the best match—and those that use a kidney from an unrelated donor who has died. Blood and other cellular material must match to ensure the greatest potential for success in a transplant. The most important complication that occurs after transplant is rejection of the kidney by the recipient. Medication must be taken every day to prevent rejection. Results of transplantation are improving steadily with research advances. Successful kidney transplants provide a chance for patients to enjoy a better quality of life and are often associated with increased energy levels and a less restricted diet.

One BODY How the Urinary System Interacts with Other Body Systems

INTEGUMENTARY SYSTEM

- Produces the precursor to vitamin D. The kidney in turn makes the active form of vitamin D, which is necessary for the absorption of calcium.

SKELETAL SYSTEM

- The pelvic cavity protects the urinary bladder.

MUSCULAR SYSTEM

- The waste product of muscle metabolism, creatinine, is excreted by the kidneys.
- Peristalsis moves urine from the bladder to the urethra.
- Sphincter muscles control the voluntary act of urination.

NERVOUS SYSTEM

- The osmoreceptors in the hypothalamus react to the osmotic pressure of blood. An increase in pressure causes release of ADH to increase reabsorption of water in the distal kidney tubules.
- Stretch receptors in the bladder transmit nerve impulses, which inform a person of the need to urinate.

ENDOCRINE SYSTEM

- The hormones aldosterone and ADH help regulate reabsorption of water and electrolytes.

CIRCULATORY SYSTEM

- Blood is constantly filtered through the kidneys to remove the waste products of metabolism. The necessary electrolytes are reabsorbed to maintain the acid-base balance of the body.
- The enzyme renin helps maintain blood pressure.
- Blood pressure influences the glomerular filtration rate.
- The hormone erythropoietin stimulates the production of red blood cells.

LYMPHATIC SYSTEM

- Kidneys filter bacteria and waste products of the inflammation process.

RESPIRATORY SYSTEM

- The lungs and kidneys maintain the proper pH of the blood.

DIGESTIVE SYSTEM

- The liver converts the end products of protein metabolism to urea to be excreted by the kidneys.
- It uses the active form of vitamin D to absorb calcium.

REPRODUCTIVE SYSTEM

- The male urethra is the common passageway for semen and urine.

Medical Terminology

afferent	conducting toward a part
arteriole	little artery
afferent arteriole	little artery going to a part
an-	without
-uria	urine
an/uria	without urine
cyst	bladder
-itis	inflammation of
cyst/itis	inflammation of the bladder
dia-	across or through
-lysis	breaking down
dia/lysis	breaking down through a semipermeable membrane
dys-	painful
dys/uria	painful urination
efferent	conducting outward from a part
efferent arteriole	little artery going away from a part
filtr	filter
-ate	relating to
filtr/ate	fluid passing through a filter
glomerulo	resembles a little ball of yarn
nephr	kidney
glomerulo/nephr/itis	inflammation of the glomeruli of the kidney
hemat	blood
hemat/uria	blood in the urine
hemo/dia/lysis	blood breaking down through a semipermeable membrane
hydro	water
-osis	abnormal condition of
hydro/nephr/osis	abnormal condition of water on the kidney
olig	scanty
olig/uria	scanty urine
pyelo	renal pelvis
pyelo/nephr/itis	inflammation of the renal pelvis of the kidney
ren	kidney
-al	relating to
ren/al	relating to the kidney
ur	urine
-emia	blood condition
ur/emia	accumulation in the blood of the normal constituents of urine
urina/lysis	chemical analysis of urine

Study Tools

Workbook	Activities for Chapter 21
Online Resources	• PowerPoint presentations • Animations

REVIEW QUESTIONS

Select the letter of the choice that best completes the statement.

1. The kidneys are responsible for excreting
   a. carbon dioxide and water.
   b. solid wastes and water.
   c. nitrogenous wastes and water.
   d. perspiration.

2. In addition to the kidneys, the organ(s) responsible for excretion of carbon dioxide and water is (are) the
   a. lungs.
   b. kidneys.
   c. skin.
   d. large intestine.

3. The kidneys are located in the
   a. abdominal area.
   b. pelvic area.
   c. peritoneal area.
   d. retroperitoneal area.

4. A ball of capillaries is called the
   a. Bowman's capsule.
   b. cortex.
   c. glomerulus.
   d. medulla.

5. The process of plasmalike fluid passing through the glomerulus to Bowman's capsule is called
   a. filtration.
   b. reabsorption.
   c. secretion.
   d. excretion.

6. The hormone ADH affects reabsorption in the
   a. glomerulus.
   b. proximal convoluted tubule.
   c. loop of Henle.
   d. distal convoluted tubule.

7. The pathway of urine formation is
   a. kidney, ureter, urethra, bladder.
   b. ureter, pelvis, urethra, bladder.
   c. kidney, urethra, bladder, ureter.
   d. kidney, ureter, bladder, urethra.

8. The average normal daily urinary output is
   a. 600 mL.
   b. 800 mL.
   c. 1500 mL.
   d. 2400 mL.

9. Inflammation of the urinary bladder is called
   a. nephritis.
   b. cystitis.
   c. pyelitis.
   d. urethritis.

10. Involuntary urination is known as
    a. polyuria.
    b. anuria.
    c. incontinence.
    d. frequency.

FILL IN THE BLANKS

1. Pus in the urine is called _______________.
2. Painful urination is called _______________.
3. Frequency of urination is called _______________.
4. Scanty production of urine is known as _______________.
5. Absence of urination is known as _______________.

COMPARE AND CONTRAST

Compare and contrast the following terms:

1. Proximal convoluted tubule and distal convoluted tubule
2. Pyelitis and pyelonephritis
3. Hydronephrosis and glomerulonephritis
4. Kidney failure and kidney stone
5. Hemodialysis and peritoneal dialysis

APPLYING THEORY TO PRACTICE

1. Keep a log for 24 hours. Measure your liquid intake and urinary output. Answer the following questions: Are you taking in enough fluid to maintain your body in good fluid balance? How much liquid did you intake during the 24 hours? Did you adjust your fluid intake for exercise?
2. You have just run a mile and sweated profusely. When you urinate, you notice there is only a small amount or urine, and it is concentrated. Explain what has happened.
3. Your physician has prescribed an antibiotic for you. The instructions say to take it every six hours. Why is it necessary to maintain this over 24 hours?
4. Alycia, a 42-year-old female, comes to the emergency health center complaining of burning during urination and frequent urination. What diagnostic technology would you use to test Alycia? Explain the purpose of these tests to Alycia.

   After the test, a diagnosis of cystitis is made. Explain cystitis to Alycia and why it is more common in females. What instructions should you give Alycia regarding her treatment?
5. In kidney failure, dialysis may be necessary. Define dialysis. What type do you think would be best for a vision-impaired 70-year-old patient? What type do you think would be best for a mother with children ages 2, 6, and 10? If dialysis does not work and the affected kidney has to be removed, how does the structure and function of the remaining kidney change?
6. Judy Ann, a 75-year-old female, was brought into the emergency room after falling off a stepstool. A physician suspects kidney damage from blunt force trauma. What symptoms would make a physician suspect kidney damage? What effects would you expect to see in the body systems? What diagnostic tests should be run to see if she suffered kidney damage?

CASE STUDY

Ken comes to the physician's office with a complaint of severe back pain located on the right side, lateral to the vertebrae and superior to the buttocks. The physician orders tests to rule out kidney stones.

1. What diagnostic tests might be ordered?
2. How does the location of the back pain indicate a problem with the kidneys?

After the test is done, the physician reports that Ken has a stone in the renal pelvis of his right kidney. The physician schedules an appointment to have the stone removed.

3. Describe the function of the kidney and the parts of the nephron. Name the parts of the kidney and their function.
4. What other body systems do diseases of the kidney affect?
5. Ken wants to know more about kidney stones and why they need to be removed. Find a picture showing the shape of a kidney stone and explain why it has to be removed given its location.
6. Explain the procedures for kidney stone removal. Which procedure do you think will be scheduled for Ken?
7. Ken inquires if he will have this problem in the future. What will the physician tell him?

LAB ACTIVITY 21-1 Kidney

- ***Objective:*** To observe the structure of the kidney
- ***Materials needed:*** sheep kidney, dissection tray and instruments, disposable gloves, textbook, paper, pencil

Step 1: Put on disposable gloves.

Step 2: Examine the sheep kidney. Identify the renal capsule, ureter, renal artery, and renal vein. Describe and record your observations.

Step 3: Using a scalpel, carefully make a longitudinal incision into the sheep kidney, dividing the kidney into two sections.

Step 4: Locate the renal cortex, renal medulla, and renal pelvis. Trace the renal artery to the glomeruli structure of the kidney. Compare with Figure 21-2. Sketch and label your observations.

Step 5: Dispose of the kidney in the appropriate manner.

Step 6: Remove gloves, dispose in the appropriate manner, and wash hands.

LAB ACTIVITY 21-2 Nephron Structure

- ***Objective:*** To observe the structure of the nephron
- ***Materials needed:*** prepared slide of kidney tissue, microscope, textbook, paper, pencil

Step 1: Examine Figure 21-3 in the textbook. Draw and label the nephron, then trace the process of filtration, reabsorption, and secretion of urine.

Step 2: Examine the slide of kidney tissue; locate and describe the glomerulus. Record your observations.

LAB ACTIVITY 21-3 Urinalysis

- ***Objective:*** To test a simulated urine sample for pH (acidic, basic, or neutral) to check for acetone or glucose
- ***Materials needed:*** simulated urine sample, disposable gloves, cup, red and blue litmus paper, bottle of dipsticks, charts to test for acetone and glucose, textbook, paper, pencil

Step 1: Wash hands, put on disposable gloves, and obtain simulated urine sample.

Step 2: Check the color clarity of the urine sample.

Step 3: Record your findings regarding color and clarity.

Step 4: Place blue litmus paper in the urine sample. Remove litmus paper and check color. Record results and dispose of litmus paper.

Step 5: Place red litmus paper in the urine sample. Remove litmus paper and check color. Record results and dispose of litmus paper.

Step 6: Take dipstick for acid-base balance and dip into urine sample; remove dipstick and compare findings with acid-base chart. Record your findings and dispose of dipstick.

Step 7: Take dipstick for acetone and dip into urine sample; remove dipstick and compare findings with acetone chart. Record your findings and dispose of dipstick.

Step 8: Take dipstick for glucose and dip into urine sample; remove dipstick and compare findings with glucose chart. Record your findings.

Step 9: Dispose of urine sample, dipstick, and gloves in the appropriate manner. Wash hands.

Step 10: Compare your results with the normal value in the textbook.

Step 11: Compare your results with those of classmates.

CHAPTER 22

Reproductive System

Objectives

- Compare somatic cell division (mitosis) with germ cell division (meiosis).
- Explain the process of fertilization.
- Identify the organs of the female reproductive system and explain their functions.
- Describe the stages and changes that occur during the menstrual cycle.
- Explain menopause and the changes that occur during this time.
- Identify the organs of the male reproductive system and explain their functions.
- List some common disorders of the reproductive system.
- Describe the stages of human growth and development.
- Define the key words that relate to this chapter.

Key Words

amenorrhea
anorchism
areola
artificial insemination
bacterial vaginosis
Bartholin's glands
benign prostatic hypertrophy (BPH)
breast cancer
breast tumors
breasts
bulbourethral glands (Cowper's glands)
cervical cancer
cervix
chlamydia
circumcision
clitoris
coitus
corona radiata
corpus luteum
cryptorchidism
cystoscopy
ductus deferens
dysmenorrhea
ectoderm
ectopic pregnancy
ejaculatory duct
endoderm
endometrial cancer
endometriosis
endometrium
epididymis
epididymitis
fallopian tubes
fertilization
fibroid tumors
fimbriae
foreskin
fundus

(*continues*)

Key Words *continued*

genital herpes
genital warts
germ cells (gametes)
glans penis
gonorrhea
graafian follicles
human papillomavirus
hymen
hysterectomy
imperforate hymen
impotence
in vitro fertilization (IVF)
infertility
labia majora
labia minora
laparoscopy
lumpectomy
mammography
mastectomy
menarche
menopause
menorrhagia
menstrual cycle
menstruation
mesoderm
missing vagina
mons pubis
mycoplasma genitalium
myometrium
oogenesis
orchitis
ova
ovarian cancer
ovarian cysts
ovaries
oviducts
ovulation
Papanicolaou (Pap) smear
pelvic inflammatory disease (PID)
penile shaft
penis
perimetrium
perineum
polycystic ovary syndrome (PCOS)
premenstrual syndrome (PMS)
prepuce
prostate cancer
prostate gland
prostatectomy
prostatitis
puberty
relaxin
salpingitis
scrotum
seminal vesicles
seminiferous tubules
spermatogenesis
spermatozoa
sterile
syphilis
testes
testicular cancer
toxic shock syndrome
trichomoniasis vaginalis
uterus
vagina
vaginal blockage
vas deferens
vestibule
vulva
yeast infections
zygote

All living organisms, whether unicellular or multicellular, small or large, must reproduce to continue their species. Humans and most multicellular animals reproduce new members of their species by sexual reproduction.

Doctors typically assign gender at birth based on the external genitalia. Most individuals identify as that gender throughout their lives, but some individuals may not. This chapter will address the anatomy and physiology of what is assigned at birth from a clinical viewpoint in order to prepare for practice as a health care provider.

FUNCTIONS OF THE REPRODUCTIVE SYSTEM

The reproductive system carries out multiple functions.

1. Has the necessary organs capable of accomplishing reproduction, the creation of a new individual
2. Manufactures hormones necessary for the development of the reproductive organs and secondary sex characteristics:
    - Females—estrogen, progesterone, and relaxin
    - Males—testosterone

Sexual Reproduction

Specialized sex cells or **germ cells (gametes)** must be produced by the gonads of both male and female sex organs before sexual reproduction can take place. The female gonads, called ovaries, produce egg cells, or ova. The male gonads, or testes, produce sperm. Normal cell division is known as mitosis. In the formation of the germ cells, a special process of cell division occurs called meiosis (my-**OH**-sis). In the female, the specific meiotic process is called **oogenesis** (**oh**-oh-**JEN**-eh-sis); in the male, it is **spermatogenesis** (**sper**-mah-toh-**JEN**-eh-sis).

In humans, the somatic (body) cells, including skin, fat, muscle, nerve, and bone cells, contain 46 chromosomes in the nucleus. Forty-four of these are autosomes, or nonsex chromosomes. The remaining two are sex chromosomes. Each chromosome has a partner of the same size and shape so that they can be paired (**Figure 22-1**). In the female, the somatic cells contain 22 pairs of autosomes, and a single pair of sex chromosomes; in the female, the sex chromosomal pair consists of two X chromosomes. In the male, the combination is also 22 autosomal pairs and a single pair of sex chromosomes; however, the male sex chromosomal pair consists of one X and one Y chromosome.

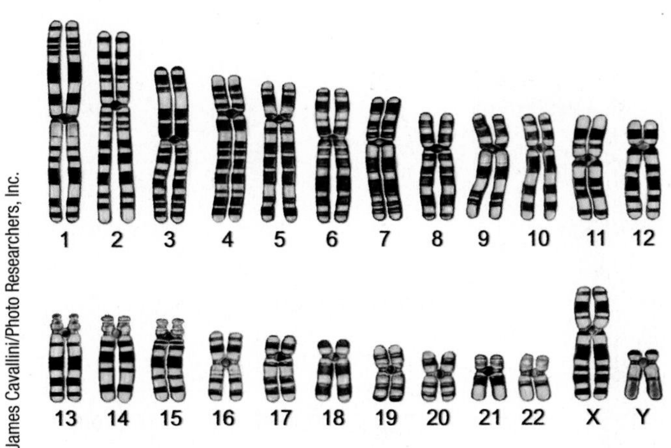

Figure 22-1 *Human karyotype: A karyotype is the arrangement of chromosome pairs according to shape and size.*

Oogenesis and spermatogenesis reduce the chromosome number from 46 to 23 in the gametes, or germ cells. All multicellular organisms start from the fusion of two gametes: the sperm (spermatozoon) from the male and the ovum from the female. **Figure 22-2** shows the structure of a spermatozoon and an ovum.

Fertilization

During sexual intercourse, or **coitus** (**KOH**-ih-tus), sperm from the testes is deposited into the vagina. Spermatozoa entering the female reproductive tract live for only a day or two at the most, though they may remain in the tract up to two weeks before degenerating. Approximately 100 million spermatozoa are contained in 1 mL of ejaculated seminal fluid. They are fairly uniform in shape and size. If the count is less than 20 million per milliliter, the male is considered **sterile**. These millions of sperm cells swim toward the ovum that has been released from the ovary. The large quantity of sperm is necessary because a great number are destroyed before they even approach the ovum. Many die from the acidity of the secretions in the male urethra or the vagina. Some cannot withstand the high temperature of the female abdomen, while others lack the propulsion ability to progress from the vagina to the upper uterine (fallopian) tube.

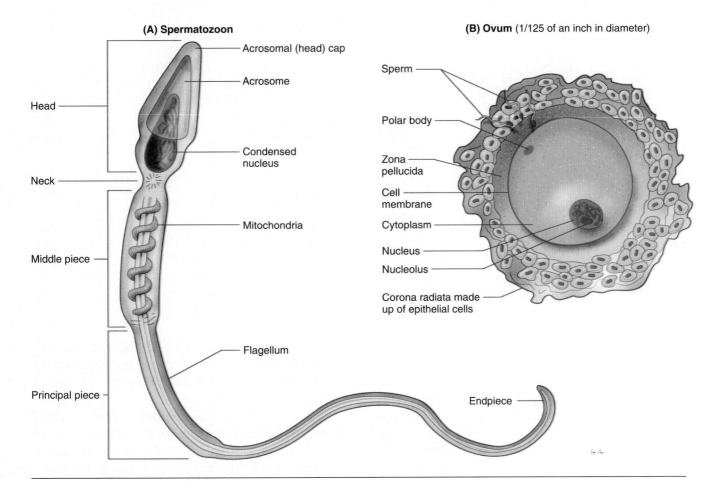

Figure 22-2 *Structure of gametes: (A) a spermatozoon, and (B) an ovum*

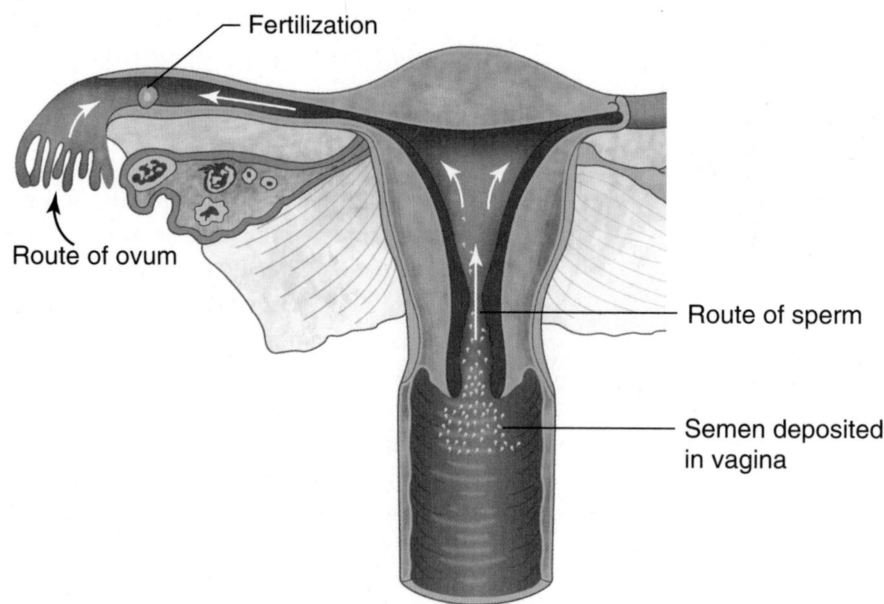

Figure 22-3 *The route of the sperm and the ovum*

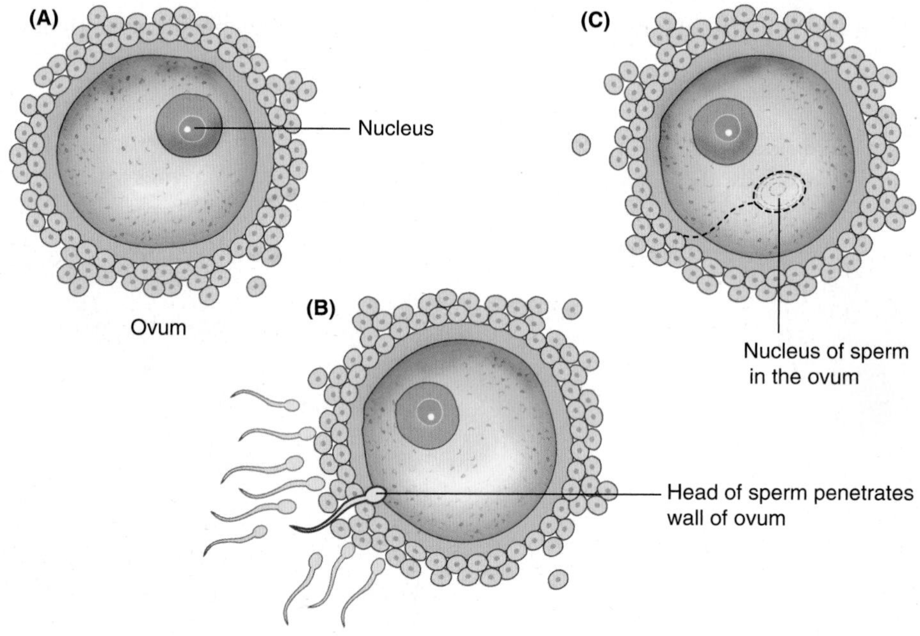

Figure 22-4 *Fertilization*

For a sperm to penetrate and fertilize an ovum, the **corona radiata** (koh-**ROH**-nah ray-dee-**AY**-tah) must first be penetrated. This is the layer of epithelial cells surrounding the zona pellucida (pah-**LOO**-sih-dah) (**Figure 22-2**). Eventually, only one sperm cell penetrates and fertilizes an ovum. To accomplish this successfully, the sperm head produces an enzyme called hyaluronidase. Hyaluronidase acts on hyaluronic acid, a chemical substance that holds together the epithelial cells of the corona radiata. As a result of the action of the hyaluronidase, the epithelial cells fall away from the ovum. This exposes an area of the plasma membrane for sperm penetration. **Figure 22-3** illustrates the route of the sperm and the ovum.

True **fertilization**, or conception, occurs when the sperm nucleus combines with the egg nucleus to form a fertilized egg cell, or **zygote** (**ZYE**-goht) (**Figure 22-4**). The type of fertilization that occurs in humans is referred to as internal fertilization because fertilization takes place within the female's body.

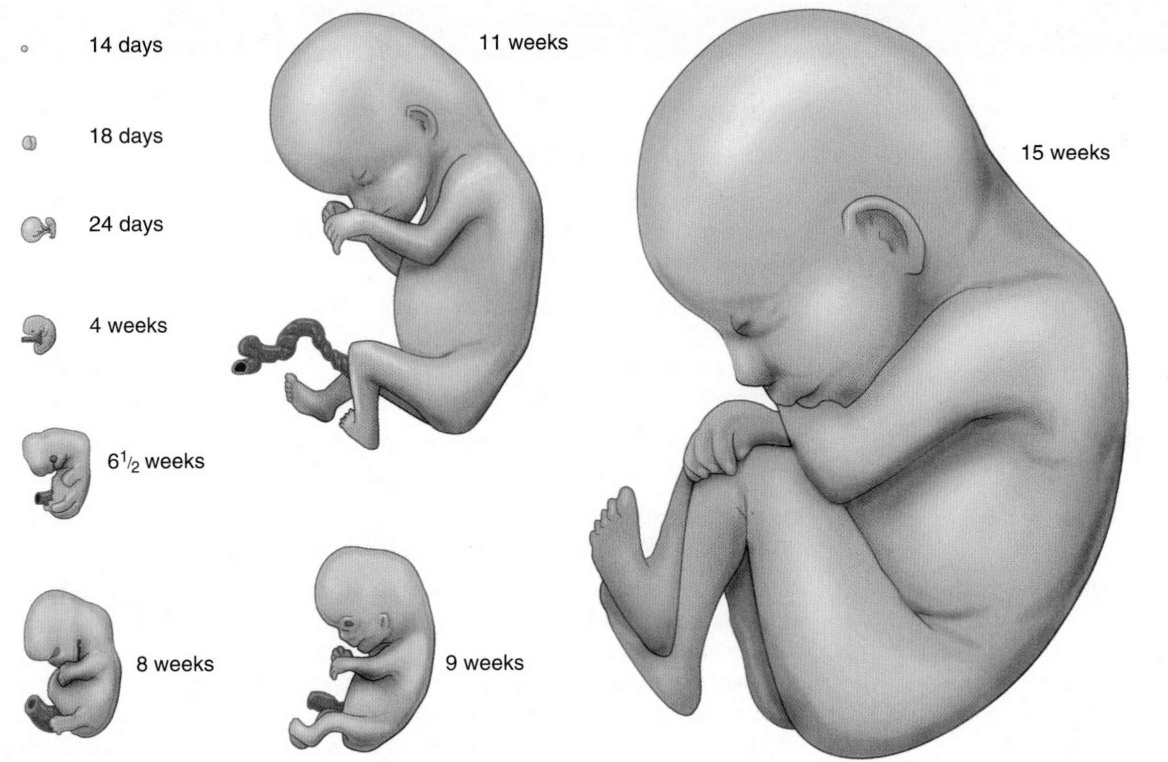

Figure 22-5 *Growth of an embryo into a fetus once fertilization has occurred*

Fertilization restores the full complement of 46 chromosomes possessed by every human cell, each parent contributing one chromosome to each of the 23 pairs. The combination of these chromosomes produces an offspring that is similar to its mother and father but is not identical to either.

Deoxyribonucleic acid (DNA) is found in the chromosomes. It contains the genetic code that is replicated and passed on to each cell as the zygote divides and redivides to form the embryo.

All of the inherited traits possessed by the offspring are established at the time of fertilization. The sex chromosomes of the male parent determine the sex of the child, whereas other characteristics are a combination of both parents. If the embryo receives an X chromosome from the father and an X chromosome from the mother during fertilization, it will be female; if the embryo receives a Y chromosome from the father and an X chromosome from the mother, it will be male.

FETAL DEVELOPMENT

If fertilization occurs, the zygote travels down the fallopian tube and is implanted in the endometrial wall of the uterus. The zygote rapidly grows into an embryo and then a fetus; see **Figure 22-5** for the growth of the embryo into a fetus and **Figure 22-6** for a view of the fetus.

Fetal development is complete after a 9-month process as outlined in **Table 22-1**.

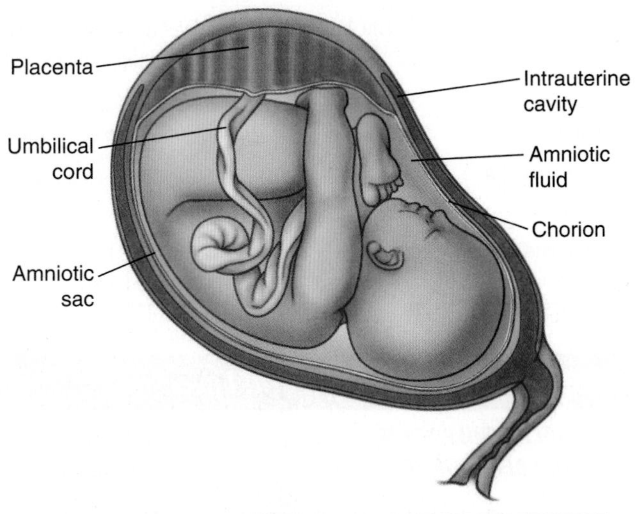

Figure 22-6 *A normal pregnant uterus with a view of the fetus*

Table 22-1 *Fetal Development*

TIME PERIOD	PHASE OF FETAL DEVELOPMENT	DEVELOPMENT
Ovulation	Phase 1	Fertilization occurs; zygote forms.
1–5 days after ovulation	Phase 2	The zygote now begins to divide. When cell division reaches about 16 cells, the zygote becomes a morula (mulberry shaped). Three to five days after fertilization, the morula leaves the fallopian tube and enters the uterine cavity.
4–6 days after ovulation	Phase 3	Cell division continues, and a cavity known as the blastocele forms in the center of the morula. The entire structure is now called a blastocyst. The presence of the blastocyst indicates that two cell types are forming: the embryoblast (inner cell mass on the inside of the blastocele) and the trophoblast (the cells on the outside of the blastocele).
5–6 days after ovulation	Phase 4	The trophoblast cells secrete an enzyme, which erodes the epithelial uterine lining and creates an implantation site for the blastocyst. This implantation site becomes swollen with new capillaries.
7–12 days after ovulation	Phase 5	Implantation is complete, and placental circulation begins. The top layers of cells will become the embryo and amniotic cavity; the lower cells will become the yolk stalk.
13–28 days after ovulation	Phase 6	In the forming of the placenta, chorionic villi "fingers" anchor the site to the uterine wall. By the end of this phase, the embryo is attached by a connecting stalk (which later will become part of the umbilical cord) to the developing placenta. A narrow line of cells appears on the embryonic disc, marking the beginning of gastrulation, the process that gives rise to all three germ layers of the embryo, which carry the pluripotent stem cells for organ development: the **ectoderm** (**EK**-toh-durm), **mesoderm** (**MEZ**-oh-durm), and **endoderm** (**EN**-doh-durm). See **Table 22-2**. The embryo is about 1/10-inch long (0.3 cm), the heart is forming, and the eyes begin to develop. The heart begins to beat on about day 24. The neural tube forms, which is the foundation of the brain, spinal cord, and nervous system. Muscles are developing. Arms and legs are budding.
At the end of 2 months	Phase 7	The embryo is about 1-inch long (2.5 cm), veins are visible, and the heart has divided into two chambers. The brain has human proportions, blood flows in fetus veins, the skeleton is formed, and reflex responses have begun.
At the end of 3 months	Phase 8	The fetus is 2.5–3-inches long (6.4–7.6 cm) and has begun swallowing and kicking. All organs and muscles are formed.
At the end of 4 months	Phase 9	The fetus is covered with a layer of thick hair called lanugo. The heartbeat can be heard; the mother may feel the fetus's first movements.
At the end of 5 months	Phase 10	A protective covering called vernix caseosa begins to form on the skin. By the end of this month, the fetus will be nearly 8 inches long (20.3 cm) and weigh almost one pound.
At the end of 6 months	Phase 11	Eyebrows and eyelids are visible. The lungs are filled with amniotic fluid and the fetus has started breathing motions. The mother's voice is heard and recognized.
At the end of 7 months	Phase 12	The fetus weighs about 3.5 pounds and is 12 inches long (30.5 cm). The body is well formed. Fingernails cover the fingertips. A fetus born at this age can live outside the uterus.
At the end of 8 months	Phase 13	The fetus is gaining about 0.5 pounds per week, and layers of fat are piling on. The fetus is normally turned head down in preparation for birth.
At the end of 9 months	Phase 14	The fetus is 6–9 pounds and measures 19–22 inches (48.3–55.9 cm). As the area becomes more crowded, movement may be limited. The fetus and/or placenta triggers labor, and birth occurs.

Table 22-2 *Embryonic Germ Layers*

ECTODERM WILL FORM (OUTER GERM LAYER)	MESODERM WILL FORM (MIDDLE GERM LAYER)	ENDODERM WILL FORM (INNER GERM LAYER)
Skin, hair, nails, lens of the eye, lining of the internal and external ear, nose sinuses, mouth, anus, tooth enamel, pituitary gland, mammary glands, and all parts of the nervous system	Muscles, bones, lymphatic tissue, spleen, blood cells, heart, lungs, and reproductive and urinary systems	Lining of the lungs, tongue, tonsils, urethra, associated glands, bladder, and digestive tract

DIFFERENTIATION OF REPRODUCTIVE ORGANS

Reproductive organs are the only organs in the human body that differ between the male and female, and yet there is still a significant similarity. This likeness results from the fact that female and male organs develop from the same group of embryonic cells. For approximately the first 10 weeks, the embryo develops without defined sexual characteristics. Then the influence of the X or Y chromosome begins to make a difference.

The gonads, or sexual organs, of the female begin to evolve at about the 10th or 11th week of pregnancy. The ovaries of the female embryo develop from the same type of tissue as the testes of the male embryo. However, the testes evolve from the medulla of the gonad, whereas the ovary develops from the cortex of the gonad. **Figure 22-7** illustrates how the undifferentiated external genitalia develop into fully differentiated structures. In the male, the tubercle becomes the **glans penis** (**GLANZ PEE**-nis), the folds become the **penile shaft** (**PEE**-nile), and the swelling develops into the **scrotum** (**SKROH**-tum). In the female, the tubercle becomes the **clitoris** (**KLIT**-oh-ris), the folds the **labia minora** (**LAY**-bee-ah mih-**NOR**-ah), and the swelling the **labia majora** (mah-**JOR**-ah). Internally, differentiation develops from initially similar structures. In the male, the embryonic Müllerian ducts degenerate, and the Wolffian ducts become the **epididymis** (**ep**-ih-**DID**-ih-mis), **vas deferens** (vas-**DEF**-er-enz), and the **ejaculatory duct** (ee-**JAK**-yoo-luh-tor-ee). In the female, the Wolffian ducts degenerate, and the Müllerian ducts develop into the fallopian tubes (fal-**LOH**-pee-an), the **uterus** (**YOU**-ter-us), and the upper portion of the vagina. It is believed that the presence of

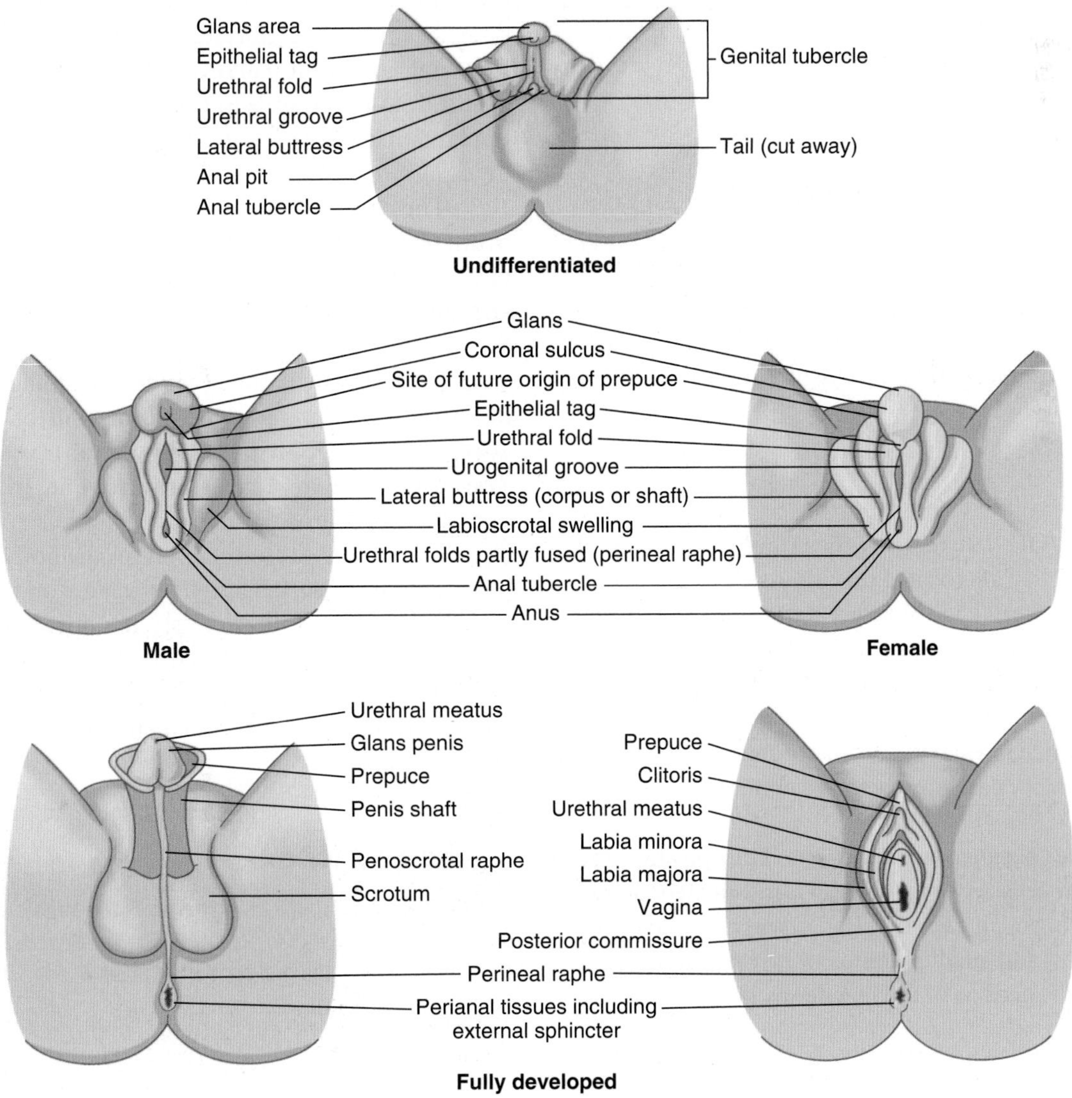

Figure 22-7 *Development of undifferentiated external genitalia into fully differentiated structures*

testes in the male is the differentiating factor in development. Without androgens (male hormones) from the testes, a female will develop. With androgens, a male develops. Another substance, the Müllerian inhibitor, works with androgen to produce the sex differentiation.

Organs of Reproduction

The function of the reproductive system is to provide for continuity of the species. In the human, the female reproductive system is composed of two ovaries, two fallopian tubes, the uterus, and the vagina. The male reproductive system is made up of two testes, ductus deferens, glands, and the penis. The principal male organs are located outside the body, in contrast to the female organs, which are largely located within the body.

The reproductive organs mature usually around the age of **puberty** (**PYU**-ber-tee), which is the period of sexual maturation and achievement of fertility. The time when puberty begins varies greatly among individuals, girls between the ages of 10 to 14 and boys between the ages of 12 to 16. Both genetic and environment factors are involved in the timing.

FEMALE REPRODUCTIVE SYSTEM

The female reproductive system consists of two ovaries, two fallopian tubes, the uterus, the vagina, and the external female genitalia. The breasts are accessory organs.

Ovaries

The **ovaries** (**OH**-vah-rees) are the primary sex organs of the female. They are located on either side of the pelvis, lateral to the uterus, in the lower part of the abdominal cavity. Each ovary is about the shape and size of a large almond, measuring about 3 cm (1.1 inches) long and from 1.5 to 3 cm (0.59–1.18 inches) wide. An ovarian ligament, which is a short, fibrous cord within the broad ligament, attaches each ovary to the upper lateral part of the uterus.

Ovaries perform two functions. They produce the female germ cells, or **ova** (**OH**-vah), and the female sex hormones *estrogen* (**ES**-troh-jen), *progesterone* (pro-**JES**-ter-ohn), and small amounts of **relaxin**. **Table 22-3** outlines the functions of the female sex hormones.

Each ovary contains thousands of microscopic, hollow sacs called **graafian follicles** (**GRAF**-ee-an **FOL**-ih-kuls) in varying stages of development. An ovum slowly develops inside each follicle. In addition, the graafian follicle produces the hormone estrogen.

As the follicle enlarges, it migrates to the outside surface of the ovary and breaks open, releasing the ovum from the ovary. This process is called **ovulation** (**ov**-yoo-**LAY**-shun); it occurs about two weeks before the menstrual period begins. The time of ovulation may vary depending on emotional and physical health, state of mind, and age. The reproductive years begin at the time of puberty and the **menarche** (meh-**NAR**-kee), which is the initial menstrual discharge of blood. During a female's reproductive years, she produces about 400 ova.

An ovum consists of cytoplasm and some yolk. This yolk is the initial food source for the growth of the early embryo. After ovulation, the ovum travels down one of the fallopian tubes, or oviducts. Fertilization of the ovum takes place only in the outer third of the oviduct. The time of fertilization is limited to a day or two following ovulation. Following fertilization, the zygote, or fertilized egg, travels to the well-prepared uterus and implants itself in the wall of the endometrium (uterus lining).

The development of the follicle and release of the ovum occur under the influence of two hormones of the pituitary gland: the follicle-stimulating hormone (FSH) and the luteinizing hormone (LH). FSH also promotes the secretion of estrogen by the ovary.

Following ovulation, the ruptured follicle enlarges, takes on a yellow fatty substance, and becomes the **corpus luteum** (**KOR**-pus **LOO**-tee-um) (yellow body). The corpus luteum secretes progesterone, which maintains the growth of the uterine lining. Corpus luteum also produces relaxin, which relaxes the body's muscles, joints, and ligaments. This effect centers on the joints of the pelvis, allowing them to stretch during delivery. If the egg is not fertilized, the corpus luteum degenerates, progesterone production stops, the thickened glandular endometrium sloughs off, and menstruation (men-stroo-**AY**-shun) begins.

Did You Know?

The largest cell in the human body is the female ovum. It is about 1/180-inch in diameter. The smallest cell in the human body is the male sperm. It takes about 175,000 sperm cells to weigh as much as one egg cell.

Table 22-3 *Functions of Estrogen, Progesterone, and Relaxin*

HORMONE	FUNCTION
Estrogen	1. Affects the development of the fallopian tubes, ovaries, uterus, and vagina 2. Produces secondary sex characteristics: • Broadening of the pelvis, making the outlet broad and oval to permit childbirth • The epiphysis (growth plate) becomes bone and growth ceases • Development of softer and smoother skin • Development of pubic and axillary hair • Deposits of fat in the breasts and development of the duct system • Deposits of fat in the buttocks and thighs • Sexual desire 3. Prepares the uterus for the fertilized egg
Progesterone	1. Develops the excretory portion of mammary glands 2. Thickens the uterine lining so it can receive the developing embryo egg 3. Decreases uterine contractions during pregnancy
Relaxin	1. Produced by the ovaries and placenta during pregnancy 2. Helps relax the ligaments in the pelvis 3. Helps soften and widen the cervix

Fallopian Tubes

The **fallopian tubes** (fal-**LOH**-pee-an), or **oviducts** (**OH**-vih-ducts), are about 10 cm (4 inches) long and are not attached to the ovaries (**Figure 22-8**). The outer end of each fallopian tube curves over the top edge of each ovary and opens into the abdominal cavity. This portion of the fallopian tube, nearest the ovary, is the infundibulum. Because the infundibulum is not attached directly to the ovary, it is possible for an ovum to accidentally slip into the abdominal cavity and be fertilized there. If the fertilized egg implants in the fallopian tube instead of the uterus it is called an **ectopic pregnancy** (eck-**TOP**-ick). An ectopic pregnancy can also occur outside the uterine cavity.

The area of the infundibulum over the ovary is surrounded by a number of fringelike folds called **fimbriae** (**FIM**-bree-eh). Each fallopian tube is lined with mucous membrane, smooth muscle, and ciliated epithelium. The combined action of the peristaltic contractions of the smooth muscles and the beating of the cilia helps propel the ova down the fallopian tube into the uterus. Conception, or fertilization, takes place in the outer third of the fallopian tube.

Uterus

The uterus is a hollow, thick-walled, pear-shaped, and highly muscular organ. The nongravid, or nonpregnant, uterus measures about 7.5 cm (3 inches) in length and is 5 cm (2 inches) wide and 2.75 cm (1 inch) thick. The uterus lies behind the urinary bladder and in front of the rectum. The broad ligaments support the uterus, extending from each side of the organ to the lateral body wall. The ovaries are suspended from the broad ligaments. The fallopian tubes lie within the upper borders of the broad ligaments. The uterine cavity is extremely small and narrow. During pregnancy, however, the uterine cavity greatly expands to accommodate the growing embryo and a large amount of fluid.

The uterus is divided into three parts: (1) the **fundus** (**FUN**-dus), the bulging, rounded upper part above the entrance of the two oviducts into the uterus; (2) the body, or middle portion; and (3) the **cervix** (**SER**-vicks), the cylindrical, lower, narrow portion that extends into the vagina (**Figure 22-8**). A short cervical canal extends from the lower uterine cavity—the internal orifice, or os of the uterus—to the external os at the end of the cervix. The uterine wall is comprised of the following three layers:

- **Perimetrium**—the outer layer
- **Myometrium** (my-oh-**MEE**-tree-um)—the extremely thick, smooth, muscular middle layer
- **Endometrium** (en-do-**MEE**-tree-um)—an inner mucous layer

The endometrium, which lines the oviducts and the vagina, is also lined with ciliated epithelial cells, numerous uterine glands, and many capillaries.

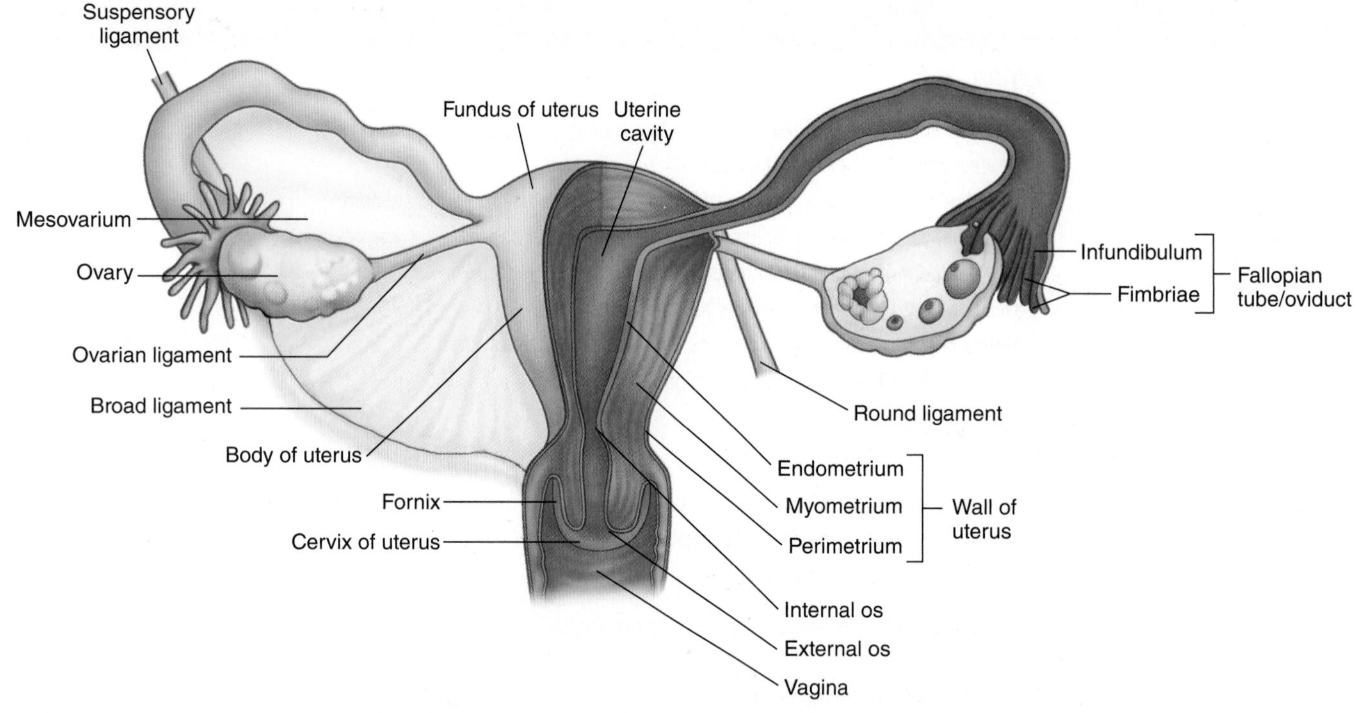

Figure 22-8 *The position of the ovaries, fallopian tubes, uterus, and vagina of the female reproductive system*

During development of the embryo-fetus, the uterus gradually rises until the top part is high in the abdominal cavity, pushing on the diaphragm. This may cause the expectant mother some difficulty in breathing during the late stages of pregnancy.

Vagina

The **vagina** (vah-**JIGH**-nah) is the short canal that extends from the cervix of the uterus to the vulva. The vagina consists of smooth muscle with a mucous membrane lining. This type of muscle tissue allows the vaginal canal to accommodate the penis during sexual intercourse; it also permits a baby to pass through the vaginal canal during the birthing process (**Figure 22-9**).

EXTERNAL FEMALE GENITALIA/VULVA

The external female genitalia, or **vulva** (**VUL**-vah), contains the external organs of the reproductive area (**Figure 22-10**). The large pad of fat that is covered with coarse hair on the mature female and overlies the symphysis pubis is known as the **mons pubis** (**MONZ PYOU**-bis). From the mons pubis, extending posteriorly and inferiorly are two longitudinal folds of hair-covered skin called the labia majora (large lips). These folds of skin have adipose tissue and sweat glands. Medial to the labia majora are two delicate folds of skin called the labia minora. The labia minora have no hair and many sebaceous glands. The opening or region between the labia minora is called the **vestibule**.

Within the vestibule is a thin fold of tissue called the **hymen** (**HIGH**-men), which partially closes the distal end of the vagina. The hymen has some openings that allow for the flow of blood during menstruation. During the first act of sexual intercourse, the openings in the hymen are enlarged, and there may be some slight bleeding. This tissue may also be torn during sports activity or the insertion of a tampon. Also located in the vestibule area are the vaginal orifice and the urethral opening. Above the urethral opening is a small structure called the clitoris, which contains many nerve endings and blood vessels. When stimulated, the blood vessels engorge, and the highly sensitive clitoris provides sexual pleasure for the female. After the clitoris is properly stimulated, the vagina is fully lubricated and ready for full insertion of the penis.

The two round glands located on either side of the vaginal orifice are the **Bartholin's glands**, which produce mucus. This mucus lubricates the distal end of the vagina during intercourse.

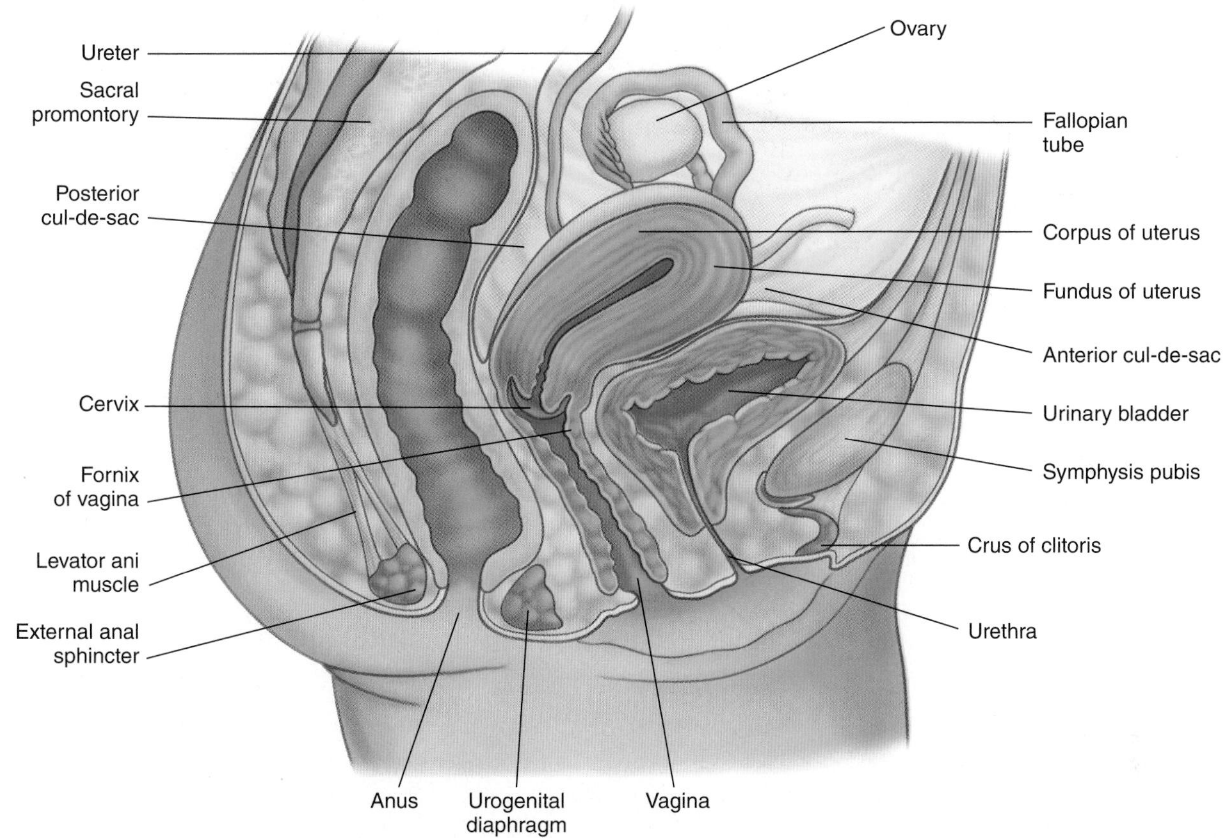

Figure 22-9 *Structures of the female reproductive system*

The **perineum** (pehr-ih-**NEE**-um) is the area between the posterior vaginal opening and the anus.

Breasts

The **breasts** are made up of fat, connective tissue, and mammary glands. Mammary glands are present in both males and females, but they normally only function in females. They are accessory organs to the female reproductive system (**Figure 22-11**). They consist of numerous lobes, which contain the milk-secreting cells. Clusters of these secreting cells surround mammary ducts. Prolactin from the anterior lobe of the pituitary gland stimulates the mammary glands to secrete milk following childbirth.

A single duct extends from each lobe to an opening in the nipple. The **areola** (ah-**REE**-oh-lah) is the darker area that surrounds the nipple. Sebaceous glands surround the areola and produce sebum, which prevents the nipple from chapping and cracking. The autonomic nervous system controls the smooth muscle fibers in the areola and nipple. The nipple becomes erect when stimulated by tactile sensation, cold, or during breast-feeding.

Media Link

View the **Female Reproduction** animation on the Online Resources.

THE MENSTRUAL CYCLE

In females, a mature egg develops and is ovulated from one of the two ovaries about once every 28 days through a complex series of actions between the pituitary gland and the ovary. Before the mature egg is released from the ovary, a series of events occurs to thicken the uterine lining, or endometrium. This is necessary to receive and hold a fertilized egg for embryonic development. If the egg is not fertilized, the endometrium starts to

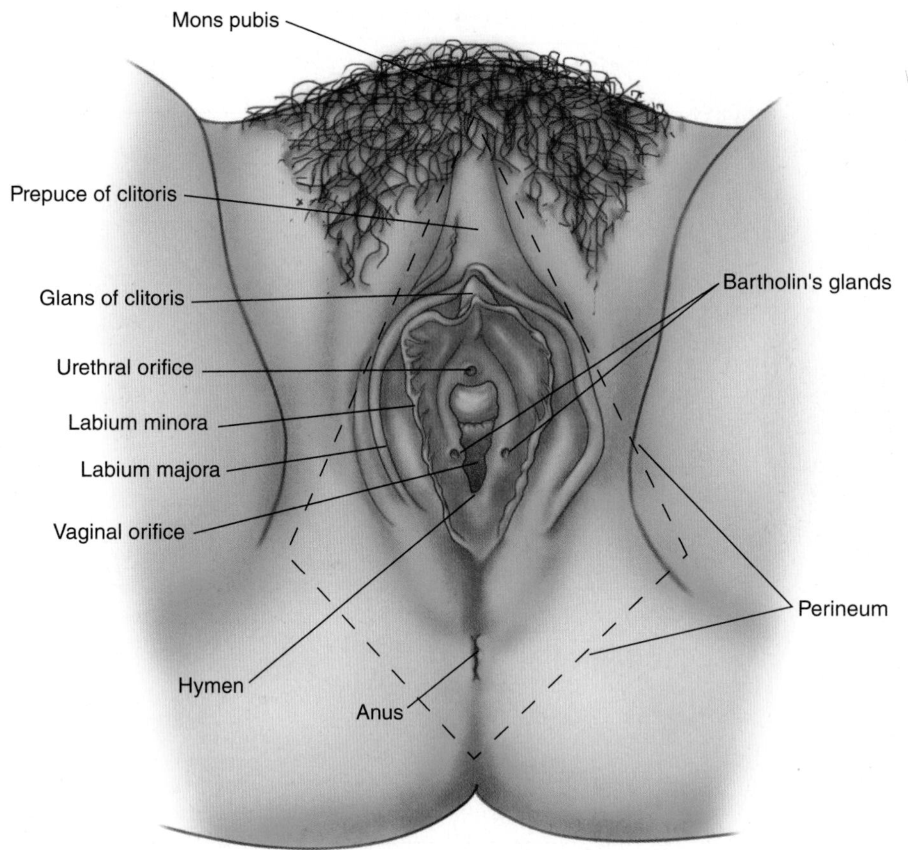

Figure 22-10 *External female genitalia, or vulva*

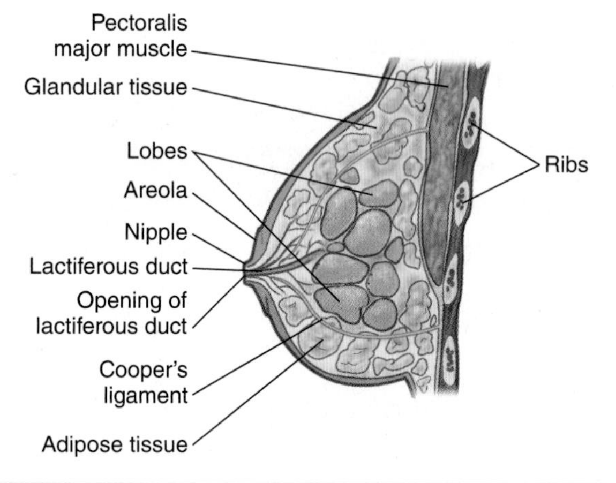

Figure 22-11 *Sagittal section of the female breast*

break down. Eventually the old, unfertilized egg and the degenerated endometrium are discharged out of the female reproductive tract in a process called **menstruation** (**men**-stroo-**AY**-shun). The cycle then starts all over again with the development of another ovum and the buildup of the endometrium.

This cycle is called the **menstrual cycle**. The menstrual cycle starts at puberty. It can start as early as 9 years of age to as late as 17 years of age. Generally, the age range is between 10 and 14. The changes that occur during the menstrual cycle involve hormones from the pituitary gland and the ovaries.

Stages of the Menstrual Cycle

The menstrual cycle is divided into four stages: the follicle stage, the ovulation stage, the corpus luteum stage, and the menstruation stage. See **Figure 22-12** for a diagram of the menstrual cycle.

FOLLICLE STAGE

Follicle-stimulating hormone (FSH) is secreted from the anterior lobe of the pituitary gland on day five of the menstrual cycle. FSH is then circulated to an ovary via the bloodstream. When FSH reaches an ovary, it will stimulate several follicles. Usually, a single follicle matures every 28 days throughout the reproductive years of a female. Occasionally two or more follicles

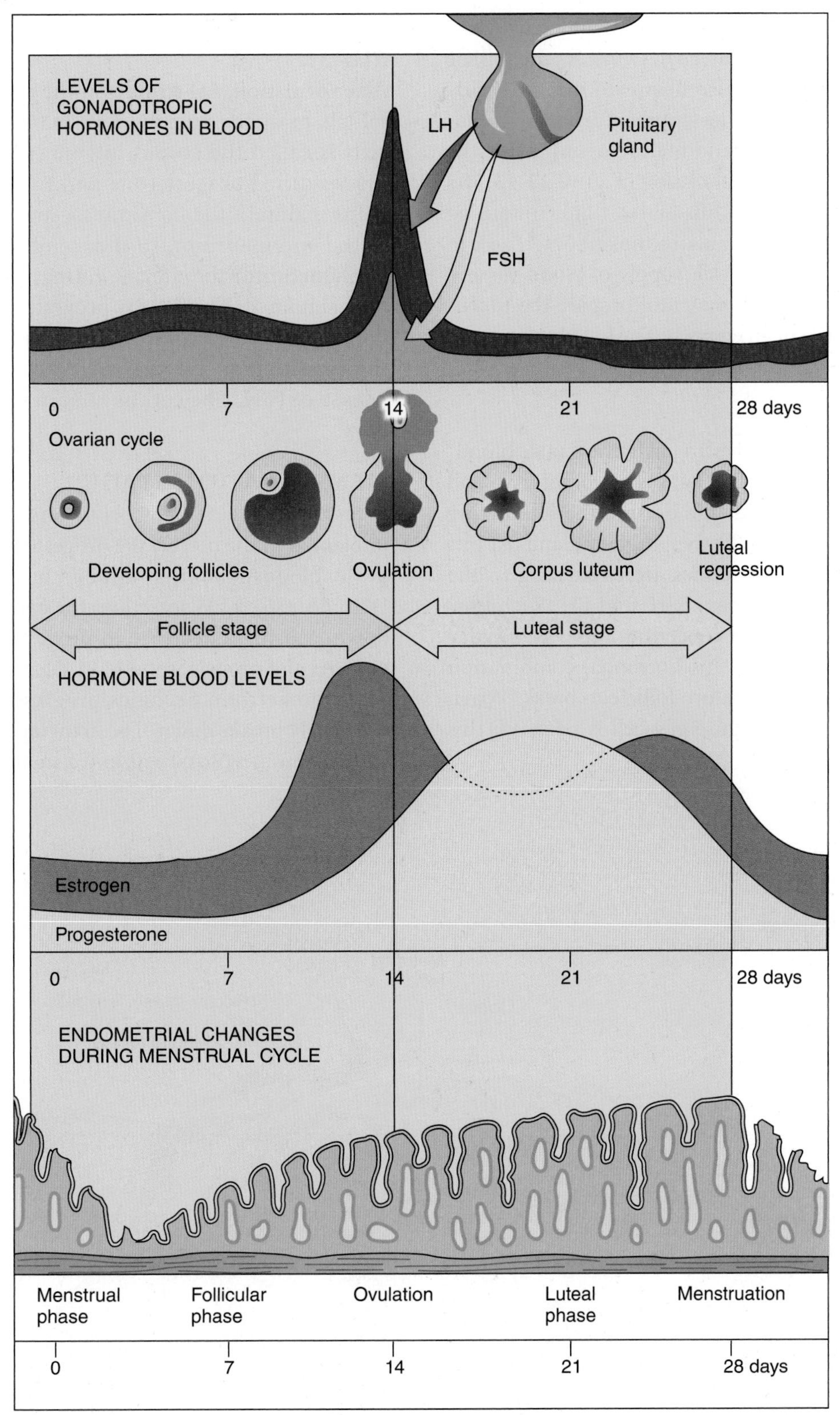

Figure 22-12 *The menstrual cycle*

may mature, releasing more than one ovum, which may result in nonidentical, or fraternal, twins. As the follicle grows in size, an egg cell also begins to mature inside the follicle. The process of development from an immature ovum to a functional and mature ovum inside the graafian follicle is called maturation (**Figure 22-13**). As the follicle grows in size, it fills with a fluid containing estrogen. The estrogen stimulates the endometrium to thicken with mucus and a rich supply of blood vessels. These changes to the endometrium prepare the uterus for the implantation of an embryo. The follicle stage lasts about 10 days.

OVULATION STAGE

When the concentration of estrogen in the female bloodstream reaches a high level, it causes the pituitary gland to stop FSH secretion. As this occurs, the luteinizing hormone (LH) is secreted by the pituitary gland. At this point, three different hormones are circulating in the female bloodstream: estrogen, FSH, and LH. Each hormone is present in different concentrations. Around day 14 of the menstrual cycle, this hormonal combination somehow stimulates the mature follicle to break. When the follicle ruptures, a mature egg cell is released; this event is called ovulation.

CORPUS LUTEUM STAGE, OR LUTEAL PHASE

After ovulation, LH stimulates the cells of the ruptured follicle to divide quickly. This mass of reddish-yellow cells is called the corpus luteum, which secretes hormones called progesterone and relaxin. Progesterone helps maintain the continued growth and thickening of the endometrium, so if an embryo happens to be implanted into the uterine lining, the pregnancy can be maintained. That is why progesterone is often called the "pregnancy hormone." Progesterone also prevents the formation of new ovarian follicles by inhibiting the release of FSH. The corpus luteum stage lasts about 14 days.

MENSTRUATION STAGE

If fertilization does not occur and an embryo is not implanted in the uterus, the progesterone reaches a level in the bloodstream that inhibits further LH secretion. With decreased LH secretion, the corpus luteum breaks down, causing a decrease in progesterone secretion as well. As the progesterone level decreases, the lining of the endometrium becomes progressively thinner and eventually breaks down. The extra layers of the endometrium, the unfertilized egg, and a small quantity of blood

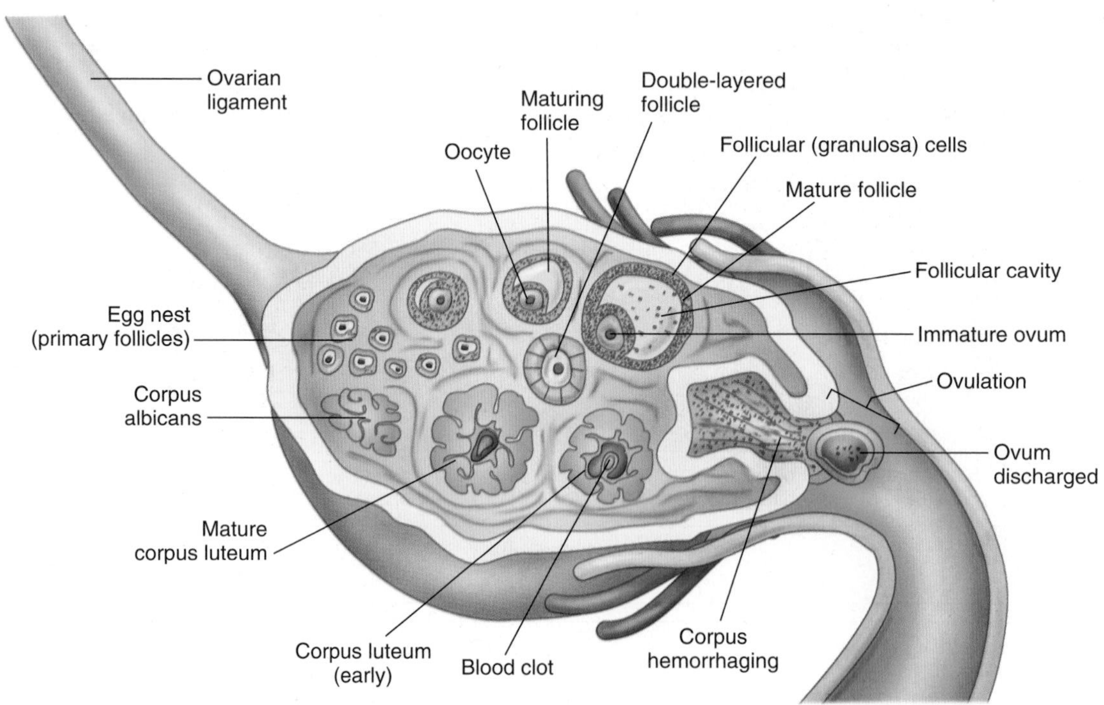

Figure 22-13 *An ovary showing the development of an ovum in a graafian follicle*

that comes from the ruptured capillaries as the endometrium peels away from the uterus are discharged from the female's body through the vagina. This causes the characteristic menstrual blood flow. The menstruation stage starts around day 28 of the cycle and lasts about four days. While menstruation is occurring, the estrogen level in the bloodstream is decreasing. The anterior lobe of the pituitary gland is now stimulated to secrete FSH; consequently, a new follicle starts to grow, and the menstrual cycle starts again.

The relationship between the pituitary gland hormones and the ovarian hormones is one of feedback. That means pituitary hormones control the functioning of the ovaries; in turn, the ovaries secrete hormones that control pituitary functioning. This is another example of the automatic regulation, or homeostasis, of many of the body's processes.

Media Link

View the **Ovulation** animation on the Online Resources.

Menopause

Menopause (**MEN**-oh-pawz) or "change in life" is the time in a female's life when the monthly menstrual cycle comes to an end. It frequently occurs between ages 45 and 55. Menopause signals the end of follicle growth, leading to a decrease in ovulation and the hormones estrogen and progesterone. Consequently, no ovulation means the end of childbearing. However, a normal sex drive usually remains.

The menopausal female will experience the following anatomical changes:

1. Atrophy of the external genitalia and of the internal reproductive structures: uterus, fallopian tubes, and ovaries
2. Vagina becomes conical-shaped
3. Atrophy of the vaginal mucous membranes, causing vaginal dryness
4. Reduction of the secretory activity of glands associated with the reproductive organs
5. More facial and chin hair

These changes do not occur overnight; they happen gradually over a period of years. Pronounced physiological changes may also occur, including "hot flashes," a brief sensation of heat that may be accompanied by a flushed face, sweating, and chill after the flash; dizziness; headaches; insomnia; rheumatic pains in joints; sweating; and susceptibility to fatigue. Sometimes, these physiological changes are also accompanied by psychological changes. These include abnormal fears, depression, excessive irritability, and a tendency to worry. Menopausal symptoms may be treated with estrogen. Menopause can be induced prematurely, creating artificial menopause, by the removal of ovarian tissue.

MALE REPRODUCTIVE SYSTEM

The male reproductive organs (**Figure 22-14**) consist of the following structures:

1. The two testes produce the male gametes, **spermatozoa** (**sper**-mat-oh-**ZOH**-ah), and the male sex hormone *testosterone* (tes-**TOS**-teh-rohn). They are suspended from the body wall by a spermatic cord and encased in a pouch called the scrotum.
2. A system of ducts carries the sperm cells out of the testes through the epididymis, two ductus deferens called vas deferens, two ejaculatory ducts, and the urethra.
3. Accessory glands include the two seminal vesicles, two bulbourethral glands, and a prostate gland. These glands add a viscous fluid to the sperm cells to form seminal fluid.
4. The penis is a copulatory structure that will transfer sperm cells to the female reproductive system.

Testes and Epididymis

The two **testes** (**TES**-teez) are the primary male reproductive organs (**Figure 22-15**). They are found in a pouch outside the male body called the scrotum. Each testis is about the size and shape of a small egg, approximately 4 cm (1.57 inches) long, 2.5 cm (0.98 inches) wide, and 2 cm (0.78 inches) thick. The testes are attached to an overlying structure called the epididymis. A fibrous tissue called the tunica albuginea (al-byoo-**JEN**-ee-ah) covers the testes and sends incomplete partitions into the body of each testis. Each partition is called a lobule, and each testis contains approximately 250 lobules.

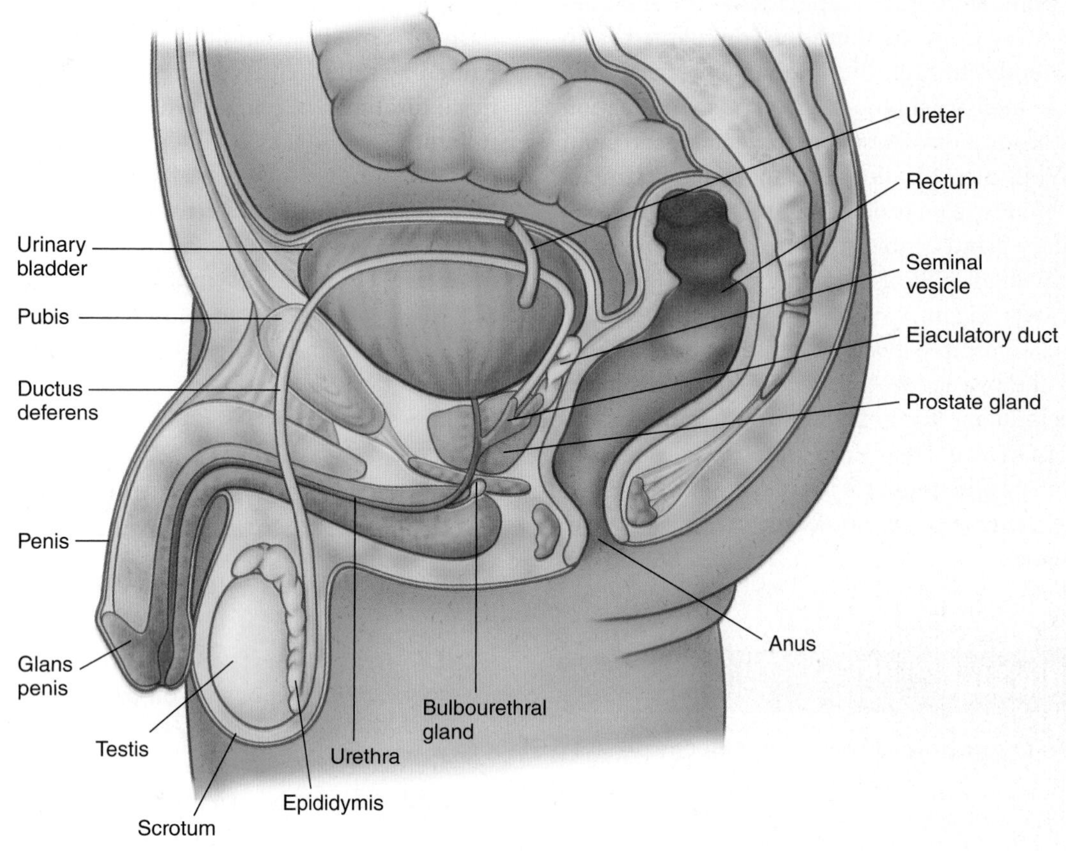

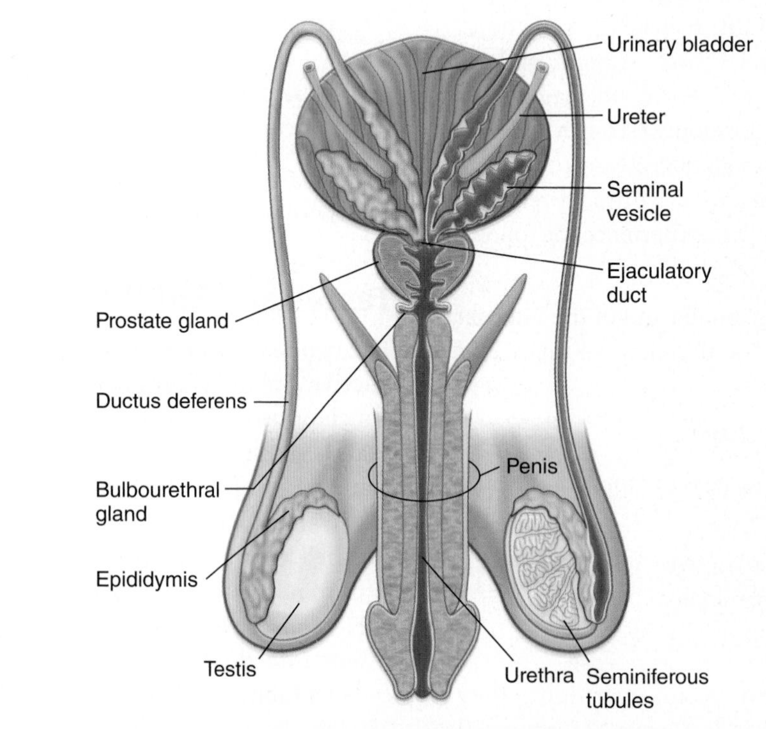

Figure 22-14 *(A) Sagittal view of the male reproductive system; and (B) internal and external structures of the male*

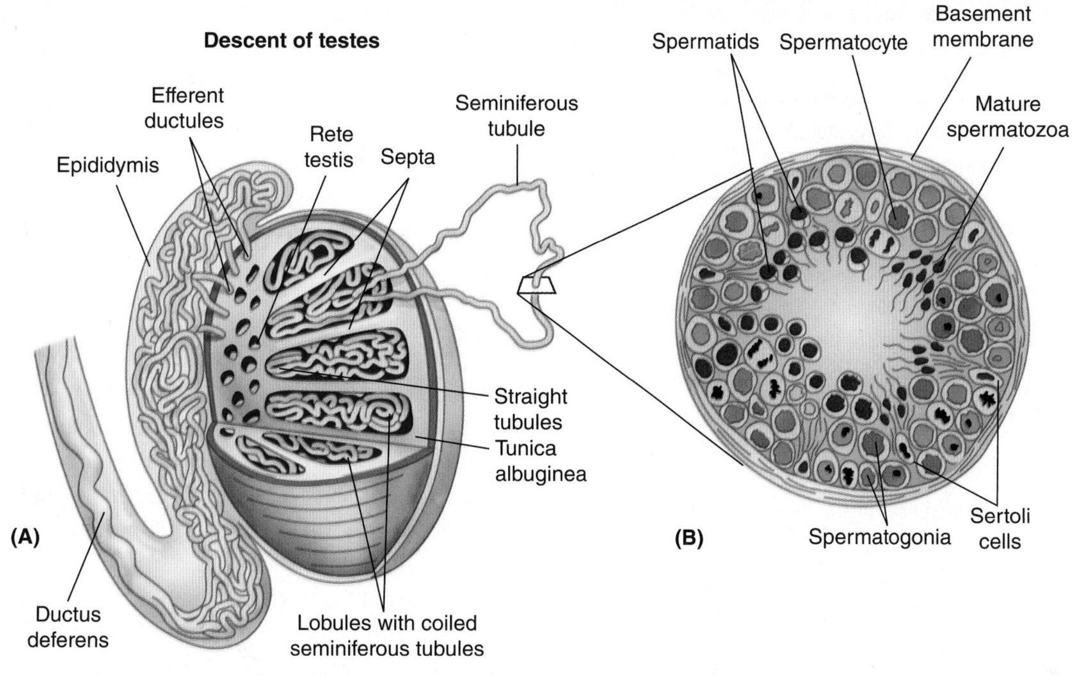

Figure 22-15 *(A) Sagittal section of the testis; and (B) saggital section of the seminiferous tubule*

Each testicular lobule contains one to four minute and highly convoluted (twisted) **seminiferous tubules** (sem-in-**IF**-er-us **TOO**-byouls). FSH from the anterior pituitary stimulates the production of sperm in the cells that line the tubules. As the sperm develop, they are released into the tubules. In males, mature sperm formation requires about 74 days. The function begins at about age 12, and the first mature sperm are ejaculated at about age 14. All of the seminiferous tubules intertwine and join to form a small, mesh-like network of tubules called the rete testis. The rete (**REE**-tee) testis unites to form the epididymis. The seminiferous tubules are supported by a type of tissue called interstitial tissue. The LH, also known as interstitial cell-stimulating hormone of the anterior pituitary, stimulates the interstitial cells to produce testosterone. Testosterone is secreted in relatively steady amounts during the adult life of the male. **Table 22-4** outlines the functions of testosterone.

The epididymides connect the testes with the ductus deferens and help in the final development of the sperm cells.

DESCENT OF THE TESTES

In the embryo, the testes are formed and developed in the abdominal wall slightly below the kidneys. During the last three months of fetal development, the testes will migrate downward through the ventral abdominal wall into the scrotum. In its descent, each testis carries with it the ductus deferens, blood and lymphatic vessels, and autonomic nerve fibers. These structures and their fibrous tissue covering form the spermatic cord.

Occasionally, the testes do not descend in premature babies. If the testes do not descend, this condition is known as **cryptorchidism** (krip-**TOR**-kih-dizm); if one testis does not descend, it is called unilateral cryptorchidism. For two testes, it is called bilateral cryptorchidism. If the testes stay inside the abdomen after puberty, spermatogenesis will be affected. The body temperature will destroy any sperm cells. A simple surgical procedure done before puberty can correct this condition.

Ductus Deferens, Seminal Vesicles, and Ejaculatory Ducts

The right and left **ductus deferens** (**DUK**-tus **DEF**-er-enz), or vas deferens, are continuations of the epididymides. The ductus deferens have a dual function. They serve as a storage site for sperm cells and as the excretory ducts for the testes.

Table 22-4 Function of Testosterone

HORMONE	FUNCTION
Testosterone	1. Develops the male reproductive organs: testes, ducts, glands, and penis 2. Produces sperm 3. Produces secondary sex characteristics: • Pubic, auxiliary, and facial hair • Enhanced hair growth on chest or other body areas • Deepening of the voice, enlarged larynx • Thickening of skin, which becomes oilier • Thickening of the bones of the skeletal system • Increases in size and mass of skeletal muscle • Growth and increased density of bones 4. Closes the epiphyseal plate and terminates skeletal growth in height 5. Is the basis for the sex drive (libido)

Each ductus runs from the epididymis up through the inguinal canal. It continues to run downward and backward to the side of the urinary bladder. Each ductus then curves around the ureter and goes down to meet with the seminal vesicle duct on the posterior side of the bladder.

The **seminal vesicles** (**SEM**-ih-nal **VES**-ih-kuls) are two highly convoluted membranous tubes. A duct leads away from a seminal vesicle that joins the ductus deferens to form the ejaculatory duct on either side. The seminal vesicles produce secretions that help nourish and protect sperm on their journey through the female reproductive system. At the precise moment of ejaculation, the seminal fluid is added to the sperm cells as they leave the ejaculatory ducts.

The ejaculatory ducts are short and very narrow. They begin where the ductus deferens and the seminal vesicle ducts join. They then descend into the prostate gland to join with the urethra, into which they discharge their contents (**Figure 22-14**).

Scrotum and Penis

The external organs are the scrotum and the penis. Internally, the scrotum is divided into two sacs, each containing a testis, epididymis, and lower part of the ductus deferens. The **penis** contains erectile tissue that becomes enlarged and rigid during sexual stimulation. Loose-fitting skin, called the **foreskin** or **prepuce** (**PRE**-pyous), covers the end of the penis. The foreskin can be removed in a simple operation known as **circumcision** (**ser**-kum-**SIZH**-un).

Prostate Gland

The **prostate gland** (**PROS**-tayt) is located in front of the rectum and just under the urinary bladder. It surrounds the opening of the bladder leading into the urethra. It also surrounds the beginning portion of the urethra that is called the prostatic urethra. The prostate gland is about the shape and size of a chestnut. It is covered by a dense fibrous capsule and contains glandular tissue surrounded by fibromuscular tissue that contracts during ejaculation.

The contraction of the prostate gland closes off the prostatic urethra during ejaculation, preventing the passage of urine through the urethra. This contraction of the muscular tissue also aids in the expulsion of semen during ejaculation. The prostate gland secretes a thin, milky, alkaline fluid that enhances sperm motility. It also gives semen its characteristic strong, musky odor. The alkaline prostatic fluid probably neutralizes the acidic vaginal secretions of the female, which enhances the viability and motility of the sperm cells.

Bulbourethral Glands

The **bulbourethral glands (Cowper's glands)** (**bul**-boh-you-**REE**-thral) are located on either side of the urethra below the prostate gland. They add an alkaline secretion to the semen. Their ducts connect with the spongy urethra. The secretion is the first to move down the urethra when a male becomes sexually aroused and develops an erection. It functions both as a lubricant for sexual intercourse and as an agent to clean the urethra of any traces of acidic urine.

Semen

Semen, also known as seminal fluid, is a mixture of sperm cells and the secretions of the seminal vesicles, prostate, and the bulbourethral glands. The fluid is milky in color and sticky due to the presence of fructose sugar, which provides the energy for the beating of the flagellum of each sperm cell. Semen is alkaline, which neutralizes the acidity of the female vagina and the male urethra. This helps protect the sperm cell. The semen is a transport medium for the swimming sperm cells.

Semen also contains enzymes that activate sperm after ejaculation. Semen contains an antibiotic called seminalplasmin that has the ability to destroy certain bacteria. Because the female reproductive tract may contain bacteria, the seminalplasmin helps keep these bacteria under control.

Media Link

View the **Male Reproductive System** animation on the Online Resources.

Erection and Ejaculation

The urethra extends down the length of the penis, opening at the urinary meatus of the glans. The urethra serves two purposes: to empty urine from the bladder and to expel semen. During ejaculation, the prostate gland contracts and the smooth muscle sphincter at the base of the urinary bladder closes. This ensures that urine is not expelled during ejaculation and that semen does not enter the urinary bladder. Sexual intercourse becomes possible due to the columns of erectile tissue in the penis. When a male is sexually aroused, nerve impulses cause the erectile tissue to engorge with blood, which makes the erectile tissue increase in size and become firm. Blood entering the dilated arteries squeezes the veins against the penile structures, prohibiting venous return.

Once stimulation of the glands results in maximum stimulation of the seminal vesicles, impulses are sent to the ejaculatory center and orgasm occurs. Muscular contractions from the ductus deferens, ejaculatory ducts, and prostate glands result in orgasm. Secretions stored in these structures along with the sperm are forcibly expelled through the urethra, after which the engorgement gradually subsides.

Media Link

View the **Sperm Formation** animation on the Online Resources.

IMPOTENCE/ERECTILE DYSFUNCTION

Impotence (**IM**-poh-tens) is the inability to have or sustain an erection during intercourse. Primary impotence refers to the male who has never had an erection. Secondary impotence refers to the male who is currently impotent but has had erections in the past. Transient periods of impotence are not considered a dysfunction and probably occur in half the adult male population between the ages of 40 and 70; the incidence increases with age.

The majority of impotence involves organic causes such as diabetes, multiple sclerosis, atherosclerosis, Parkinson's disease, or stroke. Psychological factors, such as anxiety and stress, can also be causes of impotence. Certain drugs can also cause impotence such as beta blockers, chemotherapy, central nervous system depressants, diuretics, and serotonin uptake inhibitors such as Prozac.

The type of therapy chosen depends on the specific cause of the dysfunction. Treatment may be sexual therapy if the cause is thought to be related to psychological factors. At the present time, penile implants, injection therapy, and oral medications are being used to treat impotence.

CONTRACEPTION

Some religious and ethnic groups oppose birth control; this text does not ignore that issue. This subject matter is presented factually, from a clinical viewpoint, as information required for practice as a health care provider. As the word implies, *contraception* is literally "against conception." There are many reasons why a person may choose to use contraception:

- *Avoiding health risks*. The person is in poor health and may not survive a pregnancy, or the person may have a health condition that contraceptives help to treat.
- *Spacing pregnancies*. The person is very fertile and could conceive every year or less. The infant death rate is reported to be 50% higher at one-year intervals than at two or more years.
- *Avoiding having babies with birth defects*. Some people have chromosome defects, are genetic disease carriers, or their partners are carriers and choose not to risk pregnancy.
- *Delaying pregnancy early in a partnership*. This is done to allow time for adjustment, to avoid additional stress in a new relationship, and establish a strong partnership.
- *Limiting family size*. It is sometimes a personal decision and other times a reality of limited resources.

- *Avoiding pregnancy among unmarried couples.* Single parenthood is difficult.
- *Curbing population growth.* The concern over worldwide food supply and supportive environment prompts some to promote contraception.

Several methods to prevent conception and their relative percentage of effectiveness are listed in **Table 22-5**. Selection is usually made by the person carrying the child or as a couple in consultation with a physician. The cost, ease of use, degree of effectiveness, and likelihood of side effects must be taken into consideration when selecting a method. One choice in contraception is abstinence. Abstinence is the voluntary restraint of sexual intercourse. Abstinence is a positive, healthy choice many people make. Douching is not recommended as a form of birth control because it is unreliable. It only takes a few minutes for sperm to enter the cervix; douching cannot be accomplished quickly enough to prevent impregnation.

Table 22-5 *Different Methods of Preventing Conception*

% EFFECTIVE	METHOD	DESCRIPTION/COMMENTS
100%	Abstinence	Refraining from sexual intercourse; most effective.
100%	Sterilization TYPES: Tubal ligation Hysteroscopic sterilization Vasectomy	*Tubal ligation* is a procedure in which a physician cuts, ties, or seals the fallopian tubes in the female. This blocks the passage between the ovaries and the uterus so the sperm cannot reach the egg. The surgical procedure is done through a laparoscope. *Hysteroscopic sterilization* is a nonsurgical procedure in which a physician inserts a 4-cm coil into each of the fallopian tubes. This procedure is done by way of a scope passed through the cervix, uterus, and into the fallopian tubes. During the next three months, scar tissue forms around the inserts and blocks the fallopian tubes. After three months, tests are conducted to be certain the fallopian tubes are completely blocked. A *vasectomy* in the male cuts, ties, or blocks the vas deferens, blocking the path between the testes and urethra. It takes three months for the procedure to be totally effective. Sperm production is usually significantly decreased in time.
95%–99%	Hormonal methods TYPES: Pill Skin patch Injection Vaginal ring	Many different kinds are available. The *pill* is a combination of hormones (estrogen and/or progesterone) that prevent ovulation: no ovum, no pregnancy. It is taken usually on a daily basis. The *skin patch* must be replaced on a weekly basis. The patch is used for three weeks, with no patch in the fourth week to allow for menstruation. *Injections* must be given every three months. The *vaginal ring* remains in place for three weeks and is removed for one week to allow for menstruation. *In all types of hormonal methods, instructions must be followed regarding their use. These methods are available only by prescription and require regular visits to a physician.*
93%–99%	Intrauterine device (IUD)	An IUD is a small, T-shaped device inserted into the uterus. A *copper IUD* releases a small amount of copper, causing an inflammation that prevents the sperm from reaching the egg. If fertilization does occur, the physical presence of the IUD prevents implantation. The copper IUD can remain in place for 12 years. A *hormonal IUD* releases progestin, causing thickening of the cervical mucus, which inhibits sperm from reaching or fertilizing the egg. The hormonal IUD is effective for five years. Side effects are that IUDs bother some people.

(continues)

Table 22-5 (Continued)

% EFFECTIVE	METHOD	DESCRIPTION/COMMENTS
90%–99%	Diaphragm	A diaphragm is a thin piece of dome-shaped rubber with a firm ring that is inserted into the vagina to cover the cervix and provide a barrier to sperm. It is most effective when used in combination with a contraceptive cream placed into the dome before insertion. Failure usually results from improper insertion; a defect in the rubber, such as a hole; failure to insert before any penile penetration; or failure to maintain in place for at least six hours following intercourse. There are no side effects. It requires cleaning and inspection after each use.
85%–97%	Condom	A condom is a thin sheath of rubber or latex that fits over an erect penis to catch the semen. It must be unrolled onto an erect penis *before* any penetration occurs. It is important to leave about 1/2 inch of free air space at the tip to catch the semen; otherwise, the force of the ejaculation may burst the condom. It must also remain in place throughout intercourse. After ejaculation has occurred, care must be taken to withdraw the penis with the condom in place. This is the only contraceptive that also provides a level of protection against sexually transmitted diseases. It is relatively inexpensive, easy to use, and readily available. Only a latex condom is effective against the AIDS virus.
70%–75%	Spermicides	Contraceptive foams, jellies, and creams with *sperm-killing* ingredients are inserted by applicator deep into the vagina before intercourse. The spermicide must remain for at least 6–8 hours afterward. Each application is good for only one act of intercourse. It should not be relied on alone as an effective contraceptive. Combined with a diaphragm, cervical cap, cervical sponge, or condom, a spermicide is effective. It has few side effects, is easily used, and is readily available.
70%–80%	Withdrawal	This method has been practiced since ancient times. It simply requires that the penis be withdrawn and ejaculation occur outside the vagina. It is not very effective because some sperm are deposited in the vagina before ejaculation occurs. In addition, the male may not be able to withdraw in time. This type of contraception is also not advised because it may lead to sexual dysfunction if practiced for a prolonged period of time.
65%–85%	Rhythm	The rhythm method is the practice of abstinence during an 8-day period from days 10–17 of the menstrual cycle, when conception is theoretically possible. The method works fairly well for females who are extremely regular in their cycles and couples who can practice strong self-control. However, it requires a careful assessment of at least six months of cycles to establish ovulation days. If cycles vary in length, the period of abstinence must be increased to cover the longest possible period of time.
	Emergency contraception pills	These are hormonal pills taken as a single dose or two doses, 12 hours apart, after unprotected sexual intercourse. If taken prior to ovulation, the pills delay ovulation for five days, allowing the sperm to become inactive. Pregnancy can occur if the pill is taken after ovulation.

INFERTILITY

Infertility (in-fer-**TIL**-ah-tee) is when conception does not occur despite trying for one year. Causes of infertility may be blockage of the female fallopian tubes or the male epididymis or ductus deferens, a low sperm count, hormonal imbalance, or swollen veins in the scrotum called varicoceles. In some cases, there is no known cause for infertility.

Urologists use a variety of tools and techniques to correct many of the male infertility problems, including hormone replacement (FSH and LH) to raise testosterone levels; **artificial insemination**; medications to counter retrograde ejaculation, which is when semen is redirected back into the urinary bladder; and microsurgery to correct varicoceles.

Gynecologists may order fertility drugs such as Clomid® and Pergonal® for the female to promote ovulation by stimulating the hormones from the pituitary to

Career Profile

MEDICAL ASSISTANT

The medical assistant is an important allied health provider. The medical assistant may perform both administrative and clinical tasks under the direction of licensed medical providers. Medical assistants are skilled communicators who act as liaisons between the physician, patient, hospital staff, and other health care providers.

The medical assistant may specialize in an administrative or clinical role, or both, and perform a wide variety of duties, including receiving patients, answering the telephone, maintaining medical records, handling insurance and billing tasks, educating patients, taking vital signs, and performing a variety of tests and procedures on patients.

Education for this career may be in a vocational school or college program leading to a degree. Certified Medical Assistant (CMA) or Registered Medical Assistant (RMA) organizations provide certification in this area. To be eligible for certification, the candidate must pass certain criteria. With the exception of a few states, both credentials are voluntary in that neither the federal government nor the states require a medical assistant to be certified, although certification can be expected to enhance employment opportunities. Job opportunities in this area are excellent.

prepare an egg or several eggs for ovulation each month. Laparoscopic surgery may be performed for blocked fallopian tubes, endometriosis, or ovarian cysts, all of which are implicated in fertility problems.

Infertility Treatments

If the preceding methods are not successful, the couple may try the following treatments:

1. **In vitro fertilization (IVF)**—After taking fertility drugs, mature eggs, as determined by ultrasound and hormonal blood levels, are removed from the ovaries using a needle inserted through the vaginal wall or by laparoscopy. The eggs are then combined with the sperm. When an egg is fertilized (zygote) and reaches the four- or eight-cell stage, the zygote is transferred to the uterus.
2. *Gamete intrafallopian transfer (GIFT)*—This procedure is similar to IVF, with the exception that the eggs are combined with the sperm and are placed into the fallopian tubes using a laparoscope.
3. *Zygote intrafallopian transfer*—This procedure is similar to GIFT, except technicians monitor the eggs carefully to be certain they are fertilized before placing them in the fallopian tubes.
4. *Donor egg and embryo*—In this case, a person may use both egg and sperm donors or their partner's egg or sperm. They must take a fertility drug to prepare their uterus for the implantation of the zygote. After the donor egg and sperm are fertilized in the laboratory, the procedure is similar to *IVF*.

FEMALE REPRODUCTIVE DISORDERS

Female reproductive disorders may be apparent at birth, emerge with puberty, occur in adulthood, or appear later in life.

Congenital Female Reproductive Disorders

Some female congenital disorders are not apparent at birth but may be seen when puberty occurs.

- **Imperforate hymen** blocks the vaginal opening and prevents the passage of menstrual blood; this leads to pain during puberty. Treatment may be surgical opening.
- The uterus may be absent or a double uterus may be present.
- There may be a **vaginal blockage** or **missing vagina**. Vaginal blockage is treated by surgery; a missing vagina is treated by the use of a dilator. This instrument is used to stretch or widen the area where the vagina is supposed to be. It takes 4–6 months to create a new vagina.

General Female Reproductive Disorders

Amenorrhea (ah-men-oh-**REE**-ah) defines the absence of the menstrual cycle. This is normal if the female is pregnant. Psychological factors, anorexia, and hormonal imbalance can also cause this condition.

Premenstrual syndrome (PMS) is a group of symptoms that are exhibited just prior to the beginning of the menstrual cycle. They are caused by water retention in the body tissue. Irritability, nervousness, mood swings, and weight gain are some of the symptoms of PMS. PMS is no longer considered a myth; it is treated with medication such as nonsteroidal anti-inflammatory drugs (NSAIDs), hormonal drugs, diuretics, and diet to reduce water retention.

Dysmenorrhea (dis-men-oh-**REE**-ah) is a term used to describe painful menstruation. Dysmenorrhea is characterized by cramps, which may be caused by excessive production of an inflammatory substance such as prostaglandins. NSAID substances, which block the action of prostaglandin, are helpful.

Menorrhagia (men-oh-**RAY**-jee-ah) is abnormal excessive or prolonged menstrual bleeding and is one of the most common gynecological problems. Bleeding is considered abnormal when a female is soaking enough sanitary products to require a change more often than every 1–2 hours, having a period that lasts longer than seven days, or passing blood clots with the menstrual flow for more than one day.

This condition can lead to iron deficiency and anemia, and it may also cause distress and discomfort. A pelvic examination is conducted to determine the cause of the menorrhagia, Pap smear and blood tests are done to check for any underlying cause, and an ultrasound may be done to check for abnormalities.

Common causes of menorrhagia are hormone imbalance, polyps, uterine fibroids, and an IUD. To reduce the bleeding, medications such as iron supplements, NSAID drugs, oral contraceptives, and other drugs may be prescribed to modify hormone levels. Treatment may also be a myomectomy to remove fibroids or an endometrial ablation. In endometrial ablation, the endometrial lining of the uterus is partially destroyed. It may be done under general or local anesthesia. The most common of these techniques are radio-frequency ablation, thermal balloon ablation, and electrocautery. Ninety percent of females who receive endometrial ablation have reduced menstrual bleeding one year after treatment.

It is common to experience a vaginal discharge of clear fluid for about a month after the procedure. Endometrial ablation may allow recovery and a return to normal activities in as few as two days. Only females who no longer desire to bear children are treated with this procedure because most females become infertile.

Endometriosis (en-doh-**mee**-tree-**OH**-sis) is a disease that affects females during their reproductive years. In this condition, endometrial tissue is found outside the uterus around the ovaries and other organs in the abdominopelvic cavity.

The Effects of Aging on the Reproductive System

Menopause begins between the ages of 45 and 55 and denotes the end of menstruation and the childbearing period. Estrogen and progesterone production markedly declines. These changes produce physical changes, such as narrowing of the vaginal opening, loss of tissue elasticity, and a decrease in vaginal secretions. Atrophic changes are seen in the uterus, vagina, external genitalia, and breasts. There may be a decline in sexual activity.

For the male, the changes occur at a more gradual pace, varying from person to person. Phases of the sexual response in males are slower, obtaining and maintaining an erection becomes more difficult, and impotence may result. The prostate gland increases in size, testes decrease, sperm level decreases, testosterone level decreases, and the viscosity of seminal fluid diminishes.

Age-related physical changes experienced by both sexes do not prevent sexual function and do not alter the pleasure of sex or inhibit desire. Medication regimens typically used by older adults are common causes of sexual dysfunction.

Every month, the tissue, like the lining of the uterus, responds to hormonal changes. Endometrial tissue outside the uterus allows no way for the blood to leave the body. The result is internal bleeding, inflammation of the surrounding areas, and formation of scar tissue. This condition causes pain before and during menstruation, during or after sexual activity, infertility, and heavy or irregular bleeding. The cause is unknown, but different theories exist. One theory is that during menstruation, some of the tissue backs up through the fallopian tubes, implants in the abdomen, and grows.

Diagnosis is made by **laparoscopy**. The surgeon can also remove endometrial tissue through laparoscopic surgery. Another treatment is the use of hormonal drugs to stop ovulation and force endometriosis into remission during the time of treatment. Physicians also suggest taking over-the-counter NSAIDS to relieve pain. Menopause generally ends the activity of mild or moderate endometriosis.

Fibroid tumors are usually benign growths that occur in the uterine wall. Fibroids may enlarge to cause pressure on other organs or may cause excessive bleeding (**Figure 22-16**). To treat fibroids, a **hysterectomy** (**hiss-teh-RECK-toh-mee**) (removal of the uterus) or myomectomy may be done.

Ovarian cysts are fluid-filled sacs or pockets in an ovary or on its surface. They are common, usually present little or no discomfort, and are harmless. Ovarian cysts that rupture can cause sudden, severe abdominal or pelvic pain. Treatment is watchful waiting to see if symptoms subside. Drugs such as hormonal contraceptives may prevent the recurrence of cysts, and surgery may be necessary.

Polycystic ovary syndrome (PCOS) is a set of symptoms due to elevated androgenic hormones in females. Symptoms include irregular or no menstrual cycle, excess body or facial hair, acne, difficulty in getting pregnant, and obesity. Causes may be environmental or genetic factors. Treatment includes birth control pills to regulate the menstrual cycle, metformin to prevent diabetes, statins to control high cholesterol, hormones to increase fertility, and procedures to remove hair. Treatment may help, but the condition cannot be cured.

Breast tumors are either benign or malignant. Benign tumors are usually fluid-filled cysts that enlarge during the premenstrual period.

Breast cancer involving malignant tumors is the second-most common cancer among females (other than skin cancer), striking more than 230,000 each year and killing more than 40,000. About one in eight females will develop breast cancer over the course of a lifetime. The latest government statistics show that there has been a 40% decline in deaths between 1989 and 2007. This represents a steady decline, largely due to advances in screening and treatment. For screening, see *Medical Highlights—Breast Self-Examination.*

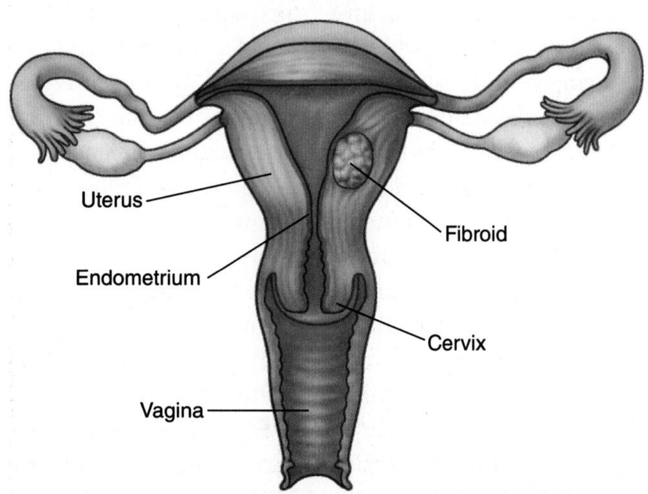

Figure 22-16 *A uterine fibroid is a benign growth in the uterine wall.*

Signs of breast cancer include the following:

- A breast lump or thickening that feels different than the surrounding tissue
- A discharge from the breast other than breast milk
- Redness, scaling, or thickening of the nipples or breast skin
- Changes in size or shape of the breast
- Breast or nipple pain, or nipple turning inward
- A lump in the axillary area (under the arm)

Individuals who have any of these signs or symptoms should see their physician immediately.

A diagnostic test for breast cancer is mammography. **Mammography** (mam-**OG**-rah-fee) is radiographic examination of the breast to detect the presence of tumors or precancerous cells. The resulting record is a mammogram. This is recommended annually for all females over the age of 45 by the American Cancer Society; after the age of 54, is it recommended that they have mammograms every two years. Some may choose to have a mammogram before the age of 45 after discussion with their physician. The United States Preventive Services Task Force recommends stopping at age 75 as there is limited data on the survival benefits. Positive or questionable mammograms are followed by a breast biopsy.

If a family history of breast cancer exists, genetic testing may be done to look for abnormal genes (BRCA1, BRAC2, or PALB2) that are linked to breast cancer.

Breast cancer can be treated in several ways, depending on the kind of breast cancer and how far it has spread. Individuals with breast cancer often get more than one kind of treatment. The patient and physician select the most appropriate treatment:

- *Surgical treatment* for breast cancer includes a **lumpectomy**, which is the removal of the tumor only, or a **mastectomy**, which is the removal of the breast. Radiation and chemotherapy may follow.
- *Chemotherapy* uses drugs to shrink or kill the cancer.
- *Radiation* uses high-energy rays to kill cancer cells.
- *Hormonal therapy* is used when the patient has a cancer that needs certain hormones to grow. This treatment is used to block cancer cells from getting the hormones they need to grow.
- *Immunological therapy* works with the body's immune system to help fight cancer or to control the side effects from other cancer treatments.
- *Targeted therapy* is the use of drugs that attack specific abnormalities like cancer cells. This is an ongoing active area of cancer research.

Medical Highlights

BREAST SELF-EXAMINATION

Breast self-examination (BSE) is recommended for all females. It should be done every month. The best time for the BSE is three to five days after the menstrual cycle has begun. At this time, breasts are not as tender or lumpy. Performing this examination on a monthly basis can help you be aware of how your breasts normally look and feel.

Steps for performing a breast self-examination are as follows:

1. Stand in front of a mirror.
   a. Examine your breasts with hands at your side (**Figure 22-17A**). Look at your breasts in the mirror, noting their usual size, shape, and color. Breasts may differ in size from right to left. Check the nipples for inversion, or when nipples turn inward. Squeeze both nipples between thumb and forefinger to check for any discharge.
   b. Clasp your hands behind your head and check for any changes; this allows you to check the underside of the breasts (**Figure 22-17B**).
   c. Place your hands on your hips and push elbows forward to tighten chest muscles. Look for dimpling; puckering; redness; or changes in the shape, size, texture, or skin color of the breasts (**Figure 22-17C**).
   d. Now, with your left hand on your waist and rolling your left shoulder forward, use the pads of three or four of your fingers to check your left underarm for enlarged lymph nodes (**Figure 22-17D**). Also check the area above and below your collarbone, and then repeat this step on the right side.
2. Examine in the shower. This part of a BSE is easier when your skin is wet and slippery.
   a. Raise your right arm over your head. Gently use the pads of the three middle fingers of your left hand and glide your fingers over the entire right breast and axilla (armpit), feeling for any lumps or thickening (**Figure 22-17E**). Move your fingers up and down over the entire breast area from the collarbone to below the breast and side-to-side from your breastbone to your axilla.
   b. Repeat the procedure on your left side by raising your left hand over your head and repeating the pattern above, using your right hand to check your left breast.
3. Lie down flat on your back. This technique helps to flatten the breast tissue by firmly pressing it against the chest wall.
   a. Place a small pillow under your left shoulder and raise your left arm over your head. Using the pads of three or four fingers of your right hand, move the pads of your fingers around the left breast in a circular motion, moving your fingers up and down over the entire breast area (**Figure 22-17F**). Use both light-to-medium and firm pressure to feel for any lumps or thickening.
   b. Repeat the procedure, placing a pillow under your right shoulder and using your left hand to examine your right breast, feeling for any lumps or thickening.

If any changes are detected during the BSE, notify your medical provider immediately.

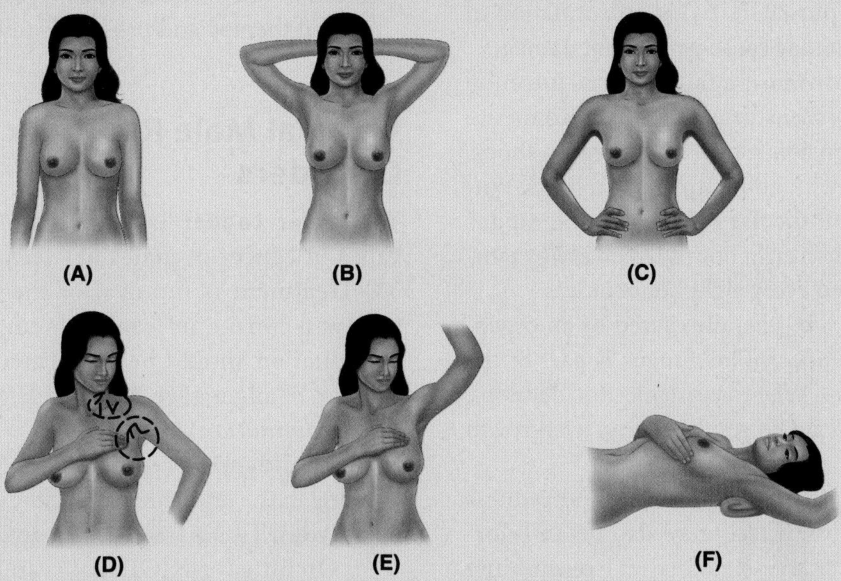

Figure 22-17 *The correct positions and motions for a BSE*

Endometrial cancer is the most common type of uterine cancer. It usually affects females after menopause. Females are advised to immediately report any vaginal bleeding that occurs after menopause to their physician. Hysterectomy and irradiation are the usual types of treatment.

Ovarian cancer ranks fifth in cancer deaths among females, but it causes more deaths than any other cancers of the female reproductive system. Early diagnosis is difficult because of vague symptoms and the lack of definitive screening methods. Treatment is aggressive surgery to remove all reproductive organs followed by chemotherapy and radiation.

Cervical cancer is frequently seen in females between the ages of 30 and 50. The human papillomavirus (HPV) is the main cause of cervical cancer. Cervical cancer can be prevented by the HPV vaccine. One test to detect cervical cancer is called the **Papanicolaou (Pap) smear**. Recommendations are for females to have a PAP smear beginning at age 21 and then every three years until age 65. A sample of cell scrapings is taken from the cervix and cervical canal for study. The HPV test looks for the virus that can cause precancerous cell changes. Treatment is surgery, chemotherapy, and/or radiation. Early detection and treatment are vital to a good prognosis.

INFECTIONS OF FEMALE REPRODUCTIVE ORGANS

Pelvic inflammatory disease (PID) may be due to infections that occur in the reproductive organs and spread to the fallopian tubes and peritoneal cavity. This disease may also be secondary to another infection, such as gonorrhea. The inflammation causes pain, high temperatures, and possible scarring of the fallopian tubes. Treatment consists of antibiotics and analgesics.

Salpingitis (sal-pin-**JIGH**-tis) is inflammation of the fallopian tubes that may cause permanent damage.

Toxic shock syndrome is a rare bacterial infection caused by a *Staphylococcus* organism. Symptoms are fever, rash, and hypotension, which may result in shock. The patient is treated with antibiotics. Toxic shock syndrome has been tied to the use of super-plus tampons being left in for more than eight hours, which is why tampons should be changed every eight hours or less.

Yeast infections are generally caused by an organism called *Candida albicans*. This fungus is part of the body's natural organisms. When a yeast infection develops, the vagina becomes less acidic, resulting in an overgrowth of candida organisms.

Symptoms include itching, burning, and redness in the vagina and vulva. There may also be an odorless, thick, white discharge (leukorrhea) resembling cottage cheese. Some of the common risk factors for yeast infections include stress; the use of antibiotics, which alter the normal flora of the vagina; problems with the immune system; wearing tight-fitting clothing or synthetics that hold in warmth and moisture such as nylon, spandex, or Lycra®; the use of feminine hygienic sprays, talc, or perfumes in the vaginal area; and douching. Treatment is the use of a fungicidal vaginal cream or insert that destroys the organism.

Bacterial vaginosis is a very common disorder caused by an imbalance of naturally occurring normal vaginal flora. Symptoms include vaginal discharge with foul odor, vaginal itching, and burning on urination. Treatment includes antibiotics and topical vaginal creams.

MALE REPRODUCTIVE DISORDERS

Male reproductive disorders can be the result of fetal development, emerge after birth, or emerge at other life stages.

Congenital Male Reproductive Disorders

The following are a couple of the more common congenital disorders of the male reproductive system:

- *Hypospadias*—A condition that affects the structure and function of the male reproductive organs; this condition is when the urethral opening is on the underside of the penis. See Chapter 21 for more details.
- **Anorchism**—This is the complete absence of development of one or both testes. Treatment includes hormonal therapy and prosthetic testes.

General Male Reproductive Disorders

Testicular cancer is rare. It is the most common cause of cancer of males between the ages of 15 and 35. Treatment is removal of the testes, chemotherapy, and hormonal replacement. A testicular self-examination should be performed monthly to check for any changes (see *Medical Highlights—Testicular Self-Examination*).

Epididymitis (**ep**-ih-did-ih-**MY**-tis) is a painful swelling in the groin and scrotum due to infection of the epididymis. This is treated with antibiotic therapy.

Orchitis (or-**KIGH**-tis) is an inflammation of the testes. It may be a complication of mumps, flu, or another infection. Symptoms are swelling of the scrotum, fever, and pain. This disease is treated with antibiotic therapy, pain relievers, and cold compresses.

Prostatitis (pros-tah-**TYE**-tis) is an infection of the prostate gland. The prostate gland lies below the urinary bladder and the prostatic urethra passes through the gland. Urinary symptoms are often the first indication there is a prostatic problem. The patient will complain of difficulty with urination. Treatment with antibiotics is effective.

Medical Highlights

TESTICULAR SELF-EXAMINATION

Testicular self-examination (TSE) should be performed monthly. The TSE is best performed after a warm bath or shower; the heat helps to relax the scrotum, making it easier to find abnormalities.

Steps for performing a testicular self-examination are as follows:

1. Stand in front of a mirror. Look for swelling on the skin of the scrotum.
2. Gently feel the scrotal sac until you locate the right testicle. Using both hands, place your index finger and middle fingers under the right testicle with your thumbs on top (**Figure 22-18A**). Gently but firmly roll the right testicle between your thumb and fingers (**Figure 22-18B**) and carefully check for any changes, including hard lumps; smooth, rounded bumps; or changes in the size, shape, or consistency of the testicle (**Figure 22-18C**).
3. As you examine the testicle, you may find a soft, ropelike structure at the base of the testicle (**Figure 21-18D**). This is the epididymis that leads upward from the top of the back part of the testicle.
4. Repeat the procedure for the left testicle. One testicle may be larger than the other, which is normal.

If any changes are detected during the TSE, notify your medical provider immediately.

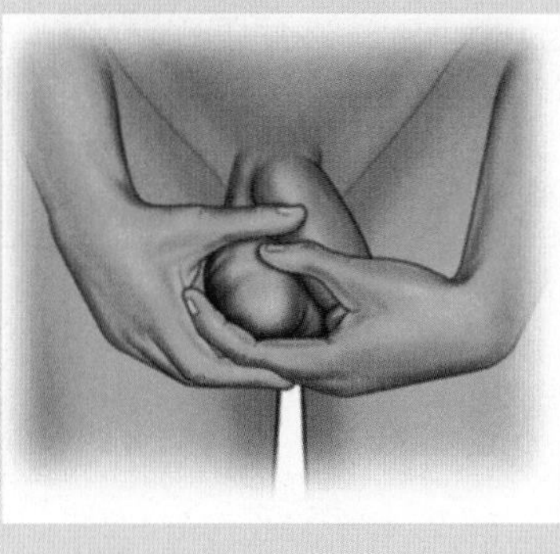

(A)

(B)

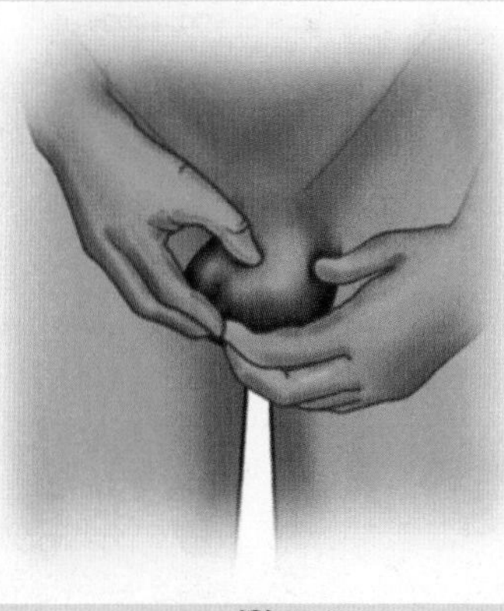

(C)

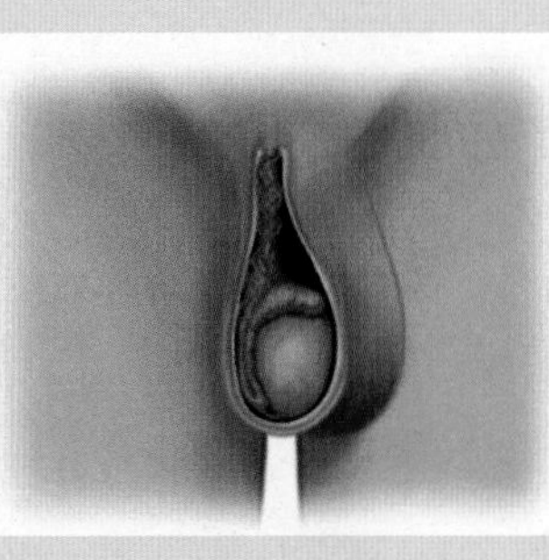

(D)

Figure 22-18 *Steps of the TSE*

Benign prostatic hypertrophy (BPH) indicates an enlarged prostate. More than half of males in their 60s and as many as 90% in their 70s have some symptoms of BPH. As the prostate enlarges, the capsule around the prostate does not, which causes the prostate to press up against the urethra like a clamp around a tube. This affects the structure and function of the bladder. First, the bladder becomes thick and irritable. Then the bladder begins to contract, even when it contains only small amounts of urine, causing frequent urination. As the bladder weakens, it loses the ability to empty itself, and urine remains in the bladder. The narrowing of the urethra may cause retention of urine, and an infection may occur (**Figure 22-19**).

Diagnosis is made by rectal exam, ultrasound, and cystoscopy. A **cystoscopy** (sis-**TOS**-koh-pee) is a process in which a flexible tube with a lens and a light system is inserted into the urethra. This enables the physician to see the inside of the urethra and the bladder (**Figure 22-20**). Treatment may depend on the extent of the symptoms. For treatment options, see *Medical Highlights—Treatment for Benign Prostatic Hypertrophy and Prostate Cancer*.

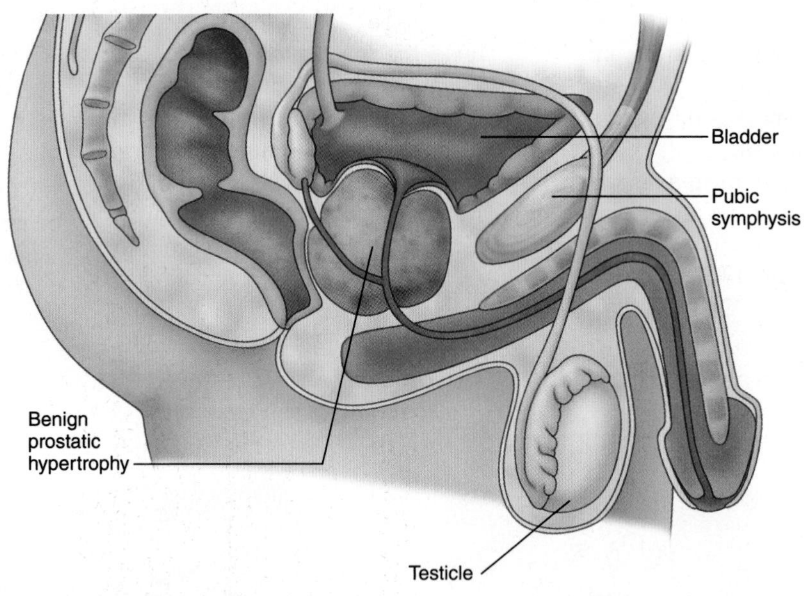

Figure 22-19 *In benign prostatic hypertrophy, the enlarged prostate presses against the bladder.*

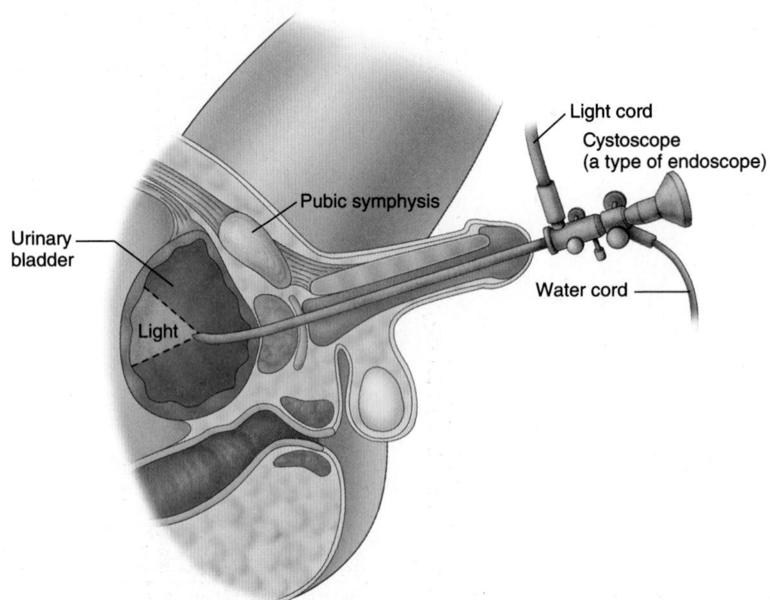

Figure 22-20 *A cystoscope is a flexible tube with a light source that is inserted into the urethra to examine the interior of the bladder.*

Medical Highlights

TREATMENT FOR BENIGN PROSTATIC HYPERTROPHY AND PROSTATE CANCER

BENIGN PROSTATIC HYPERTROPHY

A number of recent studies have questioned the need for early treatment for BPH when the prostate gland is just mildly enlarged. These studies report that early treatment may not be needed because the symptoms of BPH clear up without any treatment in as many as one-third of all mild cases. Instead of immediate treatment, the study suggests regular checkups to watch for early problems. If the condition poses a threat to health or is a major inconvenience, treatment is then recommended. Treatment for BPH includes

- watchful waiting;
- medications, such as alpha blockers that relax bladder neck muscles and muscle fibers in the prostate, making urination easier;
- medications, such as alpha reduction inhibitors that shrink the prostate by preventing hormonal changes;
- Cialis®, often used to treat erectile dysfunction, can also treat an enlarged prostate;
- transurethral resection of the prostate (TURP), which removes part of the inside portion of the prostate, making urination easier;
- transurethral incision of the prostate (TUIP), in which the surgeon makes one or two incisions in the gland, making it easier to urinate;
- transurethral microwave thermotherapy (TUMT), in which a special electrode is passed through the urethra into the prostate area. Microwave energy from the electrode destroys the inner portion of the enlarged prostate;
- transurethral needle ablation (TUNA), in which a scope is passed through the urethra, placing needles in the prostate. Radio waves pass through, heating and destroying excess prostate tissue;
- laser therapy, which removes overgrown prostate tissue through ablative procedures; and
- prostatic urethral lift (PUL) special tags, which are used to compress the prostate gland.

PROSTATE CANCER

Some prostate cancers grow very slowly and never cause problems. Treatment in these cases is watchful waiting.

Other prostate cancers are aggressive and deadly. In these cases, the benefits of treatment outweigh the risks of side effects. There are different types of treatment for prostate cancer. Types of standard treatment include the following:

- *Surgery*—A radical **prostatectomy** is performed. It involves the removal of the prostate and surrounding tissue along with nearby lymph nodes.
- *Radiation therapy*—External beam therapy uses high-energy X-rays or other types of radiation to kill cancer cells.
- *Brachytherapy*—This involves placing rice-sized radioactive seeds inside the prostate, delivering a low dose of radiation over a long period of time.
- *Hormone therapy*—This treatment removes hormones or blocks their action and stops cancer cells from growing. Luteinizing hormone-releasing hormone agonists can prevent the testicles from producing testosterone, which may accelerate the growth of cancerous tissue.
- *Cryosurgery*—This treatment uses an instrument to freeze and destroy cancer cells.
- *Chemotherapy*—Drugs are used to stop the growth of cancer cells, either by killing the cells or by stopping the cells from dividing.
- *Immunotherapy*—This uses the body's own immune system to fight cancer cells.
- *Targeted drug therapy*—This attacks special weaknesses in the cancer cells.

Note: Early detection is still the key to fighting prostate cancer.

Source: http://www.cancer.gov/types/prostate/patient/prostate-treatment-pdq

Prostate cancer is the most common cancer in males over the age of 50. Males over the age of 40 should start to have annual rectal examinations, which can detect enlargement of the prostate. Controversy exists regarding the prostate-specific antigen (PSA) blood screening test, which detects an abnormal substance released by cancer cells. American Cancer Society states that males should not be screened for PSA unless they have received information on the uncertain risks and potential benefits of PSA screening. Memorial Sloan Kettering, a leading cancer institute, recommends that males between the ages of 50 and 59 should have a PSA level check. If the PSA is 3 nanograms per milliliter (ng/ml) or higher, a biopsy may be done. Symptoms of prostate cancer include frequency of urination; dysuria, or painful urination; urgency; nocturia, or night voiding; and in some cases, hematuria, or blood in the urine. For treatment of prostate cancer, see *Medical Highlights—Treatment for Benign Prostatic Hypertrophy and Prostate Cancer.*

SEXUALLY TRANSMITTED DISEASES

Sexually transmitted diseases (STDs), also known as venereal diseases, are transmitted through the exchange of body fluids such as semen, vaginal fluid, and blood. STDs can be serious and painful and can cause long-term complications, including sterility, chronic infection, scarring of the fallopian tubes, ectopic pregnancy, cancer, and death. The most common of these diseases are chlamydia, genital herpes, genital warts, and trichomoniasis vaginalis.

Most of these diseases have no symptoms. Symptoms, if present, include the following:

- In females, an unusual discharge from the vagina, pain in the pelvic area, burning or itching around the vagina, unusual bleeding, and vaginal pain during intercourse
- In males, a discharge from the penis
- In both females and males, sores or blisters near the mouth or genitalia, burning and pain during urination or a bowel movement, flulike symptoms, painful urination, and swelling in the groin area

The patient who is at a physician's office or a health care center to be checked for an STD may feel some embarrassment. It is critical for the health care provider to treat the person in a nonjudgmental manner, because every day that the disease is untreated, it causes more severe health problems. Some STDs are diagnosed by physical examination; others require blood or other laboratory tests. For bacterial diseases such as gonorrhea, chlamydia, and syphilis, treatment is with antibiotics. Viral infections usually cannot be cured, but the symptoms can be relieved.

Protection from STDs includes abstinence and practicing safe sexual behavior. Safe sex means using condoms and looking for any signs of venereal disease *before* sexual activity occurs. Once a person is aware of the disease, they must notify any current sexual partners so they can also be checked for the disease. It is often necessary for previous sexual partners to also be notified. All STDs need to be treated. A high incidence of STDs is leading to an increase in sterility in young females.

Chlamydia (klah-**MID**-ee-ah) is caused by the *Chlamydia trachomatis* organism and is the most common curable STD in the United States. It is the major cause of nongonococcal urethritis, bacterial vaginitis, and PID. Eighty percent of females and twenty-five percent of males have no symptoms. If symptoms do appear, they may include abnormal genital discharge and burning with urination. Treatment is with antibiotics.

Human papillomavirus, or **genital warts**, is another common STD. The wart can appear on the shaft of the penis or on the vagina. It is usually asymptomatic, meaning there are no symptoms. In many cases, the warts are not visible to the naked eye. In other cases, they look like small, hard, round spots resembling a cauliflower. Although genital warts are usually painless, they become sore and itchy and may burn if rubbed or irritated. Diagnosis is made primarily by examination. Treatment involves the use of an acid to destroy wart tissue, or cryosurgery. Cryosurgery uses liquid nitrogen, which is placed on the wart and a small area of the surrounding skin. The liquid nitrogen freezes the skin, causing ice crystals, which results in the sloughing off of the wart. There is a vaccine available for HPV. It should be noted that several strains of HPV do not present with warts, so it is possible to have HPV without genital warts. See *Medical Highlights—Human Papillomavirus Vaccine.*

Gonorrhea (**gon**-oh-**REE**-ah) is a bacterial infection caused by *Neisseria gonorrhoeae.* The symptoms in the male may be painful urination and the discharge of pus from the penis. In the female, the early stages of the disease may be asymptomatic. This disease is treated with antibiotic therapy; however, some strains of the organism have become resistant to the usual treatment.

Complications may occur if the inflammation spreads to the male epididymis or the female fallopian tubes. The tubes may become scarred and blocked, resulting in sterility. In addition, if a pregnant person contracts gonorrhea and has a vaginal delivery, they run the risk of transmitting the disease to the newborn. This may result in the newborn being diagnosed with neonatal conjunctivitis.

Genital herpes is a viral infection that is sexually transmitted. The herpes lesion may cause a burning sensation, and small, blister-like areas may appear on the genitalia. Other symptoms of the herpes virus may be painful irritation and discomfort in the genitalia while sitting or standing. Herpes symptoms may simply disappear after two weeks; however, the symptoms may continue to reappear throughout the lifetime of the individual. Females who are diagnosed with herpes must consult with their physician about whether to have a cesarean section to prevent herpes infection of the newborn during childbirth. Treatment is with antiviral medications, which reduces the symptoms and makes it less likely the disease will spread to a sex partner.

Syphilis (**SIF**-ih-lis) (chancroid) is a potentially life-threatening STD caused by the bacteria *Treponema*

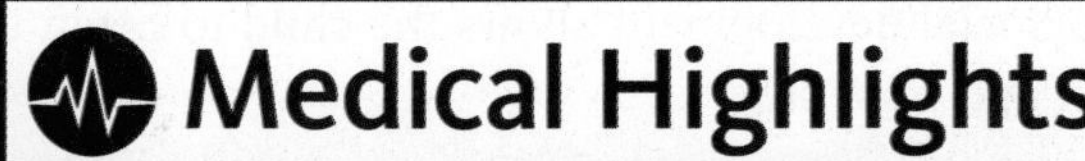

HUMAN PAPILLOMAVIRUS VACCINE

Human papillomavirus (HPV) is common and infects at least 50% of all people who have had sex. It can also be spread from skin to skin contact. There are at least 100 related viruses in this group. The word *papilloma* refers to a kind of wart that results from some HPV types. About 40 types of HPV are sexually transmitted.

The first vaccine for the human papillomavirus was approved in 2006. HPV vaccine offers the best protection to give young people the time to develop an immune response before being sexually active with another person. An HPV vaccine is recommended for preteens between the ages of 11 and 12. The vaccine is given through a series of two or three injections over a 6-month period. The vaccine is recommended for everyone through the age of 26 if not previously vaccinated; the vaccine is not recommended for people over the age of 26.

The vaccine is most effective in girls and females who have not yet acquired any of the four types of HPV covered by the vaccine. Females who have not been infected with any of these types get the full benefit of the vaccine.

Females who are sexually active may also benefit from the vaccine, but they may have already acquired one or more HPV types covered by the vaccine. They would still get protection from those types they may not have acquired.

The length of the vaccine's protection is not known. To date, studies have followed females for 10 years and found that they were still protected.

Studies also have found the vaccine to be 100% effective in preventing diseases caused by the four types of HPV, including genital warts and precancers of the cervix, vulva, and vagina.

pallidum. In the early stages of syphilis, a genital sore called a chancre (**SHANG**-ker) develops shortly after infection and eventually disappears on its own. If the disease is not treated, it can progress on its own over years. A transient rash may appear; eventually there is serious involvement of the vertebrae, brain, and heart, resulting in meningitis, lack of coordination, and stroke. The full course of the disease can take years. Penicillin is the most effective treatment for syphilis.

Trichomoniasis vaginalis (trick-oh-moh-**NYE**-ah-sis) is an STD caused by infection with the protozoan *Trichomonas vaginalis*. It causes vaginitis—inflammation of the vagina causing burning, itching, and discomfort. In males, trichomoniasis may cause similar problems in the urethra, called urethritis. Treatment is usually with a single dose of antibiotics.

Mycoplasma genitalium was acknowledged as an STD by the Centers for Disease Control in 2015. It is a small, pathogenic bacterium that lives on the skin of the urinary and genital tract. Most cases are asymptomatic. Females may experience vaginal itching and discharge, burning on urination, and painful intercourse. Males may develop urethritis. Treatment is with the antibiotic azithromycin, which is safe and effective, but there is evidence of increasing resistance to this drug.

HIV/AIDS—See Chapter 16.

TRAUMA TO THE REPRODUCTIVE ORGANS

Trauma is any injury that causes damage to the reproductive organs. Injury can be caused by bike accidents, car accidents, burns, penetrating trauma, blunt trauma, sports injuries, genital mutilation, and so on. It is important to receive treatment immediately to try to avert long-term damage that can cause infertility and issues with the bladder and urethra.

HUMAN GROWTH AND DEVELOPMENT

Even though individuals differ greatly, each person passes through certain stages of growth and development from birth to death. As a person passes through these stages, four main types of growth and development occur: physical, mental, emotional, and social. Physical growth and development refers to changes occurring in the body, body systems, and organs. Mental growth and development relates to changes occurring in the mind and the ability to solve problems and make judgments. Emotional growth and development relates to the ability to deal with feelings.

Social growth and development relates to changes in the way a person interacts with other people.

Tasks must be mastered at each stage of development before a person can progress to the next stage. Tasks build on one another as they progress from the simple to the complex. The rate at which an individual progresses through each stage varies. These stages are generally grouped by age: infant, toddler, preschool child, school-age child, adolescent, early adult, middle adult, and older adult.

Infant

Infancy lasts from birth to one year of age. This is the time in which the most dramatic changes in growth and development occur. Weight generally triples during this time, and height usually increases to 29 to 30 inches. The muscular and nervous systems undergo rapid change during this stage. Teeth develop and eyesight improves. The other senses become more defined. Mental development also occurs rapidly through the increase of verbal skills. Emotional development begins at this stage, and events occurring at this age related to emotions can have a strong impact on emotional behavior in adulthood. Social development occurs in the infant's ability to recognize others and respond to familiar people.

Toddler

The toddler stage is from 1–3 years of age. At this age, the child is very mobile. Physical growth progresses at a slower pace. Coordination is becoming fine-tuned, allowing the child to walk, run, and climb. Mental development continues to progress rapidly. The toddler learns to use more words, remembers details, and begins to understand basic concepts. Emotional development also progresses rapidly. The toddler begins to become self-aware and more independent. Social development shows the toddler progressing from being self-centered to socializing more with other adults and children.

Preschool Child

The preschool stage is from 3–5 years of age. Motor development continues, and the child is able to write and use utensils. Children also begin to achieve control over their bladder and bowels. Mental development continues as the preschooler begins to ask more questions about their surroundings and make decisions based on logic. Emotional development progresses, and the child may become frustrated as they try to do more than is within their ability. Children at this stage understand right from wrong. Social development allows the child to begin to interact and trust others.

School-Age Child

The school-age child is from 6–12 years of age. Physical development is slow but steady. Weight gain averages five to seven pounds per year and height usually increases 2–3 inches per year. Physical activities and game playing become more complex because muscular coordination is well developed at this age. The primary teeth are lost, and the permanent teeth erupt. Sexual maturity may begin. Mental development increases rapidly because a great deal of time at this stage is spent in school. Speech, reading, and writing skills develop, and school-age children are able to solve more complex problems. Children at this stage begin to understand more abstract concepts, such as honesty and values. Emotional development continues, allowing children greater independence and the development of their own personalities. Social activities change from wanting to do things on one's own to wanting to be involved in group activities.

Adolescent

Adolescence spans from 12 to 20 years of age. In early adolescence, a growth spurt may occur, and muscle coordination may not grow at the same rate. This may make the adolescent awkward. The most obvious changes are those related to puberty. Secondary sex organs and sexual characteristics become more prominent. Mental development involves increasing knowledge and sharpening skills. Adolescents make decisions and learn responsibility for their actions. Emotional development is often in conflict as adolescents try to establish their own identities and yet feel uncertain and insecure. This age group responds to peer pressure. Social development sees this age group spending more time with friends and less time with family. Many problems can develop at this stage, including eating disorders and substance abuse.

Early Adult

Early adulthood spans the ages of 20 to 40 years old. Physical development at this stage is complete. This is the prime time for childbearing. Mental development continues as education is furthered and careers are chosen. Emotional development revolves around preserving the stability founded earlier in life as stresses increase in one's life. Social development usually involves moving away from one's peer groups and developing relationships with others who share similar goals and interests.

Middle Adult

Middle adulthood is from 40 to 65 years of age. Physical changes begin, such as graying hair, formation of wrinkles, loss of hearing and vision, and weight gain. Mental ability can continue to increase; decision making and problem solving are done with more confidence. Emotionally, middle adulthood is usually a period of contentment and satisfaction. Socially, family relationships decline and work relationships grow.

Older Adult

Older adulthood ranges from 65 years and up. Physical development begins to decline, and body systems are usually affected. Mental abilities may vary; in some, these may also begin to decline. Emotional abilities vary as well and are related to the individual's ability to cope with stress and loss. Social adjustment is often required due to retirement, loss of loved ones, and physical limitations.

One BODY How the Reproductive System Interacts with Other Body Systems

INTEGUMENTARY SYSTEM

- Hormones activate oil glands that lubricate the skin and hair.
- Female hormones affect body hair growth in pubic and axillary area.
- Male hormones affect body hair growth in pubic, axillary, facial, and chest areas.
- Sensory skin receptors stimulate sexual arousal and pleasure.

SKELETAL SYSTEM

- Bones of the pelvic cavity protect the organs of reproduction.
- Hormones promote skeletal growth and bone density.
- Hormones affect epiphyseal plate closure, and skeletal growth stops.

MUSCULAR SYSTEM

- Testosterone causes an increase in muscle mass.
- Skeletal muscle contractions aid in the erection of the penis and clitoris.
- Smooth muscle contractions of the uterus expel the fetus during childbirth.

NERVOUS SYSTEM

- The hypothalamus regulates the timing of puberty.
- Sensory and motor neurons play a major role in sexual response and pleasure.

ENDOCRINE SYSTEM

- The anterior pituitary gland secretes follicle-stimulating hormone, which stimulates the growth of the graafian follicle, the production of estrogen in females, and the growth of sperm in males.
- The anterior pituitary gland secretes the luteinizing hormone responsible for the formation of the corpus luteum, which produces progesterone in females and is necessary for the production of testosterone in males.

CIRCULATORY SYSTEM

- Blood vessels of the penis and clitoris engorge during sexual activity.

LYMPHATIC SYSTEM

- The female immune system does not reject the male sperm cell, thus ensuring fertilization.
- The immune system does not reject the developing fetus during pregnancy.

RESPIRATORY SYSTEM

- Testosterone increases the size of the larynx in males, thus producing a voice change.
- Interaction between the respiratory system and the placenta provides the fetus with oxygen and removes carbon dioxide.

DIGESTIVE SYSTEM

- Breast-feeding provides nourishment for the baby.
- Digestive organs may become crowded during pregnancy, which can result in heartburn and constipation.

URINARY SYSTEM

- The male urethra serves as the common passageway for urine and semen.
- Pressure on the urinary bladder during pregnancy can cause frequent urination.

Medical Terminology

a-	without
-men	monthly
-orrhea	flow or discharge
a/men/orrhea	without a monthly flow
circumcise	cutting around
-ion	process of
circumcis/ion	process of cutting around
clitor	gatekeeper
-is	presence of
clitor/is	presence of a gatekeeper
coit	sexual intercourse
-us	presence of
coit/us	presence of sexual intercourse
crypt	hidden
orchid	testes
-ism	abnormal condition of
crypt/orchid/ism	abnormal condition of hidden testes; undescended testicles
dys-	painful
dys/men/orrhea	painful monthly flow
ectopic	out of place
ectopic pregnancy	a pregnancy that occurs outside the uterus
endo-	within
metri	uterus
-um	presence of
endo/metri/um	pertaining to the lining within the uterus
hyster	uterus
-ectomy	removal of
hyster/ectomy	removal of the uterus
leuko	white
leuk/orrhea	white discharge
mammo	breast
-gram	X-ray record
mammo/gram	X-ray record of the breasts
mast	breast
mast/ectomy	removal of the breast
myo	muscle
myo/metri/um	presence of uterine muscle
o/o	egg
-genesis	development
o/o/genesis	development of ova (egg)
ovul/a/	releasing a little egg
-tion	process of
ovul/a/tion	process of releasing a little egg
prostate	prostate gland
prostat/ectomy	removal of prostate gland
salping	fallopian tube
-itis	inflammation of
salping/itis	inflammation of the fallopian tube
spermato	seed
spermato/genesis	development of seed

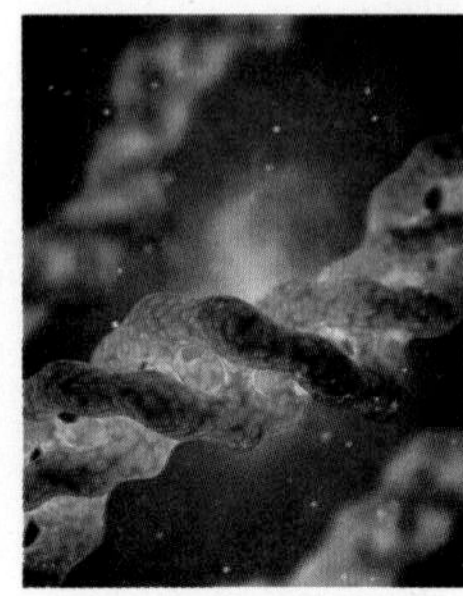

Study Tools

Workbook	Activities for Chapter 22
Online Resources	• PowerPoint presentations • Animations

REVIEW QUESTIONS

Select the letter of the choice that best completes the statement.

1. The male hormone is
   a. progesterone.
   b. luteinizing hormone.
   c. follicle-stimulating hormone.
   d. testosterone.
2. Ovulation usually occurs
   a. the day before the menstrual period begins.
   b. one week before the menstrual period begins.
   c. three weeks before the menstrual period begins.
   d. two weeks before the menstrual period begins.
3. The ovaries contain
   a. 30 graafian follicles.
   b. thousands of graafian follicles.
   c. hundreds of graafian follicles.
   d. 6 graafian follicles.
4. The development of the follicle and release of the ovum are under the influence of
   a. the follicle-stimulating hormone and the luteinizing hormone.
   b. estrogen and corpus luteum.
   c. progesterone and the follicle-stimulating hormone.
   d. estrogen and the luteinizing hormone.
5. Which one of the following statements is *not* correct?
   a. The fallopian tubes are about 4 inches long.
   b. The fallopian tubes serve as ducts for the ovum on its way to the uterus.
   c. The fallopian tubes are also called oviducts.
   d. The fallopian tubes are attached to the ovaries.

MATCHING

Match each term or stage in Column I with its correct description in Column II.

	COLUMN I	COLUMN II
______	1. skin, hair, pituitary	a. formed by mesoderm
______	2. implantation, complete ovulation	b. on or about the 24th day
______	3. zygote becomes morula	c. formed by mesoderm germ layer
______	4. vernix caseosa forms on skin	d. at the end of second month
______	5. heart begins to beat	e. at the end of four months
______	6. mother's voice heard and recognized	f. formed by ectoderm layer
______	7. lining of lungs and digestive tract	g. at the end of five months
______	8. mother feels fetal movements	h. at the end of seven months
______	9. muscles, bones, and spleen	i. 1–5 days after ovulation
______	10. brain has human proportions	j. 7–10 days after ovulation

MATCHING

Match each term in Column I with its correct description in Column II.

	COLUMN I	COLUMN II
______________	**1.** scrotum	a. secondary sex characteristics
______________	**2.** testosterone	b. external sac that holds the testes
______________	**3.** facial and pubic hair	c. accessory gland
______________	**4.** seminal vesicle	d. formed in the seminiferous tubules
______________	**5.** spermatozoa	e. male hormone produced in the testes

FILL IN THE BLANK

1. Painful or difficult menstruation is known as ______________.

2. Amenorrhea is normal when a person is ______________.

3. Gonorrhea is a sexually transmitted disease. The male complains of ______________, and the female complains of ______________.

4. The test done to detect breast tumors is called ______________.

5. Sterility results from inflammation of the fallopian tube, which can be caused by ______________.

6. A group of symptoms that occur before the menstrual cycle is called ______________.

7. The onset of ovulation is known as ______________, and the cessation of ovulation is known as ______________.

8. A sexually transmitted disease that has small, blisterlike areas is known as ______________.

9. An enlarged prostate may cause problems with ______________.

10. The best method for preventing sexually transmitted diseases is to practice ______________.

APPLYING THEORY TO PRACTICE

1. A young pregnant female comes into the physician's office and states, "I told my husband it will be a boy, because in my family, I was the only girl and I have four brothers." Is this a valid statement? Explain to the expectant mother how the sex of a newborn is determined.

2. If a male has a sperm count of 20 million, he is considered sterile. Fertilization only requires the union of one egg and one sperm; why then are so many sperm necessary for fertilization to occur?

3. You are asked to describe the fertilization process. Explain how the sperm travels from the testes and arrives at the fallopian tube in time to meet the ova.

4. You are invited to a middle school to address 11- to 14-year-old adolescents and discuss puberty. Plan a program to describe how females are affected by estrogen and progesterone and how males are affected by testosterone.

5. A 50-year-old female patient tells you it has gotten very hot in the waiting room and requests that you put on the air conditioner. The temperature outdoors is −1°C (30°F). She further states that the physician told her about changes that usually occur at midlife. Explain the physiological and psychological changes that are associated with menopause to the patient.

6. Sexually transmitted diseases are a major health concern, especially in the teenage population. These conditions can lead to pelvic inflammatory disease and sterility. Describe at least three sexually transmitted diseases; explain their symptoms and treatment.

7. As a health care provider, knowledge of the various stages of human growth and development helps to understand the needs of the patient. State the stage of development and the average age of the individual at which the following tasks are accomplished:
   a. Right from wrong is understood
   b. Group activity involvement
   c. Others are recognized and responded to
   d. Peer pressure is most important
   e. Words are learned
   f. Secondary sex characteristics become more pronounced
   g. Decision making and problem solving
   h. Education is furthered and careers chosen
   i. Decision making and problem solving are done with confidence
   j. Mental abilities may begin to decline

8. Mike, a 70-year-old male, has just been diagnosed with prostate cancer. Provide him with answers to the following question: How does prostate cancer affect the structure and function of organs and body systems?

LABELING

Study the following diagram of the male reproductive system and label the numbered structures.

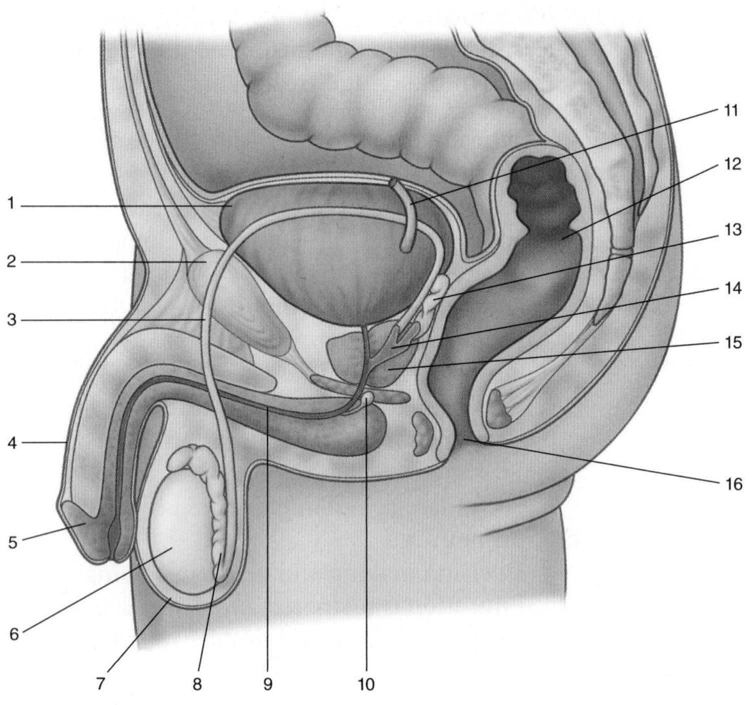

1. ______________
2. ______________
3. ______________
4. ______________
5. ______________
6. ______________
7. ______________
8. ______________
9. ______________
10. ______________
11. ______________
12. ______________
13. ______________
14. ______________
15. ______________
16. ______________

CASE STUDY

Allison, 46 years old, has found a lump in her breast during a breast self-examination. She goes to her physician, who finds other signs of breast cancer. The physician orders diagnostic tests, which confirm that Allison has breast cancer.

1. List the signs and symptoms of breast cancer.
2. Explain what diagnostic tests would be used to diagnose Allison's cancer.
3. What therapeutic techniques are available to treat Allison's cancer?
4. Who will make the decision regarding her treatment?
5. What are some of the statistics regarding breast cancer?

LAB ACTIVITY 22-1 Organs of the Reproductive System

- ***Objective:*** To observe the location and difference between the male and female reproductive organs
- ***Materials needed:*** charts of female reproductive organs and male reproductive organs, textbook, paper, pencil

Step 1: Locate and identify the female reproductive organs on the anatomical charts. Describe and record their locations and functions.

Step 2: Locate and identify the male reproductive organs on the anatomical charts. Describe and record their locations and functions.

Step 3: Compare the organs and functions of the female and male reproductive organs. What are their differences? Record your answer. Compare your answer with the textbook.

LAB ACTIVITY 22-2 Examination of an Ovary

- ***Objective:*** To observe the structure of the ovaries
- ***Materials needed:*** cross-section slide of ovarian tissue, microscope, textbook, paper, pencil

Step 1: Identify and describe a mature graafian follicle with an oocyte, and a maturing graafian follicle. Compare with Figure 22-13. Record your observations. What do the cells of the graafian cells produce? Record your answer.

Step 2: Identify and describe a corpus luteum. Compare with Figure 22-13. Record your observation. What do the cells of the corpus luteum produce?

LAB ACTIVITY 22-3 Examination of the Testis

- ***Objective:*** To observe the structure of the testis
- ***Materials needed:*** cross-section slide of the testis, microscope, textbook, paper, pencil

Step 1: Identify and locate the seminiferous tubules. Record your description.

Step 2: Observe the cells in the middle of the tubule wall of the seminiferous tubule. Observe the spermatocyte. Record your description.

Step 3: Identify the interstitial cells that are located outside the seminiferous tubules. What is their function? Record your answer.

CHAPTER 23

Genetics and Genetically Linked Diseases

Objectives

- Define *genetic mutation*.
- Identify the basic types of genetic mutations.
- Differentiate between hereditary mutations and genetic mutations.
- Name three human genetic disorders and describe the cause and symptoms of each.
- Explain genetic counseling.
- Define the key words that relate to this chapter.

Key Words

amniocentesis
chorionic villi sampling
chromosomal mutations
congenital disorder
cystic fibrosis (CF)
designer babies
Down syndrome
Duchenne's muscular dystrophy
gene mutations
genes
genetic counseling
genetic disorder
genetic engineering
genetics
genotype
hereditary mutations
Huntington's disease (HD)
lethal gene
mutagenic agents
mutation
phenotype
phenylketonuria (PKU)
recombinant DNA
sickle cell anemia
somatic cell mutations
Tay-Sachs disease
trisomy 21

GENETICS

In sexual reproduction, a new individual is created from the union of the sperm cell and the egg cell. This process is called fertilization. Contained in the nucleus of each gamete are structures called chromosomes. The chromosomes contain deoxyribonucleic acid (DNA), the hereditary material referred to in Chapter 3. The DNA is packaged in small, functional units found along the length of a chromosome called **genes**. A gene is an area of DNA that carries information for the cellular synthesis of a specific protein. These genes are transmitted to the zygote and will then control the development and characteristics of the embryo as it grows and matures. Due to the combined influence of all of the genes on all of the chromosomes, a new individual is formed. The new individual possesses all the necessary characteristics or traits needed for survival. Additionally, because the genes come from two parents, the offspring resembles both parents in some ways; however, it is also different from each parent. It has, for instance, all the characteristics of its species. Concurrently, it possesses its own unique traits that set it apart from all other members of its species.

A **genotype** is the genetic makeup of an individual; the DNA that is inherited from both parents. The **phenotype** is the external expression of the genotype, but with external influence, such as the environment.

Genetics is the branch of biology that studies how the genes are transmitted from parents to their offspring. Occasionally, a gene or chromosome is changed, or mutated. This mutated gene or chromosome is inherited by the offspring. The inheritance of such a mutated gene or chromosome will cause the appearance of a new and different trait, called a **mutation**. Sometimes the mutation is beneficial or harmless to an organism, but most inherited mutated genes are not beneficial. Still, it must be emphasized that mutations in the genetic material are responsible for biological evolution on this planet.

Scientists have completed the first working plan of the human genome. With this information, medical researchers can identify sites where mutations can occur.

TYPES OF MUTATIONS

Gene mutations occur occasionally at random in all cells of the human body. **Somatic cell mutations** occur in individual body (somatic) cells and are acquired in one's lifetime. This type of mutation can occur in any body cell. If the mutation happens in the sperm or egg, this type of mutation may be passed down to the offspring. If the mutation occurs in any other body cell, the mutation will not be passed down.

There are two types of **chromosomal mutations**: numerical and structural. Numerical mutations involve the number of chromosomes, either too many, meaning more than 46, or too few, meaning less than 46. Structural mutations occur from a deletion, inversion, translocation, or duplication of a section of the chromosome, which results in a change in the DNA. Some chromosomal mutations affect the structure of tissues and organs and can have an effect on the overall body systems. For example, the congenital disorder Proteus syndrome causes asymmetric and disproportional overgrowth of bones, muscles, skin, and blood vessels.

Hereditary mutations are passed down from generation to generation. This occurs because the mutation is present in all cells, including the gametes. Most inherited genetic diseases are recessive, which means that the mutation must be present in both the sperm and egg at the time of fertilization. Diseases caused by one defective gene are rare but do occur.

Lethal Genes

Inherited mutations generally have negative results for an individual. At times, they might even result in the formation of lethal genes. A **lethal gene** is a gene that results in death.

The time at which lethal genes exert their deadly influence varies. Some genes interfere with mitosis of the zygote, and life ends before the zygote divides. Some lethal genes interfere with implantation of the fertilized egg in the uterus; in these cases, death would occur so early that a female would never know that conception had even occurred. A lethal gene that prevents normal formation of the heart or normal blood production causes death about three weeks after fertilization because this is the time when circulating blood becomes vital for continued existence. Others may result in death at various times during development, depending on the time their products become vital for survival. Other lethal genes causing neonatal deaths involve abnormalities of the lungs and shifts in the circulatory system, which must channel blood from the heart to the lungs instead of to the umbilical cord.

Some lethal genes do not exert their effects until later in life. Tay-Sachs disease causes death several years after birth. Duchenne's muscular dystrophy causes death in a person's late 20s to early 30s. Huntington's disease usually brings about death at around 40 to 50 years of age.

Each person is estimated to carry two or three different recessive lethal genes. Two similar recessive genes must be present in an individual for the gene to be expressed. Because of the numerous kinds of lethal genes, one's chance of reproducing with someone with even one matching lethal gene is small. Statistically, should this happen, the lethal gene would be expressed in only one-fourth of the offspring.

When close relatives have offspring, the chance of the offspring inheriting two similar lethal genes increases. People who have a common ancestry are more likely to share genes than nonrelatives. As a result, spontaneous abortions, also known as miscarriages; stillbirths; and neonatal deaths are higher among children of people sharing similar gene pools.

HUMAN GENETIC DISORDERS

Some diseases caused by gene mutations in humans are phenylketonuria (PKU) (**fen**-il-**kee**-toh-**NEW**-ree-ah), sickle cell anemia, Tay-Sachs disease, Duchenne's muscular dystrophy (doo-**SHENZ**), Huntington's disease, cystic fibrosis, thalassemia, and hemophilia.

It is important to note that there is a difference between genetic disorders and congenital disorders. A hereditary or **genetic disorder** is caused by a variation in the genetic pattern; a **congenital disorder** is something that evolves during fetal development and is present at birth, regardless of the cause. Congenital disorders can affect enzymes, cells, tissues, organs, and entire body systems. Individuals can have one disorder or multiple disorders. Not all birth defects are detected at birth. Some may not present until later in life.

Phenylketonuria

Phenylketonuria (PKU) is a human metabolic disorder caused by an enzyme deficiency. The individual with the trait cannot break down the amino acid phenylalanine and, consequently, there is a buildup of this substance in the body. Excess phenylalanine disrupts the normal development of the brain. If a child born with the disorder eats proteins containing phenylalanine during childhood, intellectual disability results. Newborns are tested for this disorder, and, if the test is positive, a phenylalanine-restricted diet is prescribed. In most cases, this diet can be diminished as the child grows older and brain development and maturation are completed.

Sickle Cell Anemia

Sickle cell anemia is a blood disorder common in individuals of African descent. It is caused by a gene mutation resulting in an abnormal hemoglobin molecule in a red blood cell (**Figure 23-1**). Especially in times of low oxygen availability, the shape of a red blood cell changes from that of a biconcave disc to a crescent shape. This is referred to as sickling. The sickle shape causes the cells to clump together, thus clogging small blood vessels and capillaries. Because a sickle cell has an abnormal

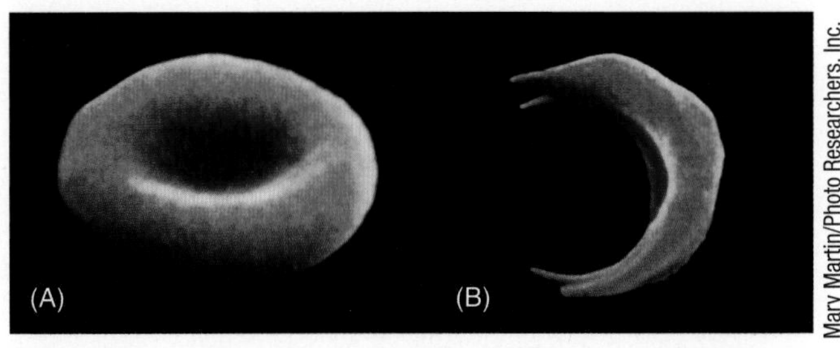

Figure 23-1 *(A) Side view of normal red blood cells, which have a concavity (or dent) on the two sides; (B) sickle-shaped (crescent-shaped) red blood cells*

hemoglobin (the pigment that combines with oxygen), it also carries less oxygen to the tissues, resulting in fatigue and listlessness. Breakage of these cells is also very common because the membranes are very fragile. **Figure 23-2** shows tissue damage and physiological effects caused by sickle cell anemia. See Chapter 13 for more information.

Tay-Sachs Disease

Tay-Sachs disease is a genetic disorder caused by a mutation resulting in a deficiency of a lysosomal enzyme. The missing enzyme functions in breaking down lipid molecules in the brain. Without the enzyme, lipids accumulate in the brain cells and destroy them. This results in severe mental and motor deterioration leading to death several years after birth. This disorder is found most frequently among Jewish people of central and Eastern European ancestry.

Duchenne's Muscular Dystrophy

In **Duchenne's muscular dystrophy**, the muscles suffer a loss of protein and the contractile fibers are eventually replaced by fat and connective tissue, rendering skeletal muscle useless. As the weakening process of the disease continues, the teen or young adult is confined to a wheelchair. Until recently, individuals would die before the age of 20 from respiratory or heart failure. Life expectancy has increased due to advances in cardiac and respiratory care.

Huntington's Disease

Huntington's disease (HD) is characterized by the degeneration of the central nervous system, which ultimately results in abnormal movements and mental deterioration. In this disorder, the product of an abnormal gene interferes with normal metabolism in nerve tissue. Treatment is aimed at the symptoms. Typically, symptoms appear between ages 30 and 50; the average life expectancy after diagnosis is 10–30 years. The HD gene is expressed in all cells, but it is expressed in higher concentrations in the brain and testes.

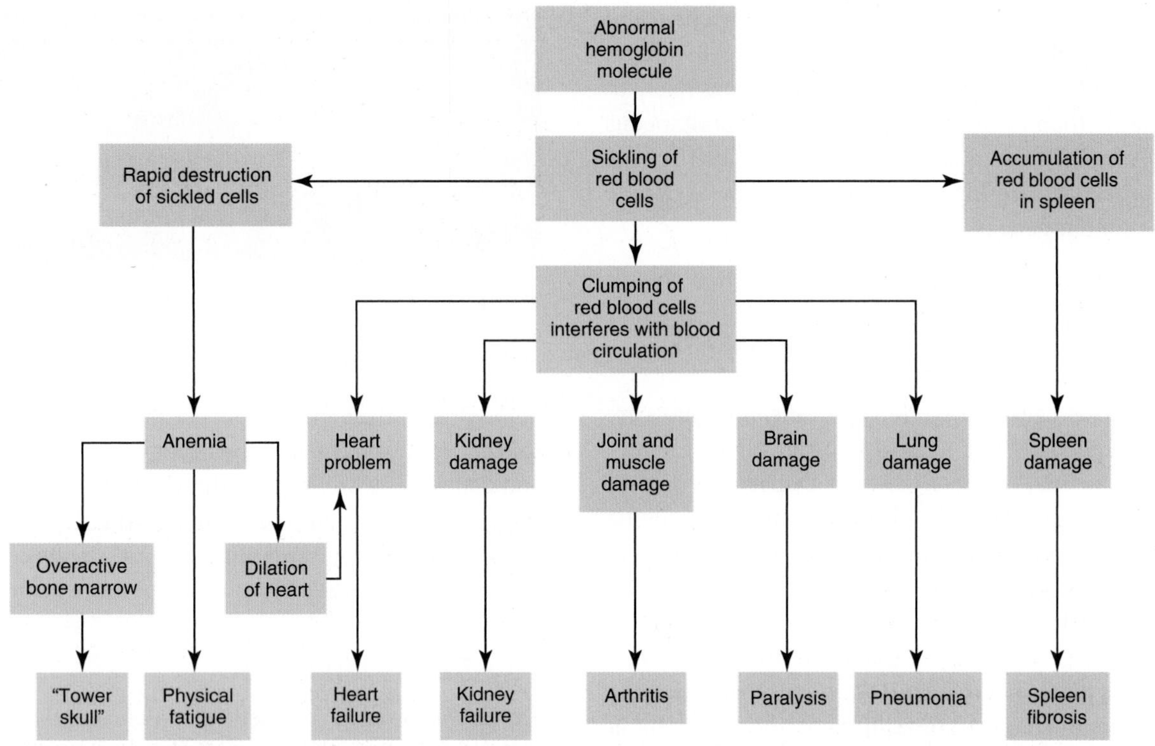

Figure 23-2 *A series of tissue damage and physiological effects caused by sickle cell anemia*

Cystic Fibrosis

Cystic fibrosis (CF) is a disease of the exocrine gland. The lining of the digestive tract, the ducts of the pancreas, and the respiratory tract produce thick mucus that blocks the passageways. Nutritional malabsorption is severe due to a deficiency of pancreatic enzymes and other contributory factors. The blockage of the respiratory passages causes chronic bronchitis and pneumonia. Pulmonary therapy consists of a procedure called "cupping and clapping," which helps dislodge the thick mucus from the respiratory tract. Treatment also includes antibiotics for respiratory infections, bronchodilators, anti-inflammatory drugs, and supplemental pancreatic enzymes. Mucus-thinning medications may be prescribed to reduce the stickiness and loosen the mucus.

Most infants are now screened for CF at birth, and most cases are identified before the age of two. As treatments for CF continue to improve, so does life expectancy—some people with CF are living into their 40s, 50s, or later. Science has discovered the gene responsible for this disease and may soon be able to treat and prevent this illness. The involvement of the lungs is what causes cystic fibrosis to be one of the most fatal hereditary disorders.

Thalassemia

Thalassemia (Cooley's anemia) is a blood disease found among people of Mediterranean descent. Symptoms are the same as those associated with any anemia; there is also enlargement of the spleen and possible congestive heart disease. This disease is treated with blood transfusions to replace the defective hemoglobin molecules. See Chapter 13 for more information.

Hemophilia

Hemophilia is a sex-linked genetic disorder, which means it is only transmitted on the X chromosome. In this disease, the person is unable to produce factor VIII, which is necessary for blood clotting. Persistent bleeding may occur as a result of an injury or spontaneously. Treatment consists of giving the person factor VIII. See Chapter 13 for more information.

Chromosomal Abnormalities

Some mutations are caused by chromosomal abnormalities. Some involve entire chromosomes, and others involve parts of chromosomes. During meiosis, the cell division that occurs during the formation of the gametes, a pair of chromosomes may adhere to each other and not pull apart at metaphase. As a result of this nondisjunction, duplicate chromosomes go to one daughter cell and none of this type of chromosome to the other. Nondisjunction of certain chromosomes referred to as sex chromosomes causes various differences in sexual development, such as Turner's syndrome in females and Klinefelter's syndrome in males.

One of the most common chromosomal abnormalities involves an extra chromosome designated as chromosome 21, meaning every cell has an extra copy of chromosome 21. This disorder is referred to as **trisomy 21** or **Down syndrome**. Risk factors for bearing a child with Down syndrome significantly increase if the mother is older than 35 or if there is a sibling with Down syndrome. Many physicians therefore recommend amniocentesis for all females who become pregnant after age 35. Cells from the amniotic fluid will show trisomy 21—as well as other chromosomal defects—if present.

The extra chromosome in the cellular structure causes dysfunction of the cells, effecting the function of tissues, organs, and body systems overall. Children with the congenital disorder Down syndrome have characteristic features including low muscle tone; almond-shaped eyes; a flat face, especially at the bridge of the nose; and a short neck. Often, the cardiovascular system is affected. There is also some degree of intellectual disability. The goal of treatment is to assist the child in learning skills to become a more independent individual.

MUTAGENIC AGENTS

Although most gene or chromosomal mutations occur spontaneously, the rate or frequency of mutations can be increased. This happens when a cell, a group of cells, or an entire organism is exposed to certain chemicals or radiations. Agents that increase the occurrence of mutations are called **mutagenic agents**. Mutagenic agents can be radiations such as cosmic rays, ultraviolet (UV) rays from the sun, X-rays, and radioactive elements. Some mutagenic chemicals are benzene, formaldehyde, phenol, and nitrous acid.

In recent times, the accelerated use of various chemical and physical agents with mutagenic properties has caused concern among some geneticists, who fear possible significant alterations to genes and chromosomes that will be passed on to future generations. The increased use of ionizing radiation in medical diagnosis and the problem of the disposal of nuclear waste from reactors are examples. Certain chemical pollutants in the environment, such as herbicides and insecticides, are also suspected to cause genetic defects.

Because people are being exposed to more and more new substances, and because changes in genes are irreversible, caution should be the rule with regard to any unnecessary exposure to substances suspected of being mutagens.

GENETIC COUNSELING

Genetic counseling involves talking to parents or prospective parents about the possibility of genetic disorders.

People who may be especially interested in genetic counseling or testing include the following:

- Those who think they have a birth defect or who have a family member with a genetic disorder
- Pregnant females who have received abnormal prenatal tests
- Females who are pregnant after age 35
- Couples who have had one child born with a genetic defect
- Females who have had two or more miscarriages or whose babies died in infancy
- Couples who need information about genetic disorders that occur frequently in their ethnic group
- Couples who are first cousins or other blood relatives

The counseling team usually is made up of members of the health care team, a genealogist, a nurse, laboratory personnel, and social service providers. In genetic counseling, a family history called a pedigree is obtained. Any and all facts that pertain to the parents or prospective parents and family members are considered. After careful analysis, a genotype is determined; this analysis is used to predict the possibility of a genetic disorder.

The prospective parents are made aware of diagnostic tests available during pregnancy that may indicate a problem. The **chorionic villi sampling** test can be done as early as 8–10 weeks into the pregnancy. A sample of fetal cells is removed from the fetal side of the placenta and examined. An **amniocentesis** is the withdrawal of amniotic fluid during week 16 of the pregnancy. An examination of the fluid is able to pick up as many as 200 possible genetic disorders. Prior to performing an amniocentesis, a sonogram is done to determine where the fetal structures are located so as to prevent any injury to the fetus.

Genetic counseling helps parents or prospective parents make informed decisions regarding children.

GENETIC ENGINEERING

Genetic engineering, **recombinant DNA** technology, and genetic modification are terms applied to manipulation of genes, generally using a process outside the organism's natural reproductive process. It involves the isolation, manipulation, and reintroduction of DNA into the cells or model organisms, usually to express a specific protein. The aim is to introduce new characteristics or attributes.

Human insulin, human growth hormone, vaccines, monoclonal antibodies, antibiotics, and human *interferon* are now being produced using the sophisticated technology of recombinant DNA.

Designer babies is a relatively new term that describes embryos that have been genetically modified to remove unfavorable traits, such as genetic disorders. One of the most common methods is preimplantation genetic diagnosis (PGD), where only embryos that lack the adverse traits are implanted. CRISPR-CAS9 modifies DNA fragments that can prevent and correct destructive traits as well as add specific genes. This process is done after fertilization.

There is some controversy regarding the ethics of genetic engineering, but most scientists believe that genetic engineering is essential to future discoveries to help alleviate certain diseases.

Gene Therapy

Gene therapy is a technique for correcting defective genes responsible for disease development. Researchers may use several approaches for correcting faulty genes:

- A normal gene may be inserted into a nonspecific location within the genome to replace a nonfunctional gene.
- An abnormal gene could be swapped for a normal gene through homologous (structures alike because of shared ancestry) recombinant DNA.
- The abnormal gene could be repaired through selective reverse mutation, which returns the gene to normal function.
- The regulation of a particular gene, or the degree to which a gene is turned on or off, could be altered.

Factors that have kept gene therapy from becoming a more effective treatment for genetic disease include the short-lived nature of gene therapy, problems with integrating therapeutic DNA into the genome, and the rapidly dividing nature of many cells. Another factor is the body's immune response. Any time a foreign object is introduced to human tissues, the immune system is designed to attack the invader. Some of the most commonly occurring disorders, such as heart disease, Alzheimer's disease, arthritis, and diabetes, are caused by the combined effect of variations in genes.

Recent developments in gene therapy include repairs to errors in messenger RNA derived from a defective gene. This technique has the potential to treat cystic fibrosis and some cancers. The Food and Drug Administration has yet to approve any human gene therapy product.

Stem Cell Treatment

Stem cell treatment is being studied to treat congenital diseases while in utero or immediately after birth. Stem cells can be retrieved from umbilical cord blood or directly from the amniotic fluid. The focus of stem cell treatment is to regenerate specific damaged cells or tissues.

Medical Terminology

amnion	closed sac surrounding fetus
-centesis	surgical puncture
amnio/centesis	surgical puncture of the amnion
congenit	dating from birth
-al	pertaining to
congenit/al	pertaining to existing at birth
chromo	colored
-some	body
muta	basic alteration
-tion	process of
genet	producing protein (DNA)
-ics	pertaining to
genet/ics	pertaining to producing protein (DNA)
hemo	bleeding
-phil	attraction for; tendency toward
-ia	abnormal condition of
hemo/phil/ia	abnormal condition of a tendency toward bleeding

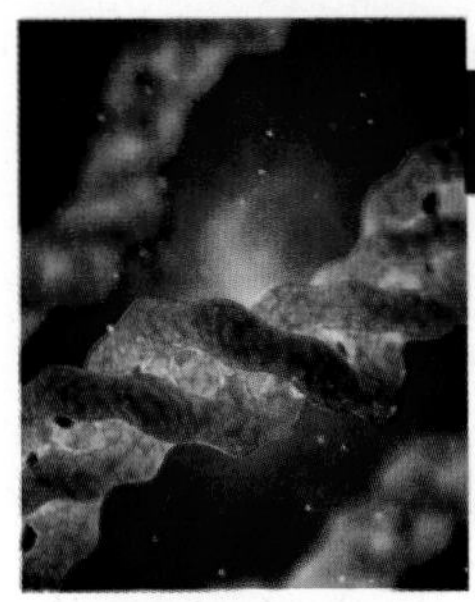

Study Tools

Workbook	Activities for Chapter 23
Online Resources	PowerPoint presentations

REVIEW QUESTIONS

Select the letter of the choice that best completes the statement.

1. The branch of science that deals with how human traits are passed down is called
   a. genetic engineering.
   b. biology.
   c. genetics.
   d. genetic counseling.
2. When a change takes place in a gene, it is called a
   a. mutation.
   b. lethal gene.
   c. congenital defect.
   d. mutant.
3. A deficiency in breaking down fat molecules is characteristic of
   a. sickle cell anemia.
   b. Tay-Sachs disease.
   c. PKU.
   d. Down syndrome.
4. An extra chromosome can cause a defect known as trisomy 21 or
   a. Cooley's anemia.
   b. Huntington's disease.
   c. Down syndrome.
   d. PKU.
5. The nutritional malabsorption the occurs with cystic fibrosis is due to
   a. an altered structure of the digestive system.
   b. the inability to absorb fats.
   c. the loss of appetite.
   d. a deficiency of pancreatic enzymes.

APPLYING THEORY TO PRACTICE

1. Mrs. Kibbee tells you she has been advised to have an amniocentesis; she is 16 weeks pregnant. The thought of having someone stick a needle in her belly and what may happen to her baby is frightening. Explain the test to her.
2. PKU is a genetic disorder that can be detected in the hospital nursery. Explain the disease and the special dietary restrictions.
3. Sickle cell anemia trait is diagnosed with a simple blood test. Some people are afraid of blood tests. Describe the steps of a blood draw.
4. Pretend you are going to participate in a debate on the issue of designer babies. One side must present legal, scientific, and ethical issues for the limited use of this science; the other side will support unrestricted use of the technology. Prepare arguments for both sides.
5. Adam was born with the congenital disorder Down syndrome. Explain to his parents the cellular differences that cause Down syndrome; the physical manifestations; and the possible way that it may affect the structure and function of tissues, organs, and body systems.

CASE STUDY

A couple comes into the family counseling center and wants information on genetic counseling. They are thinking of starting a family, but one of them has a family history of Tay-Sachs disease. Keisha is a genetic counselor who will explain the services available at the center. Keisha's discussion will include genetic mutation, tests that may be done during pregnancy, and advances being made in genetic therapy.

1. What information can Keisha give the couple regarding the services of the genetic counseling center?
2. Define genetic mutation and identify the genetic possibility of the couple having a baby with Tay-Sachs disease.
3. What types of genetic testing can be done to give the couple more information?
4. What information can be given to the couple regarding advances being made in stem cell therapy?

23-1 Phenotypes

- ***Objective:*** Identify phenotypes that are passed down in families
- ***Materials needed:*** paper, pencil

Step 1: Research and identify common phenotypes passed down from parent to child, such as attached or unattached earlobes, ability to roll tongue, and so on.

Step 2: Make a list of these inherited traits.

Step 3: Pair up with a partner and take turns recording whether you observe each of the inherited traits on your list in your partner.

Step 4: Continue to change partners until you have surveyed most or all of your classmates, keeping a running tally of the number of times a trait appeared. Graph results using a bar graph.

APPENDIX A

METRIC CONVERSION TABLES

LENGTH	CENTIMETERS	INCHES	FEET
1 centimeter	1.00	0.394	0.0328
1 inch	2.54	1.00	0.0833
1 foot	30.48	12.00	1.00
1 yard	91.4	36.00	3.00
1 meter	100.00	39.40	3.28

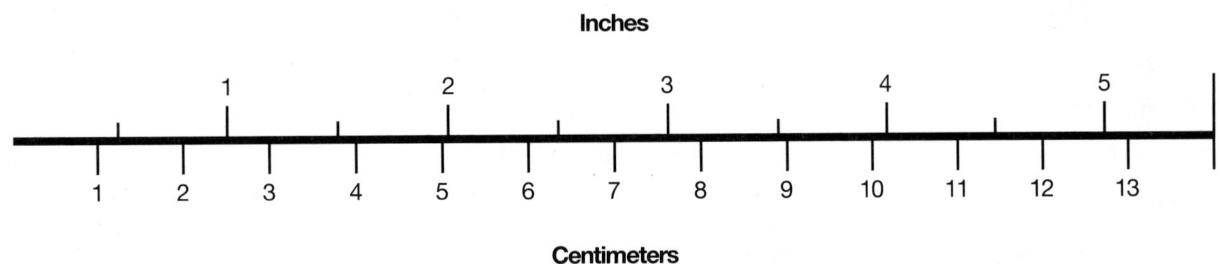

Comparison of Centimeters and Inches

VOLUME	CUBIC CENTIMETERS	FLUID DRAMS	FLUID OUNCES	QUARTS	LITERS
1 cubic centimeter	1.00	0.270	0.033	0.0010	0.0010
1 fluid dram	3.70	1.00	0.125	0.0039	0.0037
1 cubic inch	16.39	4.43	0.554	0.0173	0.0163
1 fluid ounce	29.6	8.00	1.00	0.0312	0.0296
1 quart	946.00	255.00	32.00	1.00	0.946
1 liter	1000.00	270.00	33.80	1.056	1.00

WEIGHT	GRAINS	GRAMS	APOTHECARY OUNCES	POUNDS
1 grain (gr)	1.00	0.064	0.002	0.0001
1 gram (gm)	15.43	1.00	0.032	0.0022
1 apothecary ounce	480.00	31.1	1.00	0.0685
1 pound	7000.00	454.00	14.58	1.00
1 kilogram	15432.00	1000.00	32.15	2.205

RULES FOR CONVERTING ONE SYSTEM TO ANOTHER	
Volumes	
Grains to grams	divide by 15
Drams to cubic centimeters	multiply by 4
Ounces to cubic centimeters	multiply by 30
Minims to cubic millimeters	multiply by 63
Minims to cubic centimeters	multiply by 0.06
Cubic millimeters to minims	divide by 63
Cubic centimeters to minims	multiply by 16
Cubic centimeters to fluid ounces	divide by 30
Liters to pints	divide by 2.1
Weight	
Milligrams to grains	multiply by 0.0154
Grams to grains	multiply by 15
Grams to drams	multiply by 0.257
Grams to ounces	multiply by 0.0311
Temperature	
Multiply centigrade (Celsius) degrees by 9/5 and add 32 to convert Fahrenheit to Celsius	
Subtract 32 from the Fahrenheit degrees and multiply by 5/9 to convert Celsius to Fahrenheit	

COMMON HOUSEHOLD MEASURES AND WEIGHTS	
1 teaspoon	= 4–5 mL or 1 dram
3 teaspoons	= 1 tablespoon
1 dessert spoon	= 8 mL or 2 drams
1 tablespoon	= 15 mL or 3 drams
4 tablespoons	= 1 wine glass or 1/2 gill
16 tablespoons (liquid)	= 1 cup
12 tablespoons (dry)	= 1 cup
1 cup	= 8 fluid ounces or 1/2 pint
1 tumbler or glass	= 8 fluid ounces or 240 mL
1 wine glass	= 2 fluid ounces or 60 mL
16 fluid ounces	= 1 pound
4 gills	= 1 pound
1 pint	= 1 pound

Glossary

A

abdominal cavity area of the body that contains the stomach, liver, gallbladder, pancreas, spleen, small intestine, appendix, and part of the large intestine

abdominal pain visceral pain that may be coming from the abdominal cavity

abdominopelvic cavity area below the diaphragm, with no separation between the abdomen and pelvis

abduction movement away from the midline or axis of body; opposite of adduction

abrasion an injury in which superficial layers of the skin are scraped or rubbed away

abscess pus-filled cavity

absorption passing of a substance into body fluids and tissues

accommodation as it applies to vision, the process by which the ciliary muscle of the eye controls the shape of the lens for vision at near and far distances

acetabulum area where the three bones of the hip unite to form a deep socket into which the head of the femur fits to form the hip joint

acetylcholine chemical released when a nerve impulse is transmitted

acid chemical compound that ionizes to form hydrogen ions (H^+) in aqueous solution

acidosis a disruption in the acid-base balance where the body becomes too acidic

acini cells cells in the pancreas that produce the digestive juices

acne vulgaris chronic disorder of sebaceous gland

acquired immunity immunity as a result of exposure to a disease

acquired immunodeficiency syndrome (AIDS) a potentially fatal disease causing suppression of the immune system

acromegaly excess of growth hormone in adults: overdevelopment of bones of hand, face, and feet

action potentials the ability to respond to stimuli with electrical impulses

active acquired immunity two types—natural and artificial acquired immunity

active transport process by which solute molecules are transported across a membrane to a higher concentration gradient, from an area of low concentration to one of high concentration

acute glomerulonephritis inflammation of the glomerulus of the nephron due to bacterial infection

acute kidney failure sudden loss of kidney function

Addison's disease hypofunction of the adrenal gland

adduction movement of part of the body or a limb toward the midline of body; opposite of abduction

adenoids pair of glands composed of lymphoid tissue, found in the nasopharynx; also called *pharyngeal tonsils*

adenosine triphosphate (ATP) chemical compound consisting of one molecule of adenine, one of ribose, and three of phosphoric acid; high-energy fuel a cell requires to function

adipose tissue fatty or fatlike

adrenal glands endocrine gland that sits on top of kidney; consists of cortex and medulla

adrenaline hormone produced by the adrenal gland; a powerful cardiac stimulant; epinephrine

adrenocorticotropic hormone (ACTH) hormone that stimulates the growth and secretion of the adrenal cortex

afferent arteriole arteriole that takes blood from the renal artery to the Bowman's capsule of the kidney

afferent neurons *see* sensory neuron

Age of Enlightenment the period that lasted until the late seventeenth century, which gave rise to scientific societies and academies and saw advancements in medicine, chemistry, taxonomy, mathematics, and physics; also known as the Age of Reason

agranulocytes nongranular white blood cells; known as *agranular leukocyte*

airborne transmission transfer of an agent to a susceptible host through droplet nuclei or dust particles suspended in the air

albumin plasma protein; maintains osmotic pressure

aldosterone hormone secreted by the adrenal cortex; regulates salt and water balance in the kidney

alimentary canal entire digestive tube from mouth (ingestion) to anus (excretion)

alkali a substance that, when dissolved in water, ionizes into negatively charged hydroxide (OH) ions and positively charged ions of a metal

alkalosis a disruption in the acid-base balance where the body becomes too alkaline

allergen substance that causes an allergic reaction

all or none law law that states a muscle cell, when stimulated, contracts all the way or not at all

alopecia loss of hair; baldness

alveolar sac saclike cluster of alveoli at the end of the alveolar duct; works to exchange carbon dioxide and oxygen

alveoli a globular shape with an outer layer of epithelial tissue; inner surface covered with surfactant; has a network of blood capillaries

Alzheimer's disease progressive disease with degeneration of nerve endings in the cortex of the brain that block the signals that pass between nerve cells

amblyopia dimness of vision

amenorrhea absence of menstruation

amino acids small molecular units that make up protein molecules

amniocentesis withdrawal of amniotic fluid for testing

amphiarthroses partially movable joint (e.g., symphysis pubis)

amyotrophic lateral sclerosis (ALS) a motor neuron disease that causes muscle weakness in a limb or the muscles of the mouth and throat; gradually all the muscles under voluntary control are affected

anabolism building up of complex materials in metabolism from simpler ones

analgesics drugs that reduce pain

anaphase phase 4 in mitosis; chromatid pairs fully separated

anaphylaxis (anaphylactic shock) severe and sometimes fatal allergic reaction

anatomical position body standing erect, face forward, arms at the sides, and palms forward

anatomy the study of the structure of an organism

androgens male precursor hormone converted into estrogen in females and other male hormones in males

anemia blood disorder characterized by reduction in red blood cells or hemoglobin

aneurysm a widening, or sac, formed by dilation of a blood vessel

angina pectoris severe chest pain caused by lack of blood supply to heart

angioplasty balloon surgery to open blocked blood vessels

anorchism the complete absence of development of one or both testes

anorexia nervosa an illness in which a person refuses to eat

anoxia a lack of oxygen to cellular structures

antagonist a muscle whose action opposes the action of another muscle

anterior front or ventral; in front of

anterior chamber space between cornea and iris

anterior pituitary lobe area of the pituitary gland that is responsible for the secretion of growth hormones

anthrax a disease-causing organism that has three forms: cutaneous, intestinal, and inhalation, the last of which is the most deadly form

antibody substance produced by the body that inactivates a specific foreign substance that has entered the body; formed as a reaction to an antigen

anticoagulants chemical substance that prevents or slows blood clotting (e.g., heparin)

antidiuretic hormone (ADH) hormone secreted by the posterior pituitary gland that prevents or suppresses urine excretion

antigen substance stimulating formation of antibodies against itself

antiprothrombin chemical substance, such as heparin, that directly or indirectly reduces or slows the action of prothrombin

antithromboplastin chemical substance that inhibits the clot-accelerating effect of thromboplastins

anuria absence of urine

anus outlet from rectum

anvil (incus) middle ear bone, or ossicle, in a chain of three ossicles of the middle ear

aorta largest artery in body, rising from the left ventricle of the heart; takes blood from the heart to the rest of the body

aortic semilunar valve made up of three half-moon-shaped cups, located between the junction of the aorta and the left ventricle of the heart

aphasia loss of ability to speak or understand speech; may be accompanied by loss of verbal comprehension

aplastic anemia anemia caused by a suppression of the bone marrow

apnea temporary stoppage of breathing movements

aponeurosis flattened sheet of white, fibrous connective tissue; holds one muscle to another or to the periosteum

apoptosis an orderly process by which cells intentionally die. The term is used interchangeably with the term *programmed cell death (PCD)*

appendicitis inflammation of the appendix; *see* vermiform appendix

appendicular skeleton part of skeleton consisting of pectoral and pelvic girdles and limbs

aqueous humor watery fluid found in anterior chamber of the eye

arachnid cysts benign lesions that occur as a result of the splitting of the arachnoid layer

arachnoid mater weblike middle membrane of meninges

areola pigmented ring around nipple; any small space in tissue

areolar tissue a type of connective tissue that surrounds various organs and supports both nerve and blood vessels

arrector pili muscle a smooth muscle on the side of each hair follicle; when cold, it stimulates the skin to pucker around the hair

arrhythmia absence of a normal heartbeat rhythm

arteries blood vessel that carries blood away from the heart

arterioles small branches of an artery

arteriosclerosis hardening of arteries, resulting in thickening of walls and loss of elasticity

arthritis inflammation of a joint

articular cartilage thin layer of cartilage over the ends of long bones

artificial acquired immunity type of immunity acquired as a result of an injection of vaccine, antigen, or toxoid

artificial insemination procedure in which semen is placed in vagina by means of cannula or syringe

asbestosis a respiratory disease caused by inhaling asbestos fibers

ascites accumulation of fluid in the peritoneal cavity

asepsis being free of living pathogenic microorganisms

associative neurons carries messages from sensory neuron to motor neuron; also called *interneuron*

asthma condition in which the airways are obstructed as a result of an inflammatory reaction to a stimulus

astigmatism irregular curvature of cornea or lens

atelectasis condition in which the lungs fail to expand normally

atherosclerosis hardening of arteries due to deposits of fatlike material in the lining of the arteries

athlete's foot fungal infection of the foot or fingers

atlas first cervical vertebra; articulates with the axis and occipital skull bone

atom smallest piece of an element

atresia the congenital absence or the pathological closure of an opening, passage, or cavity

atrial fibrillation occurs when abnormal impulses from the atria bombard the AV node. This action causes the ventricles to beat faster and a person can experience tachycardia

atrial septal defect a birth defect in which there is an opening between both atria

atrioventricular bundle conducting fibers in the septum; also known as the *bundle of His*

atrioventricular (AV) node small mass of interwoven conducting tissue located between the atria and ventricles

atrium upper chambers of the heart

atrophy wasting away of tissue or cells

auricle (1) pinna, or ear flap, of external ear; (2) atrium of the heart

autoimmune disorder a condition that causes destruction of the body's own tissues

autoimmunity action of antibodies against one's own body

autonomic nervous system collection of nerves, ganglia, and plexuses through which visceral organs, heart, blood vessels, glands, and smooth (involuntary) muscles receive stimulation

avascular without blood vessels

axial skeleton skeleton of head and trunk

axis second cervical vertebra; forms a pivot for the atlas

axons nerve cell structure that carries impulses away from cell bodies to dendrites

B

B lymphocytes white blood cells synthesized in the bone marrow; help form antibodies

bacteria small, one-celled microorganisms capable of causing disease or infection

bacterial vaginosis common infection caused by an imbalance of the body's normal vaginal flora

bactericidal causing the death of bacteria

ball-and-socket joints joint that connects a ball-shaped bone with a concave bone that allows the greatest freedom of movement

balloon surgery *see* angioplasty

Barrett's syndrome a condition in which the mucosa of the esophagus becomes damaged by acid reflux inflaming the lining

Bartholin's glands mucous glands at opening of vagina

basal cell carcinoma most common and least malignant type of skin cancer

basal metabolic rate (BMR) the measure of the total energy utilized by the body to maintain the body processes necessary for life

base chemical compound yielding hydroxide (OH^-) in an aqueous solution, which will react with acid to form a salt and water

basophils leukocyte cells that are activated during an allergic reaction or inflammation; produce histamine and heparin

Bell's palsy disorder that affects the facial nerve

belly the central part of a muscle

benign nonmalignant

benign prostatic hypertrophy (BPH) an enlarged prostate

biceps muscle on front part of upper arm

bicuspids premolars of the adult teeth, they have two ridges or cusps, used for grinding food

bicuspid (mitral) valve atrioventricular valve between the left atrium and left ventricle; allows blood to flow from left atrium to left ventricle

bile substance produced by liver; emulsifies fat

binge eating disorder when a person repeatedly eats unusually large amounts of food in a short period of time

biofeedback a measurement of physiological responses that yields information about the relationship between the mind and the body and helps people manipulate those responses through mental activity

biological agents living organisms that invade a host

biology the study of all forms of life

biomechanics the study of structure, function, and motion

biomarkers a normal substance found in the blood or tissue in small amounts

blepharospasm involuntary muscle contraction of the eyelid, causing blinking

blood–brain barrier choroid plexus capillaries in the brain differ in their selective permeability; thus, drugs carried in the bloodstream may not penetrate the brain tissue

body mass index (BMI) a measurement of the amount of body fat in an individual

boils bacterial infection of sebaceous gland

bolus rounded mass; food prepared by mouth for swallowing

Bowman's capsule double-walled capsule around the glomerulus of a nephron

brachial artery artery located at the crook of the elbow, along the inner biceps muscle

bradycardia abnormally slow heartbeat: less than 60 beats per minute

brain stem portion of brain other than cerebral hemispheres and cerebellum

brain tumors a mass or growth of abnormal cells in the brain

breast cancer the most common cancer in women other than skin cancer; signs include lump in breast or axillary area, changes in size, shape of breast, or discharge from nipples

breast tumors either benign or malignant; benign tumors are fluid-filled cysts that enlarge during the premenstrual period

breasts mammary glands in front of the chest; secrete milk after childbirth

bronchioles one of the small subdivisions of a bronchus

bronchitis inflammation of the bronchial tubes

bronchogenic cysts cysts found in the trachea or the lower lobes of the lung

bronchoscopy lighted, tubular instrument used to inspect the interior of the bronchial tubes

bronchus one of two primary branches of the trachea

buccal cavity mouth cavity bounded by the inner surface of the cheek

buffer a compound that maintains the chemical balance in a living organism

bulbourethral glands (Cowper's glands) located on either side of urethra in males; adds alkaline substance to semen

bulimia nervosa episodic binge eating and purging

bundle of His *see* atrioventricular bundle

bursae closed sacs with a synovial membrane lining, found in the spaces of connective tissue between muscles, tendons, ligaments, and bones

bursitis inflammation of a bursa

C

C. diff short form of *Clostridium difficile*, a highly contagious bacterial disease and the number one cause of diarrhea in the health care setting

calcaneus heel bone

calcified to deposit mineral salts

calcitonin hormone secreted by thyroid gland that controls calcium ion concentration in the body

calorie a unit that measures the amount of energy

calyces cup-shaped parts of the renal pelvis

cancellous bone *see* spongy bone

cancer the presence of a malignant tumor, which may affect all body parts

cancer of the larynx disease in which malignant cells form in the tissue of the larynx; curable if early detection is made

cancer of the lung a malignant tumor found in the lungs may be small cell or adenocarcinoma

cancer of the stomach also called gastric cancer, can develop in any part of the stomach and spread throughout the stomach

canines sharp teeth for tearing between incisors and premolars

canker sores small, painful ulcers that appear periodically on the tongue or mouth

capillaries microscopic blood vessels that connect arterioles with venules

carbohydrates an organic compound of carbon, hydrogen, and oxygen as sugar or starch

carbon monoxide (CO) poisoning a condition in which an odorless gas combines rapidly with hemoglobin and crowds out oxygen

cardiac arrest syndrome resulting from failure of the heart as a pump

cardiac catheterization a diagnostic test in which a catheter is inserted into the femoral artery or vein and fed up into the heart

cardiac muscle involuntary muscle that makes up the walls of the heart

cardiac output the total volume of blood ejected from the heart per minute

cardiac sphincter circular muscle fibers between the esophagus and stomach

cardiac stents device inserted into an artery to open a clog or plaque buildup

cardiopulmonary circulation the system of carrying blood from the heart to the lungs and back

cardiopulmonary resuscitation (CPR) life saving technique that keeps oxygenated blood flowing to the brain and other vital organs

cardiotonic drug to slow and strengthen the heart

carotid artery artery that supplies blood to the neck and head; *see* common carotid artery

carpals bones of the wrist

carpal tunnel syndrome a condition that affects the median nerve and the flexor tendons that attach to the bones of the wrist

carriers in medical terminology, a person who may have an infectious agent present in his or her body but is symptom free; a carrier may spread the disease to others

cartilage white, semiopaque, nonvascular connective tissue

catabolism the breaking down and changing of complex materials into simple ones with the release of energy; a process in metabolism

cataract condition in which the eye lens becomes opaque

caudal refers to direction toward the tail end of the body

cecum pouch at the proximal end of the large intestine

celiac disease genetic autoimmune disease in which the ingestion of gluten leads to damage to the lining of the small intestine

cell basic unit of structure and function of all living things

cell membrane structure that encloses the cell; also known as plasma membrane

cellular immunity the body's ability to resist infections, provided by T lymphocytes

cellular respiration process by which oxygen is used to release energy from a cell

central nervous system consists of the structures of the brain and spinal cord

centrioles two cylindrical organelles found near the nucleus in a tiny body called the centrosome; they are perpendicular to each other

centrosomes tiny area near the nucleus of an animal cell; it contains two cylindrical structures called centrioles

cephalic directional term used to mean toward the head

cerebellum structure of the brain behind the pons and below the cerebrum

cerebral aqueduct a narrow canal connecting the third and fourth ventricles of the brain

cerebral cortex a layer of gray matter covering the upper and lower surfaces of the cerebrum

cerebral hemorrhage bleeding from blood vessels in the brain

cerebral palsy (CP) a disturbance in voluntary muscle action due to brain damage

cerebral vascular accident (CVA) sudden interruption of blood flow to the brain; also called *stroke*

cerebral ventricles four lined cavities within the brain filled with cerebrospinal fluid

cerebrospinal fluid a substance that forms within the four brain ventricles from the blood vessels of the choroid plexus; this serves as a shock absorber to protect the brain and spinal cord

cerebrum the largest part of the brain

cerumen a waxy or oily substance produced by ceruminous (sebaceous) glands in the lining of the outer ear; also known as earwax

cervical cancer cancer of the narrow portion of the uterus (the cervix) that extends into the vaginal canal

cervical vertebrae first seven bones of the spinal column

cervix narrow end of the uterus

Charcot-Marie-Tooth (CMT) disease the most common inherited neurological condition, which is caused by a gene defect; symptoms include hammer toes, high foot arch, and foot drop; also called hereditary motor and sensory neuropathy

chemical agents substances that interact with a host, causing diseases

chemical bond bond formed when atoms share or combine their electrons with atoms of other elements

chemistry study of the structure of matter and the composition of substances, their properties, and their chemical reactions

chemokines a class of cytokines that plays an important role in introducing chemokine receptors in antitumor immune response and autoimmune disease

Chiari malformation a condition in which brain tissue extends into the spinal cord

chlamydia a sexually transmitted disease caused by the *Chlamydia trachomatis* organism

cholecystitis inflammation of the gallbladder

cholesterol a steroid normally synthesized in the liver and also ingested in egg yolks, animal fats, and tissues

chordae tendineae small, fibrous strands connecting the edges of the tricuspid valve to the papillary muscles that are projections of the myocardium

chorionic villi sampling a test done early in pregnancy to detect genetic problems

choroid coat the middle layer of the eye

choroid plexus the network of blood vessels of the pia mater found in the brain's ventricles

chromatid each strand of a replicable chromosome

chromatin DNA and protein material in a loose and diffuse state; during mitosis, chromatin condenses to form the chromosomes

chromosomal mutation a mutation that involves a change in the number of chromosomes in the organism's nucleus or a change in the structure of a whole chromosome

chromosomes short, rod-like structures that determine hereditary characteristics

chronic glomerulonephritis diminished function of the kidney due to damage to the filtration membrane, the glomerulus

chronic obstructive pulmonary disease (COPD) chronic lung condition such as emphysema or bronchitis

chronic renal failure gradual loss of function of the nephrons

chyme food that has undergone gastric digestion

cilia tiny, hairlike projections of protoplasm that extend from the cell or body surface and help move things along; they may also work like a filter

ciliary body ligaments that suspend the eye

circumcision removal of the foreskin of the penis

circumduction circular movement at a joint

cirrhosis chronic, progressive inflammatory disease of the liver characterized by the formation of fibrous connective tissue

clavicles collarbone

cleansing removal of soil or organic material from instruments and equipment used in providing patient care

cleft lip when the tissue that forms the baby's mouth or lips does not close properly during the early stages of fetal development

cleft palate when the palatine bones do not close properly during early fetal development

clitoris small structure over female urethra; has many nerve endings and blood vessels

clotting time the time it takes for blood to clot

coagulation process of blood clotting

coarctation of aorta when a part of the aorta is narrower than usual

coccyx tailbone

cochlea spiral cavity of the internal ear containing the organ of Corti

cochlear duct an endolymph-filled, triangular canal containing the spiral organ of Corti

coitus act of intercourse

collagen fibrous protein occurring in bone and cartilage

collecting tubule structure in nephron that collects urine from distal convoluted tubule

colon known as the large intestine; about 5 feet in length and 2 inches in diameter; divided into ascending, transverse, descending, and sigmoid colon

colon cancer abnormal cell growth in the colon

colorectal cancer involves the large intestine, while rectal cancer involves the rectum; collectively they are known as colorectal cancer

color blindness inability to distinguish colors

colostomy artificial opening from the colon onto the surface of the skin

common bile duct formed by the union of the hepatic duct and cystic duct; brings bile to the duodenum

common carotid artery artery found in the neck; *see* carotid artery

common cold highly contagious virus

communication dissemination of information

compact bone also known as hard bone; strong and light

comparative anatomy when different body parts and organs of humans are studied with regard to similarities to and differences from others in the animal kingdom

comparative investigation collecting data on two or more groups for comparison under different conditions with a focus on patterns or trends

complement cascade a complex series of reactions that activate more than 20 proteins that are usually inactive unless activated by a pathogen; these proteins cause the breakdown or lysis of microorganisms, attack the pathogen's cell membrane, and enhance the inflammatory process

complete proteins proteins that contain all the essential amino acids; they enable an animal to grow and carry on fundamental life activities

compounds elements combined in definite proportion by weight to form a new substance

compromised host person whose normal defense mechanisms are impaired and who is therefore susceptible to infection

conclusion/results discusses the results of an experiment

concussion severe blow to the head; may cause temporary loss of consciousness

conduction defect a defect in the electrical impulse system of the heart muscle

cones structures in the retina of the eye sensitive to bright light and responsible for color vision

congenital anomalies of kidney and urinary tract (CAKUT) a birth defect in which the baby is born with kidney or urinary tract structures that develop differently

congenital cystic adenomatoid malformation (CCAM) a birth defect in which a baby's lung tissue grows more than normal

congenital disorder present at birth

congenital hearing loss a birth defect in which there is hearing loss caused by prematurity, cytomegalovirus, congenital rubella, infection, neurodegenerative disorder, or family history of deafness

congenital heart disease (CHD) a birth defect that adversely affects the heart and its functions

congenital insensitivity to pain a birth defect in which there is insensitivity to pain; anhidrosis, or the inability to sweat; and intellectual disability

congenital lung disorder a birth defect that adversely affects the lungs and their functions; also known as cystic lung disorder or congenital lung malformations

congenital malformation of the diaphragm a birth defect in which the diaphragm fails to close properly during prenatal development

congenital myasthenia gravis *see* myasthenia gravis

congestive heart failure in right-sided heart failure, fluid builds up througout the body, edema is first noticed in lower extremities; in left-sided heart failure, fluid accumulates in the lungs

conjunctivitis an inflammation of the conjunctival membranes in the front of the eye

connective tissue cells whose intercellular secretions (matrix) support and connect the organs and tissues of the body

constipation difficulty defecating or lack of defecation

contact transmission physical transfer of an agent from an infected person to a host through direct contact with that person, indirect contact with an infected person through a fomite, or close contact with contaminated secretions

contractility the capacity of a muscle to shorten in response to a stimulus

contrecoup a head injury that occurs beneath the skull opposite to the area of impact; also called a counterblow

control group the group that does not receive treatment during an experiment

covalent bond a bond where the atoms share electrons to fill their outermost levels or shells

convalescent stage time period in which acute symptoms of an infection begin to disappear until the client returns to the previous state of health

Cooley's anemia anemia caused by defect in hemoglobin formation; also known as *thalassemia major*

cornea a clear, circular area in the very front of the sclerotic coat

corona radiata layer of epithelial cells around ova

coronal (frontal) plane imaginary line at a right angle to the sagittal plane; divides the body into anterior and posterior segments

coronary artery the first branch of the aorta

coronary artery disease (CAD) a condition in which the arteries are narrowed, affecting the supply of oxygen and nutrients to heart muscle

coronary bypass a shunt to go around an area of blockage in the coronary arteries to provide blood supply to the myocardium

coronary circulation brings oxygenated blood to the heart muscle

coronary sinus pocket in posterior of right atrium into which the coronary vein empties

corpus luteum yellow body in the ovary, formed from ruptured graafian follicle; produces progesterone

cortex the part of hair that consists of keratinized, nonliving cells; outer part of an internal organ

coughing deep breath followed by forceful exhalation from the mouth

coup a head injury that occurs within the skull near the point of impact

covalent bond chemical bond in which atoms share electrons to fill their outermost shell

cranial cavity area of the body containing the brain

cranial nerves 12 nerves originating from the brain and brain stem that each have a specific function

cretinism congenital and chronic condition due to the lack of thyroid hormone

crown pertains to the part of a tooth that is visible

cryptorchidism failure of testes to descend into the scrotal sac

Cushing's syndrome hyperfunction of adrenal cortex

cyanosis bluish, grayish, or greenish color of the skin due to insufficient oxygen in the blood

cystic duct duct from gallbladder to common bile duct

cystic fibrosis (CF) a disease of the exocrine gland that produces thick mucus that block the passageways of the respiratory and digestive systems

cystitis or urinary tract infection inflammation of the mucous membrane of the urinary bladder

cytokines in the class of interleukins (part of the immune response), it is a protein produced by damaged tissue and white blood cells; interferon is a type of cytokine

cystoscopy a process in which a flexible tube with a lens is inserted into the urethra to visualize the urethra and urinary bladder

cytology study of cells

cytoplasm protoplasm outside the nucleus of a cell

cytoskeleton internal framework of the cell consisting of microtubules, intermediate filaments, and microfilaments

D

deciduous temporary teeth one starts to lose around the age of six

decubitus ulcer a deterioration of the skin due to constant pressure on the area

deductive reasoning the reasoning process that goes from a general idea to focus on a specific outcome

deep directional term used to describe an internal organ within the body, such as the stomach

defecation elimination of waste material from the rectum

defibrillator an electrical device used to discharge an electrical current to shock the pacemaker of the heart back to a normal rhythm

dehydrated an abnormal depletion of body fluids

dehydration synthesis occurs when water is removed from a molecule: the molecule fuses together and a new substance is formed during the anabolic process

dementia loss in at least two areas of complex behavior

dendrites nerve cell processes that carries nervous impulses toward the cell body

dentin main part of the tooth, located under the enamel

deoxyribonucleic acid (DNA) a nucleic acid containing the elements of carbon, hydrogen, oxygen, nitrogen, and phosphorus; genetic material

dependent variable a variable that is measured or tested

dermatitis inflammation of the skin

dermatology study of the physiology and pathology of the skin

dermis true skin; lying immediately beneath the epidermis

descriptive investigation a process used to answer the questions of "what," "how," "where," and "why" and describes and quantifies nature and phenomena using observations and measurements to create data

designer babies embryos that have been genetically modified to remove unfavorable traits, such as genetic disorders

detached retina condition that results when the vitreous fluid contracts as it ages, pulling on the retina and causing a tear; detachment of the retina may also occur as the result of an injury

developmental anatomy the study of the growth and development of an organism from fertilization to maturity; also called *embryology*

deviated nasal septum condition in which there is a bend in the cartilage structure of the septum

diabetes insipidus condition that results when there is a drop in the amount of ADH hormone secreted by the posterior lobe of the pituitary, which causes excessive loss of water and electrolytes

diabetes mellitus condition that results when the pancreas is unable to produce insulin or is unable to produce enough insulin for the cells to use glucose

diabetic retinopathy a condition caused by changes in the blood vessels in the retina of the eye, such as blood vessels that swell and leak, or abnormal blood vessels that grow on the retina; it is the leading cause of blindness in American adults

dialysis the separation of smaller molecules from larger molecules in a solution by selective diffusion through a semipermeable membrane

dialyzer a device used to perform dialysis; a kidney machine

diapedesis passage of blood cells through unruptured vessel wall into tissues

diaphragm a dome-shaped sheet of muscle that separates the thoracic cavity from the abdomen

diaphysis shaft of a long bone

diarrhea excessive elimination of watery feces

diarthroses movable joint (e.g., elbow, knee)

diastolic blood pressure pressure measured when the ventricles are relaxed

diencephalon posterior part of the brain; contains the thalamus and hypothalamus

diffusion molecules move from higher concentration to lower concentration

digestion the complex process of the breaking down of food to be utilized by the body

dilator muscles a muscle that opens or closes an orifice

diphtheria infectious disease of the respiratory system; rarely seen because of DTaP vaccine

diplopia double vision

disaccharide double sugar formed from two monosaccharide molecules

disease any abnormal changes in the structure and function of an organism that produces symptoms

disinfection elimination of pathogens, with the exception of spores, from inanimate objects

dislocation displacement of one or more bones of a joint or organ from the original position

distal farthest from the point of attachment or origin of a structure

distal convoluted tubule distal tubular process of the nephron that ascends to the cortex of the nephron from the loop of Henle

diuretics drugs that reduce the amount of fluid in the body

diverticulitis inflammation of the wall of the colon

diverticulosis numerous diverticula (little sacs in the wall of the colon)

dorsal pertaining to the back

dorsal cavity posterior cavity of the body that houses the brain and spinal cord

dorsalis pedis artery artery located at the ankle joint

Down syndrome a disorder characterized by the presence of an extra chromosome; also known as *trisomy 21*

droplet transmission the spread of disease that occurs when bacteria or viruses travel on respiratory droplets

dry eye a scratchy feeling in the eyes that occurs when the tears fail to keep the eye surface adequately lubricated

Duchenne's muscular dystrophy a disorder in which muscles suffer from a loss of protein and the contractile fibers are replaced with fat and connective tissue, rendering skeletal muscle useless; *see* muscular dystrophy

ductus arteriosus fetal structure that permits blood to flow from the pulmonary artery to the aorta

ductus deferens the part of the duct system of the testes that runs from the epididymis to the ejaculatory duct; also known as *vas deferens*

ductus venosus fetal structure that connects the umbilical vein to the inferior vena cava in the fetus; this vessel usually closes within 30 minutes after birth

duodenum first part of small intestine, beginning at the pyloric sphincter of the stomach

dura mater fibrous membrane forming the outermost covering of the brain and spinal cord

dysmenorrhea difficult or painful menstruation

dysphasia impairment of speech and verbal comprehension

dysplasia change in size, shape, or organization of cells

dyspnea labored breathing or difficult breathing
dystonia characterized by involuntary muscle contractions that cause repetitive movements or abnormal posture
dysuria painful urination

E

ectoderm outer germ layer of embryonic layers
ectopic pregnancy implantation of a fertilized egg outside of the uterus
eczema acute or chronic noncontagious inflammation of the skin; very itchy
edema excessive fluid in tissues
effectors the responding organs to a stimulus, which may be in the form of movement
efferent arteriole arteriole that carries blood from the glomerulus
efferent neurons *see* motor neuron
ejaculatory duct short and narrow duct that begins where the ductus deferens and the seminal duct join
elasticity capable of returning to original form after being compressed or stretched
elastin elasticlike fibers found in connective tissue
electrocardiogram (ECG or EKG) device used to measure the electric conduction system of the heart
electrolytes broken down parts of a compound that help determine acid–base balance
electromyography (EMG) device used to measure electrical muscle activity
electrons a subatomic particle of an atom that is arranged around the nucleus in orbital zones or electron shells; an electron has a negative (−) charge
element made up of like atoms; substance that can neither be created nor destroyed
embolism obstruction of a blood vessel by a circulating blood clot, fat globule, air bubble, or piece of tissue
embryology study of the formation of an organism from fertilized egg to birth
empathy the ability to recognize the emotions and actions of others
emphysema lung disorder in which the alveoli of the lung become overdilated, lose their elasticity and cannot rebound, inspired air becomes trapped, and it is difficult to expire
empirical evidence data that prove that a hypothesis is true
enamel hard calcium substance that covers the teeth
encephalitis inflammation of the brain
encephalocele a condition in which the brain is exposed to the outside instead of being covered by the skull and skin
endocarditis an inflammation of the membrane that lines the heart and covers the valves
endocardium membrane lining the interior of the heart
endocrine glands organized groups of tissues that use materials from blood and lymph to make new compounds called hormones that are directly secreted into the bloodstream
endocrinology study of the physiology and pathology of the hormonal system
endoderm the inner germ layer of the embryonic layers
endometrial cancer cancer of the inner lining of the uterus, it is the most common type of uterine cancer
endometriosis the presence of endometrium, which is normally confined to the uterine cavity, in other areas of the pelvic cavity
endometrium mucous membrane lining the uterus
endoplasmic reticulum transport system of a cell; can be smooth or rough
endosteum lining of the medullary cavity in the long bone
energy ability to do work
enteritis inflammation of the small intestine
enzymes organic catalysts that initiate and accelerate a chemical reaction
eosinophils white blood cells that increase in great numbers in allergic conditions; phagocytize the remains of the antibody-antigen reaction
epicardium the thin, innermost layer covering the heart
epidermis outermost layer of skin
epididymis portion of the seminal duct lying posterior to the testes; connected by the efferent duct of each testis
epididymitis inflammation of epididymis
epigastric upper region of the abdominal cavity, located just below the sternum
epiglottis structure made of elastic cartilage that prevents food from entering the trachea
epilepsy seizure disorder
epinephrine adrenaline; secretion of the adrenal medulla, which prepares the body for energetic action (fight or flight response)
epiphysis the end of a long bone
epistaxis bloody nose usually due to a rupture of the small vessels overlying the anterior portion of the nasal system
epithelial tissue protects the body by covering external and internal surfaces
equilibrium a state of balance
erythema redness of the skin due to dilation of the capillary network
erythroblastosis fetalis hemolytic disease of the newborn
erythrocytes red blood cells
erythropoiesis formation or development of red blood cells
erythropoietin hormone produced by the kidney to accelerate the production of red blood cells
eschar blackened skin that is usually a result of third-degree burns
esophageal sphincter *see* cardiac sphincter
esophagus a muscular tube that takes food from the pharynx to the stomach
essential amino acids amino acids that are necessary for normal growth and development and are not made in the human body
essential tremor a nerve disorder causing tremors to occur in a person who is moving or trying to move; not usually associated with Parkinson's disease
estrogen secretion of the ovary; female hormone
ethmoid bone of the cranium located between the eyes
eukaryote any cell that possesses a clearly defined nucleus
eupnea normal or easy breathing with the usual quiet inhalations and exhalations
eustachian tube passageway from throat to middle ear; equalizes pressure
excitability ability to respond to stimuli
exocrine gland secretions from these glands must go through a duct to reach the body surface or organ
exophthalmos abnormal protrusion of the eyes
experiment a process that allows scientists to prove or disprove a hypothesis
experimental group the group that receives treatment during an experiment
experimental investigation a process conducted to support, refute, or validate a hypothesis
expiration the act of breathing out or expelling air from the lungs
expiratory reserve volume (ERV) amount of air a person can exhale over and above the tidal volume
extensibility the ability to lengthen (stretch) and hence increase the distance between two parts
extension act of increasing the angle between two bones

external superficial; at or near the surface of the body
external nares the nostrils of the nose
external respiration breathing; act of inspiration and expiration
extracellular fluid fluid that bathes the cell and transports nutrients in and out of the cell
extracorporeal shock wave lithotripsy (ESWL) procedure used to reduce kidney stones to sandlike particles to enable them to pass through the urinary tract
extrinsic muscles muscles responsible for moving the eye within the orbital socket
eyestrain discomfort when viewing something

F

fallopian tubes uterine tubes that carry the egg from the ovaries to the uterus; also called *oviducts*
fascia band or sheet of fibrous membranes covering or binding and supporting muscles
fats compound made up of glycerol and fatty acids
feces waste material from the digestive system
feedback the process of receiving communication from another
femoral artery artery located in the groin area
femoral nerve found in the lumbar plexus; it stimulates the hip and leg
femur longest and strongest bone in the body; thighbone
fertilization process of the union of the egg and sperm
fetal circulation brings blood to the fetus
fiber the indigestible portion of food from a plant source
fibrin an insoluble protein necessary for the clotting of blood
fibrinogen a protein that is converted into fibrin by the action of thrombin
fibroid tumors benign tumors of smooth muscle, especially in the uterus
fibromyalgia chronic muscle pain
fibula slender bone at outer edge of the lower leg
filtrate plasmalike fluid filtered from the blood in the glomerulus into the Bowman's capsule
filtration movement of water and particles across a semipermeable membrane by a mechanical force such as blood pressure
fimbriae fringelike projections of the fallopian tubes that fall over the ovary
first-degree (superficial) burn burn that affects only the epidermal layer
fissure a groove or cracklike break in the skin
fissures the deep furrows within the brain matter
flagella long, hairlike projections from the cell membrane found on sperm
flatfeet weakening of the leg muscles that support the arch of the foot; also called *talipes*
flatulence the presence of excessive gas in the digestive tract
flexion the act of bending a limb or decreasing the angle between two bones
flora microorganisms that occur or have adapted to live in a specific environment
follicle-stimulating hormone (FSH) an adenohypophyseal hormone that stimulates follicular growth in the ovary, and stimulates production of sperm in the testis
fomites objects contaminated with an infectious agent
fontanel unossified areas in the infant skull; soft spot
foramen ovale an opening in the septum between the right and left atria of the fetus
force the pull of the muscles on the joints to move a bone
foreskin loose-fitting skin around the end of the penis; also called *prepuce*
fourth ventricle a structure of the brain situated below the third ventricle, in front of the cerebellum, and behind the pons and medulla oblongata
fovea centralis structure of the eye in the retina that contains the cones for color vision
fracture a break in a bone
frontal bone of the skull that forms the forehead
frontal lobe in the cerebral cortex; it controls the motor function
functional residual capacity in lung capacity, refers to the sum of the expiratory reserve volume plus the residual volume
fundus part farthest from opening of an organ
fungal infection infections that grows in warm, moist environments that commonly affect the nails
fungi microscopic organisms that grow in single cells or in colonies

G

gallbladder a small, pear-shaped organ under the right lobe of the liver that stores bile
gallstones crystallized cholesterol that forms in the gallbladder
gamma globulin fractionated part of globulin used to treat infectious diseases
ganglia a mass of nerve cell bodies outside the central nervous system
gangrene death of body tissue due to insufficient blood supply
gastric mucosa mucosa that lines the stomach
gastritis inflammation of the stomach
gastroenteritis inflammation of the stomach and small intestines
gastroesophageal reflux disease (GERD) condition in which stomach contents flow back into the esophagus
gene part of the chromosome that transmits a specific hereditary trait
gene mutations production of a new or altered gene
genetic counseling discussions with parents or prospective parents about genetic disorders
genetic disorder a disorder caused by variation in the genetic pattern
genetic engineering the ability to snip, rearrange, edit, or program DNA
genetics the branch of biology that studies the science of heredity and the differences and similarities between parents and offspring
genital herpes a sexually transmitted recurrent disease caused by a virus that may result in blisters in the genital area
genital warts sexually transmitted disease; also called *human papillomavirus*
genotype the genetic makeup of an individual; the DNA that is inherited from both parents
germ cells (gametes) mature reproductive cells
germinal center area found within the lymph node that produces the lymphocytes
ghrelin hormone produced by the stomach that stimulates appetite
gigantism hypersecretion of the growth hormone; overgrowth of long bones
gingivae gums
gingivitis inflammation of the gums
glans penis the head or tip of the penis
glaucoma increase in intraocular eye pressure
glial cells nerve cell; sometimes referred to as "nerve glue"; *see* neuroglia
gliding joints joint in which the nearly flat surfaces of the bones glide across each other (e.g., vertebrae)
globulin plasma protein made in liver; helps in synthesis of antibodies
glomerulonephritis inflammation of the glomerulus of the kidney
glomerulus part of the nephron; tuft of capillaries situated within Bowman's capsule
glottis space within the vocal cords of the larynx
glucagon a hormone that stimulates the liver to change glycogen into glucose

glucocorticoids hormones of the adrenal cortex, namely, cortisone and cortisol

glycogen the form of glucose that is stored in liver and muscle cells

goiter enlargement of the thyroid gland

Golgi apparatus a membranous network that resembles a stack of pancakes; it stores and packages secretions to be secreted by the cell

gonads sex glands (ovaries or testes)

gonorrhea a sexually transmitted infectious disease of the genitourinary tract caused by gonococcus bacteria

gout increase in uric acid crystals in the bloodstream, which are deposited in joint cavities, especially the great toe

graafian follicle a follicle in the ovary that stores immature ova

grafts to transplant tissue into a body part to replace damaged tissue

granulation tiny red granules that are visible in the base of a healing wound; granules consist of newly formed capillaries and fibroblasts

granulocytes granular white blood cells

greater omentum double fold of peritoneum that hangs down over the abdominal organs like an apron

gross anatomy the study of large and easily observable structures on an organism

growth hormone (GH) hormone responsible for growth and development; also known as *somatotropin*

Guillain-Barré syndrome an autoimmune disease that attack the nerves that control the muscles in the legs, or sometimes in the arms and upper body

gyri convolutions in the brain between the sulci

H

hair follicle inpocketing of the epidermis that holds the hair root

hammer (malleus) a tiny bone found in the middle ear

hammertoe a toe that is curled due to a bend in the middle joint of the toe

haustra the longitudinal muscle layer of the colon made of three bands called tenae coli; the colon is gathered too fit these bands, giving it a puckered appearance; these little puckers or pockets are called haustra and provide more surface area in the colon

HDL high-density lipoprotein; removes excess cholesterol from walls of the artery

heart block interruption of the SA node message to the AV node, resulting in a lack of coordination between the atria and the ventricles

heart failure condition in which heart ventricles do not contract effectively

heartburn a burning sensation in the esophagus and stomach

heel spur a calcium deposit in the plantar fascia, near its attachment to the calcaneus bone

helminths parasitic worms; the most common are roundworms and pinworms

hematocrit blood test that measures the percentage of the volume of whole blood that is made up of red blood cells, which depends on the number and size of the red blood cells

hematoma localized clotted mass of blood formed in an organ, tissue, or space

hematopoiesis formation of blood cells

hematuria presence of blood in urine

hemiplegia paralysis of one side of the body

hemoccult strip test for hidden blood

hemodialysis a procedure for removing waste products in the circulating blood of patients with kidney failure

hemoglobin oxygen-carrying pigment of the blood

hemolysis the bursting of red blood cells

hemophilia sex-linked, hereditary bleeding disorder occurring mostly in males but transmitted by females; characterized by a prolonged clotting time and abnormal bleeding

hemorrhoids enlarged and varicose condition of the veins in the lower part of the anus or rectum and the tissues of the anus

hemostasis process of controlling or stopping bleeding

heparin substance obtained from the liver that slows blood clotting

hepatic duct structure from the liver to the common bile duct; carries bile

hepatic vein vein that drains blood from the liver into the inferior vena cava

hepatitis inflammation of the liver

herd immunity a form of indirect protection from infectious disease that occurs when a sufficient percentage of a population has become immune to an infection through vaccination or previous infection, making the spread of the disease from person to person unlikely; also called community immunity

hereditary hemochromatosis a genetic disorder that causes the body to absorb too much iron from food

hereditary lymphedema a genetic disorder that causes swelling in a certain part of the body

hereditary mutation a genetic mutation passed down from generation to generation

hereditary spherocytosis an inherited blood disorder characterized by a mutation that causes red blood cells to take on a spherical shape

Hering-Breuer reflex a reflex that prevents overstretching of the lungs

hernia protrusion of part of an organ through abnormal opening

herpes contagious viral infection in which small blisters appear

herpes simplex a viral infection that is usually seen as a blister around the face or mouth; also called cold sores or fever blisters

hiatal hernia disorder that occurs when the stomach pushes through the diaphragm

hiccoughs spasms of the diaphragm and spasmodic closures of the glottis

hilum indentation along the medial border of the kidney

hinge joints a joint that moves in one direction or plane

Hirschsprung's disease when nerve cells stop growing and are missing at the end of the colon; also called congenital megacolon

histamine a substance that increases gastric secretions; a substance released by mast cells when tissue is injured or in an allergic or inflammatory reaction

histology study of tissues and organs

Hodgkin's disease specific type of cancer of the lymph nodes

homeostasis state of balance; the ability of the healthy body to regulate the internal environment within narrow limits

hormones chemical secretion, usually from an endocrine gland

hospital-acquired infection (HAI) infection that is acquired in a hospital or health care setting; the infection was not present at the time of admission of the patient; *see* nosocomial infection

host simple or complex organism that can be affected by an agent

human immunodeficiency virus (HIV) the causative agent of AIDS

human papilloma virus *see* genital warts

humerus the bone of the upper arm

humoral immunity type of immunity donated by antibodies

Huntington's disease (HD) genetic disorder characterized by degeneration of the central nervous system

hyaline type of cartilage that forms the skeleton of the embryo

hydrocephalus increase in the volume of cerebrospinal fluid within the cerebral ventricles; may occur in fetal development

hydrogen bond bond that holds water molecules together by forming a bridge between the negative oxygen atom of one water molecule and the positive hydrogen atom of another molecule

hydrolysis occurs when water is added to the molecule to break down larger molecules in a catabolic reaction

hydronephrosis condition in which the renal pelvis and calyces become distended due to the accumulation of fluid

hydroxide one atom of hydrogen and one atom of oxygen

hymen membrane at the opening of the vagina

hyoid a U-shaped bone found in the neck to which the tongue is attached

hyperglycemia high concentration of glucose in the blood

hyperopia (farsightedness) a condition in which the focal point is beyond the retina; the eyeball is shorter than normal

hyperplasia excessive proliferation of normal cells

hyperpnea increase in the depth and rate of breathing accompanied by abnormal exaggerated respiratory movements

hypersensitivity an abnormal response to a drug or allergen

hypertension abnormally high blood pressure

hyperthermia condition in which the body temperature rises above normal

hyperthyroidism condition of overactivity of the thyroid gland

hypertonic solution a solution in which water molecules are moving out of a cell, causing it to shrink

hypertrophy an increase in the size of a muscle cell

hyperventilation rapid breathing and rapid loss of carbon dioxide; sometimes causes dizziness or fainting

hypochondriac the right and left abdominopelvic cavity regions located below the ribs

hypogastric lower region of the abdominal area

hypoglycemia low concentration of glucose in the blood

hypoperfusion inadequate flow of blood carrying oxygen to the organs and body systems; leads to shock

hypospadias characterized by the location of the urethral opening on the underside of the penis

hypotension reduced or abnormally low blood pressure

hypothalamus part of the diencephalon; lies below the thalamus and is the "brain" of the brain

hypothermia a condition in which the body temperature drops below normal

hypothesis an idea from an observation that can be tested and proven true or false

hypothyroidism condition of underactivity of the thyroid gland

hypotonic solution a solution in which water molecules are moving into a cell, causing it to swell

hypoxia decreased blood flow to cellular structures

hysterectomy partial or total surgical removal of the uterus

I

ileocecal valve an opening in the sidewall of the large intestine that permits passage of chyme from ileum to large intestine and prevents the backflow of chyme into the ileum

ileum the lower part of the small intestine, extending from the jejunum to the large intestine

ilium the broad, blade-shaped bone that forms the back and side of the hip bone

illness stage time period when the patient is manifesting specific signs and symptoms of an infectious agent

immunity ability to resist a disease

immunization process of increasing resistance to disease

immunoglobulins proteins that act like antibodies

imperforate hymen hymen that blocks the vaginal opening and blocks menstrual blood from passing

impetigo acute and contagious skin disease with distinct yellow crusts

impotence inability to sustain an erection; also known as *erectile dysfunction (ED)*

in vitro fertilization (IVF) process of fertilization outside the living organism

incisors cutting teeth

incomplete proteins proteins that lack some or most of the essential amino acids

incontinence loss of self-control, especially of urine, feces, or semen

incubation stage time interval between the entry of an infectious agent in the host and the onset of symptoms

independent variable a variable that is controlled and manipulated by the researcher

inductive reasoning the reasoning process that begins with a specific observation to reach a broad conclusion

infectious agent an entity capable of causing disease

infectious mononucleosis contagious disease caused by Epstein-Barr virus; sometimes called the "kissing disease"

inferior below another or lower

inferior concha bones that make up the sidewalls of the nasal cavity

infertility incapable of reproduction

inflammation occurs when tissue reacts to chemical or physical trauma or invasion of pathogenic microorganisms; characterized by pain, heat, redness, and swelling

inflammatory bowel disease (IBD) a disorder affecting the digestive system that is characterized by chronic diarrhea

influenza viral infection causing inflammation of the mucous membrane of the respiratory tract

ingrown nails a condition where a nail curves downward into the skin

insertion part of a muscle that is attached to a movable part

inspiration drawing in of air; inhalation

inspiratory reserve volume (IRV) amount of air you can force a person to take in over and above tidal volume

insulin hormone produced by the pancreas; necessary for glucose metabolism

integumentary system all organs and structures that make up the skin

interferon protein that interferes with virus replication

interleukins a class of cytokine that regulates cell growth, differentiation, and motility

internal term used to refer to body cavities and hollow organs

internal respiration the exchange of carbon dioxide and oxygen between the cells and the lymph surrounding them, plus the oxidative process of energy in the cells

international normalized ratio (INR) *see* prothrombin time (PT)

interneurons *see* associative neuron

interphase the resting phase in the process of mitosis

interstitial cell-stimulating hormone (ICSH) a hormone that stimulates the growth of the graafian follicle and the production of estrogen in females and the production of sperm in males

interstitial fluid fluid between tissues

interventricular foramen the area that connects the third ventricle of the brain to the two lateral ventricles

intestinal mucosa lines the small and large intestines

intracellular fluid fluid within the cell

intramuscular into the muscle

intrinsic muscles muscles that help the iris control the amount of light entering the pupil

investigative resources items such as chemicals, glassware, and thermometers necessary to conduct an investigation

ion an electrically charged atom

ionic bond bond in which one atom gives up an electron to another atom

iris colored muscular layer surrounding the pupil of the eye

iron-deficiency anemia condition that results from lack of adequate amounts of iron in the diet

irritable bowel syndrome (IBS) is a group of symptoms that affects the large intestine, including cramping, abdominal pain, bloating, gas, and diarrhea or constipation or both

irritability ability to react to a stimulus; excitability

ischium forms the lower posterior portion of the hipbone; bears the weight of the body when sitting

islets of Langerhans specialized cells in the pancreas that produce insulin

isometric muscle tension increases, but muscle does not shorten

isotonic muscle contracts and shortens

isotonic solution a solution in which movement of water molecules into and out of a cell is the same

isotopes atoms of a specific element that have the same number of protons but a different number of neutrons

J

jaundice yellowish coloring of the skin

jejunum section of small intestine between duodenum and ileum

joints place where two bones meet

K

keratin protein that keeps the skin dry and protects against UV rays, bacteria, abrasions, and some chemicals

kidneys organs of the urinary system that function to rid the body of nitrogenous wastes

kidney dysplasia when cysts grow inside one or both kidneys

kidney stones (renal calculi) clumping together of calcium phosphate crystals, uric acid, and other substances in the kidneys

kilocalorie a large calorie; equal to 1,000 calories

kinetic energy work resulting in motion

Kussmaul respiration rapid, shallow breaths caused by ketoacidosis, or when there is too much glucose (energy) and ketones in the blood; associated with diabetes

kyphosis hunchback; humped curvature of the spinal column

L

labia majora large lips or folds of skin that lie on either side of the vaginal opening; they have adipose tissue and sweat glands

labia minora delicate folds of skin that lie just inside of the vaginal opening; they have sebaceous glands

laceration a tear or jagged wound in the skin

lacrimals bones that make up the inner corner of eye orbits; contain tear ducts

lacteals specialized lymph vessels in the villi of the small intestine that absorb digested fats and transport them to the circulatory system

laparoscopy a minor surgical procedure done under anesthesia to visualize internal organs

laryngitis an inflammation of the larynx or voice box

larynx voice box, found between the trachea and base of the tongue; contains the vocal cords

lateral toward the side

lateral ventricles the two largest cerebral ventricals

LDL low-density lipoprotein; lipoprotein that carries fat to the cells

left ventricle one of the lower chambers of the heart

lens crystal structure for refraction of light rays

leptin hormone produced by adipose tissue that acts on the hypothalamus to suppress appetite

lethal gene a gene that results in death

leukemia a cancerous condition in which there is a great increase in the number of white blood cells

leukocytes white blood cells

leukocytosis an increase in the white blood cell count above 10,000 cells per cubic millimeter (mm^3)

leukopenia a decrease in the normal number of white blood cells (leukocytes)

lice parasitic insects found usually in the hair; highly contagious

life functions a series of highly organized and related activities that allow living organisms to live, grow, and maintain themselves

ligaments a band of fibrous tissue connecting bones or supporting organs

limbic lobe located in the center of the brain beneath the other four cerebral lobes; influences unconscious, instinctive behavior

lingual tonsils found at the back of the tongue

lingual frenulum band of tissue that attaches the tongue to the floor of the mouth, thus limiting the motion of the tongue

lipase enzyme that changes fats into fatty acids and glycerol

lipids fatty molecules

listening being on the receiving end of communication and processing the information

liver large organ of the digestive system, located in upper right quadrant of the abdominal cavity

logical reasoning the reasoning process in which something is evaluated using both deductive and inductive reasoning

loop of Henle the proximal convoluted tubule descends into the medulla, forming the loop of Henle in the nephron

lordosis forward curvature of lumbar region of spine

lubb dupp sounds sounds made by the heart valves when they close

lumbar puncture removal of cerebrospinal fluid for diagnostic purposes by insertion of a needle between the third and fourth lumbar vertebrae

lumbar vertebrae five vertebrae associated with the lower part of the back

lumpectomy surgical removal of an abnormal cellular growth

lupus a chronic inflammatory autoimmune disease

luteinizing hormone (LH) hormone that stimulates ovulation and the production of progesterone in females

lymph watery fluid in the lymphatic vessels

lymph nodes structures that produce lymphocytes and filter out harmful bacteria

lymph vessels structures that transport excess tissue fluid back into the circulatory system

lymphadenitis inflammation of the lymph nodes and glands

lymphatic malformations the formation of nonmalignant masses of fluid-filled channels or spaces because of abnormal development of the lymph system

lymphatic system system of vessels and nodes that supplement the blood circulatory system and carry lymph; nodes and organs produce lymphocytes to destroy invading bacteria

lymphatics large lymph vessels

lymphedema swelling of the tissue due to an abnormal collection of lymph

lymphocytes type of white blood cells

lymphokines chemicals released by T lymphocytes during the immune response; help cellular immunity by stimulating activities of monocytes and macrophages

lymphoma a tumor of the lymphatic system; usually malignant

lysosomes cytoplasmic organelle containing digestive enzymes

M

macrophages a cell that removes dead organisms and foreign substances by phagocytosis

macula lutea a yellow disc in the retina in which the fovea centralis is located

macular degeneration thinning of the retinal layer of the eye or leakage that can develop under the retina, disturbing sharp central vision

malignant melanoma a type of tumor that develops in the pigmented cells of the skin called melanocytes and spreads rapidly

mammography an X-ray of the breast

mandible lower jawbone

manubrium forms the upper region of the sternum

Marfan syndrome a connective tissue disorder; people with this condition tend to be tall and thin, with long arms, legs, and fingers, and they may also have flexible joints and scoliosis

mastectomy removal of a breast

masticated food has been chewed or prepared for swallowing and digestion

matter anything that has weight and occupies space

maxillae bone of the upper jaw

means of transmission the process that bridges the gap between the portal of exit of the infectious agent from the reservoir and the portal of entry of the "new" host

Meckel's diverticulum an outpouching, or bulge, in the lower part of the small intestine caused by leftover umbilical cord

medial directional term, toward midline of the body

mediastinum intrapleural space separating the sternum in front and the vertebral column behind

medulla innermost layer of hair; inner portion of an organ

medulla oblongata part of the brain stem; contains the nuclei for vital functions

medullary canal center of the shaft of a long bone

meiosis cell division of gametes or cells; reduces the number of chromosomes

melanin pigment that gives color to hair, skin, and eyes; protects from UV rays

melanocytes cells that make the protein melanin to protect against ultraviolet rays

melanocyte-stimulating hormone (MSH) hormone produced by intermediate pituitary lobe; stimulates melanin cells in the skin

melatonin hormone produced by the pineal gland

membrane two thin layers of tissue that cover a surface or divide an organ

membrane excitability ability of nerves to carry impulses by creating electric charges

memory process by which the brain stores old information and packages and stores new information

menarche time when menstruation begins

Ménière's disease condition affecting the semicircular canals of the inner ear, causing marked vertigo (dizziness)

meninges any of three linings enclosing the brain and spinal cord

meningitis inflammation of the lining of the brain and spinal cord

menisci the medial meniscus and the lateral meniscus of the knee

menopause physiological termination of menstruation; generally between 45 and 55 years of age

menorrhagia abnormal excessive or prolonged menstrual bleeding

menstrual cycle recurring series of changes that take place in the ovaries, uterus, and accessory sexual structures during menstruation

menstruation monthly shedding of endometrial lining of the uterus if ova are not fertilized

mesentery peritoneum attached to posterior wall of the abdominal cavity

mesoderm the middle germ layer of the embryonic layers

metabolism the functional activities of cells that result in growth, repair, secretions, and the release of energy by the cells

metacarpal bone of the wrist

metaphase phase 3 in the process of mitosis; nuclear membrane disappears

metastasis transfer of malignant cells from an original site to a distant one through the circulatory system or lymph vessels

metatarsal bone of the foot similar to the metacarpal, five metatarsals forms the arch of the foot.

metric system a decimal system based on the power of 10. The medical community uses this system to determine length (measured in cm = centimeters), weight (measured in g = gram, mg = milligram, kg = kilogram), and volume (measured in L = liter, mL = milliliter)

microscopic anatomy study of small tissues, organs, and cells that cannot be seen with the naked eye

micturition the act of emptying the bladder; urination

midsagittal plane an imaginary line dividing the body into equal right and left halves

mineral an inorganic, solid chemical compound found in nature

mineralocorticoids hormones of the adrenal cortex, mainly aldosterone

miotic pertaining to or causing contraction of the pupil

missing vagina congenital abnormality of absent vagina; treated with a dilator

mitochondria organelle that supplies energy to the cell

mitosis cell division involving two distinct processes: (1) division of a nucleus and (2) cytoplasmic division, when the cytoplasm is divided into two approximately equal parts

mitral valve prolapse condition in which the valve between the left atrium and the left ventricle does not close properly

mixed nerve nerve composed of both afferent (sensory) fibers and efferent (motor) fibers

molars teeth designed for crushing and tearing

molecule the smallest unit of a compound that still has the properties of the compound

moles benign growths that occur when melanocytes grow in a cluster with tissue surrounding them

monocytes leukocytes manufactured in bone marrow and the spleen; wall off infected areas

monosaccharides simple sugar that cannot be broken down anymore

mons pubis fatty tissue overlying the genital region; usually covered by coarse hair

motor nerve (efferent nerve) nerve fiber that carries impulses from the brain or spinal cord to muscles, organs, or glands

motor neurons carries messages from brain and spinal cord to muscles and glands; also called *efferent neuron*

motor neuron diseases (MND) conditions that cause the nervous system to lose function over time

motor unit a motor nerve plus all the muscle fibers it stimulates

MRSA form of methicillin-resistant *Staphylococcus aureus*, a type of staph bacteria that is resistant to the methicillin antibiotic drug class

mucosa mucous membrane

mucous membranes type of tissue that lines surfaces and spaces that lead to the outside of the body

multicellular many celled

multiple myeloma malignant neoplasm of plasma cells or B lymphocytes

multiple sclerosis (MS) chronic inflammatory disease in which the immune cells attack the myelin sheath of a nerve

murmurs gurgling or hissing sounds made by the heart valves that are failing to close properly

muscle fatigue caused by an accumulation of lactic acid in the muscle

muscle spasm sustained muscle contraction

muscle strain overstretching or tearing of the muscle

muscle tissue contains cell material that has the ability to contract and move the body

muscle tone muscles always in a state of partial contraction

muscular dystrophy muscle disease in which the muscle cells deteriorate

mutagenic agent any substance causing a genetic mutation

mutation the appearance of a new and different organic trait caused by the inheritance of a mutated gene or chromosome

myalgia muscular pain

myasthenia gravis disease in which there is abnormal weakness and eventual paralysis of muscles

mycoplasma genitalium a small, pathogenic bacterium that lives on the skin of the urinary and genital tract

myelin sheath covering of the axon of the neuron, providing electrical insulation and increasing the velocity of impulse transmission

myeloblasts cells that synthesize granulocytes in bone marrow

myocardial infarction a heart attack caused by a blockage of blood flow to the heart muscle

myocarditis inflammation of muscular tissue of the heart

myocardium muscle of the heart

myometrium uterine muscular structure

myopia (nearsightedness) condition in which the focal point is in front of the retina

myringotomy opening into the tympanic membrane

myxedema hypofunction of the thyroid gland

N

nasal bone that forms the bridge of the nose

nasal cavity one of the pair of cavities between the anterior nares and the nasopharynx

nasal polyps an abnormal growth that occurs in the sinus cavity

nasal septum partition between the two nasal cavities

natural acquired immunity immunity to a disease that results from having the disease and recovering

natural immunity immunity with which a person is born

nausea feeling of sickness in the stomach accompanied by a loathing for food and an impulse to vomit

neck the part of the tooth at the gumline

necrosis cell death; *see* apoptosis

negative feedback occurs in the hormonal system; when there is a drop in the blood level of a specific hormone, the drop triggers a chain reaction of responses to increase the amount of hormone in the blood

negative feedback loop a reaction that reverses disturbances in response to a stimulus to reach homeostasis

neoplasia uncontrolled growth pattern in a cell; may result in a neoplasm

neoplasms a tumor; can be benign or malignant

nephron unit of structure of the kidney; contains glomerulus, Bowman's capsule, proximal distal tubule, loop of Henle, and distal tubule

nervous tissue composed of two types of cells, neuroglia and neurons, that react to stimuli and conduct an impulse

neuralgia severe, stabbing pain along the pathway of a nerve

neuritis inflammation of a nerve

neurogenic bladder condition caused by damaged nerves that control the bladder

neuroglia a network of cells that insulate, support, and protect the nerves of the central nervous system

neurohormones a hormone produced by the neurons in the hypothalamus that are secreted into the bloodstream

neurology study of the physiology and pathology of the nervous system

neuromuscular junction point between the motor nerve axon and the muscle cell membrane

neuron nerve cell, including its processes

neurotransmitters a chemical substance that makes it possible for messages to cross the synapse of a neuron to a target receptor

neutralization process in which an acid and a base combine to form a salt and water

neutrons a subatomic particle of an atom that, with a proton, makes up the nucleus of the atom; a neutron has no electric charge

neutrophils many-lobed white blood cells that phagocytize bacteria; sometimes called "polys"

night blindness a condition that makes it difficult to see at night; the rod cells in the retina are affected in this condition

nocturia excessive urination during the night

nonverbal communication information conveyed though body language or physical behavior

norepinephrine hormone that acts as a vasoconstrictor

nosocomial infection infection that is acquired in the hospital or other health care facility and was not present or incubating at the time of the patient's admission; *see* hospital-acquired infection (HAI)

nuclear membrane double-layered membrane that surrounds the nucleus

nucleic acids organic compound containing carbon, hydrogen, oxygen, nitrogen, and phosphorus (i.e., DNA, RNA)

nucleolus small, spherical structure within the cell nucleus

nucleoplasm protoplasm of the nucleus; also called *nuclear sap* or *karyolymph*

nucleus organelle that controls cell activities and cell division; also the center of an atom

nutrients food for the body that afford nutrition

nutrition the process by which the body assimilates food and uses it for energy, growth, and repair of tissue

nystagmus rapid, involuntary movement of the eyeball

O

obesity increase of body weight due to fat accumulation of 10% to 20% above normal range for the specific age, height, and sex

observation acquisition of information through the senses

observational investigation a process used to investigate cause-and-effect relationships

obturator foramen large opening between the pubic bone and the ischium that allows for the passage of blood vessels, nerves, and tendons

occipital bone that forms the base of the skull and contains the foramen magnum

occipital lobe part of the cerebrum that houses the visual area

olfactory nerves nerves that supply the nasal mucosa, responsible for the sense of smell

oliguria diminished production of urine

oogenesis process of origin, growth, and formation of ovum in ovary during preparation for fertilization

optic disc (blind spot) an area of the eye devoid of visual reception

oral cavity encloses the teeth and tongue

oral thrush creamy-white lesions on the tongue and/or the inner lining of the mouth caused by *Candida albicans*

orbital cavity contains the eye and its external structures

orchitis inflammation of the testes

organ of Corti hearing organ

organ system organs that are grouped together because more than one is needed to perform a function

organelles microscopic structure within the cell having a special function or capacity

organic catalyst a substance that affects the rate of speed of a chemical reaction without itself being changed

organic compounds compound that contains the element carbon

organic food food produced without using most conventional pesticides, fertilizers made with synthetic ingredients or sewage sludge, bioengineering, or ionizing radiation

organs groups of tissues organized according to structure and function

origin part of the skeletal muscle that is attached to the fixed part of the bone

orthopnea difficult or labored breathing in a horizontal position

osmoreceptors structures found in the hypothalamus that are sensitive to changes in the osmotic blood pressure and control the release of the antidiuretic hormone (ADH)

osmosis passage of fluid through a membrane

osmotic pressure the pressure exerted by the flow of water through a semipermeable membrane at equilibrium

osseous (bone) tissue bony; composed of or resembling bone

ossification process of bone formation

osteoarthritis degenerative joint disease

osteoblasts cells involved in formation of bony tissue

osteoclasts cells involved in resorption of bony tissue

osteocytes bone cell

osteogenesis imperfecta a bone tissue disorder in which bones are easily fractured

osteomyelitis inflammation of the bone

osteoporosis loss of calcium in bone, causing brittleness; occurs mainly in females after menopause

osteosarcoma bone cancer

otitis media an infection of the middle ear

otosclerosis chronic, progressive ear disorder in which the bone in the region of the oval window first becomes spongy and then hardened, causing the stirrup or stapes to become fixed or immobile

ova female reproductive cells

oval window the membrane that separates the middle ear from the inner ear

ovarian cancer cancer of the ovaries; early diagnosis is difficult and treatment is to remove all reproductive organs

ovarian cysts fluid-filled sacs or pockets in an ovary or on its surface

ovaries female reproductive organs that produce ova, estrogen, and progesterone

overactive bladder condition when urination occurs eight or more times daily with nocturia; this happens when the nerve signals between the bladder and the brain are not coordinated

oviducts *see* fallopian tubes

ovulation second stage of the menstrual cycle; when a ripe egg cell is released from an ovarian follicle cell

oxyhemoglobin hemoglobin combined with oxygen

oxytocin hormone released during childbirth to cause strong contractions of the uterus

P

pacemaker *see* sinoatrial (SA) node, *see* artificial pacemaker

palatines 1) tonsils located on the side of the soft palate; 2) bones that form hard palate of mouth

palpitations irregular, rapid pulsations of the heart

pancreas organ of digestion that lies behind the stomach; produces digestive juices, insulin, and glucagon

pancreatitis inflammation of the pancreas

Papanicolaou (Pap) smear diagnostic test to detect cancer of the cervix, a sample of cell scrapings is taken from the cervix and cervical canal for microscopic study

papillae ridges formed in the lower edge of the stratum germinativum; they raise the skin into permanent ridges that form fingerprints

papilloma a type of tumor of the epithelial tissue; also known as a *wart*

paraplegia paralysis of the lower extremities caused by severe injury to the spinal cord in the thoracic or lumbar region, resulting in the loss of sensory or motor control below the area of injury

parasites organisms that live off of another organism to survive

parasympathetic system division of the autonomic nervous system that inhibits or opposes the effects of the sympathetic nervous system

parathormone hormone that controls the concentration of calcium in the bloodstream

parathyroid glands four small endocrine glands embedded in the thyroid gland; they secrete parathormone

paresthesia sensation of tingling, crawling, or burning of the skin

parietals two bones that form the roof and sides of the skull

parietal lobe division of the cerebrum that lies beneath the parietal bone; interprets nerve impulses from sensory receptors

parietal membrane outer part of a serous membrane that lines a body cavity

Parkinson's disease condition characterized by marked tremors; may be due to a decrease in the amount of the neurotransmitter dopamine

parotid glands largest of the salivary glands

passive acquired immunity borrowed immunity; has a temporary effect (i.e., gamma globulin)

passive transport the process of moving materials across a cell membrane without using energy, such as diffusion, osmosis, or filtration

patella kneecap

pathogenic disease causing

pathogenicity ability of a microorganism to produce disease

pelvic cavity area of the body containing the urinary bladder, reproductive organs, rectum, and remainder of the large intestine

pelvic inflammatory disease (PID) a disease resulting from infections of the reproductive organs

pelvis any basin-shaped structure or cavity

penile shaft erectile tissue that becomes rigid during intercourse

penis male reproductive organ

peptic ulcers sores or lesions that form in the lining of the stomach or duodenum, usually caused by *H. pylori*

pericardial membrane lining of the heart cavity

pericarditis inflammation of the outer membrane covering the heart

pericardium double layer of serous and fibrous tissue surrounding the heart

perimetrium outer layer of the uterine wall

perineum area between the vagina and the rectum

periodontal membrane membrane that anchors a tooth in place

periodontitis a chronic bacterial infection of the gums and surrounding tissue that causes oral bacteria by products to enter the bloodstream; also called periodontal disease

periosteum fibrous tissue covering the bone

peripheral nervous system made up of 12 pairs of cranial nerves and 31 pairs of spinal nerves

peripheral neuropathy term used to describe damage to the peripheral nerves

peripheral vascular disease (PVD) blockage of arteries, usually in the legs

peristalsis progressive wave of contraction in tubular structures provided with longitudinal and transverse muscular fibers, as in esophagus, stomach, and small and large intestines; moves material along the structures

peritoneal dialysis filtering of a patient's blood through the patient's own peritoneal lining

peritoneal membrane serous membrane lining of abdominal cavity; also referred to as peritoneum

peritonitis inflammation of the membrane lining the abdominal cavity

pernicious anemia caused by decrease of vitamin B12 or lack of intrinsic factor in the stomach

peroxisomes membranous sacs that contain oxidase enzymes

pertussis (whooping cough) highly contagious disease caused by bacterium *Bordatella pertussis*; characterized by repeated coughing attacks that end in a "whooping" sound; to prevent whooping cough, it is recommended that all children and adolescents receive the DTaP vaccine

Peyer's patches aggregated lymphatic follicles found in the walls of the small intestines; they produce macrophages

pH scale a measure of the acidity or alkalinity of a solution

phagocytosis ingestion of foreign or other particles by certain cells

phalanges bones of the fingers and toes

pharyngitis a red, inflamed throat caused by a bacteria or virus

pharynx throat

phenotype the external expression of a genotype, but with external influence, such as the environment

phenylketonuria (PKU) a metabolic disorder; the body cannot make an enzyme needed for normal metabolism or breakdown of the amino acid phenylalanine

pheochromocytoma tumor of the adrenal glands that causes excessive secretion of epinephrine, which may be fatal

phlebitis inflammation of a vein, with or without infection and thrombus formation

phospholipids fats that contain carbon, hydrogen, oxygen, and phosphorus

phrenic nerve nerve that controls the diaphragm and intercostal muscles

physical agents factors in the environment capable of causing disease in a host

physiology study of the functions of living organisms and their parts

physiotherapy treatment of disease and injury by physical means using light, heat, cold, water, electricity, massage, and exercise

pia mater innermost vascular covering of the brain and spinal cord

pineal gland located in the third ventricle of the brain; produces melatonin

pinocytic vesicles formed by having the cell membrane fold inward into a pocket

pinocytosis process of engulfing large molecules in solution and taking them into the cell

pituitary gland a small gland located in the sphenoid bone in the cranium; its hormones affect all other glandular activity; called the *master gland*

pituitary short stature a condition where growth of the long bones is abnormally decreased due to an inadequate production of growth hormone; formerly known as *dwarfism*

pivot joints joint in which an extension of one bone rotates in a second, arch-shaped bone

planes imaginary, anatomical dividing lines useful in separating body structures

plantar fasciitis inflammation of the plantar fascia on the sole of the foot

plasma liquid part of blood without its cellular elements

pleura serous membrane protecting the lungs and lining the internal surface of the thoracic cavity; also called *pleural membrane*

pleural fluid serous fluid that fills the pleural cavity

pleural membrane serous membrane protecting the lungs and lining the internal surface of thoracic cavity; also called *pleura*

pleurisy inflammation of pleura

plexus a network of spinal nerves

pneumonia infection of the lung

pneumothorax total collapse of a lung

poliomyelitis contagious viral disease of the nerve pathways of the spinal cord; rarely seen because of polio vaccines

polycystic ovary syndrome (PCOS) a set of symptoms due to elevated androgenic hormones in females; symptoms include irregular or no menstrual cycle, excess body or facial hair, acne, difficulty in getting pregnant, and obesity

polycythemia too many red blood cells

polydipsia excessive thirst

polyphagia excessive hunger

polysaccharides a complex sugar made of many bonded glucose molecules

polyuria excessive urination

pons part of the brain stem; pathway for nerve impulses between the brain and the rest of the nervous system

popliteal artery artery located behind the knee

portal circulation brings blood from the organs of digestion through the portal vein to the liver

portal of entry route by which an infectious agent enters a host

portal of exit route by which an infectious agent leaves a reservoir

portal vein major vein leading to the liver

positive feedback the body's ability to increase the level of an event that has already started; an example of positive feedback is blood clotting

posterior located behind or at the back; opposite to anterior

posterior chamber chamber of the eye filled with vitreous humor

posterior pituitary lobe lobe that stores the hormones produced by the hypothalamus

potential energy energy stored in cells waiting to be released

premature contractions an arrhythmia disorder that occurs when an ectopic (abnormal place) pacemaker, rather than the sinoatrial node, sparks and stimulates a contraction of the myocardium

premature ventricular contractions (PVCs) a heart contraction that originates in the ventricles and causes contractions ahead of the next anticipated beat

premenstrual syndrome (PMS) a group of symptoms exhibited just prior to the menstrual cycle; caused by water retention in body tissue

prepuce *see* foreskin

presbycusis a condition that causes deafness due to damage to the inner ear or auditory nerves

presbyopia farsightedness due to loss of elasticity in the lens of the eye

primary cilia dyskinesia a condition characterized by structural defects of the cilia that result in ineffective clearing of mucous particles, including bacteria

primary repair repair of epithelial tissue when no infection is present; new epithelial cells push themselves up toward the skin surface to repair the damage

prime mover muscle that provides movement in a single direction

prodromal stage time interval from the onset of nonspecific symptoms of a disease until specific symptoms of the infectious process begin to manifest

progesterone steroid hormone secreted by the ovaries from the corpus luteum to help maintain pregnancy

prolactin hormone (PR) hormone that develops breast tissue and stimulates the production of milk after childbirth

pronation turning of the palm of the hand downward

prone refers to lying flat, with face and chest down

prophase phase 2 in the process of mitosis; when the two centrioles start to separate

prostaglandins hormones secreted by various tissues; their function depends on which tissue they are excreted from

prostate cancer cancer of the prostate; most common cancer in males over 50

prostate gland gland located just under the urinary bladder; secretes a thin, milky alkaline fluid that enhances sperm motility

prostatectomy surgical removal of all or part of the prostate

prostatitis an infection of the prostate gland

protease pancreatic juice that breaks down protein into amino acids

proteins an organic compound containing the elements of carbon; hydrogen; oxygen; nitrogen; and most times, phosphorus and sulfur. Protein is necessary to build and repair body tissue

protein synthesis production of protein by the cells that are essential to life

prothrombin a globulin that helps blood coagulate

prothrombin time (PT) blood test done to determine clotting time of blood

protons a subatomic particle of an atom with a positive (+) charge; with neutrons, it makes up the nucleus of the atom

protoplasm an aqueous solution of carbohydrates, proteins, lipids, nucleic acids, and inorganic salts surrounded by a cell membrane

protozoa single-celled parasitic organisms with the ability to move

proximal located toward the point of attachment or trunk of the body

proximal convoluted tubule twisted tubular branch off the Bowman's capsule of the nephron of the kidney

prune belly disorder a condition in which one or more layers of abdominal muscles are missing at birth, giving the belly a wrinkled appearance

psoriasis chronic inflammatory skin disease with silvery scales

ptyalin found in salivary amylase; it converts starches into simple sugars

puberty age when reproductive organs become functional

pubis pubic bone; portion of hipbone forming front of pelvis

pulmonary artery structure that takes blood from the right ventricle to the lungs

pulmonary embolism occurs when a blood clot travels to the lung

pulmonary semilunar valve heart valve at the opening of the pulmonary artery that allows blood to flow from the right ventricle into the pulmonary artery

pulmonary veins structures that take blood from the lungs to the left atrium

pulp cavity inside of the tooth that contains blood vessels and nerves

pulse a measurement of the number of times the heart beats per minute

pulse pressure the difference between systolic and diastolic blood pressure

pupil opening in the iris of the eye for passage of light

Purkinje fibers conduction fibers that conduct impulses through the ventricles of the heart

pus product of inflammation; a cream-colored liquid that is a combination of dead tissue, dead and living bacteria, dead white blood cells, and blood plasma

pyelitis inflammation of the renal pelvis

pyelonephritis inflammation of the kidneys and the pelvis of the ureter

pyloric sphincter valve that regulates entrance of food from the stomach to the duodenum

pyloric stenosis narrowing of the pyloric sphincter

pylorospasm vomiting of undigested food due to failure of the pyloric sphincter to relax

pyrexia fever

pyrogens chemical released when there is inflammation; pyrogens circulate to the hypothalamus and affect the temperature control center

pyuria the presence of pus in the urine

Q

quadrants a term used in reference to the abdominal area by dividing it into four areas, or quadrants

quadriplegia follows severe trauma to the spinal cord below the C1–C4 vertebrae in which there is loss of movement of all four extremities with the accompanying loss of bowel, bladder, and sexual function

qualitative data data that can be observed or recorded, not through numbers or mathematical equations, but through direct observations, interviews, focus groups, or similar methods

quantitative data data that are defined by a numerical value, such as information that can be used in a mathematical calculation or statistical analysis

question the issue or problem that determines the focus of the investigation

R

radial artery artery located at the wrist

radial nerve nerve found in the brachial plexus that stimulates the wrist and hand

radioactive capable of emitting energy in the form of radiation

radius bone on the thumb side of the forearm

rales raspy-sounding breathing

receptors sensory nerves that receive a stimulus and transmit it to the central nervous system

recombinant DNA replication and artificial manipulation of DNA

rectum portion of the colon that opens into the anus

reflex involuntary action; automatic response

rehabilitation the process of restoring function through therapeutic exercise

relaxin hormone produced by the ovaries and placenta during pregnancy; helps relax the ligaments of the pelvis and soften and widen the cervix

remission a long-term disappearance of symptoms of a disease

renal columns support structures between the renal pyramids

renal fascia tough, fibrous tissue covering the kidney

renal papilla apex of the renal pyramid

renal pelvis funnel-shaped structure at the beginning of the ureter

renal pyramids striated cones that make up the medulla of the kidney

renin enzyme produced by the kidney

replication occurs when an exact copy of each nuclear chromosome is made during the early part of the first stage of mitosis (early interphase)

reservoir place where the agent can survive

resident flora microorganisms that are always present, usually without altering the patient's health

residual volume the amount of air that cannot be voluntarily expelled in the lungs

respiratory distress syndrome a condition occurring in premature infants in whom there is a deficiency of surfactant to keep the lung's alveoli open

respiratory mucosa lining of the respiratory passages

respiratory syncytial virus (RSV) a very common and contagious virus that leads to mild cold-like symptoms

retina innermost layer of the eye; contains the rods and cones

retroperitoneal located behind the peritoneum

reverse isolation barrier protection designed to prevent infection in patients who are severely compromised and highly susceptible to infection

Rh factor antigen found in red blood cells

rheumatic heart disease a disease of the lining of the heart, especially the mitral valve; thought to be caused by frequent strep throat infections that lead to rheumatic fever

rheumatoid arthritis chronic inflammatory disease that affects connective tissue and joints

rhinitis inflammation of the lining of the nose

ribonucleic acid (RNA) type of nucleic acid

ribosomes submicroscopic particle attached to endoplasmic reticulum; site of protein synthesis in the cytoplasm of a cell

rickets disease in which bones soften due to a lack of vitamin D

rickettsia intercellular parasites that need to be in living cells to reproduce

right lymphatic duct the lymphatic duct that receives lymph from the right side of the body

right ventricle one of the lower chambers of the heart

ringworm contagious fungal infection with raised circular patches

rods cells in the retina; they are sensitive to dim light

root (1) the part of a tooth that is embedded into the alveolar process of the jaw; (2) the part of a hair that is implanted in the skin

rosacea a common inflammatory disorder characterized by chronic redness and irritation to the face

rotation type of movement in which a bone moves around a central axis

rotator cuff injury an inflammation of a group of tendons that surround the shoulder joint

roughage the coarse parts of certain foods that are indigestible and stimulate peristalsis

rugae wrinkles or folds

rule of nines pertains to patients with burns; measurement of the percentage of body burned: the body is divided into 11 areas, with each area accounting for 9% of the total body surface

S

sacrum wedge-shaped bone below the lumbar vertebra at the end of the spinal column

Safety Data Sheets information on the hazards of the chemical, the personal protective equipment needed when handling the chemical, and the parts of the body that could be affected by exposure

sagittal plane directional term that divides the body into left and right parts

salivary amylase found in saliva; converts starches into simple sugars

salivary glands infection swollen glands commonly caused by salivary stones, which are buildups of crystallized saliva deposits

salpingitis an inflammation of the fallopian tubes

sarcolemma muscle cell membrane

sarcoplasm the hyaline or finely granular interfibrillar material of muscle tissue

scab dried capillary fluid that seals a wound

scapulae large, flat, triangular bone that forms the back of the shoulder

sciatic nerve largest nerve in the body; originates in the sacral plexus and runs through the pelvis and down the leg

sciatica neuritis of the sciatic nerve

science the study of what is observable and what can be tested through experimentation

scientific method a method of inquiry created by Frances Bacon in 1620 based on three main concepts: observation, experimentation, and the development of theories or natural laws

scientific models three types of models that represent concepts, objects, and phenomena: physical, conceptual, and mathematical

Scientific Revolution the period toward the end of the Renaissance that continued though the late seventeenth century, which saw a major transformation in scientific thought and ideas across multiple disciplines that led to the emergence of modern science

scientific theories theories based on natural and physical phenomena that are proven to be true

sclera tough, white covering; part of the external coat of the eye

scleroderma disease that results in thickening of the skin and blood vessels

scoliosis lateral curvature of the spine

scrotum pouch that contains the testicles

sebaceous glands gland that secretes sebum, a fatty material

sebum secretion of sebaceous glands that lubricate the skin

secondary repair repair of a wound with small or large tissue loss

second-degree (partial-thickness) burn affects the epidermis and dermis layers

section a cut made through the body in the direction of a certain plane

sedimentation rate a blood test that measures the time it takes red blood cells to settle to the bottom in an upright tube; an elevated rate indicates if there is an inflammatory condition present

segmented movement single segments of the intestine alternate between contraction and relaxation

semicircular canals structures in the inner ear involved with equilibrium

seminal vesicles two highly convoluted membranous tubes that produce substances found to help nourish and protect sperm

seminiferous tubules highly twisted tubules within the testicles

sensory nerve (afferent nerve) nerve fibers that carry messages from the sense organs to the brain and spinal cord

sensory neurons nerve that carries nerve impulses from the periphery to the central nervous system; also known as *afferent neuron*

septicemia presence of pathogenic organisms in the blood

septum partition; dividing wall between two spaces or cavities, such as the septum between the left and right sides of the heart or nose

serosa name given to specific double-walled serous membranes

serous fluid (1) normal lymph fluid; (2) thin, watery body fluid

serous membranes double-walled membrane that produces serous fluid

shaft (1) the part of the hair that extends from the skin surface; (2) the diaphysis of a long bone

shin splints injury to the muscle and tendon in front of the shins

shingles (herpes zoster) virus infection of the nerve endings

shock inadequate blood supply carrying oxygen to organs and body systems; can be caused by excessive loss of blood or fluids, severe infection, allergic reaction, or loss of muscle control; can result in serious damage or death

sickle cell anemia blood disorder; the shape of the red blood cell is a sickle shape, which makes the red blood cells clump together

sigmoid colon distal, S-shaped part of the colon

silicosis lung condition caused by breathing dust containing silicon dioxide; lungs become fibrotic

sinoatrial (SA) node dense network of fibers of conduction at the junction of the superior vena cava and right atrium; it sends out an electrical impulse that begins and regulates the heart rate; also called *pacemaker*

sinuses cavities in the skull that surround the nasal region and are filled with air

sinusitis infection of the mucous membrane lining the sinus cavities

Sjögren's syndrome the most prevalent of the autoimmune diseases, it affects the moisture glands of the body; symptoms include dry eyes or mouth or other body systems

skeletal muscle muscle attached to a bone or bones of the skeleton that aids in body movements; also known as *voluntary* or *striated muscle*

skin cancer a tumor that develops on the skin

smooth muscle nonstriated, involuntary muscle

sneezing deep breath followed by exhalation from the nose

solutes dissolved substance in a solution

somatic cell mutation alteration that occurs within individual body cells

somatic nervous system a division of the peripheral nervous system; it conducts impulses from the brain and spinal cord to the skeletal muscles, thereby causing responses to changes in our external environment

somatotropin growth hormone

spastic quadriplegia spastic paralysis of all four limbs

spermatogenesis the process of forming sperm

spermatozoa male gametes

sphenoid the key bone that connects all the skull bones

sphincter muscles circular muscle, such as the anus

spina bifida a malformation of the vertebral bones and skin surrounding the spine that leads to serious infections, bladder and bowel dysfunction, hydrocephalus, and paralysis

spinal cord part of the central nervous system within the spinal column; begins at the foramen magnum of occipital bone and continues to the second lumbar vertebra

spinal nerves 31 pairs; they originate in the spinal cord

spirometer a device that measures the volume and flow of air during inspiration and expiration

spleen lymph organ situated below and behind the stomach

spongy bone the result of hard bone when it is broken down

spores bacteria in a resistant stage that can withstand unfavorable environments

sprain wrenching of a joint that produces a stretching or laceration of ligaments

squamous cell carcinoma cancer of the epidermis that grows rapidly; often found on scalp or lower lip

stem cells primal cells common to all multicellular organisms

sterile incapable of reproducing

sterilization total elimination of all microorganisms, including spores

sternum flat, narrow bone in the median line in front of the chest; it is composed of three parts: manubrium, body, and xiphoid process

steroids lipids or fats that contain cholesterol

stethoscope instrument used for detection and study of sounds arising within the body

stimulus any change in the environment

stirrup (stapes) a tiny, stirrup-shaped bone in the middle ear

stomach a major organ of digestion; a pouchlike structure located in the upper left quadrant of the abdominal cavity, between the esophagus and the duodenum

stomatitis inflammation of the mucous membrane of the mouth

strabismus (crossed eyes) condition in which the muscles of the eyeball do not coordinate their action

strain tear in a muscle

stratum corneum the surface layer of the skin made up of dead cells that slough off daily

stratum germinativum the deepest epidermal layer of the skin

stratum granulosum the epidermal layer where keratinocyte cells change their shape, lose their nucleus, lose most of their water, and become mainly hard protein or keratin

stratum lucidum the epidermal layer found only in the palms of the hands or soles of the feet; the cells in this layer appear clear

stratum spinosum the epidermal layer that under a microscope looks prickly; contains melanocytes, keratinocytes, and Langerhans cells

strength capacity to do work

stroke *see* cerebral vascular accident (CVA)

stroke volume the amount of blood ejected from the ventricles with each heartbeat

sty (hordeolum) infection of the gland along the eyelid

sublingual glands salivary glands located under the sides of the tongue

submandibular gland salivary gland located near the angle of the lower jaw

sudden infant death syndrome (SIDS) death of an infant due to a stoppage of breathing while the infant sleeps

sulci fissures or grooves separating cerebral convolutions

superficial on or near the surface of the body

superior in anatomy, higher; denoting upper of two parts, toward vertex

supination turning palm of hand upward

supine refers to lying on one's back, with face and torso facing up

surfactant lipid material covering the inner surfaces of the alveoli

susceptible host person who lacks resistance to an agent and is thus vulnerable to disease

suspensory ligaments the ligaments that hold the lens of the eye in place

sutures (1) a line of connection or closure between bones, as in a cranial suture; (2) in surgery, a fine, threadlike catgut or silk used to repair or close a wound

sweat glands sudoriferous gland that produces sweat or perspiration

sympathetic system division of autonomic nervous system

synapse space between adjacent neurons through which an impulse is transmitted

synaptic cleft space between the axon of one neuron and the dendrite of another

synarthroses immovable joints connected by fibrous connective tissue

synergists muscles that help steady a joint

synovial fluid a lubricating substance produced by the synovial membrane

synovial membrane double layer of connective tissue that lines joint cavities and produces synovial fluid

syphilis infectious disease transmitted by sexual contact

systemic anatomy study of the structure and function of various organs or parts comprising a particular organ system

systemic circulation circulation that takes blood from the heart to the tissues and cells of the body and back to the heart

systolic blood pressure pressure measured at the moment of ventricle contraction

T

T lymphocytes cell synthesized in the thymus gland

tachycardia abnormally rapid heartbeat

tachypnea abnormally rapid rate of breathing

talus anklebone that articulates with bones of the leg

tarsal anklebone

taste buds cells on the papillae of the tongue that can distinguish salt, bitter, sweet, umami, and sour qualities of dissolved substances

Tay-Sachs disease genetic mutation caused by lack of a particular enzyme needed for the breakdown of lipid molecules in the brain

telophase final stage in the mitosis process in which the chromosomes migrate to the poles of the cell

temporals bones that form side of head and house ears

temporal artery artery located slightly above the outer edge of the eye

temporal lobe part of the cerebral hemisphere associated with the perception and interpretation of sound

tendon cord of fibrous connective tissue that attaches a muscle to a bone or other structure

tennis elbow inflammation of the tendon that connects the arm muscles to the elbow

testes male reproductive organs that produce sperm and testosterone

testicular cancer cancer of the testes is rare; most common cancer of males between 15 and 35

testosterone male sex hormone responsible for male secondary sex characteristics

tetanus infectious disease, usually fatal, characterized by spasm of voluntary muscles and convulsions caused by a toxin from tetanus bacillus

tetany a condition in which severely decreased levels of calcium affect the normal function of nerves

tethered cord syndrome when the spinal cord is abnormally attached to the surrounding tissue

tetralogy of Fallot a congenital heart disease consisting of four anatomical defects: pulmonary stenosis, ventricular septal defect, right ventricular hypertrophy, and enlarged aortic valve

thalamus part of the diencephalon; relays sensory stimuli to the cerebral cortex

thalassemia *see* Cooley's anemia

third-degree (full-thickness) burn burn that involves complete destruction of the epidermis, dermis, and subcutaneous layers of the skin

third ventricle a cavity within the brain filled with cerebrospinal fluid; located behind and below the lateral ventricles

thoracentesis aspiration of the chest cavity for removal of fluid, usually for emphysema

thoracic cavity contains the mediastinum and two other cavities: the left pleural cavity contains the left lung and the right pleural cavity contains the right lung

thoracic duct (left lymphatic duct) lymphatic duct that receives lymph from the left side of the body

thoracic vertebrae the 12 bones of the spine located in the chest area

threshold limit of reabsorption in the urinary system

thrombin enzyme found in blood; produced from an inactive precursor, prothrombin, inducing clotting by converting fibrinogen to fibrin

thrombocytes platelet; part of the megakaryocyte cells necessary for blood clotting

thrombocytopenia decrease in the number of platelets

thromboplastin substance secreted by platelets when tissue is injured; necessary for blood clotting

thrombosis formation of a clot in a blood vessel

thrombus blood clot formed in a blood vessel

thymus endocrine gland located under the sternum; produces T lymphocytes

thyroid gland endocrine gland located on anterior portion of the neck that produces thyroxine, triiodothyronine, and calcitonin

thyroid-stimulating hormone (TSH) hormone that stimulates the growth and secretion of the thyroid gland

thyroxine (T_4) hormone secreted by thyroid gland or prepared synthetically

tibia larger, inner bone of the leg below the knee; shinbone

tidal volume amount of air that moves in and out of the lungs with each breath

tinnitus ringing sensation in one or both ears

tissues cells grouped according to size, shape, and function; epithelial, connective, muscle, and nerve tissues are examples

tonsillitis infection and swelling of the tonsils

tonsils mass of lymph tissue in the back of the throat that produces lymphocytes

torque the application of force on a lever to cause rotation on an axis

torticollis a contracted state of the neck muscles that produces an unnatural position of the head; also called *wryneck*

total lung capacity measurement that includes tidal volume, inspiratory reserve, expiratory reserve, and residual air

toxic shock syndrome a bacterial infection caused by the body's reaction to a *Staphylococcus* organism

trabeculae needlelike, bony spicules within cancellous bone that contribute to the spongy appearance; their distribution along lines of stress adds to the strength of the bone

trace elements substances found in the body in very small amounts

trachea a thin-walled tube between the larynx and the bronchi; conducts air to the lungs

trans fat type of fat made when hydrogen is added to vegetable oil, a process called hydrogenation

transient flora microorganisms that attach to the skin for a brief time but do not continuously live on the skin

transient ischemic attacks (TIAs) temporary interruptions in blood flow to the brain

transmyocardial laser revascularization (TMR) the use of lasers to puncture holes in the heart muscle to improve blood flow

transverse imaginary horizontal line that divides the body into upper and lower portions

triceps three-headed muscle on the back of the upper arm

trichomoniasis vaginalis a sexually transmitted disease caused by infection with the protozoan *Trichomonas vaginalis*

tricuspid atresia a condition in which the tricuspid valve between the right atria and right ventricle does not form

tricuspid valve valve between right atrium and right ventricle; allows blood to flow from the right atrium to the right ventricle

trigeminal neuralgia painful condition affecting the fifth cranial nerve; also known as *tic douloureux*

triglycerides also called fats, they consist of glycerol and fatty acids/ and make up 95% of fats in the human body

triiodothyronine (T_3) hormone that serves to regulate body systems

trisomy 21 *see* Down syndrome

tuberculosis infectious disease caused by the tubercle bacillus; mainly affects the lungs

tumor abnormal and uncontrolled growth of a cell

tumor necrosis factor (TNF) a cytokine that stimulates macrophages and causes cell death in cancer cells

tunica adventitia (externa) the outer layer of the arterial walls

tunica intima the inner arterial layer

tunica media the middle arterial layer

turbinates shaped like a spiral; the three bones situated on the lateral side of the nasal cavity

tympanic membrane membrane that separates the external ear from the middle ear

U

ulcer inflammation that occurs on the mucosal skin surface

ulna bone on inner forearm

umami one of the five basic tastes; a savory taste

umbilical area located around the navel; the right and left lumbar region

umbilicus navel

unicellular composed of one cell

universal donor individual with type O Rh-negative blood; has no A or B antigens; can be donated to all blood types

universal recipient individual belonging to the AB blood group

uremia the presence of urea and excess waste products in the blood

ureters the long, narrow tubes that convey urine from the kidney to the urinary bladder

urethra the tube that takes urine from the bladder to the outside of the body

urinalysis the chemical analysis of urine

urinary bladder a muscular, membrane-lined sac situated in the anterior part of the pelvic cavity; used to hold urine

urinary meatus the opening to the urethra

urine at the completion of filtration, reabsorption, and secretion, the filtrate is known as urine

urticaria (hives) skin condition characterized by itching wheals or welts and usually caused by an allergic reaction

uterus hollow, thick-walled, muscular organ that houses the fetus during pregnancy

uvula projection hanging from the soft palate in the back of the throat

V

vacuole clear space in cell

vagina sheathlike structure; tube in females, extending from the uterus to the vulva

vaginal blockage obstruction of the vagina; can be fixed with surgery

valves structures that permit flow of blood in only one direction

varicose veins veins that have become abnormally dilated and tortuous due to interference with venous drainage or weakness of their walls

vas deferens a continuation of the epididymis; *see* ductus deferens

vascular malformations a type of birthmark or growth often present at birth composed of blood vessels

vasopressin hormone secreted by the posterior pituitary gland; has an antidiuretic effect; also called antidiuretic hormone (ADH)

vastus lateralis muscle found on anterior of thigh, used as an injection site for intramuscular injections

vector-borne transmission transfer of an agent to a susceptible host by animate means such as mosquitoes, fleas, and ticks

vehicle transmission transfer of an agent to a susceptible host by contaminated inanimate objects such as food, milk, and drugs

veins vessels that carry blood toward the heart

vena cava large blood vessel that returns blood to the right atrium; there are two: superior and inferior

venipuncture (phlebotomy) is a method of drawing blood using a needle to access a vein for intravenous therapy, or obtain a sampling of venous blood for testing

ventilation another term for moving air in and out of the lungs; breathing

ventral front, in front of, or anterior

ventricular fibrillation the rhythm of the heart breaks down and muscle fibers contract at random without coordination

ventricular septal defect a birth defect in which there is an opening between both ventricles

venules small veins

verbal communication information expressed in spoken words

vermiform appendix fingerlike projection protruding into the abdominal cavity; plays a role in immunity

vertebral (spinal) cavity area of the body containing the spinal cord

vertigo sensation of dizziness

vesicoureteral reflux (VUR) the abnormal upward flow of urine (reflux) from the bladder to the ureters

vestibule a small cavity at the beginning of a canal

villi hairlike projections, as in the intestinal mucous membrane

virulence frequency with which a pathogen causes disease

viruses disease-causing agents

VISA form of vancomycin-intermediate *Staphylococcus aureus*, a type of staph bacteria that has some limited resistance to the vancomycin family of antibiotics

visceral membrane the part of a serous membrane that covers each organ in a body cavity

vital lung capacity total amount of air involved with tidal volume, inspiratory reserve volume, and expiratory reserve volume

vitamin any of a group of organic compounds found in very small amounts in natural food; needed for the normal growth and maintenance of an organism

vitreous humor transparent, gelatinlike substance filling the greater part of the eyeball

VLDL very low density lipoprotein

vomer flat, thin bone that forms the lower part of the nasal septum

vomiting the expelling of undigested food or fluid through the mouth; also called *emesis*

VRSA form of vancomycin-resistant *Staphylococcus aureus*, a type of staph bacteria that is resistant to the vancomycin family of antibiotics

vulva external female genitalia

W

wart a type of tumor of the epithelial tissue; also known as a *papilloma*

West Nile virus (WNV) a mosquito-borne virus that most often has no symptoms or mild, flulike symptoms

wheezing sound produced by a rush of air through a narrowed passageway

whiplash injury trauma to cervical vertebrae

white-coat hypertension increase in a patient's blood pressure that occurs only when a medical professional, traditionally in a white coat, takes the blood pressure

written communication information expressed in written words, such as emails, manuals, reports, memos, and so on

wisdom teeth third molar teeth in the adult mouth

X

xiphoid process structure made of cartilage that forms the lower portion of the sternum

Y

yawning deep, prolonged breath that fills the lungs

yeast infections infections caused by the *Candida albicans* organism

Z

zygomatics bone that forms the prominence of the cheek

zygote organism produced by the union of two gametes

Index

Note: Page numbers in **bold** indicate tables and figures.

B

C

D

E

F

M

N

O

T

U